Emerging Technologies and Management of Crop Stress Tolerance

Emerging Technologies and Management of Crop Stress Tolerance
Biological Techniques

Volume 1

Edited by

Parvaiz Ahmad

Saiema Rasool

AMSTERDAM • BOSTON • HEIDELBERG • LONDON
NEW YORK • OXFORD • PARIS • SAN DIEGO
SAN FRANCISCO • SINGAPORE • SYDNEY • TOKYO
Academic Press is an imprint of Elsevier

Academic Press is an imprint of Elsevier
525 B Street, Suite 1800, San Diego, CA 92101-4495, USA
32 Jamestown Road, London NW1 7BY, UK
225 Wyman Street, Waltham, MA 02451, USA

Notice
No responsibility is assumed by the publisher for any injury and/or damage to persons, or property as a matter of products liability, negligence or otherwise, or from any use or, operation of any methods, products, instructions or ideas contained in the material herein. Because of rapid advances in the medical sciences, in particular, independent verification of diagnoses and drug dosages should be made.

British Library Cataloguing-in-Publication Data
A catalogue record for this book is available from the British Library

Library of Congress Cataloging-in-Publication Data
A catalog record for this book is available from the Library of Congress

ISBN: 978-0-12-810171-1

For information on all Academic Press publications
visit our website at elsevierdirect.com

14 15 16 17 18   10 9 8 7 6 5 4 3 2 1

Dedication

This book is dedicated to

Hakim Abdul Hameed
(1908–1999)
Founder of Jamia Hamdard (Hamdard University)
New Delhi, India

Contents

CHAPTER 3   Transcription Factors and Environmental Stresses in Plants   57

Loredana F. Ciarmiello, Pasqualina Woodrow, Pasquale Piccirillo,
Antonio De Luca and Petronia Carillo

CHAPTER 4   Plant Resistance under Cold Stress: Metabolomics, Proteomics, and Genomic Approaches   79

Saiema Rasool, Shailza Singh, Mirza Hasanuzzaman, Muneeb U. Rehman,
M.M. Azooz, Helal Ahmad Lone and Parvaiz Ahmad

Preface

More than 450 million small-scale producers around the world are reliant on agriculture for their way of life. A "right to food" movement has built momentum all over the developing world. "Food Security" has become a buzz word — meaning when people have enough food to feed themselves. Recent articles on the "Global Food Crisis", "Warming Planet Struggles to Feed Itself", and "Cutting Food Waste to Feed the World" reveal that roughly one third of the food produced for human consumption each year is wasted. This amounts to nearly 1.3 billion tonnes of food. The effects of a rapidly changing climate on the availability of food add to this grave situation. Climatic conditions are changing day by day and have an negative impact on the growth and dvelopment of the crops and ultimately reduced the crop yield. Environmental stress is responsible for the huge crop loss worldwide and this loss is increasing in coming years. The adverse impact of climate change in the form of declining rainfall and rising temperatures, and thus, increased severity of drought and flooding, is bound to threaten food security and livelihoods in the economy in many developing countries. To mitigate the effect of the environmental stress, plant biologists are trying to develop plants with higher levels of resistance to these stresses. They are employing new technologies to understand the physiobiochemical responses of the plants. Different approaches like genomics, proteomics, metabolomics, and micromics are being used to study plant metabolism, and up- and downregulation of genes under such stress. Advanced technologies have widened our genetic and molecular understanding of plant responses espacially under environmental stress. *Emerging Technologies and Management of Crop Stress Tolerance: Volume 1 Biological Techniques* will focus on the latest technologies used in plant stress resistance and development.

The book has 21 chapters; all of which deal with the alleviation of environmental stress through a different new technology that can be employed to increase the crop yield. Chapter 1 deals with the genomic approaches in abiotic stress tolerance in plants. Here the authors have clearly mentioned physiological, cellular and biochemical mechanisms of abiotic stress in plants. The authors have also highlighted functional genomic approaches in alleviating abiotic stress. Chapter 2 discusses the role of metabolomics in crop improvement and various techniques involved in metabolomics. Chapter 3 explains the role of transcription factors in mitigating environmental stress in plants. The author also highlights NAC, WRKY and Zinc Finger transcription factors. Chapter 4 describes the impact of cold stress in plants and the role of proteomics, genomics and metabolomics in alleviating the cold stress. Chapter 5 deals with genetic engineering of crop plants for abiotic stress tolerance. Here the overexpression of genes for transcriptional regulation, over expression of genes for osmoprotectants, and engineering of ion transport is meticulously explained.

Chapter 6 deals with Bt crops as a sustainable approach towards biotic stress tolerance. In this chapter the author discusses the transformation of crops with Bt genes, molecular analysis of transgenic plants and biosafety and risk assessment studies. Chapter 7 is about modern tools for enhancing crop adaptation to climate changes. Here the author describes diferent stresses caused by climatic changes. Conventional and molecular breeding methods, genetic engineering and GM crops is also well documented.

Chapter 8 is about interaction of nano-particles (NPs) with plants, an emerging prospective in agriculture industry. In this chapter classification and applications of nano-particles are observed. Modes of nanoparticles internalization by plants, the influence of NPs as growth promoter in plants and the influence of NPs as biological control in plants is discussed. Chapter 9 highlights the role of miRNAs in abiotic and biotic stresses in plants. Discussions in this chapter cover miRNA as a significant player in gene regulation, mechanisms of miRNA biogenesis and function, miRNA-mediated functions in plants, genome-wide miRNA profiling under stresses, and the involvement of miRNAs in plant stresses.

Chapter 10 discusses gene silencing, a novel cellular defense mechanism improving plant productivity under environmental stresses. Here the authors explain the elements of RNAi and RNAi under environmental stresses. Chapter 11 discusses the role of carbohydrates in plant resistance to abiotic stresses. The chapter highlights the importance of carbohydrates in osmotic balance and signaling.

The role of glucosinolates in plant stress tolerance is discussed in Chapter 12. The author talks about glucosinolate structure, isolation and analysis, biosynthesis of GSLs, role of glucosinolates in stress alleviation, and signaling networks. Chapters 13 and 14 deal with the trace elements and their deficiency symptoms and physiological roles is described. Also the alleviation of the different environmental stresses by these trace elements is also discussed in detail.

Chapter 15 discusses nutritional stress in dystrophic savanna soils of the Orinoco Basin, looking at the biological responses to low N and P availabilities. Here the chapter describes the climate in the Orinoco basin, nutritional stress, enhancing nitrogen-fixation and phosphorus acquistion and uptake. Chapter 16 looks at silicon and selenium, two vital trace elements that confer abiotic stress tolerance to plants. Here the author discusses silicon and selenium uptake and transport in plants, the involvement of Si and Se in plant growth, development, and physiology, and the protective roles of Si and Se under abiotic stress.

Chapter 17 highlights the mechanisms of herbicide/pesticide tolerance in plants. Chapter 18 is about the effects of humic materials on plant metabolism and agricultural productivity. Here the authors discuss the characteristics of humic materials and their functions and use of humic materials for sustainable plant production.

Chapter 19 discusses climate change and potential impacts on quality of fruit and vegetable crops. Impacts of climate change on vegetable production systems in Brazil, harvest and postharvest, and the effects of abiotic stress on different aspects of fruit is well explained. Chapter 20 regards the interplay of plant circadian clock and abiotic stress response networks. The chapter looks at the molecular basis of the circadian clock function in plants, the interaction between the clock components and cold response, and crosstalk between the circadian clock and ABA transcriptional networks. Finally, Chapter 21 throws light on the development of water-saving techniques for sugarcane (*Saccharum officinarum* L.) in the arid environment of Punjab in Pakistan.

We fervently believe that this volume comprises a wealth of knowledge, especially for plant biologists, environmentalists, agriculturists and students of different science colleges and universities. We have tried our level best to provide this compilation, but there may be some errors creeping in the volume for which we seek readers' indulgence and feedback.

We are very much thankful to our contributors who have devoted their valuable time in preparing their chapters and bearing the editors' corrections and suggestions as well. We owe our gratidude to Patricia Osborn (Acquisitions Editor, Elsevier), Carrie Bolger (Editorial Project Manager, Life Sciences, Elsevier), Melissa Read (Freelance Project Manager, Elsevier) and all the other staff members of Elsevier who were directly or indirectly associated with the project, for their constant help, valuable suggestions and timely publication of this volume.

Parvaiz Ahmad
Saiema Rasool

Acknowledgments

We acknowledge all the contributors of this volume for their valuable contributions. Parvaiz Ahmad also acknowledges the Higher Education Department, Government of Jammu and Kashmir, India for their support.

About the Editors

Dr. Parvaiz Ahmad is Senior Assistant Professor in the Department of Botany at Sri Pratap College, Srinagar, Jammu and Kashmir, India. He has completed his postgraduated study in Botany in 2000 at Jamia Hamdard, New Delhi, India. After receiving a Doctorate degree from the Indian Institute of Technology, Delhi, India, he joined the International Centre for Genetic Engineering and Biotechnology, New Delhi, in 2007. His main research area is Stress Physiology and Molecular Biology. He has published more than 35 research papers in peer-reviewed journals and 29 book chapters. He is also an editor of 12 volumes (1 with Studium Press Pvt. India Ltd., New Delhi, India, 8 with Springer USA and 3 with Elsevier USA). He is a recipient of the Junior Research Fellowship and Senior Research Fellowship by CSIR, New Delhi, India. Dr. Ahmad has been awarded the Young Scientist Award under the Fast Track scheme in 2007 by the Department of Science and Technology, Government of India. Dr. Ahmad is actively engaged in studying the molecular and physio-biochemical responses of different agricultural and horticultural plants under environmental stress.

Dr. Saiema Rasool is currently teaching plant science in the Education department, Government of Jammu and Kashmir India. Dr. Rasool completed her Masters in Botany at Jamia Hamdard New Delhi, India in 2007, specializing in plant stress physiology. She has eight research publications to her credit, published in various international and national journals of repute. She has also published 7 book chapters in international published volumes from publishers such as Springer, Elsevier and Wiley. At present, her research interests are mainly focused on the development of abiotic stress tolerant plants and the physiological and biochemical responses of crop plants to a range of biotic and abiotic stresses.

List of Contributors

Fakiha Afzal
Atta-Ur-Rahman School of Applied Biosciences, National University of Sciences and Technology, Islamabad, Pakistan

Parvaiz Ahmad
Department of Botany, S.P. College, Srinagar, Jammu and Kashmir, India

Fizza Akhter
Department of Microbiology, Quaid-I-Azam University, Islamabad, Pakistan

Jincy J. Akkarakaran
Nuclear Agriculture and Biotechnology Division, Bhabha Atomic Research Center, Trombay, Mumbai, India

Muhammad Arif
Agricultural Officer, Pesticide Quality Control Laboratory, Institute of Soil Chemistry and Environmental Science, Ayub Agricultural Research Institute, Department of Agriculture, Faisalabad, Pakistan

Muhammad Ashraf
Atta-Ur-Rahman School of Applied Biosciences, National University of Sciences and Technology, Islamabad, Pakistan

Muhammad Aslam
Arid Zone Agricultural Research Institute, Bhakkar, Punjab, Pakistan

M.M. Azooz
Department of Biological Sciences, Faculty of Science, King Faisal University, Al Hassa, Saudi Arabia

Imam Bakhsh
Faculty of Agriculture, Department of Agronomy, Gomal University, D.I. Khan Kyber, Pakhtunkhwa, Pakistan

Shagun Bali
Department of Botanical and Environmental Sciences, Guru Nanak Dev University, Amritsar, Punjab, India

A. Banerjee
Food Technology Division, Bhabha Atomic Research Center, Trombay, Mumbai, India

Ricardo Luis Louro Berbara
Federal Rural University of Rio de Janeiro, Soils Department, Seropédica, Brazil

Renu Bhardwaj
Department of Botanical and Environmental Sciences, Guru Nanak Dev University, Amritsar, Punjab, India

Muhammad Bilal
Department of Environmental Sciences, COMSATS Institute of Information Technology, Abbottabad, Pakistan

Shazia Anwar Bukhari
Department of Chemistry, GC University, Allama Iqbal Road, Faisalabad, Pakistan

Natalia A. Burmistrova
Institute of Plant Physiology Russian Academy of Science, Botanicheskaja, Moscow, Russia

Petronia Carillo
Dipartimento di Scienze e Tecnologie Ambientali Biologiche e Farmaceutiche, Seconda Università di Napoli, Caserta, Italy

Theocharis Chatzistathis
Laboratory of Pomology, Department of Horticulture, School of Agriculture, Aristotle University of Thessaloniki, Greece

Loredana F. Ciarmiello
Consiglio per la Ricerca e la Sperimentazione in Agricoltura- Unità di Ricerca per la Frutticoltura di Caserta (Fruit Tree Research Unit, CRA-FRC), Caserta, Italy

Jagoda Czarnecka
Institute of Plant Genetics, Polish Academy of Sciences, Strzeszyńska, Poznań, Poland

Antonio De Luca
Consiglio per la Ricerca e la Sperimentazione in Agricoltura—Unità di Ricerca per la Frutticoltura di Caserta (Fruit Tree Research Unit, CRA-FRC), Caserta, Italy

Anupam Dikshit
Biological Product Laboratory, Department of Botany, University of Allahabad, Allahabad, India

Mariana R. Fontenelle
Laboratory of Soil Science, Embrapa Vegetables, Brazil

Masayuki Fujita
Laboratory of Plant Stress Responses, Department of Applied Biological Science, Faculty of Agriculture, Kagawa University, Miki-cho, Kita-gun, Kagawa, Japan

Andrés Calderín García
Federal Rural University of Rio de Janeiro, Soils Department, Seropédica, Brazil; Agricultural University of Havana, Chemistry Department, Tapaste, San José de Las Lajas, Mayabeque, Cuba

Surbhi Goel
Department of Biochemical Engineering and Biotechnology, Indian Institute of Technology, Delhi, India

Ghader Habibi
Department of Biology, Payame Noor University, Tehran, Iran

Abdul Hannan
Office of Assistant Land Reclamation Officer, Factory Area, Near Madni Masjid, Sargodha, Directorate of Land Reclamation, Irrigation Department, Pakistan

Mirza Hasanuzzaman
Department of Agronomy, Faculty of Agriculture, Sher-e-Bangla Agricultural University, Sher-e-Bangla Nagar, Dhaka, Bangladesh

Rosa Mary Hernández-Hernández
Centro de Agroecología Tropical, Universidad Simón Rodríguez, Caracas, Venezuela

Ismael Hernández-Valencia
Instituto de Zoología y Ecología Tropical, Facultad de Ciencias, Universidad Central de Venezuela, Caracas, Venezuela

Tayyab Husnain
Center of Excellence in Molecular Biology, University of the Punjab, Thokar Niaz Baig, Lahore, Pakistan

Amir Hussain
Agriculture Officer (Technical) Office of District Officer Agriculture, Bhakkar, Punjab, Pakistan

Syed Sarfraz Hussain
Australian Centre for Plant Functional Genomics, University of Adelaide, Waite Campus, Urrbrae Glen Osmond, South Australia, Australia

Shabina Iram
Atta-Ur-Rahman School of Applied Biosciences, National University of Sciences and Technology, Islamabad, Pakistan

Fernando Guridi Izquierdo
Agricultural University of Havana, Chemistry Department, Tapaste, San José de Las Lajas, Mayabeque, Cuba

Sumira Jan
Centre of Research for Development, Kashmir University, Jammu and Kashmir, India

Dhriti Kapoor
Department of Botanical and Environmental Sciences, Guru Nanak Dev University, Amritsar, Punjab, India

Parminder Kaur
Department of Botanical and Environmental Sciences, Guru Nanak Dev University, Amritsar, Punjab, India

Ravinderjit Kaur
Department of Zoology, Guru Nanak Dev University, Amritsar, Punjab, India

Alvina Gul Kazi
Atta-Ur-Rahman School of Applied Biosciences, National University of Sciences and Technology, Islamabad, Pakistan

Anjali Khajuria
Department of Zoology, Guru Nanak Dev University, Amritsar, Punjab, India

Ejaz Ahmed Khan
Faculty of Agriculture, Department of Agronomy, Gomal University, D.I. Khan Kyber, Pakhtunkhwa, Pakistan

Agnieszka Kiełbowicz-Matuk
Institute of Plant Genetics, Polish Academy of Sciences, Strzeszyńska, Poznań, Poland

Marina S. Krasavina
Institute of Plant Physiology Russian Academy of Science, Botanicheskaja, Moscow, Russia

Ghulam Kubra
Atta-ur-Rahman School of Applied Biosciences, National University of Sciences and Technology, Islamabad, Pakistan

Carlos Eduardo P. Lima
Laboratory of Soil Science, Embrapa Vegetables, Brazil

Helal Ahmad Lone
Department of Botany, A.S. College, Srinagar, Jammu and Kashmir, India

Danilo López-Hernández
Instituto de Zoología y Ecología Tropical, Facultad de Ciencias, Universidad Central de Venezuela, Caracas, Venezuela

Bhawna Madan
Department of Biochemical Engineering and Biotechnology, Indian Institute of Technology, Delhi, India

Qaisar Mahmood
Department of Environmental Sciences, COMSATS Institute of Information Technology, Abbottabad, Pakistan

Leonora M. Mattos
Laboratory of Soil Science, Embrapa Vegetables, Brazil

Rohit K. Mishra
Biological Product Laboratory, Department of Botany, University of Allahabad, Allahabad, India

Vani Mishra
Nanotechnology Application Centre, University of Allahabad, Allahabad, India

Celso L. Moretti
Laboratory of Soil Science, Embrapa Vegetables, Brazil

Kamrun Nahar
Laboratory of Plant Stress Responses, Department of Applied Biological Science, Faculty of Agriculture, Kagawa University, Miki-cho, Kita-gun, Kagawa, Japan; Department of Agricultural Botany, Faculty of Agriculture, Sher-e-Bangla Agricultural University, Sher-e-Bangla Nagar, Dhaka, Bangladesh

Puja Ohri
Department of Zoology, Guru Nanak Dev University, Amritsar, Punjab, India

Avinash C. Pandey
Nanotechnology Application Centre, University of Allahabad, Allahabad, India; Bundelkhand University, Jhansi, India

Pasquale Piccirillo
Consiglio per la Ricerca e la Sperimentazione in Agricoltura—Unità di Ricerca per la Frutticoltura di Caserta (Fruit Tree Research Unit, CRA-FRC), Caserta, Italy

Muhammad Qasim
Department of Bioinformatics and Biotechnology, GC University, Allama Iqbal Road, Faisalabad, Pakistan

Mahmood-ur-Rahman
Department of Bioinformatics and Biotechnology, GC University, Allama Iqbal Road, Faisalabad, Pakistan

Galina N. Raldugina
Institute of Plant Physiology Russian Academy of Science, Botanicheskaja, Moscow, Russia

Bushra Rashid
Center of Excellence in Molecular Biology, University of the Punjab, Thokar Niaz Baig, Lahore, Pakistan

Saiema Rasool
Department of Botany, Jamia Hamdard, New Delhi, India

Amandeep Rattan
Department of Botanical and Environmental Sciences, Guru Nanak Dev University, Amritsar, Punjab, India

Muneeb U. Rehman
Molecular Biology Laboratory, Division of Veterinary Biochemistry, Faculty of Veterinary Sciences and Animal Husbandry, Sheri Kashmir University of Agricultural Science and Technology, Shuhama, Alastang, Srinagar, Jammu and Kashmir, India

Saleha Resham
Atta-ur-Rehman School of Applied Biosciences, National University of Sciences and Technology, Islamabad, Pakistan

Sheikh Riazuddin
Allama Iqbal Medical College, University of Health Sciences, Lahore, Pakistan

Abdul Gaffar Sagoo
MLL Land Reclamation Research Station Chak No. 37 TDA, Bhakkar, Punjab, Pakistan

Steven A. Sargent
Horticultural Sciences Department, University of Florida, Gainsville, Florida, USA

Tayyaba Shaheen
Department of Bioinformatics and Biotechnology, GC University, Allama Iqbal Road, Faisalabad, Pakistan

Bujun Shi
Australian Centre for Plant Functional Genomics, University of Adelaide, Waite Campus, Urrbrae Glen Osmond, South Australia, Australia

Ravinder Singh
Department of Botanical and Environmental Sciences, Guru Nanak Dev University, Amritsar, Punjab, India

Shailza Singh
National Centre for Cell Science, NCCS Complex, Ganeshkhind, Pune University Campus, Pune, India

P. Suprasanna
Nuclear Agriculture and Biotechnology Division, Bhabha Atomic Research Center, Trombay, Mumbai, India

Orooj Surriya
Atta-Ur-Rahman School of Applied Biosciences, National University of Sciences and Technology, Islamabad, Pakistan

Marcia Toro
Instituto de Zoología y Ecología Tropical, Facultad de Ciencias, Universidad Central de Venezuela, Caracas, Venezuela

P.S. Variyar
Food Technology Division, Bhabha Atomic Research Center, Trombay, Mumbai, India

Kinza Waqar
Atta-Ur-Rahman School of Applied Biosciences, National University of Sciences and Technology, Islamabad, Pakistan

Muhammad Waqas
Directorate General of Agricultural Extension, Office of Agricultural Officer, Kahna Nau, Punjab, Pakistan

Pasqualina Woodrow
Dipartimento di Scienze e Tecnologie Ambientali Biologiche e Farmaceutiche, Seconda Università di Napoli, Caserta, Italy

Genomic Approaches and Abiotic Stress Tolerance in Plants

Bushra Rashid, Tayyab Husnain and Sheikh Riazuddin

1.1 Introduction

The main goals of agricultural plant science for many decades have been to increase the yield and improve the quality of agricultural products. To attain these goals, improvement in the protection of crops against different types of abiotic stresses is important. Since the plants are sessile and complete their life cycle in a single location, they are afflicted by environmental challenges such as abiotic stresses, which include light, cold, heat, nutrition, water, salinity, and toxic concentrations of metals. As much as 80% of the crop harvest can be destroyed by these stresses. Abiotic stresses in crop plants negatively influence the whole plant and ultimately reduce the yield, whether it is for domestic use or for industrial purposes (Munns, 2002; Ashraf, 2002). Another drawback is the restricted use and further extension of the land for crop cultivation, which limits farmers to grow enough food to cope with the increased demand of the growing population. Other environmental factors and agricultural practices, like poor drainage, restricted rainfall, and higher vapor transpiration rate in combination with poor water and soil quality, may also contribute to enhance the problem in arid and semi-arid areas (Ashraf, 2004; Bao et al., 2009; Bhattarai and Midmore, 2009), and once the level is beyond the threshold then it will be more challenging and expensive to recover (Pisinaras et al., 2010).

Many agronomically important crops are affected by the abiotic stresses at different developmental stages, such as germination, leaf area and size, shoot and root length and weight, stem thickness, plant height, fruit initiation, setting, and maturity (Zhu, 2001; Akram et al., 2009b; Rodriguez-Uribe et al., 2011). Primary processes in the plants affected by abiotic stresses are photosynthesis (Munns et al., 2006; Chaves et al., 2009), osmotic potential (Hasegawa et al., 2000; Bor et al., 2003), stomatal conductance (Xue et al., 2004; Bao et al., 2009), and/or a combination of all these dynamics. These effects may ultimately influence the morphological, physiological, biochemical, cellular, and molecular mechanisms of the whole plant. Alteration in these processes may reduce the plants' fresh and dry biomass and reduce the yield (Azevedo-Neto et al., 2004; Higbie et al., 2010).

Plants have developed tolerance to abiotic stresses to some extent by evolving defense systems such as adjusting osmotic regulation and controlled uptake of ions (Senadheera and Maathuis, 2009), but this system is very complex and not completely understood. There is relative interaction of the mechanisms involved at the physiological, biochemical, morphological, and molecular levels. There are a number of means of coping with these problems such as land reclamation through hydrological and chemical means but these are expensive (Corbishley and Pearce, 2007).

P. Ahmad (Ed): Emerging Technologies and Management of Crop Stress Tolerance, Volume 1.
DOI: http://dx.doi.org/10.1016/B978-0-12-800876-8.00001-1

Conventional breeding technology has restricted success in developing stress-tolerant cultivars due to variant germplasm in order to exploit natural or artificially induced diversity and, subsequently, to select for desired properties. The problem with traditional plant breeding is that it is time consuming and laborious; it is difficult to modify single traits; and it relies on existing genetic variability (Yamaguchi and Blumwald, 2005; Ashraf et al., 2008; Zhang et al., 2008). Currently, it is well understood that these complex mechanisms generally involve interactions of a number of genes at the molecular level (Flowers, 2004). Classification of candidate genes for stress tolerance and their expression is required to understand the metabolic phenomena. Success in breeding for better adapted varieties to abiotic stresses depends upon the concerted efforts of various research domains including plant and cell physiology, molecular biology, and genetics.

Genome-based studies have the potential to endorse persistent and improved plant genetic development. The progress made since the last decade related to the studies of functional genomics is becoming increasingly important as the genome sequences of model crop species have been released. Therefore, alterations to the genes' behavior, such as overexpression or silencing related to the fabrication of particular plant components, are a possibility. This would be helpful to reveal regulatory mechanisms linked with the biosynthesis and catabolism of metabolites in crop plants (Gambino and Gribaudo, 2012). Therefore, for future progress in this area, efforts are required to develop genomic resources and tools for basic and applied genetics, genomics, and breeding research. This will pave the way to understand the molecular and metabolic pathways involved in the adaptation of plants to environmental challenges.

Nevertheless, significant progress in genomics studies has made this more valuable as this technology is one among other tools that have been exploited to recognize stress responsive genes in several species of plants (Rabbani et al., 2003; Arpat et al., 2004; Micheletto et al., 2007). This chapter presents reviews related to genomics and molecular processes associated with the developments for abiotic stress responses and tolerance in plants with significant features of the effects of stress on different crop plants.

1.2 Physiological, cellular, and biochemical mechanisms of abiotic stress in plants

Crop production is severely affected ultimately reducing yield by abiotic stress. According to their performance in extreme environmental conditions, plants have been classified into two groups, i.e., glycophytes (susceptible) and halophytes (tolerant), and most crops are considered glycophytes. These plants are not able to survive in extreme environments and accumulation of excessive metabolites in the growing medium hampers the different plants' developmental stages (Hasegawa et al., 2000). By and large, stressful environments affect metabolism, which ultimately disturbs the physiological, biochemical, cellular, and morphological processes in plants. High salt accumulation or dehydration reduces plant growth by disturbing the osmosis or the ion toxicity. Higher concentrations of salts are toxic to plants, preventing the uptake of ions leading to ionic homeostasis and disrupting osmoregulation (Senadheera and Maathuis, 2009). This will reduce the potassium and calcium ions, and sodium and chloride will be increased with an increased ionic effect. Reactive oxygen species (ROS) are accumulated at cellular level by the induced salt stress, which may produce damaging or toxic reactions like lipid peroxidation, degraded proteins, and mutations in the nucleic acids (Pitman and Läuchli, 2002; Mittler, 2002; Bor et al., 2003). These factors negatively affect development, and change the physiological mechanism including deposition of inorganic ions and organic solutes, water relations, or photosynthesis of different plant species (Greenway and Munns, 1980; Niknam and McComb, 2000; Murphy and Durako, 2003).

1.2.1 Organic and inorganic solutes

Specific types of proteins work as the essential component of cell membranes, which is porous for the ions. The membranes act as the channels to bind the ions to those proteins through diffusion and form electrochemical potential gradient. This process requires energy, which is used in the form of stored adenosine triphosphate (ATP) or ATP and pyrophosphate. This gradient across the membrane will produce differences in pH and electric potential. Therefore, the modification in electric potential forces the inward movement of cations through channels and the variation in pH regulates the movement of ions through carriers, which leads to binding of the protons and ions (Flowers and Flowers, 2005). Therefore, the efficiency of cell membranes to control the rate of ion movement in and out of the plant cell is used as an indicator of damage to a great range of tissues (Farkhondeh et al., 2012).

Ion toxicity results after the increased accumulation of salts into the crop through growing medium. This mostly includes increased levels of Na^+ and Cl^-, which form ionic effects and whose level is different in various crop species depending upon the severity and duration of the stress (Abrol et al., 1988). This leads to surpassing the capability of the cell to compartmentalize ions into the vacuole, or they may pass onto the cell wall and dehydrate the cell, which may result in interruption of the water relations and effective use of essential nutrients (Lacerda et al., 2003; Munns, 2005). The inclusion of Na^+ into the plant cell taken up through growing medium will result in cytoplasmic toxicity. Therefore, uptake, accumulation, and long-distance transport of Na^+ throughout the plant system is a critical step for plants to be affected by ion toxicity (Munns et al., 2000; Ashraf, 2004). There are reports that higher accumulation of Cl^- causes leaf injury in different plant species (Greenway and Munns, 1980; Flowers and Hajibagheri, 2001; Mansour et al., 2005). The toxicity of Cl^- causes necrosis and finally leaf senescence (Nawaz et al., 2010). The roots absorb the Cl^- and move through the xylem to shoots and leaves where its accumulation beyond the toxic level causes cell injury. Higher concentrations of salt increase the deposition of Na^+ in the growing region of the root and hence decrease the selectivity for K^+ (Ashraf et al., 2012). Therefore, proper uptake through roots and justified cellular movement, translocation, and compartmentalization in the cell are necessary for proper functioning of the plant metabolism in saline, dehydrated, and higher temperature environments (Akram et al., 2007).

Higher accumulation of Na^+ causes a higher proportion of Na^+/K^+ and Na^+/Ca^+ in stress growth medium/conditions. The concentration of K^+ is indirectly proportional to the Na^+ in the cell, i.e., if Na^+ is increasing then K^+ and Ca^{2+} are decreasing in maize and *Limonium perezii* as reported by Suarez and Grieve (1988) and Carter et al. (2005), respectively, while K^+ and Ca^{2+} both take part in the maintenance for integration and proper functioning of cell membranes, i.e., cell wall stabilization, triggering enzymes, and directing ion translocation (Carden et al., 2003; Wenxue et al., 2003).

Altered or lower levels of calcium sodium ratios in a saline environment will inhibit plant development and uptake of various nutrients in *Atriplex griffithii* (Khan et al., 2000), cotton (Higbie et al., 2010), and *Agrostitis stolonifera* (Majeed et al., 2010), and significantly change the morphological and anatomical structure (Cramer, 1992). Various nutrient deficiency with higher NaCl stress has also been reported in different crops, as can be found with K^+ in spinach (Chow et al., 1990), artichoke (Graifenberg et al., 1995), and tomato (Lopez and Satti, 1996), and with nitrogen deficiency in cucumber (Cerda and Martinez,1988) and lettuce and cabbage (Feigin et al., 1991), whereas phosphorus and potassium are deficient in tomato (Adams, 1991) and cucumber (Sonneveld and de Kreiji, 1999). However, artificial or foliar application of potassium in the form of KH_2PO_4, KCl, KOH, K_2CO_3, KNO_3, and K_2SO_4 improved the plant growth in wheat, sunflower, and cotton for various physiological parameters (Sherchand and Paulsen, 1985; Akram et al., 2007, 2009a,b; Jabeen and Ahmad, 2009).

The organic solutes protect the plant metabolism and play an important role in adjusting the osmotic potential. Osmotic adjustment in a plant is altered due to inorganic ion or organic solute accumulation in the cell. Net increase in the organic solutes in a cell is due to the decrease in the water potential. Besides inorganic ion accumulation, the role of the accumulation of organic solutes is also important. Cell membrane is important in maintaining the proper concentration of these solutes in the cell. Proline, trehalose, sucrose, polyols, glycinebetaine, prolinebetaine, alaninebetaine, hydroxylprolinebetaine, pipecolatebetaine, and choline O-sulfate are among the common organic solutes that play an important role for osmotic adjustment in plants under salt stress, dehydration, and some other abiotic stresses (Rhodes and Hanson, 1993; Hasegawa et al., 2000). Therefore, higher concentrations of these solutes under salt stress are favorable to plants as they participate in ROS scavenging activity, act as a $^1O_2$ quencher, and maintain the OP (Marcelo-Pedrosa and Queila-Souza, 2013).

1.2.2 **Role of abscisic acid under abiotic stress**

Under salinity, dehydration, extreme temperatures, and other abiotic stresses, plants respond in relation to metabolic and progressive variations. Different tissues exposed to variant stresses behave in a coordinated manner to implement these responses. Salts would not be accumulated in higher concentrations in actively dividing and growing cells. Plants' responses under stress are initiated by some indicators such as primary osmotic stress and/or secondary metabolites (Chaves et al., 2003). Secondary metabolites signals comprise growth hormones like abscisic acid, ethylene, and cytokinins, and reactive oxygen species and intracellular phospholipids or sugars. It was observed that the ability of isopentenyl transferase (IPT) to increase cytokinin biosynthesis, hence delaying senescence and the ability of CBL-interacting protein kinase to sense and respond to the calcium concentrations, has been used to make cotton plants more robust under drought conditions (Kuppu et al., 2013; He et al., 2013). Abscisic acid (ABA) is initiated to produce in roots, translocates to shoots through xylem, and signals the stomata to close, which in the long run limits the cellular division and expansion and reduction in stomatal conductance in response to stress (Aldesuquy and Ibrahim, 2001). Munns et al. (2000) conducted experiments on shoot water relations and verified that the osmotic effects under salt stress induce the hormonal signals outside the roots to regulate cell expansion. Artificial application of ABA to common bean plants induced higher K^+/Na^+ ratios and so restricted the translocation of sodium to shoots (Khadri et al., 2007; Chaves et al., 2009). It is evident that along with root cells, ABA may also be synthesized in leaf cells and distributed from there throughout the whole plant (Wilkinson and Davies, 2002). Flexas et al. (2006) also proved that stomatal conductance is reduced when the leaves are cut off the plants, but the same has been recovered very quickly after the foliar application of ABA to the plants. They also reported the response of stomatal conductance to alterations in photoperiod, temperature, and availability of CO_2 (Flexas et al., 2008).

The role of other growth hormones like GA_3 has been reported to compensate for the adverse effect of stress on different plant tissues (Naqvi, 1999; Chakraborti and Mukherji, 2003). Shah (2007) reported on the foliar application of GA_3 under salt-stressed plants and observed the improvements in the dry mass, leaf area expansion, photosynthetic rate, and stomatal conductance. The accumulation of salts under dehydrated conditions induces the production of ABA to close the stomata and regulate the plants' growth under stress conditions which accounted for the restricted production of ABA during the process of conjugation. The reduced stomatal conductance and rescuing of productivity was aggravated by the foliar application of GA_3 as these results were inconsistent with those of Afroz et al. (2005). Similarly, Guo et al. (2011) guessed that GhWRKY3—a transcription factor from *Gossypium* sp.—will possibly play a role towards ABA- and GA_3-intermediated pathways that signal and regulate plant growth and development and generate defense responses against pathogens.

Late embryogenesis abundant (LEA) genes have been identified that are responsive to ABA, and a number of ABA responsive elements (ABREs) are present on the promoters that work together with other nuclear protein factors in different crops. These LEA genes are mostly present in mature embryos and some other vegetative tissues under abiotic stresses (Luo et al., 2008). LEA proteins are known for their role in protecting cells under osmotic stress due to hydrophilic characteristics and as they accumulate under the abiotic stress in plants. Advances in this research indicate that the pH of xylem and leaf apoplast affects ABA production and subsequently translocates to the stomata. Under salt stress or more alkaline pH, exclusion of ABA from xylem and leaf apoplast may be decreased and hence more ABA will reach the guard cells, which will facilitate the variation of stomatal aperture (Jia and Davies, 2007).

1.2.3 Reactive oxygen species

Higher amounts of salts taken up from the soil by plants at higher temperatures restrict the availability of water and cause drought stress as well. Consequently, plants close the stomata to retain water and this will also limit the entrance of carbon dioxide into the leaves. This will decrease the process of photosynthesis, or if the salts are at too high a concentration or at toxic levels, then photosynthesis will be inhibited directly. In this situation, plants suffer from oxidative stress, which is an additional significant feature of this stressful condition.

The formation of reactive oxygen species (ROS) is the outcome of sunlight absorption by plants. This mainly takes place in the chloroplast through superoxide anion, hydrogen peroxide, hydroxyl radical, and singlet excited oxygen as a routine function of plants' aerobic respiratory system. ROS are the reason to why cell membranes, nucleic acids degradation, and proteins in the cell are damaged because they are extremely reactive. Photorespiration, oxidation of fatty acids, and mitochondrial and chloroplast electron transport systems (PSI and PSII) are the various categories of functions performed in the cell that generate ROS (Hernàndez et al., 2001; Foyer and Noctor, 2003). A number of reports are available for the effects of abiotic stress comprising ROS (Bor et al., 2003; Stepien and Klobus, 2005), but the limit of these processes stimulated under specific conditions as well as the magnitude of the damage or the differences in their capability acquired by the plants under stress has not been well understood. Irregular concentrations of antioxidants in chloroplast and changes to enzymatic activities may regulate the ROS to damage the plant systems (Asada, 2000), but transgenic plant development is another way to control the damage (Apse et al., 1999; Kumar et al., 2012).

1.3 Effects of abiotic stresses on physiological, cellular, and biochemical processes in plants

Flexibility of the cell wall regulates the cell growth rate and maintains the turgor pressure, which actually determines the cell growth and elongation (Peters et al., 2001). Therefore, cell wall elasticity is considered to be the prerequisite for cell growth and expansion. Cell growth is affected by the osmotic adjustment regulated by the contribution of roots and leaves to maintain the water absorption, uptake, transportation, and turgor pressure. This will lead to maintaining the primary physiological processes such as stomatal opening and photosynthesis, and cell expansion, up-regulation of antioxidants, accumulation of organic solutes like amino acids, polyamines and carbohydrates (Lockhart, 1965; Serraj and Sinclair, 2002; Munns et al., 2006; Stepien and Klobus, 2006; Yue et al., 2012). This is caused by the decrease in availability of CO_2 due to decrease in the stomatal conductance (Flexas et al., 2004).

Water relations or maintenance of relative water contents is also an important physiological criterion for stress tolerance in plants. Relative water contents are the estimation of water uptake and the leaf turgidity, whereas water potential is the maintenance of water−soil−plant−atmosphere under continuous status. Therefore, osmotic adjustment is related to the leaf relative water contents as it maintains the leaf water potential, but the water potential is not related to the osmotic adjustments (Suriya-Arunroj et al., 2004). Seed germination is also one of the physiological processes affected by the inhibitory effects of temperature, salts, and dehydration stress and this may be relieved by plant growth hormones including gibberellic acid, ethylene, cytokinin, etc. (Xu et al., 2011).

Photosynthesis is one of the main contributing factors in abiotic stresses that induce reduction of plant growth and yield (Mi et al., 2012). It is one of the major physiological processes during plant development and reduction under abiotic stresses—as water uptake and CO_2 availability are reduced. Photosynthesis will reduce the intercellular carbon dioxide availability and other non-stomatal functions. During photorespiration in C_3 plants, ribulose-1,5-bisphosphate carboxylase/ oxygenase (RUBISCO) catalyzes the absorbed carbon dioxide. The same step in C_4 plants is regulated by phosphoenol pyruvate carboxylase (PEPC). The higher accumulation of salts increases the oxygenase activity of RUBISCO as the CO_2 availability is reduced and therefore carboxylation is reduced or ceased (Sivakumar et al., 2000). Transpiration efficiency and stomatal conductance in combination with water use efficiency are the essential components of photosynthetic machinery in plants. Stressful environmental conditions decrease the stomatal conductance, which limits the transpiration rate and leads to restriction of water use efficiency, and this whole cycle will reduce the photosynthesis (Gamma et al., 2007; Akram et al., 2009b).

Chlorophyll is the green pigment essential for photosynthesis, having absorbed the spectra of visible light. Estimation of chlorophyll contents is the essential criterion to measure the pigments in the leaves and in turn the estimation of nutrient status from time to time (Gao et al., 2008). The older leaves are affected earlier and leaf dropping starts more quickly as compared to the new and actively growing leaves (Parida and Das, 2005). This is interconnected with the prolonged stress period. The accumulation of ions is increased in the chloroplast, which affects the electron transport system during photosynthesis (Sudhir and Murthy, 2004). This will affect the chlorophyll contents, i.e., an increase in susceptible plants (Hamada and El-Enany, 1994) and a decrease in tolerant plants (Singh et al., 1990). Mg^+ has been reported for its important role in the structure of chlorophyll, enzyme co-factor, and distribution of photosynthesis. Increased degradation of chlorophyll in Mg^+-deficient plants has been observed, which leads to the enhancement of oxidation of RUBISCO (Ramoliya et al., 2004; Saleh, 2011). It has been reported that the tolerant species of plants may not be affected by the saline medium, but the salt-sensitive species has great negative effects on the chlorophyll contents. This was proven when *Arabidopsis* showed sensitivity to 150 mM NaCl, in comparison to *Thellungiella*, which tolerated the same concentration of NaCl very satisfactorily and chlorophyll contents had not been altered (Stepien and Johnson, 2009).

1.4 Conventional breeding technology to induce abiotic stress tolerance in plants

Traditional breeding techniques have been the routine practice to study the genetic variability among crop species. Genetic variation has been identified among sexually compatible species and desirable agronomic traits have been introduced to the agriculturally important crops. Many of the crop cultivars such as wheat (Villareal et al., 1994; Valkoun, 2001; Zaharieva et al., 2001), rice (Mackill et al., 1993; Mishra et al., 1996), soybean (VanToai et al., 1994), and maize (Bänziger et al., 2004) have been introduced through different procedures of conventional breeding programs. Therefore, conventional breeding technology has yielded improvements to the quality of crops at morphological, biochemical, physiological, and cellular levels against various abiotic stresses. But the limitations related to this technology are worth mentioning as it is laborious, time consuming, inefficient, and costly. Hence, it takes several years to develop a new cultivar with improved traits through this technology. Moreover, some of the genetic material with irrelevant or unwanted characteristics may also be introduced, which might be difficult to remove. Another important factor for selection may be the availability of germplasm or the low genetic variation (Ashraf, 2010).

Therefore, breeding is difficult due to the complexity of the plant's adaptation under stress. So, this suggests that a number of traits may combine and contribute to adapt the plants for tolerance against a range of abiotic stresses. Thus, there is a need to seek more efficient approaches for genetically tailoring crops for enhanced drought tolerance.

1.5 Functional genomics approaches to induce abiotic stress tolerance in plants

Fundamental processes involved in abiotic stress tolerance mechanisms are water transporters and relevant practices, cell signaling modules (heat shock proteins, specific transcription factors, molecular chaperones, and late embryogenesis proteins), reactive oxygen species, osmolyte adjustment, and ion accumulation, all of which are common within plant species. These processes are regulated by a number of genes simultaneously and offer goals to conduct research in crop plants. Therefore, currently, genomic tools are suggested to use resources to generate the comprehensive datasets making contribution to adapt the genes' expression, protein synthesis, and modifications to secondary metabolites after exposure to abiotic stresses (Figure 1.1).

1.5.1 Advances in phenology assist the genomic studies of abiotic stress tolerance

An abiotic stress mechanism is complex to understand at the molecular level, but the advances in phenotypic studies have helped to quantify the components contributing to tolerate the abiotic stress at the genetic level. Therefore, phenomic developments along with genomic methods in the crops with more complex genomes are becoming increasingly amenable because the genomic evidences in non-model crops are now available (Roy et al., 2011).

Plant morphological or phenotypic studies are commonly difficult, time consuming, and destructive as most require removing the whole plant biomass. But now there are imaging tools that are non-destructive and are used to take several images of the same plant at different time intervals and wavelengths. These advances have made it possible to offer non-destructive approaches to obtain computable data in a number of crops at different growth stages under different types of abiotic stresses like drought, salt, cold, and heat (Morison et al., 2008; Rajendran et al., 2009; Sirault et al., 2009; Berger et al., 2010). Most of the phenotypic studies have concentrated upon the shoots or traits related to the stems or leaves and have ignored the roots. But in spite of the contribution of the roots to tolerate abiotic stress, the studies related to the roots are fewer. This might be because it is difficult to phenotype the roots without damaging or donating the whole plant (Richards et al., 2010; Fleury et al., 2010; Zhu et al., 2011b). Some of these reports are known for the important root traits that tolerate abiotic stresses, like metal toxicity (Jefferies et al., 1999; Ma et al., 2005) and nutrient deficiency (Laperche et al., 2006; Walk et al., 2006; Zhu et al., 2011a). Therefore, for further explanations of abiotic stress tolerance through roots, it is desirable to make improvements to the methods developed for phenotyping of roots, and environmental factors affecting plants growth may also be considered (Berger et al., 2010). Comparison of the crops growing under controlled conditions, i.e., in a greenhouse and in the field conditions, is important to elucidate the complexity of the crops' tolerance mechanism for superimposed environmental variations or the naturally occurring climatic disorders.

With the introduction of high-throughput studies, a number of phenomics assays are now accessible online to speed up the gene identification and molecular marker-assisted genomics studies related through phenomics. One of these assays is rapid elemental analysis of plant tissue, which is helpful to estimate the ionic accumulation in different plant tissues under particular abiotic stress conditions (Baxter et al., 2007). A number of developed countries have plant growth chambers/facilities furnished with mechanized systems that bring the plants for experimental analyses to, for example, irrigation, imaging, and/or weighing stations. Data obtained are analyzed with computational software that includes: The Plant Accelerator1 (http://www.plantaccelerator.org.au) in Australia, Crop Design (http://www.cropdesign.com) in Belgium, and the Leibniz Institute of Plant Genetics and Crop Plant Research in Germany (http://www.ipk-gatersleben.de/Internet). Results of

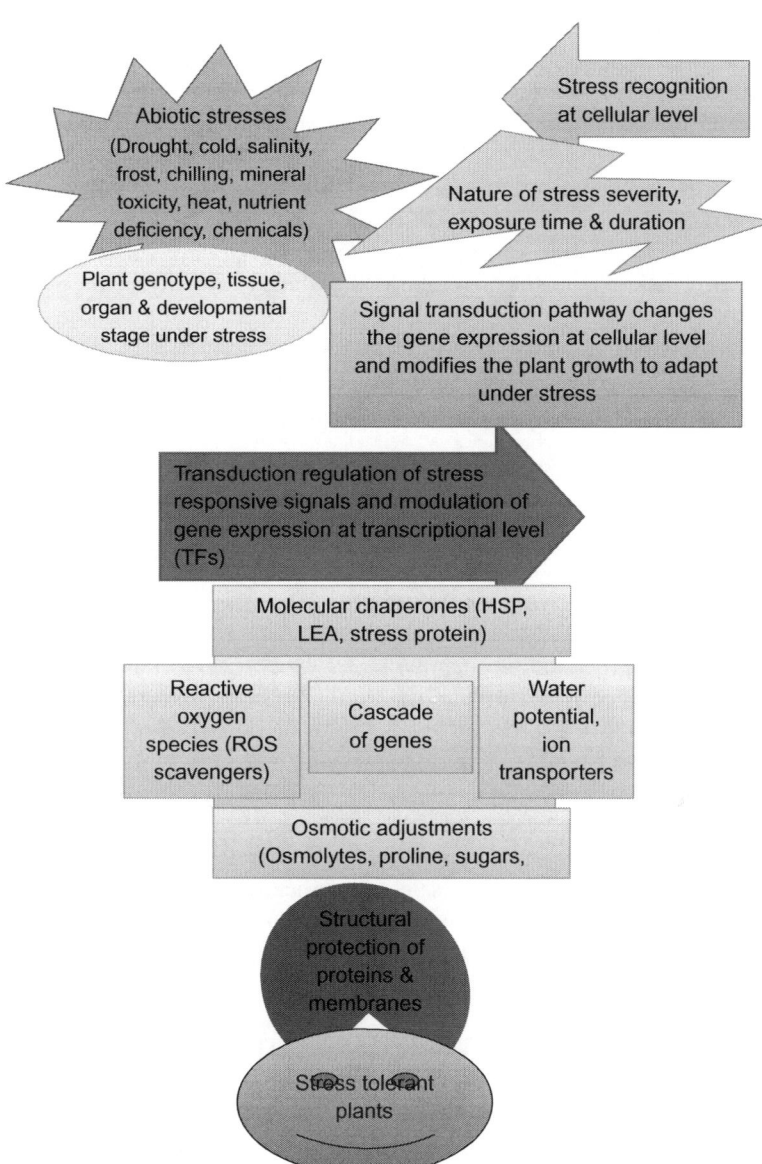

FIGURE 1.1

Plant stress response is originated at the cellular level and signal transduction is stimulated. Genes protect the cells as: accumulation of osmolytes, ionic movement across membranes, water transport potential, and homeostasis. Regulatory proteins promote to regulate the transduction of the stress signal and moderate gene expression.

these analyses in combination with the genomics approaches will help to discover the genes contributing to the known abiotic stress tolerance mechanism in crop plants. To minimize the experimental errors, the plant populations tested in greenhouses are required to be compared/tested under different field conditions several times and the abiotic stress-related components must be studied carefully.

The latest advances in phenomics for measurement of shoot-related parameters without damaging the plant aerial parts are digital RGB images and infrared thermography (Jones et al., 2009; Berger et al., 2010; Munns et al., 2010). Similarly, minirhizotrons, ground penetrating radar, and electrical resistivity imaging have been introduced to analyze the roots non-invasively (Zhu et al., 2011b). Genomics studies for plants grown in field conditions became more accurate with high-resolution EM38 mapping for characterization of defined growth conditions (Furbank, 2009; Robinson et al., 2009). Thus, genetic approaches in combination with phenotypic studies to expose the molecular mechanism of abiotic stress tolerance in crops are becoming more easily attainable.

1.5.2 Molecular mapping

Correlation of molecular markers with phenotypic studies of plants is important to know the exact point of qualitative and quantitative genetic variation harboring the dynamics manipulating the plant's response under abiotic stress. Yield or associated secondary traits are main components for crop improvement, and identification of quantitative trait loci (QTLs) for these traits is very strategic to improve stress tolerance in plants through marker-assisted selection. The loci for variation may be identified at structural or expression level and further tested in other germplasms, or alleles may be discovered to be used for breeding and selection into high yielding and elite cultivars of agriculturally important crops (Langridge et al., 2006). To date, QTL mapping is an exciting option, rather than using previously listed DNA markers such as RFLP, AFLP, RAPD, SSR, and SNP (Ashraf et al., 2008). So, gene pyramiding at two or more loci is possible through molecular marker-based selection (Asins, 2002). The selection procedure is mainly dependent on specific environmental conditions, which are major limitations encountered in the conventional breeding of the traits affected by abiotic stresses. Once a marker-related featured link has been marked undoubtedly, then selection could be minimized to a great extent (Humphreys and Humphreys, 2005; Tuberosa and Salvi, 2006).

Several mapping studies have identified numerous QTLs associated with salinity, drought, extreme temperatures, and other abiotic stress tolerance loci relevant to performance in less yielding areas in a number of crop species (such as that found at http://www.gramene.org/qtl/). The major crops mapped for different traits through QTL-related abiotic stresses are for cold, heat, salinity, mineral toxicities, nutrient deficiencies, and drought in cotton (Saranga et al., 2001), wheat (Quarrie et al., 1994), maize (Feng-Ling et al., 2008), soybean (Cornelious et al., 2005), barley (Teulat et al., 1997), sorghum (Sanchez et al., 2002), and rice (Lafitte et al., 2004; Mizoi and Yamaguchi-Shinozaki, 2013), among others (Table 1.1). Although the application and achievements through marker-assisted selection seems to be simple and easy-to-induce stress tolerance, the constraint is the precise and accurate identification of QTL and the proficient capability relevant to this research application. Broad diversity has been identified in the complexity of stress tolerance through the mapping program, and its association with breeding has expanded the mapping studies.

Table 1.1 QTLs Identified in Different Crop Species for Abiotic Stress-related Parameters

Crop	Relevant Stress	QTL Trait Identified	Trait Improved	References
Cotton	Pathogens, water potential. Osmotic potential	Fiber related, cell membrane stability	Seed cotton, panicles, disease and insect resistance, plant biomass, fiber quality and yield	Wright et al. (1998); Jiang et al. (2000); Ulloa et al. (2006)
Wheat	Drought, cold, salinity, metal toxicity, nutrient deficiency, heat	Osmotic potential	Plant biomass, early flower and maturity, yield	Yang et al. (2007); Baga et al. (2007); Balint et al. (2009); Laperche et al. (2008); Ma et al. (2007)
Barley	Drought, salinity, water logging, metal toxicity	Water potential, osmotic potential	Grain	Teulat et al. (1997); Xue et al. (2009); Li et al. (2008); Navakode et al. (2009)
Rice	Drought	Osmotic potential, water potential, cell membrane stability	Yield, grain, plant biomass	Lafitte et al. (2004)
Pearl millet	Drought	Osmotic potential, water-related attributes, cell membrane stability	Grain and yield	Serraj et al. (2004)
Maize	Drought	Ionic balance, osmotic adjustment	Yield and grain	Feng-Ling et al. (2008)
Sorghum	Drought	Water-related attributes, ionic balance, osmotic adjustment	Delayed leaf senescence, yield and grain	Sanchez et al. (2002)

Most of these studies are based on the field evaluation of crops for abiotic stress tolerance (Ahmed et al., 2013). This identified diversity in the stress tolerance mechanism may prove a key source to validate the candidate genes and will be a helpful mechanism to communicate the findings of genomics to an efficient/practical plant breeding program (Ismail et al., 2007).

Since the abiotic stress tolerance mechanism has a multi-gene trait, then wide mapping of some specific traits in different populations of the same plant species may identify the common loci (Langridge et al., 2006). Large segregating populations are needed for positional cloning through high-resolution linkage maps, and combining tasks such as identification of loci and positional cloning may be the alternative easier way in the future for plant breeding to induce abiotic stress tolerance (Roy et al., 2011).

1.5.3 Expression sequence tags

Although the genetic sequencing was found to be a good source of genome information, the large size of the genome of some plant species or the non-coding part of the genome sequence was still a limitation related to the genetic information. Plants with larger genomes are unpopular or are difficult to sequence because they contain copies of large repeats of genes, which do not provide information of coding genes; the purpose of sequencing is to get the information for discovery and characterization of the genes for proteomics. This has led to the progression to mass mode to anticipate the actual regions of protein coding constituents in a genome. Therefore, an expression sequence tags (EST)-based study was developed as the genomics approach to overcome the problems associated with the large genome sequence or noncoding part of the genetic sequence (Rudd, 2003).

The early 1980s saw the initial use of cDNA as a method to explore gene discovery (Putney et al., 1983), which in 1990 was further extended by supporting the implication of a high-throughput approach for transcript profiling as characterization of the coding region of human genome would involve messengers from the expressed genes (Brenner, 1990). Continuing this study, Adams et al. (1991) described the term EST, correlated with gene discovery and the human genome project. Since these reports, more than 10 million ESTs have been sequenced from more than 500 distinctly annotated species (fungi, plants, animals) (Rudd, 2003).

Langridge et al. (2006) considered ESTs and cDNA libraries as "Electronic Northern" and John and Spangenberg, (2005) considered them as "*in silico* Northern analysis," producing a large number of sequences appraising gene expression and being a useful method for prevalence of transcript abundance at the primary level. Analysis of EST data shows that the homologous gene shows variation among the genes and also that they are tissue specific (Mochida et al., 2004). There are a large number of data produced by the ESTs and there is a report of 449,101 ESTs for drought, 312,353 ESTs for salt, 103,898 ESTs for low temperature, 252,595 ESTs for high temperature, 19,384 ESTs for nutrient deficiency, and 135,578 ESTs for light stress on the National Center for Biotechnology Information browser (see http://www.ncbi.nlm.nih.gov/).

EST is a simple and easy way to obtain the sequencing of complicated and targeted plant species under abiotic stress conditions. It provides the mRNA with abundant expression profile of specific sequences from cDNA under specific conditions. To observe the expression of clones differentially expressing certain genes, libraries are constructed with cDNA that contain 10,000 clones acquired from targeted plant species under specific abiotic stress compared with the control (Pariset et al., 2009). Ubiquitously expressing housekeeping genes within the cells may also be considered for validation studies. It depends upon sampling of the plant tissues or organs taken from roots, leaves, etc., under salt, drought, frost, or other abiotic stresses.

Complete analysis of the core gene assembly is a complicated issue and difficult to resolve. Sampling of all transcribing genes is only possible if mRNA is gathered from all available types of cells at each and every growing stage of the plant under as many combinations of biological and environmental trials as possible. This is very complex and can only be possible through extensive experimental planning and comprehensive collection of cell types under a wider variety of environmental threats. These assemblies of cDNA and ESTs are expected to comprise the differentially expressed genes, which are stress induced and expected to play key a role in plants' tolerance mechanisms.

Demonstration of the host gene within a library and the quality of sequence are the limiting factors associated with EST technology and this is difficult to overcome due to the difficulty in observing any transcript from a tissue under some specific stress, and the sequencing may contain a background of incorrect (partially or incomplete) sequences. Initially, the EST sequencing was preferred for the 5′ end due to the coding or conserved regions, but with the advancements in sequencing technology, EST sequencing at both ends is becoming prevalent due to high-throughput sequencing of plants' ESTs. It may also provide more distinctive sequences and hence be the possible solution to sequencing problems within ESTs. Hence, this is the link to the genome if the complete sequence of the genome is lacking.

1.5.4 Microarray

Limitations for improvements or development of salt-tolerant crops are the partial understanding of physiological criterion that reveals the genetic prospective and/or hereditary restrictions in combination with conventional breeding under abiotic stress-growing medium. But the turning point in understanding the key traits responsible for reducing productivity under abiotic stress is the meaningful approach to assimilate the plant's responses at the physiological as well as molecular level. Combining this approach with traditional breeding will boost the improvements in productivity (Araus et al., 2002). This will also help to produce crops suitable for our environment and for sustainable agriculture. "Omics" (metabolomics, transcriptomics, proteomics, and genomics) is involved throughout stress responses in plants (Saeed et al., 2012). A number of abiotic stress responsive genes with known structure and functions have been identified after the advent of extensive projects on genomics and helped to unravel the complexity of many associated genes (Hamel et al., 2006). In addition to identifying genes, high-throughput studies are aimed at the regulation of genes at the transcriptional level (Yanik et al., 2013). Proteome analyses reveal several proteins as showing peculiarity in their expression patterns under drought conditions (Zadraznik et al., 2013). Careful interpretation of these analyses has brought to light various mechanisms that operate during drought and other environmental challenges at both the molecular and cellular level. Recent advances in genomics research have progressively contributed to enhance the genetic improvements of the agriculturally important crops and a number of other promising plant species.

Microarray technology has been used over the past 20 years to identify the expression of high-throughput genes or to measure the expression of thousands of genes in a single experiment. These studies may be of cDNA inserts, fragments of short 20-25mers, or long oligonucleotides of 60-80mers (Lockhart and Winzeler, 2000). The model plant *Arabidopsis* played a key role in the identification of many stress-induced genes. But now this technology is being used to identify other abiotic stress-related genes in a number of plant species such as cotton (Arpat et al., 2004; Udall et al., 2006; Zhang et al., 2008; Payton et al., 2011; Barozai and Husnain, 2012), model plant *Arabidopsis* (Bray, 2002), rice (Gorantla et al., 2007; Hadiarto and Tran, 2011), beans (Micheletto et al., 2007; Sanchez et al., 2011), alfalfa (Kersey, 2004), and many others (Oh et al., 2010).

A variety of microarray platforms is available, i.e., oligonucleotide and cDNA. Oligonucleotide microarray consists of an abundance of mRNA transcripts in some specific tissue or cell for differential expression of several genes at one time. The "array or grid" of linearized complementary DNA from genes under investigation is plotted onto the glass slides. A single slide may have an array of thousands of spots. Private companies (Affymetrix, Agilent, and NimbleGen) are also available to design and spot the arrays. The in-house printing may be cheaper than from the private companies. Spotted slides are hybridized with the mRNA extracted from the sample to be investigated. After hybridization, the expression of the differentially expressed genes will be investigated by correlating the spots of complementary DNA and abundance of mRNA transcript from the labeled sample (Lipshutz et al., 1999). Complex and extensive computational and statistical analysis is compulsory for the exploration of data from such experiments.

Several batches of cDNA- or EST-based microarray data have developed in cotton for fiber, plant growth and development, abiotic stresses, and pathogen-related genes (Zhang et al., 2008; Rodriguez-Uribe et al., 2011). Cotton fiber genes have been studied at different developmental stages such as 10 and 24 days post-anthesis and it was found that gene expression was altered progressively from initiation to elongation. Microarray results showed the developmental differences at primary to secondary cell wall metabolism. The number of genes up- or down-regulated confirmed that this process is associated with the plant developmental stage (Arpat et al., 2004). A number of ESTs have been collected to generate cDNA microarray from different plant species subjected to a number of abiotic stresses such as freezing, salt, water deficit, and metal toxicity. Microarray was developed by spotting unigene, and a transcript profile was done to classify the genes expressing under subjected abiotic stresses. Functional categorization of a number of candidate genes expressing have been performed and the method was found to be a useful tool for sequence annotation and transcriptome analysis (John and Spangenberg, 2005).

Identification and characterization of the genes involved in different metabolic processes have facilitated understanding of the basic molecular mechanism involved and thus breeders can improve the genetic capability of the crops (Parida and Das, 2005; Yokotani et al., 2013). Certain regulatory pathways are common and shared across the species, such as a microarray of certain species of wheat showing expression of enzymes and osmolytes under water deficit, which suggests that these mechanisms are genotype, duration, severity, and type of stress exposure dependent (Mohammadi et al., 2007). Insufficient identification of transcripts sometimes means that there is scarce selection of a cDNA array based on EST selection. For this solution, large-scale EST selection and sequencing from different tissues at different developmental growth stages are required (Cushman and Bohnert, 2000). There is evidence that many of the genes identified through microarray have confirmed the tolerance of abiotic stress in transgenic plants (Bar et al., 2013). For those species where representative array is not available for analysis or comparison, cross-species hybridization and analysis are now available on the publicly available repositories. Model species such as *Arabidopsis*, rice, etc., play an important role in this limitation (Pariset et al., 2009), as whole genome of *Arabidopsis* was complete in 2000, so all the predicted genes could be monitored for expression profiles in a single microarray experiment at one time (Hirayama and Shinozaki, 2010). In a broader sense, the abiotic stress-induced genes identified through microarray can be classified as the first group of proteins (heat shock proteins and late embryogenesis abundant proteins) transcribed directly under abiotic stress. The second group consists of the regulatory proteins related to signal transduction pathways (MAPK, enzymes, and transcription factors). These functional groups of genes competently stimulate the cellular machinery of plants to respond to abiotic stress.

There are public repositories, like GEO (Gene Expression Omnibus) and Array Express, where microarray data can be deposited. These databases are easily available and accessible to the public, which is helpful for further hypothesis and research findings that may lead to major conclusions (Mah et al., 2004; Sanchez et al., 2011; Tseng et al., 2012). Table 1.2 summarizes the different

Table 1.2 Summary of the Record of GEO Datasets for Different Crops under Abiotic Stresses

GDS record	Organism	Tissue Type	Stress
GDS1914	*Hordeum vulgare*	shoot tissue	salinity
GSE3097	Barley	shoot tissue	salinity
GSE13921	*Medicago truncatula*	seedling	salt stress
GSE14029		seedling	salt stress
GSE13907		NM*	salt stress
GSE10942	*Bruguiera gymnorhiza*	roots	salt stress
GSE7228		NM	salt/osmotic/ionic
GSE16401	Tomato	NM	salt stress
GSE8883		cotyledon/shoot	salt stress
GSE4961		seedling	salt stress
GSE31594	*Vitis vinifera*	NM	salt stress
GSE2981	*Thellungiella*	NM	drought/cold/high salinity
GSE4681	*Chlamydomonas reinhardtii*	NM	salt stress
GSE671	Maize	NM	UV-B
GSE33494		NM	location change
GSE28209	Rice	NM	salinity
GSE2415		NM	salt/drought
GSE32065		leaves	abiotic
GSE31974		NM	drought
GSE11175		NM	salt stress
GSE14403		roots	salt stress
GSE8380		NM	pathogen
GSE7530		shoot/seed/panicle	physiological
GSE6600		leaves	salt stress
GSE6901		seedling	drought/salt/temperature
GSE6533		leaf, shoot, panicle	drought/salt stress
GSE8064	Wheat	seedlings	salt stress
GSE8060		NM	salt stress
GSE8158	*Solanaceous* spp.	NM	salt stress
GSE10338	Potato	tuber	heat necrosis
GSE18053		leaves	salt stress
GSE8205		root/leaves	cold/heat/salt
GSE2416		NM	heat/cold and stress
GSE8554	Chickpea	leaf/root/shoot	drought/cold/salinity
GSE7504		NM	drought/cold/salinity
GSE7418		leaf/root/shoot	salt stress
GSE9748	Poplar	NM	salt stress
GSE10874		NM	ozone
GSE10932		NM	ozone
GSE27861	*Arabidopsis thaliana*/rice	NM	drought/salt stress
GSE13165	Rice/*Arabidopsis*	NM	multiple (salt/heat/light/UV)

(Continued)

Table 1.2 (Continued)

GDS record	Organism	Tissue Type	Stress
GSE29941	*Arabidopsis thaliana*	seeds/seedlings	hypoxia
GSE26983		NM	salt stress
GSE16765		NM	salt stress
GSE21553		roots	iron
GSE13803		NM	wound
GSE18217		NM	salt stress
GSE19893		NM	salt stress
GSE15063		NM	osmotic
GSE10349		NM	drought/salt
GSE10384		NM	drought/salt stress
GSE10670		NM	drought
GSE8556		seedling	salt stress
GSE8787		roots	salt stress
GSE10420		seedling	salt stress
GSE9459		NM	salt stress
GSE8936		NM	drought
GSE7942		NM	xenobiotic
GSE7743		NM	irradiance
GSE5623		NM	salt stress

*NM: not mentioned.

plant species under salt and/or in combination with multiple abiotic stresses as reported on the GEO Data Set record (http://www.ncbi.nlm.nih.gov/gds). There are certain standards for data submission like experimental design, data input, normalization, and data annotation, so that scientists from different backgrounds can access the data in the same pattern and genes have been identified related to responsiveness to abiotic stresses alone or in combination with different plant species. The significance of microarray technology can be further strengthened due to the publication and availability of 88,696 free full text articles over the last 5 years (http://www.ncbi.nlm.nih.gov/gquery/?term = microarray) compared with only 4800 articles in 2000–2003 (Mah et al., 2004).

The use of microarrays in future functional genomics tools will probably lead to gene regulation at the transcriptional level for different tissues in cotton and other agriculturally important crops. However, these microarray-based methods are expected to be innovative when they become an essential part of the research/experimental activities of a routine molecular biology laboratory.

1.5.5 Proteomics and metabolomics

Plants will always be open to environmental threats as they are exposed to various abiotic stresses. Adaptation to these stresses is a complex process at the cellular and physiological level, and the acquired modifications in gene expression monitor the alterations in protein profiles (Parker et al., 2006).

Proteomics and metabolomics are comparatively new research areas to emerge from the post-genomic era and there are only a few published reports on these applications to abiotic stress tolerance (Bahrman et al., 2004). Hence, these innovative "omics" perspectives have provoked curiosity among genomics scientists and opened up new horizons in plant stress biology (Khan et al., 2007). While transcriptomic studies are in progress to unravel gene transcription and regulation, gene functions and expression profiles can certainly be accomplished by proteomic study, as the primary protein sequences undergo proteolysis after post-translational modification. Accordingly, protein-level quantification of expressed genes is essentially done to measure the plant's response to abiotic stress.

Proteomics techniques such as GC-MS and 2D gel electrophoresis are well known for protein profiling. A number of key proteins could be identified and quantified in a single gel for a specific subject area. For example, Salekdeh et al. (2002) identified 3000 proteins from rice drought and salt-stressed plants from a single gel: over 1000 were quantified among those and 42 were found worthy of further study related to their abundance or position in response to the subjected stress. Defined tissues of wheat resolved significant proteins through proteomic studies. Those proteins identified by proteomic methods have proven their functional involvement in abiotic stress tolerance mechanisms (Skylas et al., 2005; Woo et al., 2003).

Abiotic stresses impact the plant's physiological processes such as photosynthesis—the effects on the Calvin cycle and results in reduction of plant growth and development. To adapt or to cope with these conditions, protein profiles regulating the mechanism of glycolysis and amino acid biosynthesis are up-regulated for ATP synthesis and osmotic adjustments. Molecular chaperones and C_4 photosynthesis encoding proteins are up-regulated as defense mechanisms against salt in maize, and such genes may be the valuable source for genetic transformation to induce salt tolerance to other agronomically important crops. Soybean and potatoes are relatively salt sensitive and up- or down-regulation of certain protein profiles suggests that these crops are sensitive to photosynthesis and protein biosynthesis-related proteins, but tolerant to mechanisms like water potential, osmotic adjustments, and antioxidants (Sobhanian et al., 2011).

Plasma membrane plays a key role in regulating several cellular processes; therefore, it is considered to be a prime location of damage after abiotic stress such as cold. Freezing or cold is also a serious threat to crops especially to the cold-sensitive/susceptible species that do not adapt to severe cold temperatures due to limited functions to modify the cellular machinery. Tolerant species may have the adaptation mechanism as they may alter the plasma membrane, and hence many key proteins may be involved in the pathway. So, proteomic skills are useful to identify the regulatory and functional protein in the freezing mechanism, and identification of those candidate genes may help to alter the plasma membrane functions under low temperature stress. Hence, this research application is of key importance for plant molecular breeding strategies to induce the low temperature tolerance mechanism in crop plants (Takahashi et al., 2013). To induce abiotic stress tolerance to crops, different plant species should be identified for different types of stresses, and the approaches for tolerance levels should be listed. This would be accomplished with rigorous research on each plant species' response to specific stresses. The evaluation of susceptible and tolerant cultivars/species should be done for proteome profiling in order to reveal the distinctive mechanisms involved in this challenge.

Plant growth and development is affected by exposure to salts and plants may experience various symbiotic restraints; these include water deficit, ionic toxicity, and oxidation stress, which lead to metabolic and nutrient discrepancy that culminates in a complex physiological disorder.

Metabolomics is another area in the genomics approach that deals with the metabolic changes in plant responses to abiotic stresses, which suggests that thorough profiling of metabolites may offer enormous understanding for stress tolerance pathways. This is a relatively innovative area and there are not too many reports on its application. However, one of the available reports found 88 main metabolites from the extract of rice leaves, most of them covered pathways of sugar and amino acid metabolism (Sato et al., 2004). This was further confirmed by Sanchez et al. (2010), in that polyols, sugars, and specific amino acids are accumulated and usually termed as compatible solutes or osmolytes for osmotic adjustments. Further, these metabolites are also involved to regulate physiological responses such as ROS, cell membrane stability, protein protection from degradation, and cell signaling. Accumulation or decreased metabolites is also specific for distinct species, so it may help to identify the stress-susceptible or -tolerant species (Szabados and Savouré, 2010).

Carbohydrates and secondary metabolites play key roles during the post-harvest of fruit crops. Understanding the molecular pathways involved in the shelf-life of fruit can help to improve fruit quality (softening and ripening). Sorbitol concentration was reduced due to down-regulation of sorbitol dehydrogenase, and sugar was increased due to up-regulation of sucrose synthase in the transgenic apples harboring the aldose-6-phosphate reductase (*A6PR*) gene with antisense orientation, which led to improvement in photosynthesis and CO_2 that produced good vegetative growth (Zhou et al., 2006). Sucrose contents were reduced with increased polymerization of pro-anthocyanidins with improved volatile compounds, such as carotenoid and shikimate-derived volatiles in grapevines-overexpressed *Adh* alcohol dehydrogenase, known to be regulated progressively under environmental threats. Strawberry, due to its small genome and faster growth, is a preferred model among fruit crops for functional genomics. A small subunit of ADP-glucose pyrophosphorylase (AGPase) was regulated under the promoter ascorbate peroxidase (APX), which overexpressed the pyrophosphate mechanism. Assessing the metabolic pathway reveals that fructose-6-phosphate 1-phosphotransferase (PFP) is a cytosolic enzyme that catalyzes the glycolysis, which leads to the accumulation of sugars and degrades the starch contents in transgenic plants (Basson et al., 2011).

The flavanoid compound family contains one of the water soluble pigments—anthocyanin—which are responsible for the color trait in fruits and flowers. These flavonoids are regulated and expressed through MYB well-known transcription factors. Strawberry transgenic plants harboring *FaMYB10* showed enhanced levels of anthocyanins in different plant parts, but the silencing of anthocyanidin synthase (MdANS) in red-leaved apple produced necrotic leaves, and flavonoids/anthocyanins-altered metabolic pathway was observed (Szankowski et al., 2009b; Lin-Wang et al., 2010). Transgenic plants have been used to analyze the flavonoids and other metabolic profiling, up- or down-regulated genes, and associated metabolic pathways, and this study might be helpful to unravel the unknown genes and metabolites involved during the plants' stress challenges. Hence, the genome sequences of many plant species are available and some of them are important as they contain genes involved for the production of particular components completely or partially known or unknown. Therefore this may likely to expose regulatory or functional mechanisms relevant to functions of metabolites in whole or plant parts.

1.5.6 Transgenic plants for improved salinity tolerance

Abiotic stress causes significant plant yield losses and hence lowers crop production (Orsini et al., 2010). In addition to desiccation, ionic toxicity and ROS stress plants also face metabolic and nutrient mismanagement and all these lead to different physiological responses (Ashraf, 2004; Sanchez et al., 2011). Therefore, production of crops tolerant to stress is the ultimate and important solution for sustainable agriculture to meet the demands of the growing population (Aquino et al., 2011). Molecular biology, genomics, or biotechnology applications in combination with plant breeding are being successively used to produce transgenic crops/plants to overcome this crop productivity threat. Genomics approaches have provided insights into the functional regulation of the key genes participating in the stress tolerance mechanism of plants. New attempts have been reported by using genomics approaches such as marker-free selection of plants, tissue-specific promoters, and induction of disease resistance (fungal, bacterial, and viral) (Gambino and Gribaudo, 2012). Table 1.3 summarizes a number of genes identified individually or in combination with abiotic and biotic stress tolerance in different plants.

Plants are adapted to saline environmental conditions in a number of ways and among these is the compartmentalization of Na^+ into the vacuole, which helps to maintain the osmotic balance. Transgenic legume forage (*Medicago sativa*) was developed with *AVP1*, a vacuolar H^+-pyrophosphatase (H^+-PPase) gene from *Arabidopsis thaliana*, to adapt to saline and arid soils (Bao et al., 2009). Transgenic plants were developed which expressed the *AtNHX1* gene (a vacuolar Na^+/H^+ antiporter). This gene has been reported to reduce the cytosolic Na^+ by sequestering Na^+ in the vacuole (Asif et al., 2011; Jha et al., 2011). The transgenic plants were produced with better germination rates and showed improved fresh and dry biomass in severe saline conditions (Xue et al., 2004). Another plasma membrane Na^+/H^+ antiporter, i.e., the SOS1 gene from *Arabidopsis thaliana*, was tested into the model plant tobacco and showed improved germination, vegetative growth, and chlorophyll contents in transgenic plants as compared to the non-transgenic (Pons et al., 2011; Yue et al., 2012). By reducing the expression of the sodium/proton antiporter SOS1 through genetic engineering affects the numerous pathways, indicating a role for SOS1 that exceeds its known function as an antiporter (Oh et al., 2009, 2010). Tissue-specific expressions for the SOS genes may contribute a significant role to Na^+ regulation (Kumar et al., 2009). The higher amount of Na^+ and Cl^- separately under NaCl treatment restricts the plants' growth and productivity through different but simultaneous mechanisms. It has been documented that a higher level of Na^+ hampers K^+ and Ca^{2+} nutrients whereas a high concentration of Cl^- degrades the chlorophyll and decreases the photosynthesis (Tavakkoli et al., 2010).

Cyclophilin (Cyp) genes belong to the ubiquitous proteins family and are capable of peptidyl-prolyl isomerase activity, which helps to catalyze the *cis/trans* isomerize, the peptide bond at proline residues and protein folding. These Cyps have been reported to regulate gene functions at cell division, signaling, transcriptional regulation, and pre-mRNA splicing under different environmental stresses including temperature, salt, and drought (Zhu et al., 2011a). There is the possibility that some of the genes perform vital roles to acclimatize and tolerate the stressful environment, but this is not likely to conclude the stress transcriptome of a genus on the basis of only one species (Sanchez et al., 2011). Proline and trehalose accumulation has been reported in response to plant adaptation, and related genes expression has been found increased and this could be applied to help the plants cope under abiotic stresses (Nounjan et al., 2012). The MtSAP1 gene isolated from *Medicago truncatula* has been found overexpressing in tobacco transgenic seedlings and showed tolerance to extreme temperature, osmotic, and salt stresses (Charrier et al., 2013).

Transcription factors are considered as the potent application for the plants to adapt under different abiotic stresses (Martin et al., 2012; Mizoi and Yamaguchi-Shinozaki, 2013; Naika et al., 2013).

Table 1.3 Abiotic Stress-tolerant Genes Identified in Different Crops

Gene Symbol	Gene Description	Stress Tolerance	Gene Response
Oryza sativa			
Os01g0798500	Os01g0798500	Salt	Malate synthase-like family protein—contains InterPro domain(s)
Triticum Aestivum			
LOC543151	NADPH oxidase	Salt	NADPH oxidase
HO1	Heme oxygenase 1	Salt	Endogenous signaling system in animals
Solanum Lycopersicum			
AGP-S1	ADP-glucose pyrophosphoryl as large subunit	Salt	Glyco_tranf_GTA_type; Glycosyltransferase family A (GT-A)
AIM1	ABA-induced MYB transcription factor	Salt	MYB transcription factor
Zea Mays			
LOC100283283	Serine/threonine-protein kinase SAPK8	Salt	Serine/threonine-protein kinase SAPK8
Arabidopsis Thaliana			
MPK4	Mitogen-activated protein kinase 4	Salt, pathogens	MAP kinase activity
MPK3	Extracellular signal-regulated kinase 1/2	Touch, salt, cold	Encodes a mitogen-activated kinase
SOS4	SOS4 pyridoxal kinase	Salt	Pyridoxine biosynthetic process
RSR4	Pyridoxal biosynthesis protein PDX1.3	Salt	Pyridoxine biosynthesis
GSK1	Shaggy-related protein kinase	Salt, ABA	GSK3/shaggy-like protein kinase
ddf2	Dehydration responsive element-binding protein 1E	Drought, salt	Encodes DREB subfamily A-1 of ERF/AP2 transcription factor
DDF1	Dehydration responsive element-binding protein 1F		
P5CS2	Gamma-glutamyl phosphate reductase	Drought, salt, ABA	Encodes delta 1-pyrroline-5-carboxylate synthetase B
SLT1	Protein sodium and lithium tolerant 1	Salt	Molecular function unknown
P5CS1	Gamma-glutamyl phosphate reductase	Drought, ABA, salt	Encodes a delta1-pyrroline-5-carboxylate synthase
CBL10	Calcineurin B-like protein 10	Salt	Encodes calcineurin B-like calcium sensor
RCI2A	Hydrophobic protein RCI2A	Cold, ABA, drought, salt	Encodes a small highly hydrophobic protein

(Continued)

Table 1.3 (Continued)

Gene Symbol	Gene Description	Stress Tolerance	Gene Response
BZIP24	Basic leucine zipper 24	Salt	bZIP transcription factor family
BZIP17	Basic helix-loop-helix		
SOS3	Calcineurin B-like protein 4	Salt	Calcium sensor, essential for K^+
NCED3	9-cis-epoxycarotenoid dioxygenase NCED3	Drought, salt	Encodes 9cis-epoxycarotenoid dioxygenase
VSP2	Vegetative storage protein 2	ABA, salt wound, drought	Acid phosphatase and anti-insect activity
CESA1	Cellulose synthaseA catalytic subunit1UDP forming	Salt	Encodes a cellulose synthase isomer
DELTA-Oat	Ornithine-delta-aminotransferase	Salt	Encodes ornithine-delta-aminotransferase
ZF1	Zinc finger protein 1	ABA, salt, drought, cold	Zinc finger protein, mRNA levels are up-regulated
ZF2	Zinc finger protein 2		
ZF3	Zinc finger protein 3		
HAL3A	Phosphopantothenoyl cysteine decarboxylase	Salt	Encodes for phosphopantothenoyl cysteine decarboxylase
ATHAL3B	phosphopantothenoyl cysteine decarboxylase		
RCI3	Peroxidase 3	Drought, cold, salt	RCI3 accumulates in the aerial plant parts and roots
ASN2	Asparagine synthetase 2	Salt	Asparagine biosynthetic process
CP1	Ca^{2+}-binding protein 1	Salt	Novel calcium ion binding
HVA22A	HVA22-like protein a	ABA, cold drought, salt	Molecular function is unknown but protein expression is ABA and stress inducible
HVA22B	HVA22-like protein b		
HVA22D	HVA22-like protein d		
HVA22E	HVA22-like protein e		
AT4G30650	Low temperature and salt responsive protein family	Cold, salt, fungus	Low temperature and salt responsive protein
AT4G30660			
AT2G38905			
AT2G24040			
AT1G57550			
CEST	Hypothetical protein	Multiple stress	Encodes chloroplast protein
PLDDELTA	Phospholipase D delta	Drought, cold, salt	Phospholipase D activity
GAI	DELLA protein GAI	Salt	GA-mediated signaling
RGA1	DELLA protein RGA		
RGL1	DELLA protein RGL1		
RGL2	DELLA protein RGL2		
GH9A1	Endoglucanase 25	Salt	Membrane-bound endo-1,4-betaD-glucanase

(Continued)

Table 1.3 (Continued)

Gene Symbol	Gene Description	Stress Tolerance	Gene Response
TIL	Outer membrane lipoprotein Blc	Temperature, salt	Temperature-induced lipocalin
SIP3	CBL-interacting serine/threonine-protein kinase 6	salt	Encodes CBL-interacting protein kinase 6 (CIPK6)
PTR3	Peptide transporter PTR3-A	Salt	Dipeptide transporter activity
UGT74E2	Uridine diphosphate glycosyltransferase 74E2	Multiple stresses	Encodes UDP-glucosyltransferase
AT1G18260	Hr.3-like protein	Salt	HCP-like superfamily protein
S1P	SITE-1 protease	Salt	S1P functions as a Golgi-localized subtilase
LTL1	GDSL esterase/lipase LTL1	Salt	Encodes Li-tolerant lipase1
NIG1	Transcription factor bHLH28	Salt	NaCl-inducible gene1
AT4G30996	Hypothetical protein	Salt	Protein of unknown function
SAT32	Interferon-related developmental regulator domain protein	Salt	Encodes a protein with similarity to human interferon-related developmental regulator
HKT1	Sodium transporter HKT1	Salt	Expressed in xylem parenchyma
AOX1A	Alternative oxidase 1A	Cold, salt	Encodes AOX1
THI1	Thiazolebiosynthetic enzyme	Salt	Encodes a thiamine biosynthetic gene
MBF1C	Multiprotein-bridging factor 1c	Multiple stress	Encoding multiprotein bridging factor 1
NPC4	Phospholipase C	Salt, drought	Phosphoesterase family protein
NAC069	NAC domain-containing protein 69	Salt	NAC domain-containing protein 69
NAC3	NAC domain-containing protein 55	Salt, drought	ATAF-like NAC domain transcription factor
CSD1	Superoxide dismutase Cu/Zn	Multiple stresses	Encodes a cytosolic copper/zinc superoxide dismutase CSD1
GRP2	Glycine-rich protein 2	Salt, cold	Encodes a glycine-rich protein
GA2OX7	Gibberellin 2-beta-dioxygenase 7	Salt	Protein with gibberellin 2-oxidase activity
HAK5	Potassium transporter 5	Salt	Protein of the KUP/HAK/KT, K^+ channel
MSS1	Sugar transport protein 13	Salt	Hexose-specific/H^+ symporter activity
CSD2	Superoxide dismutase [Cu/Zn]	ROS	Chloroplastic copper/zinc superoxide
AT5G62040	Protein BROTHER of FT and TFL1	Salt	PEBP (phosphatidylethanolamine-binding protein)
PLC1	Phosphoinositide phospholipase C1	Salt, cold, drought	Phosphatidylinositol-specific phospholipase C is induced under environmental stresses

(Continued)

Table 1.3 (Continued)

Gene Symbol	Gene Description	Stress Tolerance	Gene Response
AT3G12630	Zinc finger A20/AN1 domain containing stress protein 5	Salt, cold, drought	A20/AN1-like zinc finger family protein
CNBT1	Cyclic nucleotide gated channel	Salt pathogens	Cyclic nucleotide-binding transporter 1
CNGC19	Cyclic nucleotide gated channel 19		Member of cyclic nucleotide gated channel
ZFHD1	Zinc finger homeodomain 1	Drought, salt, ABA	Zinc finger homeodomain transcription factor

Transcription factors genes of the basic leucine zipper (bZIP) and zinc finger protein regulate the pathways responsible for plant growth under abiotic stress and metal toxicity. *ZmbZIP72*, a transcription factor from maize bZIP, has showed improved tolerance in *Arabidopsis* under different types of abiotic stresses (Ying et al., 2012). ABI5 is a transcription factor of the bZIP family, which binds with the ABA response element (ABRE) present in AtEm6 gene, hence regulating its expression. AtEm6 is reported to increase salt tolerance by manipulating calcium-dependent protein kinases (Tang and Page, 2013). The NAC is found to be one of the largest transcription factor families and its proteins are characterized by a highly conserved DNA-binding domain. These are abundant only in plants and perform key roles during different plant developmental stages. TaNAC2, an NAC transcription factor, was isolated from wheat and then overexpression was observed in the model plant *Arabidopsis* under different abiotic stresses including salt, drought, ABA, etc. (Mao et al., 2012). The DREB-1A transcription factor gene has enriched the drought tolerance in transgenic wheat (Shen et al., 2003). Another illustration is the overexpression of ornithine aminotransferase for salt- and water-deficit tolerance in *Arabidopsis*, which further has been transformed into wheat to induce the tolerance mechanism (Roosens et al., 1998). Another transcription factor gene from the ethylene responsive factor *W6* gene from wheat showed overexpression in tobacco and improved the tolerance mechanism under saline stress condition.

The changes in biochemical and physiological parameters observed in transgenic plants compared with non-transgenic plants found the up-regulation of transgene in transgenic plants (Yan et al., 2008). The Stress responsive Transcription Data Base (STIFDB) is a collection of identified stress-related signals, responsive genes, and transcription factors. It catalogues all the available information on putative binding sites of the stress responsive transcription factors in the model species *Arabidopsis* (Shameer et al., 2009). The updated version of this database is STIFDB2, which contains supplementary information related to other agriculturally important crop species, i.e., maize, sorghum and soybean, novel stress-related signals, newly identified transcription factors and their regulatory sites, and addition of the stress responsive genes from microarray-based experiments (Naika et al., 2013). All this information will be helpful for plant and computational biologists to further understand plants' stress-related mechanisms.

Induction for production of certain enzymes like glutamate synthase, proline, or regulation of organic solute such as proline and glycinebetain and antioxidant enzymes under salt and other abiotic stresses is regulated by signal transduction pathway (Misra and Saxena, 2009; Yang et al., 2013). Salicylic acid is reported to be part of the signaling pathway and considered to take part in improving the functions related to photosynthesis, which ultimately would help to improve the contrary effects of salts and thus help to improve the plant defense mechanism by improving the physiological, metabolic, and biochemical pathway (Idrees et al., 2011). The combined effect of cassava Cu/Zn superoxide dismutase (MeCu/ZnSOD) and catalase (MeCAT1) improved cytosolic expression and maintained the ROS scavengers by showing improved shelf-life of cassava roots to combat abiotic stress. SOD, CAT, proline accumulation, and water-related attributes were improved and lowered the malendialdehyde in transgenic cassava under cold stress (Xu et al., 2013).

Ethylene production has a key function at post-harvest and affects the shelf-life of fruits. This induces several genes altogether, which transcribes and regulates the functional genes affecting the storage of fruits. *ACS* (ACC synthase; ACC-1-aminocyclopropane-1-carboxylic acid) and *ACO* (ACC oxidase) are the basic enzymes known for ethylene biosynthesis. Transgenic apple plants silenced with either of these enzymes produced less ethylene due to suppression of volatile ester synthesis, and fruit maturity features were observed (Dandekar et al., 2004; Johnston et al., 2009). *ACO* suppression was done by RNAi in kiwi and papaya fruits and volatile production was reduced, which led to reduced ethylene production and improved shelf-life of the fruits thereby extending and fruit storage (López-Gómez et al., 2009; Atkinson et al., 2011)

Plant growth and development under abiotic stresses is also regulated by the hormonal activation and in this regard brassinosteroid (BR) is reported to interact with ABA; various BR receptive genes are also responsive to ABA (Cui et al., 2012). These BRs are steroidal compounds, ubiquitous, distributed in free form, and conjugated to starch and lipids and thus help the plants to adapt under salt and other abiotic stresses (Hayat et al., 2012). ABA is considered a stress regulating plant hormone and plays a significant role in changing gene expression profile and cellular responses, as this is the most studied plant hormone under abiotic stress. NCEDs (9-cis-epoxycarotenoid di-oxygenases) and P450 CYP707As are the two enzymes involved in ABA biosynthetic and catabolic pathways, respectively, and their genes are activated after abiotic stresses. When the plants are growing under normal conditions, ABA exists within vacuoles and apoplasts as ABA glucosyl ester, which is an inactive form released by the action of β-glucosidase when the plants are dehydrated (Hirayama and Shinozaki, 2010).

There are other reports suggesting that some of the enzyme-related genes like betaine aldehyde dehydrogenase (BADH), pyrroline-5-carboxylate synthetase (P5CS), mannitol-1-phosphodehydrogenase (mtlD), 6-sorbitol dehydrogenase phosphatase (gutD), and late embryogenesis LEA, HVA1, and ME-leaN4 have been overexpressing among different crops under different stress condition (Swire-Clark and Marcotte, 1999; Prabhavathi et al., 2002; Sawahel and Hassan, 2004; Oraby et al., 2005; Park et al., 2005; Yan et al., 2008).

Even though inducing abiotic stress tolerance in plants through genetic engineering is quite successful in many agriculturally important crops, this approach will simply be authenticated after validation of the results of successive generations produced in field trials. When plants are grown as part of the plant community under field conditions and are subjected to multiple stresses, then various essential abiotic stress reactions induced in plants relate to their performance as a crop. Thus, the initial outcomes of the genes from model species are still to be studied to reach factual consideration.

1.6 **Conclusion and future perspectives**

Increased abiotic stresses are the major threats to agricultural crops, which limit plant growth and productivity. Plants respond to these stresses at molecular and cellular levels, which makes modifications to the morphological, physiological, biochemical, and molecular behavior of the plants. The complex nature of abiotic stress responses studied so far advocate that further expansion of discrete stresses at the individual level may not be the appropriate approach. Traditional breeding has introduced crop varieties with desirable characteristics, but the limitations associated with this restrict agricultural scientists to adopt this further. Genomics approaches now empower the study of plant responses to abiotic stresses.

The marker-assisted breeding approach is a potential substitute to conventional breeding with regard to time, labor, and cost effectiveness. A number of QTLs have successfully been identified for qualitative and quantitative characteristics in many crop species. This has improved the knowledge related the stress mechanisms and some crops have exhibited excellent performance for specific traits. However, the challenges related to this technology are the preciseness of the QTL identification, the genetic environment interaction, reproducibility of the results, numerous yield-related genes, and moreover germplasm restrictions, which may restrict scientists' vigilance when using this approach. However, positional cloning may be the key challenge as QTL cloning can offer a more meaningful and reliable marker for breeding of superior allelic variation in crop species.

Plant species with longer genomes have not been sequenced due to the presence of copies of large repeats of genes lacking information of coding genes, while the purpose of sequencing is to obtain the information for discovery and characterization of the genes for proteomics. Expressed sequence tags (ESTs) helped to overcome this limitation and presently this is most abundantly sequenced nucleotide product from the plant genomes. A number of sequences from the nucleotides of models as well as several other plant species are accessible on publicly available databases—hence supporting the high-throughput technologies for transcriptome analysis. Therefore, at present, ESTs offer a high-throughput robust sequence resource to exploit gene discovery. ESTs are the key components for the cDNA microarray—the high-throughput experimentation for gene identification from certain tissues at specific developmental stages under abiotic stress. The limitations of this technology are the general demonstration of representative genes in the cDNA library and the specific features of different sequences within the collection of clones. The solution for real transcriptome analysis is the complete genome sequence, but ESTs with appropriate bioinformatics studies help to overcome the limitations.

Microarray is a high-throughput technology for gene expression analysis that consists of probes demonstrating several different genes organized on a glass slide in a systematic pattern. Several plant stress-related genes have been identified for expression profiles and functionally categorized under specific conditions. Probes are used as cDNA or oligonucleotides and the several thousands of array results are available on databases. The limitation associated with the microarray is its cost in terms of materials, design, expertise, and reproducibility of results; therefore, microarray experiments are not appropriate for small-scale funded projects. A great deal of expertise is required to design the experiment and for data analysis. Bioinformatics applications are required for *in silico* experimentation and conclusions. Gene expression analysis compared with other techniques such as real-time PCR would be more specific, which may be due to non-specific hybridization. However, the microarray is very sophisticated technology and the applications are vast, so the limitations may be overcome by considering the experimentation more accurately.

Succeeding transcriptome analysis or gene identification, significant efforts are now in progress to explore the gene expression process or their response under various abiotic stresses. So the postgenomic era in crop research now uses proteomics approaches to identify the proteins that are regulated in response to environmental threats. These studies found that proteins are expressing to change the physiological mechanisms, and plants from different species showed adaptive strategies under abiotic stresses including carbohydrate synthesis, protein synthesis, ATP formation, C_4 photosynthesis, and heat shock protein profiles for ROS scavenging, etc. Proteomic studies should be comprised of assessments of susceptible and tolerant cultivars to reveal the specific mechanisms involved in stress tolerance. This will lead to the engineering of stress tolerance into agricultural crops.

Metabolic changes upon abiotic stress response adaptation involve accumulation of or increase in the levels of particular stress responsive metabolites such as amino acids, total soluble sugars, and polyols, which are commonly termed compatible solutes. The physiological role of stress responsive metabolites which are precisely altered remain to be examined. The extent of metabolic modifications within a genotype reveals the intensity of stress experienced by the plant. Several enzymes are key components for metabolic changes and further exploration of the genes expressing for these adaptations is under way.

Genetic transformation has made real progress to understand how plants cope with the stresses as well as alterations to the signaling and response pathways. Transgenic plants harboring genes related to organic solutes, antioxidant enzymes, molecular chaperones, and some of the transcription factors have the best specific stress tolerance under controlled conditions. Although some transgenic lines consist of the fully characterized genes and are performing best, most of them are transformed with a single stress-related gene whose function may not be properly understood. When the plants are grown in natural conditions, they have to combat a number of stress threats, so it is difficult for a transgenic plant transformed with a single gene and growing in controlled conditions to cope with multiple stresses for multiple exposures with specific severity. Therefore, a meaningful future approach for effective utilization of transgenic technology is to use gene pyramiding to stack multi-genic transformation technology. This should be further progressed with fully characterized genes' known functions and the transgenic lines should be field tested. Thus, the complete system of complex cell signaling mechanisms and vital interrelating practices that hint at multi-defense responses might be elaborated.

Acknowledgments

We are thankful to Mr. Sajjad Saddique, Ms. Sameera Hassan, and Miss Fatima Batool for their help in the literature survey.

References

Abrol, I.P., Yadov, J.S.P., Massiud, F.I., 1988. Salts affected soils and their management. Soil Res Manage Conserv Ser. FAO Land and Water Dev Div Bul 39.

Adams, M.D., Kelley, J.M., Gocayne, J.D., Dubnick, M., Polymeropoulos, M.H., Xiao, H., et al., 1991. Complementary DNA sequencing: expressed sequence tags and human genome project. Science 252, 1651−1656.

Adams, P., 1991. Effect of increasing the salinity of the nutrient solution with major nutrients or sodium chloride on the yield quality and composition of tomato grown in Rockwool. J. Hort. Sci. 66, 201−207.

Afroz, S., Firoz, M., Hayat, S., Siddiqui, M.H., 2005. Exogenous application of gibberellic acid counteracts the ill effect of sodium chloride in mustard. Turk. J. Biol. 29, 233−236.

Ahmed, F., Rafi, M.Y., Ismail, M.R., Juraimi, A.S., Rahim, H.A., Asfaliza, R., Latif, M.A., 2013. Waterlogging tolerance of crops: breeding, mechanism of tolerance, molecular approaches, and future prospects. Biomed. Res. Int. 2013, 963525.

Akram, M.S., Ashraf, M., Akram, N.A., 2009a. Effectiveness of potassium sulfate in mitigating salt-induced adverse effects on different physio-biochemical attributes in sunflower (*Helianthus annuus* L.). Flora 204, 471−483.

Akram, M.S., Ashraf, M., Shahbaz, M., Aisha, N.A., 2009b. Growth and photosynthesis of salt-stressed sunflower (*Helianthus annuus*) plants as affected by foliar-applied different potassium salts. J. Plant. Nutr. Soil. Sci. 172, 884−893.

Akram, M.S., Athar, H.U.R., Ashraf, M., 2007. Improving growth and yield of sunflower (*Helianthus annuus* L.) by foliar application of potassium hydroxide (KOH) under salt stress. Pak. J. Bot. 39, 769−776.

Aldesuquy, H.S., Ibrahim, A.H., 2001. Interactive effect of seawater and growth bio-regulators on water relations, abscisic acid concentration, and yield of wheat plants. J. Agron. Crop. Sci. 187, 185−193.

Apse, M.P., Aharon, G.S., Snedden, W.A., Blumwald, E., 1999. Salt tolerance conferred by overexpression of a vacuolar Na^+/H^+ antiporter in Arabidopsis. Science 285, 1256−1258.

Aquino, R.S., Grativol, C., Mourão, P.A.S., 2011. Rising from the sea: correlations between sulfated polysaccharides and salinity in plants. PLoS ONE 6, e18862.

Araus, J.L., Slafer, G.A., Reynolds, M.P., Royo, C., 2002. Plant breeding and drought in C_3 cereals: what should we breed for? Ann. Bot.—Lond. 89, 925−940.

Arpat, A., Waugh, M., Sullivan, J.P., Gonzales, M., Frisch, D., Main, D., 2004. Functional genomics of cell elongation in developing cotton fibers. Plant Mol. Biol. 54, 911−929.

Asada, K., 2000. The water-water cycle as alternative photon and electron sinks. Philos. Trans. R. Soc. Lond. B. Biol. Sci. 355, 1419−1430.

Ashraf, M., 2002. Salt tolerance of cotton: some new advances. Crit. Rev. Plant Sci. 2, 1−32.

Ashraf, M., 2004. Some important physiological selection criteria for salt tolerance in plants. Flora 199, 361−376.

Ashraf, M., 2010. Inducing drought tolerance in plants: recent advances. Biotechnol. Adv. 28, 169−183.

Ashraf, M., Athar, H.R., Harris, P.J.C., Kwon, T.R., 2008. Some prospective strategies for improving crop salt tolerance. Adv. Agron. 97, 45−110.

Ashraf, M., Qasim, A., Ashraf, M.A., 2012. Assessment of variation in drought tolerance using some key physiological criteria in potential wheat (*Triticum aestivum* L.) cultivars of different geographic origins. Arch. Agron. Soil Sci. 7, 437−454.

Asif, M.A., Zafar, Y., Iqbal, J., Iqbal, M.M., Rashid, U., Ali, G.M., 2011. Enhanced expression of AtNHX1, in transgenic groundnut (*Arachis hypogaea* L.) improves salt and drought tolerance. Mol. Biotechnol. 49, 250−256.

Asins, M.J., 2002. Present and future of quantitative trait locus analysis in plant breeding. Plant Breed 121, 281−291.

Atkinson, R.G., Gunaseelan, K., Wang, M.Y., 2011. Dissecting the role of climacteric ethylene in kiwifruit (Actinidia chinensis) ripening using a 1-aminocyclopropane-1-carboxylic acid oxidase knockdown line. J. Exp. Bot. 62, 3821−3835.

Azevedo-Neto, A.D., Prisco, J.T., Enéas-Filho, J., de Lacerda, C.F., Silva, J.V., PHA, da Costa, 2004. Effects of salt stress on plant growth, stomatal response and solute accumulation of different maize genotypes. Braz. J. Plant Physiol. 16, 31−38.

Baga, M., Chodaparambil, S.V., Limin, A.E., Pecar, M., Fowler, D.B., Chibbar, R.N., 2007. Identification of quantitative trait loci and associated candidate genes for low-temperature tolerance in cold-hardy winter wheat. Funct. Integr. Genom. 7, 53−68.

Bahrman, N., Negroni, L., Jaminon, O., 2004. Wheat leaf proteome analysis using sequence data of proteins separated by two-dimensional electrophoresis. Proteomics 4, 2672−2684.

Balint, A.F., Szira, F., Roder, M.S., Galiba, G., Borner, A., 2009. Mapping of loci affecting copper tolerance in wheat: the possible impact of the vernalization gene Vrn-A1. Environ. Exp. Bot. 65, 369−375.

Bänziger, M., Setimela, P.S., Hodson, D., Vivek, B. (2004) Breeding for improved drought tolerance in maize adapted to southern Africa. Proc 4th Int. Crop. Sci. Cong. Brisbane, Australia 26 Sep−01 Oct. Published on CDROM, https://www.cropscience.org.au.

Bao, A., Wang, S., Wu, G., Xi, J., Zhang, J., Wang, C., 2009. Overexpression of the Arabidopsis H^+PPase enhanced resistance to salt and drought stress in transgenic alfalfa (*Medicago sativa* L.). Plant Sci. 176, 232−240.

Bar, M., Leibman, M., Schuster, S., Pitzhadza, H., Avni, A., 2013. EHD1 functions in endosomal recycling and confers salt tolerance. PLoS One 8, e54533.

Barozai, M.Y.K., Husnain, T., 2012. Identification of biotic and abiotic stress up-regulated ESTs in *Gossypium arboreum*. Mol. Biol. Rep. 39, 1011−1018.

Basson, C.E., Groenewald, J.-H., Kossmann, J., Cronjé, C., Bauer, R., 2011. Upregulation of pyrophosphate: fructose 6-phosphate 1-phosphotransferase (PFP) activity in strawberry. Trans. Res. 20, 925−931.

Baxter, I., Ouzzani, M., Orcun, S., Kennedy, B., Jandhyala, S.S., Salt, D.E., 2007. Purdue ionomics information management system. An integrated functional genomics platform. Plant Physiol. 143, 600−611.

Berger, B., Parent, B., Tester, M., 2010. High-throughput shoot imaging to study drought responses. J. Exp. Bot. 61, 3519−3528.

Bhattarai, S.P., Midmore, D.J., 2009. Oxygation enhances growth, gas exchange and salt tolerance of vegetable soybean and cotton in a saline vertisol. J. Integ. Plant Biol. 51, 675−688.

Bor, M., Ozdemir, F., Turkan, I., 2003. The effect of salt stress on lipid peroxidation and antioxidants in leaves of sugar beet *Beta vulgaris* L. and wild beet *Beta maritima* L. Plant Sci. 164, 77−84.

Bray, E.A., 2002. Classification of genes differentially expressed during water-deficit stress in *Arabidopsis thaliana*: an analysis using microarray and differential expression data. Ann. Bot. 89, 803−811.

Brenner, S., 1990. The human genome: the nature of the enterprise. CIBA Found Symp 149, 6−17.

Carden, D.E., Walker, D.J., Flowers, T.J., Miller, A.J., 2003. Single cell measurement of the contributions of cytosolic Na^+ and K^+ to salt tolerance. Plant Physiol. 131, 676−683.

Carter, C.T., Grieve, C.M., Poss, J.A., 2005. Salinity effects on emergence, survival, and ion accumulation of *Limonium perezii*. J. Plant Nutr. 28, 1243−1257.

Cerda, A., Martinez, V., 1988. Nitrogen fertilization under saline conditions in tomato and cucumber plants. J. Hort. Sci. 63, 451−458.

Chakraborti, N., Mukherji, S., 2003. Effect of phytohormone pretreatment on nitrogen metabolism in *Vigna radiata* under salt stress. Biol. Plant 46, 63−66.

Charrier, A., Lelièvre, E., Limami, A.M., Planchet, E., 2013. *Medicago truncatula* stress associated protein 1 gene (MtSAP1) overexpression confers tolerance to abiotic stress and impacts proline accumulation in transgenic tobacco. J. Plant Physiol. 170, 874−877.

Chaves, M.M., Maroco, J.P., Pereira, J.S., 2003. Understanding plant responses to drought—from genes to the whole plant. Funct. Plant Biol. 30, 239−264.

Chaves, M.M., Flexas, J., Pinheiro, C., 2009. Photosynthesis under drought and salt stress: regulation mechanisms from whole plant to cell. Ann. Bot. ¡03, 551−560.

Chow, W.S., Ball, M.C., Anderson, J.M., 1990. Growth and photosynthetic responses of spinach to salinity: implication of K nutrition for salt tolerance. Aust. J. Plant Physiol. 17, 563–578.

Corbishley, J., Pearce, D. (2007) Growing trees on salt-affected land. ACIAR Impact Assessment Series 2007; Report No. 51 (www.aciar.gov.au).

Cornelious, B., Chen, P., Chen, Y., De Leon, N., Shannon, J.G., Wang, D., 2005. Identification of QTLs underlying waterlogging tolerance in soybean. Mol. Breed 6 (2), 103–112.

Cramer, G.R., 1992. Kinetics of maize leaf elongation. II. Responses of a Na-excluding cultivar and a Na-including cultivar to varying Na/Ca salinities. J. Exp. Bot. 43, 857–864.

Cui, F., Liu, L., Zhao, Q., Zhang, Z., Li, Q., Lin, B., Wu, Y., Tang, S., Xie, Q., 2012. Arabidopsis ubiquitin conjugase UBC32 is an ERAD component that functions in brassinosteroid mediated salt stress tolerance. Plant Cell 24, 233–244.

Cushman, J.C., Bohnert, H.J., 2000. Genomic approaches to plant stress tolerance. Curr. Opin. Plant Biol. 3, 117–124.

Dandekar, A.M., Teo, G., Defilippi, B.G., Uratsu, S.L., Passey, A.J., Kader, A.A., et al., 2004. Effect of down-regulation of ethylene biosynthesis on fruit flavor complex in apple fruit. Trans. Res. 13, 373–384.

Farkhondeh, R., Nabizadeh, E., Jalilnezhad, N., 2012. Effect of salinity stress on proline content, membrane stability and water relations in two sugar beet cultivars. Int. J. Agri. Sci. 2, 385–392.

Feigin, A., Pressman, E., Imas, P., Milta, O., 1991. Combined effects of KNO_3 and salinity on yield and chemical composition of lettuce and Chinese cabbage. Irrig. Sci. 12, 223–230.

Feng-Ling, F.U., Zhi-Lei, F., Shi-Bing, G., Shu-Feng, Z., Wan-Chen, L., 2008. Evaluation and quantitative inheritance of several drought-relative traits in maize. Agric. Sci. China 7, 280–290.

Fleury, D., Jefferies, S., Kuchel, H., Langridge, P., 2010. Genetic and genomic tools to improve drought tolerance in wheat. J. Exp. Bot. 61, 3211–3222.

Flexas, J., Bota, J., Loreto, F., Cornic, G., Sharkey, T.D., 2004. Diffusive and metabolic limitations to photosynthesis under drought and salinity in C_3 plants. Plant Biol. 6, 269–679.

Flexas, J., Ribas-Carbo, M., Diaz-Espejo, A., Galmés, J., Medrano, H., 2008. Mesophyll conductance to CO_2: current knowledge and future prospects. Plant Cell Environ 31, 602–612.

Flexas, J., Ribas-Carbó, M., Bota, J., Galmés, J., Henkle, M., Martínez-Cañ ellas, S., 2006. Decreased Rubisco activity during water stress is not induced by decreased relative water content but related to conditions of low stomatal conductance and chloroplast CO_2 concentration. New Phytol. 172, 73–82.

Flowers, T.J., 2004. Improving crop salt tolerance. J. Exp. Bot. 55, 1–13.

Flowers, T.J., Flowers, S.A., 2005. Why does salinity pose such a difficult problem for plant breeders? Agric. Water Manage 78, 15–24.

Flowers, T.J., Hajibagheri, M.A., 2001. Salinity tolerance in *Hordeum vulgare*: ion concentration in root cells cultivars differing in salt tolerance. Plant Soil 231, 1–9.

Foyer, C.H., Noctor, G., 2003. Redox sensing and signaling associated with reactive oxygen in chloroplasts, peroxisomes and mitochondria. Physiol. Plant 119, 355–364.

Furbank, R.T., 2009. Plant phenomics: from gene to form and function. Funct. Plant Biol. 36, 5–6.

Gambino, G., Gribaudo, I., 2012. Genetic transformation of fruit trees: current status and remaining challenges. Transgenic Res. 21, 1163–1181.

Gamma, P.B., Inanaga, S., Tanaka, K., Nakazawa, R., 2007. Physiological response of common bean (*Phaseolus Vulg.* L.) seedlings to salinity stress. Afric. J. Biotechnol. 6, 79–88.

Gao, Y.H., Chen, L.F., Zhou, X., Li, L., Liu, Q.H., Tian, G.L., 2008. Analysis on optimal bands for retrieval of mixed canopy chlorophyll content based on remote sensing. Remote Sensing Spatial Infor. Sci. 37, 1391–1396.

Gorantla, M., Babu, P.R., Reddy, V.B., Lachagari, A.M., Reddy, M., Wusirika, R., 2007. Identification of stress-responsive genes in an Indica Rice (*Oryza sativa* L.) using ESTs generated from drought-stressed seedlings. J. Exp. Bot. 58, 253–265.

Graifenberg, A., Giustiniani, L., Temperini, O., di Paola, M.L., 1995. Allocation of Na, Cl, K and Ca within plant tissues in globe artichoke (*Cynara scolimus* L.) under saline-sodic conditions. Sci. Hort. 63, 1−10.

Greenway, H., Munns, R., 1980. Mechanisms of salt tolerance in non halophytes. Ann. Rev. Plant Physiol. 31, 149−190.

Guo, R., Yu, F., Gao, Z., An, H., Cao, X., Guo, X., 2011. GhWRKY3, a novel cotton (*Gossypium hirsutum* L.) WRKY gene is involved in diverse stress responses. Mol. Biol. Rep. 38, 49−58.

Hadiarto, T., Tran, L.S.P., 2011. Progress studies of drought-responsive genes in rice. Plant Cell Rep. 30, 297−310.

Hamada, A.M., El-Enany, A.E., 1994. Effect of NaCl salinity on growth, pigment and mineral element contents, and gas exchange of broad bean and pea plants. Biol. Plant 36, 75−81.

Hamel, L.P., Nicole, M., Sritubtim, S., Morency, M.-J., Ellis, M., Ehlting, J., 2006. Ancient signals: comparative genomics of plant MAPK and MAPKK gene families. Trends Plant Sci. 11, 192−198.

Hasegawa, P.M., Bressan, R.A., Zhu, J.K., Bohnert, H.J., 2000. Plant cellular and molecular response to high salinity. Ann. Rev. Plant Physiol. Plant Mol. Biol. 51, 463−499.

Hayat, S., Maheshwari, P., Wani, A.S., Irfan, M., Alyemeni, M.N., Ahmad, A., 2012. Comparative effect of 28 homobrassinolide and salicylic acid in the amelioration of NaCl stress in *Brassica juncea* L. Plant Physiol. Biochem. 53, 61−68.

He, L., Yang, X., Wang, L., Zhu, L., Zhou, T., Deng, J., Zhang, X., 2013. Molecular cloning and functional characterization of a novel cotton CBL-interacting protein kinase gene (GhCIPK6) reveals its involvement in multiple abiotic stress tolerance in transgenic plants. Biochem. Biophys. Res. Commun. 435, 209−215.

Hernàndez, J.A., Ferrer, M.A., Jimenez, A., Ros-Barcelo, A., Sevilla, F., 2001. Antioxidant system and O_2/H_2 O_2 production in the apoplast of *Pisum sativum* L. leaves: its relation with NaCl induced necrotic lesions in minor veins. Plant Physiol. 127, 817−831.

Higbie, S.M., Wang, F., Stewart, J.McD., Sterling, T.M., Lindemann, W.C., Hughs, E., 2010. Physiological response to salt (NaCl) stress in selected cultivated tetraploid cottons. Int. J. Agron. 2010, 1−12.

Hirayama, T., Shinozaki, K., 2010. Research on plant abiotic stress responses in the post-genome era: past, present and future. The Plant J. 61, 1041−1052.

Humphreys, M.O., Humphreys, M.W., 2005. Breeding for stress resistance: general principles. In: Ashraf, M., Harris, P.J.C. (Eds.), Abiotic stresses: plant resistance through breeding and molecular approaches. Hawarth, New York, London, Oxford, pp. 19−46.

Idrees, M., Naeem, M., Khan, M.N., Aftab, T., Khan, M.M.A., Moinuddin, 2011. Alleviation of salt stress in lemongrass by salicylic acid. Protoplasma 249, 709−720.

Ismail, A.M., Heuer, S., Thomson, M.J., Wissuwa, M., 2007. Genetic and genomic approaches to develop rice germplasm for problem soils. Plant Mol. Biol. 65, 547−570.

Jabeen, R., Ahmad, R., 2009. Alleviation of the adverse effects of salt stress by foliar application of sodium antagonistic essential minerals on cotton (*Gossypium hirsutum*). Pak. J. Bot. 41, 2199−2208.

Jefferies, S.P., Barr, A.R., Karakousis, A., Kretschmer, J.M., Manning, S., Chalmers, K.J., et al., 1999. Mapping of chromosome regions conferring boron toxicity tolerance in barley (*Hordeum vulgare* L.). Theor. Appl. Genet. 98, 1293−1303.

Jha, A., Joshi, M., Yadav, N.S., Agarwal, P.K., Jha, B., 2011. Cloning and characterization of the *Salicornia brachiata* Na^+/H^+ antiporter gene SbNHX1 and its expression by abiotic stress. Mol. Biol. Rep. 38, 1965−1973.

Jia, W., Davies, W.J., 2007. Modification of leaf apoplastic pH in relation to stomatal sensitivity to root sourced abscisic acid signals. Plant Physiol. 143, 68−77.

Jiang, C., Wright, R., Woo, S., Delmonte, T., Paterson, A., 2000. QTL analysis of leaf morphology in tetraploid *Gossypium* (Cotton). Theor. Appl. Genet. 100, 409−418.

John, U.P., Spangenberg, J.C., 2005. Xenogenomics: genomic bioprospecting in indigenous and exotic plants through EST discovery, cDNA microarray-based expression profiling and functional genomics. Comp. Funct. Genom. 6, 230−235.

Johnston, J.W., Gunaseelan, K., Pidakala, P., Wang, M., Schaffer, R.J., 2009. Coordination of early and late ripening events in apples is regulated through differential sensitivities to ethylene. J. Exp. Bot. 60, 2689−2699.

Jones, H.G., Serraj, R., Loveys, B.R., Xiong, L., Wheaton, A., Price, A.H., 2009. Thermal infrared imaging of crop canopies for the remote diagnosis and quantification of plant responses to water stress in the field. Funct. Plant Biol. 36, 978−989.

Kersey, R.K., (2004) Microarray expression analysis to identify drought responsive genes involved in carbohydrate, and lipid metabolism in *Medicago sativa* leaves. PhD thesis, New Mexico State University.

Khadri, M., Tejera, N.A., Lluch, C., 2007. Sodium chloride−ABA interaction in two common bean (*Phaseolus vulgaris*) cultivars differing in salinity tolerance. Environ. Exp. Bot. 60, 211−218.

Khan, M.A., Ungar, I.A., Showalter, A.M., 2000. Effects of salinity on growth, water relations and ion accumulation in the sub tropical perennial halophyte, *Atriplex griffithii* var. stocksii. Ann. Bot. 85, 225−232.

Khan, PSSV, Hoffman, L., Renaut, J., Housman, J.F., 2007. Current initiatives in proteomics for analysis of plant salt tolerance. Curr. Sci. 93, 807−817.

Kumar, G., Purtya, R.S., Sharmac, M.P., Singla-Pareekb, S.L., Pareek, A., 2009. Physiological responses among Brassica species under salinity stress show strong correlation with transcript abundance for SOS pathway related genes. J. Plant Physiol. 166, 507−520.

Kumar, R., Goswami, S., Sharma, S., Singh, K., 2012. Protection against heat stress in wheat involves change in cell membrane stability, antioxidant enzymes, osmolyte, H_2O_2 and transcript of heat shock protein. Int. J. Plant Physiol. Biochem. 4, 83−91.

Kuppu, S., Mishra, N., Hu, R., Sun, L., Zhu, X., Shen, G., et al., 2013. Water-deficit inducible expression of a cytokinin biosynthetic gene IPT improves drought tolerance in cotton. PLoS One 8 (5), e64190.

Lacerda, C.F., Cambraia, J., Cano, M.A.O., Ruiz, H.A., Prisco, J.T., 2003. Solute accumulation and distribution during shoot and leaf development in two sorghum genotypes under salt stress. Environ. Exp. Bot. 49, 107−120.

Lafitte, H.R., Price, A.H., Courtois, B., 2004. Yield response to water deficit in an upland rice mapping population: associations among traits and genetic markers. Theor. Appl. Genet. 109, 1237−1246.

Langridge, P., Paltridge, N., Fincher, G., 2006. Functional genomics of abiotic stress tolerance in cereals. Brief Funct. Genom. Proteom. 4, 343−354.

Laperche, A., Devienne-Barret, F., Maury, O., Le Gouis, J., Ney, B., 2006. A simplified conceptual model of carbon/nitrogen functioning for QTL analysis of winter wheat adaptation to nitrogen deficiency. Theor. Appl. Genet. 113, 1131−1146.

Laperche, A., Le Gouis, J., Hanocq, E., Brancourt-Hulmel, M., 2008. Modelling nitrogen stress with probe genotypes to assess genetic parameters and genetic determinism of winter wheat tolerance to nitrogen constraint. Euphytica 161, 259−271.

Li, H.B., Vaillancourt, R., Mendham, N., Zhou, M.X., 2008. Comparative mapping of quantitative trait loci associated with waterlogging tolerance in barley (*Hordeum vulgare* L.). BioMed. Cent. Genom. 9, 401.

Lin-Wang, Bolitho K, Grafton, K., Kortstee, A., Karunairetnam, S., McGhie, T.K., et al., 2010. An R2R3 MYB transcription factor associated with regulation of the anthocyanin biosynthetic pathway in Rosaceae. BMC Plant Biol. 10, 50.

Lipshutz, R.J., Fodor, S.P., Gingeras, T.R., Lockhart, D.J., 1999. High density synthetic oligonucleotide arrays. Nat. Genet. 21, 20−24.

López-Gómez, R., Cabrera-Ponce, J.L., Saucedo-Arias, L.J., Carreto-Montoya, L., Villanueva-Arce, R., Díaz-Perez, J.C., et al., 2009. Ripening in papaya fruit is altered by ACC oxidase cosuppression. Trans. Res. 18, 89−97.

Lockhart, D.J., Winzeler, E.A., 2000. Genomics, gene expression and DNA arrays. Nature 405, 827–836.

Lockhart, J.A., 1965. Analysis of irreversible plant cell elongation. J. Theor. Biol. 8, 264–275.

Lopez, M.V., Satti, S.M.E., 1996. Calcium and potassium enhanced growth and yield of tomato under sodium chloride stress. Plant Sci. 114, 19–27.

Luo, K., Zhang, G., Deng, W., Luo, F., Qiu, K., Pei, Y., 2008. Functional characterization of a cotton late embryogenesis abundant D113 gene promoter in transgenic tobacco. Plant Cell Rep. 27, 707–717.

Ma, H.X., Bai, G.H., Carver, B., Zhou, L.L., 2005. Molecular mapping of a quantitative trait locus for aluminum tolerance in wheat cultivar Atlas 66. Theor. Appl. Genet. 112, 51–57.

Ma, L.Q., Zhou, E.F., Huo, N.X., Zhou, R.H., Wang, G.Y., Jia, J.Z., 2007. Genetic analysis of salt tolerance in a recombinant inbred population of wheat (*Triticum aestivum* L.). Euphytica 153, 109–117.

Mackill, D.J., Amante, M.M., Vergara, B.S., Sarkarung, S., 1993. Improved semi-dwarf rice lines with tolerance to submergence of seedlings. Crop. Sci. 33, 749–775.

Mah, N., Thelin, A., Lu, T., Nikolaus, S., Kühbacher, T., Gurbuz, Y., 2004. A comparison of oligonucleotide and cDNA based microarray systems. Physiol. Genom. 16, 361–370.

Majeed, A., Nisar, M.F., Hussain, K., 2010. Effect of saline culture on the concentration of Na^+, K^+ and ClG in *Agrostis tolonifera*. Curr. Res. J. Biol. Sci. 2, 76–82.

Mansour, M.M.F., Salama, K.H.A., Ali, F.Z.M., AbouHadid, A.F., 2005. Cell and plant responses to NaCl in *Zea mays* cultivars differing in salt tolerance. Gen. Appl. Plant Physiol. 31, 29–41.

Mao, X., Zhang, H., Qian, X., Li, A., Zhao, G., Jing, R., 2012. TaNAC2, a NAC-type wheat transcription factor conferring enhanced multiple abiotic stress tolerances in Arabidopsis. J. Exp. Bot. 63, 2933–2946.

Marcelo-Pedrosa, G., Queila-Souza, G., 2013. Reactive oxygen species and seed germination. Biologia 68, 110–115.

Martin, R.C., Glover-Cutter, K., Baldwin, J.C., Dombrowski, J.E., 2012. Identification and characterization of a salt stress inducible zinc finger protein from Festuca arundinacea. BMC Res. Notes 5, 66.

Mi, Z., Ji, C., Zhang, S., Zhao, X., 2012. Effects of climate change on water use efficiency in rain-fed plants. Intl. J. Plant Prod. 6, 513–520.

Micheletto, S., Rodriguez-Uribe, L., Hernandez, R., Richins, R.D., Curry, J., O'Connell, M.A., 2007. Comparative transcript profiling in roots of *Phaseolus acutifolius* and *P. vulgaris* under water deficit stress. Plant Sci. 73, 510–520.

Mishra, S.B., Senadhira, D., Manigbas, N.L., 1996. Genetics of submergence tolerance in rice (*Oryza sativa L.*). Field Crop Res. 46, 177–181.

Misra, N., Saxena, P., 2009. Effect of salicylic acid on proline metabolism in lentil grown under salinity stress. Plant Sci. 177, 181–189.

Mittler, R., 2002. Oxidative stress, antioxidants and stress tolerance. Trends Plant Sci. 7, 405–410.

Mizoi, J., Yamaguchi-Shinozaki, K., 2013. Molecular approaches to improve rice abiotic stress tolerance. Methods Mol. Biol. 956, 269–283.

Mochida, K., Yamazaki, Y., Ogihara, Y., 2004. Discrimination of homologous gene expression in hexaploid wheat by SNP analysis of contigs grouped from a large number of expressed sequence tags. Mol. Genet. Genom. 270, 371–377.

Mohammadi, M., Kav, N.N., Deyholos, M.K., 2007. Transcriptional profiling of hexaploid wheat (*Triticum aestivum* L.) roots identifies novel, dehydration-responsive genes. Plant Cell Environ. 30, 630–645.

Morison, J.I.L., Baker, N.R., Mullineaux, P.M., Davies, W.J., 2008. Improving water use in crop production. Phil. Trans. Biol. Sci. 363, 639–658.

Munns, R., 2005. Genes and salt tolerance: bringing them together. New Phytol. 167, 645–663.

Munns, R., 2002. Comparative physiology of salt and water stress. Plant Cell Environ. 25, 239–250.

Munns, R., Hare, R.A., James, R.A., Rebetzke, G.J., 2000. Genetic variation for salt tolerance of durum wheat Aust. J. Agric. Res. 51, 69–74.

Munns, R., James, R.A., Lauchli, A., 2006. Approaches to increasing the salt tolerance of wheat and other cereals. J. Exp. Bot. 57, 1025–1043.

Munns, R., James, R.A., Sirault, X.R.R., Furbank, R.T., Jones, H.G., 2010. New phenotyping methods for screening wheat and barley for beneficial responses to water deficit. J. Exp. Bot. 61, 3499–3507.

Murphy, K.S.T., Durako, M.J., 2003. Physiological effects of short-term salinity changes on *Ruppia maritima*. Aquat. Bot. 75, 293–309.

Naika, M., Shameer, K., Mathew, O.K., Gowda, R., Sowdhamini, R., 2013. STIFDB2: an updated version of plant stress-responsive transcription factor database with additional stress signals, stress-responsive transcription factor binding sites and stress-responsive genes in arabidopsis and rice. Plant Cell Physiol. 54 (2), e8(1–15).

Naqvi, S.S.M., 1999. Plant hormones and stress phenomena. In: Pessarakli, M. (Ed.), Handbook of plant and crop. stress. Marcel Dekker, New York, pp. 709–730.

Navakode, S., Weidner, A., Varshney, R.K., Lohwasser, U., Scholz, U., Borner, A., 2009. A QTL analysis of aluminium tolerance in barley, using gene-based markers. Cereal Res. Comm. 37, 531–540.

Nawaz, K., Hussain, K., Majeed, A., Khan, F., Afghan, S., Ali, K., 2010. Fatality of salt stress to plants: morphological, physiological and biochemical aspects. Afric. J. Biotechnol. 9, 5475–5480.

Niknam, S.R., McComb, J., 2000. Salt tolerance screening of selected Australian woody species—a review. Forest Ecol. Manage 139, 1–19.

Nounjan, N., Nghia, P.T., Theerakulpisut, P., 2012. Exogenous proline and trehalose promote recovery of rice seedlings from salt stress and differentially modulate antioxidant enzymes and expression of related genes. J. Plant Physiol. 169, 596–604.

Oh, D., Dassanayake, M., Haas, J.S., Kropornika, A., Wright, C., d'Urzo, M.P., 2010. Genome structures and halophyte-specific gene expression of the extremophile *Thellungiella parvula* in comparison with *Thellungiella salsuginea* (*Thellungiella halophila*) and Arabidopsis. Plant Physiol. 154, 1040–1052.

Oh, D., Zahir, A., Yun, D., Bressan, R.A., Bohnert, H.J., 2009. SOS1 and halophytism. Plant Signal Behav. 4, 1081–1083.

Oraby, H.F., Ransom, C.B., Kravchenko, A.N., Sticklen, M.B., 2005. Barley HVA1 gene confers salt tolerance in R3 transgenic oat. Crop Sci. 45, 2218–2227.

Orsini, F., D'Urzo, M.P., Inan, G., Serra, S., Oh, D., Mickelbart, M.V., 2010. A comparative study of salt tolerance parameters in 11 wild relatives of *Arabidopsis thaliana*. J. Exp. Bot. 61, 3787–3798.

Parida, A.K., Das, A.B., 2005. Salt tolerance and salinity effects on plants: a review. Ecotoxicol. Environ. Safety 60, 324–349.

Pariset, L., Chillemi, G., Bongiorni, S., Spica, Valentini A, 2009. Microarrays and high-throughput transcriptomic analysis in species with incomplete availability of genomic sequences. New Biotechnol. 25, 272–279.

Park, B.J., Liu, Z., Kanno, A., Kameya, T., 2005. Increased tolerance to salt- and water-deficit stress in transgenic lettuce (*Lactuca sativa* L.) by constitutive expression of LEA. Plant Growth Regul. 45, 165–171.

Parker, R., Flowers, T.J., Moorem, A.L., Harpham, N.V.J., 2006. An accurate and reproducible method for proteome profiling of the effects of salt stress in the rice leaf lamina. J. Exp. Bot. 57, 1109–1118.

Payton, P., Kottapalli, K.R., Kebede, H., Mahan, J.R., Wright, R.J., Allen, R.D., 2011. Examining the drought stress transcriptome in cotton leaf and root tissue. Biotechnol Lett. 33, 821–828.

Peters, W.S., Farm, S.M., Kopf, J.A., 2001. Does growth correlate with turgor induced elastic strain in stems? A re-evaluation of de Vries' classical experiments. Plant Physiol. 125, 2173–2179.

Pisinaras, V., Tsihrintzis, V.A., Petalas, C., Ouzounis, K., 2010. Soil salinization in the agricultural lands of Rhodope District, northeastern Greece. Environ. Monit Assess 166, 79–94.

Pitman, M.G., Läuchli, A., 2002. Global impact of salinity and agricultural ecosystems. In: Läuchli, A., Lüttge, A.E. (Eds.), Salinity: environment-plants-molecules. Kluwer Academic, Netherlands, p. 32.

Pons, R., Cornejo, M., Sanz, A., 2011. Differential salinity induced variations in the activity of H^+-pumps and Na^+/H^+ antiporters that are involved in cytoplasm ion homeostasis as a function of genotype and tolerance level in rice cell lines. Plant Physiol. Biochem. 49, 1399–1409.

Prabhavathi, V., Yadav, J.S., Kumar, P.A., Rajam, M.V., 2002. Abiotic stress tolerance in transgenic egg plant (*Solanum melongena* L.) by introduction of bacterial mannitol phosphodehydrogenase gene. Mol. Breed 9, 137–147.

Putney, S.D., Herlihy, W.C., Schimmel, P., 1983. A new troponin T and cDNA clones for 13 different muscle proteins, found by shotgun sequencing. Nature 302, 718–721.

Quarrie, S.A., Gulli, M., Calestani, C., Steed, A., Marmiroli, N., 1994. Location of a gene regulating drought-induced abscisic acid production on the long arm of chromosome 5A of wheat. Theor. Appl. Genet. 89, 794–800.

Rabbani, M.A., Maruyama, K., Abe, H., Khan, M.A., Katsura, K., Ito, Y., 2003. Monitoring expression profiles of rice genes under cold, drought and high-salinity stresses and abscisic acid application using cDNA microarray and RNA get blot analyses. Plant Physiol. 133, 1755–1767.

Rajendran, K., Tester, M., Roy, S.J., 2009. Quantifying the three main components of salinity tolerance in cereals. Plant Cell Environ. 32, 237–249.

Ramoliya, P.J., Patel, H.M., Pandey, A.N., 2004. Effect of salinization of soil on growth and macro and micronutrient accumulation in seedlings of *Salvadora persica* (Salvadoraceae). Forest Ecol Manage 202, 181–193.

Rhodes, D., Hanson, A.D., 1993. Quaternary ammonium and tertiary sulfonium compounds in higher plants. Annu. Rev. Plant Physiol. Plant Mol. Biol. 44, 357–384.

Richards, R.A., Rebetzke, G.J., Watt, M., Condon, A.G., Spielmeyer, W., Dolferus, R., 2010. Breeding for improved water productivity in temperate cereals: phenotyping, quantitative trait loci, markers and the selection environment. Funct. Plant Biol. 37, 85–97.

Robinson, N.J., Rampant, P.C., Callinan, A.P.L., Rab, M.A., Fisher, P.D., 2009. Advances in precision agriculture in south-eastern Australia. II. Spatio-temporal prediction of crop yield using terrain derivatives and proximally sensed data. Crop. Pasture Sci. 60, 859–869.

Rodriguez-Uribe, L., Higbiea, S.M., Stewart, J.M., Wilkins, T., Lindemann, W., Sengupta-Gopalana, C., 2011. Identification of salt responsive genes using comparative microarray analysis in Upland cotton (*Gossypium hirsutum* L.). Plant Sci. 180, 461–469.

Roosens, N.H., Thu, T.T., Iskandar, H.M., 1998. Isolation of the ornithine-delta-aminotransferase cDNA and effect of salt stress on its expression in *Arabidopsis thaliana*. Plant Physiol. 117, 263–271.

Roy, S.J., Tucker, E.J., Tester, M., 2011. Genetic analysis of abiotic stress tolerance in crops. Curr. Opin. Plant Biol. 4, 232–239.

Rudd, S., 2003. Expressed sequence tags: alternative or complement to whole genome sequences? Trends Plant Sci. 8, 321–329.

Saeed, M., Dahab, A.H., Wangzhen, G., Tianzhen, Z., 2012. A cascade of recently discovered molecular mechanisms involved in abiotic stress tolerance of plants. OMICS 16, 188–199.

Saleh, B., 2011. Ion partitioning and Mg^{2+}/Na^+ ratio under salt stress application in cotton. J. Stress Physiol. Biochem. 7, 292–300.

Salekdeh, G.H., Siopongco, J., Wade, L.J., Ghareyazie, B., Bennet, J., 2002. A proteomic approach to analyzing drought and salt responsiveness in rice. Field Crops Res. 76, 199–219.

Sanchez, A.C., Subudhi, P.K., Rosenow, D.T., Nguyen, H.T., 2002. Mapping QTLs associated with drought resistance in sorghum (*Sorghum bicolor* L. Moench). Plant Mol. Biol. 48, 713–726.

Sanchez, D.H., Pieckenstain, F.L., Escaray, F., Erban, A., Kraemer, U., Udvardi, M.K., 2011. Comparative ionomics and metabolomics in extremophile and glycophytic Lotus species under salt stress challenge the metabolic pre-adaptation hypothesis. Plant Cell Environ. 34, 605–617.

Sanchez, D.H., Szymanski, J., Erban, A., Udvardi, M.K., Kopka, J., 2010. Mining for robust transcriptional and metabolic responses to long-term salt stress: a case study on the model legume Lotus japonicus. Plant Cell Environ. 33, 468−480.

Saranga, Y., Menz, M., Jiang, C.X., Wright, R.J., Yakir, D., Paterson, A.H., 2001. Genomic dissection of genotype × environment interactions conferring adaptation of cotton to arid conditions. Genome Res. 11, 1988−1995.

Sato, S., Soga, T., Nishioka, T., Tomita, M., 2004. Simultaneous determination of the main metabolites in rice leaves using capillary electrophoresis mass spectrometry and capillary electrophoresis diode array detection. The Plant J. 40, 151−163.

Sawahel, W.A., Hassan, A.H., 2004. Generation of transgenic wheat plants producing high levels of the osmoprotectant proline. Biotechnol. Lett. 24, 712−725.

Senadheera, P., Maathuis, F.J.M., 2009. Differentially regulated kinases and phosphatases in roots may contribute to inter-cultivar difference in rice salinity tolerance. Plant Signal Behav. 4, 1163−1165.

Serraj, R., Krishnamurthy, L., Kashiwagi, J., Kumar, J., Chandra, S., Crouch, J.H., 2004. Variation in root traits of chickpea (*Cicer arietinum* L.) grown under terminal drought. Field Crops Res. 88, 115−127.

Serraj, R., Sinclair, T.R., 2002. Osmolyte accumulation: can it really help increase crop yield under drought conditions? Plant Cell Environ. 25, 333−341.

Shah, H.H., 2007. Effects of salt stress on mustard as affected by gibberellic acid application. Gen. Appl. Plant Physiol. 33, 97−106.

Shameer, K., Ambika, S., Varghese, S.M., Karaba, N., Udayakumar, M., Sowdhamini, R., 2009. STIFDB— Arabidopsis Stress-responsive Transcription Factor DataBase. Int. J. Plant Genom. 2009, 583429.

Shen, Y.G., Zhang, W.K., He, S.J., 2003. An EREBP/AP2-type protein in *Triticum aestivum* was a DRE-binding transcription factor induced by cold, dehydration and ABA stress. Theor. Appl. Genet. 106, 923−930.

Sherchand, K., Paulsen, G.M., 1985. Response of wheat to foliar phosphorus treatments under field and high temperature regimes. J. Plant Nutr. 12, 1171−1181.

Singh, M.P., Pandey, S.K., Singh, M., Ram, P.C., Singh, B.B., 1990. Photosynthesis, transpiration, stomatal conductance and leaf chlorophyll content in mustard genotypes grown under sodic conditions. Photosynthetica 24, 623−627.

Sirault, X.R.R., James, R.A., Furbank, R.T., 2009. A new screening method for osmotic component of salinity tolerance in cereals using infrared thermography. Funct. Plant Biol. 36, 970−977.

Sivakumar, P., Sharmila, P., Saradhi, P.P., 2000. Proline alleviates salt-stress-induced enhancement in ribulose-1, 5-biphosphate oxygenase activity. Biochem. Biophys. Res. Commun. 279, 512−515.

Skylas, D., Van-Dyk, D., Wrigley, C.W., 2005. Proteomics of wheat grain. J. Cereal. Sci. 41, 165−179.

Sobhanian, H., Aghaei, K., Komatsu, S., 2011. Changes in the plant proteome resulting from salt stress: toward the creation of salt-tolerant crops? J. Proteom. 74, 1323−1337.

Sonneveld, C., de Kreiji, C., 1999. Response of cucumber (*Cucumis sativus* L.) to an unequal distribution of salt in the root environment. Plant Soil 209, 47−56.

Stepien, P., Johnson, G.N., 2009. Contrasting responses of photosynthesis to salt stress in the glycophyte Arabidopsis and the halophyte Thellungiella: role of the plastid terminal oxidase as an alternative electron sink. Plant Physiol. 149 (2), 1154−1165.

Stepien, P., Klobus, G., 2005. Antioxidant defense in the leaves of C_3 and C_4 plants under salinity stress. Physiol. Plant 125, 31−40.

Stepien, P., Klobus, G., 2006. Water relations and photosynthesis in *Cucumis sativus* L. leaves under salt stress. Biol. Plant 50, 610−616.

Suarez, D.L., Grieve, C.M., 1988. Prediction cation ratios in corn from saline solution composition. J Exp. Bot. 39, 605−612.

Sudhir, P., Murthy, S.D.S., 2004. Effects of salt stress on basic processes of photosynthesis. Photosynthetica 42, 481–486.

Suriya-Arunroj, D., Nopporn, S., Theerayut, T., Apichart, V., 2004. Relative leaf water content as an efficient method for evaluating rice cultivars for tolerance to salt stress. Sci. Asia 30, 411–415.

Swire-Clark, G.A., Marcotte Jr, W.R., 1999. The wheat LEA protein Em functions as an osmoprotective molecule in *Saccharomyces cerevisiae*. Plant Mol. Biol. 39, 117–128.

Szabados, L., Savouré, A., 2010. Proline: a multifunctional amino acid. Trends Plant Sci. 15, 89–97.

Szankowski, I., Flachowsky, H., Li, H., Halbwirth, H., Treutter, D., Regos, I., et al., 2009b. Shift in polyphenol profile and sublethal phenotype caused by silencing of anthocyanidin synthase in apple (Malus sp.). Planta 229, 681–692.

Takahashi, D., Li, B., Nakayama, T., Kawamura, Y., Uemura, M., 2013. Plant plasma membrane proteomics for improving cold tolerance. Front Plant Sci. 17 (4), 90.

Tang, W., Page, M., 2013. Overexpression of the Arabidopsis AtEm6 gene enhances salt tolerance in transgenic rice cell lines. Plant Cell Tiss. Org. Cult. 114, 339–350.

Tavakkoli, E., Rengasamy, P., McDonald, G.K., 2010. High concentrations of Na^+ and Cl^- ions in soil solution have simultaneous detrimental effects on growth of faba bean under salinity stress. J. Exp. Bot. 61, 4449–4459.

Teulat, B., Monneveux, P., Wery, J., Borriès, C., Souyris, I., Charrier, A., 1997. Relationships between relative water content and growth parameters in barley: a QTL study. New Phytol. 137, 99–107.

Tseng, G.C., Ghosh, D., Feingold, E., 2012. Comprehensive literature review and statistical considerations for microarray meta-analysis. Nucl. Acids Res. 40, 3785–3799.

Tuberosa, R., Salvi, S., 2006. Genomics approaches to improve drought tolerance in crops. Trends Plant Sci. 11, 405–412.

Udall, J.A., Swanson, J.M., Haller, K., Rapp, R.A., 2006. A global assembly of cotton ESTs. Genome Res. 16, 441–4450.

Ulloa, M., Hutmacher, R.B., Davis, R.M., Wright, S.D., Percy, R., Marsh, B., 2006. Breeding for *Fusarium wilt* race 4 resistance in cotton under field and greenhouse conditions. J. Cotton Sci. 10, 114–127.

Valkoun, J.J., 2001. Wheat pre-breeding using wild progenitors. Euphytica 119, 17–23.

VanToai, T.T., Beuerlein, J.E., Schmitthenner, A.F., Martin, S.K., 1994. Genetic variability for flooding tolerance in soybeans. Crop. Sci. 34, 1112–1115.

Villareal, R.L., Mujeeb-Kazi, A., Rajaram, S., Toro, E.D., 1994. Morphological variability in some synthetic hexaploid wheats derived from *Triticum turgidum* × *Triticum tauschii*. J. Genet Breed 48, 7–16.

Walk, T., Jaramillo, R., Lynch, J., 2006. Architectural tradeoffs between adventitious and basal roots for phosphorus acquisition. Plant and Soil 279, 347–366.

Wenxue, W., Bilsborrow, P.E., Hooley, P., Fincham, D.A., Lombi, E., Forster, B.P., 2003. Salinity induced differences in growth, ion distribution and partitioning in barley between the cultivar Maythorpe and its derived mutant Golden Promise. Plant Soil 250, 183–191.

Wilkinson, S., Davies, W.J., 2002. ABA-based chemical signaling: the co-ordination of responses to stress in plants. Plant Cell Environ. 25, 195–210.

Woo, S.H., Kimura, M., Higa-Nishiyama, A., 2003. Proteome analysis of wheat lemma. BioSci. Biotech. Biochem. 67, 2486–2491.

Wright, R.J., Thaxton, P.M., El-Zik, K.M., Paterson, A.H., 1998. D-subgenome bias of Xem resistance genes in tetraploid *Gossypium* (cotton) suggests that polyploidy formation has created novel avenues for evolution. Genet. 149 (4), 1987–1996.

Xu, J., Duan, X., Yang, J., Beeching, J.R., Zhang, P., 2013. Coupled expression of Cu/Zn-superoxide dismutase and catalase in cassava improves tolerance against cold and drought stresses. Plant Signal Behav. 19, e24525.

Xu, S., Hu, B., He, Z., Ma, F., Feng, J., Shen, W., 2011. Enhancement of salinity tolerance during rice seed germination by presoaking with hemoglobin. Int. J. Mol. Sci. 12, 2488−2501.

Xue, Z.Y., Zhi, D.Y., Xue, G.P., Zhang, H., Zhao, Y.X., Xia, G.M., 2004. Enhanced salt tolerance of transgenic wheat (*Triticum aestivum* L.) expressing a vacuolar Na^+/H^+ antiporter gene with improved grain yields in saline soils in the field and a reduced level of leaf Na^+. Plant Sci. 167, 849−859.

Xue, D.W., Huang, Y.Z., Zhang, X.Q., Wei, K., Westcott, S., Li, C.D., et al., 2009. Identification of QTLs associated with salinity tolerance at late growth stage in barley. Euphytica 169, 187−196.

Yamaguchi, T., Blumwald, E., 2005. Developing salt tolerant crop plants: challenges and opportunities. Trends Plant Sci. 10, 615−620.

Yan, L., Pei, L., Zhao-Shi, X., Rui-Yue, Z., Li, L., Lian-Cheng, L., 2008. Overexpression of W6 gene increases salt tolerance in transgenic tobacco plants. Acta Agron. Sin. 34, 984−990.

Yang, D.L., Jing, R.L., Chang, X.P., Li, W., 2007. Identification of quantitative trait loci and environmental interactions for accumulation and remobilization of water-soluble carbohydrates in wheat (*Triticum aestivum* L.) stems. Genetics 176, 571−584.

Yang, Y., Lu, X., Yan, B., Li, B., Sun, J., Guo, S., 2013. Bottle gourd rootstock-grafting affects nitrogen metabolism in NaCl-stressed watermelon leaves and enhances short-term salt tolerance. J. Plant Physiol. 170, 653−661.

Yanik, H., Turktas, M., Dundar, E., Hernandez, P., Dorado, G., Unver, T., 2013. Genome-wide identification of alternate bearing-associated microRNAs (miRNAs) in olive (*Olea europaea* L.). BMC Plant Biol. 13, 10.

Ying, S., Zhang, D., Fu, J., Shi, Y., Song, Y., Wang, T., 2012. Cloning and characterization of a maize bZIP transcription factor, ZmbZIP72, confers drought and salt tolerance in transgenic Arabidopsis. Planta 235, 253−266.

Yokotani, N., Ichikawa, T., Kondou, Y., Iwabuchi, M., Matsui, M., Hirochika, H., 2013. Role of the rice transcription factor JAmyb in abiotic stress response. J. Plant Res. 126, 131−139.

Yue, Y., Zhang, M., Zhang, J., Duan, L., Li, Z., 2012. SOS1 gene overexpression increased salt tolerance in transgenic tobacco by maintaining a higher K^+/Na^+ ratio. J. Plant Physiol. 169, 255−261.

Zadraznik, T., Hollung, K., Egge-Jacobsen, W., Meglic, V., Sustar-Vozlic, J., 2013. Differential proteomic analysis of drought stress response in leaves of common bean (*Phaseolus vulgaris* L.). J. Proteom. 78, 254−272.

Zaharieva, M., Gaulin, E., Havaux, M., Acevedo, E., Monneveux, P., 2001. Drought and heat responses in the wild wheat relative *Aegilops geniculata* Roth: potential interest for wheat improvement. Crop Sci. 41, 1321−1329.

Zhang, H., Li, Y., Wang, B., Chee, P.W., 2008. Recent advances in cotton genomics. Int. J. Plant Genom. 742304, 20.

Zhou, R., Cheng, L., Dandekar, A.M., 2006. Down-regulation of sorbitol dehydrogenase and up-regulation of sucrose synthase in shoot tips of the transgenic apple trees with decreased sorbitol synthesis. J Exp Bot 57, 3647−3657.

Zhu, C., Wang, Y., Li, Y., Bhatti, K.H., Tian, Y., Wu, J., 2011a. Overexpression of a cotton cyclophilin gene (GhCyp1) in transgenic tobacco plants confers dual tolerance to salt stress and *Pseudomonas syringae* pv. tabaci infection. Plant Physiol. Biochem. 49, 1264−1271.

Zhu, J., Ingram, P.A., Benfey, P.N., Elich, T., 2011b. From lab to field, new approaches to phenotyping root system architecture. Curr. Opin. Plant Biol. 14, 310−317.

Zhu, J.K., 2001. Plant salt tolerance. Trends Plant Sci 6, 66−72.

Metabolomics Role in Crop Improvement

Saleha Resham, Fizza Akhter, Muhammad Ashraf and Alvina Gul Kazi

2.1 Introduction

Metabolomics is a technology that deals with comprehensive analysis followed by identification and quantification of all the metabolites of an organism at a given time (Fiehn, 2002). Being aware of the current scenario, the following question comes to mind: Would we ever be able to meet the increasing demand for food to feed a continuously increasing population? The world's population will increase to 10 billion by 2050, as estimated by Evans in 1998 (Evans, 1998). Would the rate of increase in crop yield and the rate of its consumption ever be balanced by appropriate stats and economic planning?

Such are the questions that circulate in the minds of biotechnologists, plant scientists, agriculturalists, and economists. Looking at the potential of the field of metabolomics, it deals with the study of all low-molecular-mass compounds synthesized and modified by a living cell or organism. Metabolic profiling (metabolomics/metabonomics) is the measurement in biological systems of the complement of low-molecular-weight metabolites and their intermediates that reflects the dynamic response to genetic modification and physiological, pathophysiological, and developmental stimuli. Plant metabolomics is relatively young and still very much in development, but is now being widely applied and is already considered as a technology—a "maturing science," which is "established and robust" (Dixon et al., 2006; Schaur and Fernie, 2006). The use of metabolomics offers us information about the fundamental biochemical basis of the food we eat. The potential of the field of metabolomics has been explored in a study by Chandna (Chandna et al., 2013). According to the review, different metabolomics techniques/strategies can be employed to detect changes at different levels (Chandna et al., 2013). The metabolic changes that occur when plants are in a stressed condition can be evaluated through metabolite fingerprinting and metabolite profiling (Chandna et al., 2013). Accurate analysis through identification and quantification of a stressed sample was possible until the change occurred at the transcriptome or proteome level. It certainly would help to eliminate or somewhat lessen the two highly contrasting nutrition-related problems: the malnutrition/undernourishment of the developing world and the overnutrition/overconsumption of the developed world. So, the wealth of information generated from metabolomics studies would help us to design certain strategies to generate nutritious food products that would not only improve the general health of the population but also help in decreasing the cost of the food products when produced in abundance (Hall et al., 2008). Metabolomics serves as an important tool in the

P. Ahmad (Ed): Emerging Technologies and Management of Crop Stress Tolerance, Volume 1.
DOI: http://dx.doi.org/10.1016/B978-0-12-800876-8.00002-3

Table 2.1 Terms and Their Definitions Used in Metabolomics Technology (Hall, 2006)

Term	Definition
Metabolome	The complete complement of small molecules present in an organism.
Metabolomics	The technology geared towards providing an essentially unbiased, comprehensive qualitative and quantitative overview of the metabolites present in an organism.
Metabolic fingerprinting	High-throughput qualitative screening of the metabolic composition of an organism or tissue with the primary aim of sample comparison and discrimination analysis. Generally no attempt is initially made to identify the metabolite present. All steps from the sample preparation, separation and detection should be rapid and as simple as feasible. It is often used as a forerunner to metabolic profiling.
Metabolic profiling	Identification and quantification of metabolites present in the organism. For practical reasons, this is only feasible for a limited number of components which are generally chosen on the basis of discriminant analysis or on the molecular relationships based on molecular networks/pathways.
Targeted analysis	Following broad-scale metabolomics analysis or based upon previous knowledge, biochemical profiling can be performed in a greater detail on a selected group of metabolites by using optimized extraction and dedicated separation/detection techniques.

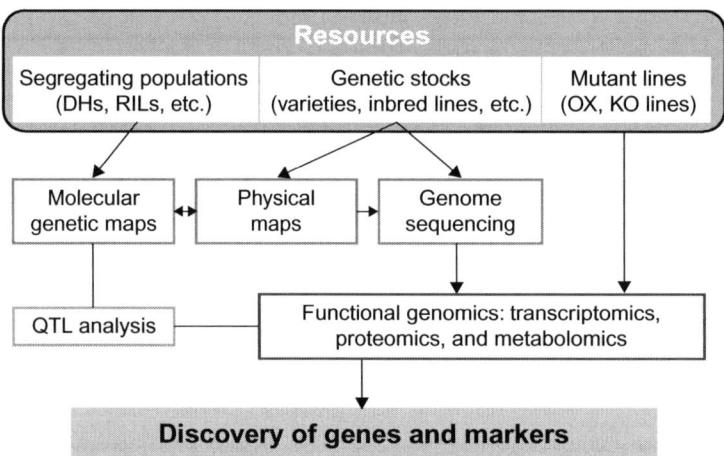

FIGURE 2.1

An overview of gene discovery and markers for crop improvement based on genetic and genomic strategies (Fukushima et al., 2009).

genomic-assisted selection for crop improvement. See Table 2.1 for a list of terms and their definitions used in metabolomics technology.

The integrated analysis of transcriptomics and metabolomics enables the depiction of genes and the discovery of molecular networks involved in biological processes of *Arabidopsis* (Figure 2.1) (Fukushima et al., 2009; Kusano and Saito, 2012). Soil loss and degradation; water logging, drought, and salinity; the co-evolution of pests, pathogens, and hosts; and the impact of climate change are the leading resources and environmental constraints faced by the world's farmers today (Tilman et al., 2001).

Once the potential new genes are narrowed down through correlation analysis, the functions of these genes are confirmed by approaches of reverse genetics or reverse biochemistry (Fukushima et al., 2009; Kusano and Saito, 2012).

Understanding the plants' metabolism acts as a driving force for understanding and interpreting food-based metabolomics.

2.2 Techniques involved in metabolomics

Metabolite detection has been carried out for a number of years, but the technical landmark was contributed by the development of chromatographic methods that act by identifying and detecting individual metabolites (Fritz, 2004; Unger, 2004). The following are the major and frequently used techniques involved in metabolomics analysis:

- Liquid chromatography-mass spectrometry (LC-MS)
- Gas chromatography mass spectrometry (GCMS)
- Nuclear magnetic resonance (NMR)
- Fourier transform-infrared (FT-IR) spectroscopy

2.2.1 Metabolite target analysis

Metabolite target analysis is a technique that involves the preparation and analysis of samples of one or a small number of compounds from complex mixtures. It is a widely used technique applied to the monitoring of phytohormones and to directly study the primary effect of genetic alteration (Fiehn, 2002).

2.2.2 Metabolite profiling

Metabolite profiling involves the measurement of hundreds or thousands of metabolites, and requires streamlined extraction, separation, and analysis in a high-throughput manner, so as to measure large numbers of metabolites in a complex mixture of chemicals found in cellular extracts (Kopka et al., 2004).

2.2.3 Metabolite fingerprinting

Metabolite fingerprinting involves looking into metabolites to help differentiate samples according to their genotype, phenotype, or biological relevance (Shanks, 2005).

LC-MS and GCMS are now being exploited to obtain a more detailed insight into the variation in composition of food in the context of both quality and nutrition. The quality of crop plants, nutritionally or otherwise, is a direct function of metabolite content (Memelink, 2004).

The metabolites are the end products of cellular regulatory processes, and their levels can be viewed as the response of biological systems to environmental or genetic manipulations (Maloney, 2004). The metabolome is very diverse and includes lipid soluble chemicals normally found in

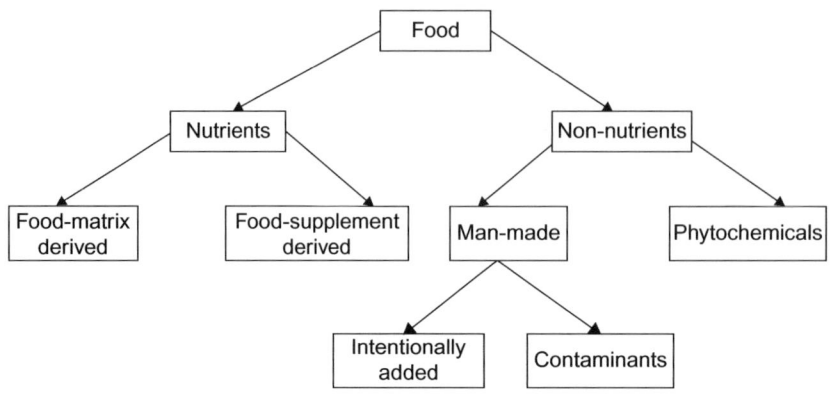

FIGURE 2.2

Nutrients and non-nutrients in the human food supply (Gibney et al., 2005).

membranes, polar chemicals for aqueous parts of the cell, acidic and basic ions, stable structures, and structures that oxidize at the slightest mistreatment (Maloney, 2004).

There is a huge opportunity for improving crops in a great variety of ways as has recently been demonstrated in studies on genetic metabolomics (linking metabolic profiles with genomic information) using *Arabidopsis* and tomato (Keurentjes et al., 2006; Schauer and Fernie, 2006).

2.2.4 Bio-resource databases

There is a wide variety of bio-resource databases available such as TAIR (http://arabidopsis.org; Swarbreck et al., 2008), Phenome (Kuromori et al., 2006), and the FOX hunting system (Ichikawa et al., 2006). Data in TAIR is derived in large part from manual curation of the *Arabidopsis* research literature and direct submissions from the research community. See Figure 2.2 for nutrients and non-nutrients in human food supply.

Feihn (2009) thoroughly reviewed metabolomics as a tool to establish a link between genotype and phenotype. According to him "metabolome" is the complete set of metabolites synthesized by a biological system. Feihn (2009) mentioned careful consideration of the methods employed for tissue extraction, sample preparation, data acquisition, and data mining as the prerequisites for the analysis of metabolomics data. When compared to proteome and transcriptome, metabolome can be defined on all levels of complexity, such as organisms, tissues, cells, or cell compartments. Figure 2.3 displays how multiparallel metabolite and transcript profiling will help delineate future breeding strategies (Fernie and Schauer, 2009).

2.3 Metabolomics and nutrigenomics—a link

Nutrigenomics is a science that links nutrition to genome and is growing speedily these days. It is the study of the bidirectional interactions between genes and diet. Metabolomics as introduced previously is the integrated study of many small molecules produced by metabolism (Zeisel, 2007).

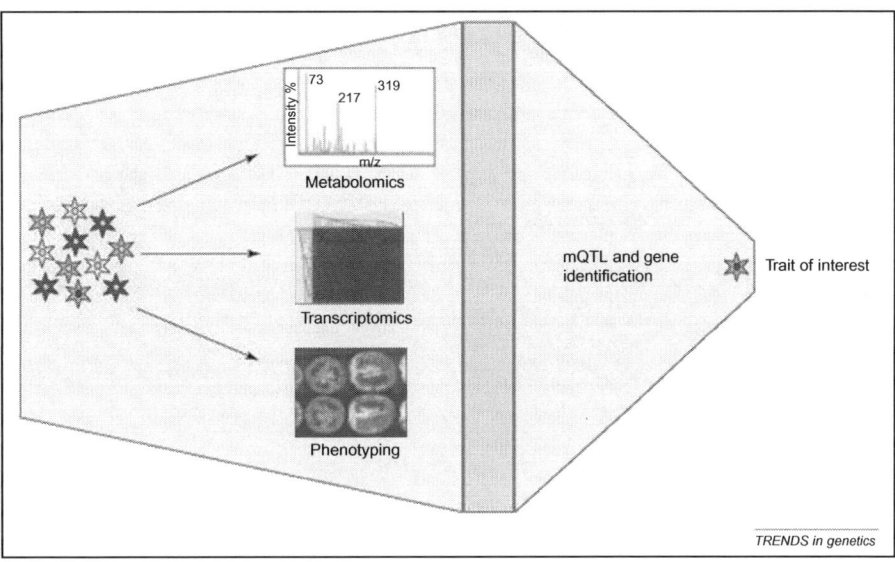

FIGURE 2.3

Profiling large populations to define novel metabolic QTL. Combining metabolomics, transcriptomics analysis, and extensive phenotyping of large, genetically diverse populations (e.g., tomatoes) with an integrated bioinformatics platform will facilitate the identification of novel mQTL and the underlying genetics of the trait of interest.

Both are rapidly emerging and developing fields that help contribute to improving human nutrition, which is the ultimate aim of such research initiatives.

Zeisel (2007) has predicted that the relations between diet and nutrigenomic, metabolomic profiles and health have become important components of research that could change clinical practice in nutrition. There are two contrasting demands of both worlds: improved crop varieties with better nutritional content (essential amino acids, vitamins and other micronutrients) to alleviate hunger and malnutrition-associated diseases are needed for the developing world, whereas for the developed world, overnutrition is a concern (Rist et al., 2006). Nutrigenomic and metabolomic profiling can enhance nutrition epidemiology and nutrition intervention research (Zeisel, 2007).

Figure 2.4 shows how the process of metabolomics research can be initiated. Sample extraction is the first step, followed by analysis and data acquisition using techniques mentioned previously. Data processing is then performed using normalizing and transforming techniques. Statistical analysis is the next step, followed by data visualization and ultimately biological interpretation of data obtained from which useful conclusions can be drawn. Figure 2.5 shows a scheme of the metabolomics workflow.

2.4 Applications of metabolomics in crop improvement

Metabolomics has been found useful in studying responses of plants to stressed conditions, as has been performed by Bowne and his colleagues (2012). Drought responses of leaf tissues from wheat

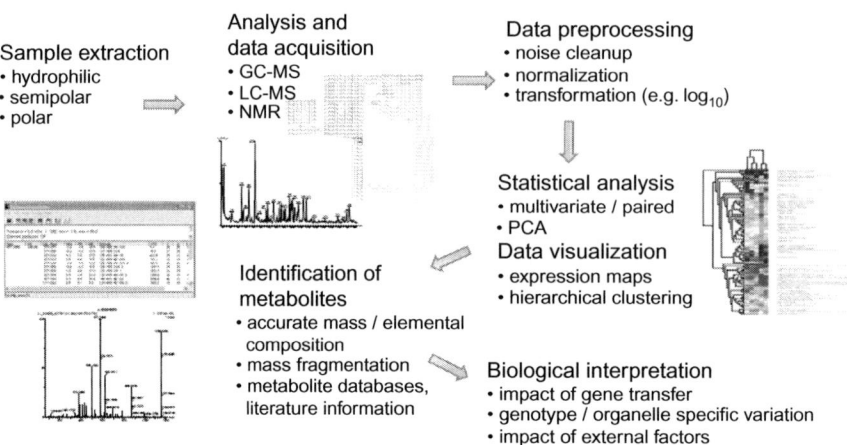

FIGURE 2.4

Steps involved in metabolomics research (www.weizmann.ac.il/plants/aharoni/PDFs/b4.pdf).

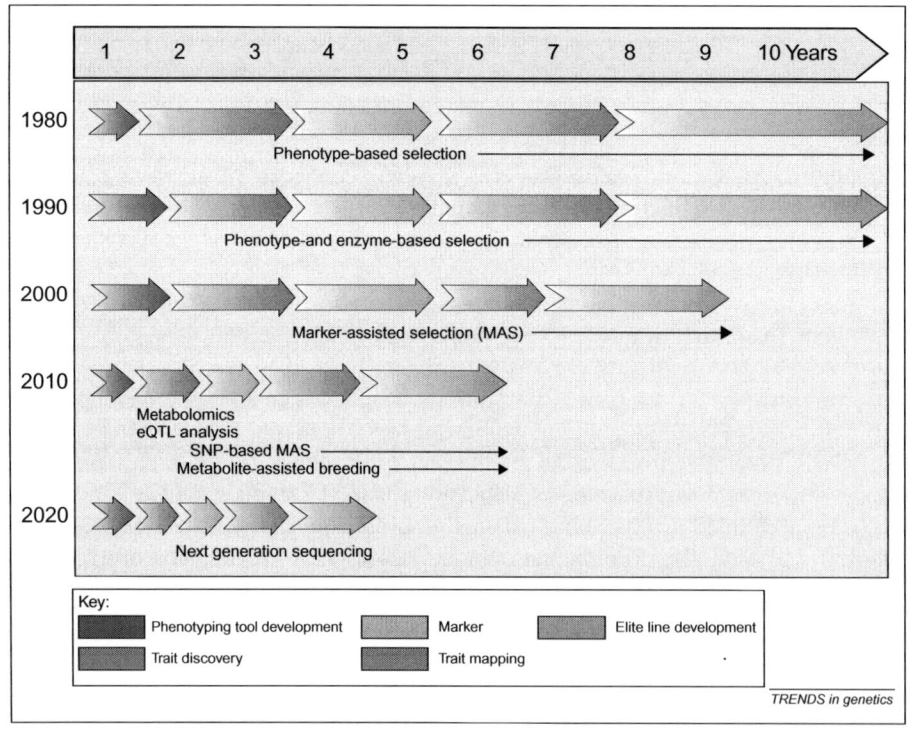

FIGURE 2.5

A scheme of the metabolomics workflow.

cultivars of differing drought tolerance at the metabolite level was performed using metabolomics techniques (Bowne et al., 2012). What they found was the increased level of certain amino acids, proline, tryptophan, and the branched chain amino acids leucine, isoleucine, and valine under drought stress in all the cultivars.

Metabolomics of crop plants highlights the metabolic profile at various stages of crop development. NMR and MS techniques have benefits and limitations in the metabolite profiling of plants. Several metabolomic techniques have been employed in order to estimate the metabolic content of plants of commercial importance (Wolfender et al., 2013). Phenotype is intimately connected to a particular set of metabolites termed the metabolome (Wishart et al., 2009).

2.4.1 Improvement of tomato quality by metabolomics

Alteration in the physiological condition of metabolites affects tomato fruit development, thus influencing tomato fruit quality. The balance between organic acids, free amino acids, and sugars influences tomato fruit flavor (Petró-Turza, 1986). Changes in metabolic balance depict variation in different stages of tomato fruit development (Carrari and Fernie, 2006).

Tomato fruit extract constitutes various metabolites including compounds that are good anti-oxidants. These anti-oxidants are listed in Table 2.2.

The polar metabolites present in the peel or surface tissue of tomato are given in Table 2.3.

Various volatile metabolites are produced in response to toxigenic fungi infecting tomato. Gas chromatography and mass spectrometry have been used to profile volatile metabolites in response to the toxigenic fungi. Oleic acid amide is one of the most significant metabolites in ripe normal tomato fruits, whereas stable production of octadecenoic acid is found in the case of infection with *Aspergillus niger*, *Aspergillus flavus*, and *Fusarium oxysporum* (Ibrahim et al., 2011).

The pericarp composition of the tomato fruit has been analyzed using gas chromatography-mass spectrometry (GC-MS) (Roessner-Tunali et al., 2003). Oms-Oliu et al. (2011) analyzed tomato fruit

Table 2.2 Anti-oxidants in Tomato Fruit Extract (Engelhard et al., 2006)

Anti-Oxidants in Tomato Fruit Extract
Lycopene
β-Carotene
Vitamin E

Table 2.3 Polar Metabolites in Surface Tissue of Tomato Fruit (Mintz-Oron et al., 2008)

Polar Metabolites in Tomato Fruit Surface Tissue
Flavonoids
Glycoalkaloids
Amyrin-type pentacyclic triterpenoids

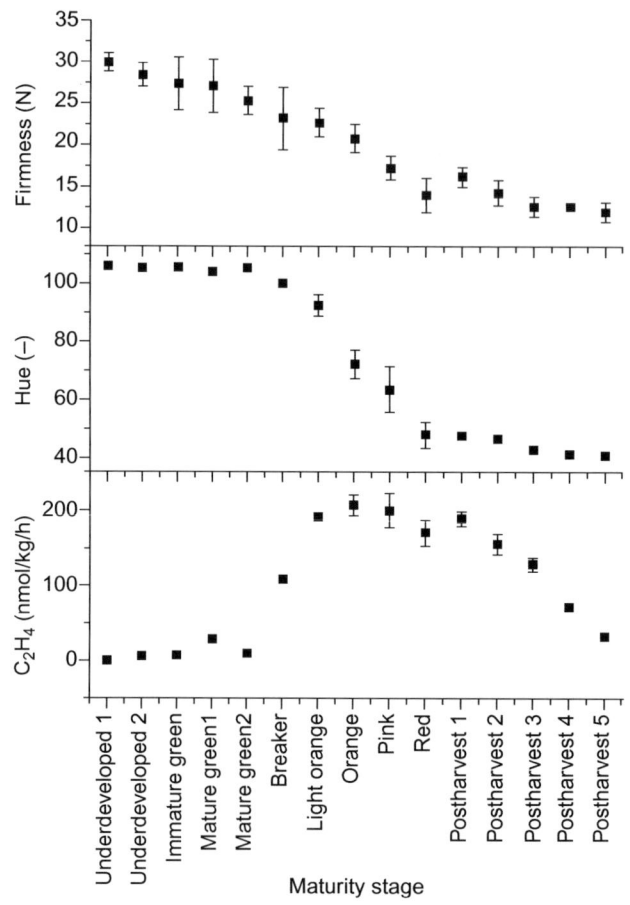

FIGURE 2.6

The level of stiffness, hue, and ethylene during ripening stage (Oms-Oliu et al., 2011).

ripening by measuring the ethylene content, stiffness, and color. The production of ethylene increases during tomato fruit ripening and tomato color and stiffness decrease during the ripening stage (see Figure 2.6).

2.4.1.1 Organic acid content

Upon maturation, citric acid content increases according to the GC-MS analyses (see Table 2.4). Malic acid composition decreases as the tomato fruit ripens. During the post-harvest stage, fumaric acid and lactic acid are measurable. Fumaric acid is not measurable at the post-harvest stage and its level decreases during the maturation stage.

The level of certain organic acids needs to be balanced as it influences the pH of the tomato fruit. The pH of tomato fruit above a specific threshold leads to decomposition by microbes.

Table 2.4 Metabolites Identified by GC-MS as Components of a Methanol Extract from Tomato Tissue (Oms-Oliu et al., 2011)

Amino Acids	Organic Acids	Sugars	Others
Glutamic acid (glu)	Citric acid (cit)	Fructose (frc)	Palmitic acid (pal)
Pyroglutamic acid (pyglu)	Malic acid (malic)	Glucose (glc)	Linoleic acid (lin)
γ-Aminobutyric acid (GABA)	Isocitric acid (icit)	Galactose (gal)	Phosphoric acid (P)
Aspartic acid (asp)	Tartaric acid (tar)	Sucrose (suc)	Adenosine (aden)
Glutamine (gln)	Saccharic acid (sacc)	Myo-inositol (ino)	Guanosine (guan)
Serine (ser)	Maleic acid (maleic)	Galacturonic acid (galactu)	Chlorogenic acid (chlor)
Asparagine (asparag)	Succinic acid (succ)	Mannose (man)	
Methionine (met)	Lactic acid (lac)	Maltose (mal)	
Threonine (thr)	Citramalic acid (citmal)	Mannitol (manol)	
Leucine (leu)	Fumaric (fum)	Galactinol (galol)	
β-Alanine (β-ala)	Shikimic acid (shik)	Glucose6p (G6P)	
Phenylalanine (phe)	2-Keto-ʟ-gulonic acid (kgul)	Raffinose (raff)	
Glycine (gly)	Gluconic acid lactone (gluc)	Sorbitol/galactitol (galol)	
Valine (val)	Ascorbic acid (AA)		
Isoleucine (ile)	Dehydroascorbic acid (DHA)		

During ripening, the level of malate concentration falls. Subsequently, NADP-ME activity falls with the decrease in citric acid production (Knee and Finger, 1992).

2.4.1.2 Sugar content

During fruit ripening, glucose and fructose are present in equal quantities according to the GC-MS analysis. At maturation, the sucrose level decreases by more than 90% (Richardson et al., 1990). The level of mannose, galactose, and glucuronic acid is elevated at the maturation stage, whereas the levels of galactinol, raffinose, and myo-inositol decrease during tomato fruit development. Raffinose family oligosaccharides (RFO) are also important in tomato fruit development (Zhao et al., 2004). GC-MS analysis aids in analyzing the sugar content at different stages of tomato fruit development.

2.4.1.3 Free amino acid content

During maturation and ripening, the level of phenylalanine, threonine, glutamic acid, and aspartic acid is noticeable. There is a decline in the level of gamma-aminobutyric acid (GABA) and β-alanine during maturation. The overall results using GC-MS revealed that the amount of GABA, aspartate, and glutamine is maximum at the maturation stage (Oms-Oliu et al., 2011).

Moco et al. (2007) recorded a profile for the level of secondary metabolites in tomato tissue and tomato peel. The tomato metabolome database (MoToDB) was developed later on. Various tomato metabolite databases have thus far been developed. The databases contain information from liquid chromatography-mass spectrometry (LC-MS) analysis of metabolites. The metabolites from tomato fruit extract relate to the metabolites present in the fruit extract of transgenic as well as wild cultivated tomatoes. The scheme for finding the required metabolite present in tomato using the tomato metabolome database includes the entry of the accurate mass value of the metabolite as well as its form of ionization. The result indicated on the pane is exhibited as the CAS number, estimated formula, accurate mass of the compound, and its alternative name (Grennan, 2009).

Bino (Bino et al., 2005) used GC-MS for the analysis of changes in the metabolite composition in light hyper-responsive high pigment tomato mutant plant. Metabolic profiling using GC-MS was combined with phenotypic study of the tomato introgression lines. This revealed that yield quality traits were associated with 50% of the quantitative trait loci (QTL) (Schauer and Fernie, 2006).

Capillary electrophoresis-mass spectrometry has been used to identify polar tomato metabolites such as sugar phosphates, organic acids, and nucleotides (William et al., 2007). Nuclear magnetic resonance spectroscopy has been used as an analytic tool to assess the changes in the metabolite composition of transgenic tomato overexpressing maize transcription factors LC and C1. It has been depicted by the NMR study that the level of flavonols increases as the tomato fruit ripens and matures. The level of citric acid, phenylalanine, sucrose, and trigonelline increases significantly in transgenic tomato line as compared to the wild-type tomatoes (Gall et al., 2003). Nuclear magnetic resonance spectroscopy-based metabolite profiling of transgenic tomato fruit has been used to record metabolites accumulated in transgenic tomatoes. The metabolites recorded were accumulated spermidine and spermine. The results indicated that choline accumulation in tomato was enhanced by polyamines (Mattoo et al., 2006).

Citrus exocortis viroid (CEVd) or *Pseudomonas syringae* infection-related metabolites have been characterized using NMR-based metabolomics (López-Gresa et al., 2010). The interaction of tomato mosaic virus and tomato host plant has been studied with the help of nuclear magnetic resonance. The metabolomics approach has been successful in analyzing the metabolite balance in tomato mosaic virus-infected tomato plant. The metabolites include flavonoids, sugars, amino acids, and organic acids (López-Gresa et al., 2010). The NMR-based metabolomics led to the analysis of metabolites involved in the hypersensitive response shown by tobacco plants in the case of tobacco mosaic virus infection. The NMR-based analysis gives special information regarding the physiological conditions and selection of virus-free varieties of plants for metabolic profiling (Choi et al., 2006).

Ultra performance liquid chromatography combined with the mass spectrometry technique using time-of-flight (TOF) gives high sensitivity in the detection of tomato metabolites (Wilson et al., 2005).

2.4.2 Improvement of rice quality by metabolomics

To date, several techniques have aimed at investigating the loci crucial in the production of good quality traits in the rice genotypes (Wright et al., 2010). The metabolomics approach to investigating the key metabolites involved in improved rice quality is of prime importance for scientists. The most applied technique is GC-MS (gas capillary-mass spectroscopy) combined with LC-MS and NMR (Jellum, 1997). GC-TOF/MS (gas chromatography-time-of-flight/mass spectrometry) is beneficial in analyzing the unfamiliar metabolites. Seventy rice cultivars with various metabolic compositions leading to genetic diversity have been analyzed using 2D GC-MS (Kusano et al., 2007).

The gain- and loss-of-function approach in rice provides a way of altering the metabolic profile of the rice crop, thus improving quality. The FOX hunting system was developed to express the full length cDNA of rice ectopically in heterologous plants for functional gene analysis (Kondou et al., 2009). Insertional mutagenesis of Tos-17 retrotransposon and Ds-transposon in rice has been achieved as a loss-of-function approach. In the case of the gain-of-function approach, the random insertion of cauliflower 35S transcriptional enhancer into the rice genome led to the production of improved rice crops. The total number of rice genes calculated amounts to approximately 32,000. Elucidation of the genes important in crop development needs to be done. The analysis of loss-of-function phenotypes is difficult to achieve. The FOX hunting system provides the best alternative to achieve good quality rice (Kusano and Saito, 2012).

2.4.2.1 Strategy of the fox hunting system

This technology finds its usage in plants in which *Agrobacterium*-mediated transformation is feasible (Toki et al., 2006).

The Full length cDNA Over-eXpressor (FOX) gene hunting system is one of the remarkable techniques in the study of transgenic plants overexpressing full length complementary DNA of known or unknown function. The technique is used in the analysis of gain-of-function by the expression of full length complementary DNA in a large population of transgenic plants. The overexpression of the complementary DNA is achieved by placing the complementary DNA under the control of maize *Ubiquitin-1* promoter. So far, the FOX hunting system is one of the most feasible and effective transformation systems in rice. The technique involves the construction of a rice/FOX *Agrobacterium* library followed by PCR amplification of the transgene. The construction of a binary vector for the overexpression of full length complementary DNA is required. The level of expression of the full length complementary DNA can be analyzed using Southern blot analysis (Nakamura et al., 2007) (see Figures 2.7 and 2.8).

2.5 Improvement of strawberry quality by metabolomics

A major part of the Mediterranean diet consists of strawberries. Anthocyanins and ellagitannins are the two most beneficial components of strawberry fruit (Giampieri et al., 2012). Anthocyanin is broken down into phenolic acid of a smaller size (Aura et al., 2005). Amino acids and sugars play an important role in imparting nutritional value to the strawberry fruit.

Associated with the sugars are phenolic compounds that are also critical in determining the characteristics of the strawberry fruit. Ellagitannin profiles along with phenolic compounds significantly increase as a result of benzothiadiazole treatment in strawberry fruit. Ellagitannins are polyphenols having chemopreventive effects in humans (Cerda et al., 2005). Phenolic compounds that are produced with enhanced metabolism during benzothiadiazole treatment are listed in Table 2.5 (Hukkanen et al., 2007).

Ellagitannins found in strawberry are beneficial to human health. Ellagic acid is produced as a result of ellagitannin hydrolysis. The final product is a compound related to urolithin-like metabolites. These metabolites are active against colon carcinoma. GC-MS analysis of strawberry fruit analysis along with LC-MS analysis provides significant details regarding improvement in the production of high ellagitannin-producing strawberry fruit (Sharma et al., 2010).

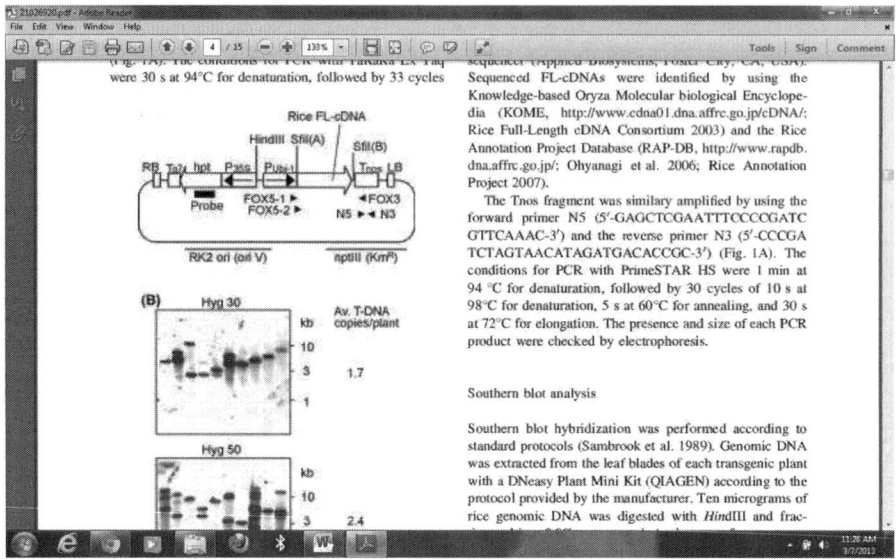

FIGURE 2.7

Full length complementary DNA overexpressing plant binary vector in *Arabidopsis*, pBIG2113SF. The binary vector contains the RIKEN *Arabidopsis* full length complementary DNA between SfiI sites and under the influence of 35S promoter (Ichikawa et al., 2006).

Ellagitannins found in the 100 μM concentration range starting from 10 μM are effective against thrombosis and angiogenesis. Inflammation is blocked as a result of ellagic acid production. Urolithin present in the blood plasma after strawberry consumption is known to exert strong vascular effects in humans (Larrosa et al., 2010). Ellagitannin content in strawberries ranges from 77 to 85 mg/100 g f.w. (Koponen et al., 2007). The overall phenolic composition of the strawberry fruit is up to 50% of ellagitannins and ellagitannin-like compounds. Another beneficial aspect of these compounds is their activity against type 2 diabetes. Alpha-amylase has been known to be inhibited by ellagitannins in the purified extract (Marcia et al., 2010). Seventy milligrams of ellagitannins are served per glass of strawberry juice (Tomas-Barberan et al., 2009).

Strawberry fruit belonging to the family Rosaceae has its own nutritional value. The metabolite composition of the strawberry fruit at different stages of development determines the quality of strawberry fruit at the particular phase of growth. The variation in metabolite composition is dependent on the alteration in the physiological parameters of fruit development. The composition of strawberry metabolites has been studied in detail. Zhang et al. (2011) analyzed the metabolic profiles of the strawberry fruit at different stages of development. Profiling of metabolites was done using GC-MS analysis. Targeted profiling of metabolites was achieved via LC-MS. Today, polar as well as non-polar metabolites in strawberry are analyzed by employing principal component analysis. GC-MS analysis determines the exact chemical content of the fruit at different stages of

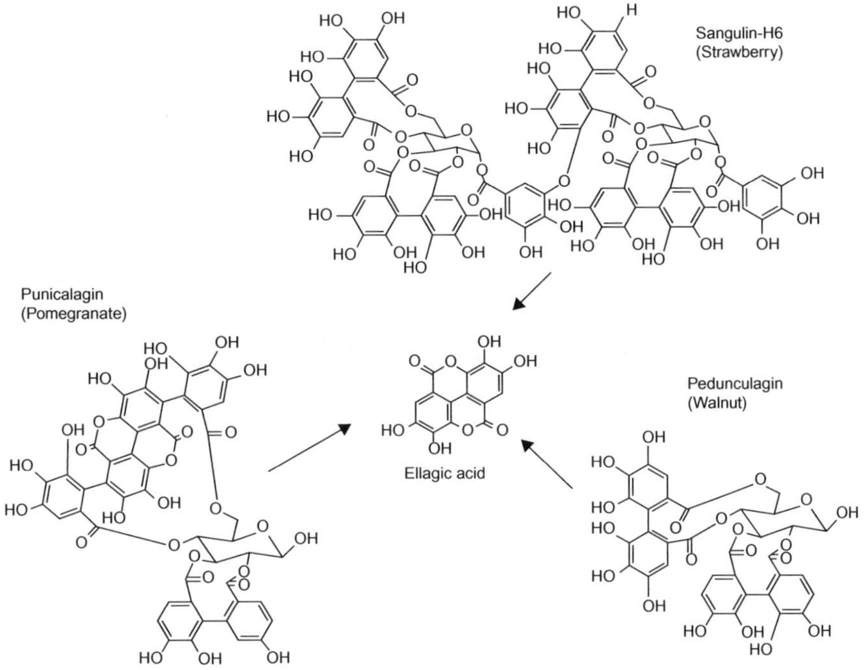

FIGURE 2.8

The construction of a binary vector for the overexpression of FL-cDNA. The FL-cDNA has been placed in the forward orientation between two *Sfi*I sites. For the purpose of FOX hunting in rice, the pRice-FOX binary vector has been constructed from pBIG2113-S. The full length complementary DNA has been placed under the control of PUb-i promoter as it is a strong promoter in rice (Nakamura et al., 2007).

Table 2.5 A Scheme of the Metabolomics Workflow

Ellagic acid deoxyhexose	Abundant in leaves
Agrimoniin-like ellagitannins	
Sanguiin H-10	
Lambertianin C-like ellagitannins	
Ellagic acid	Abundant in cell-bound fraction of the strawberry leaves
p-Coumaric acid	
Gallic acid	
Kaempferol hexose	
Kaempferol malonylglucoside	Abundant in the strawberry fruit

development including the polar and non-polar metabolites. Combined with the LC-MS data, the overall amino acid content as well as anthocyanin composition of fruit is obtained. In red-ripped strawberry fruit, certain amino acids are of primary importance in imparting flavor to the strawberry. The amino acids are listed in Table 2.6 (Zhang et al., 2011).

Table 2.6 Amino Acids Critical in Flavor Development in Red-ripped Strawberry Fruit (Zhang et al., 2011)

n-Hexadecanoic acid

9, 12-(Z,Z)-Octadecadienoic acid

3, 7, 11-Trimethyl-1, 6, 10-dodecatrien-3-ol

9, 12, 15-(Z, Z, Z)-Octadecatrien-1-ol

2.6 Conclusion and future prospects

Metabolomics is a current field of interest among all the "omics" technologies in order to design a more tailor-made approach to crop improvement. To meet the increased demand for food, plant scientists, agriculturalists, and biotechnologists are always seeking strategies to increase yield and the availability of better varieties/strains of crops. Metabolic profile changes can be used as a marker for stress physiology and metabolic movements, and factors can be analyzed in combination with other "omic" techniques such as transcriptomics. There is therefore great interest in using a metabolomics approach to understand better what is happening during crop domestication in order to design new concepts for more targeted crop improvement in ways more tailor-made to current needs.

Metabolomics is a viable option for crop improvement as it provides a basis for sustainable analysis of both primary and secondary metabolites. So far, different metabolomics approaches have been used to statistically assess the metabolite balance in wild as well as in transgenic plants. Tomato crops have been widely studied for their metabolic profile in diseased and normal conditions. This ultimately provided data regarding the important metabolites involved in improved fruit quality. Similarly, in the case of rice it has been a viable option to express full length complementary DNA for the improved version of transgenic rice crop. Cotton crop improvement via the application of metabolomics needs to be further enhanced and efforts in this area are forthcoming.

References

Aura, A.M., Lopez, P.M., O'Leary, K.A., Williamson, G., Oksman-Caldentey, K.M., Poutanen, K., et al., 2005. In vitro metabolism of anthocyanins by human gut micro flora. Eur. J. Nutr. 133−142.

Bino, R.J., de Vos, C.H.R., Lieberman, M., Hall, R.D., Bovy, A., Jonker, H.H., et al., 2005. The light-hyperresponsive high pigment-2[dg] mutation of tomato: alterations in the fruit metabolome. New Phytol. 166, 427−438.

Bowne, J.B., Erwin, T.A., Juttner, J., Schnurbusch, T., Langridge, P., Bacic, A., et al., 2012. Drought responses of leaf tissues from wheat cultivars of differing drought tolerance at the metabolite level. Mol. Plant 5, 418−429.

Carrari, F., Fernie, A., 2006. Metabolic regulation underlying tomato fruit development. J. Exp. Bot. 57, 1883−1897.

Cerda, B., Tomas-Barberan, F.A., Espin, J.C., 2005. Metabolism of antioxidant and chemopreventive ellagitannins from strawberries, raspberries, walnuts and oak-aged wine in humans: identification of biomarkers and individual variability. J. Agric. Food Chem. 53, 227−235.

Chandna, Ruby, Azooz, M.M., Ahmad, P., 2013. Recent advances of metabolomics to reveal plant response during salt stress. In: Ahmad, P., Azooz, M.M., Prasad, M.N.V. (Eds.), Salt Stress in Plants: Omics, Signalling and Adaptations. Springer, Dordrecht, pp. 1−14.

Choi, Y.H., Kim, H.K., Linthorst, H.J.M., Hollander, J.G., Lefeber, A.W.M., Erkelens, C., 2006. NMR metabolomics to revisit the tobacco mosaic virus infection in *Nicotiana tabacum* leaves. J. Nat. Prod. 69, 742−748.

Dixon, R.A., Gang, D.R., Chariton, A.J., Fiehn, O., Kuiper, H.A., Reynolds, T.L., et al., 2006. Applications of metabolomics in agriculture. J. Agric. Food Chem. 54, 8984−8994.

Engelhard, Y.N., Gazer, B., Paran, E., 2006. Natural anti-oxidants from tomato extract reduce blood pressure in patients with grade-1 hypertension: a double blind, placebo-controlled pilot study. Am. Heart J. 151, 100.

Evans, L.T., 1998. Feeding the Ten Billion: Plants and Population Growth. Cambridge University Press.

Fernie, R., Schauer, N., 2009. Metabolomics-assisted breeding: a viable option for crop improvement? Trends Gen. 25, 39−48.

Fiehn, O., 2002. Metabolomics—the link between genotypes and phenotypes. Plant Mol. Biol. 48, 155−171.

Fritz, J.S., 2004. Early milestones in the development of ion-exchange chromatography: a personal account. J. Chromatogr. A 1039, 3−12.

Fukushima, A., et al., 2009. Integrated omics approaches in plant systems biology. Curr. Opin. Chem. Biol. 13, 532−538.

Gall, G.L., Colquhoun, I.J., Davis, A.L., Collins, G.J., Verhoeyen, M.E., 2003. Metabolite profiling of tomato (*Lycopersicon esculentum*) using 1H NMR spectroscopy as a tool to detect potential unintended effects following a genetic modification. J. Agric. Food Chem. 51, 2447−2456.

Giampieri, F., Tulipani, S., Alvarez-Suarez, J.M., Quiles, J.L., Mezetti, B., Battino, M., 2012. The strawberry: composition, nutritional quality and impact on human health. Nutrition 28, 9−19.

Gibney, M.J., Walsh, M., Brennan, L., Roche, H.M., Berman, J.B., van Ommen, B., 2005. Metabolomics in human nutrition: opportunities and challenges. Am. J. Clin. Nutr. 82, 497−503.

Grennan, A.K., 2009. MoTo DB: a metabolic database for tomato. Plant Physiol. 151, 1701−1702.

Hall, R.D., Brouwer, I.D., Fitzgerald, M.A., 2008. Plant metabolomics and its potential application for human nutrition. Physiol. Plant 132, 162−175.

Hukkanen, A.T., Kokko, H.I., Buchala, A.J., McDougall, G.J., Stewart, D., Karenlampi, S.O., et al., 2007. Benzothiadiazole induces the accumulation of phenolics and improves resistance to powdery mildew in strawberries. J. Agric. Food Chem. 55, 1862−1870.

Ibrahim, A.D., Husnaini, H., Sani, A., Aleiro, A.A., Yakubu, S.E., 2011. Volatile metabolites profiling to discriminate diseases of tomato fruits inoculated with three toxigenic fungal pathogens. Res. Biotechnol. 2, 14−22.

Ichikawa, T., Nakazawa, M., Kawashima, M., Iizumi, H., Kuroda, H., Kondou, Y., et al., 2006. The FOX hunting system: an alternative gain-of-function gene hunting technique. Plant J. 48, 974−985.

Jellum, E., 1997. Profiling of human body fluids in healthy and diseased states using gas chromatography and mass spectrometry, with spatial reference to organic acids. J. Chromatogr. 143, 427−462.

Keurentjes, J.J.B., Fu, Y., De Vos, C.H.R., Lommen, A., Hall, R.D., Bino, R.J., et al., 2006. The genetics of plant metabolism. Nat. Gen. 38, 842−849.

Knee, M., Finger, F.L., 1992. NADP$^+$-Malic enzyme and organic acid levels in developing tomato fruits. J. Am. Soc. Horticul. Sci. 117, 799−801.

Kondou, Y., Higuchi, M., Takahashi, S., Sakurai, T., Ichikawa, T., Kuroda, H., et al., 2009. Systematic approaches to using the FOX hunting system to identify useful rice genes. Plant J. 57, 883−894.

Kopka, J., Fernie, A., Weckwerth, W., 2004. Metabolite profiling in plant biology: platforms and destinations. Genome Biol. 5, 109–127.

Koponen, J.M., Happonen, A.M., Mattila, P.H., Torronen, A.R., 2007. Contents of anthocyanins and ellagitannins in selected foods consumed in Finland. J. Agric. Food Chem. 55, 1612–1619.

Kuromori, T., Wada, T., Kamiya, A., Yuguchi, M., Yokouchi, T., Imura, Y., et al., 2006. A trial of phenome analysis using 4000 Ds-insertional mutants in gene-coding regions of *Arabidopsis*. Plant J. 47, 640–651.

Kusano, M., Saito, K., 2012. Role of metabolomics in crop improvement. J. Plant Biochem. Biotechnol. 21 (Suppl. 1), S24–S31.

Kusano, M., Fukushima, A., Kobayashi, M., Hayashi, N., Jonsson, P., Moritz, T., et al., 2007. Application of a metabolomics method combining one-dimensional and two-dimensional gas chromatography-time-of-flight/mass spectrometry to metabolic phenotyping of natural variants in rice. J. Chromatogr. B 855, 71–79.

Larrosa, M., Garcia-Conesa, M.T., Espin, J.C., Tomas-Barberan, F.A., 2010. Ellagitannins, ellagic acid and vascular health. Mol. Asp. Med. 31, 513–539.

López-Gresa, M.P., Maltese, F., Bellés, J.M., Conejero, V., Kim, H.K., Choi, Y.H., 2010. Metabolic response of tomato leaves upon different plant–pathogen interactions. Phytochem. Anal. 21, 89–94.

Maloney, V., 2004. Plant metabolomics. Biotechnol. J. 12, 92–99.

Marcia, S.P., Joao, E.C., Lajolo, F.M., Genovese, M.I., Shetty, K., 2010. Evaluation of Ant proliferative, anti-type 2 diabetes, and antihypertension potentials of ellagitannins from strawberries (Fragaria × ananassa Duch.) using in vitro models. J. Med. Food 13, 1027–1035.

Mattoo, A.K., Sobolev, A.P., Neelam, A., Goyal, R.K., Handa, A.K., Segre, A.L., 2006. Nuclear magnetic resonance spectroscopy-based metabolite profiling of transgenic tomato fruit engineered to accumulate spermidine and spermine reveals enhanced anabolic and nitrogen-carbon interactions. Plant Physiol. 142, 1759–1770.

Memelink, J., 2004. Tailoring the plant metabolome without a loose stitch. Trends Plant Sci. 10, 305–307.

Mintz-Oron, S., Mandel, T., Rogachev, I., Feldberg, L., Lotan, O., Yativ, M., et al., 2008. Gene expression and metabolism in tomato fruit surface tissues. Am. Soc. Plant Biol. 147, 823–851.

Moco, S., Capanoglu, E., Tikunov, Y., Bino, R.J., Boyacioglu, D., Hall, R.D., et al., 2007. Tissue specialization at the metabolite level is perceived during the development of tomato fruit. J. Exp. Bot. 58, 4131–4146.

Nakamura, H., Hakata, M., Amano, K., Miyao, A., Toki, N., Kajikawa, M., et al., 2007. A genome-wide gain-of function analysis of rice genes using the FOX-hunting system. Plant Mol. Biol. 65, 357–371.

Oms-Oliu, G., Hertog, M.L.A.T.M., Van de Poel, B., Ampofo-Asiama, J., Geeraerd, A.H., Nicolaï, B.M., 2011. Metabolic characterization of tomato fruit during preharvest development, ripening, and postharvest shelf-life. Postharvest Biol. Technol. 62, 7–16.

Petró-Turza, L., 1986. Flavor of tomato and tomato products. Food Rev. Int. 309–351.

Richardson, D.L., Davies, H.V., Ross, H.A., Mackay, G.R., 1990. Invertase activity and its relation to hexose accumulation in potato tubers. J. Exp. Bot. 41, 95–99.

Rist, M.J., Wenzel, U., Daniel, H., 2006. Nutrition and food science go genomic. Trends Biotechnol. 24, 172–178.

Roessner-Tunali, U., Hegemann, B., Lytovchenko, A., Carrari, F., Bruedigam, C., Granot, D., et al., 2003. Metabolic proofing of transgenic tomato plants overexpressing hexokinase reveals that the influence of hexose phosphorylation diminishes during fruit development. Plant Physiol. 133, 84–99.

Schauer, N., Fernie, A.R., 2006. Plant metabolomics: towards biological function and mechanism. Trends Plant Sci. 11, 508–516.

Shanks, J.V., 2005. Phytochemical engineering: combining chemical reaction engineering with plant science. AICHE J. 51, 2−7.

Sharma, M., Li, L., Celver, J., Killian, C., Kovoor, C., Seeram, N.P., 2010. Effects of fruit ellagitannin extracts, ellagic acid, and their colonic metabolite, urolithin A, on Wnt signaling. J. Agri. Food Chem. 58, 3965−3969.

Swarbreck, D., Wilks, C., Lamesch, P., Berardini, T.Z., Garcia-Hernandez, M., Foerster, M., et al., 2008. The Arabidopsis Information Resource (TAIR): gene structure and function annotation. Nucleic Acids Res. 36 (Database Issue), D1009−D1014.

Tilman, D., Fargione, J., Wolf, B., 2001. Forecasting agriculturally driven global environmental change. Science 292, 281−284.

Toki, S., Hara, N., Ono, K., Onodera, H., Tagiri, A., Oka, S., et al., 2006. Early infection of scutellum tissue with *Agrobacterium* allows high speed transformation of rice. Plant J. 47, 969−976.

Tomas-Barberan, F.A., Espin, J.C., Garcia-Conesa, M., 2009. Bioavailability and metabolism of ellagic acid and ellagitannins. In: Quideau, S. (Ed.), Chemistry and Biology of Ellagitannins. World Scientific, London, pp. 273−297.

Unger, K.K., 2004. Scientific achievements of Jack Kirkland to the development of HPLC and in particular to HPLC silica packings—a personal perspective. J. Chromatogr. A 1060, 1−7.

William, B.J., Cameron, C.J., Workman, R., Broekling, C.D., Summer, L.W., Smith, J.T., 2007. Amino acid profiling in plant cell cultures: an inter-laboratory comparison of CE-MS and GC-MS. Electrophoresis 28, 1371−1379.

Wilson, I., Nicholson, J., Castro-Perez, J., Granger, J., Johnson, K., Smith, B., et al., 2005. High resolution "ultra-performance" liquid chromatography coupled to oa-TOF mass spectrometry as a tool for differential metabolic pathway profiling in functional genomic studies. J. Proteome Res. 4, 591−598.

Wishart, D.S., Knox, C., Guo, A.C., Eisner, R., Young, N., Gautman, B., et al., 2009. HMDB: a knowledge-base for the human metabolome. Nucleic Acids Res. 37, D603−D610.

Wolfender, J.L., Rudaz, S., Choi, Y.H., Kim, H.K., 2013. Plant metabolomics: from holistic data to relevant biomarkers. Curr. Med. Chem. 20, 1056−1090.

Wright, M., Tung, C.-W., Zhao, K., Reynolds, A., McCouch, S.R., Bustamante, C.D., 2010. ALCHEMY: a reliable method for automated SNP genotype calling for small batch sizes and highly homozygous populations. Bioinformatics 26, 2952−2960.

Zeisel, S.H., 2007. Nutrigenomics and metabolomics will change clinical nutrition and public health practice: insights from studies on dietary requirements for choline. Am. J. Clin. Nutr. 86, 542−548.

Zhang, J., Wang, X., Yu, O., Tang, J., Gu, X., Wan, X., et al., 2011. Metabolite profiling of strawberry (*Fragaria* x *ananassa* Duch.) during fruit development and maturation. J. Exp. Bot. 62, 1103−1118.

Zhao, T.Y., Thackera, R., Corum, J.W., Snyder, J.C., Meeley, R.B., Obendorf, R.B., et al., 2004. Expression of the maize galactinol synthase gene family. I. Expression of two different genes during seed development and germination. New Phytol. 121, 634−646.

Transcription Factors and Environmental Stresses in Plants

3

Loredana F. Ciarmiello, Pasqualina Woodrow, Pasquale Piccirillo, Antonio De Luca and Petronia Carillo

3.1 Introduction

Plants, being sessile organisms, must cope with a wide range of unfavorable environmental conditions. Therefore, they act within a broad spectrum of response mechanisms, among which their expression of stress-protective proteins and metabolism changes. In fact, the integration of signals and their processing from various sources is regulated by complex networks of genes—proteins and/or protein—protein interactions, which in turn synergistically controls multiple cellular processes allowing adaptation and survival of plants (Dufour et al., 2010). Tolerance and susceptibility to abiotic stresses are very complex phenomena regulated by intricate networks of signal cascades and responses, which change according to plant development stages and environmental conditions. Therefore, the perception of abiotic stresses and signal transduction in plants exposed to adverse environments are crucial steps to determine plant growth, reproduction, and survival (Chinnusamy et al., 2003). Recently, transcriptomic approaches have shown that different environmental stresses cause similar responses. The overlap between responses to different stresses is known as cross-tolerance, an ability of plants to decrease the damage inflicted by a stress when they are exposed to the same and/or a different stress.

Numerous transcription factor families are known to translate stress signals into changes in gene expression. The main families involved in plant abiotic stress response are: CBF/DREB regulon; the NAC (NAM, ATAF, and CUC) and ZF-HD (zinc-finger homeodomain) regulon; the AREB/ABF (ABA-responsive element binding protein/ABA binding factor) regulon; and the MYC (myelocytomatosis oncogene)/MYB (myeloblastosis oncogene) regulon.

3.2 Transcription factors activate stress responsive genes

Transcription factors (TFs) and kinases are the main proteins involved in gene expression and signal transduction; other proteins, as well as late embryogenesis-abundant (LEA) proteins and chaperones, osmolyte biosynthesis, and reactive oxygen species (ROS) detoxification enzymes, are involved in stress tolerance (Yamaguchi-Shinozaki and Shinozaki 2005; Carillo et al., 2011). The genes are mainly induced at the transcriptional level; the spatiotemporal regulation of their expression is a fundamental component of plant response to stress (Rushton et al., 2010). Recently,

P. Ahmad (Ed): Emerging Technologies and Management of Crop Stress Tolerance, Volume 1.
DOI: http://dx.doi.org/10.1016/B978-0-12-800876-8.00003-5

significant advances have been made in order to unravel the mechanisms of transcriptional changes induced by environmental stresses and to identify signaling proteins (among them TFs) involved in the regulation of gene expression under stress. Usually, genes' transcriptional regulation is directly controlled by a network of TFs and transcription factor binding sites (TFBS) (Chaves and Oliveira, 2004).

The families of the plant-specific TFs are defined by their characteristic TFBS. After completion of the *Arabidopsis thaliana* (*Arabidopsis*) genome sequence, approximately 1500 probable TFs belonging to approximately 30 TF families were identified. Approximately half of the families were considered plant specific, because no members have been identified in other eukaryotic lineages. In particular, a plant TFBS three-dimensional structure analysis done by Yamasaki and co-workers (Todaka, 2012) has demonstrated that they are likely to have originated from endonucleases associated with transposable elements. After the TFBSs have been established in unicellular eukaryotes, they experienced extensive plant-specific expansion, by acquiring new functions and complexity. Indeed, to adapt a specific response to a given abiotic stress stimulus, complex pathways are present in plants, which involve the interplay of specific additional cofactors and signaling molecules. In particular, TFs are specific proteins that regulate stress responses by modifying gene expression. They interact with specific target genes' *cis*-elements in their promoter regions and form a complex of transcription initiation on the TATA box (core promoter) upstream of the transcription start sites (Figure 3.1). This transcriptional regulatory system is known as "regulon."

There are numerous regulons involved in plant responses to abiotic stresses. The dehydration-responsive element binding protein 1 (DREB1)/C-repeat binding factor (CBF) regulon plays an important role in plant acclimation to low temperature, whereas the DREB2 regulon is involved in plant response to high temperature and osmotic stress and can modify gene expression independently of abscisic acid (ABA). On the contrary, the ABA-responsive element (ABRE) binding protein (AREB)/ABRE binding factor (ABF) regulon is dependent on ABA and modulates gene expression under osmotic stress. There are also regulons involved in minor pathways, such as NAC (or NAM, no apical meristem) and myeloblastosis–myelocytomatosis (MYB/MYC), that control ABA and jasmonic acid (JA)-inducible gene expression under cold and osmotic stresses (Saibo et al., 2009; Todaka et al., 2012; Yamasaki et al., 2012) (Figure 3.2).

We will discuss the different TFs and their regulatory networks by focusing on the main regulons and therefore on the set of their target genes in order to understand the different mechanisms of action upon exposures to abiotic stress.

3.3 APETALA 2/ethylene-responsive element-binding factor

A unique class of plant proteins that regulates biotic and abiotic stress responses is the APETALA 2/ethylene-responsive element binding factor (AP2/ERF) (Agarwal et al., 2006a; Shoji et al., 2013; Wang et al., 2013b). This TF's family is characteristic because it contains an AP2/ERF-type DNA binding site (DBDs). Four major subfamilies belong to the AP2/ERF family: AP2, ERF, dehydration-responsive element binding protein (DREB), and related to AB13/VP1 (RAV) (Sakuma et al., 2002). AP2/ERF TFs have a conserved DBD of about 60 amino acids binding two GCC box *cis*-elements, found in many pathogen-related (PR) gene promoters, which is ethylene responsive (Gu et al., 2000).

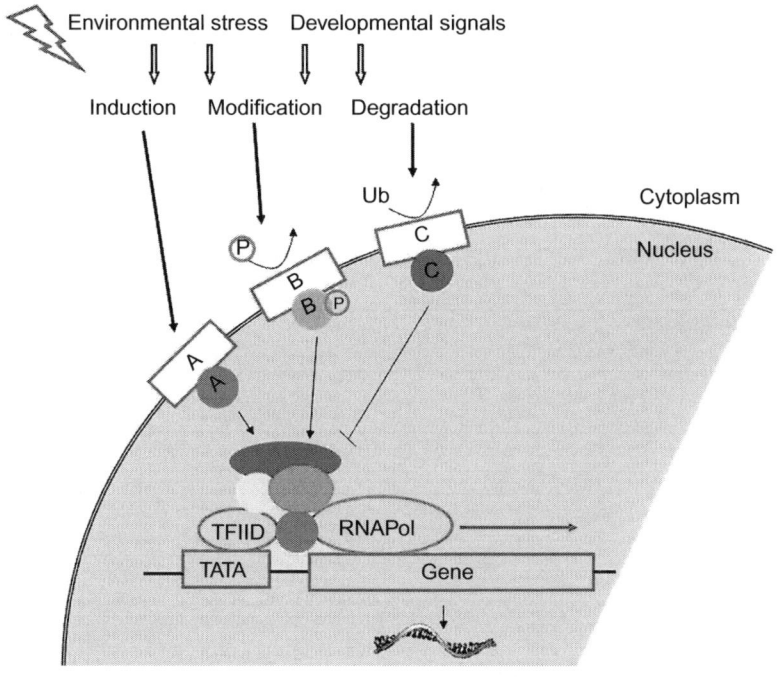

FIGURE 3.1

Transcriptional regulatory networks. Environmental stress and/or development signals active transcriptional factors. The rectangular boxes named A, B, and C represent the *cis*-acting factors and circles labeled A, B, and C represent the transcription factors.

Abbreviations: Ub, ubiquitin; P, phoshorylation.

Moreover, they share a C repeat CRT/dehydration-responsive element (DRE) domain related to gene expression under cold and dehydration (Agarwal et al., 2006b). TFs families share a homologous DBD. Their DBDs recognize and bind DNA sequences that differ among the subfamilies. Furthermore, individual TFs belonging to a family are able to act differently in response to diverse stresses (Dietz et al., 2010; Akhtar et al., 2012). Mutational and deletion analyses of gene promoters responsive to UV radiation, light, and JA demonstrated that mutations in the G-box affect the promoter capacity to respond to specific stimuli. At least a *cis*-acting element is required in addition to the G-box within promoter regions for the proper transcriptional activation.

An AP2/ERF binding domain was found for the first time in the *Arabidopsis* homeotic gene APETALA 2 (Jofuku et al., 1994; Mizoi et al., 2012), while an analogue ethylene-responsive domain was found in *Nicotiana tabacum* (EREBPs) (Ohme-Takagi and Shinshi, 1995). Because bacterial and viral endonucleases show sequences similar to the AP2/ERF domain, it has been suggested that a cyanobacterium may have transferred this domain by endosymbiosis, or a bacterium or a virus has done the same by lateral gene transfer events (Magnani et al., 2004).

The structure of the *Arabidopsis* ERF1 protein complexed with a target DNA molecule was analyzed by NMR showing that the AP2/ERF element has an N-terminal three-strand antiparallel

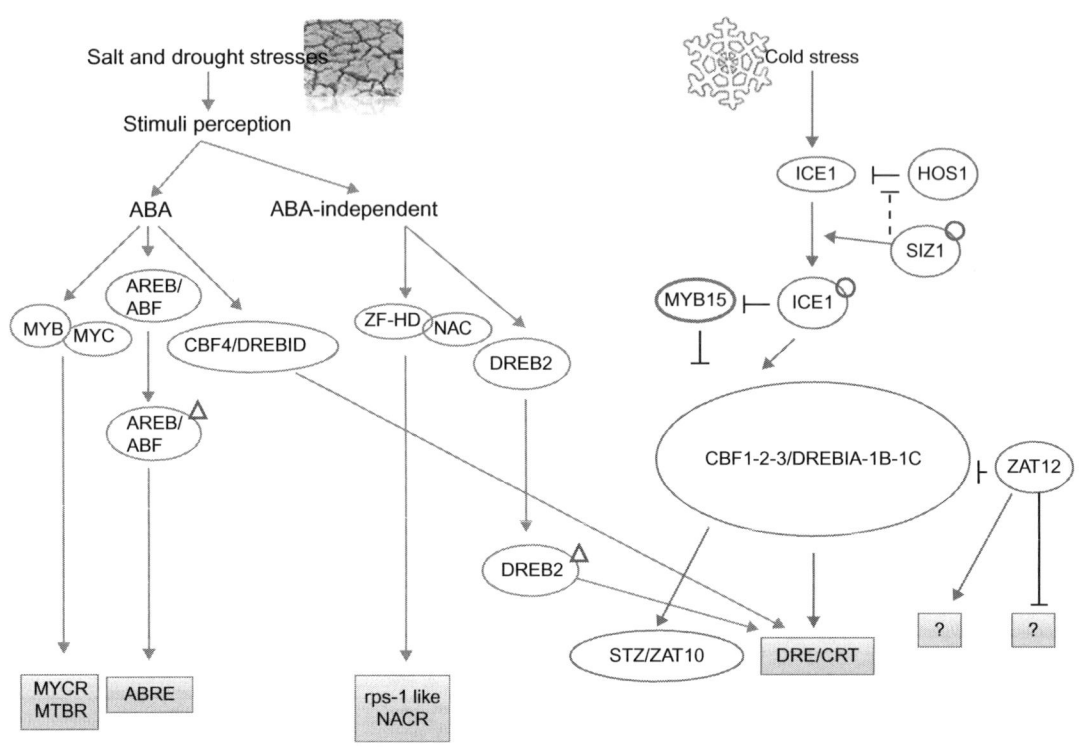

FIGURE 3.2

Transcriptional network of abiotic stress responses. Transcription factors are shown in ovals. Transcription factor-modifying enzymes are shown in circles. The small triangles correspond to post-translational modifications. Sky-blue squares with question marks represent putative MYC ICE1-like transcription factors that may activate CBF1/DREB1B and CBF2/DREB1C. The sky-blue boxes represent the *cis*-elements present in stress-responsive genes. The red dot corresponds to the sumoylation modification by SIZ1 of the ICE1 transcription factor. The dashed black line from SIZ1 to HOS1 represents competition for binding sites on the ICE1 transcription factor. SIZ1 blocks the access of HOS1 to the ubiquitination sites on the ICE1.

Adapted from Woodrow et al., 2012b.

β-sheet able to recognize a target sequence connected by loops and a C-terminal α-helix, arranged almost parallel between them (Allen et al., 1998; Figure 3.3).

DREBs, known also as CBF (C-repeat binding factor) proteins, members of the ERF subfamily, are essential for activating abiotic stress-responsive genes conferring resistance to plants (Dubouzet et al., 2003). As previously mentioned, DREB TFs could be divided into DREB1 and DREB2, which take part in two separate signal transduction pathways, respectively responsive to low temperature and dehydration. The AP2 family genes possess two repeats of the AP2/ERF domain (14 members in *Arabidopsis*) and the ERF family proteins contain a single AP2/ERF domain, whereas RAV subfamily proteins (six members) have an additional B3 DNA binding domain (Kagaya et al., 1999; Rashid et al., 2011). In the ERF/AP2 domain, the specific DRE *cis*-element DNA binding is

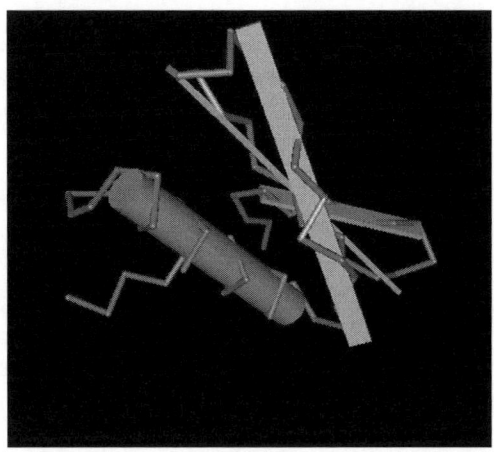

FIGURE 3.3

AP2; DNA-binding domain found in transcription regulators in plants such as APETALA2 and ethylene-responsive element binding protein (EREBP). In EREBPs, the domain specifically binds to the 11 bp GCC box of the ethylene response element (ERE), a promoter element essential for ethylene responsiveness. EREBPs and the C-repeat binding factor CBF1, which is involved in stress response, contain a single copy of the AP2 domain.

Modified from Allen et al., 1998.

principally due to two amino acids, valine (14) and glutamic acid (19) (Sakuma et al., 2002). Arginine (6, 8, and 25), tryptophan (10 and 27), and glutamic acid (16) are among the amino acids that allow the DNA binding (Allen et al., 1998). DREB protein phosphorylation is probably allowed by a conserved Ser/Thr-rich region, adjacent to the ERF/AP2 domain, while a conserved alanine (37) is capable of binding to the DRE element (Liu et al., 1998; Magnani et al., 2004).

The entry of DREB proteins is mediated by one or two nuclear localization signals (NLS) (Akhtar et al., 2012).

A basic amino acid-rich stretch with a consensus sequence PKRPAGRTKFRETRHP is typical of the DREB1/CBF1-type proteins and sets them apart from others belonging to the ERF/AP2 family. Ethylene-responsive TF RAV1 binds specifically to bipartite recognition sequences composed of two unrelated motifs, 5′-CAACA-3′ and 5′-CACCTG-3′. This TF contains two distinct DNA binding domains (Kagaya et al., 1999; Sohn et al., 2006): one is located in the N-terminal region and binds the 5′-CAACA-3′ motif and the other one is located in the C-terminal region and binds the 5′-CACCTG-3′ motif (Kagaya et al., 1999). RAV1, in fact, interacts in a selective and non-covalent manner with the two previously mentioned DNA sequences, in order to modulate transcription. *RAV1* expression is down-regulated by brassinosteroid and zeatin (Hu et al., 2004). RAV proteins have important roles in several physiological processes such as lateral root development and leaf and flower development (Ikeda and Ohme-Takagi, 2009). Moreover, they are involved in plant response to both biotic and abiotic stresses. For example, RAV1 is expressed in pepper and in *Brassica napus* where it has functional roles in bacterial disease resistance and drought and salt stress tolerance (Sohn et al., 2006; Zhuang et al., 2011). In *Arabidopsis*, members of the RAV

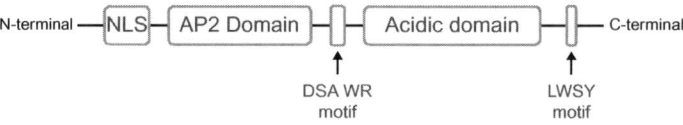

FIGURE 3.4

Schematic structure of typical DREB1/CBF domains.

family as RAV1 and RAV2 exhibit transcriptional repressive activity when they bind a plant-specific repression domain that contains the conserved amino acid sequence LXLXL (Ikeda and Ohme-Takagi, 2009). The AP2 domain includes two regions: YRG and RAYD. YRG has an N-terminal stretch of 20 amino acids rich in basic and hydrophobic amino acids that confers it a basic nature probably responsible in part for the DNA binding activity. RAYD is approximately 40 amino acids long and influences the YRG element conformation, being responsible for the regulation of DREB proteins (Okamuro et al., 1997).

Two motifs are strongly conserved in most of the DREB1-type proteins: the DSAW motif (A (A/V)xxA (A/V)xxF) and the LWSY motif, which show a consensus sequence of (L/Y)(L/Y)x (N/S)(M/L)A(E/Q)G(M/L)(L/M)xxPP, located at the end of the ERF/AP2 domain and at the end of the C-terminal, respectively (Figure 3.4). Protein—protein interaction is mediated by an 18 amino acid amphipathic α-helix present in the C-terminal region.

In the eudicot plants, the LLxNM motif is located at the beginning of the domain, while in the monocot CBF homologues there is a YYxSL motif (Akhtar et al., 2012).

3.3.1 Cis-acting regulatory element

Various TFs form a complex on the TATA box (core promoter) upstream of transcription start sites by interacting with promoter regions' *cis*-regulatory elements.

In *Arabidopsis*, promoter analysis of RD29A/COR78/LT178 genes induced by drought, high salinity, and cold highlighted a 9 bp conserved sequence (TACCGA-CAT) (Akhtar et al., 2012). Low-temperature-responsive element (LTRE) and C-repeat (CRT), containing an A/GCCGAC element forming the DRE sequence core and similar to the previous *cis*-acting motif, regulate cold-responsive genes (Stockinger et al., 1997; Thomashow, 1999). In the promoter regions of several ABA-inducible genes was found a conserved *cis*-acting element, PyACGTGGC, identified for the first time as an ABA-responsive element (ABRE) in the wheat Em gene (Guiltinan et al., 1990) and then in the rice RAB16 gene expressed in both dehydrated vegetative tissues and maturating seeds (Mundy et al., 1990).

DREB1A and DREB2A TFs show the highest affinity for the DRE core sequence A/GCCGAC (Sakuma et al., 2002). Accurate analyses of up-regulated gene promoters in transgenic plants that overexpress DREBs have demonstrated that DREB1A binds preferentially A/GCCGACNT, while DREB2A binds preferentially ACCGAC (Sakuma et al., 2006). In *DREB2A* and its homologue *DREB2B* expression is induced by extreme salt stress and dehydration (Liu et al., 1998). These two proteins need to be activated by post-translational modification but the activation mechanism has not yet been unraveled. Some studies have demonstrated that abiotic stresses can induce at different time periods the expression of diverse *DREB* genes. In particular, drought/dehydration or salt stress (250 mM NaCl) can activate *AtDREB2A* in 10 min (Liu et al., 1998) and *Os-DREB2A* in 24 h, while this latter responds weakly to cold stress and ABA (Dubouzet et al., 2003). Information on the tissue-specific expression of DREBs is scant at the moment. But it has been found in soybean seedlings that drought, salt, and cold promote *GmDREBa* and *GmDREBb* transcription in leaves, while the same abiotic stresses plus ABA treatments induce high expression levels of *GmDREBc* in roots (Li et al., 2005; Chen et al., 2007). Five ERF subfamily proteins show the highest affinity for the core sequence of the ERE, the AGCCGCC motif (Fujimoto et al., 2000), while the preferred binding motif for barley DREB2 (A-2) subgroup proteins (HvDRF1) is TT/AACCGCCTT (Xue and Loveridge, 2004).

Based on the above-mentioned studies, DREB and ERF are TFs deeply involved in the regulation of abiotic stress-responsive genes, which have a key role in conferring plant stress tolerance.

3.4 The MYC/MYB transcriptional factors

MYC/MYB TFs take part in the ABA-dependent stress-signaling pathway up-regulating the abiotic stress-responsive genes.

The N-terminal region of the plant MYC-like basic helix−loop−helix TFs (bHLH TFs) contains two domains; one around 190 amino acids is located at the N-terminus that comprises a domain of interaction with MYB-like TFs. It is an activation domain liable for transactivation. Indeed, the activation domain binds the RNA polymerase II complex and afterwards starts the transcription (Pattanaik et al., 2008). MYC-like bHLH TFs show multiple interfaces of protein dimers and one of these, located at the C-terminus, is an ACT-like domain. The bHLH proteins are part of a large transcriptional regulatory family characterized by a domain of about 60 amino acids. The key region domain binds an E-box (CANNTG) sequence and the HLH region plays an important role in homo- and heterodimerization (Li et al., 2006). The MYB domain is characterized by the presence of one to three repeats, each one of about 52 amino acids, showing in the major groove a helix-turn-helix conformation intercalated to DNA (Yanhui et al., 2006). Three major groups of MYB proteins are present in plants: group I (R2R3-MYB) which shows two adjacent repeats; group II (R1R2R3-MYB) with three adjacent repeats; and group III (MYB-related proteins) which has a single MYB repeat (Stracke et al., 2001; Lata et al., 2011).

MYB TFs function as important components in the complex signaling pathways of plant defense against abiotic stress (i.e., low and high temperature, light and water and nutrient deficiency) and they also play important roles in several developmental and physiological processes (i.e., cell morphogenesis, meristem formation, floral and seed development, cell cycle control, and hormone signaling) (Kirik et al., 1998; Higginson et al., 2003; Araki et al., 2004; Newman et al., 2004: Wang et al., 2013a). ABA increase can induce MYC and MYB TF accumulation. MYB and MYC together with the transcriptional repressor family of Cys2/His2 (C2H2)-type zinc-finger proteins respond to ABA increase under drought.

MYC, in cooperation with MYB, is involved in the regulation of ABA-inducible genes under drought stress conditions. In fact, in *Arabidopsis*, drought-inducible gene responsive to dehydration

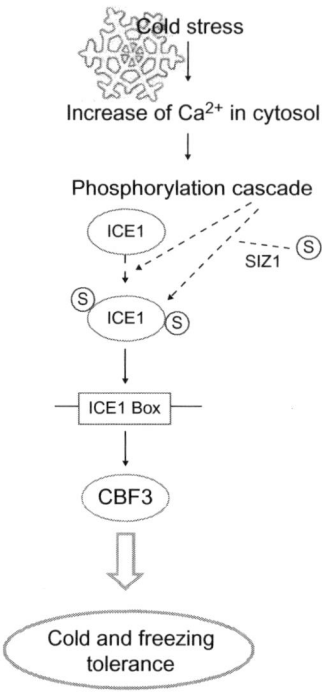

FIGURE 3.5

Cold-responsive transcriptional network in *Arabidopsis*. Constitutive expressed ICE1 is activated by cold stress after sumoylation and phosphorylation post-transcriptional modifications. Active ICE1 induces the transcription of CBFs. Broken arrows indicate post-translational regulation; solid arrows indicate activation.

22 (RD22) responds to ABA. RD22 contains in its promoter region the *cis*-element recognition sites MYC (CANNTG) and MYB (C/TAACNA/G). An enhanced ABA sensitivity and tolerance to drought can be obtained by overexpressing these TFs (Abe et al., 2003).

AtMYB102 is able to integrate drought, salt, and osmotic stresses or wound-signaling and ABA pathways (Denekamp and Smeekens, 2003). AtMYB60 and AtMYB61 are involved in stomata opening induced by light (Cominelli et al., 2005) and their closure induced by dark (Liang et al., 2005). These TFs can operate as a negative regulator of blue light-mediated photomorphogenic growth and blue and far red light-regulated gene expression (Yadav et al., 2005; Novillo et al., 2012). Moreover, studies have shown that AtMYB15 expression is up-regulated by cold and salt stresses (Agarwal et al., 2006a). Inducer of CBF Expression1 (ICE1), a MYC-type bHLH TF, interacts with the CBF3 promoter MYC recognition elements, allowing CBF3 expression in *Arabidopsis* during acclimation to cold. ICE1 is localized in the nucleus and constitutively expressed, but it promotes the expression of CBFs only during low temperature stress (Figure 3.5). Therefore, a post-translational modification induced by cold is necessary for ICE1-dependent activation of stress-responsive genes in plants (Chinnusamy et al., 2003).

3.5 **NAC transcriptional factors**

The plant-specific NAC [no apical meristem (NAM); *Arabidopsis* transcription activation factor (ATAF); cup-shaped cotyledon (CUC)] proteins constitute a major TF superfamily renowned mostly for their important functions in plant development and stress responses (Fujita et al., 2004; Puranik et al., 2011; Todaka et al., 2012). NAC TFs were originally identified due to their role in diverse developmental processes (Souer et al., 1996); a growing body of evidence suggests that they also play a significant role in both biotic and abiotic stress responses. These hypotheses have been made mainly based on observations of rapid induction of NAC genes in response to stress stimuli and enhanced stress tolerance in transgenic plants overexpressing these genes (Lata et al., 2011 and references therein). It has been proved that at least one type of abiotic stress signal (drought, cold, salt stress, or ABA) can induce NAC genes (Jensen et al., 2010). In addition to rapid gene expression in response to stress stimuli, NAC transcription factors can be regulated at the mRNA and/or protein level.

The NAC family is the largest family of TFs in plants (Puranik et al., 2012). Its members have been identified and characterized in model plants like *Arabidopsis* (Ooka et al., 2003; Fujita et al., 2004; Tran et al., 2004), crops such as rice (*Oryza sativa*) (Ooka et al., 2003; Hu et al., 2006, 2008), soybean (*Glycine max*) (Hao et al., 2011; Le et al., 2011), and wheat (*Triticum* sp.) (Kawaura et al., 2008; Xia et al., 2010), but also in poplar (*Populus trichocarpa*) (Hu et al., 2010), citrus (*Citrus* sp.) (Liu et al., 2009; de Oliveira et al., 2011), and grape (*Vitis vinifera*) (Wang et al., 2013b). One hundred and seventeen NAC genes have been identified in *Arabidopsis* (Jensen et al., 2010), 151 in rice (Nuruzzaman et al., 2010), 163 in poplar, 152 in tobacco (*Nicotiana tabacum*) and soybean, and 79 in grape (*Vitis vinifera*) (Rushton et al., 2008; Hu et al., 2010; Nuruzzaman et al., 2010; Le et al., 2011), which makes them one of the largest family of TFs in plants (Hao et al., 2010; Puranik et al., 2012).

3.5.1 **NAC TFs' role in abiotic stress response**

NAC TFs are implicated in abiotic stress responses to salt and ABA (He et al., 2005).

Some NAC TFs associated with abiotic stresses of subgroup III-3, called ATAF (Jensen et al., 2010), are also indicated as stress-responsive NAC (SNAC) (Nuruzzaman et al., 2010). ANAC019 and ANAC055, which are members of the SNAC subgroup, and RESPONSIVE TO DEHYDRATION 26 (RD26) (ANAC072), are induced by drought, high salinity, or cold stress, and are under the control of the central ABA perception and signaling network (Tran et al., 2004; Jensen et al., 2010). Transformed plants that overexpress SNAC genes under drought up-regulate several genes acquiring stress tolerance (Tran et al., 2004; Naika et al., 2013). These TFs bind the EARLY RESPONSIVE TO DEHYDRATION STRESS 1 (ERD1) promoter (Tran et al., 2004; Nakashima et al., 2012), which is not only induced by dehydration but is also up-regulated during senescence and dark-induced etiolation. ERD1 up-regulation depends on the cooperative overexpression of two *cis*-acting elements: the ZF homeodomain transcriptional activator and the NAC TFs (Lindemose et al., 2013). ATAF1 overexpression affects drought tolerance in plants even if it is not yet clear if ATAF1 acts negatively or positively (Lu et al., 2007; Wu et al., 2009). What is clear is that the NAC TFs subgroup III-3 is certainly involved in perception of abiotic stresses and the different NAC members show overlapping responses to stress (Figure 3.6).

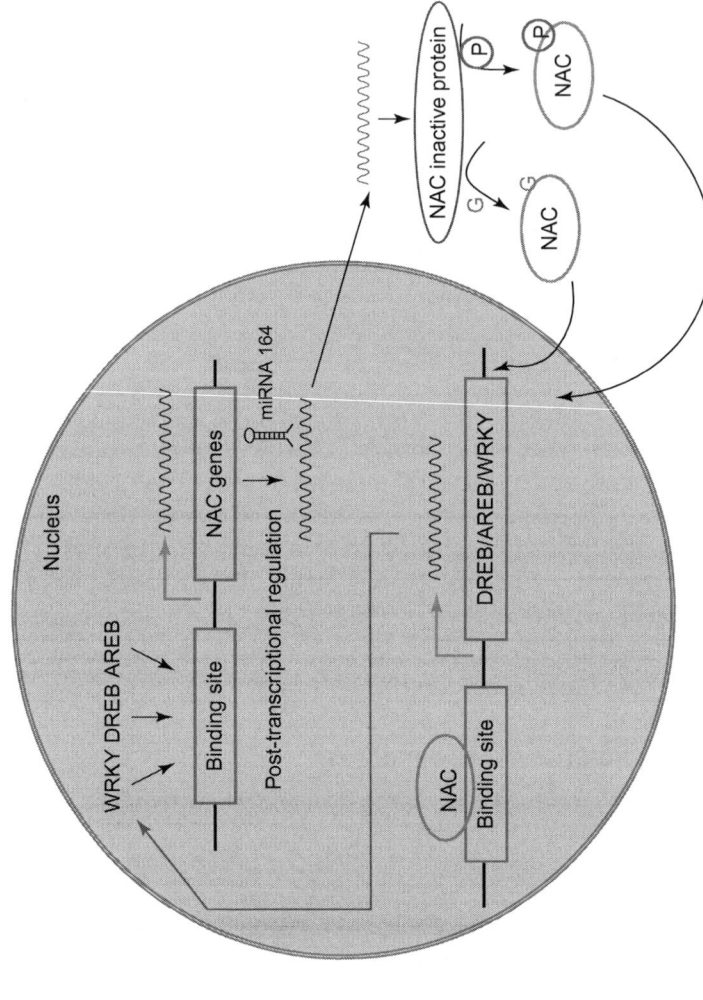

Plasma membrane

Cytoplasm

FIGURE 3.6

A model representing the regulation of NAC TFs at different levels during stress. Upstream TFs like DREBs, WRKYs or ABREs may bind to stress-related *cis*-regulatory elements in the promoter of regulated NAC genes and influence their transcription. Post-transcriptional control of pre-NAC mRNAs by miRNA164 and alternate splicing may also regulate their expression. The translated protein may be subjected to activation by phosphorylation or glycosylation. Upon their nuclear import, the activated NAC proteins homo- or heterodimerize or bind to promoters of their target genes to control their expression. Abbreviations: AREB, ABA-responsive element binding protein; DREB, dehydration-responsive element binding protein; G, N-acetyl glucosamine; miRNA, microRNA; and P, phosphate.

Adapted from Puranik et al., 2012.

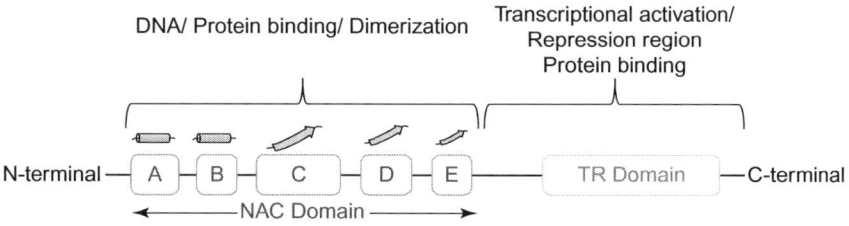

FIGURE 3.7

Schematic structure of a typical NAC domain protein with a highly conserved NAC domain at the N-terminal, which is further divided into five conserved subdomains (indicated with the letters A–E). This region holds DNA binding (DB) ability and/or is responsible for protein binding and dimerization. Secondary elements are shown at the top of subdomains (A and B α-helices; C–E β-strands). The C-terminal region is more diverged and serves as a potential transcriptional regulatory (TR) domain (activator or repressor); it may occasionally have protein binding activity.

Cis-acting elements present in the stress-responsive NACs are responsible for their own expression. Among these *cis*-acting elements are elements responsive to ABA (ABREs), dehydration (DREs), low temperature (LTREs), salicylic and jasmonic acids, but also myelocytomatosis (MYC) and myeloblastosis (MYB) binding sites and W-box (Nakashima et al., 2012).

3.5.2 Structural features of NAC TFs

The DNA sequences that are recognized show weak differences among the family members and possess palindromic properties with different distances, such as $AN_5TCN_7ACACGCATGT$ with a consecutive pseudo-palindromic core (underlined), for abscisic acid-responsive NAC (ANAC) proteins (Tran et al., 2004) and (T/A)NN(C/T)(T/C/G)TN_7A(A/C)GN(A/C/T)(A/T) with a pseudo-palindromic sequence with a seven-base gap for SND1 and VND7 (Yamasaki et al., 2012). A conserved N-terminal DNA binding domain is a common feature to all NAC domain proteins. This DNA binding domain is about 150 amino acids long and initially it was found in other four genes: ATAF1, ATAF2, CUC2, and NAM (Christianson et al., 2010; Wang and Culver, 2012). Five conserved subdomains are present in the N-terminal domain of NAC proteins, which is responsible for protein binding and dimerization as well as for DNA binding activity (Figure 3.7). The C-terminal, which can present a transmembrane motif, appears more divergent and has a putative transcriptional positive or negative regulatory function, but also has protein binding activity (Figure 3.7) (Olsen et al., 2005).

The presence of diverse isoforms of NAC proteins differing in their functional domains and mechanisms of action highlights the complexity of defense systems in plants (Puranik et al., 2012).

3.6 WRKY transcriptional factors

Members of plant WRKY TF families are among the best-characterized classes of plant TFs. They participate in disease resistance (Pandey and Somssich, 2009) and in a number of different physiological processes such as embryogenesis, trichome and maternal seed coat development, hormone signaling, and senescence (Song et al., 2010; Zhou et al., 2011; Li et al., 2013). Furthermore, WRKY TFs have an important role in the complex signaling networks in plant responses to abiotic stresses, such as water, nutrient deficiency, and cold. WRKY-GCM1 is a superfamily of ZF TFs, originated from Mutator or Mutator-like (MULE) transposase evolution (Marquez and Pritham, 2010). WRKY proteins are a class of DNA binding proteins that recognize the TTGAC(C/T) W-box elements found in the promoters of a large number of plant defense-related genes. The DNA binding domain shows a conserved WRKY N-terminus amino acid sequence (Eulgem et al., 2000). Only in a few WRKY proteins, the WRKY amino acid sequences have been replaced by WRRY, WSKY, WKRY, WVKY, or WKKY (Xie et al., 2005).

Three groups of WRKY TFs are known: two domains are present in group I, one domain in group II, and one domain plus one ZF structure in group III (C2-H/C or C2-H2 ZF motif) (Chen et al., 2012). Moreover, group II genes, not being monophyletic on the base of their amino acid sequence, can be divided into five subgroups (IIa − IIe) (Zhang and Wang, 2005; Rushton et al., 2008).

3.6.1 The WRKY domain and the W-Box

The WRKY protein family shows a highly conserved region consisting of a peptide 60 amino acids long, called the WRKY domain (Maeo et al., 2001; Ciolkowski et al., 2008). In this domain, an invariant heptapeptide WRKYGQK at the N-terminus, which shows slight variations in a few WRKY proteins (Yang et al., 2009), partly protrudes from a side of the protein allowing DNA binding. In fact, it has been well established that the b-strand containing the WRKYGQK motif makes contact with an approximately 6-bp region, and this is consistent with the length of the W-box (Rushton et al., 2010).

At the C-terminus of the WRKY domain there is also a novel ZF-like motif that contributes to the strength of DNA binding, and probably the N-terminal domain might participate in the binding process, increasing the affinity or specificity of these proteins for their target sites (Figure 3.8). Therefore, proper DNA binding of the protein is allowed by the ZF motif and not by the WRKYGQK residues (Rushton et al., 2010).

The conservation of the WRKY domain depends on the W-box (TTGACC/T) binding site's high conservation (Eulgem et al., 2000). Several experimental assays made on many different WRKY motifs showed that the W-box is the minimal consensus for DNA binding (Ciolkowski et al., 2008). It has been shown that some WRKY proteins, such as OsWRKY13, instead of binding the W-box, bind the PRE4 element (TGCGCTT) (Cai et al., 2008) or SUSIBA2 (a sucrose-regulated barley WRKY transcription factor), which bind both W-box and SURE (a sugar-responsive element, TAAAGATTACTAATAGGAA) (Sun et al., 2003).

3.6.2 Role of WRKY genes under abiotic stresses

The WRKY TFs are induced in response to different abiotic stresses and so they could be a useful tool for conferring stress tolerance. The role of WRKY TFs under environmental stimuli is not clearly unraveled. Microarray analysis on *Arabidopsis* transcriptome under salt stress reveals up-regulation of 18 WRKY and down-regulation of 8 WRKY genes (Jiang and Deyholos, 2006). The WRKY TFs can be induced by ABA, NaCl, PEG, cold, and heat treatment (Xie et al., 2005; Ramamoorthy et al., 2008) (Figure 3.9).

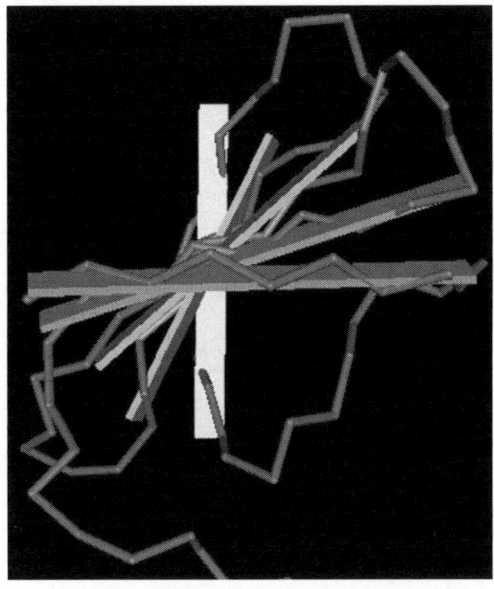

FIGURE 3.8

Crystal structure of the Arabidopsis C-terminal WRKY domain.

Adapted from Eulgem et al., 2000.

WRKY TFs regulate the plant stress response in two ways: by interacting with other TFs or directly regulating some stress-responsive genes. WRKY TFs are also in turn early response genes, and their target genes, encoding enzymes such as chitinase, glutathione S-transferase, and strictosidine synthase, are late response genes (Cheong et al., 2002) (Figure 3.10).

GmWRKY21 and GmWRKY54 transgenic *Arabidopsis* plants show a good tolerance to environmental stimuli such as cold, salt, and drought stresses. This ability is probably due to WRKY TFs, which regulate TFs like DREB2A and STZ/Zat10 (Wei et al., 2008; Zhou et al., 2008). This example shows that factors like WRKY are fully involved in signaling cascades associated with TF regulation of gene expression under abiotic stresses.

3.7 CYS2HIS2 zinc-finger (C2H2 ZF) TFs

Zinc-finger proteins (ZFPs) contain a specific catalytic domain made up of cysteines and/or histidines coordinating zinc atom(s) (C2H2) (Persikov and Singh, 2014), which can be associated with a transcriptional domain EAR (ERF-associated amphiphilic repression), modulating biotic and abiotic stress responses in plants (Gommans et al., 2005; Persikov and Singh, 2014).

Several ZAT proteins, a family of putative ZF-containing TFs, are involved in abiotic stress responses. This is confirmed by several studies on plant response to abiotic stresses. For example, in *Arabidopsis*, as reported by Lee et al., (2002), Zat10/STZ can repress the *RD29A* gene, while the *Zat10* constitutive expression enhances plants' tolerance to salt, heat, and osmotic stress. By

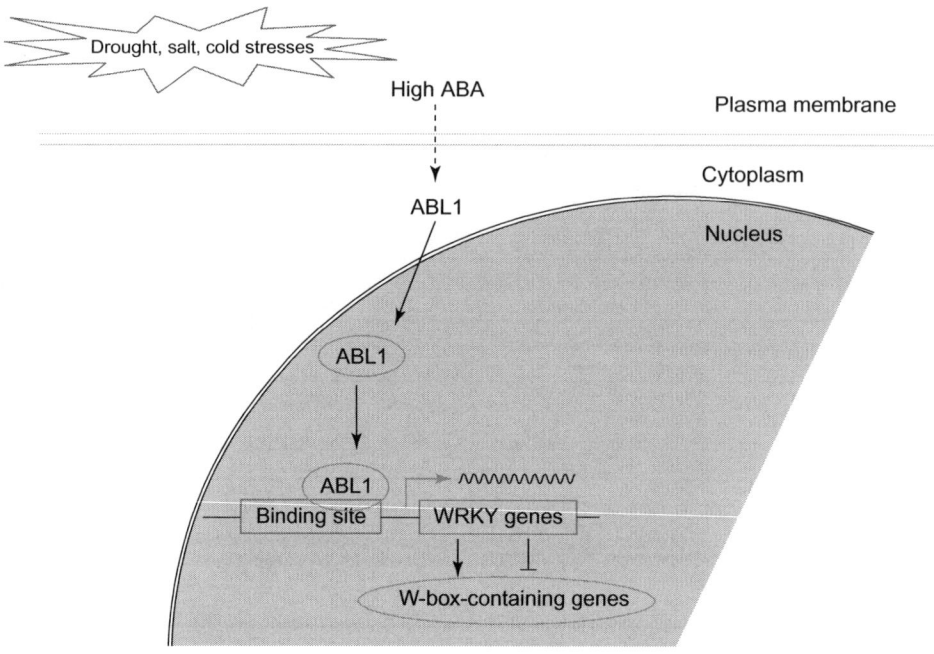

Abiotic responses; ABA responsive gene expression

FIGURE 3.9

Role of *Arabidopsis* WRKY transcriptional factors in abiotic stress response. ABA signaling pathway leads to the activation of ABL1 expression. This latter stimulates the expression of WRKY genes, which differentially regulate the downstream W-box containing genes to regulate abiotic responses and ABA signaling.

contrast, overexpression of *Zat7* also results in increased tolerance to salinity stress (Ciftci-Yilmaz et al., 2007; Cushman et al., 2007). In *Arabidopsis*, ZAT12 is involved in response signaling to oxidative stress (Rizhsky et al., 2004). During this stress, ZAT12 induces the expression of other TFs and the up-regulation of the H_2O_2-scavenging enzyme, the cytosolic ascorbate peroxidase (Woodrow et al., 2012a). ROS accumulation is a common consequence of both biotic and abiotic stresses, including hypoxia and anoxia. ZAT12 plays an important role in oxidative stress acclimation as established by several independent analyses that show ZAT12 transcription is up-regulated in response to oxygen deprivation (Branco-Price et al., 2005). ABA, IAA, and osmotic stress induce *AZF1* and *AZF2* expression, which in turn represses some genes, such as SAUR63 and SAUR20, binding their promoter region. The overexpression of *AZF1* and *AZF2* can affect plant growth and survival (Kodaira et al., 2011).

3.7.1 **C2H2 ZF DBD and overall structure of Zat7**

In rice under salt and drought stresses, C2H2 ZFP TF *DROUGHT AND SALT TOLERANCE (DST)* is involved in stomatal opening regulation mediated by H_2O_2 homeostasis (Huang et al., 2009).

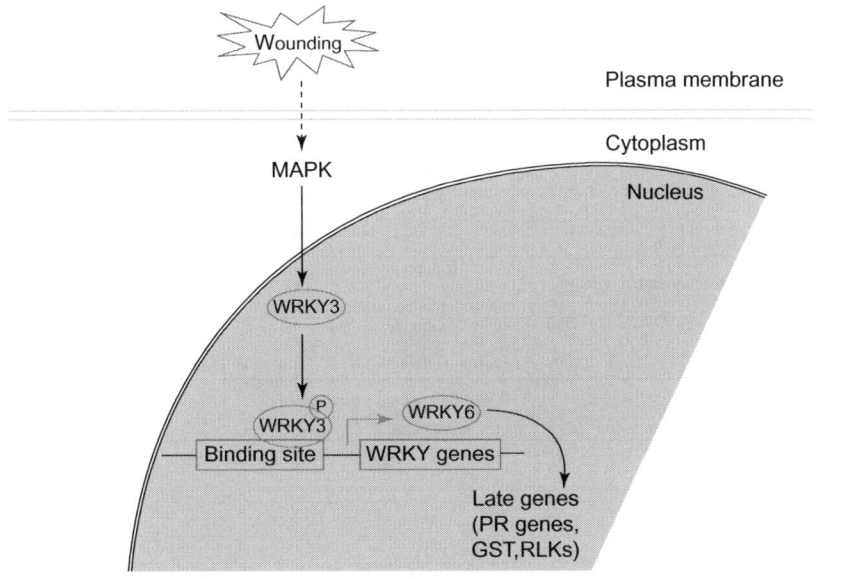

FIGURE 3.10

Role of *Arabidopsis* WRKY transcriptional factors in wounding response pathways. After activation of MAPK phosphorylation cascade, WRKY3 regulate the expression of WRKY genes. The WRKY6 activates the expression of late genes, such as chitinase and glutathione S-transferase, which allow plants to defend.

C2H2-type ZFP's (Ciftci-Yilmaz and Mittler, 2008) structure shows two fingers, in ZAT7, each formed by two β-strands and one α-helix separated by long flexible spacers (Figure 3.11).

Moreover, ZAT7 protein contains two EAR repression domains (Lindemose et al., 2013). C2H2-type ZF TF ZAT7 consists of 168 amino acids containing two substructure C2H2 ZFs with a QALGGH motif showing DNA binding activity (Ciftci-Yilmaz and Mittler, 2008).

3.8 Conclusion and future perspectives

Extremes of environmental conditions, such as, drought, salinity, heat, cold, and wounding, induce stress and decrease plant growth and crop productivity. A large number of genes have been reported as abiotic stress responsive genes and the expression of these genes is crucial for providing stress tolerance to plants. The use of several diverse methodological approaches (i.e., physiological, chemical, genetical, molecular) has allowed rapid progress in plant signal transduction and gene regulation comprehension (Chen et al., 2012), which helps us to unravel the complex mechanisms of abiotic stress responses in plants and to develop tolerant cultivars better able to cope with the increasing environmental constraints.

The interaction between TFs and genes mediated by the binding of *cis*-elements in the promoter regions activates the ABA-dependent and -independent responses to multiple stresses, increasing in

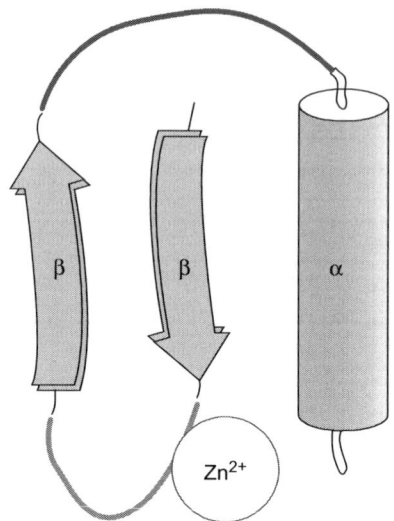

FIGURE 3.11

Schematic structure of a Cys2-His2 type ZF. This zinc-finger motif consists of two β-strands and one α-helix separated by a long spacer. Amino acid residues within the α-helix recognize the target DNA sequence.

general plant tolerance. Their further characterization and in particular their functional analysis could help to explain the complex abiotic stress regulatory pathways and the correlations between diverse stress adaptation signaling networks. The ability of TFs to induce or repress the expression of stress-responsive genes may be a useful key point to develop improved crop plants resistant or tolerant to multiple abiotic stresses. Therefore, TF coding genes should be considered as a target in breeding programs, not only to unravel the intricate pathways of plant stress signal transduction and responses, but principally to improve and facilitate the development of crop varieties superior in stress tolerance.

References

Abe, H., Urao, T., Ito, T., Seki, M., Shinozaki, K., Yamaguchi-Shinozaki, K., 2003. Arabidopsis AtMYC2 (bHLH) and AtMYB2 (MYB) function as transcriptional activators in abscisic acid signalling. Plant Cell 15, 63–78.

Agarwal, M., Hao, Y., Kapoor, A., Dong, C.H., Fuji, H., Zheng, X., et al., 2006a. A R2R3 type MYB TF is involved in the cold regulation of CBF genes and in acquired freezing tolerance. J. Biol. Chem. 281, 37636–37645.

Agarwal, P.K., Agarwal, P., Reddy, M.K., Sopory, S.K., 2006b. Role of DREB transcription factors in abiotic and biotic stress tolerance in plants. Plant Cell Rep. 25, 1263–1274.

Akhtar, M., Jaiswal, A., Taj, G., Jaiswal, J.P., Qureshi, M.I., Singh, N.K., 2012. DREB1/CBF transcription factors: their structure, function and role in abiotic stress tolerance in plants. J. Genet. 91, 385–395.

Allen, M.D., Yamasaki, K., Ohme-Takagi, M., Tateno, M., Suzuki, M., 1998. A novel mode of DNA recognition by a β-sheet revealed by the solution structure of the GCC-box binding domain in complex with DNA. EMBO J. 17, 5484−5496.

Araki, S., Ito, M., Soyano, T., Nishihama, R., Machida, Y., 2004. Mitotic cyclins stimulate the activity of c-Myb-like factors for transactivation of G2/M phase-specific genes in tobacco. J. Biol. Chem. 279, 32979−32988.

Branco-Price, C., Kawaguchi, R., Ferreira, R., Bailey-Serres, J., et al., 2005. Genome-wide analysis of transcript abundance and translation in Arabidopsis seedlings subjected to oxygen deprivation. Ann. Bot. 96, 647−660.

Cai, M., Qiu, D., Yuan, T., Ding, X., Li, H., Duan, L., et al., 2008. Identification of novel pathogen-responsive cis-elements and their binding proteins in the promoter of OsWRKY13, a gene regulating rice disease resistance. Plant Cell Environ. 31, 86−96.

Carillo, P., Parisi, D., Woodrow, P., Pontecorvo, G., Massaro, G., Annunziata, M.G., et al., 2011. Salt induced accumulation of glycine betaine is inhibited by high light in durum wheat. Funct. Plant Biol. 38, 139−150.

Chaves, M.M., Oliveira, M.M., 2004. Mechanisms underlying plant resilience to water deficits: prospects for water-saving agriculture. J. Exp. Bot. 55, 2365−2384.

Chen, L., Song, Y., Li, S., Zhang, L., Zou, C., Yu, D., 2012. The role of WRKY transcription factors in plant abiotic stresses. Biochim. Biophys. Acta. 1819, 120−128.

Chen, M., Wang, Q.Y., Cheng, X.G., Xu, Z.S., Li, L.C., Ye, X.G., et al., 2007. GmDREB2, a soybean DRE-binding transcription factor, conferred drought and high-salt tolerance in transgenic plants. Biochem. Biophys. Res. Commun. 353, 299−305.

Cheong, Y.H., Chang, H.-S., Gupta, R., Wang, X., Zhu, T., Luan, S., 2002. Transcriptional profiling reveals novel interactions between wounding, pathogen, abiotic stress, and hormonal responses in Arabidopsis. Plant Physiol. 129, 661−677.

Chinnusamy, V., Ohta, M., Kanrar, S., Lee, B., Hong, X., Agarwal, M., et al., 2003. ICE1: a regulator of cold-induced transcriptome and freezing tolerance in Arabidopsis. Gene Dev. 17, 1043−1054.

Christianson, J.A., Dennis, E.S., Llewellyn, D.J., Wilson, I.W., 2010. ATAF NAC transcription factors: regulators of plant stress signalling. Plant Signal Behav. 5, 428−432.

Ciftci-Yilmaz, S., Mittler, R., 2008. The zinc finger network of plants. Cell. Mol. Life. Sci. 65, 1150−1160.

Ciftci-Yilmaz, S., Morsy, M.R., Song, L., Coutu, A., Krizek, B.A., Lewis, M.W., et al., 2007. The EAR-motif of the Cys2/His2-type zinc finger protein Zat7 plays a key role in the defense response of Arabidopsis to salinity stress. J. Biol. Chem. 282, 9260−9268.

Ciolkowski, I., Wanke, D., Birkenbihl, R.P., Somssich, I.E., 2008. Studies on DNA-binding selectivity of WRKY transcription factors lend structural clues into WRKY domain function. Plant Mol. Biol. 68, 81−92.

Cominelli, E., Galbiati, M., Vavasseur, A., Conti, L., Sala, T., Vuylsteke, M., et al., 2005. A guard-cell-specific MYB transcription factor regulates stomatal movements and plant drought tolerance. Curr. Biol. 15, 1196−1200.

Cushman, J., Connolly, E.L., Mittler, R., 2007. The EAR-motif of the Cys2/His2-type zinc fingerprotein Zat7 plays a key role in the defense response of Arabidopsis to salinity stress. J. Biol. Chem. 282, 9260−9268.

de Oliveira, T.M., Cidade, L.C., Gesteira, A.S., Coelho Filho, M.A., Soares Filho, W.S., Costa, M.G.C., 2011. Analysis of the NAC transcription factor gene family in citrus reveals a novel member involved in multiple abiotic stress responses. Tree Genet. Genomes 7, 1123−1134.

Denekamp, M., Smeekens, S.C., 2003. Integration of wounding and osmotic stress signals determines the expression of the AtMYB102 transcription factor gene. Plant Physiol. 132, 1415−1423.

Dietz, K.-J., Vogel, M.O., Viehhauser, A., 2010. AP2/EREBP transcription factors are part of gene regulatory networks and integrate metabolic, hormonal and environmental signals in stress acclimation and retrograde signalling. Protoplasma 254, 3−14.

Dubouzet, J.G., Sakuma, Y., Ito, Y., Kasuga, M., Dubouzet, E.G., Miura, S., et al., 2003. *OsDREB* genes in rice, *Oryza sativa* L. encode transcription activators that function in drought-, high-salt- and cold-responsive gene expression. Plant J. 33, 751–763.

Dufour, Y.S., Kiley, P.J., Donohue, T.J., 2010. Reconstruction of the core and extended regulons of global transcription factors. PLoS. Genet. 6, e1001027.

Eulgem, T., Rushton, P.J., Robatzek, S., Somssich, I.E., 2000. The WRKY superfamily of plant transcription factors. Trends. Plant Sci. 5, 199–206.

Fujimoto, S.Y., Ohta, M., Usui, A., Shinshi, H., Ohme-Takagi, M., 2000. Arabidopsis ethylene-responsive element binding factors act as transcriptional activators or repressors of GCC box-mediated gene expression. Plant Cell 12, 393–404.

Fujita, M., Fujita, Y., Maruyma, K., Seki, M., Hiratsu, K., Ohmetakagi, M., et al., 2004. A dehydration-induced NAC protein, RD26, is involved in a novel ABA-dependent stress-signalling pathway. Plant J. 39, 863–876.

Gommans, W.M., Haisma, H.J., Rots, M.G., 2005. Engineering zinc finger protein transcription factors: the therapeutic relevance of switching endogenous gene expression on or off at command. J. Mol. Biol. 354, 507–519.

Gu, Y.Q., Yang, C., Thara, V.K., Zhou, J., Martin, G.B., 2000. Pti4 is induced by ethylene and salicylic acid, and its product is phosphorylated by Pto kinase. Plant Cell 12, 771–786.

Guiltinan, M.J., Marcotte Jr, W.R., Quatrano, R.S., 1990. A plant leucine zipper protein that recognizes an abscisic acid response element. Science 250, 267–271.

Hao, Y.J., Song, Q.X., Chen, H.W., Zou, H.F., Wei, W., Kang, X.S., et al., 2010. Plant NAC-type transcription factor proteins contain a NARD domain for repression of transcriptional activation. Planta 232, 1033–1043.

Hao, Y.J., Wei, W., Song, Q.X., Chen, H.W., Zhang, Y.Q., Wang, F., et al., 2011. Soybean NAC transcription factors promote abiotic stress tolerance and lateral root formation in transgenic plants. Plant J. 68, 302–313.

He, X.J., Mu, R.L., Cao, W.H., Zhang, Z.G., Zhang, J.S., Chen, S.Y., 2005. AtNAC2, a transcription factor downstream of ethylene and auxin signalling pathways, is involved in salt stress response and lateral root development. Plant J. 44, 903–916.

Higginson, T., Li, S.F., Parish, R.W., 2003. AtMYB103 regulates tapetum and trichome development in *Arabidopsis thaliana*. Plant J. 35, 177–192.

Hu, H., Dai, M., Yao, J., Xiao, B., Li, X., Zhang, Q., et al., 2006. Overexpressing a NAM, ATAF, and CUC transcription factor enhances drought resistance and salt tolerance in rice. PNAS 103, 12987–12992.

Hu, H., You, J., Fang, Y., Zhu, X., Qi, Z., Xiong, L., 2008. Characterization of transcription factor gene SNAC2 conferring cold and salt tolerance in rice. Plant Mol. Biol. 67, 169–181.

Hu, R., Qi, G., Kong, Y., Kong, D., Gao, Q., Zhou, G., 2010. Comprehensive analysis of NAC domain transcription factor gene family in *Populus trichocarpa*. BMC Plant Biol. 10, 145.

Hu, Y.X., Wang, Y.H., Liu, X.F., Li, J.Y., 2004. Arabidopsis RAV1 is down-regulated by brassinosteroid and may act as a negative regulator during plant development. Cell. Res. 14, 8–15.

Huang, X.Y., Chao, D.Y., Gao, J.P., Zhu, M.Z., Shi, M., Lin, H.X., 2009. A previously unknown zinc finger protein, DST, regulates drought and salt tolerance in rice via stomatal aperture control. Genes Dev. 23, 1805–1817.

Ikeda, M., Ohme-Takagi, M., 2009. A novel group of transcriptional repressors in Arabidopsis. Plant Cell Physiol. 50, 970–975.

Jensen, M.K., Kjaersgaard, T., Nielsen, M.M., Galberg, P., Petersen, K., O'Shea, C., et al., 2010. The *Arabidopsis thaliana* NAC transcription factor family: structure–function relationships and determinants of ANAC019 stress signalling. Biochem. J. 426, 183–196.

Jiang, Y., Deyholos, M.K., 2006. Comprehensive transcriptional profiling of NaCl-stressed Arabidopsis roots reveals novel classes of responsive genes. BMC Plant Biol. 6, 25.

Jofuku, K.D., Boer, B., Montagu, M.V., Okamuro, J.K., 1994. Control of Arabidopsis flower and seed development by the homeotic gene APETALA2. Plant Cell 6, 1211−1225.

Kagaya, Y., Ohmiya, K., Hattori, T., 1999. RAV1, a novel DNA-binding protein, binds to bipartite recognition sequence through two distinct DNA-binding domains uniquely found in higher plants. Nucleic. Acids. Res. 27, 470−478.

Kawaura, K., Mochida, K., Ogihara, Y., 2008. Genome-wide analysis for identification of salt-responsive genes in common wheat. Funct. Integr. Genom. 8, 277−286.

Kirik, V., Kölle, K., Miséra, S., Baümlein, H., 1998. Two novel MYB homologues with changed expression in late embryogenesis-defective Arabidopsis mutants. Plant Mol. Biol. 37, 819−827.

Kodaira, K.S., Qin, F., Tran, L.S., Maruyama, K., Kidokoro, S., Fujita, Y., et al., 2011. Arabidopsis Cys2/His2 zinc-finger proteins AZF1 and AZF2 negatively regulate abscisic acid-repressive and auxin-inducible genes under abiotic stress conditions. Plant Physiol. 157, 742−756.

Lata, C., Yadav, A., Prasad, M., 2011. Role of plant transcription factors in abiotic stress tolerance. In: Shanker, A., Venkateswarlu, B. (Eds.), Abiotic stress response in plants- physiological, biochemical and genetic perspectives. InTech.

Le, D.T., Nishiyama, R., Watanabe, Y., Mochida, K., Yamaguchi-Shinozaki, K., Shinozaki, K., et al., 2011. Genome-wide survey and expression analysis of the plant-specific NAC transcription factor family in soybean during development and dehydration stress. DNA. Res. 18, 263−276.

Lee, H., Guo, Y., Ohta, M., Xiong, L., Stevenson, B., Zhu, J.K., 2002. LOS2, a genetic locus required for cold-responsive gene transcription encodes a bi-functional enolase. EMBO J. 21, 2692−2702.

Li, S., Zhang, P., Zhang, M., Fu, C., Yu, L., 2013. Functional analysis of a WRKY transcription factor involved in transcriptional activation of the DBAT gene in Taxus chinensis. Plant Biol. (Stuttg) 15, 19−26.

Li, X., Tian, A.G., Luo, G.Z., Gong, Z.Z., Zhang, J.S., Chen, S.Y., 2005. Soybean DRE-binding transcription factors that are responsive to abiotic stresses. Theor. Appl. Genet. 110, 1355−1362.

Li, X., Duan, X., Jiang, H., Sun, Y., Tang, Y., Yuan, Z., et al., 2006. Genome-wide analysis of basic/helix−loop−helix transcription factor family in rice and Arabidopsis. Plant Physiol. 141, 1167−1184.

Liang, Y.K., Dubos, C., Dodd, I.C., Holroyd, G.H., Hetherington, A.M., Campbell, M.M., 2005. AtMYB61, an R2R3-MYB transcription factor controlling stomatal aperture in *Arabidopsis thaliana*. Curr. Biol. 15, 1201−1206.

Lindemose, S., O'Shea, C., Krogh Jensen, M., Skriver, K., 2013. Structure, function and networks of transcription factors involved in abiotic stress responses. Int. J. Mol. Sci. 14, 5842−5878.

Liu, Q., Kasuga, M., Sakuma, Y., Abe, H., Miura, S., Yamaguchi-Shinozaki, K., et al., 1998. Two transcription factors, DREB1 and DREB2, with an EREBP/AP2 DNA binding domain separate two cellular signal transduction pathways in drought- and low-temperature-responsive gene expression, respectively, in Arabidopsis. Plant Cell 10, 1391−1406.

Liu, Y.Z., Baig, M.N.R., Fan, R., Ye, J.L., Cao, Y.C., Deng, X.X., 2009. Identification and expression pattern of a novel NAM, ATAF, and CUC-like gene from *Citrus sinensis* Osbeck. Plant Mol. Biol. Rep. 27, 292−297.

Lu, P.L., Chen, N.Z., An, R., Su, Z., Qi, B.S., Ren, F., et al., 2007. A novel drought-inducible gene, ATAF1, encodes a NAC family protein that negatively regulates the expression of stress-responsive genes in Arabidopsis. Plant Mol. Biol. 63, 289−305.

Maeo, K., Hayashi, S., Kojima-Suzuki, H., Morikami, A., Nakamura, D., 2001. Role of conserved residues of the WRKY domain in the DNA-binding of tobacco WRKY family proteins. Biosci. Biotechnol. Biochem. 65, 2428−2436.

Magnani, E., Sjölander, K., Hake, S., 2004. From endonucleases to transcription factors: evolution of the AP2 DNA binding domain in plants. Plant Cell 16, 2265–2277.

Marquez, C.P., Pritham, E.J., 2010. Phantom, a new subclass of Mutator DNA transposons found in insect viruses and widely distributed in animals. Genetics 185, 1507–1517.

Mizoi, J., Shinozaki, K., Yamaguchi-Shinozak, K., 2012. AP2/ERF family transcription factors in plant abiotic stress responses. Biochem. Biophys. Acta 1819, 86–96.

Mundy, J., Yamaguchi-Shinozaki, K., Chua, N.-H., 1990. Nuclear proteins bind conserved elements in the abscisic acid-responsive promoter of a rice rab gene. PNAS 87, 1406–1410.

Naika, M., Shameer, K., Mathew, O.K., Gowda, R., Sowdhamini, R., 2013. STIFDB2: An updated version of plant stress-responsive transcription factor database with additional stress signals, stress-responsive transcription factor binding sites and stress-responsive genes in *Arabidopsis* and rice. Plant Cell. Physiol. 54, e8, 1–15.

Nakashima, K., Takasaki, H., Mizoi, J., Shinozaki, K., Yamaguchi-Shinozaki, K., 2012. NAC transcription factors in plant abiotic stress responses. Biochim. Biophys. Acta. 1819, 97–103.

Newman, L.J., Perazza, D.E., Juda, L., Campbell, M.M., 2004. Involvement of the R2R3-MYB, AtMYB61, in the ectopic lignification and dark-photomorphogenic components of the det3 mutant phenotype. Plant J. 37, 239–250.

Novillo, F., Medina, J., Rodríguez-Franco, M., Neuhaus, G., Salinas, J., 2012. Genetic analysis reveals a complex regulatory network modulating CBF gene expression and Arabidopsis response to abiotic stress. J. Exp. Bot. 63, 293–304.

Nuruzzaman, M., Manimekalai, R., Sharoni, A.M., Satoh, K., Kondoh, H., Ooka, H., et al., 2010. Genome-wide analysis of NAC transcription factor family in rice. Gene 465, 30–44.

Ohme-Takagi, M., Shinshi, H., 1995. Ethylene-inducible DNA binding proteins that interact with an ethylene-responsive element. Plant Cell 7, 173–182.

Okamuro, J.K., Caster, B., Villarroel, R., Van Montagu, M., Jofuku, K.D., 1997. The AP2 domain of APETALA2 defines a large new family of DNA binding proteins in Arabidopsis. PNAS 94, 7076–7081.

Olsen, A.N., Ernst, H.A., Leggio, L.L., Skriver, K., 2005. NAC transcription factors: structurally distinct, functionally diverse. Trends. Plant Sci. 10, 79–87.

Ooka, H., Satoh, K., Doi, K., Nagata, T., Otomo, Y., Murakami, K., et al., 2003. Comprehensive analysis of NAC family genes in *Oryza sativa* and *Arabidopsis thaliana*. DNA. Res. 10, 239–247.

Pandey, S.P., Somssich, I.E., 2009. The role of WRKY transcription factors in plant immunity. Plant Physiol. 150, 1648–1655.

Pattanaik, S., Xie, C.H., Yuan, L., 2008. The interaction domains of the plant Myc-like bHLH transcription factors can regulate the transactivation strength. Planta 227, 707–715.

Persikov, A.V., Singh, M., 2014. De novo prediction of DNA-binding specificities for Cys2His2 zinc finger proteins. Nucl. Acid. Res. 42, 97–108.

Puranik, S., Bahadur, R.P., Srivastava, P.S., Prasad, M., 2011. Molecular cloning and characterization of a membrane associated NAC family gene, SiNAC from foxtail millet [*Setaria italica* (L.) P. Beauv.]. Mol. Biotech. 49, 138–150.

Puranik, S., Sahu, P.P., Srivastava, P.S., Prasad, M., 2012. NAC proteins: regulation and role in stress tolerance. Trends. Plant Sci. 17, 369–381.

Ramamoorthy, R., Jiang, S.Y., Kumar, N., Venkatesh, P.N., Ramachandran, S., 2008. A comprehensive transcriptional profiling of the WRKY gene family in rice under various abiotic and phytohormone treatments. Plant Cell Physiol. 49, 865–879.

Rashid, M., Guangyuan, H., Guangxiao, Y., Hussain, J., Xu, Y., 2011. AP2/ERF transcription factor in rice: genome-wide convas and syntenic relationships between monocots and eucots. Evol. Bioinformatics 8, 321–355.

Rizhsky, L., Davletova, S., Liang, H., et al., 2004. The zinc finger protein Zat12 is required for cytosolic ascorbate peroxidase 1 expression during oxidative stress in Arabidopsis. J. Biol. Chem. 279, 11736–11743.

Rushton, P.J., Bokowiec, M.T., Han, S., Zhang, H., Brannock, J.F., Chen, X., et al., 2008. Tobacco transcription factors: novel insights into transcriptional regulation in the solanaceae. Plant Physiol. 147, 280–295.

Rushton, P.J., Somssich, I.E., Ringler, P., Shen, Q.J., 2010. WRKY transcription factors. Trends. Plant Sci. 15, 247–258.

Saibo, N.J., Lourenço, T., Oliveira, M.M., 2009. Transcription factors and regulation of photosynthetic and related metabolism under environmental stresses. Ann. Bot. 103, 609–623.

Sakuma, Y., Liu, Q., Dubouzet, J.G., Abe, H., Shinozaki, K., Yamaguchi-Shinozaki, K., 2002. DNA-binding specificity of the ERF/AP2 domain of Arabidopsis DREBs, transcription factors involved in dehydration- and cold-inducible gene expression. Biochem. Biophys. Res. Commun. 290, 998–1009.

Sakuma, Y., Maruyama, K., Osakabe, Y., Qin, F., Seki, M., Shinozaki, K., et al., 2006. Functional analysis of an Arabidopsis transcription factor, DREB2A, involved in drought-responsive gene expression. Plant Cell 18, 1292–1309.

Shoji, T., Mishima, M., Hashimoto, T., 2013. Divergent DNA-binding specificities of a group of ETHYLENE RESPONSE FACTOR transcription factors involved in plant defense. Plant. Physiol. 162, 977–990.

Sohn, K.H., Lee, S.C., Jung, H.W., Hong, J.K., Hwang, B.K., 2006. Expression and function roles of the pepper pathogen-induced transcription factor RAV1 in bacterial disease resistance, and drought and salt stress tolerance. Plant Mol. Biol. 61, 897–915.

Song, Y., Ai, C.R., Jing, S.J., Yu, D.Q., 2010. Research progress on function analysis of rice WRKY gene. Rice Sci. 17, 60–72.

Souer, E., van Houwelingen, A., Kloos, D., Mol, J., Koes, R., 1996. The *No Apical Meristem* gene of petunia is required for pattern formation in embryos and flowers and is expressed as meristem and primordia boundaries. Cell 85, 159–170.

Stockinger, E.J., Gilmour, S.J., Thomashow, M.F., 1997. *Arabidopsis thaliana* CBF1 encodes an AP2 domain-containing transcription activator that binds to the C-repeat/DRE, a cis-acting DNA regulatory element that stimulates transcription in response to low temperature and water deficit. PNAS 94, 1035–1040.

Stracke, R., Werber, M., Weisshaar, B., 2001. The R2R3-MYB gene family in *Arabidopsis thaliana*. Curr. Opin. Plant Biol. 4, 447–456.

Sun, C., Palmqvist, S., Olsson, H., Borén, M., Ahlandsberg, S., Jansson, C., 2003. A novel WRKY transcription factor, SUSIBA2, participates in sugar signalling in barley by binding to the sugar responsive elements of the iso1 promoter. Plant Cell 15, 2076–2092.

Thomashow, M.F., 1999. Plant cold acclimation: freezing tolerance genes and regulatory mechanisms. Annu. Rev. Plant Physiol. Plant Mol. Biol. 50, 571–599.

Todaka, D., Nakashima, K., Shinozaki, K., Yamaguchi-Shinozaki, K., 2012. Toward understanding transcriptional regulatory networks in abiotic stress responses and tolerance in rice. Rice 5, 6.

Tran, L.S., Nakashima, K., Sakuma, Y., Simpson, S.D., Fujita, Y., Maruyama, K., et al., 2004. Isolation and functional analysis of Arabidopsis stress-inducible NAC transcription factors that bind to a drought-responsive *cis*-element in the early responsive to dehydration stress 1 promoter. Plant Cell 16, 2481–2498.

Wang, X., Culver, J.N., 2012. DNA binding specificity of ATAF2, a NAC domain transcription factor targeted for degradation by Tobacco mosaic virus. BMC Plant Biol. 12, 157.

Wang, P., Du, Y., Zhao, X., Miao, Y., Song, C.P., 2013a. The MPK6-ERF6-ROS-responsive cis-acting Element7/GCC box complex modulates oxidative gene transcription and the oxidative response in Arabidopsis. Plant Physiol. 161, 1392–1408.

Wang, N., Zheng, Y., Xin, H., Fang, L., Li, S., 2013b. Comprehensive analysis of NAC domain transcription factor gene family in *Vitis vinifera*. Plant Cell. Rep. 32, 61–75.

Wei, W., Zhang, Y., Han, L., Guan, Z., Chai, T., 2008. A novel WRKY transcriptional factor from *Thlaspi caerulescens* negatively regulates the osmotic stress tolerance of transgenic tobacco. Plant Cell. Rep. 27, 795–803.

Woodrow, P., Fuggi, A., Pontecorvo, G., Kafantaris, I., Annunziata, M.G., Massaro, G., et al., 2012a. cDNA cloning and differential expression patterns of ascorbate peroxidase during post-harvest in *Brassica rapa* L. Mol. Biol. Rep. 39, 7843–7853.

Woodrow, P., Pontecorvo, G., Ciarmiello, L.F., Annunziata, M.G., Fuggi, A., Carillo, P., 2012b. Transcription factors and genes in abiotic stress. In: Bandi, V., Shanker, A.K., Shanker, C., Mandapaka, M. (Eds.), Crop stress and its Management: Perspectives and Strategies. Springer, The Netherlands.

Wu, Y., Deng, Z., Lai, J., Zhang, Y., Yang, C., Yin, B., et al., 2009. Dual function of Arabidopsis ATAF1 in abiotic and biotic stress responses. Cell. Res. 19, 1279–1290.

Xia, N., Zhang, G., Liu, X.Y., Deng, L., Cai, G.L., Zhang, Y., et al., 2010. Characterization of a novel wheat NAC transcription factor gene involved in defense response against stripe rust pathogen infection and abiotic stresses. Mol. Biol. Rep. 37, 3703–3712.

Xie, Z., Zhang, Z.L., Zou, X., Huang, J., Ruas, P., Thompson, D., et al., 2005. Annotations and functional analyses of the rice WRKY gene superfamily reveal positive and negative regulators of abscisic acid signalling in aleurone cells. Plant Physiol. 137, 176–189.

Xue, G.P., Loveridge, C.W., 2004. HvDRF1 is involved in abscisic acid-mediated gene regulation in barley and produces two forms of AP2 transcriptional activators, interacting preferably with a CT-rich element. Plant J. 37, 326–339.

Yadav, V., Mallappa, C., Gangappa, S.N., Bhatia, S., Chattopadhyay, S., 2005. A basic helix-loop-helix transcription factor in Arabidopsis, MYC2, acts as a repressor of blue light-mediated photomorphogenic growth. Plant Cell 17, 1953–1966.

Yamaguchi-Shinozaki, K., Shinozaki, K., 2005. Organization of cis-acting regulatory elements in osmotic- and cold-stress-responsive promoters. Trends. Plant Sci. 10, 88–94.

Yamasaki, K., Kigawa, T., Seki, M., Shinozaki, K., Yokoyama, S., 2012. DNA-binding domains of plant-specific transcription factors: structure, function, and evolution. Trends. Plant Sci. S1360–S1385.

Yang, B., Jiang, Y., Rahman, M.H., Deyholos, M.K., Kav, N.N., 2009. Identification and expression analysis of WRKY transcription factor genes in canola (*Brassica napus* L.) in response to fungal pathogens and hormone treatments. BMC Plant Biol. 9, 68.

Yanhui, C., Xiaoyuan, Y., Kun, H., Meihua, L., Jigang, L., Zhaofeng, G., et al., 2006. The MYB transcription factor superfamily of Arabidopsis: expression analysis and phylogenetic comparison with the rice MYB family. Plant Mol. Biol. 60, 107–124.

Zhang, Y., Wang, L., 2005. The WRKY transcription factor superfamily: its origin in eukaryotes and expansion in plants. BMC Evol. Biol. 5, 1.

Zhou, Q.Y., Tian, A.G., Zou, H.F., Xie, Z.M., Lei, G., Huang, J., et al., 2008. Soybean WRKY-type transcription factor genes, GmWRKY13, GmWRKY21, and GmWRKY54, confer differential tolerance to abiotic stresses in transgenic Arabidopsis plants. Plant Biotechnol. J. 6, 486–503.

Zhou, X., Jiang, Y.J., Yu, D.Q., 2011. WRKY22 transcription factor mediates dark-induced leaf senescence in Arabidopsis. Mol. Cells. 31, 303–313.

Zhuang, J., Sun, C.C., Zhou, X.R., Xiong, A.S., Zhang, J., 2011. Isolation and characterization of an AP2/ERF-RAV transcription factor Bna RAV-1-HY15 in *Brassica napus* L. HuYou15. Mol. Biol. Rep. 38, 3921–3928.

Plant Resistance under Cold Stress: Metabolomics, Proteomics, and Genomic Approaches

Saiema Rasool, Shailza Singh, Mirza Hasanuzzaman, Muneeb U. Rehman, M.M. Azooz, Helal Ahmad Lone and Parvaiz Ahmad

4.1 Introduction

Cold stress is the most important environmental factor that restricts the agricultural output of plants. Low temperature has a massive impact on the geographical distribution and survival of plants. Cold stress includes chilling ($<20°C$) and/or freezing ($<0°C$) temperatures, and damagingly affects the growth and development of plants, notably constraining their spatial distribution. About two-thirds of the world's landmass is annually subjected to temperatures below freezing point and about half suffers from temperatures below $-20°C$ (Larcher, 2001; Baldi et al., 2011). Thus, cold stress negatively affects growth and cultivation of many ecologically and economically important plant species. Cold stress averts the expression of full genetic potential of plants due to its direct inhibition of metabolic reactions and, indirectly, through cold-induced osmotic (chilling-induced inhibition of water uptake and freezing-induced cellular dehydration), oxidative, and other stresses (Chinnusamy et al., 2007; Krasensky and Jonak, 2012). Some temperate region plants such as *Arabidopsis*, canola, wheat, *Avena nuda*, etc. are tolerant to chilling temperatures, although most are not so tolerant to freezing, but can augment their freezing tolerance by being exposed to low temperatures, a process called "cold acclimation" (Levitt, 1980; Liu et al., 2012; Takahashi et al., 2013). Cold acclimation is a phenomenon in which plants attain freezing tolerance upon previous exposure to low non-freezing temperatures. Acclimation involves the activation of numerous mechanisms that contribute to an increase in freezing tolerance together with the accumulation of low-molecular-weight cryoprotective molecules, the synthesis of cryoprotective proteins, and alterations in membrane lipid composition (Artus et al., 1996; Gilmour et al., 2000; Taji et al. 2002; Easlon et al., 2013). In contrast, the plants of tropical and subtropical regions, including several crops such as rice, maize, and tomato, are sensitive to cold stress and usually lack the capacity for cold acclimation (Zhu et al., 2007). For cold tolerance and enhanced crop winter hardiness, identification of the key genes underlying cold stress has become a top priority.

The in-depth perspective of the regulation of genes involved and their response to cold stress would allow elucidation of the behaviors in which plants adjust to cold stress. This knowledge is basically fundamental for the exploitation of gene expression in crop plants, with a view to engineering advanced levels of cold tolerance.

P. Ahmad (Ed): Emerging Technologies and Management of Crop Stress Tolerance, Volume 1.
DOI: http://dx.doi.org/10.1016/B978-0-12-800876-8.00004-7

4.2 Causes of freezing injury

Chilling-sensitive plants exposed to low temperature often show signs of water stress due to decreased root hydraulic conductance, leading to associated decreases in leaf water and turgor potential, followed by a reduction or cessation of growth (Fennell and Markhart, 1998; Aroca et al., 2001; Figure 4.1).

The phenomenon is primarily reversible, but later becomes irreversible and can result in cell death. The extent of injury associated with freezing temperatures varies with plant species, rate of change of temperature, stage of crop development (Steffen et al., 1989), duration of exposure (Rajashekar et al., 1983), irradiance, and mineral nutrition (Ercoli et al., 2004). Previously published reports indicate that the membrane systems of the cell are the primary site of freezing injury in plants (Levitt, 1980; Steponkus, 1984). One of the primary factors of membrane damage under cold stress is the association between chilling sensitivity and the degree of unsaturation of fatty acids. Membrane lipids are mainly composed of two types of fatty acids: saturated as well as unsaturated. Saturated fatty acids are fully saturated with hydrogen atoms ($-CH2-CH2-$), whereas unsaturated fatty acids have one or more double bonds between two carbon atoms ($-CH=CH-$). The relative proportion of unsaturated fatty acids in the membrane strongly influences the fluidity of the membrane (Steponkus et al., 1993). The temperature at which a membrane transforms from semi-fluid state to a semi-crystalline state is called the transition temperature. Cold-sensitive plants possess increased levels of saturated fatty acids in phosphatidylglycerol in membranes. Since saturated fatty acid has a high melting point, membranes isolated from chilling sensitive plants can undergo a phase transition from the liquid crystalline phase to the gel phase even at room temperature, while cold-tolerant plants contain a high magnitude of unsaturated fatty acids, which keep the phase transition temperature lower than the applied chilling temperature. In this way, a phase

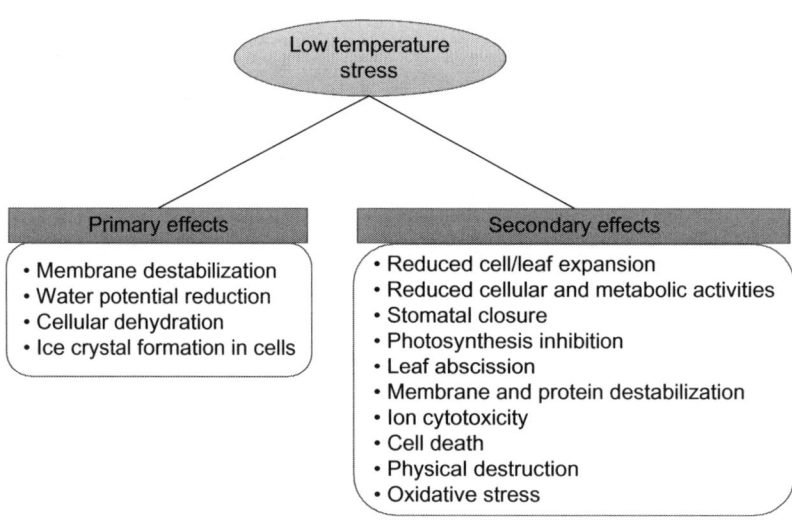

FIGURE 4.1

Major effects of cold stress in a plant.

transition is avoided. Transgenic tobacco expressing glycerol-3-phosphate acyltransferase or omega-3 fatty acid desaturase from *Arabidopsis thaliana* is able to improve cold tolerance by increasing trienoic fatty acids (Murata et al., 1992; Kodama et al., 1994).

Cellular membranes are fluid structures, and freezing temperatures can decrease their fluidity, causing increased rigidity. As the temperature goes below 0°C, ice typically forms at the intercellular spaces of plant tissues. It occurs in this location, in contrast to intracellularly (which is thought to be a fatal event), in part because the intercellular fluid generally has an increased freezing point than the intracellular fluid. The buildup of ice in the intercellular spaces can potentially result in physical damage of cells and tissues caused in part by the formation of adhesions between the intercellular ice and the cell walls and membranes (Levitt, 1980). However, the majority of injuries results from the harsh cellular dehydration that occurs with freezing (Levitt, 1980; Steponkus and Webb, 1992). At a given subzero temperature, the chemical potential of ice is less than that of liquid water. Thus, when ice forms intercellularly, there is a decrease in water potential outside the cell. Consequently, there is movement of unfrozen water down the chemical potential gradient from inside the cell to the intercellular spaces. The net amount of water movement required to bring the system into chemical equilibrium depends on both the initial solute concentration of the intracellular fluid and the subzero temperature, which directly determines the chemical potential of the ice. At $-10°C$, more than 90% of the osmotically active water will generally move out of the cells to the intercellular spaces, and the osmolality (Osm) of the remaining unfrozen intracellular and intercellular water will be in excess of 5 Osm (Thomashow, 1998).

It has been confirmed that freeze-induced dehydration can lead to multiple forms of membrane lesions (Steponkus et al., 1993). At comparatively high freezing temperatures between -2 and $-4°C$, the major injury in non-acclimated plants appears to be "expansion-induced lysis," which is caused as a result of the osmotic expansion and contraction cycle, which occurs with freezing and thawing. At decreased temperatures between about -4 and $-10°C$, the principal form of injury in non-acclimated plants is freeze-induced lamellar-to-hexagonal-II phase transitions (a non-lamellar phase that is a three-dimensional (3D) array of inverted cylindrical micelles with water in the central core of each cylinder) or interlamellar attachments (Figure 4.1). Uemura et al. (1995) reported that in protoplasts isolated from cold-acclimated leaves of *Arabidopsis*, neither freeze-induced formation of the hexagonal-II phase nor expansion-induced lysis occurred. Instead, injury was associated with the "fracture-jump lesion," which is evident as localized variation of the plasma membrane fracture plane to subtending lamellae.

Horváth et al. (1998) and Orvar et al. (2000) reported that temperature-induced change in membrane fluidity is one of the instant consequences in plants during temperature-induced stresses and may possibly signify a potential site of injury.

The significance of appropriate membrane fluidity in temperature tolerance has been explained by mutation analysis and physiological and transgenic studies.

Wu et al. (1997) reported that at low temperature, enormous membrane lipid unsaturation appears to be critical for optimum membrane function. An *Arabidopsis* fatty acid biosynthesis FAB1 mutant with more saturated membranes showed decreased quantum efficiency of photosystem II (PSII), chlorophyll content, and the amount of chloroplast glycerolipids after prolonged exposure to low temperature. The other factor causing cold injury by membrane injury is loss of function produced by lipid peroxidation and the decline in membrane fluidity (Barclay and McKersie, 1994). Cold stress enhances production of free radicals and peroxidized membranes.

This causes the loss of unsaturated fatty acids, an increase in membrane rigidity due to the formation of covalent bonds among lipid radicals, a higher lipid phase-transition temperature, and membrane degradation (Alonso et al., 1997). Roxas et al. (2000) published that transgenic tobacco overexpressing glutathione S-transferase/glutathione peroxidase shows higher metabolic activity and lower malondialdehyde (MDA) content than the wild-type plant at low temperatures.

4.3 **Freezing-tolerance mechanisms**

The mechanisms underlying freezing tolerance are not well understood. However, over the last few decades, researchers have studied and reported several mechanisms of cold stress (Figure 4.2). A decade ago, transcriptome analysis using microarray technology (Bohnert et al., 2001; Zhu et al., 2001) revealed that genes induced by stress could be categorized into two groups on the basis of their product functions. The first group constitutes functional proteins such as membrane proteins

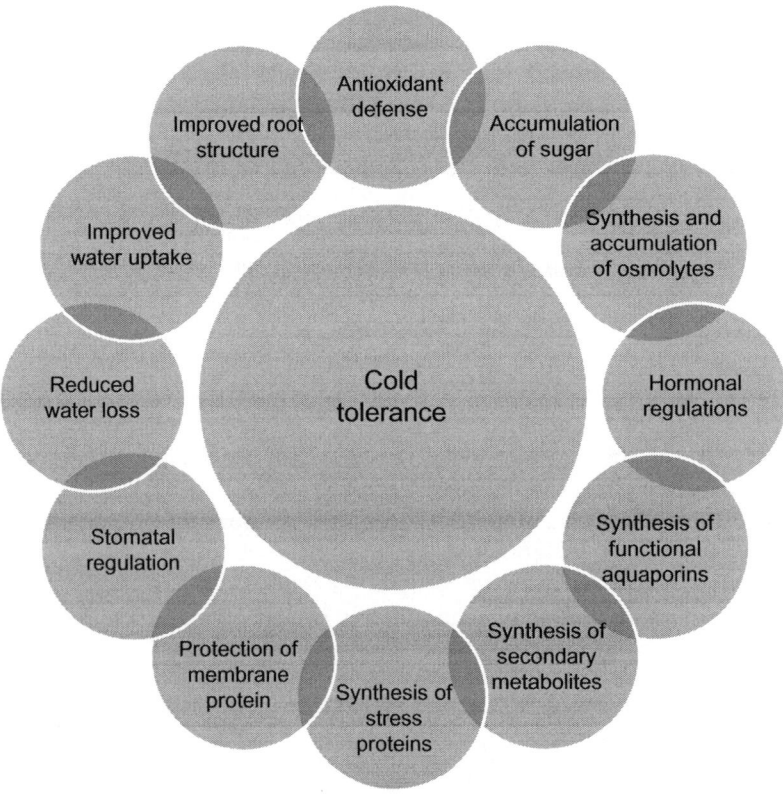

FIGURE 4.2

Different mechanisms of cold stress tolerance.

that sustain water movement through membranes (water channel proteins and membrane transporters); key enzymes for osmolyte biosynthesis (betaine, proline, sugars, etc.); the detoxification enzymes allowing biochemical, cellular, physiological, or metabolism to maintain a normal level (glutathione S-transferase, hydrolase, catalase, superoxide dismutase, ascorbate peroxidase, etc.); and other proteins for the protection of macromolecules (LEA protein, osmotin, antifreeze proteins, chaperons, mRNA binding protein, etc.).

4.3.1 Metabolomic approaches

Plants being sessile have developed sophisticated and efficient mechanisms to overcome adverse environmental challenges. Diverse "omics" approaches have enabled the active analyses of regulatory networks during abiotic stress responses such as cold. These analyses of plant responses and adaptation to stress conditions allow planning of better strategies for improved plant breeding programs. In recent times, metabolomic approaches have gained much acclaim as they monitor the complete set of metabolites that are involved in the physiological response to stress. Since metabolome (phenotype) of an organism represents the downstream result of gene expression, extensive knowledge on metabolic flows could allow the assessment of significant differences in plant species exhibiting different levels of tolerance under cold stress. Also, it is one of the easiest tools of stress marker selection (Perez-Clemente et al., 2013). The analytical tools to measure metabolites that have been recently developed to study stress responses are targeted analysis, metabolic fingerprinting, and metabolite profiling (Table 4.1).

Targeted analysis measures the concentration of a "targeted" number of known metabolites under stress conditions by using analytical techniques of gas chromatography (GC) or liquid chromatography (LC) coupled to mass spectrometry (MS) or nuclear magnetic resonance spectroscopy (NMR). Metabolic fingerprinting uses signals from hundreds to thousands of metabolites for rapid

Table 4.1 Major Approaches in Metabolomics

Approach	Advantage	Disadvantage
Targeted analysis	Quantitative	Limited number of compounds can be targeted
	Low limit of detection	Does not detect compounds that were not targeted
	High throughput	Targeted compounds must be available purified for calibration
Metabolite profiling	Global (not targeted)	Semi-quantitative
		Majority of peaks are not identifiable
		Difficult informatics
		Medium throughput
Metabolic fingerprinting	Global (not targeted)	No compound identification
	Directly applicable to pattern recognition	
	Highest throughput	

Source: Shulaev, 2006.

sample classification via statistical analysis. Metabolite profiling identifies and quantifies a specific class or classes of related metabolites that share chemical properties facilitating their simultaneous analysis (Shulaev, 2006).

Metabolomic analysis generates a large amount of data which requires specialized mathematical, statistical, and bioinformatic tools. Other than processing of microarray and proteomics data, metabolomics has unique bioinformatics needs like data and information management, raw analytical data processing, metabolomics standards and ontology, statistical analysis and data mining, data integration, and mathematical modeling of metabolic networks within the framework of systems biology. Metabolomics raw data processing is probably the most challenging and time-consuming step in data analysis. Generally, processing a set of raw chromatograms involves noise reduction, spectrum deconvolution, peak detection and integration, chromatogram alignment, compound identification, and quantification. Therefore, metabolomics requires automated data processing solutions such as the AMDIS (automated mass spectral deconvolution and identification system; http://chemdata.nist.gov/mass-spc/amdis/) software that utilizes well-described algorithms to process GC-MS data. ESI-LC-MS data can be processed using component detection algorithms (CODA) or the "windowed mass selection method" (WMSM). Metabolomics data have been analyzed with a wide range of statistical and machine-learning algorithms, classified in two major groups: unsupervised (hierarchical clustering, principal component analysis (PCA), and self-organizing maps) and supervised (ANOVA, partial least squares (PLS), and discriminant function analysis (DFA)) algorithms. These can be combined to evolutionary algorithms such as genetic algorithms or genetic programming, which helps reduce the variables in multivariate data obtained in metabolomics.

Integration of data from parallel "omics" from a systems biology framework, which creates huge data, storing, and data management, is important. A number of databases, data management, analysis, and visualization tools are currently publicly available. These include, among others, metabolic pathway databases and pathway viewers KEGG (http://www.genome.ad.jp/kegg/), MetaCyc (http://metacyc.org/), AraCyc (http://www.Arabidopsis.org/tools/aracyc/), MapMan (http://gabi.rzpd.de/projects/MapMan/), KaPPA-View (http://kpv.kazusa.or.jp/kappa-view/), the data model for plant metabolomics experiments ArMet (http://www.armet.org/), functional genomics databases MetNet (http://metnet.vrac.iastate.edu/), and DOME (http://medicago.vbi.vt.edu). DOME, a database developed by the Mendes group at the Virginia Bioinformatics Institute, provides an example of a comprehensive data management system for metabolomics as well as for other genomics data.

Guevara et al. (2012) have done a comparative metabolomics analysis of Yukon *Thellungiella* plants grown in cold and salt stress conditions. In order to gain insight into the biochemical activities of these plants, polar metabolites were profiled by gas chromatography/mass spectrometry (GC/MS). Using this approach, it was possible to simultaneously monitor over 300 mass spectral tags (MSTs) corresponding to chemically diverse compounds. AMDIS software was used to extract peak abundance and mass spectra information of the chemical diversity. Statistical analysis was performed using ANOVA between the control and stressed plant groups. Euclidean distance was used to calculate the distance matrix and a complete linkage method was used for hierarchical clustering of MSTs and transcripts using the program Cluster. Heat maps were constructed using the Java Treeview program. There is considerable overlap in the expression of metabolites under the experimental conditions (Guevara et al., 2012).

Similarly, Wang et al. (2006) used an ESI-tandem MS (MS/MS)-based approach to quantitatively profile membrane lipid molecular species in plant response to low temperatures, and the profiling analysis revealed significant and distinct lipid changes during cold acclimation and freezing. Welti et al. (2002) profiled membrane lipid molecular species in *Arabidopsis* undergoing cold and freezing stresses using ESI-MS/MS and found that freezing at a sublethal temperature induced a decline in many molecular species of phosphatidylcholine, phosphatidylethanolamine, and phosphatidylglycerol but induced an increase in phosphatidic acid and lysophospholipids, which resulted in a destabilizing membrane bilayer structure. Gray and Heath (2005) examined the effects of cold acclimation on the *Arabidopsis* metabolome using a non-targeted metabolic fingerprinting approach and found that a global reprogramming of metabolism occurs as a result of cold acclimation. They determined a comprehensive, unbiased assessment of metabolic processes relative to cold acclimation by measuring an entire spectrum of putative metabolites based on mass-to-charge (*m/z*) ratios versus an individual or group of metabolite(s). The leaves shifted to low temperature presented metabolic profiles that were constantly changing, but leaves developed at low temperature demonstrated a stable complement of components. Kaplan et al. (2004) explored the temperature—stress metabolome of *Arabidopsis* and found that cold shock influenced metabolism far more profoundly than heat shock. The steady-state pool sizes of 143 and 311 metabolites or mass spectral tags were altered in response to heat and cold shock, respectively, and comparison of heat and cold shock response patterns revealed that the majority of heat shock responses were shared with cold shock responses. Coordinate increases in the pool sizes of amino acids derived from pyruvate and oxaloacetate, polyamine precursors and compatible solutes were observed during both heat and cold shocks. However, it could be highlighted that many of the metabolites that showed increases in response to both heat and cold shocks were previously unlinked with temperature stress.

4.3.2 LEA proteins

About three decades ago, Leon Dure identified various families of low-molecular-weight (10−30 kDa) proteins that build up to elevated levels during the maturation phase of cotton (*Gossypium hirsutum*) embryogenesis (Dure and Chlan, 1981; Eriksson and Harryson, 2011), which gave rise to the name late embryogenesis abundant (LEA) proteins. The characterization of diverse representatives of cDNAs from many of these protein families revealed their common structural features, some of which were first noticed by Dure and his colleagues. These comprise increased hydrophilicity, a lack or low proportion of Cys and Trp residues, and a preponderance of certain amino acid residues such as Gly, Ala, Glu, Lys/Arg, and Thr, which later led them to be considered as a subset of hydrophilins (Dure, 1993; Garay-Arroyo et al., 2000). LEA proteins are mainly concerned with defending higher plants from injury caused by environmental stresses, especially drought. Bo et al. (2005) reported LEA biological activities, protein synthesis, and expression are regulated by many factors (e.g., hormones, ion change, developmental stages, and dehydration), signal transduction pathways, and LEA genes. There is emerging evidence that polypeptides too have a role in the stabilization of membranes against freeze-induced damage.

The dehydrins are a group of glycine-rich, heat-stable, LEA proteins thought to be important for stabilization of membrane and the defense of proteins from denaturation when the cytoplasm becomes dehydrated. Nakayama et al. (2008) have suggested that some of them, especially COR15am, function as a protectant by preventing protein aggregation. Two types of dehydrins, ERD10 (early response to dehydration) and ERD14, function as chaperones and through electrostatic forces interact with phospholipid vesicles (Kovacs et al., 2008). A number of dehydrins are significantly accumulated during times of cold stress (Allagulova et al., 2003; Kawamura and Uemura, 2003; Renaut et al., 2004). Microarray experiments revealed that the expression profile of a specific blend of dehydrin genes can present a reliable indication of low temperature and drought stress (Tommasini et al., 2008). Okawa et al. (2008) identified COR413im as a vital membrane protein directed to the inner envelop of chloroplast in response to freezing temperatures, where it adds to plant-freezing tolerance. However, the SFR2 protein, a defensive protein of the chloroplast during freezing, is localized in the outer envelope of the chloroplast membrane (Fourrier et al., 2008).

4.3.3 HSP proteins

The heat shock response refers to increased transcription of a set of genes in response to heat and/ or additional toxic agent. It is a highly conserved biological response, occurring in all organisms (Waters et al., 1996).

The response is intervened by heat shock transcription factor (HSF), which is present in a monomeric, non-DNA binding structure under unstressed cells, and on activation by stress it is changed to a trimeric form, which can bind to promoters of genes responsible for HSP. One of the major responses observed at the molecular level of organisms exposed to high temperature is induction of genes encoding heat shock proteins (Kimpel and Key, 1985; Vierling, 1991). Some genes contribute to freezing tolerance by stabilizing proteins against freeze-induced denaturation, e.g., cold responsive genes encoding molecular chaperones including a spinach hsp70 gene (Anderson et al., 1994) and a Brassica napus hsp90 gene (Krishna et al., 1995). Mitogen-activated protein (MAP) kinase (Mizoguchi et al., 1993), MAP kinase kinase kinase (MAPKKK) (Mizoguchi et al., 1996), and the calmodulin-related proteins (Polisensky and Braam, 1996) are some of the cold-responsive genes identified for encoding various signal transduction and regulatory proteins. These proteins could contribute to cold tolerance and tolerance to other stresses by regulating the activity and expression of the major stress genes and their proteins (Sun et al., 2002; Mogk et al., 2003).

4.3.4 COR genes

COR are a set of genes that encode a related family of cold-regulated (COR) proteins that are particularly induced at the time of cold acclimation (Hajela et al., 1990; Gilmour et al., 2004). Few of the COR genes are named as cold acclimation-specific (CAS), low temperature-induced (LTI), cold-induced (KIN), and responsive to drought (RD) genes (Kurkela and Franck, 1990; Monroy et al., 1993; Yamaguchi-Shinozaki and Shinozaki, 1994). A few examples of cold-responsive genes are: alfalfa Cas15 (Monroy et al., 1993), COR15a (Artus et al., 1996), and wheat WCS120 (Houde et al., 1993). Expression of COR genes has been shown to be critical for both chilling tolerance and cold acclimation in plants (Thomashow, 1999). *Arabidopsis* COR genes include: COR78/ RD29, COR47, COR6.6, COR15a, and encode LEA-like proteins (Thomashow, 1999). These genes are induced by cold, ABA, or dehydration. COR15A polypeptide is targeted to the chloroplast. Development of hexagonal II phase lipids is a chief cause of damage to membranes in non-acclimated plants. COR15a expression diminishes the propensity of the membranes to form hexagonal II phase lipids in response to freezing (Uemura and Steponkus, 1997). The promoter elements of COR genes contain CRT (C-repeats) or DRE (dehydration responsive elements) and a few of them contain ABRE (ABA-responsive element) as well (Yamaguchi-Shinozaki and Shinozaki,

1994; Stockinger et al., 1997). The COR genes are induced by overexpression of transcription factor core binding factor (CBF) (CRT/DRE binding factor) (Stockinger et al., 1997). CBF binds to the CRT/DRE elements present in the promoter of the COR genes and other cold-regulated genes. The overexpression of these dogmatic elements resulted in increased freezing tolerance as well as an increase to drought tolerance (Liu et al., 1998), highlighting to fundamental role of cold-inducible genes to defend the plant cells from cellular dehydration.

4.3.5 Osmoprotectants

Proline, a pyrrolidine ring-based secondary amino acid, has been reported to accumulate in huge amounts in cold-resistant trees before the start of deep dormancy (Kandarova, 1964). It is also reported to build up in shoot apices of wheat at the beginning of fall frosts, but vanishes when temperatures return back to normal (Telstcherova, 1967). Accumulation of simple sugars and sucrose that usually occurs with cold acclimation also possibly contributes to stabilization of membranes as these molecules are reported to defend membranes against freeze-induced damage *in vitro* (Strauss and Hauser, 1986; Anchordoguy et al., 1987). Low temperatures induce quantitative as well as qualitative changes in the carbohydrate status of the plant. Sucrose is the most abundant accumulated free sugar (Guy et al., 1992), with a 10-fold increase in cytosol but not in vacuole (Rolland et al., 2006; Koster and Lynch, 1992). A few other frequently accumulated compatible solutes in higher plants include glucose, raffinose, glycinebetaine, fructose, and proline. Walker et al. (2008) reported that in *Atriplex halimus* L. the rise in leaf sugar content is closely associated with increase in freezing tolerance, and it is not simply a low temperature response (due to reduced activity of sugar metabolic enzymes); however, detailed studies are needed to understand the cryoprotective role of sugars in plants.

4.4 **Antioxidant defense under cold stress**

During cold treatment, the enzymes of the Calvin–Benson cycle are slowed by simple thermodynamics, thus limiting the supply of $NADP^+$ for reduction, and ADP and Pi for phosphorylation. Incoming light energy continues to be channeled into ETC as long as the pigment beds remain intact and connected to the photosystems (PS I and PS II). These two factors, a slowing of the dark reactions and continuing energy absorption, overreduce the photosynthetic ETC leading to the leakage of absorbed energy in an uncontrolled manner from the thylakoid membrane. As the light-independent reaction of photosynthesis is very temperature sensitive, the energy leaked during chilling in light causes the formation of ROS (Wise, 1995), whose increased concentration causes damage to membrane lipids, proteins, and nucleic acids, leading to PCD (Apel and Hirt, 2004; Hasanuzzaman et al., 2012, 2013). Plants exposed to low temperatures use several non-enzymatic and enzymatic antioxidants to cope with the harmful effect of oxidative stress; higher contents of antioxidant defense enzymes are correlated with higher chilling tolerance (Kang and Saltveit, 2002; Huang and Guo, 2005; Table 4.2). Antioxidant enzymes have higher activity in chilling-tolerant cultivars than in susceptible ones (Guo et al., 2005). Streb et al. (1999) and Hasanuzzaman et al. (2013) found that to increase the contents of AsA and α-tocopherol in chilling-tolerant cereal leaves helped to maintain better photosynthesis levels as compared to the chilling-sensitive

Table 4.2 Modulation of Antioxidant Defense System in Some Crop Plants under Cold Stress

Crop	Temperature and Duration	Modulation of Antioxidant Defense System	References
Solanum lycopersicum (WT)	4°C, 6 h	AsA, DHA, AsA/DHA, and total ascorbate increased	Wang et al. (2013)
Avena nuda L.	1°C, 7 d	Increased SOD, POD, and CAT	Liu et al. (2013)
Trichosanthes kirilowii	4°C, 3 d	Increased CAT Decreased SOD	Yao et al. (2013)
Lycopersicon esculentum cv. Zhongshu 6 (WT)	4°C, 12 h	Increased CAT and SOD Decreased APX activity	Duan et al. (2012)
Poncirus trifoliate (L.) Raf.	4°C, 5 d	GST activity increased	Peng et al. (2012)
Cucumber (*Cucumis sativus* cv. Jingchun no. 4)	15/8°C (D/N), 2 d	MDHAR, GR, GPX, GSH-Px, and APX activities are decreased	Li et al. (2012)
Broccoli cultivar "Monaco" F1 (seedlings)	2°C, 2 wk	Increased CAT and POD	Leja et al. (2012)
Maize (single cross 704)	15°C	Increased SOD, CAT, and APX activities	Saeidnejad et al. (2012)
Zea mays L.	10/7°C (day/night), 3 d	Increased SOD, CAT, GR, and APX activities Increase AsA and GSH content Decreased POX activity	Erdal (2012)
Triticum aestivum L.	2 ± 0.5°C, 48 h	Increased CAT, SOD, and APX activities	Malekzadeh et al. (2012)
Oil palm	10° C, 7 d	Increased SOD and POD activities	Cao et al. (2011)
Ipomea batatas L.	5°C, 72 h	Increased peroxidase activity	Islam et al. (2011)
Cucurbita pepo L.	8°C, 8 d	SOD, APX, DHA, and reduced glutathione increased Decreased total and reduced ascorbate, CAT	Tartoura and Youssef (2011)
Cucumis sativus cv. Jinchun 4	15/8°C (D/N), 6 d	Increased SOD, CAT, GPX, Decreased GSH-Px MDHAR, DHAR, and GR activities Increased AsA and GSH content, GSH/total glutathione ratio Decreased AsA/total ascorbate ratio	Liu et al. (2009)
Triticum aestivum L. (Dogu-88)	5/3°C, 5 d	Increased CAT, POX, and SOD	Çakmak and Atıcı (2009)
Triticum aestivum L. (Gerek-79)		Increased POX and SOD Decreased CAT	

varieties. Fortunato et al. (2010) stated that the ROS production indicated by H_2O_2 and $OH^{\cdot}$ was reduced by the overproduction of AsA and α-tocopherol contents under LT stress in *Coffea* sp. The ratio of GSH/GSSG is also important because a higher ratio indicates a better tolerance to stress. Under stressful conditions including the cold, the higher GSH/GSSG ratio is desirable for the sufficient amount of GSH in the AsA-GSH cycle (Kocsy et al., 2000). Takáč et al. (2003) showed that activities of some antioxidant enzymes partially correlate with the chilling sensitivity of maize cultivars and possess a significant importance in the chilling tolerance of *Zea mays*. In rice, a greater efficiency of antioxidant enzymes was observed in chilling-tolerant cultivars which were far higher than chilling-susceptible cultivars (Huang and Guo, 2005). The study indicated that the activities of SOD, CAT, APX, and GR, as well as AsA content of tolerant cultivar (Xiangnuo-1), remained high, but in chilling-susceptible cultivar (IR-50), the activities decreased. Yang et al. (2011) observed that the enhanced activities of SOD, CAT, APX, and POX in cucumber plants reflected better tolerance to chilling injury. Gechev et al. (2010) reported that CAT and DHAR are most strongly affected by chilling (5°C) and may be the rate-limiting factor of the antioxidant system at low temperatures. In chickpea cultivars the activities of SOD, APX, GR, and POX increased in cold-acclimated plants and subsequent chilling stress (2°C and 4°C for 12 d), which indicated the enhanced chilling tolerance capacity of this cultivar to protect plants from oxidative damage (Turan and Ekmekçi, 2011).

Formerly regarded purely as damaging agents, reactive oxygen species (ROS) are now understood as important signal molecules vital to normal plant growth. Previously published reports suggest that plant responses to cold stress are directly linked to ROS signaling (Lee et al., 2002; Vogel et al. 2005). During response to environmental stress and developmental stimuli, ROS are key signal transduction molecules (Mittler et al., 2004). Yong et al. (2003) reported low temperature stress to trigger an increase of ROS and this early accumulation of ROS in plants might lead to the production of an antioxidant defense system (see Table 4.2).

Cold stress enhances transcript abundance and protein concentrations of ROS-scavenging enzymes, as well as ROS accumulation (O'Kane et al., 1996). Numerous genes have been implicated in both cold stress and ROS signaling. For example, overexpression of *ZAT12*, a C_2H_2 zinc-finger-type transcription factor gene, induces cold-inducible genes and confers increased cold tolerance in plants when overexpressed. Furthermore, *ZAT12* plays an essential role and controls a number of genes involved in plant responses to oxidative stress (Davletova et al., 2005; Liu et al., 2012).

In *Arabidopsis*, various cold-responsive genes such as *KIN1*, *RD29A*, *COR47*, *KIN2*, *DREB2A*, *COR15A*, *DREB1A*, and *ERD10* have been identified (Seki et al., 2002; Thomashow, 1999). Lee et al. (2002) suggested the contribution of some cold-responsive genes in controlling ROS in cold stress. *Arabidopsis* frostbite1 (fro1) mutant showed decreased expression of cold-responsive genes such as *KIN1*, *COR15A*, *RD29A*, and *COR47* and built up ROS constitutively. The FRO1 gene was shown to encode a mitochondrial complex I protein, suggesting that expression of the cold-responsive genes and ROS accumulation might be modulated by the disruption of a mitochondrial function. The double-edged nature of ROS, i.e., on the one hand the damaging toxic molecule, and on the other hand the beneficial signal transduction molecule, emphasizes the need to control the steady-state level of ROS in cells. Explaining the mechanisms that control ROS signaling in cells during heat, cold, or freezing stress could therefore provide an additional powerful strategy to enhance the resistance of crops to these environmental stress conditions.

4.5 Cold signal transducers

Understanding the mechanisms by which plants perceive environmental signals and transmit the signals to cellular machinery to activate adaptive responses is of fundamental importance to biology. The identity of the plant sensors of low temperature remains as yet unknown (Chinnusamy et al., 2006). Xiong et al. (2002) showed that multiple primary sensors may be implicated, with each perceiving a particular aspect of the cold stress, and each involved in a distinct arm of the cold signaling pathway.

4.5.1 Ca^{2+} influx channels

The increase in cytosolic Ca^{2+} from extracellular spaces is an early episode in the response to cold. This cytosolic Ca^{2+} is recommended as an important secondary messenger in low temperature signal transduction and developing cold acclimation. A positive link between cold temperature Ca^{2+} influx and accumulation of cold-induced transcripts has been shown for *Arabidopsis* (Henriksson and Trewavas, 2003) and alfalfa (Monroy and Dhindsa, 1995; Reddy and Reddy, 2004). Knight et al. (1991) developed transgenics of *Arabidopsis* and tobacco expressing the calcium-sensitive luminescent protein aequorin and showed a rise in cytosolic Ca^{2+} concentration in response to chilling temperature. Calcium chelators or Ca^{2+} channel blockers barred the calcium influx as well as the expression of low temperature-responsive cas15 gene and the development of freezing tolerance in cells (Monroy and Dhindsa, 1995; Sangwan et al., 2001). Further, Knight and Knight (2000) reported that effective Ca^{2+} signature is produced only in specific tissue or organs. During cold stress, cytosolic Ca^{2+} influx occurs in the whole plant, in contrast to drought, where it is present only in the roots. The cytosolic rise in Ca^{2+} transmits primary signal through Ca^{2+}-regulated proteins called Ca^{2+} sensors and alters the state of protein phosphorylation (Monroy et al., 1993). Calmodulin (CaM), CaM domain-containing protein kinases (CDPKs), calcineurin B-like proteins (CBLs), and CBL-interacting protein kinases (CIPKs), the key Ca^{2+} sensors in plants, are categorized into "sensor relay" and "responders." Sensor relay, which binds to Ca^{2+}, undergoes conformational changes that in turn regulates the gene expression, e.g., calmodulin (CaM) and calcineurin B-like proteins (CBLs). Responders, like protein kinase and phospholipase, transmit the message to their downstream targets through their effector domains (Reddy and Reddy, 2004; Klimecka and Muszyńska, 2007). The Ca^{2+}-dependent protein kinases could thus be activated by a rise in cytosolic Ca^{2+} alone or in connection with calmodulin. Different Ca^{2+} signatures are distinguished by different Ca^{2+} binding proteins and protein kinases, and decoding of these signals causes changes in gene expression leading to appropriate physiological responses (Yang and Poovaiah, 2003; Sathyanarayanan and Poovaiah, 2004).

4.5.2 ABA and stress signal transduction networks

ABA is identified as a stress hormone because of its quick accumulation in response to stresses and its intervention of many stress responses that aid in the survival of plants in stress conditions (Zhang et al., 2006). ABA works as a secondary signal to transduce cold signals, as verified by the LOS5 (low expression of osmotically responsive genes) mutant isolated through bioluminescent PRD29A::LUC genetic screening. The LOS5 mutant is impaired in molybdenum co-factor (MoCo) sulfurase, which synthesizes MoCo for abscisic aldehyde oxidase, and is thus defective in ABA synthesis. The LOS5 mutant demonstrated a marked decline in cold- and salt/drought-induced expression of COR genes (*RD29A*, *COR15*, *COR47*, *RD22*, and pyrroline-5-carboxylate synthetase) and is incapable of attaining freezing tolerance. Thus, ABA plays an important role in cold acclimation of plants (Xiong et al., 2001). SA and jasmonic acid (JA) also showed tolerance in various plants under stress. These hormones may work with one another in regulating stress signaling and plant stress tolerance, e.g., ethylene has been demonstrated to augment ABA action in seeds (Gazzarrini and McCourt, 2001), but may neutralize ABA effects in vegetative tissues under drought stress (Spollen et al., 2000). Nonetheless, ABA is undoubtedly the most important plant hormone involved in stress signal transduction.

4.6 Conclusion and future prospects

Since cold stress is a multigenic as well as a quantitative trait, it is far more complicated to comprehend the response of the plants towards this stress. Although many genes and their alleles have been illustrated and their functions identified, much remains to be explained about the interactions between genes, proteins, and metabolites in the area of gene networking.

By the use of molecular and genetic approaches, a number of related genes have been identified and new information repetitively emerges to develop the CBF cold-responsive pathway. However, the results of the studies related to trancriptome have confirmed the intricate nature of plant adaptation to low temperature. The fact that a large number of genes recognized by these studies are at present annotated with "unknown function" and involve novel genes and new pathways indicates that our knowledge of the transcriptional control of the low temperature response is incomplete, and the regulation of these transcriptional responses is far more complex than previously believed. Signaling pathways have to be considered as composite networks having multiple points of divergence and convergence that enable signal assimilation at different stages, and provide the molecular foundation for appropriate downstream responses that characterize these signal transduction networks. Plants not only have stress-specific signal transduction pathways such as the SOS pathway, which play a vital role in the ion stress and the ICE-CBF-COR signaling pathway of cold stress, but also the cross-talk between abiotic stresses, e.g., the ABA signaling and the MAPK cascade. With the progress in recent years towards recognizing the genes that are regulated under stress, and sequencing of the whole plant genomes, research is being taken up aggressively to understand the molecular basis of abiotic stress responses and to maneuver these processes via genetic engineering. Thus, information on the low temperature proteome, metabolome, and transcriptome is anticipated to continue to increase in the near future, which is necessary for our understanding of the intricate network of molecular changes that are vital for freezing and chilling tolerance.

References

Allagulova, C.R., Gimalov, F.R., Shakirova, F.M., Vakhitov, V.A., 2003. The plant dehydrins: structure and putative functions. Biochem. (Moscow) 68, 945−951.

Alonso, A., Queiroz, C.S., Magalhães, A.C., 1997. Chilling stress leads to increased cell membrane rigidity in roots of coffee (*Coffea arabica* L.) seedlings. Biochem. Biophys. Acta 1323, 75–84.

Anchordoguy, T.J., Rudolph, A.S., Carpenter, J.F., Crowe, J.H., 1987. Modes of interaction of cryoprotectants with membrane phospholipids during freezing. Cryobiology 24, 324–331.

Anderson, J.V., Li, Q.B., Haskell, D.W., Guy, C.L., 1994. Structural organization of the spinach endoplasmic reticulum-luminal 70-kilodalton heat-shock cognate gene and expression of 70-kilodalton heat-shock genes during cold acclimation. Plant Physiol. 104, 1359–1370.

Apel, K., Hirt, H., 2004. Reactive oxygen species: metabolism, oxidative stress and signal transduction. Annu. Rev. Plant Biol. 55, 373–399.

Aroca, R., Tognoni, F., Irigoyen, J.J., Sánchez-Diaz, M., Pardossi, A., 2001. Difference in root low temperature response of two maize genotypes differing in chilling sensitivity. Plant Physiol. Biochem. 39, 1067–1075.

Artus, N.N., Uemura, M., Steponkus, P.L., Gilmour, S.J., Lin, C., Thomashow, M.F., 1996. Constitutive expression of the cold-regulated Arabidopsis thaliana COR15a gene affects both choroplast and protoplast freezing tolerance. Proc. Natl. Acad. Sci. USA 93, 13404–13409.

Baldi, P., Pedron, L., Hietala, A.M., Porta, N.A., 2011. Cold tolerance in cypress (*Cupressus sempervirens* L.): a physiological and molecular study. Tree Genetics Genomes 7, 79–90.

Barclay, K.D., McKersie, B.D., 1994. Peroxidation reactions in plant membranes: effects of free fatty acids. Lipids 29, 877–883.

Bo, S.H., Suo, L.Z., An, S.M., 2005. LEA proteins in higher plants: structure, function, gene expression and regulation. Colliods and Surfaces B: Biointerfaces 45, 131–135.

Bohnert, H.J., Ayoubi, P., Borchert, C., Bressan, R.A., Burnap, R.L., Cushman, J.C., et al., 2001. A genomics approach towards salt stress tolerance. Plant Physiol. Biochem. 39, 1–17.

Çakmak, T., Atıcı, Ö., 2009. Effects of putrescine and low temperature on the apoplastic antioxidant enzymes in the leaves of two wheat cultivars. Plant Soil Environ. 55, 320–326.

Cao, H.X., Sun, C.X., Shao, H.B., Lei, X.T., 2011. Effects of low temperature and drought on the physiological and growth changes in oil palm seedlings. Afr. J. Biotechnol. 10, 2630–2637.

Chinnusamy, V., Zhu, J., Zhu, J.K., 2006. Gene regulation during cold acclimation in plants. Physiol. Plant 126, 52–61.

Chinnusamy, V., Zhu, J., Zhu, J.K., 2007. Cold stress regulation of gene expression in plants. Trends. Plant Sci. 12, 444–451.

Davletova, S., Schlauch, K., Coutu, J., Mittler, R., 2005. The zinc-finger protein Zat12 plays a central role in reactive oxygen and abiotic stress signaling in Arabidopsis. Plant Physiol. 139, 847–856.

Duan, M., Ma, N.-N., Li, D., Deng, Y.-S., Kong, F.-Y., Lv, W., et al., 2012. Antisense mediated suppression of tomato thylakoidal ascorbate peroxidase influences anti-oxidant network during chilling stress. Plant Physiol. Biochem. 58, 37–45.

Dure, L., 1993. Structural motifs in LEA proteins. In: Close, T.J., Bray, E.A. (Eds.), Plant Responses to Cellular Dehydration During Environmental Stress. American Society of Plant Physiologists, Rockville, MD, pp. 91–103.

Dure, L., Chlan, C., 1981. Developmental biochemistry of cotton-seed embryogenesis and germination. Purification and properties of principal storage proteins. Plant Physiol. 68, 180–186.

Easlon, H.M., Asensio, J.S.R., St-Clair, D.A., Bloom, J.A., 2013. Chilling-induced water stress: variation in shoot turgor maintenance among wild tomato species from diverse habitats. Am J. Bot. 100, 1991–1999.

Ercoli, L., Mariotti, M., Masoni, A., Arduini, I., 2004. Growth responses of sorghum plants to chilling temperature and duration of exposure. Eur. J. Agro. 21, 93–103.

Erdal, S., 2012. Androsterone-induced molecular and physiological changes in maize seedlings in response to chilling stress. Plant Physiol. Biochem. 57, 1–7.

Eriksson, S.K., Harryson, P., 2011. Dehydrins: molecular biology, structure and function. In: Luttge, et al., (Eds.), Plant Desiccation Tolerance. Springer-Verlag, Berlin, Heidelberg.

Fennell, A., Markhart, A.H., 1998. Rapid acclimation of root hydraulic conductivity to low temperature. J. Exp. Bot. 49, 879–884.

Fortunato, A.S., Lidon, F.C., Batista-Santos, P., Leitào, A.E., Pais, I.P., Ribeiro, A.I., et al., 2010. Biochemical and molecular characterization of the antioxidative system of *Coffea* sp. Under cold conditions in genotypes with contrasting tolerance. J. Plant Physiol. 167, 333–342.

Fourrier, N., Bedard, J., Lopez-Juez, E., Barbrook, A., Bowyer, J., Jarvis, P., et al., 2008. A role for sensitive to freezing in protecting chloroplasts against freeze induced damage in Arabidopsis. Plant J. 55, 734–745.

Garay-Arroyo, A., Colmenero-Flores, J.M., Garciarrubio, A., Covarrubias, A.A., 2000. Highly hydrophilic proteins in prokaryotes and eukaryotes are common during conditions of water deficit. J. Biol. Chem. 275, 5668–5674.

Gazzarrini, S., McCourt, P., 2001. Genetic interactions between ABA, ethylene and sugar signaling pathways. Curr. Opin. Plant Biol. 4, 387–391.

Gechev, T., Willekens, H., Montagu, M.V., Inzé, D., Camp, W.V., Toneva, V., et al., 2010. Different responses of tobacco antioxidant enzymes to light and chilling stress. J. Plant Physiol. 160, 509–515.

Gilmour, S.J., Sebolt, A.M., Salazar, M.P., Everard, J.D., Thomashow, M.F., 2000. Overexpression of Arabidopsis CBF3 transcriptional activator mimics multiple biochemical changes associated with cold acclimation. Plant Physiol. 124, 1854–1865.

Gilmour, S.J., Fowler, S.G., Thomashow, M.F., 2004. Arabidopsis transcriptional activators CBF1, CBF2, and CBF3 have matching functional activities. Plant Mol. Biol. 54, 767–781.

Gray, G.R., Heath, D., 2005. A global reorganization of the metabolome in *Arabidopsis* during cold acclimation is revealed by metabolic fingerprinting. Physiol. Plant 124, 236–248.

Guevara, D.R., Champigny, M.J., Tattersall, A., Dedrick, J., Wong, C.E., Li, Y., et al., 2012. Transcriptomic and metabolomic analysis of Yukon *Thellungiella* plants grown in cabinets and their natural habitat show phenotypic plasticity. BMC. Plant. Biol. 12, 175.

Guo, A., He, K., Liu, D., Bai, S., Gu, X., Wei, L., et al., 2005. A database of *Arabidopsis* transcription factors. Bioinformatics. 21, 2568–2569.

Guy, C., Haskell, D., Neven, L., Klein, P., Smelser, C., 1992. Hydration state-responsive proteins link cold and drought stress in spinach. Planta 188, 265–270.

Hajela, R.K., Horvath, D.P., Gilmour, S.J., Thomashow, M.F., 1990. Molecular cloning and expression of COR (cold-regulated) genes in *Arabidopsis thaliana*. Plant Physiol. 93, 1246–1252.

Hasanuzzaman, M., Hossain, M.A., da Silva, J.A.T., Fujita, M., 2012. Plant responses and tolerance to abiotic oxidative stress. In: Bandi, V., Shanker, A.K., Shanker, C., Mandapaka, M. (Eds.), Antioxidant Defense is a Key Factor in Crop Stress and its Management: Perspectives and strategies. Springer, Berlin, pp. 261–316.

Hasanuzzaman, M., Nahar, K., Fujita, M., 2013. Extreme temperatures, oxidative stress and antioxidant defense in plants. In: Vahdati, K., Leslie, C. (Eds.), Abiotic Stress—Plant Responses and Applications in Agriculture. InTech, Rijeka, Croatia, pp. 169–205.

Henriksson, K.N., Trewavas, A.J., 2003. The effect of short-term low-temperature treatments on gene expression in Arabidopsis correlates with changes in intra-cellular Ca^{2+} levels. Plant Cell. Environ. 26, 485–496.

Horváth, I., Glatz, A., Varasovszki, V., Torok, Z., Pali, T., Balogh, G., et al., 1998. Membrane physical state controls the signaling mechanism of the heat shock response in Synechocystis PCC 6803: identification of hsp17 as a "fluidity gene". Proc. Natl. Acad. Sci. USA 95, 3513–3518.

Houde, A., Pommier, S.A., Roy, R., 1993. Detection of the ryanodine receptor mutation associated with malignant hyperthermia in purebred swine populations. J. Anim. Sci. 71, 1414–1418.

Huang, M., Guo, Z., 2005. Responses of antioxidant system to chilling stress in two rice cultivars differing in sensitivity. Biol. Plant 49, 81–84.

Islam, S., Izekor, E., Garner, J.O., 2011. Effect of chilling stress on the chlorophyll fluorescence, peroxidase activity and other physiological activities in *Ipomoea batatas* L. genotypes. Am. J. Plant Physiol. 6, 72−82.

Kandarova, I.V., 1964. Features of the nitrogen metabolism of frost-resistant and non-frost-resistant woody plants. The Physiology of Frost Resistance of Woody Plants. Nauka, Moscow, pp. 61−63

Kang, H.M., Saltveit, M.E., 2002. Reduced chilling tolerance in elongating cucumber seedling radicles is related to their reduced antioxidant enzyme and DPPH-radical scavenging activity. Physiol. Plant 115, 244−250.

Kaplan, F., Kopka, J., Haskell, D.W., Zhao, W., Schiller, K.C., Gatzke, N., et al., 2004. Exploring the temperature-stress metabolome of Arabidopsis. Plant Physiol. 136, 4159−4168.

Kawamura, Y., Uemura, M., 2003. Mass spectrometric approach for identifying putative plasma membrane proteins of Arabidopsis leaves associated with cold acclimation. Plant J. 36, 141−154.

Kimpel, J.A., Key, J.L., 1985. Heat shock in plants. Trends. Biochem. Sci. 10, 353−357.

Klimecka, M., Muszyńska, G., 2007. Structure and functions of plant calcium-dependent protein kinases. Acta. Biochim. Pol. 54, 219−233.

Knight, H., Knight, M.R., 2000. Imaging spatial and cellular characteristics of low temperature calcium signature after cold acclimation in *Arabidopsis*. J. Exp. Bot. 51, 1679−1686.

Knight, M.R., Campbell, A.K., Smith, S.M., Trewavas, A.J., 1991. Transgenic plant aequorin reports the effect of touch and cold shock and elicitors on cytoplasmic calcium. Nature 352, 524−526.

Kocsy, G., Szalai, G., Vágújfalvi, A., Stéhli, L., Orosz, G., Galiba, G., 2000. Genetic study of glutathione accumulation during cold hardening in wheat. Planta 210, 295−301.

Kodama, H., Hamada, T., Horiguchi, G., Nishimura, M., Iba, K., 1994. The enhancement of cold tolerance by expression of a gene for chloroplast ω-desaturase in transgenic tobacco. Plant Physiol. 105, 601−605.

Koster, K.L., Lynch, D.V., 1992. Solute accumulation and compartmentation during the cold acclimation of *puma rye*. Plant Physiol. 98, 108−113.

Kovacs, D., Kalmar, E., Torok, Z., Tompa, P., 2008. Chaperone activity of ERD10 and ERD14, two disordered stress-related plant proteins. Plant Physiol. 147, 381−390.

Krasensky, J., Jonak, C., 2012. Drought, salt, and temperature stress-induced metabolic rearrangements and regulatory networks. J. Exp. Bot. 63, 1593−1608.

Krishna, P., Sacco, M., Cherutti, J.F., Hill, S., 1995. Cold-induced accumulation of Hsp90 transcripts in Brassica napus. Plant Physiol. 107, 915−923.

Kurkela, S., Franck, M., 1990. Cloning and characterization of a cold- and ABA-inducible *Arabidopsis* gene. Plant Mol. Biol. 15, 137−144.

Larcher, W., 2001. Ökophysiologie der Pflanzen. 6th edition Verlag Eugen Ulmer, Stuttgart, p. 302.

Lee, B.H., Lee, H., Xiong, L., Zhu, J.K., 2002. A mitochondrial complex I defect impairs cold-regulated nuclear gene expression. Plant Cell 14, 1235−1251.

Leja, M., Długosz-Grochowska, O., Grabowska, A., Kunicki, E., 2012. The effect of preliminary chilling of broccoli transplants on some antioxidative parameters. Folia Hortic. 24, 131−139.

Levitt, J., 1980. Responses of plants to environmental stress. Chilling, freezing, and high temperature stress. Academic Press, New York.

Li, D.M., Guo, Y.K., Li, Q., Zhang, J., Wang, X.J., Bai, J.G., 2012. The pretreatment of cucumber with methyl jasmonate regulates antioxidant enzyme activities and protects chloroplast and mitochondrial ultrastructure in chilling-stressed leaves. Sci. Hortic. 143, 135−143.

Liu, J.J., Lin, S.H., Xu, P.L., Wang, X.J., Bai, J.G., 2009. Effects of exogenous silicon on the activities of antioxidant enzymes and lipid peroxidation in chilling-stressed cucumber leaves. Agric. Sci. Chin. 8 (9), 1075−1086.

Liu, Q., Ksauga, M., Sakuma, Y., Abe, H., Miura, S., Yamaguchi-Shinozaki, K., et al., 1998. Two transcription factors, DREB1 and DREB2, with an EREBP/AP2 DNA binding domain separate two cellular signal transduction pathways in drought and low temperature-responsive gene expression respectively in *Arabidopsis*. Plant Cell 10, 1391–1406.

Liu, Q.L., Dong, F.L., Xiao, F., Wu, J., Li, Z.J., 2012. Isolation and molecular characterization of DgZFP2: a gene encoding a Cys2/His2-type zinc finger protein in *chrysanthemum*. Afri. J. Agri. Res. 7, 4499–4504.

Liu, W., Yu, K., He, T., Li, F., Zhang, D., Liu, J., 2013. The low temperature induced physiological responses of *Avena nuda* L., a cold-tolerant plant species. Scientific World Journal 658793.

Malekzadeh, P, Khara, J, Heidari, R., 2012. Effect of exogenous Gama-aminobutyric acid on physiological tolerance of wheat seedlings exposed to chilling stress. Iranian J. Plant Physiol. 3, 611–617.

Mittler, R., Vanderauwera, S., Gollery, M., Van Breusegem, F., 2004. Reactive oxygen gene network of plants. Trend Plant Sci. 9, 490–498.

Mizoguchi, T., Hayashida, N., Yamaguchi-Shinozaki, K., Kamada, H., Shinozaki, K., 1993. ATMKs: a family of plant MAP kinases in *Arabidopsis thaliana*. FEBS Lett. 336, 440–444.

Mizoguchi, T., Irie, K., Hirayama, T., Hayashida, N., Yamaguchi-Shinozaki, K., Matsumoto, K., et al., 1996. A gene encoding a mitogen-activated protein kinase is induced simultaneously with genes for a mitogen-activated protein kinase and an S6 ribosomal protein kinase by touch, cold, and water stress in *Arabidopsis thaliana*. Proc. Natl. Acad. Sci. USA 93, 765–769.

Mogk, A., Deuerling, E., Vorderwülbecke, S., Vierling, E., Bukau, B., 2003. Small heat shock proteins, ClpB and the DnaK system form a functional triad in reversing protein aggregation. Mol. Microbiol. 50, 585–595.

Monroy, A.F., Dhindsa, R.S., 1995. Low temperature signal transduction: induction of cold acclimation specific genes of alfalfa by calcium at 25°C. Plant Cell 7, 321–331.

Monroy, A.F., Castonguay, Y., Laberge, S., Sarhan, F., Vezina, L.P., Dhindsa, R.S., 1993. A new cold- induced alfalfa gene is associated with enhanced hardening at sub-zero temperature. Plant Physiol. 102, 873–879.

Murata, N., Ishizaki-Nishizawa, O., Higashi, S., Hayashi, H., Tasaka, Y., Nishida, I., 1992. Genetically engineered alteration in the chilling sensitivity of plants. Nature 356, 710–713.

Nakayama, K., Okawa, K., Kakizaki, T., Inaba, T., 2008. Evaluation of the protective activities of a late embryogenesis abundant (LEA) related protein, Cor15am, during various stresses in-vitro. Biosci. Biotechnol. Biochem. 72, 1642–1645.

O'Kane, D., Gill, V., Boyd, P., Burdon, R., 1996. Chilling, oxidative stress and antioxidant responses in *Arabidopsis thaliana* callus. Planta 198, 371–377.

Okawa, K., Nakayama, K., Kakizaki, T., Yamashita, T., Inaba, T., 2008. Identification and characterization of Cor413im proteins as novel components of the chloroplast inner envelope. Plant Cell. Environ. 31, 1470–1483.

Orvar, B.L., Sangwan, V., Omann, F., Dhindsa, R.S., 2000. Early steps in cold sensing by plant cells: the role of actin cytoskeleton and membrane fluidity. Plant J. 23, 785–794.

Peng, T., Zhu, X.F., Fan, Q.J., Sun, P.P., Liu, J.H., 2012. Identification and characterization of low temperature stress responsive genes in Poncirus trifoliata by suppression subtractive hybridization. Gene 492, 220–228.

Perez-Clemente, R.M., Vives, V., Zandalinas, S.I., Lopez-Climent, M.F., Munoz, V., Gomez-Cadenas, A., 2013. Biotechnological approaches to study plant responses to stress. Bio. Med. Res. Int. 654120.

Polisensky, D.H., Braam, J., 1996. Cold-shock regulation of the *Arabidopsis* TCH genes and the effects of modulating intracellular calcium levels. Plant Physiol. 111, 1271–1279.

Rajashekar, C.B., Li, P.H., Carter, J.V., 1983. Frost injury and heterogeneous ice nucleation in leaves of tuber-bearing *Solanum* species. Ice nucleation activity of external source of nucleants. Plant Physiol. 71, 749–755.

Reddy, V.S., Reddy, A.S.N., 2004. Proteomics of calcium-signaling components in plants. Phytochem. 65, 1745–1776.

Renaut, J., Lutts, S., Hoffmann, L., Hausman, J.F., 2004. Responses of poplar to chilling temperatures: proteomic and physiological aspects. Plant Biol. 6, 81–90.

Rolland, F., Baena-Gonzales, E., Sheen, J., 2006. Sugar sensing and signaling in plants: conserved and novel mechanisms. Annual Review Plant Biol. 57, 675–709.

Roxas, V.P., Lodhi, S.A., Garrett, D.K., Mahan, J.R., Allen, R.D., 2000. Stress tolerance in transgenic tobacco seedlings that over-express glutathione S-transferase/glutathione peroxidase. Plant Cell. Physiol. 41, 1229–1234.

Saeidnejad, A.H., Pouramir, F., Naghizadeh, M., 2012. Improving chilling tolerance of maize seedlings under cold conditions by spermine application. Notulae Scientia Biologicae 4 (3), 110–117.

Sangwan, V., Foulds, I., Singh, J., Dhindsa, R.S., 2001. Cold-activation of *Brassica napus* BN115 promoter is mediated by structural changes in membranes and cytoskeleton and requires Ca^{2+} influx. Plant J. 27, 1–12.

Sathyanarayanan, P.V., Poovaiah, B.W., 2004. Decoding Ca^{2+} signals in plants. A review. Crit. Rev. Plant Sci. 23, 1–11.

Seki, M., Narusaka, M., Ishida, J., Nanjo, T., Fujita, M., Oono, Y., et al., 2002. Monitoring the expression profiles of 7000 Arabidopsis genes under drought, cold and high-salinity stresses using a full-length cDNA micro-array. Plant J. 31, 279–292.

Shulaev, V., 2006. Metabolomics technology and bioinformatics. Brief. Bio.-info. 7, 128–139.

Spollen, W.G., LeNoble, M.E., Samuels, T.D., Bernstein, N., Sharp, R.E., 2000. Abscisic acid accumulation maintains maize primary root elongation at low water potentials by restricting ethylene production. Plant Physiol. 122, 967–976.

Steffen, K.L., Arora, R., Palta, J.P., 1989. Relative sensitivity of photosynthesis and respiration to a freeze-thaw stress: role of realistic freeze-thaw protocol. Plant Physiol. 89, 1372–1379.

Steponkus, P.L., 1984. Role of the plasma membrane in freezing injury and cold acclimation. Annu. Rev. Plant Physiol. 35, 543–584.

Steponkus, P.L., Uemura, M., Webb, M.S., 1993. A contrast of the cryo-stability of the plasma membrane of winter rye and spring oat—two species that widely differ in their freezing tolerance and plasma membrane lipid composition. In: Steponkus, P.L. (Ed.), Advances in Low- Temperature Biology, 2. JAI Press, London, pp. 211–312.

Steponkus, P.L., Webb, M.S., 1992. Freeze-induced dehydration and membrane destabilization in plants. In: Somero, G.N., Osmond, C.B., Bolis, C.L. (Eds.), Water and Life: Comparative Analysis of Water Relationships at the Organismic, Cellular and Molecular Level. Springer-Verlag, Berlin, Germany, pp. 338–362.

Stockinger, E.J., Gilmour, S.J., Thomashow, M.F., 1997. *Arabidopsis thaliana* CBF1 encodes an AP2 domain-containing transcriptional activator that binds to the C-repeat/DRE, a cis-acting DNA regulatory element that stimulates transcription in response to low temperature and water deficit. Proc. Natl. Acad. Sci. USA 94, 1035–1040.

Strauss, G., Hauser, H., 1986. Stabilization of lipid bilayer vesicles by sucrose during freezing. Proc. Natl. Acad. Sci. USA 83, 2422–2426.

Streb, P., Shang, W., Feierabend, J., 1999. Resistance of cold-hardened winter rye leaves (*Secale cereale* L.) to photo-oxidative stress. Plant Cell. Environ. 22, 1211–1223.

Sun, W., Van Montagu, M., Verbruggen, N., 2002. Small heat shock proteins and stress tolerance in plants. Biochem. Biophys. Acta 1577, 1–9.

Taji, T., Ohsumi, C., Iuchi, S., Seki, M., Kasuga, M., Kobayashi, M.K., 2002. Important roles of drought and cold-inducible genes for galactinol synthase in stress tolerance in *Arabidopsis thaliana*. Plant J. 29, 417–426.

Takáč, T., Luxová, M., Gašparíková, O., 2003. Cold induced changes in antioxidant enzymes activity in roots and leaves of two maize cultivars. Biologia Bratislava 58, 875–880.

Takahashi, D., Li, B., Nakayama, T., Kawamura, Y., Uemura, M., 2013. Plant plasma membrane proteomics for improving cold tolerance. Frontiers Plant Sci. 4, 90.

Tartoura, K.A.H., Youssef, S.A., 2011. Stimulation of ROS-scavenging systems in squash (*Cucurbita pepo* L.) plants by compost supplementation under normal and low temperature conditions. Sci. Hortic. 130, 862–868.

Telstcherova, L., 1967. On some metabolic changes in shoot apices of wheat plants at low temperatures. In: Troshin, A.S. (Ed.), The cell and environmental temperature. Proc Internatl Symposium on Cytoecology, Leningrad, 1963, 34. Pergamon Press, New York, pp. 53–58.

Thomashow, M.F., 1998. Role of cold responsive genes in plant freezing tolerance. Plant Physiol. 118, 1–7.

Thomashow, M.F., 1999. Plant cold acclimation: freezing tolerance genes and regulatory mechanisms. Annu. Rev. Plant. Physiol., Plant Mol. Biol. 50, 571–599.

Tommasini, L., Svensson, J.T., Rodriguez, E.M., Wahid, A., Malatrasi, M., Kato, K., et al., 2008. Dehydrin gene expression provides an indicator of low temperature and drought stress: transcriptome-based analysis of barley (*Hordeum vulgare* L.). Funct. Integr. Genomics. 8, 387–405.

Turan, Ö., Ekmekçi, Y., 2011. Activities of photosystem II and antioxidant enzymes in chickpea (*Cicer arietinum* L.) cultivars exposed to chilling temperatures. Acta. Physiol. Plant 33, 7849–7855.

Uemura, M., Steponkus, P.L., 1997. Effect of cold acclimation on membrane lipid composition and freeze induced membrane destabilization, plant cold hardiness. In: Li, P.H., et al., (Eds.), Plat Cold Hardiness. Plenum, New York, pp. 171–179.

Uemura, M., Joseph, R.A., Steponkus, P.L., 1995. Cold acclimation of *Arabidopsis thaliana* (effect on plasma membrane lipid composition and freeze-induced lesions). Plant Physiol. 109, 15–30.

Vierling, E., 1991. The roles of heat shock proteins in plants. Annu. Rev. Plant. Physiol. Plant Mol. Biol. 42, 579–620.

Vogel, J.T., Zarka, D.G., Van Buskirk, H.A., Fowler, S.G., Thomashow, M.F., 2005. Roles of the CBF2 and ZAT12 transcription factors in configuring the low temperature transcriptome of Arabidopsis. Plant J. 41, 195–211.

Walker, D.J., Romero, P., Hoyos, A., Correal, E., 2008. Seasonal changes in cold tolerance, water relations and accumulation of cations and compatible solutes in *Atriplexhalimus* L. Environ. Exp. Bot. 64, 217–224.

Wang, L.-Y., Li, D., Deng, Y.-S., Lv, W., Meng, Q.-W., 2013. Antisense-mediated depletion of tomato GDP-l-galactose phosphorylase increases susceptibility to chilling stress. J. Plant Physiol. 170, 303–314.

Wang, X.M., Li, W.Q., Li, M.Y., Welti, R., 2006. Profiling lipid changes in plant response to low temperatures. Physiol. Plant 126, 90–96.

Waters, E.R., Lee, G.J., Vierling, E., 1996. Evolution, structure and function of the small heat shock proteins in plants. J. Exp. Bot. 47, 325–338.

Welti, R., Li, W.Q., Li, M.Y., Sang, Y.M., Biesiada, H., Zhou, H.E., et al., 2002. Profiling membrane lipids in plant stress responses: role of phospholipase D α in freezing-induced lipid changes in *Arabidopsis*. J. Biol. Chem. 277, 31994–32002.

Wise, R.R., 1995. Chilling-enhanced photooxidation: the production, action and study of reactive oxygen species produced during chilling in the light. Photosynthesis Res. 45, 79–97.

Wu, J., Lightner, J., Warwick, N., Browse, J., 1997. Low-temperature damage and subsequent recovery of fab1 mutant Arabidopsis exposed to 28°C. Plant Physiol. 113, 347–356.

Xiong, L., Ishitani, M., Lee, H., Zhu, J.K., 2001. The Arabidopsis LOS5/ABA3 locus encodes a molybdenum co-factor sulfurase and modulates cold stress and osmotic stress responsive gene expression. Plant Cell 13, 2063–2083.

Xiong, L., Schumaker, K.S., Zhu, J.K., 2002. Cell signaling during cold, drought, and salt stress. Plant Cell 14, 165–183.

Yamaguchi-Shinozaki, K., Shinozaki, K., 1994. A novel cis-acting element in an Arabidopsis gene is involved in responsiveness to drought, low temperature, or high-salt stress. Plant Cell 6, 251–264.

Yang, H., Wu, F., Cheng, J., 2011. Reduced chilling injury in cucumber by nitric oxide and the antioxidant response. Food. Chem. 127, 1237–1242.

Yang, T., Poovaiah, B.W., 2003. Calcium/calmodulin mediated signal network in plants. Trends. Plant Sci. 8, 505–512.

Yao, G.H., Gao, P.P., Wang, Y.P., Wang, L.H., Xu, G.D., Liu, P., 2013. Abscisic acid improves chilling-induced oxidative stress in *trichosanthes kirilowii* maxim seedlings. J. Agr. Sci. Tech. 15, 583–592.

Yong, I.K., Ji, S.S., Nilda, R.B., 2003. Antioxidative enzymes offer protection from chilling damage in rice plants. Crop Science Society of America 43, 2109–2117.

Zhang, J., Jia, W., Yang, J., Ismail, A.M., 2006. Role of ABA in integrating plant responses to drought and salt stress. Field Crop. Res. 97, 111–119.

Zhu, J., Dong, C.H., Zhu, J.K., 2007. Interplays between cold responsive gene regulation, metabolism and RNA processing during plant cold acclimation. Curr. Opin. Plant Biol. 10, 290–295.

Zhu, T., Budworth, P., Han, B., Brown, D., Chang, H.S., Zou, G., et al., 2001. Toward elucidating the global expression patterns of developing Arabidopsis: parallel analysis of 8300 genes by a high-density oligonucleotide probe array. Plant Physiol. Biochem. 39, 221–242.

Genetic Engineering of Crop Plants for Abiotic Stress Tolerance

Surbhi Goel and Bhawna Madan

5.1 Introduction

Growth and survival of plants are greatly affected by unfavorable environmental factors such as high salinity, water drought, low and high temperatures, flooding, and deprivation of nutrients (Kasuga et al., 1999; Sreenivasulu et al., 2007; Bhatnagar-Mathur et al., 2008). These abiotic stress factors affect plant growth, reduce productivity, and deteriorate seed quality, thus reducing yield potential (Levitt, 1980). It has been estimated that abiotic stresses lead to more than 50% of yield reduction (Rodríguez et al., 2005; Acquaah, 2007). Plant species that are more genotypically advanced than others show better survival under abiotic stress conditions, but to sustain the growth of agriculturally important plants lacking this advantage, measures need to be taken to increase stress resistance. For this, selection of stress-resistant plants and conventional breeding programs has been used; however, no significant success was achieved (Richards, 1996; Singla-Pareek et al., 2001). Agricultural plants require maintenance of their growth and development to ensure high productivity and yield along with high quality standards. To feed the increasing population, it is required to explore new lands where agriculture is impractical due to water-drought condition, low and high temperatures, heavy metal poisoning, and high saline concentration. This can be achieved by engineering plants with improved abiotic stress tolerance (Reguera et al., 2012).

Genetic engineering approaches are one of the important ways to produce plants with increasing abiotic stress tolerance. Abiotic stress leads to up-regulation of expression of a number of genes in plants, which enables them to survive in stress conditions. These abiotic stress-inducible genes are classified into two groups: (1) genes whose products are directly involved against stress such as osmoprotectants, antifreeze proteins, chaperons, detoxification enzymes, and (2) genes that regulate expression of stress-related genes and signal transduction in stress response such as transcription factors, protein kinases, etc. (Kasuga et al., 1999). The stress-inducible genes and their regulatory genes can be altered, transferred, and expressed in different species via an *Agrobacterium*-mediated transformation system or expressed ectopically, thus inducing biochemical, molecular, and physiological alteration leading to increase in growth, development, and yield of plants in stress conditions (Bhatnagar-Mathur et al., 2008). Overexpression of stress-inducible genes is controlled under strong constitutive promoters like CaMV 35S, which improves stress tolerance, but this causes expression in an uncontrolled manner, which costs cells their energy reserve resulting in growth retardation (Kasuga et al., 2004). Recently, the use of stress-inducible promoters for expression of stress-inducible genes ensured a time-specific and optimum level expression of genes (Katiyar et al., 1999).

P. Ahmad (Ed): Emerging Technologies and Management of Crop Stress Tolerance, Volume 1.
DOI: http://dx.doi.org/10.1016/B978-0-12-800876-8.00005-9

In this chapter, the various approaches used to improve abiotic stress tolerance by genetic engineering of transcription factors, ion transport channels, overexpression of osmoprotectants and stress-signaling genes, and maintenance of osmotic balance have been discussed.

5.2 Overexpression of genes for transcriptional regulation

Transcriptional factors (TFs) are proteins that interact with the *cis*-elements of promoter regions and up-regulate the expression of stress-signaling genes, imparting enhanced tolerance towards multiple stresses (Bhatnagar-Mathur et al., 2008; Hussain et al., 2011). The various TFs involved in stress signaling, e.g., zinc-finger domains, dehydration response elements, basic leucine zippers, MYB, and NAC transcriptional factors, are described below (Figure 5.1). As the overexpression of the TFs in transgenic plants has shown increased stress tolerance (Mukhopadhyay et al., 2004; Fujita et al., 2005; Ito et al., 2006; Jung et al., 2008), it has, therefore, emerged as a powerful tool for development of drought-tolerant crops.

5.2.1 Zinc-finger domains

Cys-2/His-2-type zinc fingers, also called the classical or TFIIIA-type fingers, are DNA binding motifs found in eukaryotic transcription factors. They contain about 30 amino acids, having a consensus sequence of $Cx2-4Cx9-12Cx2Cx4Cx2Hx5HxC$, where x represents any amino acid, and two

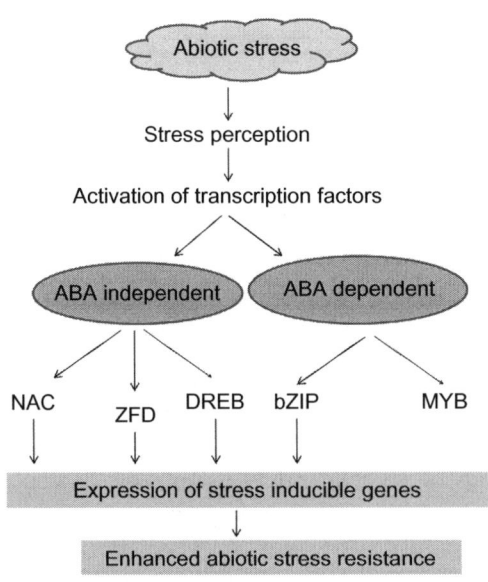

FIGURE 5.1

A schematic representation of activation of transcriptional factors leading to abiotic stress tolerance.

pairs of conserved Cys and His bound tetrahedrally to a zinc ion (Pabo et al., 2001). In multiple-fingered proteins, the adjacent fingers are separated by a long spacer that is highly variable in length and sequence from one protein to another. The EPF family is a subfamily of TFIIIA-type zinc-finger proteins of plants, which are characterized by a highly conserved sequence QALGGH in their zinc-figure motifs (Frugier et al., 2000). This family includes soybean SCOF-1 for cold tolerance (Kim et al., 2001), *Arabidopsis* STZ/ZAT10 for salt tolerance (Lippuner et al., 1996), and *Arabidopsis* RHL41/ZAT12 for light acclimatization response (Lida et al., 2000). In *Petunia hybrida* both ZPT2-2 and ZPT2-3 (EPF family zinc-finger motifs) were induced by stress, desiccation, cold, wounding, etc. (Van Der Krol et al., 1999; Sugano et al., 2003). The constitutive overexpression of ZPT2-3 in transgenic *Petunia* plants results in increased tolerance to dehydration (Sugano et al., 2003). In *Arabidopsis*, four ZPT2-related proteins (AZF1, AZF2, AZF3, and STZ) were characterized. RNA gel-blot analysis showed that expression of AZF2 and STZ was strongly induced by dehydration, high-salt and cold stresses, and abscisic acid treatment. The transgenic *Arabidopsis* plants overexpressing AZF2 and STZ show increased stress tolerance following growth retardation (Sakamoto et al., 2004). In rice, the gene *OSISAP1* encoding a zinc-finger protein is induced after different types of stresses such as cold, desiccation, salt, submergence, heavy metals as well as injury. The overexpression of this gene *OSISAP1* in transgenic tobacco conferred tolerance to cold, dehydration, and salt stress at the seed-germination/seedling stage (Mukhopadhyay et al., 2004).

5.2.2 **Dehydration-responsive element**

Dehydration-responsive element/C-repeat (DRE/CRT), a 9 bp conserved DNA sequence, TACCGACAT, a *cis*-acting element, plays an important role in both dehydration and low temperature tolerance. In *Arabidopsis*, DRE-related motifs have been reported in the promoter regions of cold- and drought-inducible genes such as *kin1*, *cor6.6*, and *rd17* (Wang et al., 1995; Iwasaki et al., 1997). In addition, two DRE binding proteins, DREB1 and DREB2, have been isolated using the yeast one-hybrid screening technique (Liu et al., 1998). DREB1 is up-regulated in response to cold stress whereas DREB2 is up-regulated in response to drought and salinity (Liu et al., 1998). Both the DREB1A and DREB2A homologues belong to the family of EREBP/AP2 transcription factors. The DREB binding proteins have also been reported in other plants such as *Triticum aestivum*, *Zea mays*, and *Oryza sativa*, and are found to be induced by low temperature, salt, and drought (Kizis et al., 2001; Shen et al., 2003). The overexpression of DREB1A under a strong constitutive promoter in transgenic *Arabidopsis* plants induced strong expression of the stress genes in unstressed conditions, which in turn imparts freezing and dehydration tolerance along with dwarf phenotype. However, no growth alterations were observed when the DREB1A was overexpressed under the control of a stress-inducible promoter (Jaglo-Ottosen, 1998; Liu et al., 1998; Kasuga et al., 1999). The transgenic plants overexpressing DREB2A proteins caused growth retardation but did not show stress resistance (Liu et al., 1998). Similarly, in *Oryzae sativa* and *Zea mays*, the overexpression of *OsDREB1* and *ZmDREB1A*, respectively, has resulted in increased tolerance to drought, salt, and freezing stress (Qin et al., 2004; Ito et al., 2006). This indicates that both *OsDREB1* and *ZmDREB1A* show functional similarity to *Arabiopsis* DREB1. The transgenic rice and *Arabidopsis* plants overexpressing *DREB1* also accumulated osmoprotectants such as proline and various other sugars (Gilmour et al., 2000; Ito et al., 2006).

5.2.3 NAC transcription factor

NAC are a class of transcription factors expressed in plants in response to stress and have shown to play an important role in development. The NAC protein is characterized by the presence of a highly conserved region, the *NAC domain* (Kikuchi et al., 2000; Nakashima et al., 2012). In *Petunia*, the first NAC gene *NAM* was reported (Souer et al., 1996) and the first cDNA of the NAC protein for the "RESPONSIVE TO DEHYDRATION 26" (RD26) gene was described in *Arabidopsis* (Yamaguchi-Shinozaki et al., 1992). In *Arabidopsis* and rice, 106 and 149 NAC genes have been predicted, respectively (Gong et al., 2004; Xiong et al., 2005). Overexpressing *SNAC1* (stress-responsive *NAC1*) in rice enhances drought resistance significantly at the reproductive stage without any changes in phenotype and yield. The transgenic rice showed increased sensitivity to abscisic acid and reduced water loss by stomata closing. Also, a number of stress-related genes were found to be up-regulated in SNAC1 overexpressing rice plants suggesting its promising role in increasing drought and salt tolerance (Hu et al., 2006). Rice *SNAC1* gene was expressed in Chinese wheat variety Yangmai 12 under maize ubiquitin promoter for improving salinity and drought tolerance. *SNAC1* expressing plant showed increased tolerance with high water and chlorophyll content in leaves and increased fresh and dry weight of roots and high yield. Also, expression of *SNAC1* increased the expression of other abiotic stress-signaling genes such as wheat 1-phosphatidylinositol-3-phosphate-5-kinase, sucrose phosphate synthase, and type 2C protein phosphatases resulting in enhanced tolerance (Saad et al., 2013).

Transgenic rice plants overexpressing rice NAC gene *ONAC045* showed enhanced drought and salt tolerance and up-regulation of other stress-responsive genes (Zheng et al., 2009). In *Arabidopsis*, a stress-responsive NAC gene, *ATAF1*, is induced by drought, salinity, infection, etc. The transgenic *Arabidopsis* plants overexpressing *ATAF1* increased the sensitivity of plants to salt, drought, and oxidative stress (Wu et al., 2009). The transgenic rice plants overexpressing rice NAC gene *ONAC045* showed enhanced drought and salt tolerance and up-regulation of other stress-responsive genes (Zheng et al., 2009). Gao et al. (2010) identified and isolated a NAC gene *OsNAC52* from *Oryza sativa L cDNA* library using rapid amplification of cDNA ends (RACE) and studied its effect on drought tolerance in *Arabidopsis*. Overexpression of *OsNAC52* in *Arabidopsis* increased drought tolerance by activation of expression of a number of downstream genes. Also, increased drought, salt, and freezing tolerance was observed in transgenic *Arabidopsis* lines overexpressing *TaNAC2*, a NAC transcription factor from wheat, which activated the expression of abiotic stress-responsive genes (Mao et al., 2012). The transgenic *Arabidopsis* overexpressing *Populus euphratica* stress-responsive gene *PeNAC1* showed enhanced tolerance to salt stress, with lower Na^+/K^+ ratios in roots and leaves and reduced expression of *AtHKT1* (Wang et al., 2013).

5.2.4 Basic leucin zipper transcription factors

ABA, a plant hormone, show increased expression under stress conditions such as drought, and induced responses essential for growth and yield. Under stress conditions, ABA is responsible for up-regulation of a number of genes containing the *cis*-element called ABA-responsive element (ABRE) with (C/T)ACGTGGC consensus sequence in the promoter region (Bray, 1994; Giraudat et al., 1994; Busk and Pagès, 1998). Many of these ABRE elements contain G-box (CACGTG) sequence and another group of ABREs known as coupling element (CE) has CGCGTG core sequence (Giuliano et al., 1988; Busk and Pagès, 1998).

The ABA-responsive element binding factors (ARBE or ABF) belong to the class of basic leucine zipper transcription factors that binds to the ABA-responsive element (ABRE) motif in the promoter region of ABA-inducible genes. Four ARBEs (ABF1−ABF4) were isolated first by using yeast one-hybrid screening of *Arabidopsis* cDNA expression library, were shown to be induced by ABA and stress, and are capable of binding to ABREs and transactivation of ABRE-containing genes (Uno et al., 2000; Choi et al., 2000). In *Arabidopsis*, the constitutive overexpression (two- to 10-folds) of *ABF3* and *ABF4* leads to ABA hypersensitivity, reduced transpiration and enhanced drought tolerance (Kang et al., 2002), suggesting their role in ABA-mediated stress response. Similarly, constitutive expression of ABF3 (*Arabidopsis* gene) in rice resulted in expression of ABA-related genes such as LEA, Hsp 70, and protein phosphatase 2C leading to stress tolerance to drought (Oh et al., 2005).

Fujita et al. (2005) showed that under normal conditions, the expression of the *AREB*1 gene is not sufficient to induce the expression of downstream genes. To overcome this, an activated form of *AREB*1 (*AREB*1ΔQT) was created. Plants overexpressing AREB1ΔQT showed ABA hypersensitivity and enhanced drought tolerance, and late embryogenesis abundant class genes and ABA and drought stress-inducible regulatory genes with two or more ABRE motifs in the promoter regions were greatly up-regulated.

Overexpression of *OsbZIP23*, a member of bZIP transcription factor in rice (*Oryza sativa*), resulted in improved drought tolerance, high-salinity stresses, and sensitivity to ABA (Xiang et al., 2008). On the other hand, a null mutant of this factor showed decreased sensitivity to ABA. Also, it was shown that expressions of hundreds of genes were altered in transgenic rice overexpressing *OsbZIP23*, indicating that *OsbZIP23* is a transcriptional regulator of stress-related genes through an ABA-dependent manner (Xiang et al., 2008; Nijhawan et al., 2008).

5.2.5 MYB transcription factors

MYB is one of the largest family of transcription factors in the plant kingdom involved in the expression of stress-induced genes. Based on the presence of repeats in DNA binding domain, MYB proteins are classified into three subfamilies: MYBR2R3, MYBR1R2R3, and MYB related (Stracke et al., 2001; Chen et al., 2006). In plants, the most common family present is MYB with R2R3 repeat (Stracke et al., 2001). In rice and *Arabidopsis*, about 126 and 109 R2R3 MYB proteins were found, respectively (Chen et al., 2006). MYB genes are known to play roles in a number of developmental and stress-responsive processes (Dubos et al., 2010).

The first plant MYB gene was isolated from *Zea mays* C1 involved in anthocyanin biosynthesis by encoding c-myb-like transcription factor (Paz-ares et al., 1987). In *Arabidopsis*, two transcription factors, *AtMYC2* and *AtMYB2*, were found to be induced by drought and salt stress and function in an ABA-dependent manner (Abe et al., 2003). In *Arabidopsis*, *AtMYB44*, an R2R3 MYB transcription factor gene, showed activation under dehydration, low temperature, and salinity. When overexpressed, transgenic *Arabidopsis* showed reduction in water loss during dehydration and enhanced salt stress (Jung et al., 2008).

Overexpression of *Osmyb4* cDNA in *Arabidopsis thaliana* showed significant increase in freeze tolerance and has also affected the expression of genes involved in cold-induced pathways suggesting its major role in cold tolerance (Vannini et al., 2006). Similarly, the constitutive expression of *Chrysanthemum CmMYB2*, an *R2R3-MYB* transcription factor gene in *Arabidopsis thaliana*, leads to increased expression of stress-tolerant genes resulting in increase drought and salinity tolerance (Shan et al., 2012). Also, overexpression of *OsMYB2* in rice imparts enhanced tolerance to salt, cold, and dehydration stress. The enhanced accumulation of soluble sugar and proline and less accumulation of H_2O_2 and malondialdehyde was observed in transgenic rice plants overexpressing *OsMYB2*. Microarray analysis also revealed the increased expression of a number of genes involved in stress tolerance suggesting the role of transcription factor in tolerance to dehydration, salt, and cold stress (Yang et al., 2012).

In wheat (*Triticum aestivum*), the function of MYB transcription factor *TaMYB73* was studied by He et al. (2011). TaMYB73 is an MYB protein of R2R3 type with transactivation activity known to be induced by NaCl, dehydration, and stress-responsive *cis*-elements. Overexpression of *TaMYB73* in *Arabidopsis* showed increased salinity tolerance to NaCl, LiCl, and KCl. Also, the transgenic plants showed superior germination ability compared to wild type (He et al., 2012). In another report, ectopic overexpression of *TaMYB33* in *Arabidopsis* showed enhanced drought tolerance and NaCl stress, but did not show tolerance towards LiCl and KCl stress (Qin et al., 2012).

5.3 Overexpression of genes for osmoprotectants

Osmoregulation involves up-regulation of osmolytes such as amino acids (proline) and quaternary amines (glycine betaine, polyamines, dimethylsulfoniopropronate, and polyol/sugars), which maintains cell turgor, regulates redox potentials and the hydroxy radical scavenger, protects macromolecules against denaturation, and reduces acidity in the cell (McNeil et al., 2001; Vinocur and Altman, 2005). Many crops lack the ability to synthesize osmoprotectants; therefore, the crops are engineered to overexpress osmoprotectants, resulting in enhanced stress tolerance (Bhatnagar-Mathur et al., 2008).

5.3.1 Glycine betaine

Betaine, an osmoprotectant, is a quaternary ammonium compound with a methyl group on every nitrogen atom. Of these, glycine betaines (GB, N, N, N-trimethyglycine) are widely distributed in higher plants and microorganisms, and their increased levels are found under various environmental stresses (Rhodes and Hanson, 1993; Chen and Murata, 2008). The halotolerant plants accumulate GB in chloroplasts and plastids in response to abiotic stress (Kishitani et al., 1994; Allard et al., 1998). *In vitro* studies have shown that GB, known as a "compatible solute," effectively stabilizes the complex enzymes and protein structures and helps in maintaining the ordered structure of membranes at high salt concentration and temperature (Papageorgiou and Murata, 1995). The synthesis of glycine betaine takes place either by dehydrogenation of choline or *N*-methylation of glycine (Chen and Murata, 2002). In higher plants, choline monooxygenase (CMO) and betaine aldehyde dehydrogenase (BADH) mediate GB synthesis in choloroplast with choline as precursor. In microorganisms, GB is synthesized by choline dehydrogenase (CDH) and BADH (Rathinasabapathi et al., 1997; Takabe et al., 1998). In some bacteria such as *Arthrobacter globiformis* and *Arthrobacter pascens*, a single enzyme choline oxidase is involved in GB synthesis (Ikuta et al., 1977). The transgenic *Arabidopsis* and rice plants overexpressing the *Arthrobacter globiformis* COD gene for glycine betaine synthesis showed a high level of salt and cold tolerance (Hayashi et al., 1997; Sakamoto et al., 1998; Huang et al., 2000; Mohanty et al., 2002). The *Oryza sativa* contains two BADH genes, but lacks a functional CMO gene, which is unable to accumulate GB. A CMO gene from spinach (*Spinacia oleracea*) is transferred in rice plants by the *Agrobacterium*-mediated transformation method. This overexpression leads to the accumulation of GB at the level of 0.29–0.43 $\mu mol/g^{-1}$ dry weight and showed increased salt and temperature stress at the seedling stage (Shirasawa et al., 2006). The transgenic tobacco (*Nicotiana tabacum*) plant after transformation with *E. coli* genes for GB biosynthesis accumulated glycine betaine and showed increased tolerance to salt stress (Holmström et al., 1994, 2000). Also, betaine can be synthesized directly from glycine by two N-methyltransferase enzymes. The transgenic *Arabidopsis* plant expressing N-methyltransferase (ApGSMT and ApDMT) showed high accumulation of betaine and increased seed yield under stress conditions (Waditee et al., 2005). Accumulation of GB was found to be low in transgenic plants even after transformation with GB synthesis genes (CMO, COD), which may be due to insufficient supply of choline. A key enzyme in choline synthesis, phosphoethanolamine N-methyltransferase, was overexpressed in tobacco and transgenic plants showed increased levels of phosphocholine (five-fold) and free choline (50-fold), which in turn led to increased production of GB via an engineered GB pathway (McNeil et al., 2001).

5.3.2 Proline

Amino acid proline is accumulated in plants under salt stress. In *Arabidopsis*, salt stress and water deficiency resulted in increased expression of Δ^1-pyrroline-5-carboxylate synthase (P5CS), a key enzyme involved in proline biosynthesis proving its role as an osmoprotectant. Proline synthesis takes place in cytosol from glutamate via glutamate semialdehyde (GSA), which undergoes cyclization to Δ^1-pyrroline-5-carboxylate (Hare et al., 1999).

Overexpression of the mothbean (*Vigna aconitifolia*) Δ^1-pyrroline-5-carboxylate synthase (*p5cs*) gene in tobacco plant resulted in 10- to 18-fold increased proline production than control plants, and showed enhanced biomass in stress conditions (Kishor et al., 1995). The transgenic tobacco plants overexpressing proline were reported to show enhanced freezing tolerance (Konstantinova et al., 2002), reduced levels of reactive oxygen species (ROS), and increased salt tolerance (Hong et al., 2000). Also, the high level of proline was achieved by down-regulating its catabolism. Proline dehydrogenase catalyzes the first step of proline degradation. In *Arabidopsis*, transformation with an antisense proline dehydrogenase cDNA resulted in enhanced proline accumulation and stress tolerance (up to 600 mM NaCl) and freezing (up to $-7°C$) (Nanjo et al., 1999).

The transgenic *Arabidopsis* plant overexpressing *LcSAIN2* (*Leymus chinensis* salt-induced 2) gene from sheepgrass (*Leymus chinensis*) showed increased salt tolerance. These plants also showed accumulation of osmolytes, especially proline, and improved expression of other stress-inducible genes such as *RD29B*, *RAB18*, and transcription factors *MYB2* and *RD26*, which are suggested to be the factors responsible for increased salt tolerance (Li et al., 2013).

5.3.3 Polyamines

Polyamines (PAs), putrescine (Put), spermidine (Spd), and spermine (Spm) are low-molecular-weight organic polycations, found ubiquitously in a wide range of organisms (Fariduddin et al., 2012).

These play crucial roles in a wide range of plant growth and developmental processes such as cell division, root formation, embryogenesis, organogenesis, floral initiation, leaf senescence, fruit development, and fruit ripening (Bagni and Tassoni, 2001; Kusano et al., 2008). The overexpression of PA biosynthetic genes, such as arginine decarboxylase (Masgrau et al., 1997; Roy and Wu, 2001; Capell et al., 2004), ornithine decarboxylase (Kumria and Rajam, 2002), S-adenosylmethionine decarboxylase (Torrigiani et al., 2005), and spermidine synthase (Franceschetti et al., 2004; Kasukabe et al., 2004, 2006) in rice, tobacco, *Arabidopsis*, and sweet potato plants leads to increased tolerance to environmental stress. The polyamines have stress protective properties and increase tolerance to stresses such as salinity (Shevyakova et al., 2006), drought (Yamaguchi et al., 2007), chilling (Nayyar, 2005), oxidative stress (Ye et al., 1997), metal toxicity (Shevyakova et al., 2011), and paraquat (PQ) (Benavides et al., 2000). Put is produced either by decarboxylation of L-ornithine to putrescine catalyzed by ornithine decarboxylase (ODC, EC 4.1.1.17) or by arginine decarboxylase (ADC, EC 4.1.1.19). Spd is synthesized by spermidine synthase (SPDS, EC 2.5.16) by addition of the aminopropyl group to ODC. The transgenic rice overexpressing ADC gene under the control of an ABA-inducible promoter resulted in the accumulation of PAs and improved biomass against salt stress than the wild-type plants and have shown more tolerance to high salinity (Roy and Wu, 2001). The transformed tobacco (*Nicotiana tabacum*) plants with human ODC gene showed that human ODC accumulated to high levels in both the cytosol and apoplast in transgenic tobacco leaves and putrescine levels increased by up to 8.5-fold ($P < 0.05$) compared to wild-type plants (Nölke et al., 2008). Also, the transgenic rice plants containing *Datura stramonium* ODC produced much higher levels of putrescine under stress, promoting spermidine and spermine synthesis and ultimately protecting the plants from drought. The transgenic tomato plants expressing *Saccharomyces cerevisiae* S-adenosyl-l-methionine decarboxylase (SAMDC) produced 1.7- to 2.4-fold higher levels of spermidine and spermine than wild-type plants under high temperature stress. Also, enhanced antioxidant enzyme activity and the protection of membrane lipid peroxidation were also observed, which in turn improved the efficiency of CO_2 assimilation and protected the plants from high temperature stress (Cheng et al., 2009).

5.3.4 **LEA proteins**

Late-embryogenesis-abundant (LEA) proteins are low molecular weight proteins synthesized during maturation of embryo and seed desiccation. Their transcription is induced in response to low temperature, osmotic stress and drought (Ingram and Bartels, 1996). Based on their amino acid sequence, mRNA homology, and expression pattern, LEA proteins are classified into six different groups (Wise, 2003).

The role of group 1, 2, and 3 LEA proteins in abiotic stress is well known, but the role of group 4 proteins LEA4-1 from *Brassica napus* was studied by expressing it in both *E. coli* and *Arabidopsis*. Overexpression of *BnLEA4-1* cDNA conferred salt, temperature, and drought tolerance displaying its role in abiotic stress (Dalal et al., 2009). Also, overexpression of *B. napus* group 3 LEA proteins in Chinese cabbage (*Brassica campestris* ssp. *pekinensis*) and lettuce (*Lactuca sativa* L.) resulted in enhanced growth under salt and drought conditions (Park et al., 2005a,b). Barley HVA1, a group 3 LEA protein, was expressed in rice (Chandra Babu et al., 2004; Xu et al., 1994), wheat (Sivamani et al., 2000), and mulberry (Lal et al., 2008), which resulted in maintenance of plant growth, delayed development of damage symptoms, and relatively high leaf water content under drought condition, as the production of HVA1 protein maintains the cell membrane stability under stress conditions. Similarly, the expression of barley HVA1 gene in a drought-intolerant grass, creeping bentgrass (*Agrostis stolonifera* var. *palustris*), resulted in high water content in leaves and delayed wilting in water-deficient conditions (Fu et al., 2007). A novel LEA gene (DQ663481) from *Tamarix androssowii* transformed in tobacco resulted in increased height and reduction in wilted leaves under drought conditions (Wang et al., 2006). Two wheat genes, *PMA80* (encoded for LEA group I protein) and *PMA1959* (encoded for LEA group 2 protein), were shown to increase in drought tolerance in transgenic rice (Cheng et al., 2002).

A cDNA clone of the LEA gene, *OsLEA3-1*, was overexpressed in drought-sensitive japonica rice Zhonghua 11 rice under the control of three different promoters: the drought-inducible promoter of OsLEA3-1 (OsLEA3-H), the CaMV 35S promoter (OsLEA3-S), and the rice Actin1 promoter (OsLEA3-A). Transgenic rice with OsLEA3-H and OsLEA3-S constructs showed higher grain yield under drought and salt stress (Xiao et al., 2007). Transgenic tobacco plants overexpressing *Rab16A*, an LEA gene from indica rice Pokkali, displayed increased salt stress tolerance with increased production of osmolytes, delayed onset of damage symptoms, and maintenance of mineral balance (RoyChoudhury et al., 2007). Two dehydration-responsive genes *BhLEA1* and *BhLEA2* from *Boea hygrometrica* when ectopically expressed in tobacco plant conferred dehydration tolerance with relatively high leaf water content and photosystem II activity after a prolonged drought period (Liu et al., 2009). These experiments clearly display the efficiency of LEA proteins in conferring water and salt stress tolerance in plants.

Dehydrins are an extensively studied group II LEA protein family and are known to provide drought stress tolerance in plants by ion sequestration (Bray, 1993), maintenance and stabilization of protein and membrane structure (Baker et al., 1988; Koag et al., 2003), chaperons (Close, 1996), and transporting specific molecules to the nucleus (Goday et al., 1994). The transgenic tobacco plants overexpressing citrus (*Citrus unshiu* Marcov.) dehydrin showed less electrolyte leakage, early germination, and seedling growth than wild type (Hara et al., 2003). To study the stress tolerance effect of dehydrins, chimeric double constructs of *RAB18* and *COR47* or *LTI29* and *LTI30* were prepared under the CaMV 35S promoter and its overexpression in *Arabidopsis* resulted in accumulation of dehydrin up to levels similar to cold-acclimated wild-type plants and increased survival rate (Puhakainen et al., 2004). Wheat (*Triticum durum*) plants containing DHN-5 (dehydrin), induced in response to salt and abscisic acid (ABA) when ectopically expressed in *Arabidopsis thaliana*, showed improved growth under high salt concentrations and drought conditions through osmotic adjustments (Brini et al., 2007).

The transgenic maize plant overexpressing the maize *Rab28* LEA gene under the constitutive maize ubiquitin promoter resulted in increased desiccation tolerance. The transgenic seedlings showed better growth than wild type with increased water content, leaf and root area, lowering chlorophyll loss, and malondialdehyde production under dehydration stress conditions (Amara et al., 2013).

5.3.5 Polyols

Polyols or sugar alcohols are reduced forms of aldose or ketose monosaccharide (Brimacombe and Webber, 1972; Williamson et al., 2002). They are present in algae, certain halophytic plants, and insects exposed to freezing (Yancey et al., 1982). The most frequently found polyols in plants are mannitol, sorbitol, galactitol, and cyclitol myo-inositol. The primary pathway for synthesis of mannitol is the reduction of fructose 6-phosphate to mannitol 1-phosphate and subsequent dephosphorylation to mannitol. In *Aspergillus niger*, mannitol has proved to be involved in conidial stress protection, particularly oxidative and high temperature stresses (Solomon et al., 2007). The transgenic tobacco containing mannitol 1-phosphate dehydrogenase (mtlD) from *Escherichia coli* resulted in mannitol production and a salinity-tolerant phenotype (Tarczynski et al., 1993; Thomas et al., 1995). The cyclitols such as myo-inositol have shown better stress protection. The myo-inositol is synthesized by myo-inositol 1-phosphate synthase and is induced by salinity (Ishitani et al., 1996). In *Mesembryanthemum crystallinum*, myo-inositol is converted to the osmoprotectants, D-ononitol, and D-pinitol (Vernon and Bohnert, 1992; Adams et al., 1992; Nelson et al., 1998). The methylation of myo-inositol forms O-methyl inositol (D-ononitol) under abiotic stress using myo-inositol methyltransferase (IMT). The transgenic *Nicotiana tabacum* plants expressing myo-inositol O-methyltransferase (IMT1) accumulated the methylated inositol O-ononitol 35 pmol/g^{-1} fresh weight. These transgenic tobacco plants have shown less inhibition on photosynthetic CO_2 fixation during salt stress and drought, and the plants recovered faster than wild type (Sheveleva et al., 1997). The transgenic *Arabidopsis* plants overexpressing *Glycine max* IMT have displayed improved tolerance to dehydration stress treatment and high salinity stress treatment (Ahn et al., 2011). The transgenic *Arabidopsis thaliana* plants expressing halophyte *Mesembryanthemum crystallinum* myo-inositol-O-methyltransferase (Imt1) in response to low temperature stress have shown elevated cold tolerance. These transgenics were confirmed by accumulation of malondialdehyde and higher levels of proline and soluble sugar contents (Zhu et al., 2012). The transgenic *Arabidopsis* plant overexpressing *SaINO1* gene, which encodes for myo-inositol 1-phosphate synthase (MIPS) from the grass halophyte *Spartina alterniflora*, showed improved growth, development, and germination under 150 mM NaCl. This transgenic plant protects its photosystem II by increased retention of chlorophyll and carorenoids, indicating the effect of *SaINO1* overexpression in increasing salt stress tolerance (Joshi et al., 2013).

5.3.6 Ectoine

Ectoine (1,4,5,6-tetrahydro-2-methyl-4-pyrimidinecarboxylic acid) is a common compatible osmolyte in halophilic bacteria (Regev et al., 1990; Wohlfarth et al., 1990; Farwick et al., 1995; Bernard et al., 1993; Del Moral et al., 1994; Malin and Lapidot, 1996). It is synthesized from L-aspartate β-semialdehyde and requires three enzymes: L-2,4-diaminobutyric acid aminotransferase (gene: ectB), L-2,4-diaminobutyric acid acetyl transferase (gene: ectA), and L-ectoine synthase (gene: ectC). The transgenic *Nicotiana tabacum* overexpressing ectB, ectA, and ectC showed increased tolerance to hyperosmotic shock (900 mOsm) (Nakayama et al., 2000). The uptake and translocation of nitrate-N in roots is impaired by salinity and the impairment can be alleviated by accumulation of ectoine in the roots of ectoine-transformed tobacco plants (Moghaieb et al., 2006). The genetically engineered tomato plants expressing the three *Halomonas elongata* genes (ectA, ectB, and ectC) have shown increased photosynthetic rates through enhancement of cell membrane stability in oxidative conditions under salt stress (Moghaieb et al., 2011).

5.4 **Engineering of ion transport**

Plants have developed a number of mechanisms to survive under high salt concentration, which involve synthesis of compatible solutes, variations in membrane structure and photosynthetic activity, sequestration of ions into vacuoles, selective uptake, and removal of ions (Parida and Das, 2005). Compartmentalization of ions (Na^+) into vacuoles enables plants to carry out their normal metabolic function under saline conditions (Figure 5.2) (Adams et al., 1992; Reddy et al., 1992). By regulating the expression of H^+ pumps, the vacuolar type H^+-ATPase and the vacuolar pyrophosphatase (V-PPase) plants maintain high concentrations of K^+ in the cytosol and exclude Na^+ (Zhu et al., 1993; Dietz et al., 2001). By overexpressing these pumps, salt tolerance can be induced in plants thereby increasing their ability to grow under saline conditions.

The tobacco seedling overexpressing apple vacuolar H^+-ATPase subunit A (MdVHA-A) showed enhance drought tolerance with extended lateral root growth and osmotic adjustments (Dong et al., 2013).

The overexpression of Na^+/H^+ antiporter gene *AtNHX1*, which encodes a tonoplast antiporter *AtNHX1* in plant salt tolerance, resulted in salt tolerance to transgenic *Arabidopsis* plants

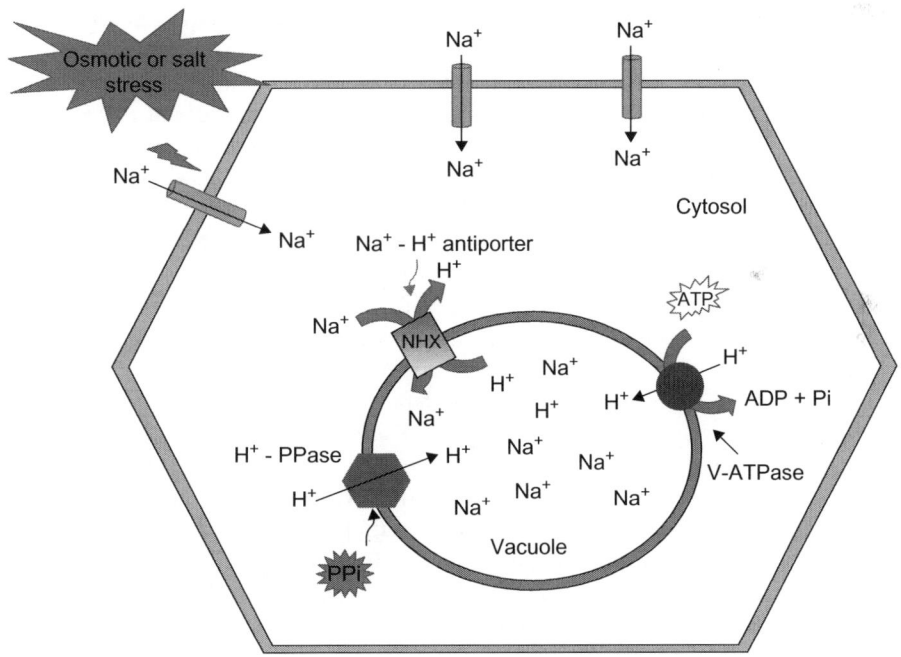

FIGURE 5.2

Compartmentalization of Na^+ ions in vacuole: Under highly saline conditions, Na^+ concentration in the plant is increased to toxic level. To overcome this toxicity, cells expel this Na^+ to the outside via plasma membrane Na^+/H^+ antiporter, but the transported Na^+ ions may create problem for neighboring cells. Therefore, the Na^+ ions are sequestered into vacuoles via Na^+/H^+ antiporter. The H^+ concentration is maintained higher inside the vacuole by H^+-ATPase and V-ATPase. (Gaxiola et al., 2002; Bayat et al., 2011).

(Apse et al., 1999). Arabidopsis vacuolar Na^+/H^+ antiporter gene was overexpressed in tomato plants and in *Brassica napus* with the assumption that these transgenic plants, with increased ability to sequester sodium in their vacuole, would show ability to grow in high salinity. At a salt concentration of 200 mM, transgenic plants showed growth with fruit and flower and seed production (Zhang and Blumwald, 2001; Zhang et al., 2001). These results strongly suggest that modification of a single Na^+ transporter on the tonoplast can significantly improve salinity tolerance.

Overexpression of the yeast *HAL1* gene in tomato showed increased fruit production and enhanced growth following salt treatment than wild-type plants. Increased water and K^+ content and decreased intracellular Na^+ in transgenic plants were found in the presence of salts (Gisbert et al., 2000; Rus et al., 2001). Overexpression of the *Schizosaccharomyces pombe* SOD2 (*Sodium2*) gene in *Arabidopsis* resulted in improved seed germination and salt tolerance. Na^+ and K^+ content analysis showed that transgenic lines accumulated less Na^+ and more K^+ in the symplast than wild-type plants. Also, the photosynthetic rate and fresh weight of transgenic plants were found to be higher after salt treatment than wild-type plants (Gao et al., 2003).

Rajgopal et al., (2007) studied that vacuolar Na^+/H^+ antiporter (*PgNHX1*) from *Pennisetum glaucum*, a glycophyte with a natural ability to withstand high salinity, drought, and heat stress, confers salt tolerance when overexpressed in *Brassica juncea*. Up-regulation of *PgNHX1* transcript was found after NaCl and ABA treatment, and transgenic *B. juncea* plants showed normal flowering and produced normal seeds when grown up to 70 mM NaCl. Similarly, this vacuolar Na^+/H^+ antiporter was also overexpressed in rice and resulted in a high level of salt tolerance. The transgenic rice plants showed increased root system, set flowering, and seeds in the presence of 150 mM NaCl (Verma et al., 2007). The overexpression of H^+-pyrophosphatase AVP1 in *Arabidopsis* showed increased salt and drought tolerance, mainly by increased accumulation of ions into vacuoles (Gaxiola et al., 2001). This Arabidopsis AVP1 (vacuolar H^+-PPase) gene was then expressed in tomato and cotton, and co-expressed in rice along with *Suaeda salsa SsNHX1* (vacuolar membrane Na^+/H^+ antiporter) where enhanced salt and drought tolerance was observed in transgenic plants with larger root systems (Park et al., 2005c; Zhao et al., 2006; Zhang et al., 2011).

Tomato, an important crop, is sensitive even to moderate salinity. To generate salt tolerant tomato plants, AVP1, a vacuolar H^+-pyrophosphatase gene from *Arabidopsis thaliana*, and *PgNHX1*, a vacuolar Na^+/H^+ antiporter gene from *Pennisetum glaucum*, were co-expressed and transformants were self-pollinated. *In vitro* and *in vivo* salt tolerances of progeny were studied. Enhanced salt tolerance was observed when *AVP1* and *PgNHX1* were co-expressed as they showed growth in the presence of 200 mM NaCl in comparison to a single gene transgenic plant. About 1.5 times higher Na^+ sequestration in leaf tissue was found in transgenic plants, thus reducing its toxic effect (Bhaskaran and Savithramma, 2011). Also, transgenic *Arabidopsis* plant overexpressing *HvNHX2*, a vacuolar Na^+/H^+ antiporter from barley, showed normal growth in the presence of 200 mM NaCl and longer roots in early seedling stage (Bayat et al., 2011).

5.5 Overexpression of genes for stress signaling

A cascade of signal perception and transduction genes is activated under abiotic stress by a variety of signaling pathways: inositol 1,4,5-trisphosphate [Ins (1,4,5) P3] pathway, protein phosphorylation involving mitogen-activated protein kinases (MAPKs) (Kyriakis and Avruch, 1996; Gustin et al., 1998; Viswanathan and Zhu, 2004). High salinity leads to increased cytosolic Ca^{2+}, which initiates the stress signal transduction pathways for stress tolerance. In the inositol 1,4,5-trisphosphate [Ins (1,4,5) P3] pathway, signals activate phosphatidylinositol-specific phospholipase C (PI-PLC), which in turn catalyzes the hydrolysis of phosphoinositide (4,5) bisphosphate [PtdIns $(4,5)P_2$] to form diacylglycerol (DAG) and Ins (1,4,5) P_3 (IP_3) (Meijer and Teun Munnik, 2003). IP_3 is converted to IP_6, which provokes Ca^{2+} release from intracellular stores (Alexandre et al., 1990; Lemtir-Chlieh et al., 2003). DAG is rapidly phosphorylated by the diacylglycerol kinase (DGK) to phosphatidic acid (PA) (Munnik et al., 1998) and this molecule is believed to function as a second messenger in plants (Testerink and Munnik, 2005). The enhanced PA levels are observed when plant cells received different kinds of stresses such as osmotic (Meijer et al., 2001; Munnik and Meijer, 2001), water deficit (Frank et al., 2000; Munnik et al., 2000), and cold stresses (Ruelland et al., 2002). Also, the transgenic maize plants overexpressing phosphatidylinositol-specific phospholipase C (PI-PLC) improved the drought tolerance under drought stress conditions. These transgenic plants in drought stress conditions had higher relative water content, better osmotic adjustment, increased photosynthesis rates, lower percentage of ion leakage, less lipid membrane peroxidation, and higher grain yield than the wild type (Wang et al., 2008). The yeast calcineurin (CaN) is a Ca^{2+}- and calmodulin-dependent protein phosphatase (PP2B), which modulates the K^+ and Na^+ uptake system in order to have higher affinity for K^+ and restrict uptake of Na^+. The calcineurin also maintains cytosolic Ca^{2+} homeostasis. The functional expression of yeast CaN in transgenic tobacco cells resulted in substantial salt tolerance of transgenic plants, which suggests that essential components of Ca^{2+} and calmodulin-dependent CaN salt stress signal pathway is present in plants and functions in conjunction with yeast CaN to facilitate salt adaptation (Pardo et al., 1998). In *Arabidopsis thaliana*, overexpression of cysteine-rich, receptor-like protein kinase CRK45 leads to enhanced tolerance to drought. Also, these plants show enhanced expression of stress-responsive genes in response to salt stress (Zhang et al., 2013).

In yeast, animals, and plants, MAPK pathways induce the production of osmolytes and antioxidants (Hasegawa et al., 2000). These MAPK pathways are activated by receptors/sensors such as protein tyrosine kinases, G-protein-coupled receptors, and two-component histidine kinases. H_2O_2 is reported to activate a specific *Arabidopsis* mitogen-activated protein kinase, ANP1, which initiates a phosphorylation cascade involving two stress MAPKs, AtMPK3 and AtMPK6, which in turn induce stress-responsive genes and enhanced tolerance to multiple stresses (Sharma et al., 1996; Schweizer et al., 1997; Alvarez et al., 1998; Chamnongpol et al., 1998). The transgenic tobacco plants, expressing a constitutively active tobacco ANP1 orthologue, NPK1, display enhanced tolerance to multiple environmental stress conditions (Kovtun et al., 1999).

5.6 Quenching of reactive oxygen species

Reactive oxygen species (ROS) include superoxide anion ($O_2^{\cdot-}$), hydrogen peroxide (H_2O_2), hydroxylic free radical ($OH^{\cdot}$), singlet oxygen ($^1O_2$), methyl radical ($CH_3^{\cdot}$), and lipid peroxidation free radicals ($LOO^{\cdot}$, $ROO^{\cdot}$). $O_2^{\cdot-}$ induces a series of electron transfer processes in plants, such as the spontaneous or enzyme-assisted superoxide dismutase (SOD) dismutation to H_2O_2. H_2O_2 reacts with Fe^{3+} (or Cu^{2+}) to generate Fe^{2+} (or Cu^+), which will then react with H_2O_2 to generate hydroxyl radical (OH·) (which is highly active and affects the cell membrane). Abiotic stresses result in generation of ROS, which are detoxified either directly by non-enzymatic antioxidants (reduced glutathione (GSH), ascorbate (ASH), tocopherols, carotenoids, etc.) or by antioxidative enzymes (superoxide dismutase, ascorbate peroxidase (APx), glutathione peroxidase (GPx), and glutathione-S-transferase (GST), etc). The transgenic tobacco plants overexpressing glyoxalase pathway enzymes resist an increase in the level of methyl glyoxal (MG) that increased to over 70% in wild-type plants under salinity stress. These plants showed increased activity of glutathione-related antioxidative enzymes (Yadav et al., 2005). These transgenic potato plants (*Solanum tuberosum*) overexpressing the strawberry *GalUR* gene under the control of the CaMV 35S promoter with increased ascorbic acid levels. These transgenic lines retained higher chlorophyll compared to control plants and, therefore, better tolerance to abiotic stresses (Upadhyaya et al., 2009).

The manganese stabilizing protein (MSP, 33 kDa) represents a key component of the oxygen evolving complex (OEC) of photosynthetic machinery of plants. Fischer et al. (2008) demonstrated lack of MSP isoforms in mutant potato lines showing delayed senescence, reduced rooting, and enhanced tuberization. The transgenic potato plants with reduced expression of MSP accumulated increased levels of proline and low-molecular-weight metabolites such as ascorbate and tocopherol, and have a significant increase in stress tolerance as compared to wild-type plants (Gururani et al., 2013). The ectopic expression of plant helicase, *M. sativa* helicase 1 (MH1) in *Arabidopsis*, resulted in improved seed germination and plant growth under drought, salt, and oxidative stresses. In these transgenic plants, the capacity for osmotic adjustment, superoxide dismutase (SOD), ascorbate eroxidase (APX) activities, and proline content were also elevated (Luo et al., 2013).

Selenium results in decreased levels of ROS subjected to various stresses such as in grain sorghum (*Sorghum bicolor*) exposed to high temperature ($O_2^{·-}$ and H_2O_2) (Na_2SeO_4) (Djanaguiraman et al., 2010), wheat (*Triticum aestivum*) seedlings under UV-B radiation stress ($O_2^{·-}$) (Na_2SeO_3) (Yao et al., 2010, 2011) and cold stress ($O_2^{·-}$) (Na_2SeO_3) (Chu et al., 2010), rapeseed seedlings under salt and drought stress (H_2O_2) (Na_2SeO_4) (Hasanuzzaman et al., 2011; Hasanuzzaman and Fujita, 2011), and *Trifolium repens* under water stress (H_2O_2) (Na_2SeO_4) (Wang, 2011).

5.7 Conclusion and future perspectives

Abiotic stress factors severely affect growth, development, and productivity of crops. Many plants in the course of their evolution develop physiological, metabolic, and biochemical mechanisms to withstand these stresses, which includes activation of specific stress signaling pathways followed by expression of stress signaling genes, accumulation of compatible salts, and stress proteins. With a better understanding of underlying mechanisms responsible for stress tolerance, genetic engineering of crop plants, overexpressing stress signaling pathway genes, and stress tolerant genes have been achieved. Genetic approaches have been tremendously successful in obtaining transgenic plants with increased abiotic stress tolerance, which uses the constitutive or stress-induced overexpression of stress-inducible genes and transcription factors, engineering of pathways, which leads to accumulation of compatible solutes, sequestering of salts, and quenching of ROS. Extensive studies of gene expression under different abiotic stress conditions will help to choose the suitable regulatory sequence, promoter transcription factor, and gene for generating transgenic stress-tolerant plants. Also, the co-expression of various stress-tolerant genes can lead to genetically superior crop plants with increased abiotic stress resistance.

References

Abe, H., Urao, T., Ito, T., Seki, M., Shinozaki, K., Yamaguchi-Shinozaki, K., 2003. *Arabidopsis* AtMYC2 (bHLH) and AtMYB2 (MYB) function as transcriptional activators in abscisic acid signaling. Plant Cell 15, 63−78.

Acquaah, G., 2007. Principles of Plant Breeding. Blackwell Publishing, USA, pp. 387−391.

Adams, P., Thomas, J.C., Vernon, D.M., Bohnert, H.J., Jensen, R.G., 1992. Distinct cellular and organismic responses to salt stress. Plant Cell Physiol. 33, 1215−1223.

Ahn, C., Park, U., Park, P.B., 2011. Increased salt and drought tolerance by D-ononitol production in transgenic *Arabidopsis thaliana*. Biochem. Biophys. Res. Commun. 415, 669−674.

Alexandre, J., Lassales, J., Kado, R., 1990. Opening of Ca^{2+} channels in isolated beet root vacuole membrane by inositol 1,4,5 trisphosphate. Nature 343, 567−570.

Allard, F., Houde, M., Krol, M., Ivanov, A., Huner, N.P., Sarhan, F., 1998. Betaine improves freezing tolerance in wheat. Plant Cell Physiol. 39, 1194−1202.

Alvarez, E., Pennell, R.I., Meijer, P., Ishikawa, A., Dixon, R.A., Lamb, C., 1998. Reactive oxygen intermediates mediate a systemic signal network in the establishment of plant immunity. Cell 92, 773−784.

Amara, I., Capellades, M., Ludevid, M.D., Pages, M., Goday, A., 2013. Enhanced water stress tolerance of transgenic maize plants over-expressing LEA Rab28 gene. J. Plant Physiol. 170, 864−873.

Apse, M.P., Aharon, G.S., Snedden, W.A., Blumwald, E., 1999. Salt tolerance conferred by overexpression of a vacuolar Na+/H+ antiport in *arabidopsis*. Science 285, 1256−1258.

Bagni, N., Tassoni, A., 2001. Biosynthesis, oxidation and conjugation of aliphatic polyamines in higher plants. Amino Acids 20, 301−317.

Baker, J., Steele, C., Dure III, L., 1988. Sequence and characterization of 6 Lea proteins and their genes from cotton. Plant Mol. Biol. 291, 277−291.

Bayat, F., Shiran, B., Belyaev, D.V., 2011. Overexpression of *HvNHX2* , a vacuolar Na + /H + antiporter gene from barley, improves salt tolerance in *Arabidopsis thaliana*. Aust. J. Crop. Sci. 5, 428−432.

Benavides, P., Gallego, S.M., Comba, E., Tomaro, L., 2000. Relationship between polyamines and paraquat toxicity in sunflower leaf discs. Plant Growth Regul. 31, 215−224.

Bernard, T., Jebbar, M., Rassouli, Y., Himdi-Kabbab, S., Hamelin, J., Blanco, C., 1993. Ectoine accumulation and osmotic regulation in *Brevibacterium linens*. J. Gen. Microbiol. 139, 129−136.

Bhaskaran, S., Savithramma, D.L., 2011. Co-expression of *Pennisetum glaucum* vacuolar Na^+/H^+ antiporter and *Arabidopsis* H^+-pyrophosphatase enhances salt tolerance in transgenic tomato. J. Exp. Bot. 62, 5561−5570.

Bhatnagar-Mathur, P., Vadez, V., Sharma, K.K., 2008. Transgenic approaches for abiotic stress tolerance in plants: retrospect and prospects. Plant Cell Rep. 27, 411−424.

Bray, E.A., 1993. Molecular responses to water deficit. Plant Physiol. 103, 1035−1040.

Bray, E.A., 1994. Alterations in Gene Expression in Response to Water Deficit. In: Basra, A.S. (Ed.), Stress Induced Gene Expression in Plants. Hardwood Academic Publishers, Chur, Switzerland, pp. 1−23.

Brimacombe, J.S., Webber, J.M., 1972. Alditols and derivatives, in the carbohydrates. In: Pigman, W., Horton, D. (Eds.), Chemistry and Biochemistry. Academic Press, New York and London, p. 495.

Brini, F., Hanin, M., Lumbreras, V., Amara, I., Khoudi, H., Hassairi, A., et al., 2007. Overexpression of wheat dehydrin DHN-5 enhances tolerance to salt and osmotic stress in *Arabidopsis thaliana*. Plant Cell Rep. 26, 2017−2026.

Busk, P.K., Pagès, M., 1998. Regulation of abscisic acid-induced transcription. Plant Mol. Biol. 37, 425−435.

Capell, T., Bassie, L., Christou, P., 2004. Modulation of the polyamine biosynthetic pathway in transgenic rice confers tolerance to drought stress. Proc. Nat. Acad. Sci. USA 101, 9909−9914.

Chamnongpo, S., Willekens, H., Moeder, W., Langebartels, C., Sandermann, H., Van Montagu, M., et al., 1998. Defense activation and enhanced pathogen tolerance induced by H_2O_2 in transgenic tobacco. Proc. Nat. Acad. Sci. USA 95, 5818−5823.

Chandra Babu, R., Zhang, J., Blum, A., Ho, T.H.D., Wu, R., Nguyen, H.T., 2004. HVA1, a LEA gene from barley confers dehydration tolerance in transgenic rice (Oryza sativa L.) via cell membrane protection. Plant Sci. 166, 855—862.

Chen, T.H.H., Murata, N., 2002. Enhancement of tolerance of abiotic stress by metabolic engineering of betaines and other compatible solutes. Curr. Opin. Plant Biol. 5, 250—257.

Chen, T.H.H., Murata, N., 2008. Glycinebetaine: an effective protectant against abiotic stress in plants. Trends Plant Sci. 13, 499—505.

Chen, Y., Zhang, X., Wu, W., Chen, Z., Gu, H., Qu, L.-J., 2006. Overexpression of the wounding-responsive gene AtMYB15 activates the shikimate pathway in Arabidopsis. J. Integr. Plant Biol. 48, 1084—1095.

Cheng, L., Zou, Y., Ding, S., Zhang, J., Yu, X., Cao, J., et al., 2009. Polyamine accumulation in transgenic tomato enhances the tolerance to high temperature stress. J. Integr. Plant Biol. 51, 489—499.

Cheng, Z., Targolli, J., Huang, X., Wu, R., 2002. Wheat LEA genes, PMA80 and PMA1959, enhance dehydration tolerance of transgenic rice (Oryza sativa L.). Mol. Breed. 10, 71—82.

Choi, H., Hong, J.-H., Ha, J.-O., Kang, J.-Y., Kim, S.Y., 2000. ABFs, a family of ABA-responsive element binding factors. J. Biol. Chem. 275, 1723—1730.

Chu, J.Z., Yao, X.Q., Zhang, Z.N., 2010. Responses of wheat seedlings to exogenous selenium supply under cold stress. Biol. Trace Elem. Res. 136, 355—363.

Close, T.J., 1996. Dehydrins: emergence of a biochemical role of a family of plant dehydration proteins. Physiol. Plant 97, 795—803.

Dalal, M., Tayal, D., Chinnusamy, V., Bansal, K.C., 2009. Abiotic stress and ABA-inducible Group 4 LEA from Brassica napus plays a key role in salt and drought tolerance. J. Biotech. 139, 137—145.

Del Moral, A., Severin, J., Ramos-Cormenzana, A., Trüper, H.G., Galinski, E.A., 1994. Compatible solutes in new moderately halophilic isolates. FEMS Microbiol. Lett. 122, 165—172.

Dietz, K.J., Tavakoli, N., Kluge, C., Mimura, T., Sharma, S.S., Harris, G.C., et al., 2001. Significance of the V-type ATPase for the adaptation to stressful growth conditions and its regulation on the molecular and biochemical level. J. Exp. Bot. 52, 1969—1980.

Djanaguiraman, M., Prasad, P.V.V., Seppanen, M., 2010. Selenium protects sorghum leaves from oxidative damage under high temperature stress by enhancing antioxidant defense system. Plant Physiol. Biochem. 48, 999—1007.

Dong, Q.L., Wang, C.R., Liu, D.D., Hu, D.G., Fang, M.J., You, C.X., et al., 2013. MdVHA-A encodes an apple subunit A of vacuolar H + -ATPase and enhances drought tolerance in transgenic tobacco seedlings. J. Plant Physiol. 70, 601—609.

Dubos, C., Stracke, R., Grotewold, E., Weisshaar, B., Martin, C., Lepiniec, L., 2010. MYB transcription factors in Arabidopsis. Trends Plant Sci. 15, 573—581.

Fariduddin, Q., Varshney, P., Yusuf, M., Ahmad, A., 2012. Polyamines: potent modulators of plant responses to stress. J. Plant Interact. 8, 37—41.

Farwick, M., Siewe, R.M., Krämer, R., 1995. Glycine betaine uptake after hyperosmotic shift in Corynebacterium glutamicum. J. Bacteriol. 177, 4690—4695.

Fischer, L., Lipavska, H., Hausman, J.F., Opatrny, Z., 2008. Morphological and molecular characterization of a spontaneously tuberizing potato mutant: an insight into the regulatory mechanisms of tuber induction. BMC Plant Biol. 8, 117.

Franceschetti, M., Fornale, S., Tassoni, A., Zuccherelli, K., Mayer, M.J., Bagni, N., 2004. Effects of spermidine synthase over-expression on polyamine biosynthetic pathway in tobacco plants. J. Plant Physiol. 161, 989—1001.

Frank, W., Munnik, T., Kerkmann, K., Salamini, F., Bartels, D., 2000. Water deficit triggers phospholipase D activity in the resurrection plant Craterostigma plantagineum. Plant Cell 12, 111—124.

Frugier, F., Poirier, S., Satiat-Jeunemaître, B., Kondorosi, A., 2000. A Kruppel-like zinc finger protein is involved in nitrogen-fixing root nodule organogenesis. Gene. Develop. 14, 475—482.

Fu, D., Huang, B., Xiao, Y., Muthukrishnan, S., Liang, G.H., 2007. Overexpression of barley hva1 gene in creeping bentgrass for improving drought tolerance. Plant Cell Rep. 26, 467−477.

Fujita, Y., Fujita, M., Satoh, R., Maruyama, K., Parvez, M.M., Seki, M., et al., 2005. AREB1 is a transcription activator of novel ABRE-dependent ABA signaling that enhances drought stress tolerance in *Arabidopsis*. Plant Cell 17, 3470−3488.

Gao, F., Xiong, A., Peng, R., Jin, X., Xu, J., Zhu, B., et al., 2010. *OsNAC52*, a rice NAC transcription factor, potentially responds to ABA and confers drought tolerance in transgenic plants. Plant Cell Tiss. Organ Cult. 100, 255−262.

Gao, X., Ren, Z., Zhao, Y., Zhang, H., 2003. Overexpression of SOD2 increases salt tolerance of *Arabidopsis*. Plant Physiol. 133, 1873−1881.

Gaxiola, R.A., Li, J., Undurraga, S., Dang, L.M., Allen, G.J., Alper, S.L., et al., 2001. Drought- and salt-tolerant plants result from overexpression of the AVP1 H^+-pump. Proc. Nat. Acad. Sci. USA 98, 11444−11449.

Gilmour, S.J., Sebolt, A.M., Salazar, M.P., Everard, J.D., Thomashow, M.F., 2000. Overexpression of the *Arabidopsis* CBF3 transcriptional activator mimics multiple biochemical changes associated with cold acclimation. Plant Physiol. 124, 1854−1865.

Giraudat, J., Farcy, F., Bertauche, N., Gosti, F., Leung, J., Morris, P.-C., et al., 1994. Current advances in abscisic acid action and signaling. Plant Mol. Biol. 26, 1557.

Gisbert, C., Rus, A.M., Bolarín, M.C., López-Coronado, J.M., Arrillaga, I., Montesinos, C., et al., 2000. The yeast HAL1 gene improves salt tolerance of transgenic tomato. Plant Physiol. 123, 393−402.

Giuliano, G., Pichersky, E., Malik, V.S., Timko, M.P., Scolnik, P.A., Cashmore, A.R., 1988. An evolutionarily conserved protein binding sequence upstream of a plant light-regulated gene. Proc. Nat. Acad. Sci. USA 85, 7089−7093.

Goday, A., Jensen, A.B., Culiáñez-Macià, F.A., Mar Albà, M., Figueras, M., Serratosa, J., et al., 1994. The maize abscisic acid-responsive protein Rab17 is located in the nucleus and interacts with nuclear localization signals. Plant Cell 6, 351−360.

Gong, W., Shen, Y.P., Ma, L.G., Yi, P., Du, Y.L., Wang, D.H., et al., 2004. Genome-wide ORFeome cloning and analysis of *Arabidopsis* transcription factor genes. Plant Physiol. 135, 773−782.

Gururani, M.A., Upadhyaya, C.P., Strasser, R.J., Yu, J.W., Park, S.W., 2013. Evaluation of abiotic stress tolerance in transgenic potato plants with reduced expression of PSII manganese stabilizing protein. Plant Sci. 198, 7−16.

Gustin, M.C., Albertyn, J., Alexander, M., Davenport, K., Gustin, M.C., Albertyn, J., et al., 1998. MAP kinase pathways in the yeast Saccharomyces cerevisiae. Microbiol. Mol. Biol. Rev. 62, 1264−1300.

Hara, M., Terashima, S., Fukaya, T., Kuboi, T., 2003. Enhancement of cold tolerance and inhibition of lipid peroxidation by citrus dehydrin in transgenic tobacco. Planta 217, 290−298.

Hare, P.D., Cress, W.A., Staden, J.V., 1999. Proline synthesis and degradation: a model system for elucidating stress-related signal transduction. J. Exp. Bot. 50, 413−434.

Hasanuzzaman, M., Fujita, M., 2011. Selenium pretreatment upregulates the antioxidant defense and methylglyoxal detoxification system and confers enhanced tolerance to drought stress in rapeseed seedlings. Biol. Trace Elem. Res. 143, 1758−1776.

Hasanuzzaman, M., Hossain, M.A., Fujita, M., 2011. Selenium-induced up-regulation of the antioxidant defense and methylglyoxal detoxification system reduces salinity-induced damage in rapeseed seedlings. Biol. Trace Elem. Res. 143, 1704−1721.

Hasegawa, P.M., Bressan, R.A., Zhu, J., Bohnert, H., 2000. Plant cellular and molecular responses to high salinity. Annu. Rev. Plant Physiol. Plant Mol. Biol. 51, 463−499.

Hayashi, H., Alia, Mustardy, L., Deshnium, P., Ida, M., Murata, N., 1997. Transformation of *Arabidopsis thaliana* with the codA gene for choline oxidase; accumulation of glycinebetaine and enhanced tolerance to salt and cold stress. Plant J. 12, 133−142.

He, Y., Li, W., Lv, J., Jia, Y., Wang, M., Xia, G., 2012. Ectopic expression of a wheat MYB transcription factor gene, TaMYB73, improves salinity stress tolerance in *Arabidopsis thaliana*. J. Exp. Bot. 63, 1511−1522.

Holmström, K.-O., Welin, B., Mandal, A., Kristiansdottir, I., Teeri, T.H., Strom, A.R., et al., 1994. Production of the *Escherichia coli* betaine-aldehyde dehydrogenase, an enzyme required for the synthesis of the osmoprotectant glycine betaine in transgenic plants. Plant J. 6, 749−758.

Holmström, K.-O., Somersalo, S., Mandal, A., Palva, T.E., Welin, B., 2000. Improved tolerance to salinity and low temperature in transgenic tobacco producing glycine betaine. J. Exp. Bot. 51, 177−185.

Hong, Z., Lakkineni, K., Zhang, Z., Verma, D.P., 2000. Removal of feedback inhibition of delta(1)-pyrroline-5-carboxylate synthetase results in increased proline accumulation and protection of plants from osmotic stress. Plant Physiol. 122, 1129−1136.

Hu, H., Dai, M., Yao, J., Xiao, B., Li, X., Zhang, Q., et al., 2006. Overexpressing a NAM, ATAF, and CUC (NAC) transcription factor enhances drought resistance and salt tolerance in rice. Proc. Nat. Acad. Sci. USA 103, 12987−12992.

Huang, J., Hirji, R., Adam, L., Rozwadowski, K.L., Hammerlindl, J.K., Keller, W.A., et al., 2000. Genetic engineering of glycinebetaine production toward enhancing stress tolerance in plants: metabolic limitations. Plant Physiol. 122, 747−756.

Hussain, S.S., Kayani, M.A., Amjad, M., 2011. Transcription factors as tools to engineer enhanced drought stress tolerance in plants. Biotechnol. Prog. 27, 297−306.

Ikuta, S., Matuura, K., Imamura, S., Misaki, H., Horiuti, Y., 1977. Oxidative pathway of choline to betaine in the soluble fraction prepared from *Arthrobacter globiformis*. J. Biochem. 82, 157−163.

Ingram, J., Bartels, D., 1996. The molecular basis of dehydration tolerance in plants. Annu. Rev. Plant Physiol. Plant Mol. Biol. 47, 377−403.

Ishitani, M., Majumder, A., Bornhouser, A., Michalowski, C., Jensen, R., Bohnert, H., 1996. Coordinate transcriptional induction of myo-inositol metabolism during environmental stress. Plant J. 9, 537−548.

Ito, Y., Katsura, K., Maruyama, K., Taji, T., Kobayashi, M., Seki, M., et al., 2006. Functional analysis of rice DREB1/CBF-type transcription factors involved in cold-responsive gene expression in transgenic rice. Plant Cell Physiol. 47, 141−153.

Iwasaki, T., Kiyosue, T., Yamaguchi-Shinozaki, K., Shinozaki, K., 1997. The dehydration-inducible rd17 (cor47) gene and its promoter region in *Arabidopsis thaliana*. Plant Physiol. 115, 1287.

Jaglo-Ottosen, K.R., 1998. *Arabidopsis* CBF1 overexpression induces COR genes and enhances freezing tolerance. Science 280, 104−106.

Joshi, R., Ramanarao, M.V., Baisakh, N., 2013. *Arabidopsis* plants constitutively overexpressing a myo-inositol 1-phosphate synthase gene (SaINO1) from the halophyte smooth cordgrass exhibits enhanced level of tolerance to salt stress. Plant Physiol. Biochem. 65, 61−66.

Jung, C., Seo, J.S., Han, S.W., Koo, Y.J., Kim, C.H., Song, S.I., et al., 2008. Overexpression of AtMYB44 enhances stomatal closure to confer abiotic stress tolerance in transgenic *Arabidopsis*. Plant Physiol. 146, 623−635.

Kang, J.-Y., Choi, H., Im, M., Kim, S.Y., 2002. *Arabidopsis* basic leucine zipper proteins that mediate stress-responsive abscisic acid signaling. Plant Cell 14, 343−357.

Kasuga, M., Miura, S., Yamaguchi-Shinozaki, K., Shinozaki, K., 1999. Improving plant drought, salt and freezing tolerance by gene transfer of a single stress-inducible transcription factor. Nat. Biotechnol. 17, 176−186.

Kasuga, M., Miura, S., Shinozaki, K., Yamaguchi-Shinozaki, K., 2004. A combination of the *Arabidopsis* DREB1A gene and stress-inducible rd29A promoter improved drought- and low-temperature stress tolerance in tobacco by gene transfer. Plant Cell Physiol. 45, 346−350.

Kasukabe, Y., He, L., Nada, K., Misawa, S., Ihara, I., Tachibana, S., 2004. Overexpression of spermidine synthase enhances tolerance to multiple environmental stresses and up-regulates the expression of various stress-regulated genes in transgenic *Arabidopsis thaliana*. Plant Cell Physiol. 45, 712−722.

Kasukabe, Y., He, L., Watakabe, Y., Otani, M., Shimada, T., Tachibana, S., 2006. Improvement of environmental stress tolerance of sweet potato by introduction of genes for spermidine synthase. Plant Biotechnol. 23, 75−83.

Katiyar-Agarwal, S., Agarwal, M., Grover, A., 1999. Emerging trends in agricultural biotechnology research: use of abiotic stress induced promoter to drive expression of a stress resistence gene in the transgenic system leads to high level stress tolerance associated with minimal negative effects on growth. Curr. Sci. 77, 1577−1579.

Kikuchi, K., Ueguchi-Tanaka, M., Yoshida, K.T., Nagato, Y., Matsusoka, M., Hirano, H.Y., 2000. Molecular analysis of the NAC gene family in rice. Mol. Gen. Genet. 262, 1047−1051.

Kim, J.C., Lee, S.H., Cheong, Y.H., Yoo, C.M., Lee, S.I., Chun, H.J., et al., 2001. A novel cold-inducible zinc finger protein from soybean, SCOF-1, enhances cold tolerance in transgenic plants. Plant J. 25, 247−259.

Kishitani, S., Watanabe, K., Yasuda, S., Arakawa, K., Takabe, T., 1994. Accumulation of glycinebetaine during cold acclimation and freezing tolerance in leaves of winter and spring barley plants. Plant Cell Environ. 17, 89−95.

Kishor, P., Hong, Z., Miao, G.H., Hu, C., Verma, D., 1995. Overexpression of [delta]-pyrroline-5-carboxylate synthetase increases proline production and confers osmotolerance in transgenic plants. Plant Physiol. 108, 1387−1394.

Kizis, D., Lumbreras, V., Pagès, M., 2001. Role of AP2/EREBP transcription factors in gene regulation during abiotic stress. FEBS Lett. 498, 187−189.

Koag, M., Fenton, R.D., Wilkens, S., Close, T.J., 2003. The binding of maize DHN1 to lipid vesicles. Gain of structure and lipid specificity. Plant Physiol. 131, 309−316.

Konstantinova, T., Parvanova, D., Atanassov, A., Djilianov, D., 2002. Freezing tolerant tobacco, transformed to accumulate osmoprotectants. Plant Sci. 163, 157−164.

Kovtun, Y., Chiu, W., Tena, G., Sheen, J., 1999. Functional analysis of oxidative stress-activated mitogen-activated protein kinase cascade in plants. Proc. Nat. Acad. Sci. USA 97, 2940−2945.

Kumria, R., Rajam, M.V., 2002. Alteration in polyamine titres during Agrobacterium-mediated transformation of indica rice with ornithine decarboxylase gene affects plant regeneration potential. Plant Sci. 162, 769−777.

Kusano, T., Berberich, T., Tateda, C., Takahashi, Y., 2008. Polyamines: essential factors for growth and survival. Planta 228, 367−381.

Kyriakis, J.M., Avruch, J., 1996. Sounding the alarm: protein kinase cascades activated by stress and inflammation. J. Biol. Chem. 271, 24313−24316.

Lal, S., Gulyani, V., Khurana, P., 2008. Overexpression of HVA1 gene from barley generates tolerance to salinity and water stress in transgenic mulberry (*Morus indica*). Transgenic Res. 17, 651−663.

Lemtir-Chlieh, F., MacRobbie, E.A.C., Webb, A.A.R., Manison, N.F., Brownlee, C., Skepper, J.N., et al., 2003. Inositol hexakisphosphate mobilizes an endomemebrane store of calcium in guard cells. Proc. Natl. Acad. Sci. USA 100, 10091−10095.

Levitt, J., 1980. Response of Plants to Environmental Stress. Chilling, Freezing, and High Temperature Stresses, vol. I. Academic Press Inc.

Li, X., Gao, Q., Liang, Y., Ma, T., Cheng, L., Qi, D., et al., 2013. A novel salt-induced gene from sheepgrass, *LcSAIN2*, enhances salt tolerance in transgenic *Arabidopsis*. Plant Physiol. Biochem. 64, 52−59.

Lida, A., Kazuoka, T., Torikai, S., Kikuchi, H., Oeda, K., 2000. A zinc finger protein *RHL41* mediates the light acclimatization response in *Arabidopsis*. The Plant J. 24, 191−203.

Lippuner, V., Cyert, M.S., Gasser, C.S., 1996. Two classes of plant cDNA clones differentially complement yeast calcineurin mutants and increase salt tolerance of wild-type yeast. J. Biol. Chem. 271, 12859−12866.

Liu, Q., Kasuga, M., Sakuma, Y., Abe, H., Miura, S., Yamaguchi-Shinozaki, K., et al., 1998. Two transcription factors, DREB1 and DREB2, with an EREBP/AP2 DNA binding domain separate two cellular signal transduction pathways in drought- and low-temperature-responsive gene expression, respectively, in *Arabidopsis*. Plant Cell 10, 1391−1406.

Liu, X., Wang, Z., Wang, L., Wu, R., Phillips, J., Deng, X., 2009. LEA 4 group genes from the resurrection plant *Boea hygrometrica* confer dehydration tolerance in transgenic tobacco. Plant Sci. 176, 90—98.

Luo, X., Wu, J., Li, Y., Nan, Z., Guo, X., Wang, Y., et al., 2013. Synergistic effects of GhSOD1 and GhCAT1 overexpression in cotton chloroplasts on enhancing tolerance to methyl viologen and salt stresses. PloS One 8, e54002.

Malin, G., Lapidot, A., 1996. Induction of synthesis of tetrahydropyrimidine derivatives in Streptomyces strains and their effect on Escherichia coli in response to osmotic and heat stress. J. Bacteriol. 178, 385—395.

Mao, X., Zhang, H., Qian, X., Li, A., Zhao, G., Jing, R., 2012. TaNAC2, a NAC-type wheat transcription factor conferring enhanced multiple abiotic stress tolerances in *Arabidopsis*. J. Exp. Bot. 63, 2933—2946.

Masgrau, C., Altabella, T., Farr, R., Flores, D., Thompson, A.J., Besford, R.T., et al., 1997. Inducible overexpression of oat arginine decarboxylase in transgenic tobacco plants. Plant J. 11, 465—473.

McNeil, S.D., Nuccio, M.L., Ziemak, M.J., Hanson, A.D., 2001. Enhanced synthesis of choline and glycine betaine in transgenic tobacco plants that overexpress phosphoethanolamine N-methyltransferase. Proc. Nat. Acad. Sci. USA 98, 10001—10005.

Meijer, H.J.G., Munnik, T., 2003. Phospholipid-based signaling in plants. Annu. Rev. Plant Biol. 54, 265—306.

Meijer, H.J.G., Arisz, S.A., Van Himbergen, J.A.J., Musgrave, A., Munnik, T., 2001. Hyperosmotic stress rapidly generates lyso-phosphatidic acid in *Chlamydomonas*. Plant J. 25, 541—548.

Moghaieb, R.E.A., Tanaka, N., Saneoka, H., Murooka, Y., Ono, H., Morikawa, H., 2006. Characterization of salt tolerance in ectoine-transformed tobacco plants (Nicotiana tabaccum): photosynthesis, osmotic adjustment, and nitrogen partitioning. Plant Cell Environ. 29, 173—182.

Moghaieb, R.E.A., Nakamura, H., Saneoka, Fujita, K., 2011. Evaluation of salt tolerance in ectoine-transgenic tomato plants (Lycopersicon esculentum) in terms of photosynthesis, osmotic adjustment, and carbon partitioning. GM Crops 2, 58—65.

Mohanty, A., Kathuria, H., Ferjani, A., Sakamoto, A., Mohanty, P., Murata, N., et al., 2002. Transgenics of an elite indica rice variety Pusa Basmati 1 harbouring the codA gene are highly tolerant to salt stress. Theor. Appl. Genet. 106, 51—57.

Mukhopadhyay, A., Vij, S., Tyagi, A.K., 2004. Overexpression of a zinc-finger protein gene from rice confers tolerance to cold, dehydration, and salt stress in transgenic tobacco. Proc. Nat. Acad. Sci. USA 101, 6309—6314.

Munnik, T., Meijer, H.J., 2001. Osmotic stress activates distinct lipid and MAPK signalling pathways in plants. FEBS Lett. 498, 172—178.

Munnik, T., Irvine, R.F., Musgrave, A., 1998. Phospholipid signalling in plants. Biochim. Biophys. Acta 1389, 222—272.

Munnik, T., Meijer, H.J., Ter Riet, B., Hirt, H., Frank, W., Bartels, D., et al., 2000. Hyperosmotic stress stimulates phospholipase D activity and elevates the levels of phosphatidic acid and diacylglycerol pyrophosphate. Plant J. 22, 147—154.

Nakashima, K., Takasaki, H., Mizoi, J., Shinozaki, K., Yamaguchi-Shinozaki, K., 2012. NAC transcription factors in plant abiotic stress responses. Biochim. Biophys. Acta 1819, 97—103.

Nakayama, H., Yoshida, K., Ono, H., Murooka, Y., Shinmyo, A., 2000. Ectoine, the compatible solute of Halomonas elongata, confers hyperosmotic tolerance in cultured tobacco cells. Plant Physiol. 122, 1239—1247.

Nanjo, T., Kobayashi, M., Yoshiba, Y., Kakubari, Y., Yamaguchi-Shinozaki, K., Shinozaki, K., 1999. Antisense suppression of proline degradation improves tolerance to freezing and salinity in *Arabidopsis thaliana*. FEBS Lett. 461, 205—210.

Nayyar, H., 2005. Putrescine increases floral retention, pod set and seed yield in cold stressed chickpea. J. Agron. Crop Sci. 191, 340—345.

Nelson, D.E., Rammesmayer, G., Bohnert, H.J., 1998. Regulation of cell-specific inositol metabolism and transport in plant salinity tolerance. Plant Cell 10, 753—764.

Nijhawan, A., Jain, M., Tyagi, A.K., Khurana, J.P., 2008. Genomic survey and gene expression analysis of the basic leucine zipper transcription factor family in rice. Plant Physiol. 146, 333—350.

Nölke, G., Schneider, B., Agdour, S., Drossard, J., Fischer, R., Schillberg, S., 2008. Modulation of polyamine biosynthesis in transformed tobacco plants by targeting ornithine decarboxylase to an atypical subcellular compartment. Open Biotechnol. J. 2, 183—189.

Oh, S., Song, S.I., Kim, Y.S., Jang, H., Kim, S.Y., Kim, M., et al., 2005. *Arabidopsis* CBF3/DREB1A and ABF3 in transgenic rice increased tolerance to abiotic stress without stunting growth. Plant Physiol. 138, 341—351.

Pabo, C.O., Peisach, E., Grant, R.A., 2001. Design and selection of novel Cys2 His2 zinc finger domains. Annu. Rev. Biochem. 70, 313—340.

Papageorgiou, G.C., Murata, N., 1995. The unusually strong stabilizing effects of glycine betaine on the structure and function of the oxygenevolving photosystem II complex. Photosynth. Res. 44, 243—252.

Pardo, J., Reddy, M., Yang, S., Maggio, A., Huh, G.H., Matsumoto, T., et al., 1998. Stress signaling through Ca^{2+}-calmodulin-dependent protein phosphatase calcineurin mediates salt adaptation in plants. Proc. Nat. Acad. Sci. USA 95, 9681—9686.

Parida, A.H., Das, A.B., 2005. Salt tolerance and salinity effects on plants: a review. Ecotox. Environ. Safe 60, 324—349.

Park, B.-J., Liu, Z., Kanno, A., Kameya, T., 2005a. Increased tolerance to salt- and water-deficit stress in transgenic lettuce (*Lactuca sativa* L.) by constitutive expression of LEA. J. Plant Growth Regul. 45, 165—171.

Park, B.J., Liu, Z., Kanno, A., Kameya, T., 2005b. Genetic improvement of Chinese cabbage for salt and drought tolerance by constitutive expression of a B. napus LEA gene. Plant Sci. 169, 553—558.

Park, S., Li, J., Pittman, J.K., Berkowitz, G.A., Yang, H., Undurraga, S., et al., 2005c. Up-regulation of an H+ -pyrophosphatase (H+ -PPase) as a strategy to engineer drought-resistant crop plants. Proc. Nat. Acad. Sci. USA 102, 18830—18835.

Paz-ares, J., Ghosal, D., Wienand, U., Petersont, A., Saedler, H., 1987. The regulatory c1 locus of *Zea mays* encodes a protein with homology to myb proto-oncogene products and with structural similarities to transcriptional activators. EMBO J. 6, 3553—3558.

Puhakainen, T., Hess, M.W., Mäkelä, P., Svensson, J., Heino, P., Palva, E.T., 2004. Overexpression of multiple dehydrin genes enhances tolerance to freezing stress in Arabidopsis. Plant Mol. Biol. 54, 743—753.

Qin, F., Sakuma, Y., Li, J., Liu, Q., Li, Y.Q., Shinozaki, K., et al., 2004. Cloning and functional analysis of a novel DREB1/CBF transcription factor involved in cold-responsive gene expression in *Zea mays* L. Plant Cell Physiol. 45, 1042—1052.

Qin, Y., Wang, M., Tian, Y., He, W., Han, L., Xia, G., 2012. Over-expression of TaMYB33 encoding a novel wheat MYB transcription factor increases salt and drought tolerance in *Arabidopsis*. Mol. Biol. Rep. 39, 7183—7192.

Rajagopal, D., Agarwal, P., Tyagi, W., Singla-Pareek, S.L., Reddy, M.K., Sopory, S.K., 2007. *Pennisetum glaucum* Na+/H+ antiporter confers high level of salinity tolerance in transgenic *Brassica juncea*. Mol. Breed. 19, 137—151.

Rathinasabapathi, B., Burnet, M., Russell, B.L., Gage, D.A., Liao, P.C., Nye, G.J., et al., 1997. Choline monooxygenase, an unusual iron-sulfur enzyme catalyzing the first step of glycine betaine synthesis in plants: prosthetic group characterization and cDNA cloning. Proc. Nat. Acad. Sci. USA 94, 3454—3458.

Reddy, M., Sanish, S., Iyengar, E., 1992. Photosynthetic studies and compartmentation of ions in different tissues of *Salicornia brachiata* under saline conditions. Photosynthetica (Praha) 26, 173—179.

Regev, R., Peri, I., Gilboa, H., Avi-Dor, Y., 1990. ^{13}C NMR study of the interrelation between synthesis and uptake of compatible solutes in two moderately halophilic eubacteria media containing. Arch. Biochem. Biophys. 278, 106—112.

Reguera, M., Peleg, Z., Blumwald, E., 2012. Targeting metabolic pathways for genetic engineering abiotic stress-tolerance in crops. Biochim. Biophys. Acta 1819, 186—194.

Rhodes, D., Hanson, A.D., 1993. Quaternary ammonium and tertiary sulfonium compounds in higher plants. Annu. Rev. Plant Physiol. Plant Mol. Biol. 44, 357–384.

Richards, R.A., 1996. Defining selection criteria to improve yield under drought. J. Plant Growth Regul. 20, 157–166.

Rodríguez, M., Canales, E., Borrás-Hidalgo, O., 2005. Molecular aspects of abiotic stress in plants. Biotecnologia Aplicada 22, 1–10.

Roy, M., Wu, R., 2001. Arginine decarboxylase transgene expression and analysis of environmental stress tolerance in transgenic rice. Plant Sci. 160, 869–875.

RoyChoudhury, A., Roy, C., Sengupta, D.N., 2007. Transgenic tobacco plants overexpressing the heterologous lea gene Rab16A from rice during high salt and water deficit display enhanced tolerance to salinity stress. Plant Cell Rep. 26, 1839–1859.

Ruelland, E., Cantrel, C., Gawer, M., Kader, J., Zachowski, A., 2002. Activation of phospholipases C and D is an early response to a cold exposure in *Arabidopsis* suspension cells. Plant Physiol. 130, 999–1007.

Rus, A.M., Estañ, M.T., Gisbert, C., Serrano, R., Caro, M., Moreno, V., et al., 2001. Expressing the yeast HAL1 gene in tomato increases fruit yield and enhances K^+/Na^+ selectivity under salt stress. Plant Cell Environ. 24, 875–880.

Saad, A.D.I., Li, X., Li, H.P., Huang, T., Gao, C.S., Guo, M.W., et al., 2013. A rice stress-responsive *NAC* gene enhances tolerance of transgenic wheat to drought and salt stresses. Plant Sci. 203–204, 33–40.

Sakamoto, A., Alia, Murata, N., 1998. Metabolic engineering of rice leading to biosynthesis of glycinebetaine and tolerance to salt and cold. Plant Mol. Biol. 38, 1011–1019.

Sakamoto, H., Maruyama, K., Sakuma, Y., Meshi, T., Iwabuchi, M., 2004. *Arabidopsis* Cys2/His2-type zinc-finger proteins function as transcription repressors under drought, cold and high salinity conditions. Plant Physiol. 136, 2734–2746.

Schweizer, P., Buchala, A., Métraux, J.P., 1997. Gene-expression patterns and levels of jasmonic acid in rice treated with the resistance inducer 2,6-dichloroisonicotinic acid. Plant Physiol. 115, 61–70.

Shan, H., Chen, S., Jiang, J., Chen, F., Yu, C., Gu, C., et al., 2012. Heterologous expression of the *Chrysanthemum* R2R3-MYB transcription factor CmMYB2 enhances drought and salinity tolerance, increases hypersensitivity to ABA and delays flowering in *Arabidopsis thaliana*. Mol. Biotechnol. 51, 160–173.

Sharma, Y.K., Léon, J., Raskin, I., Davis, K.R., 1996. Ozone-induced responses in *Arabidopsis thaliana*: the role of salicylic acid in the accumulation of defense-related transcripts and induced resistance. Proc. Nat. Acad. Sci. USA 93, 5099–5104.

Shen, Y.G., Zhang, W.K., He, S.J., Zhang, J.S., Liu, Q., Chen, S.Y., 2003. An EREBP/AP2-type protein in Triticum aestivum was a DRE-binding transcription factor induced by cold, dehydration and ABA stress. Theor. Appl. Genet. 106, 923–930.

Sheveleva, E., Chmara, W., Bohnert, H.J., Jensen, R.G., 1997. Increased salt and drought tolerance by D-ononitol production in transgenic *Nicotiana tabacum* L. Plant Physiol. 115, 1211–1219.

Shevyakova, N.I., Shorina, M.V., Rakitin, V.Y., Kuznetsov, V.V., 2006. Stress-dependent accumulation of spermidine and spermine in the halophyte *Mesembryanthemum crystallinum* under salinity conditions. Russ. J. Plant Phys. 53, 739–745.

Shevyakova, N.I., Cheremisina, A.I., Kuznetsov VI. V., 2011. Phytoremediation potential of amaranthus hybrids: antagonism between nickel and iron and chelating role of polyamines. Russ. J. Plant Physl. 58, 634–642.

Shirasawa, K., Takabe, T., Takabe, T., Kishitani, S., 2006. Accumulation of glycinebetaine in rice plants that overexpress choline monooxygenase from spinach and evaluation of their tolerance to abiotic stress. Ann. Bot. 98, 565–571.

Singla-Pareek, S.L., Reddy, M.K., Sopory, S.K., 2001. Transgenic approach towards developing abiotic stress tolerance in plants. Proc. Ind. Nat. Sci. Acad. 67, 265–284.

Sivamani, E., Bahieldin, A., Wraith, J., Al-Niemi, T., Dyer, W., Ho, T., et al., 2000. Improved biomass productivity and water use efficiency under water deficit conditions in transgenic wheat constitutively expressing the barley HVA1 gene. Plant Sci. 155, 1–9.

Solomon, P.S., Waters, O.D.C., Oliver, R.P., 2007. Decoding the mannitol enigma in filamentous fungi. Trends Microbiol. 15, 257–262.

Souer, E., Van Houwelingen, A., Kloos, D., Mol, J., Koes, R., 1996. The no apical meristem gene of *Petunia* is required for pattern formation in embryos and flowers and is expressed at meristem and primordia boundaries. Cell 85, 159–170.

Sreenivasulu, N., Sopory, S.K., Kavi Kishor, P.B., 2007. Deciphering the regulatory mechanisms of abiotic stress tolerance in plants by genomic approaches. Gene 388, 1–13.

Stracke, R., Werber, M., Weisshaar, B., 2001. The R2R3-MYB gene family in *Arabidopsis thaliana*. Curr. Opin. Plant Biol. 4, 447–456.

Sugano, S., Kaminaka, H., Rybka, Z., Catala, R., Salinas, J., Matsui, K., et al., 2003. Stress-responsive zinc finger gene ZPT2-3 plays a role in drought tolerance in petunia. Plant J. 36, 830–841.

Takabe, T., Nakamura, T., Nomura, M., Hayashi, Y., Ishitani, M., Muramoto, Y., et al., 1998. Glycinebetaine and the genetic engineering of salinity tolerance in plants. In: Satoh, K., Murata, N. (Eds.), Stress Responses of Photosynthetic Organisms. Elsevier Science, Amsterdam, pp. 115–131.

Tarczynski, M.C., Jensen, R.G., Bohnert, H.J., 1993. Stress protection of transgenic tobacco by production of the osmolyte mannitol. Science 259, 508–510.

Testerink, C., Munnik, T., 2005. Phosphatidic acid: a multifunctional stress signaling lipid in plants. Trends Plant Sci. 10, 368–375.

Thomas, J.C., Sepahi, M., Arendall, B., Bohnert, H.J., 1995. Enhancement of seed germination in high salinity by engineering mannitol expression in Arabidopsis thaliana. Plant Cell Environ. 18, 801–806.

Torrigiani, P., Scaramagli, S., Ziosi, V., Mayer, M., Biondi, S., 2005. Expression of an antisense *Datura stramonium* S-adenosylmethionine decarboxylase cDNA in tobacco: changes in enzyme activity, putrescine-spermidine ratio, rhizogenic potential, and response to methyl jasmonate. J. Plant Physiol. 162, 559–571.

Uno, Y., Furihata, T., Abe, H., Yoshida, R., Shinozaki, K., Yamaguchi-Shinozaki, K., 2000. *Arabidopsis* basic leucine zipper transcription factors involved in an abscisic acid-dependent signal transduction pathway under drought and high-salinity conditions. Proc. Nat. Acad. Sci. USA 97, 11632–11637.

Upadhyaya, C.P., Young, K.E., Akula, N., Kim, H.S., Heung, J.J., Oh, O.M., et al., 2009. Over-expression of strawberry d-galacturonic acid reductase in potato leads to accumulation of vitamin C with enhanced abiotic stress tolerance. Plant Sci. 177, 659–667.

Van Der Krol, A.R., Van Poecke, R.M.P., Vorst, O.F.J., Voogd, C., Van Leeuwen, W., Van Borst-Vrensen, T.W.M., et al., 1999. Developmental and wound, cold, dessication, ultravoilet B-stress-induced modulations in the expression of the petunia zinc finger transcription factor gene ZPT2. Plant Physiol. 121, 1153–1162.

Vannini, C., Iriti, M., Bracale, M., Locatelli, F., Faoro, F., Croce, P., et al., 2006. The ectopic expression of the rice Osmyb4 gene in *Arabidopsis* increases tolerance to abiotic, environmental and biotic stresses. Physiol. Mol. Plant Pathol. 69, 26–42.

Verma, D., Singla-Pareek, S.L., Rajagopal, D., Reddy, M.K., Sopory, S.K., 2007. Functional validation of a novel isoform of Na + /H + antiporter from *Pennisetum glaucum* for enhancing salinity tolerance in rice. J. Biosci. 32, 621–628.

Vernon, D.M., Bohnert, H.J., 1992. A novel methyl transferase induced by osmotic stress in the facultative halophyte *Mesembryanthemum crystallinum*. EMBO J. 11, 2077–2085.

Vinocur, B., Altman, A., 2005. Recent advances in engineering plant tolerance to abiotic stress: achievements and limitations. Curr. Opin. Biotechnol. 16, 123–132.

Viswanathan, C., Zhu, J.K., 2004. Molecular perspectives on cross-talk and specificity in abiotic stress signaling in plants. J. Exp. Bot. 55, 225–236.

Waditee, R., Bhuiyan, M.N.H., Rai, V., Aoki, K., Tanaka, Y., Hibino, T., et al., 2005. Genes for direct methylation of glycine provide high levels of glycinebetaine and abiotic–stress tolerance in *Synechococcus* and *Arabidopsis*. Proc. Nat. Acad. Sci. USA 102, 1318–1323.

Wang, C.Q., 2011. Water-stress mitigation by seleniumin *Trifolium repens* L. J. Plant Nutr. Soil Sci. 174, 276–282.

Wang, C.R., Yang, A.F., Yue, G.D., Gao, Q., Yin, H.Y., Zhang, J.R., 2008. Enhanced expression of phospholipase C1 (ZmPLC1) improves drought tolerance in transgenic maize. Planta 227, 1127–1140.

Wang, H., Datla, R., Georges, F., Loewen, M., Cutler, A.J., 1995. Promoters from kin1 and cor6.6, two homologous *Arabidopsis thaliana* genes: transcriptional regulation and gene expression induced by low temperature, ABA, osmoticum and dehydration. Plant Mol. Biol. 28, 605–617.

Wang, J.Y., Wang, J.P., He, Y., 2013. A *Populus euphratica* NAC protein regulating Na(+)/K(+) homeostasis improves salt tolerance in *Arabidopsis thaliana*. Gene 521, 265–273.

Wang, Y., Jiang, J., Zhao, X., Liu, G., Yang, C., Zhan, L., 2006. A novel LEA gene from *Tamarix androssowii* confers drought tolerance in transgenic tobacco. Plant Sci. 171, 655–662.

Williamson, J.D., Jennings, D.B., Guo, W., Pharr, D.M., 2002. Sugar alcohols, salt stress, and fungal resistance: polyols—multifunctional plant protection? J. Amer. Soc. Hort. Sci. 127, 467–473.

Wise, M.J., 2003. LEAping to conclusions: a computational reanalysis of late embryogenesis abundant proteins and their possible roles. BMC Bioinform. 4, 52.

Wohlfarth, A., Severin, J., Galinski, E.A., 1990. The spectrum of compatible solutes in heterotrophic halophilic eubacteria of the family Halomonadaceae. J. Gen. Microbiol. 136, 705–712.

Wu, Y., Deng, Z., Lai, J., Zhang, Y., Yang, C., Yin, B., et al., 2009. Dual function of *Arabidopsis* ATAF1 in abiotic and biotic stress responses. Cell Res. 19, 1279–1290.

Xiang, Y., Tang, N., Du, H., Ye, H., Xiong, L., 2008. Characterization of OsbZIP23 as a key player of the basic leucine zipper transcription factor family for conferring abscisic acid sensitivity and salinity and drought tolerance in rice. Plant Physiol. 148, 1938–1952.

Xiao, B., Huang, Y., Tang, N., Xiong, L., 2007. Over-expression of a LEA gene in rice improves drought resistance under the field conditions. Theor. Appl. Genet. 115, 35–46.

Xiong, Y., Liu, T., Tian, C., Sun, S., Li, J., Chen, M., 2005. Transcription factors in rice: a genome-wide comparative analysis between monocots and eudicots. Plant Mol. Biol. 59, 191–203.

Xu, D., Duan, X., Wang, B., Hong, B., Ho, T.D., Wu, R., 1994. Expression of a late embryogenesis abundant protein gene, HVA7, from barley confers tolerance to water deficit and salt stress in transgenic rice. Plant Physiol. 110, 249–257.

Yadav, S.K., Singla-Pareek, S.L., Reddy, M.K., Sopory, S.K., 2005. Transgenic tobacco plants overexpressing glyoxalase enzymes resist an increase in methylglyoxal and maintain higher reduced glutathione levels under salinity stress. FEBS Lett. 579, 6265–6271.

Yamaguchi, K., Takahashi, Y., Berberich, T., Imai, A., Takahashi, T., Michael, A.J., et al., 2007. A protective role for the polyamine spermine against drought stress in Arabidopsis. Biochem. Biophys. Res. Commun. 352, 486–490.

Yamaguchi-Shinozaki, K., Koizumi, M., Urao, S., Shinozaki, K., 1992. Molecular cloning and characterization of 9 cDNAs for genes that are responsive to desiccation in *Arabidopsis thaliana*: sequence analysis of one cDNA clone that encodes a putative transmembrane channel protein. Plant Cell Physiol. 33, 217–224.

Yancey, P.H., Clark, M.E., Hand, S.C., Bowlus, R.D., Somero, G.N., 1982. Living with water stress: evolution of osmolyte systems. Science 217, 1214–1222.

Yang, A., Dai, X., Zhang, W.H., 2012. A R2R3-type MYB gene, OsMYB2, is involved in salt, cold, and dehydration tolerance in rice. J. Exp. Bot. 63, 2541–2556.

Yao, X., Chu, J., Ba, C., 2010. Antioxidant responses of wheat seedlings to exogenous selenium supply under enhanced ultraviolet-B. Biol. Trace Elem. Res. 136, 96–105.

Yao, X., Chu, J., He, X., Ba, C., 2011. Protective role of selenium in wheat seedlings subjected to enhanced UV-B radiation. Russ. J. Plant Physl. 58, 283−289.

Ye, B., Muller, H.H., Zhang, J., Gressel, J., 1997. Constitutively elevated levels of putrescine and putrescine-generating enzymes correlated with oxidant stress resistance in *Conyza bonariensis* and wheat. Plant Physiol. 115, 1443−1451.

Zhang, H.X., Blumwald, E., 2001. Transgenic salt-tolerant tomato plants accumulate salt in foliage but not in fruit. Nat. Biotechnol. 19, 765−768.

Zhang, H.X., Hodson, J.N., Williams, J.P., Blumwald, E., 2001. Engineering salt-tolerant Brassica plants: characterization of yield and seed oil quality in transgenic plants with increased vacuolar sodium accumulation. Proc. Nat. Acad. Sci. USA 98, 12832−12836.

Zhang, X., Yang, G., Shi, R., Han, X., Qi, L., Wang, R., et al., 2013. Arabidopsis cysteine-rich receptor-like kinase 45 functions in the responses to abscisic acid and abiotic stresses. Plant Physiol. Biochem. 67C, 189−198.

Zhang, X., Zhang, M., Takano, T., Liu, S., 2011. Characterization of an AtCCX5 gene from *Arabidopsis thaliana* that involves in high-affinity $K(+)$ uptake and $Na(+)$ transport in yeast. Biochem. Biophys. Res. Commun. 414, 96−100.

Zhao, F.Y., Zhang, X.J., Li, P.H., Zhao, Y.X., Zhang, H., 2006. Co-expression of the *Suaeda salsa* SsNHX1 and *Arabidopsis* AVP1 confer greater salt tolerance to transgenic rice than the single SsNHX1. Mol. Breed. 17, 341−353.

Zheng, X., Chen, B., Lu, G., Han, B., 2009. Overexpression of a NAC transcription factor enhances rice drought and salt tolerance. Biochem. Biophys. Res. Commun. 379, 985−989.

Zhu, B., Peng, R.H., Xiong, A.S., Xu, J., Fu, X.Y., Zhao, W., et al., 2012. Transformation with a gene for myo-inositol O-methyltransferase enhances the cold tolerance of *Arabidopsis thaliana*. Biologia Plantarum 56, 135−139.

Zhu, J.K., Shi, J., Singh, U., Wyatt, S.E., Bressan, R.A., Hasegawa, P.M., et al., 1993. Enrichment of vitronectin and fibronectin like proteins in NaCl-adapted plant cells and evidence for their involvement in plasma membrane-cell wall adhesion. Plant J. 3, 637−646.

Bt Crops: A Sustainable Approach towards Biotic Stress Tolerance

6

Mahmood-ur-Rahman, Muhammad Qasim, Shazia Anwar Bukhari and Tayyaba Shaheen

6.1 Introduction

The transgenic plants harboring genes from *Bacillus thuringiensis* (Bt) produce crystal proteins causing resistance to insect pests harmful to crops and thus help to minimize the use of chemical insecticides. Since 1996, many plants have been genetically transformed with genes derived from Bt. Using this technology, plants acquire the ability to produce toxins and perform functions of self-pest resistance and eliminating the need to apply chemical sprays/synthetic pesticides. Genetically modified (GM) crops used to produce an insecticide were first marketed in the late 1990s.

Currently, all commercial Bt crops have genes that produce various forms of insecticidal toxins of Gram-positive bacteria (Federici, 2002). GM crops produce a variety of toxins that have specific effects on specific orders of insect pests (for example, Lepidoptera, Coleoptera, Diptera, etc.). Commercial Bt crops available today are cotton transformed with Bt genes (*Cry1Ac*, *Cry2A*, *Cry1F*, etc.), maize (*Cry1Ab*, *Cry1Ac Cry1F*, *Cry3B*, *Cry9C*, etc.), and potatoes (*Cry3Aa*) (Federici, 2002; Shelton et al., 2002). Some other crops, such as apple, broccoli, cabbage, tobacco, tomato, soybean, and rice, have been developed to express Bt genes but they have not yet been marketed (Table 6.1).

The first successful commercial Bt crop, Bollgard cotton, was marketed in the United States in 1996. In 2000, it occupied approximately 1.8 million acres (Perlak et al., 2001). It has an innate resistance to lepidopteran insects. Bollgard cotton contributed significant economic benefits to US agriculture. Studies have shown that US farmers receive an average benefit return of about $50/acre (Perlak et al., 2001). The introduction of Bt cotton drastically reduced the use of conventional chemical pesticides (Smith, 1997). Transgenic Bt crops are particularly protected against flies, worms, and beetles. Other benefits associated with Bt plants are:

- No need to use synthetic chemical pesticides.
- Increased opportunities for beneficial insects; Bt proteins do not kill non-target insects.
- Reduce exposure to pesticides for farm workers.

Important crops, which are damaged by the larvae and beetles, have been genetically modified to produce Bt toxin to control these pests. Genetically, modified Bt cotton contains a natural toxin released by the bacteria. In any case, the use of these systems means that many chemical pesticides will be eliminated. This practice is good for the consumer and for the environment and farmers.

P. Ahmad (Ed): Emerging Technologies and Management of Crop Stress Tolerance, Volume 1.
DOI: http://dx.doi.org/10.1016/B978-0-12-800876-8.00006-0

Table 6.1 Economics of Bt Cotton and Maize

Crop	Country	Insecticide Reduction (%)	Increase in Yield (%)	References
Cotton	Argentina	47	33	Qaim and de Janvry (2005)
	Australia	48	0	Fitt (2003)
	China	65	24	Pray et al. (2002)
	India	41	37	Sadashivappa and Qaim (2009)
	Mexico	77	9	Traxler et al. (2003)
	South Africa	33	22	Gouse et al. (2004)
	USA	36	10	Carpenter et al. (2002)
Maize	Argentina	0	9	Brooks and Barfoot (2005)
	Philippines	5	34	Yorobe and Quicoy (2006)
	South Africa	10	11	Gouse et al. (2006)
	Spain	63	6	Gomez-Barbero et al. (2008)
	USA	8	5	Fernandez-Cornejo and Li (2005)

6.2 Bacillus thuringiensis

Bacillus thuringiensis is a Gram-positive bacterium. It lives in the soil and is a biological pesticide that has been known for more than a century. It is also found naturally in the gut of caterpillars and on dark areas of the surfaces of plants (Madigan and Martinko, 2005). Most of the strains of *B. thuringiensis* have *cry* genes which encode δ-endotoxin insecticidal proteins and are present on the plasmids (Xu et al., 2006). Previously, the presence of a plasmid in a strain of *B. thuringiensis* suggested its involvement in endospore and crystal formation (Cheng, 1984). See Table 6.2 for various milestones of Bt gene discovery.

The development of genetic transformation was an important milestone in the field of plant biotechnology. Now, scientists can introduce genes from different sources into the main crops, in order to induce resistance to various pests (Lycett and Grierson, 1990; Dhaliwal et al., 1998). *B. thuringiensis* has been widely known as a source of Bt genes for quite some time. These genes produce crystal proteins that are toxic to specific orders of various insect pests such as Lepidoptera (Whiteley and Schanept, 1986; Hoftey and Whiteley, 1989; Cohen et al., 2000), beetles (Krieg et al., 1983; Herrnstadt et al., 1986), and Diptera (Andrews et al., 1987). When insects feed on toxin crystals, the alkaline pH of their digestive system triggers the toxin. *Cry* toxin gets into the cell membrane of the insect gut and forms a pore. Pores lead to cell lysis and death of the insect (Dean, 1984). *B. thuringiensis*-based insecticides are generally used in the form of liquid sprays in plants. Insecticides are effective mainly when internalized. The toxins are soluble and are believed to produce pores in the epithelium of the midgut of the affected larvae. Some studies suggested the midgut bacteria of susceptible larvae are required for *B. thuringiensis* insecticidal activity (Broderick et al., 2006).

Table 6.2 History of *Bacillus thuringiensis* and Bt Gene Discovery

Year	Event
1901	Shigetane Ishiwatari isolated *Bacillus thuringiensis*
1911	Ernst Berliner isolated a bacteria that killed a Mediterranean flour moth and rediscovered *B. thuringiensis*
1915	Berliner reported the existence of a crystal within *B. thuringiensis*
1920	Farmers started to use Bt as a pesticide
1938	France started to make commercialized spore-based formulations of Bt called Sporine
1956	Hannay, Fitz-James, and Angus found that the main insecticidal activity against lepidoteran (moth) insects was due to the parasporal crystal
1958	Bt was started to be used commercially in USA
1961	Bt was registered as a pesticide to the EPA
1977	The first subspecies toxic to dipteran (flies) species was found
1983	First discovery of strains toxic to species of coleopteran (beetles)
1980s	Routine use of Bt as pesticide
1995	The first genetically engineered plant, corn, was registered with the EPA

6.3 Transformation of crops with Bt genes

Genetic modification of plants developed into a new era with progress in the science of recombinant DNA technology in the 1970s. With this technology, genes from different heterologous systems, namely, animals, bacteria, or insects, can be transferred to plants (Deineko et al., 2007). Genetic engineering involves the transformation of a gene responsible for a particular trait into the plant genome and the subsequent development of a complete plant from the transformed tissue. Several methods of gene transfer in plants, such as the transfer of PEG-mediated genes (Uchimiya et al., 1986), microinjection (de la Pena et al., 1987), electroporation (Fromm et al., 1985, 1986; Lorz et al., 1985; Arencibia et al., 1995) and the method of particle bombardment (Sanford, 1988), were developed.

Currently, the methods used for the transformation of genes in crops are Agrobacterium-mediated transformation and transformation by particle bombardment (Deineko et al., 2007; Gelvin, 2012; Azad et al., 2013) (Figure 6.1). Transgenic plants have become increasingly popular among farmers in developed countries and also in the developing world. Millions (16.7 million) of farmers in 29 countries grew biotech crops on 160 million hectares in 2011, of which 90% (about 15 million) were small farmers in developing countries (James, 2012).

6.3.1 Agrobacterium-mediated genetic transformation

Agrobacterium tumefaciens is a bacterium that has the ability to transfer foreign genes in host plant cells. *A. tumefaciens* causes wound infections in dicotyledonous plants, and leads to the formation of root-knot (tumors). The first evidence that this bacterium was the causative agent of crown gall

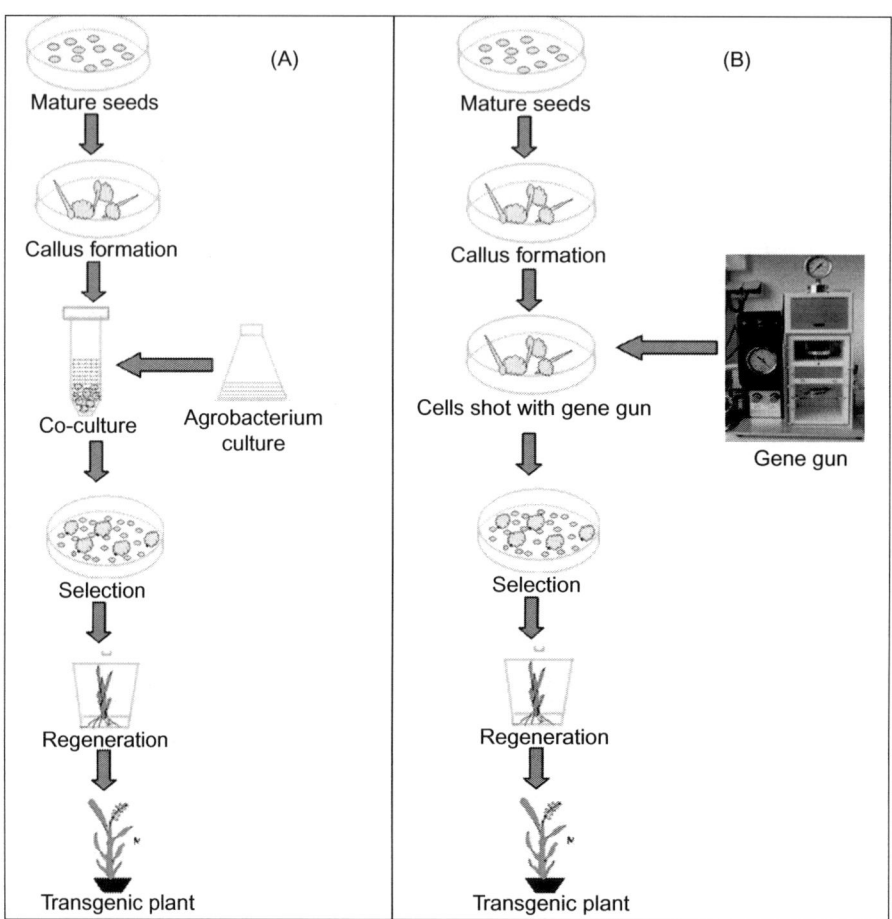

FIGURE 6.1

Transformation of crop plants by (A) Agrobacterium and (B) gene gun-mediated transformation methods.

was found over a century ago (Smith and Townsend, 1907). It has the capability of transferring a DNA fragment of the carcinogenic (Ti) plasmid into the nucleus of infected cells, causing crown gall (Nester et al., 1984). Tumor formation is a transformation of plant cells due to the transfer and integration of T-DNA and the subsequent expression of the genes of T-DNA and foreign DNA between the T-DNA borders placed in plant cells (Deblaere et al., 1985; Hooykaas and Schilperoort, 1992; Hamilton, 1997; Torisky et al., 1997; Rao et al., 2009; Kumar et al., 2012; Anami et al., 2013).

The first report of a successful plant expressing foreign genes was available at the beginning of 1980 (Herrera-Estrella, 1983). *A. tumefaciens* naturally produces tumors in dicotyledonous plants only. This method of genetic transformation has many advantages compared to those of other methods. Low number of copies of the transgene is transformed in this process, which is desirable.

The method has fewer problems of instability of the transgene and co-suppression (Koncz et al., 1994; Hansen et al., 1997). In addition, it takes a mutated gene in a single cell and not a patchwork of plants, which is very common when the direct gene transformation method is used (Enriquez-Obregon et al., 1997, 1998).

Agrobacterium-mediated transformation has been established in dicotyledous plants for many years and has recently been applied in monocotyledous plants (Hiei et al., 1994; Cheng et al., 1998). Previously, it was not possible in monocots. Now it is being used in various plants after the development of efficient and reproducible methods in banana (May et al., 1995), corn (Ishida et al., 1996), wheat (Cheng et al., 1997), sugarcane (Enriquez-Obregon et al., 1997, 1998), and chilli (Kumar et al., 2012). Now, the process has been used with success in several monocotyledonous crops (Chan et al., 1993; Hiei et al., 1994; Park et al., 1996; Rashid et al., 1996). Currently, Agrobacterium-mediated transformation is also being successfully used in rice transformation.

6.3.2 Transformation of plants by particle bombardment

Particle bombardment is a method of genetic transformation of crop plants by which the DNA is transformed directly into plant cells by physical or chemical method. The advantage of the direct transformation of genes is that each piece of DNA can be transferred and no special vectors are needed (Potter and Jones, 1997). There are many methods of direct transfer of plant-derived genes. The most promising approach is the "gene gun method" (Sanford, 1990; Jenes et al., 1992). This method involves high speed microprojectiles containing foreign DNA which penetrates through the cell wall and membrane. Thus, DNA can enter the plant genome.

Genetic transformation by particle bombardment makes use of different physical methods to obtain transformed plants. Currently, it is the most effective way to obtain transformed plants and is the only method previously used to achieve the transformation of chloroplasts and mitochondria (Altpeter et al., 2005). Successful and stable transformation of foreign genes is reported for rice, barley, and many other crops. Different types of plant tissues, such as floral tissues (microspores and anthers), tillers, and immature embryos, were successfully transformed by particle bombardment (King and Kasha, 1994; Stiff et al., 1995). Bt plants developed by both the Agrobacterium method and gene gun method, registered in the United States, are listed in Table 6.3.

6.4 Molecular analyses of putative transgenic plants

Detection of genetically modified organisms (GMOs) is possible through biochemical and molecular methods. It can be qualitative or quantitative. The following are the procedures to detect the presence and expression of Bt genes in the host genome.

6.4.1 Selection marker

A selectable marker gene in a GM organism develops a characteristic that can be used for artificial selection; this attribute is called a reporter gene/s. They are often genes that confer antibiotic resistance (Tuteja et al., 2012; Bala et al., 2013).

Table 6.3 Bt Crops Registered in USA up to 2009

Crop	Bt Toxins	Target Insects	Registered
Corn	Cry1Ab	Lepidoptera	1995
	Cry1F	Lepidoptera	2001
	Cry3Bb1	Coleoptera	2003
	Cry1Ab + Cry3Bb1	Lepidoptera and Coleoptera	2003
	Cry34Ab1 + Cry35Ab1	Coleoptera	2005
	Cry1F + Cry34Ab1 + Cry35Ab1	Lepidoptera and Coleoptera	2005
	Modified Cry3A	Coleoptera	2006
	Cry1Ab + Modified Cry3A	Lepidoptera and Coleoptera	2007
	Cry1A.105 + Cry2Ab2	Lepidoptera	2008
	Cry1A.105 + Cry2Ab2 + Cry3Bb1	Lepidoptera and Coleoptera	2008
	Vip3Aa20	Lepidoptera	2008
	Cry1Ab + Vip3Aa20	Lepidoptera and Coleoptera	2009
	Cry1Ab + Vip3Aa20 + modified Cry3A	Lepidoptera and Coleoptera	2009
	Cry1A.105 + Cry2Ab2 + Cry1F + Cry3Bb1 + Cry34Ab1 + Cry35Ab1	Lepidoptera and Coleoptera	2009
Cotton	Cry1Ac	Lepidoptera	1995
	Cry1Ac + Cry2Ab2	Lepidoptera	2002
	Cry1Ac + Cry1F	Lepidoptera	2004
	Modified Cry1Ab + Vip3Aa19	Lepidoptera	2008

Source: Tabashnik et al., 2009.

6.4.1.1 Types of selection marker
6.4.1.1.1 Antibiotic-resistant marker genes
Marker genes cause resistance patterns in a transgenic organism, an antibiotic to which the gene has been introduced. For example, neomycin phosphotransferase II (NPT-II gene) is typically being used in transgenic technology and results in resistance to the antibiotics kanamycin and neomycin (Tuteja et al., 2012).

6.4.1.1.2 The marker genes for herbicide tolerance
Transgenic plants resistant to herbicides have marker genes specific to chemical herbicides. The gene for tolerance to the herbicide glufosinate-ammonium is often used as a marker in the technology of GM plants. Those cells that survive exposure to the herbicide are selected and regenerated into a whole organism (Dale et al., 2002). The *ALS* gene is also being used as a selectable marker and is gaining popularity for several reasons. It is present in all plants, so there is no potential food safety concern (Yao et al., 2013).

6.4.1.1.3 Metabolic markers genes
Metabolic or auxotrophic marker genes allow transformed cells to synthesize an essential component, usually an amino acid, which is not produced by the application in any other way to

synthesize the cells. The surrounding medium intentionally does not have the essential component that cells need to grow. Cells that have successfully incorporated the selection marker and the rest of the gene construct will be able to produce the principal components in the cells and therefore survive. Such cells are selected and regenerated into whole organisms.

6.4.1.1.4 Screenable marker genes
Detectable marker genes encode a protein that can be identified by various laboratory tests. The presence of the protein confirms that the transformation has successfully occurred. Detectable marker is used as an alternative to a selectable marker as an aid for researchers who intend to distinguish between wanted and unwanted cells, based on the phenotype (Miki and McHugh, 2004). There are different types of methods of detection:

1. Green fluorescent protein makes the cells glow green under UV light. Yellow fluorescent protein and red fluorescent protein are also available. Using different colors, scientists can easily study multiple genes simultaneously. The technology is widely used in transgenic crop technology (e.g., maize, Kirienko et al., 2012; Coussens et al., 2012).
2. GUS assay (using β-glucuronidase) is a method for the detection of a single cell for blue stain without requiring complicated equipment.

6.4.2 Molecular analyses based on DNA
6.4.2.1 Polymerase chain reaction (PCR)
Polymerase chain reaction (PCR) is a method for the molecular amplification of a fragment of DNA. It allows the detection of specific DNA strands possessing millions of copies of a sequence of the target gene. The method works by synchronizing the sequence of the gene-specific DNA fragments with complementary primer. Specific genetic primers can be used to detect the presence of the genes that can be studied by nucleotide sequences of Bt genes in the presence of the target sequence. The primer matches the newly inserted gene with the assistance of such primer and starts a chain reaction. The process is repeated several times by sequential heating and cooling, and the target sequence is multiplied several million times. The millions of identical fragments are dyed to identify a segment of the primer-amplified-specific DNA on the gel under UV light (Figure 6.2A).

6.4.2.1.1 Qualitative detection
B. thuringiensis genes in a sample can be analyzed by qualitative PCR and multiplex PCR. Multiplex PCR uses multiple, unique primer sets within a single PCR. Amplicons of varying sizes specific to different DNA sequences that produce different transgenes are detected on agarose gel. By targeting multiple genes at once, additional information from a single test cycle is obtained, which saves time and the reagents (Figure 6.2A).

6.4.2.1.2 Quantitative detection
Quantitative PCR (Q-PCR) was used to measure the amount of PCR product. It is the preferred method to measure quantitatively the levels of transgenic DNA. Q-PCR is often used to determine the number of copies in the sample. The method is endowed with the highest accuracy of real-time quantitative PCR. Methods of QRT-PCR use fluorescent dyes such as SYBR Green or DNA probes

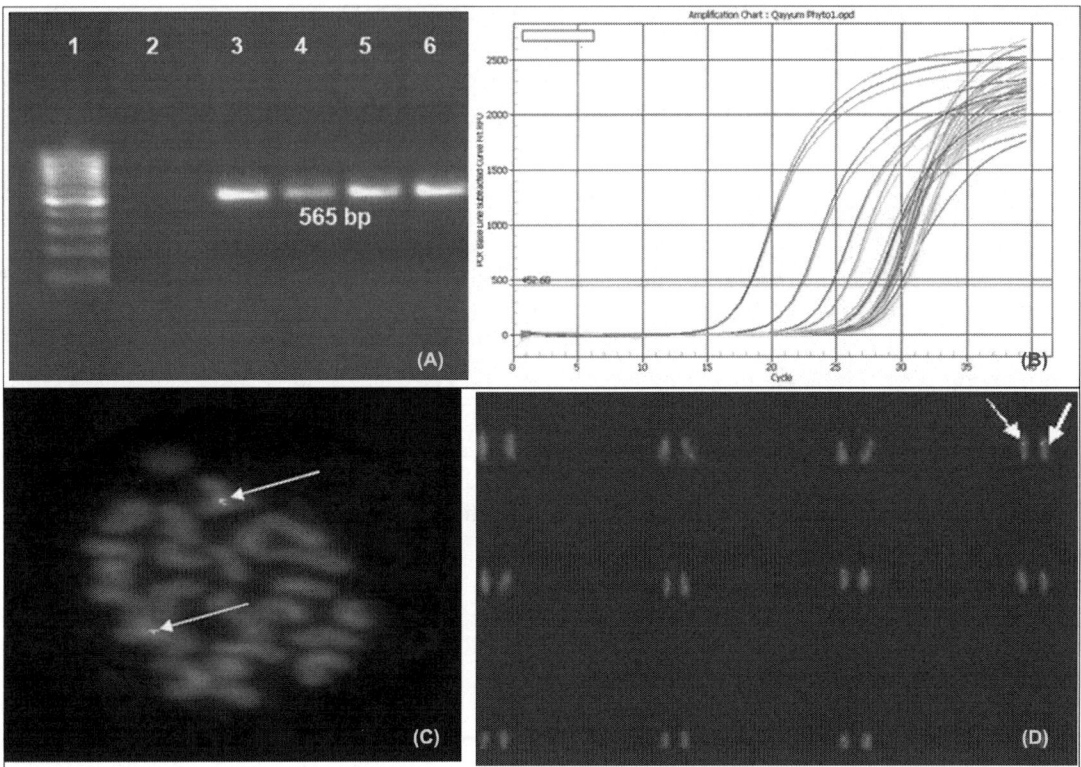

FIGURE 6.2

DNA-based detection methods of transgene in GM plants: (A) PCR amplification of Bt gene (Mahmood-ur-Rahman et al., 2010); (B) quantitative detection of Bt gene in cotton by real-time PCR; (C) detection of Bt gene on Bt rice chromosome through fluorescence *in situ* hybridization (Mahmood-ur-Rahman et al., 2010); (D) karyotype analysis of Bt gene on BT rice chromosomes (Mahmood-ur-Rahman et al., 2010).

containing a fluorophore, such as TaqMan, to measure the amount of amplified color product in real time (Figure 6.2B).

6.4.2.2 Southern blot

This technique was developed by Southern (Southern, 1975) to detect a complementary known DNA sequence used for the identification of desired DNA fragments using a specific probe. Southern blot allows a comparison of the genome of a particular organism of which a fragment of a gene or genes is available (the probe). One can tell if an organism contains a particular gene, and provide information on the organization and restriction map of the gene. The chromosomal DNA isolated from the microorganism of interest is completely digested with restriction endonucleases. The restriction fragments are separated by electrophoresis on agarose gel. The next step is the transfer of fragments of gel to a nitrocellulose filter or nylon membrane. This technique is dependent on the size and the specific activity of the probe and requires a large amount of DNA. Short probes are more accurate in general.

6.4.2.3 Fluorescent *in situ* hybridization

In situ hybridization, as the name implies, is a method for locating any mRNA or chromosomal DNA in the cell cytoplasm by hybridization of the sequence of interest in a free strand of a nucleotide probe. It is a powerful tool for the physical mapping of genes (Jiang et al., 1995) and target nucleic acid sequences that can be detected on the chromosomes, cells, or tissue sections. The technique was originally developed by Pardue and Gall (1969). They observed sequences hybridized with radioactive probes, because at that time only radioisotope-labeled nucleic acid materials were available.

Several disadvantages of isotopic hybridization inspired the development of new techniques. The dangers of using radioactive probes are: (1) the probes are unstable due to isotope decay over time, and, therefore, the specific activity of the probes is not constant; (2) sensitivity to X-rays is generally high and resolution is limited; (3) they require long exposure times to produce measurable signals on X-ray film; and (4) probes of radioactive materials are relatively expensive and dangerous, and must be transported, processed, stored, and disposed of with great care (Jeffrey and Singer, 2003).

Now, fluorochrome is used for labeling the probe for the detection of nucleic acid complementary sequence; the method is known as "fluorescent *in situ* hybridization" or FISH. The first fluorescence detection of antibody-dependent nucleic acid hybrids was by Rudkin and Stollar (1977), but this technology has been replaced with the advent of nucleic acid probes labeled with fluorescence. The technique of FISH was introduced for the first time in 1980. FISH allows (1) high sensitivity, (2) a short measurement period, and (3) to reach the manipulation of the sample (Ohmido and Fukui, 1997). Using this technique, a specific transgene was detected (Mahmood-ur-Rahman et al., 2010) (Figure 6.2C and D).

6.4.3 Molecular analyses based on protein

6.4.3.1 Enzyme-linked immunosorbent assay

Enzyme-linked immunosorbent assay (ELISA) is a method for detecting the presence of a particular protein in a sample. The technique requires a binding reagent that can be immobilized on the solid phase with a detection reagent that is specifically used for an enzyme and generates a signal that can be quantified correctly. Specifically, only the required protein and its specific interactions of antigen—antibody binding to the solid phase are detected, while the non-specific components are unbound and are washed.

6.4.3.2 Western blotting

The Western blot technique is widely used to detect proteins in a given sample. In this technique, gel electrophoresis is used to separate proteins. These proteins are transferred onto a nitrocellulose membrane, where they are examined by the use of antibodies specific for the transferred target protein.

6.5 Greenhouse and field experiments

The development of resistance in insects against Bt toxins is a major problem after prolonged exposure in the field. The most important technologies to increase the effectiveness of the Bt plant is a pyramid of two or more genes in crops (Cohen et al., 2000), observed in field conditions by evaluation over several years. Several transgenic lines of Basmati rice with Bt genes were evaluated under field conditions over several generations (Bashir et al., 2005; Mahmood-ur-Rahman et al., 2007) and sustainable resistance was monitored. These transgenic lines have shown resistance to target pests at all stages of growth and development of plants. After five years of field exposure, it has been observed that insect resistance was very stable over the years and the lines were significantly better as compared to untransformed control plants under field conditions (Bashir et al., 2005; Mahmood-ur-Rahman et al., 2007). These lines have shown resistance to advanced generations and there was no problem of gene silencing or gene segregation. Although these lines were significantly different from non-transformed control plants in height and days to maturity, these features have been observed up to the eighth generation, with trouble-free operation or segregation associated with these inherited lines.

Resistance against various lepidopteran pests in rice has been studied with the Bt genes (Bashir et al., 2004a, 2005; Breitler et al., 2004; Ye et al., 2001a,b, 2003). Zhau et al. (2003) reported that in the greenhouse, pyramidal two toxin genes with different mechanisms of cultivated plants show delays in the development of resistance. Toxin titers varied between the different lines, the plant with the same Bt gene construct (Cheng et al., 1998; Datta et al., 1998), and transformed assumes substantially in the later stages of development, particularly during the reproductive phase in rice (Alinia et al., 2000) (Table 6.4).

It is already documented that the transgenic lines could be early or late maturation (Jiang et al., 2000). There are several possible reasons for the morphological changes in transgenic plants, including somaclonal variations (Larkin and Scowcroft, 1981), disruption of indigenous genes in transgenic plants (Van et al., 1991), pleiotropy, and gene silencing (Matzke et al., 2000). Somaclonal variations may be one of the main reasons, as it took more time for the production of transgenic plants, compared to normal tissue culture techniques. The longer the time in tissue culture, the higher the frequency of somaclonal variations in plants (Kaeppler et al., 2000). The antibodies used as selectable markers can induce mutations in rice (Wu et al., 2000).

There are several studies of genetic transformation and evaluation of its effectiveness in the greenhouse and field (Bashir et al., 2004a, 2005; Breitler et al., 2004; Shu et al., 2000;

Table 6.4 Field Evaluation of Transgenic Rice

Gene	Trait	Location	References
cry1Ab and cry1Ac	Insect resistance	China	Tu et al. (2000)
cry1Ab	Insect resistance	China	Shu et al. (2000)
cry1Ab	Insect resistance	China	Ye et al. (2001a)
cry1Ab	Insect resistance	China	Shu et al. (2002)
cry1Ac and cry2A	Insect resistance	Pakistan	Bashir et al. (2004a,b)
cry1Aa and cry1B	Insect resistance	Spain	Breitler et al. (2004a)
cry2A	Insect resistance	China	Chen et al. (2005)
cry1Ac and cry2A	Insect resistance	Pakistan	Mahmood-ur-Rahman et al. (2007)
cry1Ac and CpT1	Insect resistance	China	Han et al. (2006)

Ye et al., 2001a,b, 2003). Most of the transgenic lines used in the field tests expressed Cry1Ab, Cry1Ac Cry2A, or Cry1Ab/Cry1Ac genes. With the passage of time, over 500 species of insects have acquired resistance to the conventional insecticides. So far, the history of Bt is better. In the field, only the diamondback moth is reported to have developed resistance against Bt sprays but has not acquired resistance to Bt crops (Ferre and Rie, 2002).

6.6 Biosafety and risk assessment studies

A series of problems associated with the development of resistance in target pest species, such as horizontal gene transfer, impact on non-target invertebrates, and morphological and agronomic variations, are being discussed globally. It is also important to consider if advanced generations of these lines are also resistant to insects in the field or lose their effectiveness over time, which could lead to a more rapid development of resistance in target pests. So, it is recommended that not all common varieties should be transformed with Bt genes to maintain refugia (Cohen et al., 2000). The potential of gene flow via pollen from Bt lines and other commercial varieties is an important consideration. The response of non-target insects on transgenic plants is well studied for herbivorous insects, insects, and invertebrates (Bashir et al., 2004a,b; Noor et al., 2010).

The specificity of *cry* genes to some orders of insects is its most important feature. Bt sprays do not disrupt the overall structure of a community's predators or parasitoids, or population trajectories of non-target herbivores (grasshoppers and crickets; Schoenly, 2004). Several researchers have tried to determine the number of non-target insects in transgenic and control fields (Fitt et al., 1994; Sims, 1995; Orr and Landis, 1998; Bashir et al., 2004a,b). The results showed that Bt crops are safe for non-target insects pests and visiting insects (Hendriksma et al., 2012, 2013).

There is some concern that purified Bt toxins can pose risks to the soil environment. The fate of Bt protein in earth is an important parameter (Saxena et al., 1999; Saxena and Stotzky, 2001; Stotzky, 2001). Bt concentrations could accumulate in the soil and pose a threat to earthworms, nematodes, protozoa, bacteria, and fungi. Contrasting results were obtained for evaluating the persistence of Bt protein. Donegan et al. (1995) observed immunological activity of the protein up to 28 days and for proteins Cry1Ab and Cry1Ac up to 56 days in the soil.

There are two types of gene flow: horizontal gene transfer and vertical gene flow. Horizontal gene transfer (HGT) is the transfer of genetic material from one organism to another organism, which is not sexually compatible with the donor (Gay, 2001). HGT is based on the known mechanisms in bacteria, including transduction, conjugation, and transformation. Accidental transfer of DNA between bacterial cells by bacteriophage infection has been studied (Kidambi et al., 1994; Herron, 1995). Ochman et al. (2000) have reported that it has been detected in up to 16% of the protein coding DNA in bacteria in the vicinity. Transfer of mobile sequences (plasmids, transposons, and chromosomal genes mobility) between the bacterial cells capable of mediating between the bacterial population in the rhizosphere soil and plant surfaces and water is also reported.

Non-target effects are no adverse effects of GM plants on friendly organisms in the environment (Dale et al., 2002). Non-target organisms are pollinators and herbivores feed on the affected and surrounding vegetation. It is important to remember that every human intervention to protect crops from pests has some negative effect on non-target organisms (Schuler, 2000) and on the entire community of life (Shelton et al., 2002). The impact of GM crops on the population dynamics of natural enemies depends on several factors, including the expression of the transgene, the specificity of the transgene product, and the tissue specificity of the transgene (Schuler, 2000). The populations of many non-target species were higher in Bt cotton fields as compared to the field sprayed by conventional pesticides (Xia et al., 1999; Head et al., 2001). Some insects and spiders were strongly influenced by insecticide sprays (Candolfi et al., 2004).

Some chemicals, known as allelochemicals, released from certain plants are known to have inhibitory or stimulatory effects on the growth of other crops (Torres et al., 1996). GM crops could have the potential to suppress other cultures, which decreases biodiversity in an ecosystem. Some allelochemicals isolated from the transgenic rice hull and identified as momilactone A and momilactone B have no allelopathic effects, while allelochemicals from barnyard grass have inhibitory effects on the germination and growth of plants (Kato-Noguchi, 2004; Xu et al., 2005; Chung et al., 2006).

6.7 Conclusion and future prospects

The year 2011 was the 16th year of the commercialization of GM crops (James, 2012). An increase of 94 times in the area of 1.7 million hectares in 1996 to 160×10^6 hectares in 2011 makes biotech crops the fastest adopted agricultural technology in the history of modern agriculture (James, 2012). In 2011, a record number of GM plants were grown by 16.7 million farmers. Over 90% or 15 million of these farmers were small resource-poor farmers in developing countries. Bt cotton farmers' income has increased significantly while their exposure to pesticides has decreased (James, 2012). There are 13 nations, both developed and developing, growing Bt cotton today, including the United States, Brazil, Argentina, India, China, Pakistan, and Australia.

The world needs at least 70% more food by 2050. For developing countries, food production must double by 2050, where 2.5 million poor small farmers need to survive. GM crops have had an exemplary safety record because they benefit farmers, the environment, and consumers. GM crops have been adopted faster than any other agricultural advance in the history of humanity (Alberts et al., 2013). The adoption of GM crops should be increased while keeping in view the following three factors: (1) timely implementation of appropriate, responsible, and cost/time-effective regulatory systems; (2) a strong political will and financial support; and (3) constant improvement in biotech crops in Asia, Latin America, and Africa (James, 2012), to meet the requirements of developed as well as developing countries.

Acknowledgments

The authors acknowledge the financial support of the Higher Education Commission, Government of Pakistan. The authors are also thankful to Prof. Dr. Muhammad Ibrahim Rajoka, Department of Bioinformatics and Biotechnology, GC University, Faisalabad (Pakistan) who read the manuscript critically.

References

Alberts, B., Beachy, R., Baulcombe, D., Blobel, G., Datta, S., Fedoroff, N., et al., 2013. Standing up for GMOs. Science 341, 1320.
Alinia, F., Ghareyyazie, B., Rubia, L.G., Bennett, J., Cohen, M.B., 2000. Effect of plant age, larval age and fertilizer treatment on resistance of a *cry1Ab* transformed aromatic rice to lepidopterous stem borers and foliage feeders. J. Econ. Entomol. 93, 484–493.

Altpeter, F., Baisakh, N., Beachy, R., Bock, R., Capell, T., Christou, P., et al., 2005. Particle bombardment and the genetic enhancement of crops: myths and realities. Mol. Breed. 15, 305−327.

Anami, S., Njuguna, E., Coussens, G., Aesaert, S., Lijsebettens, M.V., 2013. Higher plant transformation: principles and molecular tools. Int. J. Dev. Biol. 57, 483−494.

Andrews, R.W., Fausr, R., Wabiko, M.H., Roymond, K.C., Bulla, L.A., 1987. Biotechnology of Bt: a critical review. Biotechnol. 6, 163−232.

Arencibia, A., Molina, P., de la Riva, G., Selman-Housein, G., 1995. Production of transgenic sugarcane (*Saccharum officinarum* L.) plants by intact cell electroporation. Plant Cell Rep. 14, 305−309.

Azad, M.A.K., Rabbani, M.G., Amin, L., Sidik, N.M., 2013. Development of transgenic papaya through *Agrobacterium*-mediated transformation. Int. J. Genom. 2013, 235487.

Bala, A., Roy, A., Das, A., Chakraborti, D., Das, S., 2013. Development of selectable marker free, insect resistant, transgenic mustard (*Brassica juncea*) plants using Cre/lox mediated recombination. BMC Biotechnol. 13, 88.

Bashir, K., Husnain, T., Fatima, T., Latif, L., Mehdi, S.A., Riazzuddin, S., 2004a. Field evaluation and risk assessment of transgenic indica Basmati rice. Mol. Breed. 13, 301−312.

Bashir, K., Husnain, T., Riazuddin, S., 2004b. Response of transgenic rice expressing two Bt genes to nontarget insects. Int. Rice Res. Notes 29, 15−16.

Bashir, K., Husnain, T., Fatima, T., Riaz, N., Makhdoom, R., Riazuddin, S., 2005. Novel indica Basmati line (B-370) expressing two unrelated genes of *Bacillus thuringiensis* is highly resistant to two lepidopteran insects in the field. Crop Prot. 24, 870−879.

Breitler, J.C., Vassal, J.M., del mar Catala, M., Meynard, D., Marfa, V., Mele, E., et al., 2004. *Bt* rice harbouring *cry* genes controlled by a constitutive or a wound-inducible promoter: protection and transgene expression under Mediterranean field conditions. Plant Biotechnol. J. 2, 417−430.

Broderick, N.A., Raffa, K.F., Handelsman, J., 2006. Midgut bacteria required for Bacillus thuringiensis insecticidal activity. Proc. Natl. Acad. Sci. USA 103, 15196−15199.

Brookes, G., Barfoot, P., 2005. GM Crops: The Global Socioeconomic and Environmental Impact—The First Nine Years. PG Econ, Dorchester.

Candolfi, M.P., Brown, K., Grimm, C., Reber, B., Schmidli, H., 2004. A faunistic approach to assess potential side-effects of genetically modified Bt-corn on non-target arthropods under field conditions. Biocontrol Sci. Techn. 14, 129−170.

Carpenter, J., Felsot, A., Goode, T., Hammig, M., Onstad, D., Sankula, S., 2002. Comparative Environmental Impacts of Biotechnology-Derived and Traditional Soybean, Corn, and Cotton Crops. Counc. Agric. Sci. Technol., Ames, IA.

Chan, M., Chang, H., Ho, S., Tong, W., Yu, S., 1993. *Agrobacterium*-mediated production of transgenic rice plants expressing a chimeric α-amylase promoter/glucuronidase gene. Plant Mol. Biol. 22, 491−506.

Chen, H., Tang, W., Xu, C., Li, X., Lin, Y., Zhang, Q., 2005. Transgenic indica rice plants harboring a synthetic cry2A gene of Bacillus thuringiensis exhibit enhanced resistance against lepidopteran rice pests. Theor. Appl. Genet. 111, 1330−1337.

Cheng, M., Fry, J.E., Pang, S.Z., Zhou, H.P., Hironaka, C.M., Duncan, D.R., et al., 1997. Genetic transformation of wheat mediated by *Agrobacterium tumefaciens*. Plant Physiol. 115, 971−980.

Cheng, T.C., 1984. Pathogens of invertebrates: application in biological control and transmission mechanisms. Society for Invertebrate Pathology Meeting. 7, 159.

Cheng, X.Y., Sardana, R., Kaplan, H., Altosaar, I., 1998. *Agrobacterium*-transformed rice expressing synthetic *cry1Ab* and *cry1Ac* genes are highly toxic to striped stem borer and yellow stem borer. Proc. Natl. Acad. Sci. USA 95, 2767−2772.

Chung, I.M., Kim, J.T., Kim, S.H., 2006. Evaluation of allelopathic potential and quantification of momilactone A, B from rice hull extracts and assessment of inhibitory bioactivity on paddy field weeds. J. Agric. Food Chem. 54, 2527−2536.

Cohen, B.M., Gould, F., Benture, J., 2000. *Bt* rice: practical steps to sustainable use. Int. Rice Res. Notes 25 (2), 4−10.

Coussens, G., Aesaert, S., Verelst, W., Demeulenaere, M., de Buck, S., Njuguna, E., et al., 2012. Brachypodium distachyon promoters as efficient building blocks for transgenic research in maize. J. Exp. Bot. 63, 4263−4273.

Dale, P.J., Clarke, B., Fontes, E.M.G., 2002. Potential for the environmental impact of transgenic crops. Nat. Biotechnol. 20, 567−574.

Datta, K., Vasquez, A., Tu, J., Torrizo, L., Alam, M.F., Olivia, N., et al., 1998. Constitutive and tissue-specific differential expression of *cry1A(b)* gene in transgenic rice plants conferring resistance to rice insect pests. Theor. Appl. Genet. 97, 20−30.

de la Pena, A., Lorz, H., Schell, J., 1987. Transgenic rye plants obtained by injecting DNA into young floral tillers. Nature 325, 274−276.

Dean, D.H., 1984. Biochemical genetics of the bacterial insect-control agent Bacillus thuringiensis: basic principles and prospects for genetic engineering. Biotechnol. Genet. Eng. Rev. 2, 341−363.

Deblaere, R., Bytebier, B., De Greve, H., Deboeck, F., Schell, J., Van Montagu, M., et al., 1985. Efficient octopine Ti plasmid vectors for Agrobacterium-mediated gene transfer to plants. Nucl. Acids Res. 13, 4777−4788.

Deineko, E.V., Zagorskaya, A.A., Shumny, V.K., 2007. T-DNA induced mutations in transgenic plants. Russ. J. Genet. 43, 1−11.

Dhaliwal, H.S., Kawai, M., Uchimiya, H., 1998. Genetic engineering for abiotic stress tolerance in plants. Plant Biotechnol. 15, 1−10.

Donegan, K.K., Palm, C.J., Fieland, V.J., Porteous, L.A., Ganio, L.M., Schaller, D.L., et al., 1995. Changes in levels, species, and DNA fingerprints of soil microorganisms associated with cotton expressing the *Bacillus thuringiensis var kurstaki* endotoxin. Appl. Soil Ecol. 2, 111−124.

Enriquez-Obregon, G.A., Vazquez-Padron, R.I., Prieto-Samsonov, D.L., Perez, M., Selman-Housein, G., 1997. Genetic transformation of sugarcane by *Agrobacterium tumefaciens* using antioxidants compounds. Biotecnologia Aplicada 14, 169−174.

Enriquez-Obregon, G.A., Vazquez-Padron, R.I., Prieto-Sansonov, D.L., de la Riva, G.A., Selman-Housein, G., 1998. Herbicide resistant sugarcane (*Saccharum officinarum* L.) plants by *Agrobacterium*-mediated transformation. Planta 206, 20−27.

Federici, B., 2002. Case study: Bt crops. In: Atherton, K. (Ed.), Genetically Modified Crops: Assessing Safety. Taylor & Francis, New York, pp.

Fernandez-Cornejo, J., Li, J., 2005. The impacts of adopting genetically engineered crops in the USA: the case of Bt corn. Presented at American Agricultural Economics Association Annual Meeting, 24−27 July, Providence, RI, pp. 164−200.

Ferre, J., Rie, V., 2002. Biochemistry and genetics of insect resistance to *Bacillus thuringiensis*. Ann. Rev. Entomol. 47, 501−533.

Fitt, G., 2003. Implementation and Impacts of Transgenic Bt Cottons in Australia. ICAC Rec., December: 14−19.

Fitt, G.P., Mares, C.L., Llewellyn, D.J., 1994. Field evaluation and potential ecological impact of transgenic cotton (*Gossypium hirsutum*) in Australia. Biocontrol Sci. Techn. 4, 535−548.

Fromm, M., Taylor, L., Walbot, V., 1985. Expression of genes transferred into monocotyledonous and dicotyledonous plant cells by electroporation. Proc. Natl. Acad. Sci. USA 82, 5824−5828.

Fromm, M., Taylor, L., Walbot, V., 1986. Stable transformation of maize after gene transfer by electroporation. Nature 319, 791−793.

Gay, P., 2001. The biosafety of antibiotic resistance markers in plant transformation and the dissemination of genes through horizontal gene flow. In: Custers, R. (Ed.), Safety of Genetically Engineered Crops. Flanders Interuniversity Institute for Biotechnology, Zwijnaarde, Belgium, pp. 135−159.

Gelvin, S.B., 2012. Traversing the cell: *Agrobacterium* T-DNA's journey to the host genome. Front. Plant Sci. 3, 52.

Gomez-Barbero, M., Berbel, J., Rodriguez-Cerezo, E., 2008. Bt corn in Spain—the performance of the EU's first GM crop. Nat. Biotechnol. 26, 384–386.

Gouse, M., Pray, C., Schimmelpfennig, D., 2004. The distribution of benefits from Bt cotton adoption in South Africa. AgBioForum 7, 187–194.

Gouse, M., Pray, C., Schimmelpfennig, D., Kirsten, J., 2006. Three seasons of subsistence insect-resistant maize in South Africa: have smallholders benefited? AgBioForum 9, 15–22.

Hamilton, C.M., 1997. A binary-BAC system for plant transformation with high-molecular-weight DNA. Gene 200, 107–116.

Han, L., Wu, K., Peng, Y., Wang, F., Guo, Y., 2006. Evaluation of transgenic rice expressing Cry1Ac and CpTI against Chilo suppressalis and intrapopulation variation in susceptibility to Cry1Ac. Environ. Entomol. 35, 1453–1459.

Hansen, G., Shillito, R.D., Chilton, M.D., 1997. T-strand integration in maize protoplasts after codelivery of a T-DNA substrate and virulence genes. Proc. Natl. Acad. Sci. USA 94, 11726–11730.

Head, G., Brown, C.R., Groth, M.E., Duan, J.J., 2001. Cry1Ab protein levels in phytophagous insects feeding on transgenic corn: implications for secondary exposure risk assessment. Entomol. Exp. Appl. 99, 37–45.

Hendriksma, H.P., Hartel, S., Babendreier, D., von der Ohe, W., Steffan-Dewenter, I., 2012. Effects of multiple Bt proteins and GNA lectin on in vitro-reared honey bee larvae. Apidologie 43, 549–560.

Hendriksma, H.P., Kuting, M., Hartel, S., Nather, A., Dohrmann, A.B., Dewenter, I.S., et al., 2013. Effect of stacked insecticidal cry proteins from maize pollen on nurse bees (*Apis mellifera carnica*) and their gut bacteria. PLoS One 8, e59589.

Herrera-Estrella, L., 1983. Transfer and Expression of Foreign Genes in Plants. PhD Thesis. Laboratory of Genetics, Gent University, Belgium.

Herrnstadt, C., Soanres, G.G., Wilcox, E.R., Edward, D.L., 1986. A new strain of *Bacillus thuringiensis* with activity against coleopteran insects. Biotechnology 4, 305–308.

Herron, P.R., 1995. Phage ecology and gene exchange in soil. In: Akkermans, A.D.L., Van Elsas, J.D., de Brujin, F.J. (Eds.), Molecular Microbial Ecology Manual. Kluwer Academic Publishers, Dordrecht, The Netherlands, , 5.3.2:1–12.

Hiei, Y., Ohta, S., Komari, T., Kumashiro, T., 1994. Efficient transformation of rice (*Oryza sativa* L.) mediated by *Agrobacterium* and sequence analysis of the boundaries of the T-DNA. Plant J. 6, 271–282.

Hoftey, H., Whiteley, H.R., 1989. Insecticidal crystal proteins of *Bacillus thuringiensis*. Microbiol. Rev. 53, 242–255.

Hooykaas, P.J.J., Shilperoort, R.A., 1992. *Agrobacterium* and plant genetic engineering. Plant Mol. Biol. 19, 15–38.

Ishida, Y., Saito, H., Ohta, S., Hiei, Y., Komari, T., Kumashiro, T., 1996. High efficiency transformation of maize (*Zea mayz* L.) mediated by *Agrobacterium tumefaciens*. Nat. Biotechnol. 4, 745–750.

James, C., 2012. Global Status of Commercialized Biotech/GM Crops: 2012. ISAAA, Ithaca, NY (ISAAA Brief No. 43).

Jeffrey, M.L., Singer, R.H., 2003. Fluorescence *in situ* hybridization: past, present and future. J. Cell Sci. 116, 2833–2838.

Jenes, B., Moore, H., Cao, J., Zhang, W., Wu, R., 1992. Techniques for gene transfer. In: Kung, S.D., Wu, R. (Eds.), Transgenic Plants. Academic Press, Inc., San Diego, CA, pp. 125–146.

Jiang, J., Gill, B.S., Wang, G.L., Ronald, P.C., Ward, D.C., 1995. Metaphase and interphase fluorescence *in-situ* hybridization mapping of the rice genome with bacterial artificial chromosomes. Proc. Natl. Acad. Sci. USA 92, 4487–4491.

Jiang, J., Linscombe, S.D., Wang, J., Oard, J.H., 2000. Field evaluation of transgenic rice (*Oryza sativa* L.) produced by *agrobacterium* and particle bombardment methods. Plant and Animal Genome VIII Conference, January 9–12, 2000, Town and County Hotel, San Diego, CA.

Kaeppler, S.M., Kaeppler, H.F., Rhee, Y., 2000. Epigenetic aspects of somaclonal variation in plants. Plant Mol. Biol. 43, 179−188.

Kato-Noguchi, H., 2004. Allelopathic substance in rice root exudates: rediscovery of momilactone B as an allelochemical. J. Plant Physiol. 161, 271−276.

Kidambi, S.P., Ripp, S., Miller, R.V., 1994. Evidence for phage mediated gene transfer among Pseudomonos aeruginosa strains on the phylloplane. Appl. Environ. Microbiol. 60, 496−500.

King, S.P., Kasha, K.J., 1994. Optimising somatic embryogenesis and particle bombardment of barley (*Hordeum vulgare* L.) immature embryos. In Vitro Cell Dev. Biol. Plant 30, 117−123.

Kirienko, D.R., Luo, A., Sylvester, A.W., 2012. Reliable transient transformation of intact maize leaf cells for functional genomics and experimental study. Plant Physiol. 159, 1309−1318.

Koncz, C., Nemeth, K., Redei, G.P., Scell, J., 1994. In: Paszkowski, J. (Ed.), Homologous Recombination and Gene Silencing in Plants. Kluwer, Dordrecht, The Netherlands, pp. 167−189.

Krieg, A., Huger, A., Langenbuch, G., Schneter, W., 1983. *Bacillus thuringiensis var. tenebrionis*: a new pathotype effective against larvae of coleoptera. J. Appl. Entomol. 96, 500−508.

Kumar, R.V., Sharma, V.K., Chattopadhyay, B., Chakraborty, S., 2012. An improved plant regeneration and Agrobacterium-mediated transformation of red pepper (*Capsicum annuum* L.). Physiol. Mol. Biol. Plants 18, 357−364.

Larkin, P.J., Scowcroft, W.R., 1981. Somaclonal variation—a novel source of variability from cell culture from plant improvement. Theor. Appl. Genet. 60, 197−214.

Lorz, H., Baker, B., Schell, J., 1985. Gene transfer to cereal cells mediated by protoplast transformation. Mol. Gen. Genet. 199, 473−497.

Lycett, G.W., Grierson, D., 1990. Genetic Engineering of Crop Plants. Butterworth, London, UK.

Madigan, M., Martinko, J., 2005. Brock Biology of Microorganisms. eleventh ed. Prentice Hall, New York.

Mahmood-ur-Rahman, Noreen, S., Husnain, T., Riazuddin, S., 2010. Fast and efficient method to determine the position of alien genes in transgenic plants. Em. J. Food Agric. 22, 223−231.

Mahmood-ur-Rahman, Rashid, H., Shahid, A.A., Bashir, K., Husnain, T., Riazuddin, S., 2007. Insect resistance and biosafety studies of advanced lines of indica Basmati rice (B-370) expressing two genes of *Bacillus thuringensis*. Elect. J. Biotechnol. 10, 240−251.

Matzke, M.A., Mette, M.F., Matzke, A.J.M., 2000. Transgene silencing by the host genome defense: implications for the evolution of epigenetic control mechanism in plants and vertebrates. Plant Mol. Biol. 43, 401−415.

May, G.D., Afza, R., Mason, H.S., Wiecko, A., Novak, F.J., Arntzen, C.J., 1995. Generations of transgenic Banana (*Musa acuminata*) plants via *Agrobacterium*-mediated transformation. Biotechnology 13, 486−492.

Miki, B., McHugh, S., 2004. Selectable marker genes in transgenic plants: applications, alternatives and biosafety. J. Biotechnol. 107, 193−232.

Nester, E.W., Gordon, M.P., Amasino, R.M., Yanofsky, M.F., 1984. Crown gall: a molecular and physiological analysis. Ann. Rev. Plant Physiol. 35, 387−413.

Noor, M., Mahmood-ur-Rahman, Shahid, A.A., Husnain, T., Riazuddin, S., 2010. Risk assessment and biosafety studies of transgenic Bt rice (*Oryza sativa* L.). J. Agric. Sci. Technol. 4, 1−9.

Ochman, H., Lawrence, J.G., Groisman, E.A., 2000. Lateral gene transfer and the nature of bacterial innovation. Nature 405, 299−304.

Ohmido, N., Fukui, K., 1997. Visual verification of close disposition between rice A genome-specific DNA sequence (*TrsA*) and telomere sequence. Plant Mol. Biol. 35, 963−968.

Orr, D.B., Landis, D.L., 1998. Oviposition of European corn borer (Lepidoptera: Pyralidae) and impact of natural enemy populations in transgenic versus isogenic corn. J. Econ. Entomol. 90, 905−909.

Pardue, M.L., Gall, J.G., 1969. Molecular hybridization of radioactive DNA to the DNA of cytological preparations. Proc. Natl. Acad. Sci. USA 64, 600−604.

Park, S.H., Pinson, S.R.M., Smith, R., 1996. T-DNA integration into genomic DNA of rice following *Agrobacterium* inoculation of isolated shoot apices. Plant Mol. Biol. 32, 1135−1148.

Perlak, F., Oppenhuizen, M., Gustafson, K., Voth, R., Sivasupramnium, S., Heering, D., et al., 2001. Development and commercial use of Bollgard cotton in the USA: early promises versus today's reality. Plant J. 27, 489−501.

Potter, R.H., Jones, M.G.K., 1997. Plant gene transfer. In: Clark, M.S. (Ed.), Plant Molecular Biology. A Laboratory Manual. Springer Verlag, Berlin−Heidelberg−New York, pp. 399−426.

Pray, C.E., Huang, J., Hu, R., Rozelle, S., 2002. Five years of Bt cotton in China—the benefits continue. Plant J. 31, 423−430.

Rao, A.Q., Bakhsh, A., Kiani, S., Shahzad, K., Shahid, A.A., Husnain, T., et al., 2009. The myth of plant transformation. Biotechnol. Adv. 27, 753−763.

Rashid, H., Yokoi, S., Toriyama, K., Hinata, K., 1996. Transgenic plant production mediated by *Agrobacterium* in *Indica* rice. Plant Cell Rep. 15, 727−730.

Rudkin, G.T., Stollar, B.D., 1977. High resolution detection of DNA: RNA hybrids *in situ* by indirect immunofluorescence. Nature 265, 472−473.

Sadashivappa, P., Qaim, M., 2009. Effects of Bt cotton in India during the first five years of adoption. Presented at Int. Assoc. Agric. Econ. Triennial Conf., Beijing, China.

Sanford, J.C., 1988. The biolistic process. Trends Biotechnol. 6, 299−302.

Sanford, J.C., 1990. Biolistic plant transformation. Physiol. Plant 79, 206−209.

Saxena, D., Stotzky, G., 2001. *Bacillus thuringiensis* (Bt) toxin released from root exudates and biomass of Bt corn has no apparent effect on earthworms, nematodes, protozoa, bacteria, and fungi in soil. Soil Biol. Biochem. 33, 1225−1230.

Saxena, D., Saul, F., Stotzky, G., 1999. Transgenic plants: insecticidal toxin in root exudates from Bt corn. Nature 402, 480.

Schoenly, K.G., 2004. Effects of *Bacillus thuringiensis* on non-target herbivore and natural enemy assemblages in tropical irrigated rice. Environ. Biosafety Res. 2, 181−206.

Schuler, T.H., 2000. The impact of insect resistant GM crops on populations of natural enemies. Antenna 24, 59−65.

Shelton, A., Zhao, J., Roush, R., 2002. Economic, ecological, food safety, and social consequences of the deployment of Bt transgenic plants. Ann. Rev. Entomol. 47, 845−881.

Shu, Q.U., Ye, G.Y., Cui, H.R., Cheng, X.Y., Xiang, Y.B., Wu, D.X., et al., 2000. Transgenic rice plants with a synthetic cry1Ab gene from Bacillus thuringiensis were highly resistant to eight lepidopteran rice pest species. Mol. Breed. 6, 433−439.

Shu, Q.U., Cui, H., Ye, G., Wu, D., Xia, Y., Gao, M., et al., 2002. Agronomic and morphological characterization of Agrobacterium-transformed Bt rice plants. Euphytica 127, 345−352.

Sims, S.R., 1995. *Bacillus thuringiensis* var. kurstaki (Cry1Ac) protein expressed in transgenic cotton: effects on beneficial and other non-target insects. Southwestern Entomologist 20, 493−500.

Smith, E.F., Towsend, C.O., 1907. A plant tumor of bacterial origin. Science 25, 671−673.

Smith, R.H., 1997. An extension entomologist's 1996 observations of Bollgard (Bt) technology. Proceedings Beltwide Cotton Conferences.

Southern, E.M., 1975. Detection of specific sequences among DNA fragments separated by gel electrophoresis. J. Mol. Biol. 98, 503−517.

Stiff, C.M., Kilian, A., Zhou, H., Kurdna, D.A., Kleinhofs, A., 1995. Stable transformation of barley callus using biolistic particle bombardment and the phosphinotricin *acetyltransferase (bar)* gene. Plant Cell Tissue. Organ Cult. 40, 243−248.

Stotzky, G., 2001. Release, persistence, and biological activity in soil of insecticidal proteins from *Bacillus thuringiensis*. In: Letourneau, D.K., Burrows, B.E. (Eds.), Genetically Engineered Organisms: Assessing Environmental and Human Health Effects. CRC Press, Boca Raton, Fla, pp. 187−222.

Tabashnik, B.E., Van Rensburg, J.B.J., Carriere, Y., 2009. Field-evolved insect resistance to Bt crops: definition, theory and data. J. Econ. Entomol. 102, 2011–2025.

Torisky, R.S., Kovacs, L., Avdiushko, S., Newman, J.D., Hunt, A.G., Collins, G.B., 1997. Development of a binary vector system for plant transformation based on supervirulent *Agrobacterium tumefaciens* strain Chry5. Plant Cell. Rep. 17, 102–108.

Torres, A., Oliva, R.M., Castellano, D., Cross, P., 1996. First World Congress on Allelopathy. A Science of the Future. SAI (University of Cadiz), Spain, Cadiz, p. 278.

Traxler, G., Godoy-Avila, S., Falck-Zepeda, J., Espinoza-Arellano, J., 2003. Transgenic cotton in Mexico: a case study of the Comarca Lagunera. See Kalaitzandonakes 10, 183–202.

Tu, J., Datta, K., Khush, G.S., Zhang, Q., Datta, S.K., 2000. Field performance of Xa21 transgenic indica rice (Oryza sativa L.), IR72. Theor. Appl. Genet. 101, 15–20.

Tuteja, N., Verma, S., Sahoo, R.K., Raveendar, S., Reddy, I.N., 2012. Recent advances in development of marker-free transgenic plants: regulation and biosafety concern. J. Biosci. 37, 167–197.

Uchimiya, H., Fushimi, T., Hashimoto, H., Harada, H., Syono, K., Sugawara, Y., 1986. Expression of a foreign gene in callus derived from DNA-treated protoplasts of rice (*Oryza sativa* L.). Mol. Gen. Genet. 204, 204–207.

Van, L.M., Vanderhaeghen, R., Van, M.M., 1991. Insertional mutagenesis in *Arabidopsis thaliana*. Isolation of a T-DNA-linked mutation that alters leaf morphology. Theor. Appl. Genet. 81, 277–284.

Whiteley, H.R., Schanept, H.E., 1986. The molecular biology of parasporal crystal body formation in *Bacillus thuringiensis*. Ann. Rev. Microbiol. 40, 549–576.

Wu, G., Cui, H.R., Shu, Q., Xia, Y.W., Xiang, Y.B., Gao, M.W., et al., 2000. Stripped stem borer (*Chilo suppressalis*) resistant transgenic rice with a *cryIAb* gene from Bt (*Bacillus thuringiensis*) and its rapid screening. J. Zhejiang Univ. 19, 315–318.

Xia, J.Y., Cui, J.J., Ma, L.H., Dong, S.L., Cui, X.F., 1999. The role of transgenic cotton in integratede pest management. Acta Gossypii Sin. 11, 57–64.

Xu, J., Liu, Q., Yin, X., Zhu, S., 2006. A review of recent development of Bacillus thuringiensis ICP genetically engineered microbes. Entomol. J. East China 15, 53–58.

Xu, Z., He, Y., Zhu, C., Yu, G., 2005. Inhibitory effects of allelopathic rice materials on *Echinochloa crusgalli* and related field weeds. Ying Yong Sheng Tai Xue Bao (Chinese) 16, 726–731.

Yao, J.L., Tomes, S., Gleave, A.P., 2013. Transformation of apple (*Malus* x *domestica*) using mutants of apple acetolactate synthase as a selectable marker and analysis of the T-DNA integration sites. Plant Cell Rep. 32, 703–714.

Ye, G.Y., Shu, Q.Y., Yao, H.W., Cui, H.R., Cheng, X.Y., Hu, C., et al., 2001a. Field evaluation of resistance of transgenic rice containing a synthetic *cryIAb* gene from *Bacillus thuringiensis* Berliner. to two stem borers. J. Econ. Entomol. 94, 271–276.

Ye, G.Y., Tu, J., Datta, K., Datta, S.K., 2001b. Transgenic IR72 with fused Bt gene *cryIAb/cryIAc* from *Bacillus thuringiensis* is resistant against four lepidopteran species under field conditions. Plant Biotechnol. 18, 125–133.

Ye, G.Y., Colburn, S.M., Xu, C.W., Hajdukiewicz, P., Staub, J., 2003. Persistence of unselected transgenic DNA during a plastid transformation and segregation approach to herbicide resistance. Plant Physiol. 133, 402–410.

Yorobe Jr., J.M., Quicoy, C.B., 2006. Economic impact of Bt corn in the Philippines. Philipp. Agric. Sci. 89, 258–267.

Zhau, J., Cao, J., Li, Y., Collins, H.L., Roush, R.T., Earle, E.D., et al., 2003. Transgenic plants expressing two *Bacillus thuringiensis* toxins delay insect resistance evolution. Nat. Biotechnol. 21, 1493–1497.

Modern Tools for Enhancing Crop Adaptation to Climatic Changes

Kinza Waqar, Orooj Surriya, Fakiha Afzal, Ghulam Kubra, Shabina Iram, Muhammad Ashraf and Alvina Gul Kazi

7.1 Introduction

Over time, agriculturalists have embraced novel crop varieties and accustomed their practices according to the changes in the environment (Rotter et al., 2013). Gradual but permanent changes to the climate due to deforestation has led to haphazard rainfall patterns, extreme temperature conditions, drought, salinity, increased levels of carbon dioxide, flood, and pH, which has affected plant growth badly and in turn compromised crop yield (Redden, 2013). The rise in Earth's temperature, formally called "global warming," is one of the hottest issues of recent years, and one which has had an adverse effect on plants (Houghton, 2009). Extreme weather events are already affecting agricultural systems around the world and by 2050, the challenge is to provide food for the 9 billion estimated world population (Lal, 2013). An example of such an event is in Australia where after 10 years of drought and terrible floods in 2010 and 2011, the result has been the loss of nearly $6 billion in grain yields. These spontaneous weather changes and severe conditions are a threat to human lives such as starvation and poverty (Hatfield et al., 2011). Keeping in mind all of the above-mentioned problems, it is important to build modern agricultural tools to enhance crop adaptation to the changing environment and meet the world's need for food in the future. The weather has been in a state of fluctuation, but the present ratio of variation is much faster, and the weather variables are much wider (Lal, 2013). This chapter discusses various outcomes caused by climatic changes and methods for specific approaches to lessen its drastic effects.

7.2 Stresses caused by climatic changes

Due to climatic changes, extreme weather conditions such as high temperature, robust winds, hurricanes and tornadoes, etc., are predictable. Other abiotic stresses are also associated with climatic changes such as drought, floods, and increased levels of CO_2 or ozone. All of these can be harmful to plant growth and development (Ahmad and Prasad, 2012a,b; Ahmad et al., 2013a,b; Ahmad and Wani, 2013a,b; Hakeem et al., 2013). Furthermore, certain changes in climate lead to the proliferation of biotic pressures in farming organization including pests, weeds, bacteria, fungi, and viruses. It has been estimated that more than 50% of the crop yield is lost around the world due to environmental stresses (Bray et al., 2000), and also these lead to decreased soil fertility and increased soil

P. Ahmad (Ed): Emerging Technologies and Management of Crop Stress Tolerance, Volume 1.
DOI: http://dx.doi.org/10.1016/B978-0-12-800876-8.00007-2

erosion (Bray, 2002). Following are the major expected effects on agriculture due to climate change.

7.2.1 Drought

Farmers are already coping with drought and it has been responsible for many severe famines throughout history. It is estimated that due to drought, the yield from just over half of Earth's land will be decreased in 50 years' time (Cattivelli et al., 2008). The competition for water among cultivated and urban areas will complicate the problem of water accessibility. The usage of saline water might also help to ease Earth's water difficulties, but this is only feasible with the development of crops that are salt tolerant (Sinclair et al., 2004). In order to limit the problems of drought, there is an urgent need for crop diversity that can preserve water and maintain good harvest during phases of water shortage (Rosegrant et al., 2009). Development of such crops is rather difficult, because of the interchanging crop reaction systems to drought at the genomic, biological, metabolic, and biochemical levels. In order to make varieties that are drought tolerant and easily accessible to growers, interdisciplinary teams of scientists must cooperate (Kelly et al., 2005). Less crop yield and reduced grain size are the result of terminal drought (Vadez et al., 2011).

7.2.2 Temperature

The development and growth of all crops is influenced by temperature. The present heat range in the Midwest and southern USA is best for crop production, resulting in increased profit. Temperature ranges that are higher than ordinary are predicted to decrease grain legume and cereal yields (Hatfield et al., 2011). Raised temperatures decrease the pollen viability and grain-filling period (Boote and Sinclair, 2006). Furthermore, changes in temperature can lead to a warmer environment and less cold winters, which occasionally allow infections and pests to endure and hibernate, thus increasing the probability of subsequent reduced harvests (Ostberg et al., 2013). In areas of high latitude and elevation, where temperature is the major constraint, high temperatures will increase the growing seasons, which lead to late maturation. However, adapting the latest technologies in agriculture and research on the development of new and resistant crop varieties can help to lessen these effects.

7.2.3 Ozone

Ozone varies locally and is usually seen to have greater magnitude around industrial zones. Ozone is a significant agricultural contaminant and a greenhouse fume that keeps on increasing because of fossil petroleum ignition (Staehelin et al., 2001). Plants take in ozone during the process of photosynthesis, which leads to a drop in photosynthesis and quickens leaf expiry, disturbing crop development and output (Krupa et al., 2001). Being a very reactive molecule, ozone reacts with rubisco, which is the key enzyme of photosynthesis and thus destroys it. The frequency at which plants can take in ozone depends on the amount of ozone in the atmosphere as well as on the opening and closing of leaf pores or stomata (CCSP, 2008). In current studies, it is estimated that ozone is responsible for approximately a 10% decrease in yield of soybean and wheat, and about 3–5% of maize and rice.

New implements and diverse methods of crop propagation will help harvests to flourish in spite of contact with increasing ozone concentration in the air.

7.2.4 Increased carbon dioxide levels

Carbon dioxide (CO_2) levels increase consistently everywhere in the world, unlike ozone which varies locally. CO_2 is the major raw material of photosynthesis and its increased levels enhance photosynthesis resulting in efficient plant growth and yield (Battisti and Naylor, 2009). However, this is still debatable because plants may have low nutritional value including low protein content (Deressa and Hassan, 2009). It has also been found by researchers that poison ivy becomes more poisonous with higher levels of CO_2. Higher levels of atmospheric CO_2 make herbicides less effective, thus making plants more prone to damage caused by herbs (Taub, 2012).

7.2.5 Biotic factors

Biotic pressures on harvesting systems include pests, weeds, bacteria, fungi, and viruses. The chief element responsible for how pests and bugs affect crop manufacture and harvest is temperature (Coakley et al., 1999). For example, some populations of insect species, such as flea beetles, are showing signs of overwintering because of warmer winter temperatures, high moisture, and elevated atmospheric CO_2. Bacterial, fungal, and viral agents and certain weeds also proliferate significantly due to temperature, rainfall, and humidity. It is quiet challenging to make plants with increased resistance to such biotic factors (Juroszek and Tiedemann, 2011).

7.3 Modern tools for enhancing crop adaptation

Not all plants' responses to stresses are the same. It depends upon the degree of severity of climate change. To maintain sustainable yields, it is important for plants to have greater adaptability to the changing climate (Capiati et al., 2006). Research is ongoing in various domains such as plant physiology, cell physiology, plant breeding (both molecular breeding and conventional breeding), genetics, genomics and molecular biology in order to improve crop adaptations to climatic change for better yield and food quality (Bhatnagar-Mathur et al., 2008; Ahmad and Prasad, 2012a,b; Ahmad et al., 2013a,b; Ahmad and Wani 2013a,b). These approaches have led to the enhancement of plant genetic material with stress-tolerant and adaptive genes or alleles in the population (Redden, 2013).

Some of the methods to enhance crop adaptability are listed below.

7.3.1 Domestication bottlenecks

Domesticating crops for wild ancestral relatives is called domestication bottleneck (Abbo et al., 2003). It favors the transfer of a narrow range of genes, thus limiting genetic diversity (Xue and McIntyre, 2011). The process of domesticating bottleneck is done to increase the variability in the genome and to have more diversity (Ladizinsky, 1998). Researchers hunt for wild ancestors of a particular crop and then crosses are made with wild type. Selection is done on the basis of a crop with desirable traits as valuable genes are lost during the bottleneck process (Meyer et al., 2013). By this process we can have a variable number of traits in crops such as disease resistance, increased grain yield, and wider stress tolerance (Gur and Zamir, 2004).

7.3.2 Conventional breeding methods

Conventional breeding has been practiced for centuries, and is still frequently used today. In the past, farmers used to carry out cross-pollination or artificial mating of some plants in order to increase yield. Due to advancement in plant breeding, new and improved plant varieties were made by breeders, which not only increased plant growth efficiency but also plant worth for food, fiber, and feed. In order to have a better crop with greater adaptability to climatic changes, breeding is the best tool of the twenty-first century (Ainsworth et al., 2008). This will not only help crops to cope with stresses such as drought, high temperature, ozone, and carbon dioxide, but will also help in having greater yields (Bohnert et al., 2006) and limiting the use of chemical fertilizers both economically and ecologically (Subbarao et al., 2006).

Plant breeding can be defined as the recognition of required favorable characters, picking them, and then merging them into a discrete plant. The laws of Mendel's genetics gave technical foundation to plant breeding. As all characters of a plant are organized by genes, conventional breeding of plants is taken as chromosomal arrangement. Overall, there are three chief ways to work on plant chromosomal arrangement, namely, pure line selection, hybridization, and polyploidy. These methods and others are listed below.

7.3.2.1 Pure line selection

The oldest and most fundamental process in plant breeding is selection. It typically includes three diverse stages.

1. First, a huge number of selections are made from the novel population that is adjustable genetically.
2. Second, the development of progeny rows is made from the selection of separate plants for observational reasons. After clear eradication, varieties are matured over many years to certify observations of performance under diverse environmental circumstances for additional rejections.
3. Finally, the selected and inbred lines are associated with current profitable varieties in their yielding abilities and other features of agronomic significance (Jannink et al., 2010).

7.3.2.2 Hybridization

This is widely used in plant breeding methods. The main goal of hybridization is to bring all the desired qualities together that are found in diverse plant ranks into one plant via the process of cross-pollination. The first stage is to produce inbred lines that are homozygous; this is done by self-pollinating plants. Once a pure line is produced, it is outcrossed in order to make a progeny having a combination of the desired traits. The progeny made has desirable characteristics (e.g., disease-resistant trait) along with undesirable traits (e.g., bad taste, low nutritional value, and low yield). These disapproving characteristics must be detached by frequent crossing with the crop parent; this is called back-crossing (Holland et al., 2003). There are two kinds of hybrid plants: intergeneric and interspecific.

Heterosis is achieved by crossing highly inbred lines of crop plants. Inbreeding of most crops results in a strong decrease in size and vigor in the first generation. After six or seven generations, no additional decreasing in size or vigor is seen. The most successful and notable is the hybrid maize. The first profitable hybrid maize was available in the USA in 1919. Now, nearly all maize is hybrid, although the farmers have to buy new hybrid seed every year.

7.3.2.3 Polyploidy

Most of the existing plants are diploid. Polyploids are plants having three or more complete sets of chromosomes. This can be artificially manufactured by using the chemical colchicine, leading to doubling of the chromosomal number. Usually, the chief effect of polyploidy is increase in genetic variability and size. The disadvantages of polyploid plants are slow growth and lower fertility (Hijmans, 2011).

7.3.2.4 Mutation breeding

In the late 1920s, research showed that a large number of mutations and variations can be introduced in plants by exposing them to chemicals and X-rays. This practice was further developed after World War II. In order to check whether certain radiations cause valuable mutations, plants were irradiated with several radiations. Chemicals such as ethyl methanesulfonate and sodium azide were used to cause mutations. Some of the examples of plants that have been formed using this technique include barley, wheat, rice, potatoes, onions, and soybean (Chakraborty and Paul, 2013). In spite of solely trusting the introduction of genetic variability from wild plant genes or other cultivars, a good substitute is the mutation that is induced by radiation and chemicals. The mutants obtained are verified and are additionally nominated for wanted qualities. The place to induce mutation cannot be controlled while using chemicals or radiation because the countless mutants have undesirable characteristics. This technique has not been extensively used in breeding programs. In future, induced mutation technology will prove to be the gold standard for making crops adaptable to varying numbers of climatic changes. Almost 3000 mutants with desired genetic variance have been generated; these have many desirable traits including staple foods with reduced anti-nutrients, fruits with enhanced phytonutrients, increased protein, and high quality of starch and oil (Chakraborty and Paul, 2013).

7.3.2.5 Hybrid seed technology

The upgrading of seeds that are pollinated openly has resistance towards the infections, pests, and insects. Hybridized seeds are made by crossing pure lines. The pure lines cross to form a population, which takes several years to develop desirable characteristics. Mostly, this is made up of numerous former crossings in order to obtain necessary characters. The progeny thus formed is not mature until a genetically pure form is obtained. The qualities possessed by hybrid plants like trueness to type, good potency, high uniformity, and heavy yields make them desirable. They also have other characteristics like disease resistance, late senescence, good water holding ability, and insect resistance. However, these rewards come at an expense. An F_1 hybrid generation takes a long period of research to make pure lines. The farmer, in turn, is rewarded by greater harvests as well as restored value crops. Today, mostly all corn available and 50% of all rice are hybrids (Danida, 2002). China increased rice production from 140 million tons to 188 million tons with the help of hybrid rice technology. In terms of better plant features, vegetable breeders from tropical areas can celebrate some notable achievements:

1. Yield enhancement: Hybrids frequently yield 50–100% more when compared to old-fashioned OP ranges, leading to genetic disease-resistant plants, higher female/male flower ratios, and better-quality fruit setting under pressure.
2. Extended growing season: Hybrids often mature 15 days earlier in comparison to local OP diversities.
3. Quality improvement: Hybrids have assisted in a higher cost, more uniform and stable product.

7.3.2.5.1 Drawbacks of hybrid seed technology

Drawbacks of hybrid seed technology are that:

1. A pure line formation takes several years to develop.
2. If the first progeny hybrid seeds are being used for developing upcoming crops, the resultant floras are not able to perform well compared to the first generations, leading to less vigor or yields, thus the new F_1 seeds are needed to be purchased by the farmer every year.

7.3.3 Molecular breeding methods

Molecular and biochemical skills have shortened the time period of breeding programs from months to weeks and years to months. Breeding is merely the technique for selective coupling to blend wanted physiological, morphological, or genetic characters of the individuals of a population. This is done with the support of recognizable characters (Jacobsen et al., 2013). Naturally evident DNA markers are currently used effectively in plant breeding (Murphy et al., 2013). The sections of genome, which are used for recognition of the organism under question, are called as DNA markers (Brenner et al., 2013). They are naturally present tags that are attached to particular sections of a chromosome, which are related to particular phenotypes. It can either be situated within the desired gene or linked to a gene responsible for a characteristic of concern (Brenner et al., 2013). Thus, marker-assisted selection (MAS) is the selection of a characteristic built on a genotype using related markers (Foolad and Sharma, 2005). Selection is made possible for breeders by using markers in order to combine wanted alleles at larger amounts of loci and at earlier generations. The markers are valuable tools for pyramiding resistance genes of numerous kinds and also for multi-line cultivar development by aiming for long-lasting resistance to infection. This helps in the development of stress-tolerant crops having better yield tendency. Molecular breeding for cold tolerance has been performed by scientists on alfalfa (*Medicago sativa* L.), which is a legume, and was found to be freeze tolerant making it a better plant (Castonguay et al., 2012).

MAS is a procedure in which a marker is used for indirect selection of a genetic determinant of a feature (Mohan et al., 1997). This procedure is used in animal as well as plant breeding. Conventionally, breeders have trusted noticeable qualities to choose better-quality diversities, whereas MAS depends on recognizing marker DNA sequences inherited together with a wanted feature during the first few progenies. Subsequently, the floras carrying the characters are selected swiftly by observing the marker sequences and permitting multiple rounds of breeding (Kumar et al., 2007).

Some plants are in great danger of being attacked by diseases and pathogens. But plants have adopted certain behaviors to manage these advances, known as quantitative resistance or non-host resistance, e.g., resistance controlling genes (Silva et al., 2013). It is essential to look into the complete genome of such plants so that these various dangerous diseases can be managed. The study of genes helps us to understand the communication between different genetically controlled pathways by which we are able to know about the intricacies of resistance towards a disease. DNA markers have transformed the existing situation of plant breeds. They have a successful role in numerous areas like choice of donor parent in back-crossing, first generation choice, enhancement of difficult F_1s, choice and connection block analysis, and retrieval of recurrent parental genotype in back-crossing (Panigrahi, 2011). The following are the types of DNA markers:

- RFLP (restriction fragment polymorphism) markers
- RAPD (randomly amplified polymorphic DNA) markers
- AFLP (amplified fragment length polymorphism) markers
- SNP (single nucleotide polymorphism) markers
- SSRs (simple sequence repeats) markers

Each marker has its own usage and utility and is frequently used in research. Identification of QTLs (quantitative trait loci) underlying tolerance to stresses is an emerging technology. QTL mapping in many plants has been done against drought and salt stress (Foolad et al., 2003; Foolad, 2004; Gur and Zamir, 2004). In certain plants, tolerance to stress is developmental stage dependent. It is quantitatively inherited and its expression involves multiple genes. The environment has pronounced influence on the expression of traits that are stress tolerant. Merging the knowledge of phenotypes and genotypes, identification of such traits is done after calculating breeding values, which is specific to each marker. Selection of back-cross progeny can also be done by using this approach to check the presence of rare alleles (Rafalski, 2010). Markers can be used for the detection of the presence of various antioxidant genes in plants. Expression of these genes in plants leads to stress induced by reactive oxygen species (Ozgur et al., 2013).

7.3.4 **Plant tissue culture**

Assembly of procedures used to sustain and cultivate plant tissues, organs, or cells under sterile environment, on a culture medium of known composition, is called plant tissue culture. This is an extensively used technique to create clones of a plant using a process known as micro-propagation. The plant tissue culture has many advantages over old-style procedures, for example:

- They are able to produce exact copies of plants.
- They have a short plant growth period and mature plants are acquired early.
- They help in the redevelopment of entire plants using plant cells that are genetically improved.
- Plant production in sterile vessels significantly lessens the risks of pests, pathogens, and spreading diseases.

7.3.4.1 *Applications of tissue culture in crop improvement*
The most important applications of tissue culture are:

- Micro-propagation is extensively used in forestry. It is also used to protect rare plant species.

- It is used to produce plants that are disease free.
- Tissue culture can be used to screen cells by the plant breeder.
- Plant cells can be developed on a large scale in liquid culture in order to have valued compounds like recombinant proteins and secondary metabolites.
- Tissue culture can aid in the crossing of vaguely related species by regeneration and protoplast fusion of the original hybrid.
- Di-haploids can be made to achieve homozygous lines in breeding techniques.

7.3.5 Genetic engineering and genetically modified (GM) crops

Genetic engineering allows the introduction of foreign genes to plants with the help of techniques different from traditional breeding procedures. Gene engineering technology enhances the production of plants, as well as increases resistance to frost, pests, viruses, etc. (Ahmad et al., 2012c). Numerous genetic engineering diversities are on the market especially alfalfa, cotton, canola, soybean, squash, and maize (James, 2007). Currently, the characters introduced in these crops are effectively tolerant to pests and herbicides. Genetic engineering techniques extend the boundaries of possible profits and hazards. This technique has helped in food security and quality maintenance by making plants less susceptible to frost, drought, viruses, and insects, and also producing more competent plants that can strive against weeds to obtain soil nutrients. Food nutrient value has also been improved by altering their makeup. Nevertheless, the practice of biotechnology has increased anxieties of the possible danger to people and the environment. Commonly found fears among people are pest tolerance and issues related to health.

The United States obtained most of its ethanol from maize kernels, approximately 4 billion gallons per year (Klopfenstein et al., 2013). Currently, biofuel production is achieved by converting plant biomass to fermentable sugar using different pretreatment processes so that by disrupting the lignocellulose, lignin is removed and microbial enzymes are accessed for cellulose degradation. However, both pretreatment and enzyme production are costly. By using plant genetic engineering tools, biomass conversion costs could be reduced. Developing crop varieties with self-producing cellulase and ligninase, or plants that have increased cellulose or an overall biomass yield, could minimize costs (Sticklen, 2008).

Sugarcane has been genetically modified at the Sugarcane Research Institute in South Africa by recombinant DNA and *in vitro* culture technologies. The objective was to attain resistant forms and to look into the genetic basis of sucrose accretion. Besides this, the underlying aims are to develop genetic resources by isolating the appropriate gene along with its promoter and to optimize tissue culture engineering. The pioneering advancements include expressed sequence tag library analysis, which was made from sugarcane tissues (Deschamps and Campbell, 2010).

Plants become transgenic when their genetics is altered either by introducing a non-native gene or by bringing about a change in the internal gene sequence. This phenomenon is called recombinant DNA technology. The protein produced by the altered gene confers specific characteristics to the engineered plant. Genetic engineering has various applications, for example, to confer resistance to stresses, both biotic and abiotic (Ahmad et al., 2012c). The technology also proved useful to enhance the nutritional value of the plant; an approach particularly useful with respect to developing countries. Both food quality and health are strengthened by producing crop plants with disease-resistant genes. Development of new generation GM crops can be used for the manufacture of industrial products and recombinant medicines like vaccines, monoclonal antibodies, biofuels, and plastics (Sticklen, 2008). There are certain systemic risks associated with genetically modified organisms (GMOs) (Meyer, 2011). Table 7.1 summarizes the overview of deregulation from 1999 to 2009 in the USA.

Generally, tissue culture techniques transform the targeted plant cell into GM plants. There are some concerns regarding human health related to GM plants (Key et al., 2008):

- The usage of specific markers to recognize changed cells.
- Transmission of extraneous DNA into the genome of the plant.
- The likelihood of amplified mutations in GM plants associated with non-GM counterparts.

Table 7.1 Overview of Deregulated and Cultivated GM Traits in the USA 1992−2009

Phenotype	Plant Species	Number of Deregulated Traits	Transgenic Species in Cultivation
Herbicide tolerance		48	
Glyphosate	Canola, cotton, maize, soy, sugar beet	15	Yes
Glufosinate	Canola, cotton, maize, rice, soy, sugar beet	27	Yes, not all species
Others	Cotton, flax, maize, soy	6	Yes, not all species
Insect resistance		41	
Corn borer	Maize	26	Yes
Corn root worm	Maize	3	Yes
Colorado beetle	Potato	11	No
Other	Tomato	1	No
Altered fruit ripening		40	
Flavr Savr	Tomato	33	No
Other	Tomato	7	No
Virus resistance		12	
Papaya ring spot virus	Papaya	3	Yes
Cucumber mosaic virus, zucchini yellow mosaic virus, watermelon mosaic virus 2	Squash	2	Yes
Potato leaf roll virus, potato virus Y	Potato	5	No
Plum pox virus	Plum	1	No
Male sterility	Cichoria, canola, maize	8	No
Altered oil composition	Canola, soy	5	No
Higher lysin content	Maize	1	No
Lower nicotine content	Tobacco	1	No

Source: Meyer, 2011.

Approvals for the usage of genetic engineering for plant breeding entitle its accuracy. The assumptions regarding genetic engineering are given below that are essential for the regulation of biosafety rules:

1. Only known and specific genotypic changes are allowed in the engineered plant.
2. Only specific and known phenotypic changes are allowed in the engineered plant.

GMO plants started with the Flavr Savr tomato. Subsequently, genetic engineering agriculture really began with the making of Bt cotton. Roundup Ready soybeans are exported as an elementary constituent for the food industry and started the global debate on the usage of GM crops. Fifteen countries grow more than 50,000 hectares of GM crops today (James, 2007; Ahmad et al., 2012c).

7.3.5.1 Importance of *Arabidopsis thaliana* in transgenesis
Because of the simple genome, *Arabidopsis thaliana* has become the most important model plant in genomics. Some of the examples given below reveal its importance:

1. Expression of stress-responsive genes such as salt, freezing, and drought have been induced in various crops by binding of transcription factors, i.e., *DREB1* and *DREB2* (dehydration-responsive element binding) to their respective promoters. This led to the expression of *Arabidopsis* gene rd29A, which is a responsive gene to dehydration (Ortiz et al., 2007 and references therein).
2. Transgenic rice was produced with reduced transpiration, enhanced photosynthesis, increased water usage efficacy, and high biomass production versus water usage by the expression of the *HARDY* or *HRD* gene from *Arabidopsis thaliana* (Karaba et al., 2007). The resultant rice was drought tolerant with increased roots and shoot biomass. A list of different transgenic plants with tolerance to various stresses is summarized in Table 7.2.

7.3.6 EST sequencing and NGS
EST (expressed sequence tags) sequencing and NGS (next generation sequencing) have greatly improved the efficiency of investigating the presence of stress-tolerant genes. By using these approaches, genes with the following traits have been identified successfully (Xue and McIntyre, 2011):

- Transpiration efficacy
- Osmotic modification
- Drought tolerance
- Heat tolerance
- Photosynthesis maintenance

NGS gives information about the presence and localization of a large amount of ESTs in genomes, which have proved to be very helpful in mapping (Xue et al., 2008). NGS also discovers single nucleotide polymorphisms in the genome, which can be employed for association mapping and alien introgression (Varshney et al., 2009).

Table 7.2 List of Different Transgenic Plants with Improved Characters

Transgenic Plant	Gene Induced	Effect(s) of the Expression of Transgenesis	References
Tobacco		Suppression of drought-induced leaf senescence Drought tolerance Minimal yield loss	Rivero et al. (2007)
Maize	*Escherichia coli*'s glutamate dehydrogenase (*gdhA*) gene	Drought tolerance Increased germination and grain biomass production	Lightfoot et al. (2007)
Multiple plant species	Cold shock proteins (CSPs) from bacteria, CspA from *E. coli*, and CspB from *Bacillus subtilis*	Cold tolerance Other stress tolerance	Castiglioni et al. (2008)
Maize	Phospholipase C1 gene (*ZmPLC1*)	Drought tolerance Better osmotic adjustment Increased photosynthesis rates Lower percentage of ion leakage Less lipid membrane peroxidation Higher grain yield	Wang et al. (2008)
Arabidopsis thaliana	Dof1transcription factor induced the up regulation of genes encoding enzymes for carbon skeleton	Marked increase of amino acids Increased N assimilation Improved growth under a low N	Yanagisawa et al. (2004)

7.4 Conclusion and future prospects

Developing a new crop with improved adaptability is very laborious, expensive, and time consuming. It takes more than a decade to make a cultivar available to agriculturists. From discovering the trait to field tests involves a lot of other developmental phases. Before coming to market, all risk factors including environment, human health, safety issues, and biodiversity are assessed. Changing climate is always a hot issue for scientists in terms of crop growth as it badly affects the yield. Scientists have been busy improving the existing breeding technologies and other high-level technologies such as NGS to produce germplasm with greater adaptability to the climate. New tools in molecular biology have greatly improved the efficiency of finding underlying stress-tolerant genes. With the world's changing climate and exponential increase of population, it is very important to meet food needs and hunger in the coming years. This can be done by making crops more adaptable to the climate—otherwise people will starve.

There is a dire need to develop different programs and organizations to encourage research for the development of stress-tolerant cultivars. One such program is the CGIAR Research Program on Climate Change, Agriculture and Food Security (CCAFS), which is working efficiently in this regard. More work is needed to explore the plant genome to obtain more tolerant traits to improve their growth under various climatic stresses. Due to the use of a broad range of chemicals, pests and other disease-causing organisms have developed resistance to many pest control and disease control agents, therefore it is important to explore the natural resistance in plants by advanced immunology. Management of soil can greatly enhance plant nutritional value. In the future, our climate will become harsher and there is a strict need for extensive research work to develop more tolerant varieties.

References

Abbo, S., Berger, J., Turner, N.C., 2003. Evolution of cultivated chickpea: four bottlenecks limit diversity and constrain adaptation. Funct. Plant Biol. 30, 1081–1087.

Ahmad, P., Prasad, M.N.V., 2012a. Environmental Adaptations and Stress Tolerance in Plants in the Era of Climate Change. Springer Science + Business Media, LLC, New York, NY, USA.

Ahmad, P., Prasad, M.N.V., 2012b. Abiotic Stress Responses in Plants: Metabolism, Productivity and Sustainability. Springer Science + Business Media, LLC, New York, NY, USA.

Ahmad, P., Wani, M.R., 2013a. Physiological Mechanism and Adaptation Strategies in Plants Under Changing Environment, vol. I. Springer Science + Business Media, LLC, New York, NY, USA.

Ahmad, P., Wani, M.R., 2013b. Physiological Mechanism and Adaptation Strategies in Plants Under Changing Environment, vol. II. Springer Science + Business Media, LLC, New York, NY, USA.

Ahmad, P., Azooz, M.M., Prasad, M.N.V., 2013a. Ecophysiology and Responses of Plants Under Salt Stress. Springer Science + Business Media, LLC, New York, NY, USA.

Ahmad, P., Azooz, M.M., Prasad, M.N.V., 2013b. Salt Stress in Plants: Signalling, Omics and Adaptations. Springer + Science Business Media, LLC, New York, NY, USA.

Ahmad, P., Ashraf, M., Younis, M., Hu, X., Kumar, A., Akram, N.A., et al., 2012c. Role of transgenic plants in agriculture and biopharming. Biotechnol. Adv. 30, 524–540.

Ainsworth, E., Rogers, A., Leakey, A.D.B., 2008. Targets for crop biotechnology in a future high-CO_2 and high O_3 world. Plant Physiol. 147, 13–19.

Battisti, D.S., Naylor, R.L., 2009. Historical warnings of future food insecurity with unprecedented seasonal heat. Science 323, 240–244.

Bhatnagar-Mathur, P., Vadez, V., Sharma, K.K., 2008. Transgenic approaches for abiotic stress tolerance in plants: retrospect and prospects. Plant Cell Rep. 27, 411–424.

Bohnert, H.J., Gong, Q., Li, P., Ma, S., 2006. Unravelling abiotic stress tolerance mechanisms getting genomics going. Curr. Opin. Plant Biol. 9, 180–188.

Boote, K.J., Sinclair, T.R., 2006. Crop physiology: significant discoveries and our changing perspective on research. Crop. Sci. 46, 2270–2277.

Bray, E.A., 2002. Abscisic acid regulation of gene expression during water-deficit stress in the era of the *Arabidopsis* genome. Plant Cell Environ. 25, 153–161.

Bray, E.A., Bailey-Serres, J., Weretilnyk, E., 2000. Responses to abiotic stresses. In: Gruissem, W., Buchannan, B., Jones, R. (Eds.), Biochemistry and Molecular Biology of Plants. John Wiley and Sons, Rockville, Maryland, pp. 1158–1249.

Brenner, E.A., Beavis, W.D., Andersen, J.R., Lubberstedt, T., 2013. Prospects and limitations for development and application of functional markers in plants. Diagn. Plant Breed 329–346.

Capiati, D.A., País, S.M., Téllez-Iñón, M.T., 2006. Wounding increases salt tolerance in tomato plants: evidence on the participation of calmodulin-like activities in cross-tolerance signaling. J. Exp. Bot. 57, 2391–2400.

Castiglioni, P., Warner, D., Bensen, R.J., Anstrom, D.C., Harrison, J., Stoecker, M., et al., 2008. Bacterial RNA chaperones confer abiotic stress tolerance in plants and improved grain yield in maize under water-limited conditions. Plant Physiol. 147, 446–455.

Castonguay, Y., Dubé, M.P., Cloutier, J., Bertrand, A., Michaud, R., Laberge, S., 2012. Molecular physiology and breeding at the crossroads of cold hardiness improvement. Physiol. Plant 147, 64–74.

Cattivelli, L.F., Rizza, F.W., Badeck, E., Mazzucotelli, A.N., Mastrangelo, E., Francia, C., et al., 2008. Drought tolerance improvement in crop plants: an integrated view from breeding to genomics. Field Crops Res. 105, 1–14.

CCSP (Climate Change and Science Program) (2008) The effects of climate change on agriculture, land resources, water resources, and biodiversity. A Report by the U.S. Climate Change Science Program and the Sub-committee on Global Warming. US Climate Change and Science Program, USA.

Chakraborty, N.R., Paul, A., 2013. Role of induced mutations for enhancing nutrition quality and production of food. Intl. J. Bio-res. Stress Manag. 4, 014–019.

Coakley, S.M., Scherm, H., Chakraborty, S., 1999. Climate change and plant disease management. Annu. Rev. Phytopathol. 37, 399–426.

DANIDA, 2002. Assessment of Potentials and Constraints for Development and use of Plant Biotechnology in Relation to Plant Breeding and Crop Production in Developing Countries. Working Paper. Ministry of Foreign Affairs, Denmark.

Deressa, T.T., Hassan, R.M., 2009. Economic impact of climate change on crop production in Ethiopia: evidence from cross-section measures. J. Afr. Econ. 18, 529–554.

Deschamps, S., Campbell, M.A., 2010. Utilization of next-generation sequencing platforms in plant genomics and genetic variant discovery. Mol. Breed 25, 553–570.

Foolad, M.R., 2004. Recent advances in genetics of salt tolerance in tomato. Plant Cell Tissue Organ Culture 76, 101–119.

Foolad, M.R., Sharma, A., 2005. Molecular markers as selection tools in tomato breeding. Acta Hort. 695, 225–240.

Foolad, M.R., Zhang, L.P., Subbiah, P., 2003. Genetics of drought tolerance during seed germination in tomato: inheritance and QTL mapping. Genome 46, 536–545.

Gur, A., Zamir, D., 2004. Unused natural variation can lift yield barriers in plant breeding. Plant Biol. 2, 1610–1615.

Hakeem, K.R., Ahmad, P., Ozturk, M., 2013. Crop Improvement—New approaches and modern techniques. Springer Science + Business Media, LLC, New York, NY, USA.

Hatfield, J.L., Boote, K.J., Kimball, B.A., Ziska, L.H., Izaurralde, R.C., Ort, D., et al., 2011. Climatic impacts on agriculture: implications for crop production. Agron. J. 103, 351–370.

Hijmans, R.J., 2011. Comment on "Changes in climatic water balance drive downhill shifts in plant species' optimum elevations." Science 334, 177, author reply 177.

Holland, J.B., Nyquist, W.E., Cervantes-Martinez, C.T., 2003. Estimating and interpreting heritability for plant breeding: an update. Plant Breed Rev. 22, 9–112.

Houghton, J., 2009. Global warming: the complete briefing, 4. Cambridge University Press, Cambridge, 3–7.

Jacobsen, S.R., Sorenson, M., Pedersen, S.M., Weiner, J., 2013. Feeding the world: genetically modified crops versus agricultural biodiversity. Agron Sustain Dev. 10, 593–613.

James, C., 2007. Global Status of Commercialized Biotech/GM Crops. ISAAA Brief Summary 37. International Service for the Acquisition of Agri-Biotech Applications, Ithaca, New York.

Jannink, J.L., Lorenz, A.J., Iwata, H., 2010. Genomic selection in plant breeding: from theory to practice. Brief Funct. Gen. 9, 166–177.

Juroszek, P., Tiedemann, A.V., 2011. Potential strategies and future requirements for plant disease management under a changing climate. Plant Pathol. 60, 100–112.

Karaba, A., Dixit, S., Greco, R., Aharoni, A., Trijatmiko, K.R., Marsch-Martinez, N., et al., 2007. Improvement of water use efficiency in rice by expression of *HARDY*, an Arabidopsis drought and salt tolerance gene. Proc. Natl. Acad. Sci. USA 104, 15270–15275.

Kelly, D.L., Kolstad, C.D., Mitchell, G.T., 2005. Adjustment costs from environmental change. J. Environ. Econ. Manage 50, 468–495.

Key, S., Ma, J.K., Drake, P.M.W., 2008. Genetically modified plants and human health. J. R. Soc. Med. 8, 101–290.

Klopfenstein, T.J., Erickson, G.E., Berger, L.L., 2013. Maize is a critically important source of food, feed, energy and forage in the USA. Field Crops Res. 153, 5–11.

Krupa, S., McGrath, M.T., Andersen, C.P., Booker, F.L., Burkey, K.O., Chappelka, A.H., et al., 2001. Ambient ozone and plant health. Plant Dis. 85, 4–12.

Kumar, S., Kumar, M., Kumar, Mukesh, Y.K., Manoj, Kumar, et al., 2007. Integration of conventional and non-conventional breeding approaches for seed spices improvement. Proc National Workshop on Spices and Aromatic Plants 14–20.

Ladizinsky, G., 1998. The course of reducing and maintaining genetic diversity under domestication. In: Ladizinsky, G. (Ed.), Plant Evolution under Domestication. Kluwer Academic Publishers, The Netherlands, pp. 113–126.

Lal, R., 2013. Food security in a changing climate. Ecohydrol. Hydrobiol. 13, 8–21.

Lightfoot, D.A., Mungur, R., Ameziane, R., Nolte, S., Long, L., Bernhard, K., et al., 2007. Improved drought tolerance of transgenic *Zea mays* plants that express the glutamate dehydrogenase gene (*gdhA*) of *E. coli*. Euphytica 156, 103–116.

Meyer, H., 2011. Systemic risks of genetically modified crops: the need for new approaches to risk assessment. Env. Sci. Europe 23, 7.

Meyer, R.S., DuVal, A.E., Jensen, H.R., 2013. Patterns and processes in crop domestication: an analysis of 203 global food crops. New Phytol. 196, 29–48.

Mohan, M., Nair, S., Bhagwat, A., Krishna, T.G., Yano, M., Bhatia, C.R., et al., 1997. Genome mapping, molecular markers and marker-assisted selection in crop plants. Mol. Breed 3, 87–103.

Murphy, K.M., Carter, A.H., Jones, S.S., 2013. Evolutionary breeding and climate change. In: Kole, C. (Ed.), Genomics and Breeding for Climate-Resilient Crops. Springer, Heidelberg NY, Dordrecht, London, pp. 377–389.

Ortiz, R., Iwanaga, M., Reynolds, M.P., Wu, H., Crouch, J.H., 2007. Overview on crop genetic engineering for drought-prone environments. J. SAT. Agric. Res. 4.

Ostberg, S., Lucht, W., Schaphoff, S., Gerten, D., 2013. Critical impacts of global warming on land ecosystems. Earth Sys. Dyn. Dis. 4, 541–565.

Ozgur, R., Uzilday, B., Sekmen, A., Turkan, I., 2013. Reactive oxygen species regulation and antioxidant defence in halophytes. Funct. Plant Biol. In press.

Panigrahi, J., 2011. Molecular mapping and map based cloning of genes in plants. In: Thangadurai, D. (Ed.), Plant Biotechnology and Transgenics. Bentham Science Publishers, USA.

Rafalski, J.A., 2010. Association genetics in crop improvement. Curr. Opin. Plant Biol. 13, 174–180.

Redden, R., 2013. New approaches for crop genetic adaptation to the abiotic stresses predicted with climate change. Agronomy 3, 419–432.

Rivero, R.M., Kojima, M., Gepstein, A., Sakakibara, H., Mittler, R., Gepstein, S., et al., 2007. Delayed leaf senescence induces extreme drought tolerance in a flowering plant. Proc. Natl. Acad. Sci. USA 104, 19631–19636.

Rosegrant, et al., Building Climate Resilience in the Agriculture Sector. Asian Development Bank and International Food Policy Research Institute, Manila, Philippines and Washington, DC.

Rotter, R.P., Hohn, J.G., Fronzek, S., 2013. Projections of climate change impacts on crop production: a global and a Nordic perspective. Acta Agr. Scand. 62, 166–188.

Silva, V.M.P., Menezes-Junior, J.A.N., Carneiro, P.C.S., Carneiro, J.E.S., Cruz, C.D., 2013. Genetic improvement of plant architecture in the common bean. Genet. Mol. Res. 217–233.

Sinclair, T.R., Purcell, L.C., Sneller, C.H., 2004. Crop transformation and the challenge to increase yield potential. Trends Plant Sci. 9, 70–75.

Staehelin, J., Harris, N., Appenzeller, C., Eberhard, J., 2001. Ozone trends: a review. Rev. Geophys. 39, 231–290.

Sticklen, M.B., 2008. Plant genetic engineering for biofuel production: towards affordable cellulosic ethanol. Nature Rev. Gen. 9, 433–443.

Subbarao, G.V., Ito, O., Berry, W., Sahrawat, K.L., Rondon, M., Rao, I.M., et al., 2006. Scope and strategies for regulation of nitrification in agricultural systems—challenges and opportunities. Crit. Rev. Plant Sci. 25, 1–33.

Taub, D.R., 2012. Effects of rising atmospheric concentrations of carbon dioxide on plants. Intl. J. Bio-res. stress Manag. 4, 014–019.

Vadez, V., Kholova, J., Choudhary, S., Zindy, P., Terrier, M., Krishnamurthy, L., et al., 2011. Responses to increased moisture stress and extremes: whole plant response to drought under climate change. In: Yadav, S.S., Redden, R.J., Hatfield, J.L., Lotze-Campen, H., Hall, A.E. (Eds.), Crop Adaptation to Climate Change. Wiley-Blackwell, UK, pp. 186–197.

Varshney, R.K., Nayak, S.N., Gregory, D., May, G.D., Jackson, S.A., 2009. Next-generation sequencing technologies and their implications for crop genetics and breeding. Trends Biotechnol. 27, 522–530.

Wang, C.R., Yang, A.F., Yue, G.D., Gao, Q., Yin, H.Y., Zhang, J.R., 2008. Enhanced expression of phospholipase C1 (*ZmPLC1*) improves drought tolerance in transgenic maize. Planta 9, 22–35.

Xue, G.P., McIntyre, C.L., 2011. Wild relative and transgenic innovation for enhancing crop adaptation to warmer and drier climate. In: Yadav, S.S., Redden, R.J., Hatfield, J.L., Lotze-Campen, H., Hall, A.E. (Eds.), Crop adaptation to climate change Wiley-Blackwell, UK. pp. 522–545.

Xue, G.P., McIntyre, C.L., Jenkins, C.L.D., Glassop, D., Herwaarden, V.A.F., Shorter, R., 2008. Molecular dissection of variation in carbohydrate metabolism related to water soluble carbohydrate accumulation in stems of wheat (*Triticum. aestivum* L.). Plant Physiol. 146, 441–454.

Yanagisawa, S., Akiyama, A., Kisaka, H., Uchimiya, H., Miwa, T., 2004. Metabolic engineering with Dof1 transcription factor in plants: improved nitrogen assimilation and growth under low-nitrogen conditions. Proc. Natl. Acad. Sci. USA 101, 7833–7838.

Interactions of Nanoparticles with Plants: An Emerging Prospective in the Agriculture Industry

Vani Mishra, Rohit K. Mishra, Anupam Dikshit and Avinash C. Pandey

8.1 Introduction

Agriculture forms the backbone of the economy of most countries and is considered the fundamental contributor to their overall growth, industrialization, and modernization. Agriculture has made remarkable advances over the past decades, but its sustainable development is required to support the ever-increasing population. Among the latest trends in technological advancements, nanotechnology has opened up new avenues in the field of crop improvement and food processing. Nanoparticles (NPs) are classified as aggregates or components ranging between 1 and 100 nm (Ball, 2002; Roco, 2003), which have specific physicochemical properties such as strength, optical, and electrical features as compared to their bulk counterparts (Nel et al., 2006).

NPs with a high surface/volume ratio and high reactivity are considered as the building blocks of nanotechnology. They possess the ability to cross cell walls and plasma membranes (Stern and McNeil, 2008; Farre et al., 2011) and may bear intrinsic toxicity upon their surface (Donaldson et al., 2004; Das and Ansari, 2009). It is worth noting that these extraordinary properties differentiate them from bulk materials and bring about characteristic environmental fate and behaviors. Nanotechnology empowers agricultural researches, by a broad range of advancements in reproductive science and technology, transformation of wastes from agricultural and food products to energy and other useful byproducts through nano-enzymatic processing, and control of diseases and their treatment in plants through using various NPs (Carmen et al., 2003). Nanotechnology could provide possible solutions to many major risks in agriculture. It can enhance our understanding of various crops and their biology to potentially enhance their yield and nutritional value with control over plant diseases and pest incidences (Nair and Kumar, 2013).

8.2 Classification of nanoparticles

NPs can be classified on the basis of their dimension, morphology, content, conformity, and agglomeration (Buzea et al., 2007). Based on their characteristic features, NPs can acquire various shapes, namely, spherical, tubular, and irregular, and can also be found in fused aggregates or agglomerates (Nowack and Bucheli, 2007). Dimensionality is the shape and morphology of NPs

P. Ahmad (Ed): Emerging Technologies and Management of Crop Stress Tolerance, Volume 1.
DOI: http://dx.doi.org/10.1016/B978-0-12-800876-8.00008-4

upon which they are accredited with a number of dimensions, namely, one-dimensional (1D), two-dimensional (2D), and three-dimensional (3D). Further, on the basis of their origin, another class of distinction has been made between NSPs:

1. **Natural NPs** date back to the origin of Earth and still exist in the current environment; examples are: volcanic dust, mineral composites, lunar dust, etc.
2. **Incidental NPs**, also referred to as anthropogenic particles, are the result of human actions like industrial processes or household applications (diesel exhaust, welding fumes, coal combustion, etc.)
3. **Engineered or designed NPs** can be categorized into following types:
    a. **Carbon based particles:** these are the most abundant NPs and comprise fullerenes, single-walled carbon nanotubes (SWCNT), and multi-walled carbon nanotubes (MWCNT) (Ma et al., 2010).
    b. **Metal-based particles:** these include quantum dots (Qds), gold (Au)NPs, zinc (Zn)NPs, aluminum (Al)NPs, and metal oxides like TiO_2, ZnO, and Al_2O_3, which have been variously used for their catalytic support in the field of heterogeneous catalysis (Biener et al., 2005).
    c. **Dendrimers:** these are characterized as nano-sized polymers made up of branched subunits, and when efficiently engineered they can perform specific chemical functions.
    d. **Composites:** these are a combination of a nanoparticle of one kind and that of other NPs or with larger bulk-type materials (Lin and Xing, 2007) and present different morphologies such as spheres, tubes, rods, and prisms (Yu-Nam and Lead, 2008).

8.3 Applications of NPs

Nanomaterial technology has reached new heights by serving the fields of medicine, drug development, the information and communication sector, and many more. This could be attributed to the characteristic features of NPs that include small size, shape, and larger surface area to mass ratio. NPs have been diversely used in innumerable ways as summarized in Table 8.1 (The Royal Society, 2004; Biswas and Wu, 2005).

The extensive advancement of nanomaterial technologies has caught the attention of people from public health interventions towards the potent risk induced by NPs into the environment and human health. Thus, a stage has been set for a huge debate over the benefits and risks of engineered NPs (USEPA, 2007). The literature on the ecotoxicity of NPs and nanomaterials as well as the chemistry of both manufactured and natural NSPs are summarized in recent reports (Handy et al., 2008; Yu-Nam and Lead, 2008).

Since NPs have been excessively and routinely used by the consumers, it has become evident that NPs will pave their way into all sorts of environments (Bandyopadhyay et al., 2013), including aquatic, terrestrial, and atmospheric, where their interaction and behavior with the surroundings and their subsequent fate remain unexplored. These features, therefore, bring the closely related organisms under surveillance for studying interactions with NPs. Navarro et al. (2008) highlighted three topics: (1) sources, transformation, and fate of NPs; (2) biotransformation of NPs that includes their entrance and fate when administered to algae, fungi, or higher plants; and (3) the toxicological

Table 8.1 Nanoparticles and their Applications	
Products	**Nanoparticles**
Cosmetics and sunscreens	Zinc oxide and titanium dioxide
Lipsticks	Iron oxide
Detergents and shampoos	Alumina
Ceramic tiles and window panes	Titanium oxides
Medical implants	Selenium
Food packaging	Zinc, magnesium, silica
Biosensors	Silver and Gold
Water treatment technology	Silver
Paints	Vanadium pentoxide
Batteries and displays	Silicon
Fuel additives	Palladium and cobalt
Military battle suits	CNTs as reinforced fibers

impact of NPs on organisms, their specific target sites and mode of action, and further their transfer to other organisms though the food chain, thereby influencing the entire ecosystem. There has been an increasing amount of research on the interaction of NPs with animals, microorganisms, etc., and limited studies are available highlighting higher plant and NP interactions. This review, therefore, deals with the effect of NPs on the plant kingdom.

8.4 Plant—nanoparticle interactions: yet to reach "the state of art"

Plants are one of the fundamental components of the environment and perform essential functions in maintaining equilibrium across the ecosystem including the food chain and food web through the translocation of minerals and nutrients. Higher plants work in network with their soil, water, and environmental chambers making way for NPs through all the specified routes (Monica and Cremonini, 2009). In addition to NPs destined for interaction with plants, the latter are also subjected to extensive human manipulations that unknowingly add a spectrum of engineered NPs (ENPs). The various routes that engineered NPs can acquire to reach the plants include direct application, release via accidents, contaminations in soil, sediments or water, and atmospheric outcomes.

Even though studies have been performed over plant—NP interactions, no clear understanding has been substantiated defining their impact on the overall development of plant. Plants are reported to take up different essential as well as non-essential elements in various concentrations under specific growth conditions, above which they may pose toxic threats to plants (Ke et al., 2007). Scientists globally are divided into groups over the impact of NPs over plants. Accumulated literature indicates both beneficial as well as toxic effects of NPs. For example, Canas et al. (2008) reported that SWCNTs increased the root length in onion and cucumber, whereas reduction was observed in tomato over the same size range. Similarly, ZnO was found to play contradictory roles in soybean where it increased root length (Lopez-Moreno et al., 2010), while shrunken root tip and

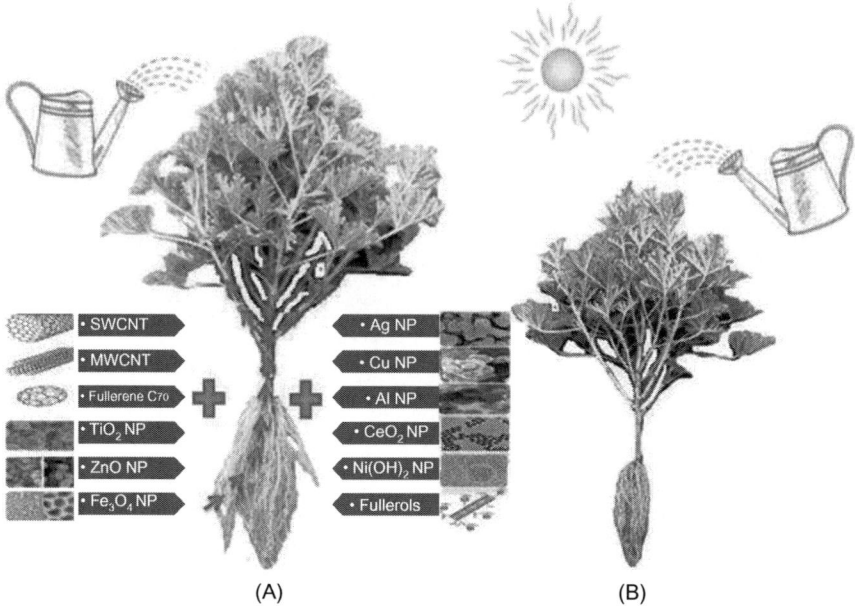

FIGURE 8.1

Effect of different NPs in plant growth. (A) Representing enhanced growth of plant treated with NPs in addition to other basic requirements in comparison to (B) plant representing growth in absence of NPs.

broken root caps were observed in ryegrass (Lin and Xing, 2007). This review will, however, focus on the NPs and the conditions that have been attributed to the positive impact over crop improvement (Figure 8.1).

As the interactions between NPs and vascular plants have become inevitable, it becomes important to resolve the highly convoluted modes of action of NPs following their uptake and translocation into plants, such that the information could be utilized to accelerate further developments in plant biotechnology.

8.5 Mode of nanoparticle internalization by plants

As already stated, shape, size, chemical entities, stability, and functionalization of NPs influence the uptake, translocation, and accumulation; properties are also found to be variably affected by plant type, species, and site facilitating internalization of NPs. The plant cell wall is the primary site of interaction with the external environment, which acts as a barrier and does not allow any alien particles including nanomaterials to access plant cells easily. The active functionalized sites among the plant cell wall components include carboxylate, phosphate, hydroxyl, amine, sulfhydryl, and imidazole functional groups (Vinopal et al., 2007) that interlink themselves to form complex biomolecules like cellulose, carbohydrates, and proteins (Knox, 1995) and facilitate the selective

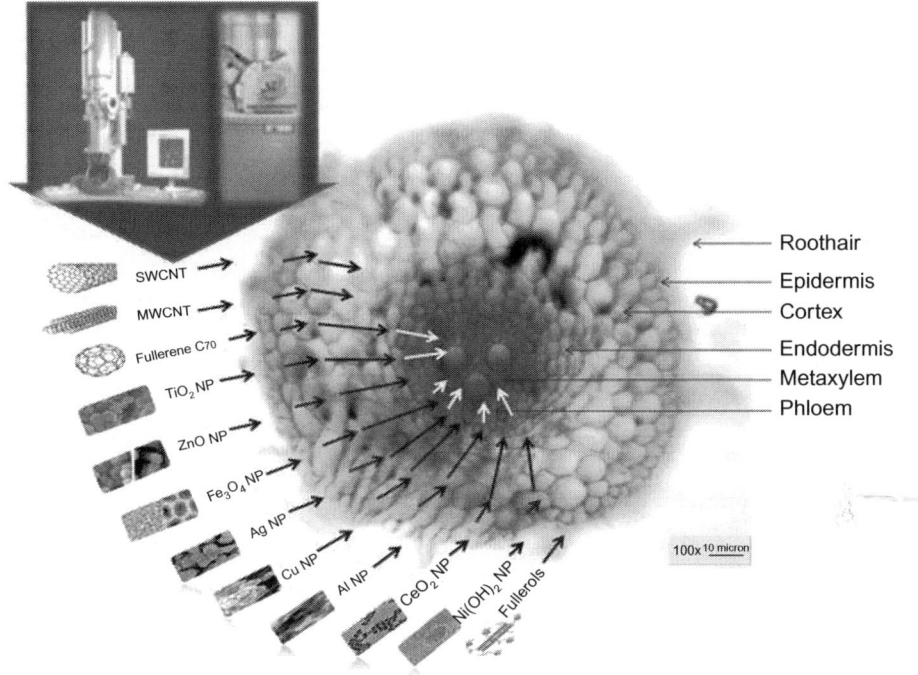

FIGURE 8.2

Uptake of differential NPs by roots. Transverse cross-section representing uptake, distribution, and accumulation of NPs in different zones of root.

uptake of NPs. The differential or selective uptake and translocation of NPs by the cell wall is due to its semipermeable property that allows the small-sized particles to move through them and sieve out the larger particles and limit their entry into the plant system. Thus, the pore size of the cell wall with a variable diameter in the range of 5 to 20 nm provides the plant with a sieving property and the exclusive basis of NP uptake (Fleischer et al., 1999). Thus, NPs with a diameter comparable to the pore size of the cell wall can penetrate through them and reach the plasma membrane (Moore, 2006; Navarro et al., 2008). Various routes of NP uptake through the root cell wall are depicted in Figure 8.2.

Formation of new cell wall pores during the reproduction or the enlargement of already existing pores under the influence of NPs has also been observed, which subsequently makes the cell wall more permeable and enhances nanoparticle uptake (Wessels, 1993; Ovecka et al., 2005). After penetrating into the cell wall, internalization further occurs through endocytosis in which the NPs are surrounded by the plasma membrane in the form of a cavity-like structure. Aquaporins, ion channels, or the organic chemicals in the environmental media that are the membrane-embedded carrier proteins responsible for transportation also bind with the NPs and facilitate their entrance into the plant. Another route of transportation includes the complex formation of NPs with root exudates or membrane transporters (Watanabe et al., 2008; Kurepa et al., 2010). Hall and Williams (2003) have identified a variety of ion transporters for diverse families of NPs. Following their entry into the cells, NPs follow either an apoplastic or symplastic mode of transportation. Cell-to-cell transportation occurs through plasmodesmata.

In the cytoplasm, the NPs interact and may bind to different organelles in various ways to further interact with the plant metabolism in both positive as well as negative ways (Jia et al., 2005). NPs when administered over the leaf surfaces percolate into various tissues via stomatal openings or bases of trichomes (Eichert et al., 2008; Uzu et al., 2010). The stomatal openings in photosynthetic areas are obstructed due to the accumulation of NPs, resulting in heating of foliar chambers thereby causing alterations in gas exchange and further modifying physiological and cellular functions of plants (Da Silva et al., 2006). Ao et al. (2013) reported novel conjugated nanospheres derived from 1-naphthylacetic acid (NNA), 3-aminopropyltriethoxysilane (APTES), and tetraethyl orthosilicate as nano-sized controlled-release formulations (CRFs) for plant growth. However, the exact mechanisms of selective uptake of NPs by different plant species is still unknown and needs to be explore.

8.5.1 Carbon-based NPs: uptake, translocation and accumulation in plants

With recent advancements in technologies involving carbon nanotubes (CNTs) as carriers for biomolecules/genes/drugs into the cells, experiments are being performed to find the mechanism behind the process of uptake and transport of CNTs into the plant tissues (Rico et al., 2011). Larger particles of single-walled carbon nanotubes (SWCNTs) fail to penetrate cell walls, but experiments on an *Arabidopsis thaliana* leaf cell report formation of an endocytosis-like structure in the plasma membrane (Shen et al., 2010). As a model plant for studying nanomaterial uptake by plants and their metabolism, the observations made in *Arabidopsis* arouse interest and encourage more relevant and significant research with edible plants. Water-soluble SWCNTs (<500 nm in length) when applied on *Nicotiana tabacum* cv. could penetrate the integral and rigid cell wall and the plasma membrane and Bright Yellow (BY-2) through fluidic phase endocytosis (Liu et al., 2009). Multi-walled carbon nanotubes (MWCNTs) uptake through the seed and root systems of various plants has been reported by many researchers (Khodakovskaya et al., 2009; Wild and Jones, 2009). It has also been suggested that the MWCNTs were able to create new pores in the seed coat to facilitate their internalization, increasing water uptake and facilitating seed germination and plant growth. Penetration of whole MWCNTs was not noticed in the cytoplasm of wheat seedlings due to limitation of their large size as compared to SWCNTs. The epidermal and root hair cells and root cap of wheat seedlings were pierced by the MWCNTs accumulated initially in the rhizosphere (Wild and Jones, 2009).

The unique hydrophobic property of the MWCNTs makes them capable of interacting with many organic substances that ultimately become sorbed by them (Yang et al., 2006b). It has been suggested that the flow of organic substances into the cytoplasm is facilitated by very low surface friction of CNTs (Whitby and Quirke, 2007). The process of phytoremediation utilizes the low surface friction property. Ma and Wang (2010) reported increased uptake of trichloroethylene by 82% in cottonwood (*Populus deltoides*) cuttings when treated with C_{60} at 15 mg/L. Thus, more potential modes of uptake of NPs along with macro- and micro-plant constituents are yet to be explored. Inversely, Canas et al. (2008) reported that roots of cucumber seedlings after 48 h of treatment with SWCNTs and functionalized SWCNTs (F-SWCNTs) showed no uptake of nanotubes. Nanotubes were functionalized by poly-3-aminobenzenesulfonic acid. However, primary and secondary roots and their external surfaces were surrounded by the SWCNTs.

Observations report that the cell does not permit the MWCNTs across the plasma membrane of rice cell suspensions (Tan and Fugetsu, 2007). They subsequently form black clumps and associate them with the cell surface by forming tightly around them (Tan et al., 2009). The size and number of clumps remain directly proportional to the concentration of NPs. This interaction leads to negative impact-like mortality over plants by interacting with protein and polysaccharide contents of the cell wall, and elevated hypersensitivity reactions are observed (Tan et al., 2007; Lin et al., 2009). These allergenic reactions were found similar to pathogenic infection in plants due to ultra-small size. Nevertheless, the negative impacts could be rationalized by optimizing the concentrations of the NPs. The conditioning and fabrication of MWCNTs into biocompatible molecules may help in preventing plants from hypersensitive reactions and facilitating their uptake into plants.

The media for growing plant material have been found to be important for the uptake of CNTs that are being modulated by the characteristic of dispersion in the media. The natural ecosystem that is a composite medium of soil, sludge, and sediments, referred to as natural organic matter (NOM), with diversity in its physical/chemical/colloidal properties and organic content, has been found to influence the uptake and effect of NPs over plants. An NOM is an assortment of organic substances heterogeneously obtained from decaying living materials, which present essential moieties that affect nanoparticle presentation to plants and ultimately influence their uptake (Chen et al., 2010). The constituents of NOM bearing hydrophobic characteristics interact with hydrophobic carbon present on the surface of NPs maintaining a dynamic equilibrium (Hyung et al., 2007).

The process of NOM-suspended fullerenes C_{70} and MWCNTs uptake, their accumulation, and translocation were studied by Lin et al. (2009) in rice plants. The observations confirmed the presence of C_{70} in the form of black aggregates in the seed and roots in large amounts as compared to stems and leaves that did not show significant presence of the particles. They suggested that the particles moved through xylem, the route for the translocation of water and nutrients. Inverse results were reported in mature plants where the particle abundance was found around the shoot vascular system and leaves. C_{70} NPs were absent from the roots indicating strong translocation of the particles from roots and rhizosphere to the aerial part of the plant. Inversely, trivial uptake of a few aggregates of NOM-MWCNT was seen in tissues and in the vascular system. Fullerene C_{70} NPs individually were proposed to interact with root cells via pores and plasmodesmata present on cell surface via the symplastic route or under the effect of forces like osmotic pressure and capillary force.

Contradicting evidence is also available regarding hydrophobic fullerenes C_{70}-NOM, which were reported to block the cell wall pores in the cell suspensions of *Allium cepa*. This caused an inappreciable intake of the NPs by the cells. However, fullerols$_{60}$(OH)$_{20}$ with small size and higher hydrophobic property percolated into the cell wall of *A. cepa* and consequentially accumulated at the boundary separating cell wall and plasma membrane. This accumulation between adjacent epidermal cell walls was attributed to the apoplastic transportation in the plant cells. The uptake, translocation, and accumulation of fullerols in bitter melon have also been reported recently using bright field imaging and Fourier transform infrared spectroscopy. Fullerol treatment enhanced the biomass yield and water content up to 54% and 24%, respectively. The fruit length, fruit number and fruit weight were also significantly increased (Kole et al., 2013). Miralles et al. (2012) studied the interaction of industrialized MWCNTs (75 w% CNTs) and their impurities on alfalfa and wheat.

Internalization of CNTs was facilitated through the adsorption on the root surfaces of the plant; however, no uptake or translocation was reported.

8.5.2 Metal oxide-based NPs: uptake, translocation, and accumulation in plants

Although TiO_2 NPs are abundantly consumed in the form of daily products, their uptake, translocation, and relevance has not yet been undertaken appreciably. *A. thaliana* seedlings showed tissue and cell-specific distribution of ultra-small TiO_2 (<5 nm) complexed with Alizarin red S nanoconjugate. Release of mucilage by *A. thaliana* and formation of a pectin hydrogel-like capsule that surrounds the roots are responsible for facilitating selective entry of TiO_2 complexed with Alizarin red S or sucrose (Kurepa et al., 2010). Other studies indicate the activation of heavy metals by the polysaccharides in mucilage that function as an adsorbent and enhance NP accumulation in a species-specific manner (Watanabe et al., 2008).

Servin et al. (2012) reported the transportation of TiO_2 from roots to leaf trichomes suggesting that trichomes might be acting as a sink or excretory system for TiO_2 in cucumber tissues. Experimental evidence also suggests that TiO_2 are not biotransformed within the tissues. TiO_2 NP (30 nm) uptake and translocation was also studied in maize with exuded roots with intact apexes. No uptake was facilitated by the root cells due to incomparable sizes of NPs with the pore diameter of the root cell wall (Asli, 2009).

Researchers investigated the soybean seedlings for the uptake and accumulation of ZnO NPs of 8 nm size. The uptake was inversely related to concentrations of NPs due to aggregation of crystals in higher concentrations. The larger agglomerates could not make way through the smaller pores of the cell wall, thereby reducing the uptake and accumulation of NPs (Lopez-Moreno et al., 2010). Cell internalization and translocation of ZnO NPs in ryegrass was studied by Lin and Xing (2008). NPs were found to adhere to the root surface and accumulation was reported in apoplast, protoplast, cytoplasmic compartment, endodermal nuclei, and the vascular cylinders. ZnO uptake has also been studied in roots of *Allium cepa* (Figure 8.3), which were found to accumulate on the inner side of the cell membrane cytoplasm resulting in a deformed nucleus (Kumari et al., 2011).

Iron oxide NPs (Fe_3O_4 NPs) of 20 nm size were studied in pumpkin and lima beans. Signals of magnetic NPs were traced in roots, stems, and leaves under hydroponic conditions, while the plants growing in soil or in sand did not show any signals of magnetic NPs indicating no particle uptake. The uptake was therefore found to be dependent on growth media that might have caused the adherence of particles on the soil and sand grains (Zhu et al., 2008). The uptake has also been reported to be species specific because no uptake was observed in lima bean plants treated with NPs. Wang et al. (2006), on the contrary, reported no uptake of the NP by the pumpkin plants. The uptake here seems to be size dependent, as the larger NPs fail to penetrate through pore sizes of 2 to 20 nm and thus no transportation was noticed across the plasma membrane.

Seedlings of soybean, alfalfa, corn, and tomato have been considered for the CeO_2 NP (7 nm) uptake process. The accumulation was found to be both concentration and species specific. Various species offered differential rhizospheric conditions for NP uptake including differences in root microstructure and the type of interactions between the NPs and root exudates (Lopez-Moreno et al., 2010). Parsons et al. (2010) investigated the uptake of $Ni(OH)_2$ NPs in mesquite. Plants were treated with uncoated and citrate-coated NPs; only the roots were found with accumulated NPs.

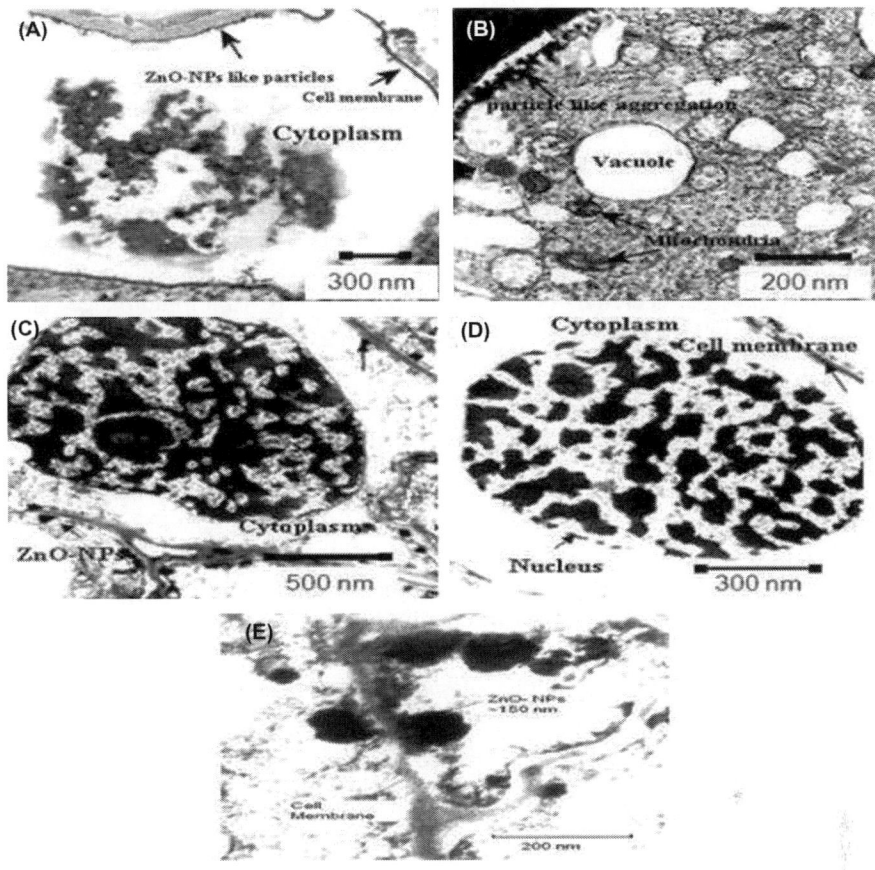

FIGURE 8.3

TEM images of *Allium cepa* root cells treated with $100\,\mu g\,ml^{-1}$ ZnO-NPs; (A) ZnO-NP-treated root cells showing the presence of NPs around the cell membranes and the cytoplasm; (B) ZnO-NP-treated root cells showing the presence of particles at the inner side of cell membrane; (C) ZnO-NP-treated root cells showing the presence of NPs around cell membrane, in the cytoplasm, and deformed nucleus; (D) control sample treated with Milli-Q water showing no NPs around cell membrane and in the cytoplasm; (E) ZnO-NP-treated root cells showing aggregation of nanoparticles around cell membrane, in the cytoplasm, with a size of 150 nm. Magnification for images (A) 15,000×, (B) 10,000×, (C) 4500×, (D) 4500×, and (E) ×45,000.

Adapted from Kumari et al. (2011).

8.5.3 Metal-based nanoparticles: uptake, translocation, and accumulation in plants

Metallic NPs like silver (Ag), gold (Au), aluminum (Al), copper (Cu), etc., have been studied for their uptake and translocation in different plants. Stampoulis et al. (2009) used *Cucurbita pepo* to compare the uptake of Ag. They treated the plants with AgNPs and their bulk counterparts. AgNP-treated plants revealed 4.7 times more Ag concentration in shoots of plants as compared to bulk powder at similar concentrations. It was hypothesized that AgNPs could release greater numbers of ions resulting in higher Ag concentration in shoots. However, no accumulation of Ag in any form was observed in *Brassica juncea* when treated with AgNPs (Haverkamp and Marshall, 2009).

Carbon-treated FeNPs are found only in epidermal cells close to the area where NPs were applied to the leaf petioles of living pumpkin plants. Faraway areas like xylem were devoid of any NPs (Corredor et al., 2009). CuNPs were studied with *Phaseolus radiate* and *Triticum aestivum*. Lee et al. (2008) reported that NPs can permeate across the cell membrane and consequently accumulate in cells. Doshi et al. (2008) investigated red kidney beans for the uptake of Al exposed to AlNPs in soil. No significant accumulation was observed in comparison to untreated plants.

8.6 Influence of nanoparticles as growth promoters in plants

Increased application of carbon-based nanotubes in the field of food and agriculture has also raised concerns regarding their impact. Various studies have produced contradictory results on the basis of size of NPs and the species of plant. Other factors including media and mode of plant part have also influenced the results of NP application. The functionalized SWCNTs (fCNTs; with poly-3-aminobenzenesulfonic for high dispersibility) and non-functionalized SWCNTs (CNTs) were administered to the root areas of six different crop species, namely, tomato (*Solanum lycopersicum*), cabbage (*Brassica oleracea*), onion (*Allium cepa*), cucumber (*Cucumis sativus*), lettuce (*Lactucasativa*), and carrot (*Daucus carota*), to study their effect on root elongation and to delineate the mode of uptake, translocation, and their ultimate effect on crops (Canas et al., 2008).

CNTs and fCNTs on interaction with the root structures formed nanosheets on the root surface, but they did not enter the roots; CNTs, however, enhanced the elongation of onion and cucumber roots while reduced the root length significantly in tomato. This contradictory effect in tomato roots was attributed to CNTs that accumulated in high concentration around the area of root elongation, also known as base of apical meristem (Canas et al., 2008). fCNTs also gave inhibitory effect over lettuce root length, while cabbage and carrot remained unaffected by all forms of nanotubes. These experiments were, however, performed for 24 and 48 h of exposures, whereas ideal exposure time is about 4 days.

MWCNTs have been reported to yield positive response over growth of tomato plants and their seed germination (Khodakovskaya et al., 2009). Moreover, MWCNTs also possess the ability to enhance the growth of tobacco cell culture (Khodakovskaya et al., 2012). Germination of seeds was favored by the presence of MWCNTs due to increased uptake of water by the seeds. Six different crop species, radish (*Raphanus sativus*), rape (*Brassica napus*), ryegrass (*Lolium perenne*), lettuce (*Lactuca sativa*), corn (*Zea mays*), and cucumber (*Cucumis sativus*), also showed positive impact of MWCNTs over root elongation and seed germination (Lin and Xing, 2007). Nair et al. (2010) also reported enhancement of root growth and germination in rice seedlings in response to both SWCNTs and MWCNTs (Table 8.2). Exposure time-based experimental findings in zucchini plants gave no negative effect of MWCNTs over seed germination and root length elongation in tested concentration range.

However, an increased exposure towards SWCNTs reduced the biomass of the plants. These results have been found to vary and present contradictory results on the basis of plants, their

Table 8.2 Positive Consequential Effects of Nanoparticles in Routinely used Food Crops

Name of Crop	Nanoparticles	Size of NPs (nm)	Consequential Effects	References
Allium cepa *Cucumis sativus*	SWCNT	8	Significantly increased root length	Canas et al. (2008)
Solanum lycopersicum	MWCNT	Internal dimension 110–170	Significant increase in germination rate, fresh biomass, and length of stem	Khodakovskaya et al. (2009)
Spinacia oleracea	Nanoanatase (TiO_2)	5	Improved spinach growth related to N_2 fixation by TiO_2	Yang et al. (2007)
Glycine max	ZnO	8	Increased root growth	Lopez-Moreno et al. (2010)
Curcurbita mixta	Fe_3O_4	20	No toxic effect	Zhu et al. (2008)
Linum usitatissimum	Ag	20	No effect on the germination	El-Temsah and Joner (2010)
Lactuca sativa	Cu	10	No effect on the germination; improved shoot/root ratio	Shah and Belozerova (2009)
Raphanus sativus *Brassica napus*	Al	1–100	Improved root growth	Lin and Xing (2007)
Cucumis sativus *Lactuca sativa*	Au	10	Positive effect on germination index	Barrena et al. (2009)
Zea mays *Medicago sativa* *Glycine max*	CeO_2	7	Significantly increased root and stem growth	Lopez-Moreno et al. (2010)
Prosopis sp.	$Ni(OH)_2$	8.7	No effect	Parsons et al. (2010)

species, and growth phases as well as the kind, size, and structure and time of exposure of NPs. Certain studies have demonstrated a hypersensitive reaction generated by rice suspension in response to MWCNTs as a defense mechanism to escape the detrimental effects of NPs (Tan and Fugetsu, 2007). Interaction of MWCNTs with protein and polysaccharide contents resulted in the thickening of the cell wall. This was the result of elevation of a signaling cascade that produces molecules necessary for cell wall thickening and is the signature step adopted by plants in biotic and abiotic stress adversely affecting plant growth. Generation of reactive oxygen species (ROS), which shows increased oxidative stress in plants and decreased cell proliferation, was found in response to MWCNTs (Tan et al., 2009). MWCNTs have also been reported to adversely affect the root and leaf morphology of spinach. However, the effect was reversed when CNTs were administered with ascorbic acid (Begum and Fugetsu, 2012). CNTs did not penetrate the plant cell in the above studies; however, their impact was very well noticed.

MWCNTs have proved to be detrimental on *Arabidopsis* T87 suspension cells when differentiated on the basis of loose agglomerate and fine agglomerates (obtained after ultrasonication) (Lin et al., 2009). The cultured cells represented decreased dry weight, less viability, lower chlorophyll content, and superoxide dismutase. The effects were more pronounced in fine agglomerates than in loose ones. This was explained due to the clump formation feature of plant cells with the surrounding cells. Loose but large agglomerates could not penetrate into such cell clumps, whereas fine and small agglomerates can easily be distributed into plant clumps due to their small size and their interactions with protein and polysaccharides of the cell wall. These size-based characteristics of NPs might be responsible for the toxic behavior of NPs. Lin et al. (2009) also suggested the presence of metallic impurities (residual metals used as catalysts for CNT synthesis) might be the cause behind MWCNT toxicity (Table 8.3).

CNTs can be stabilized through NOM on the basis of their hydrophobic behavior (Navarro et al., 2008). The uptake, translocation, and accumulation of NOM-CNTs were observed in rice plants (Lin et al., 2009), Fullerene C_{70}-NOM or MWCNTs-NOM with reduced hydrophobicity were studied in rice seedlings at different concentrations. The carryover of NPs was studied in the first generation as well as the progeny. Microscopic sections of both generations at various stages of the life cycle were carried out. The results indicated the occurrence of C_{70} NPs in both generations; however, a small amount of MWCNTs was observed inside the cell. The results consolidated the fact that the NPs are transferred to the F_1 generation through the seeds of the mother generation. This could have occurred through osmotic pressure and capillary action at the plant root tip, and NPs may have entered through the cell wall pores and translocated through plasmodesmata. The presence of NPs near the vascular bundle could interfere with the water and nutrient uptake of plants, thus posing a side effect of plant growth. Epigenetic modifications through global deacetylation of histone H3 have also been induced through SWCNT treatment in maize root, resulting in changes in gene expression and accordingly affecting root growth and development (Yan et al., 2013).

The positive impact of TiO_2 has been studied in spinach seed germination and growth. The plant showed improvement in light absorption and promoted the activity of Rubisco activase, thereby accelerating growth (Zheng et al., 2005; Hong et al., 2005a; Gao et al., 2006, 2008; Lei et al., 2007b; Xuming et al., 2008; Mingyu et al., 2008). Nano-TiO_2 (anatase) also enhanced nitrogen metabolism (Yang et al., 2006a), which is responsible for promoting nitrate absorption by the plant and which favors the formation of organic nitrogen from inorganic nitrogen, thereby increasing the fresh weight and dry weight of the plant. Gao et al. (2013) reported the effect of nano-TiO_2 on the photosynthetic characteristics of *Ulmus elongata* seedlings and found that the net photosynthetic rate had been lowered in plants treated with various concentrations of anatase as compared to control.

NP-treated spinach plant also showed improved plant growth on the basis of nitrogen photoreduction (Mingyu et al., 2007a; Yang et al., 2007). The thylakoid membrane with light harvesting complex II was enhanced upon exposure to the NPs (Hong et al., 2005b; Lei et al., 2007a). These changes then promote energy transfer and oxygen evolution in photosystem II of the plant (Mingyu et al., 2007b).

The activity of certain enzymes like superoxide dismutase, guaiacol peroxidase, ascorbate peroxidase, and catalase has been found to be up-regulated, thereby down-regulating the accumulation of superoxide radicals, malonyldialdehyde content, and hydrogen peroxide, which further up-regulated the oxygen evolution rate in spinach chloroplasts under UV-B radiation (Lei et al., 2008). Highly comparable data regarding the light harvesting complex content have been observed in

Table 8.3 Negative Effects of NPs in Routinely used Food Crops

Name of Crop	Nanoparticles (NPs)	Size of NPs (nm)	Consequential Effects	References
Oryza sativa	SWCNT	1.19 (major) 18, 722	Delayed flowering, decreased yield	Lin et al. (2009)
Curcurbita pepo	MWCNT	Diameter range 10–30	Reduced biomass (38%)	Stampoulis et al. (2009)
Lactuca sativa	MWCNT	Diameter range 10–30	Reduced root length	Lin and Xing (2007)
Zea mays	TiO_2/inorganic bentonite clay	30/1–60	Inhibited hydraulic conductivity, leaf growth, and transpiration	Asli and Neumann (2009)
Lolium perenne	ZnO	9–37 (mean: 19 ± 7)	Reduced biomass, shrunken root tips, epidermis and root cap were broken	Lin and Xing (2008)
Glycine max	ZnO	8	Decreased root growth	Lopez-Moreno et al. (2010)
Curcurbita pepo	Ag	100	Reduced transpiration (41–79%)	Stampoulis et al. (2009)
Allium cepa	Ag	<100	Decreased mitosis; disturbed metaphase; sticky chromosome; cell wall disintegration and breaks	Kumari et al. (2009)
Triticum aestivum Vigna radiate	Cu	<100	Reduced root and seedling growth	Lee et al. (2008)
Curcurbita pepo	Cu	50	Reduced biomass (90%)	Stampoulis et al. (2009)
Zea mays Lactuca sativa	Al	10	Reduced root length	Lin and Xing (2007)
Zea mays Solanum lycopersicum Glycine max Cucumis sativus	CeO_2	7	Reduced germination	Lopez-Moreno et al. (2010)

Ti—quantum dot (QD) assembly (Kongkanand et al., 2008) for the conversion of solar energy. The solar energy trapping characteristics of QDs can be exploited for the enhanced light uptake by plants for increased photosynthesis.

Accumulation of TiO_2 in maize roots was followed by noteworthy reduction in the diameter of cell pores (Asli and Neumann, 2009). Transpiration and leaf growth were suppressed as the primary roots showed reduction in hydraulic conductivity. NPs did not impart toxicological impact over plants; however, they were found to interact with cells and inhibit the apoplastic flow through the cell. However, the mixture of SiO_2 and TiO_2 was found to enhance the fertilizer and water uptake through the increase in activity of nitrate reductase. These effects on the whole stimulate the antioxidant system of the plant (Lu et al., 2002). Nano-CuO has been found to adversely affect the growth of rice seedlings. Modulation in the ascorbate–glutathione cycle, membrane damage, *in vivo* ROS detection, foliar H_2O_2, and proline accumulation has been observed under the stress of CuO-NPs (Shaw and Hossain, 2013).

ZnO-NPs in most of the studies have influenced plant growth depending on the concentration range. Data report root elongation of soybean at 500 mg/L, whereas higher concentrations have been found to be involved in the reduction of root. An even higher concentration (4000 mg/L) has shown no effect on seed germination in soybean (Lopez-Moreno et al., 2010). This effect could be ascribed to Zn ions released in excess by the NPs and their interaction with plant root systems. Zucchini seeds in a hydroponic solution system remained unaffected under ZnO-NPs (Stampoulis et al., 2009). Several other studies have demonstrated the toxic effect of ZnO over different species of plants. For example, seed germination of ryegrass and corn (Lin and Xing, 2007) and root growth of radish and rape was inhibited by ZnO-NPs; however, due to selective permeability of the seed coat such an inhibition was not observed when soaked in nano-Zn suspension (Lin and Xing, 2008). A "tunneling like effect" exhibiting deep invaginations has been reported in the primary root tip of maize overexposure to ZnO-NP while exposure to AgNO3 led to cell erosion in the root apical meristem (Pokhrel and Dubey, 2013). Milani et al. (2012) studied the solubility and dissolution kinetics of Zn and compared the characteristics in ZnO-NPs and their bulk counterpart. The particles were coated onto two selected granular macronutrient fertilizers, urea and monoammonium phosphate (MAP). Enhanced solubility and higher dissolution rates were observed in coated MAP granules in sand columns compared to coated urea granules, which could be attributed to pH differences in the solution surrounding the fertilizer granules. Au-NPs of 24 nm have been found to enhance the total seed yield of *Arabidopsis thaliana* by three times over control. Moreover, seed germination rate, vegetative growth, and free radical scavenging activity were also interestingly enhanced after treatment (Kumar et al., 2013).

Pure alumina NPs have been reported to retard the growth of various plants, like carrot (*Daucus carota*), cabbage (*Brassica oleracea*), corn (*Zea mays*), soybean (*Glycine max*), and cucumber (*Cucumis sativus*) by reducing root elongation (Yang and Watts, 2005). However, their toxicity was found to be negligible when coated with phenanthrene (a major constituent of polycyclic aromatic hydrocarbons), thereby having a negative influence on plant roots.

This study highlights the importance of surface modification of NPs to make them biocompatible and further reduce toxicity. Similar experiments studying the effect of aluminum oxide and carboxylate ligand-coated aluminum oxide on California red kidney beans and ryegrass showed no adverse effect on plant growth (Doshi et al., 2008).

8.7 Influence of NPs as biological control in plants

The random application of pesticides and fertilizers results in environmental pollution, emergence of agricultural pests and pathogens, and loss of biodiversity (Ghormade et al., 2011). Nanotechnology permits new routes for improving existing crop management techniques by controlling plant pathogens and alleviating the above-mentioned problems. The chemicals used for the control of plant pathogens remain unaffected for most of the times as the required concentration does not reach the target site. Moreover, their effects are also restricted due to various factors like leaching of chemicals, degradation by photolysis, hydrolysis, and microbial degradation. Higher concentrations, if applied, translocate into the plants and are accumulated in the food web affecting consumers. They also change soil and other environmental profiles thus affecting crops completely. Therefore, the current scenario demands effective solutions that could be employed in low doses and released in a time-controlled manner at the specific target sites rather than affecting whole plant population. NP-based agrochemicals can be designed in such a way so as to give them specific characteristics that could result in achieving effective dose concentration, stimulus and time-controlled release of NPs, safe, easy and enhanced target-based delivery and activity, and less ecotoxicity, thus avoiding repeated applications (Tsuji, 2001; Nair et al., 2010).

For the past few decades nano-drugs have gained increased attention in agriculture over chemical pesticides. The release of drugs at the specific stimulus is the most important requirement of the advanced drug delivery mechanisms. This has been mainly achieved through the encapsulation of the functional particles that diffuse through the coating material resulting in sustained and time-controlled release of drug (Li et al., 2006). In this section, therefore, we bring into account the antimicrobial effect of various NPs employed for the management of plant diseases and subsequent crop improvement.

Lamsal et al. (2010) used various concentration ranges of Ag-NPs at pre- and post-disease outbreak in plants to establish its antifungal potential in cucumber and pumpkin. The highest inhibition rate for both plant species under the diseased and normal conditions was observed with 100 ppm Ag-NPs. Moreover, a similar concentration exhibited the highest degree of inhibition regarding the growth and germination of fungal hyphae and conidia in *in vivo* models. This effect was substantiated through scanning electron micrographs (SEM) that showed detrimental effects of Ag-NPs on both mycelial growth and conidial germination.

A nano-silver—silica composite is prepared by mixing silver salt with silicate and a water-soluble polymer followed by exposing the mixture to radioactive rays. The composite has been reported to possess excellent antimicrobial effects even at low concentrations. Such NP composites could be successfully used against phytopathogens, namely, *Phytophthora* spp., *Rhizoctonia* spp. (Figure 8.4), *Colletotrichum* spp., *Botrytis* spp., and *Phythium* spp.

A dose of 3.0 ppm of NP composite has been reported to be effective against the above-mentioned pathogens causing various infections in plants (Park et al., 2006). The antifungal property of the NP composite could be attributed to Ag-NPs that exhibit a disinfecting property against fungus on absorption, while silica provides a physical barrier to pathogenic fungi.

NP composites were also found to be effective against phytopathogenic bacteria at a concentration higher than 10 ppm. Thus, it is necessary to optimize the minimal effective concentrations of such formulations, i.e., effective enough to fight against each pathogen individually as well as together. Another advantage of using NP-silver—silica is that it could provide long-term control of microorganisms very selectively depending upon its concentration with a single application. Moreover, it does not cast a negative impact on farmers as it does not cause chemical injuries and remains non-toxic to humans.

ZnO-NPs were tested for their antibacterial efficacy against their bulk counterpart. ZnO-NPs were reported to possess more antibacterial potential that ZnO powder. Gram-negative bacteria

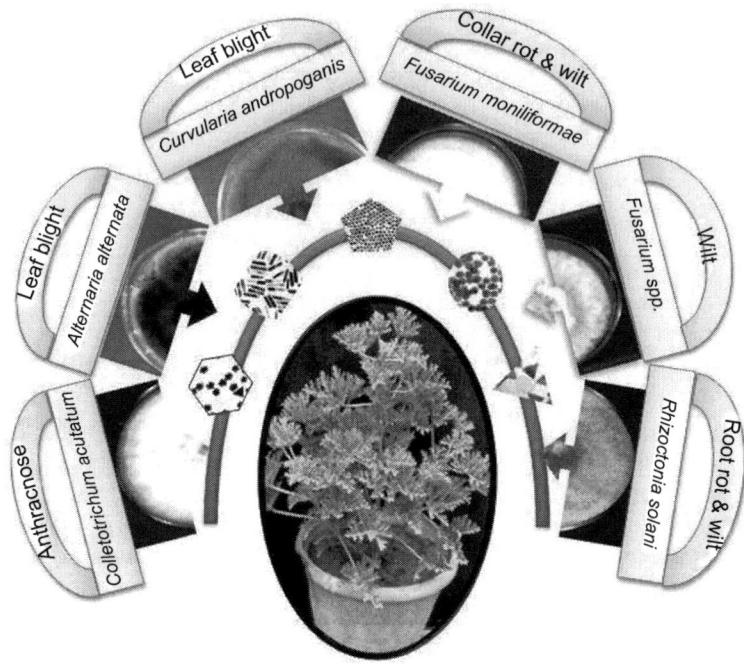

FIGURE 8.4

Schematic representation of NPs and their applications in biological control. Figure represents different phytopathogens (PDA culture plates) restricting plant growth by causing various diseases (represented in semicircles) and their therapeutic measures in the form of NPs.

were generally less sensitive to ZnO than Gram-positive bacteria. NP-treated bacteria were found to be greatly affected when observed through a scanning electron microscope. Complete lysis of the treated bacterial cells was observed. Such antibacterial efficiency of ZnO-NPs prompts further research for its risk assessment and its application in the food industry as a preservative (Tayel et al., 2011).

Titanium dioxide is another class used in NP generation, and is a non-toxic pigment widely used in the manufacture of paints, paper, ink, cosmetics, ceramics, leather, etc. It also possesses very strong disinfectant characteristics comparable to that of chlorine and ozone. TiO_2 has been approved for use in food products up to 1% of product final weight because of its non-toxic nature. It has great potential in various agricultural applications including plant protection from pathogenic microorganisms on the basis of its photo-catalyst surface. Researchers have been trying to modify the property of disinfection by preparing TiO_2 thin films by dye doping and other suitable techniques (Yao et al., 2007). Accumulated evidence has reported that TiO_2 application reduced the effects of *Curvularia* leaf spot and bacterial leaf blight in rice and maize plants and also decreased the incidences and severity of rice blast disease and tomato spray molds (Chao and Choi, 2005). Hence, nano-TiO_2 could be successfully promoted as an environmentally-friendly fungicide and/or bactericide thus inhibiting various phytopathogenic diseases.

8.8 Conclusion and future perspectives

Despite the advancement of NPs in the field of nanotechnological applications reaching state of the art, their implications in agriculture and crop improvement are still at an elementary phase. In order to exploit the promised advantages of NPs, it has become essential to enhance our understanding regarding plant—NP interactions through their characterization and relative phytotoxic aspects. Another important problem remains: how to delineate the mode of uptake and translocation of NPs by plants. However, various routes have been proposed, and results have been deviating on the basis of growth conditions, plant species, and size and concentration of NPs. Therefore, it becomes important to explore the uptake kinetics of NPs under the influence of particle size, agglomeration and compositions. Their translocation, accumulation, and biotransformation in different plant parts are another approach to be taken care of. Accumulating evidence suggests the toxic effects of NPs; however, results have been modulated to yield positive effects through the surface modification of NPs. fCNTs and phenanthrene-coated Al-NPs highlight progress in this regard. NPs in different concentrations have been reported to influence different species of plants with both positive and negative effects. This characteristic could be utilized simultaneously to promote growth in edible crops and kill the weeds or phytopathogens affecting the crops. Size and concentrations of NPs could be optimized to yield such desirable effects. Structure of the vascular bundle with significant variations in pore size and miscellaneous modes of nutrient and water transport is also an essential parameter, as it has already been determined that xylem plays a key role in maintaining the speed of water uptake. NP assimilation and its subsequent accumulation in the food web is also a major concern. Therefore, attempts should be taken to design experimental models depicting plant—NP—animal interaction and the effects studied at individual atrophic levels. The NP—plant cell interaction modifies the gene as well as protein profiles of plant cells, ultimately leading to changes in biological pathways that bring about changes in growth and development of plants. Hence, experimentation should be taken up to generate information at the molecular level caused by uptake and translocation of NPs. Antimicrobial activity also makes NPs a strong tool to use as a biological control agent to help in crop management and improvement.

Acknowledgments

The authors are thankful to the University of Allahabad, India, for providing facilities and UGC for providing financial assistance under the Dr. D.S. Kothari Post Doctoral Fellowship program.

References

Ao, M., Zhu, Y., He, S., Li, D., Li, P., Li, J., et al., 2013. Preparation and characterization of 1-naphthylacetic acid-silica conjugated nanospheres for enhancement of controlled-release performance. Nanotechnology 24, 035601.

Asli, S., Neumann, M., 2009. Colloidal suspensions of clay or titanium dioxide nanoparticles can inhibit leaf growth and transpiration via physical effects on root water transport. Plant Cell Environ. 32, 577—584.

Ball, P., 2002. Natural strategies for the molecular engineer. Nanotechnology 13, 15—28.

Bandyopadhyay, S., Peralta-Videa, J.R., Gardea-Torresdey, J.L., 2013. Advanced analytical techniques for the measurement of nanomaterials in food and agricultural samples: a review. Environ. Eng. Sci. 30, 118−125.

Barrena, R., Casals, E., Colon, J., Font, X., Sanchez, A., Puntes, V., 2009. Evaluation of the eco-toxicity of model nanoparticles. Chemosphere 75, 850−857.

Begum, P., Fugetsu, B., 2012. Phytotoxicity of multi-walled carbon nanotubes on red spinach (*Amaranthus tricolor* L.) and the role of ascorbic acid as an antioxidant. J. Hazard. Mater. 243, 212−222.

Biener, J., Farfan-Arribas, E., Biener, M., Friend, C.M., Madix, R.J., 2005. Synthesis of TiO_2 nanoparticles on the Au(111) surface. J. Chem. Phy. 123, 94705.

Biswas, P., Wu, C.Y., 2005. Critical review: nanoparticles and the environment. J. Air Waste Manage. Assoc. 55, 708−746.

Buzea, C., Pacheco, I.I., Robbie, K., 2007. Nanomaterials and nanoparticles: sources and toxicity. Biointerphases 2, MR17−MR71.

Canas, J.E., Long, M., Nations, S., Vadan, R., Dai, L., Luo, M., et al., 2008. Effects of functionalized and non-functionalized single-walled carbon-nanotubes on root elongation of select crop species. Nanomater. Environ. 27, 1922−1931.

Carmen, I.U., Chithra, P., Huang, Q., Takhistov, P., Liu, S., Kokini, J.L., 2003. Nanotechnology: a new frontier in food science. Food Technol. 57, 24−29.

Chao, S.H.L., Choi, H.S., 2005. Method for Providing Enhanced Photosynthesis, vol. 11. Korea Research Institute of Chemical Technology, Jeonju, South Korea Bull, pp. 1−34.

Chen, R., Ratnikova, T.A., Stone, M.B., Lin, S., Lard, M., Huang, G., et al., 2010. Differential uptake of carbon nanoparticles by plant and mammalian cells. Small 6, 612−617.

Corredor, E., Testillano, P.S., Coronado, M.J., Gonzalez-Melendi, P., Fernandez-Pacheco, R., Marquina, C., et al., 2009. Nanoparticle penetration and transport in living pumpkin plants: in situ subcellular identification. BMC Plant Biol. 9, 45.

Da Silva, L.C., Oliva, M.A., Azevedo, A.A., De Araujo, J.M., 2006. Responses of resting plant species to pollution from an iron pelletization factory. Water Air Soil Pollut. 175, 241−256.

Das, I., Ansari, S.A., 2009. Nanomaterials in science and technology. J. Sci. Ind. Res. 68, 657−667.

Donaldson, K., Stone, V., Tran, C.L., Kreyling, W., Borm, P.J.A., 2004. Nanotoxicology. Occup. Environ. Med. 61, 727−728.

Doshi, R., Braida, W., Christodoulatos, C., Wazne, M., O'Connor, G., 2008. Nano-aluminum: transport through sand columns and environmental effects on plants and soil communities. Environ. Res. 106, 296−303.

Eichert, T., Kurtz, A., Steiner, U., Goldbach, H.E., 2008. Size exclusion limits and lateral heterogeneity of the stomatal foliar uptake pathway for aqueous solutes and water-suspended nanoparticles. Physiol. Plant 134, 151−160.

El-Temsah, Y.S., Joner, E.J., 2010. Impact of Fe and Ag nanoparticles on seed germination and differences in bioavailability during exposure in aqueous suspension and soil. Environ. Toxicol. 27, 42−49.

Farre, M., Sanchis, J., Barcelo, D., 2011. Analysis and assessment of the occurrence, the fate and the behavior of nano-materials in the environment. Trends Analyt. Chem. 30, 517−527.

Fleischer, M.A., Neill, O., Ehwald, R., 1999. The pore size of non-graminaceous plant cell wall is rapidly decreased by borate ester cross-linking of the pectic polysaccharide rhamnogalacturon II. Plant Physiol. 121, 829−838.

Gao, F., Hong, F., Liu, C., Zheng, L., Su, M., Wu, X., et al., 2006. Mechanism of nano-anatase TiO_2 on promoting photosynthetic carbon reaction of spinach: inducing complex of rubisco-rubisco activase. Biol. Trace Elem. Res. 111, 239−253.

Gao, F., Liu, C., Qu, C., Zheng, L., Yang, F., Su, M., et al., 2008. Improvement of spinach growth by nano-TiO_2 treatment related to the changes of rubisco activase. Biometals 21, 211−217.

Gao, J., Xu, G., Qian, H., Liu, P., Zhao, P., Hu, Y., 2013. Effects of nano-TiO$_2$ on photosynthetic characteristics of *Ulmus elongata* seedlings. Environ. Pollut. 176, 63−70.

Ghormade, V., Deshpande, M.V., Paknikar, K.M., 2011. Perspectives for nano-biotechnology enabled protection and nutrition of plants. Biotechnol. Adv. 29, 792−803.

Hall, J.L., Williams, L.E., 2003. Transition metal transporters in plants. J. Exp. Bot. 54, 2601−2613.

Handy, R.D., Owen, R., Valsami-Jones, E., 2008. The ecotoxicology of nanoparticles and nanomaterials: current status, knowledge gaps, challenges and future needs. Ecotoxicology 17, 315−325.

Haverkamp, R.G., Marshall, A.T., 2009. The mechanism of metal nanoparticle formation in plants: limits on accumulation. J. Nanopart. Res. 11, 1453−1463.

Hong, F., Zhou, J., Liu, C., Yang, F., Wu, C., Zheng, L., et al., 2005a. Effect of nano-TiO$_2$ on photochemical reaction of chloroplasts of spinach. Biol. Trace Elem. Res. 105, 269−280.

Hong, F., Yang, P., Gao, F., Liu, C., Zheng, L., Yang, F., et al., 2005b. Effect of nano-TiO$_2$ on spectral characterization of photosystem particles from spinach. Chem. Res. Chin. Univ. 21, 196−200.

Hyung, H., Fortner, J.D., Hughes, J.B., Hong, K.J., 2007. Natural organic matter stabilizes carbon nanotubes in the aqueous phase. Environ. Sci. Technol. 41, 179−184.

Jia, G., Wang, H., Yan, L., Wang, X., Pei, R., Yan, T., et al., 2005. Cytotoxicity of carbon nano-materials: single-wall nanotube, multi-wall nanotube, and fullerene. Environ. Sci. Technol. 39, 1378−1383.

Ke, W., Xiong, Z.T., Chen, S., Chen, J., 2007. Effects of copper and mineral nutrition on growth, copper accumulation and mineral element uptake in two *Rumex japonicas* populations from a copper mine and an uncontaminated field sites. Environ. Exp. Bot. 59, 59−67.

Khodakovskaya, M., Dervishi, E., Mahmood, M., Xu, Y., Li, Z., Watanabe, F., et al., 2009. Carbon nanotubes are able to penetrate plant seed coat and dramatically affect seed germination and plant growth. ACS Nano 3, 3221−3227.

Khodakovskaya, M.V., de Silva, K., Biris, A.S., Dervishi, E., Villagarcia, H., 2012. Carbon nanotubes induce growth enhancement of tobacco cells. ACS Nano 6, 2128−2135.

Knox, J.P., 1995. The extra cellular-matrix in higher-plants. 4. Developmentally-regulated proteoglycans and glycoproteins of the plant-cell surface. FASEB J. 9, 1004−1012.

Kole, C., Kole, P., Randunu, K.M., Choudhary, P., Podila, R., Ke, P.C., et al., 2013. Nanobiotechnology can boost crop production and quality: first evidence from increased plant biomass, fruit yield and phytomedicine content in bitter melon (*Momordica charantia*). BMC Biotechnol. 13, 37.

Kongkanand, A., Tvrdy, K., Takechi, K., Kuno, M., Kamat, P.V., 2008. Quantum dot solar cells. Tuning photoresponse through size and shape of CdSe-TiO$_2$ architecture. J. Am. Chem. Soc. 130, 4007−4015.

Kumar, V., Guleria, P., Kumar, V., Yadav, S.K., 2013. Gold nanoparticle exposure induces growth and yield enhancement in *Arabidopsis thaliana*. Sci. Total Environ. 461−462, 462−468.

Kumari, M., Mukherjee, A., Chadrasekaran, N., 2009. Genotoxicity of silver nanoparticle in *Allium cepa*. Sci. Total Environ. 407, 5243−5246.

Kumari, M., Khan, S.S., Pakrashi, S., Mukherjee, A., Chandrasekaran, N., 2011. Cytogenetic and genotoxic effects of zinc oxide nanoparticles on root cells of *Allium cepa*. J. Hazard. Mater. 190, 613−621.

Kurepa, J., Paunesku, T., Vogt, S., Arora, H., Rabatic, B.M., Lu, J., et al., 2010. Uptake and distribution of ultrasmall anatase TiO$_2$ alizarin red S nanoconjugates in *Arabidopsis thaliana*. Nano Lett. 10, 2296−2302.

Lamsal, K., Kim, S.W., Jung, J.H., Kim, Y.S., Kim, K.S., Lee, Y.S., 2010. Inhibition effects of silver nanoparticles against powdery mildews on cucumber and pumpkin. Mycobiology 39, 26−32.

Lee, W.M., An, Y.J., Yoon, H., Kweon, H.S., 2008. Toxicity and bioavailability of copper nanoparticles to the terrestrial plants mung bean (*Phaseolus radiatus*) and wheat (*Triticum aestivum*): plant agar test for water-insoluble nanoparticles. Nanomater. Environ. 27, 1915−1921.

Lei, Z., Mingyu, S., Xiao, W., Chao, L., Qu, C., Liang, C., et al., 2007a. Effects of nano-anatase on spectral characteristics and distribution of LHC II on the thylakoid membranes of spinach. Biol. Trace Elem. Res. 120, 273−283.

Lei, Z., Mingyu, S., Chao, L., Liang, C., Hao, H., Xiao, W., et al., 2007b. Effects of nano-anatase TiO_2 on the photosynthesis of spinach chloroplasts under different light illumination. Biol. Trace Elem. Res. 119, 68−76.

Lei, Z., Mingyu, S., Xiao, W., Chao, L., Chunxiang, Q., Liang, C., et al., 2008. Antioxidant stress is promoted by nano-anatase in spinach chloroplasts under UV-beta radiation. Biol. Trace Elem. Res. 121, 69−79.

Li, Y., Leung, P., Yao, L., Song, Q.W., Newton, E., 2006. Antimicrobial effect of surgical masks coated with nanoparticles. J. Hosp. Infect. 62, 8−63.

Lin, C., Fugetsu, B., Su, Y., Watari, F., 2009. Studies on toxicity of multi-walled carbon nanotubes on *Arabidopsis* T87 suspension cells. J. Hazard. Mater. 170, 578−583.

Lin, D., Xing, B., 2007. Phytotoxicity of nanoparticles: inhibition of seed germination and root growth. Environ. Pollut. 150, 243−250.

Lin, D., Xing, B., 2008. Root uptake and phytotoxicity of ZnO nanoparticles. Environ. Sci. Technol. 42, 5580−5585.

Lin, S., Reppert, J., Hu, Q., Hudson, J.S., Reid, M.L., Ratnikova, T.A., et al., 2009. Uptake, translocation, and transmission of carbon nanomaterials in rice plants. Small 5, 1128−1132.

Liu, Q., Chen, B., Wang, Q., Shi, X., Xiao, Z., Lin, J., et al., 2009. Carbon nanotubes as molecular transporters for walled plant cells. Nano Lett. 9, 1007−1010.

Lopez-Moreno, M.L., De La Rosa, G., Hernandez-Viezcas, J.A., Castillo-Michel, H., Botez, C.E., Peralta-Videa, J.R., et al., 2010. Evidence of the differential biotransformation and genotoxicity of ZnO and CeO_2 nanoparticles on soybean (Glycine max) plants. Environ. Sci. Technol. 44, 7315−7320.

Lopez-Moreno, M.L., De La Rosa, G., Hernandez-Viezcas, J.A., Peralta-Videa, J.R., Gardea-Torresdey, J.L., 2010. X-ray absorption spectroscopy (XAS) corroboration of the uptake and storage of CeO_2 nanoparticles and assessment of their differential toxicity in four edible plant species. J. Agric. Food Chem. 58, 3689−3693.

Lu, C.M., Zhang, C.Y., Wen, J.Q., Wu, G.R., Tao, M.X., 2002. Research of the effect of nanometer materials on germination and growth enhancement of Glycine max and its mechanism. Soybean Sci. 21, 168−172.

Ma, X., Wang, C., 2010. Fullerene nanoparticles affect the fate and uptake of trichloroethylene in phytoremediation systems. Environ. Eng. Sci. 27, 989−992.

Ma, X., Geiser-Lee, J., Deng, Y., Kolmakov, A., 2010. Interactions between engineered nanoparticles (ENPs) and plants: phytotoxicity, uptake and accumulation. Sci. Total Environ. 408, 3053−3061.

Milani, N., McLaughlin, M.J., Stacey, S.P., Kirby, J.K., Hettiarachchi, G.M., Beak, D.G., et al., 2012. Dissolution kinetics of macronutrient fertilizers coated with manufactured zinc oxide nanoparticles. J. Agric. Food Chem. 60, 3991−3998.

Mingyu, S., Hong, F., Liu, C., Wu, X., Liu, X., Chen, L., et al., 2007a. Effects of nano-anatase TiO_2 on absorption, distribution of light and photoreduction activities of chloroplast membrane of spinach. Biol. Trace Elem. Res. 118, 120−130.

Mingyu, S., Wu, X., Liu, C., Qu, C., Liu, X., Chen, L., et al., 2007b. Promotion of energy transfer and oxygen evolution in spinach photosystem II by nano-anatase TiO_2. Biol. Trace Elem. Res. 119, 183−192.

Mingyu, S., Liu, J., Yin, S., Linglan, M., Hong, F., 2008. Effects of nano-anatase on the photosynthetic improvement of chloroplast damaged by linolenic acid. Biol. Trace Elem. Res. 124, 173−183.

Miralles, P., Johnson, E., Church, T.L., Harris, A.T., 2012. Multiwalled carbon nanotubes in alfalfa and wheat: toxicology and uptake. Environ. Sci. Technol. 9, 3514−3527.

Monica, R.C., Cremonini, R., 2009. Nanoparticles and higher plants. Caryologia 62, 161−165.

Moore, M.N., 2006. Do nanoparticles present ecotoxicological risks for the health of the aquatic environment. Environ. Int. 32, 967−976.

Nair, R., Kumar, D.S., 2013. Plant disease-control and remedy through nanotechnology. In: Tuteja, N., Gill, S.S. (Eds.), Crop Improvement Under Adverse Conditions. Springer, pp. 231–244.

Nair, R., Varghese, S.H., Nair, B.G., Maekawa, T., Yoshida, Y., Kumar, D.S., 2010. Nanoparticulate material delivery to plants. Plant Sci. 179, 154–163.

Navarro, E., Baun, A., Behra, R., Hartmann, N.B., Filser, J., Miao, A.J., et al., 2008. Environmental behavior and ecotoxicity of engineered nanoparticles to algae, plants, and fungi. Ecotoxicology 17, 372–386.

Nel, A., Xia, T., Madler, L., Li, N., 2006. Toxic potential of materials at the nano level. Science 311, 622–627.

Nowack, B., Bucheli, T.D., 2007. Occurrence, behavior and effects of nanoparticles in the environment. Environ. Pollut. 150, 5–22.

Ovecka, M., Lang, I., Baluska, F., Ismail, A., Illes, P., Lichtscheidl, I.K., 2005. Endocytosis and vesicle trafficking during tip growth of root hairs. Protoplasma 226, 39–54.

Park, H.J., Kim, S.H., Kim, H.J., Choi, S.H., 2006. A new composition of nanosized silica-silver for control of various plant diseases. Plant Pathol. J. 22, 295–302.

Parsons, J.G., Lopez, M.L., Gonzalez, C.M., Peralta-Videa, J.R., Gardea-Torresdey, J.L., 2010. Toxicity and biotransformation of uncoated and coated nickel hydroxide nanoparticles on mesquite plants. Environ. Toxicol. Chem. 29, 1146–1154.

Pokhrel, L.R., Dubey, B., 2013. Evaluation of developmental responses of two crop plants exposed to silver and zinc oxide nanoparticles. Sci. Total Environ. 452–453, 321–332.

Rico, C.M., Majumdar, S., Duarte-Gardea, M., Peralta-Videa, J.R., Gardea-Torresdey, J.L., 2011. Interaction of nanoparticles with edible plants and their possible implications in the food chain. J. Agric. Food Chem. 59, 3485–3498.

Roco, M.C., 2003. Broader societal issue of nanotechnology. J. Nanopart. Res. 5, 181–189.

The Royal Society, The Royal Academy of Engineering, 2004. Nanoscience and Nanotechnologies: Opportunities and Uncertainties. The Royal Society & The Royal Academy of Engineering, London, UK.

Servin, A.D., Castillo-Michel, H., Hernandez-Viezcas, J.A., Diaz, B.C., Peralta-Videa, J.R., Gardea-Torresdey, J.L., 2012. Synchrotron micro-XRF and micro-XANES confirmation of the uptake and translocation of TiO_2 nanoparticles in cucumber (*Cucumis sativus*). Plants Environ. Sci. Technol. 46, 7637–7643.

Shah, V., Belozerova, I., 2009. Influence of metal nanoparticles on the soil microbial community and germination of lettuce seeds. Water Air Soil Pollut. 197, 143–148.

Shaw, A.K., Hossain, Z., 2013. Impact of nano-CuO stress on rice (*Oryza sativa* L.) seedlings. Chemosphere 93, 906–915.

Shen, C.X., Zhang, Q.F., Li, J., Bi, F.C., Yao, N., 2010. Induction of programmed cell death in Arabidopsis and rice by single-wall carbon nanotubes. Am. J. Bot. 97, 1–8.

Stampoulis, D., Sinha, S.K., White, J.C., 2009. Assay-dependent phytotoxicity of nanoparticles to plants. Environ. Sci. Technol. 43, 9473–9479.

Stern, S.T., McNeil, S.E., 2008. Nanotechnology safety concerns revisited. Toxicol. Sci. 101, 4–21.

Tan, X.M., Fugetsu, B., 2007. Multi-walled carbon nanotubes interact with cultured rice cells: evidence of a self-defense response. J. Biomed. Nanotechnol. 3, 285–288.

Tan, X.M., Lin, C., Fugetsu, B., 2009. Studies on toxicity of multiwalled carbon nanotubes on suspension rice cells. Carbon 47, 3479–3487.

Tayel, A.A., Waelf, T., Shaaban, M.A., Mohammed, F.S., 2011. Antibacterial action of zinc oxide nanoparticle against foodborne pathogens. J. Food Saf. 31, 211–218.

Tsuji, K., 2001. Microencapsulation of pesticides and their improved handling safety. J. Microencapsul. 18, 137–147.

USEPA, 2007. Nanotechnology white paper. Document Number EPA 100-B-07001. Available at <www.epa.gov/osa>.

Uzu, G., Sobanska, S., Sarret, G., Munoz, M., Dumat, C., 2010. Foliar lead uptake by lettuce exposed to atmospheric pollution. Environ. Sci. Technol. 44, 1036−1042.

Vinopal, S., Ruml, T., Kotrba, P., 2007. Biosorption of Cd^{2+} and Zn^{2+} by cell surface-engineered Saccharomyces cerevisiae. Int. Biodeterior. Biodegr. 60, 96−102.

Wang, J.X., Sun, X.W., Wei, A., Lei, Y., Cai, X.P., Li, C.M., et al., 2006. Zinc oxide nanocomb biosensor for glucose detection. Appl. Phys. Lett. 88, 233106.

Watanabe, T., Misawa, S., Hiradate, S., Osaki, M., 2008. Root mucilage enhances aluminum accumulation in *Melastoma malabathricum*, an aluminum accumulator. Plant Signal. Behav. 3, 603−605.

Wessels, J.G.H., 1993. Wall growth, protein excretion and morphogenesis in fungi. New Phytol. 123, 397−413.

Whitby, M., Quirke, N., 2007. Fluid flow in carbon nanotubes and nanopipes. Nat. Nanotechnol. 2, 87−94.

Wild, E., Jones, K.C., 2009. Novel method for the direct visualization of in vivo nanomaterials and chemical interactions in plants. Environ. Sci. Technol. 43, 5290−5294.

Xuming, W., Fengqing, G., Linglan, M., Jie, L., Sitao, Y., Ping, Y., et al., 2008. Effects of nano-anatase on ribulose-1,5-biphosphate carboxylase/oxygenase mRNA expression in spinach. Biol. Trace Elem. Res. 126, 280−289.

Yan, S., Zhao, L., Li, H., Zhang, Q., Tan, J., Huang, M., et al., 2013. Single-walled carbon nanotubes selectively influence maize root tissue development accompanied by the change in the related gene expression. J. Hazard. Mater. 246−247, 110−118.

Yang, F., Hong, F., You, W., Liu, C., Gao, F., Wu, C., et al., 2006a. Influences of nano-anatase TiO_2 on the nitrogen metabolism of growing spinach. Biol. Trace Elem. Res. 110, 179−190.

Yang, F., Liu, C., Gao, F., Su, M., Wu, X., Zheng, L., et al., 2007. The improvement of spinach growth by nano-anatase TiO_2 treatment is related to nitrogen photoreduction. Biol. Trace Elem. Res. 119, 77−88.

Yang, K., Zhu, L., Xing, B., 2006b. Adsorption of polycyclic aromatic hydrocarbons by carbon nanomaterials. Environ. Sci. Technol. 40, 1855−1861.

Yang, L., Watts, D.J., 2005. Particle surface characteristics may play an important role in phytotoxicity of alumina nanoparticles. Toxicol. Lett. 158, 122−132.

Yao, K.S., Wang, D.Y., Chang, C.Y., Weng, K.W., Yang, L.Y., Lee, S.J., et al., 2007. Photocatylitic disinfection of phytopathogenic bacteria by dye sensitized TiO_2 thin film activated by visible leaf. Surf. Coat. Technol. 202, 1329−1332.

Yu-Nam, Y., Lead, R., 2008. Manufactured nanoparticles: an overview of their chemistry, interactions and potential environmental implications. Sci. Total Environ. 400, 396−414.

Zheng, L., Hong, F., Lu, S., Liu, C., 2005. Effects of nano-TiO_2 on strength of naturally aged seeds and growth of spinach. Biol. Trace Elem. Res. 104, 83−92.

Zhu, H., Han, J., Xiao, J.Q., Jin, Y., 2008. Uptake, translocation, and accumulation of manufactured iron oxide by pumpkin plants. J. Environ. Monit. 10, 713−717.

Role of miRNAs in Abiotic and Biotic Stresses in Plants

Syed Sarfraz Hussain and Bujun Shi

9.1 Introduction: miRNA as a significant player in gene regulation

The diversity and widespread existence of small RNAs can safely be attributed to their crucial roles as regulators of gene expression in eukaryotes (Carrington and Ambros, 2003; Lai et al., 2003; Bartel, 2004). Historically, the idea of small RNA molecules as regulators of gene expression is long standing (Jacob and Monod, 1961; Britten and Davidson, 1969). However, recent break-throughs in identification and characterization of small RNAs and associated regulatory mechanisms have been instrumental in the understanding of complex gene regulatory pathways in plants (Reyes and Chau, 2007; Schommer et al., 2008; Subramanian et al., 2008; Liu et al., 2009c). These small RNAs encompass many different classes of non-coding RNAs, which include microRNAs (miRNAs), short interfering RNAs (siRNAs), small nucleolar RNAs (snoRNAs), small nuclear RNAs (snRNAs), and small RNAs derived from other elements such as transfer RNAs, ribosome RNAs, and so on. Each class of these small RNAs has their own properties and functions. miRNAs are around 22 nucleotides in length and down-regulate gene expression at the levels of translational, transcription, and post-transcription. miRNAs have been shown to be able to regulate almost every aspect of plant development and growth (Willmann and Poethig, 2007; Mathieu et al., 2009). In plants, miRNA genes are abundant. Their intriguing expression patterns suggest that miRNAs might play pivotal roles in stress tolerances (see Table 9.1).

Many genes including transcription factors that were previously revealed to control cell division and expression (Garcia, 2008), leaf morphogenesis and polarity (Chitwood et al., 2007; Nogueira et al., 2009), floral differentiation and development (Achard et al., 2004; Ru et al., 2006; Wu et al., 2006; Zhao et al., 2007), developmental phase transition (Jones-Rhoades et al., 2006; Chuck et al., 2009), and plant hormone signaling (Reyes and Chau, 2007; Lu and Huang, 2008) turned out to be under the control of miRNAs. miRNA regulatory activity in plants has immediate implications in normal growth and development. One of the first reports providing the genetic basis for an miRNA involvement was that loss-of-function *dcl1 Arabidopsis* mutant plants impaired in miRNA production and ectopically expressing miRNA target genes exhibited a range of developmental abnormalities ranging from vegetative to reproductive phase change (Laufs et al., 2004; Mallory et al., 2004b; Bäurle and Dean, 2006). Moreover, the mutations in the regulators of miRNA biogenesis, transport, and processing genes such as *ago1*, *hen1*, and *hyl1* also lead to *dcl1*-like phenotypes (Mallory and Vaucheret, 2006).

P. Ahmad (Ed): Emerging Technologies and Management of Crop Stress Tolerance, Volume 1.
DOI: http://dx.doi.org/10.1016/B978-0-12-800876-8.00009-6

Table 9.1 A Few Pioneer Studies Exploring the Role of microRNAs in Abiotic Stresses and Nutrient Deprivation in Plants

miRNA	Plant Species	Target	Target Status	Stress				Additional Result	References
				Salt	Dehydration	Cold	ABA		
miR319	*Arabidopsis*	MYB/TCP	Confirmed				+		Sunkar and Zhu (2004); Lee et al. (2004)
miR389	*Arabidopsis*	Unknown		−	−	−	−		Sunkar and Zhu (2004)
miR393	*Arabidopsis, rice*	TIR1	Putative	+	+	+	+		Sunkar and Zhu (2004); Zhao et al. (2007); Liu et al. (2008)
miR395	*Arabidopsis, corn, P. tremula*	APS	Confirmed	+			+	Sulfate deficiency	Chiou (2007); Ding et al. (2009); Jia et al. (2009)
miR396	*Arabidopsis, corn*	GRL	Confirmed	+(−)	+	−(+)	−(+)		Sunkar and Zhu (2004); Liu et al. (2008); Ding et al. (2009); Liu et al. (2009b)
miR397	*Arabidopsis*	Laccase	Putative	+	+	+	+		Sunkar and Zhu (2004)
miR398	*Arabidopsis, P. tremula*	CSD	Confirmed	−(+)		−(UV-B)	−(+)		Sunkar et al. (2006); Yamasaki et al. (2007); Zhou et al. (2007); Jagadeeswaran et al. (2009); Jia et al. (2009)
miR399	*Arabidopsis, P. tremula*	PHO2/UBC24	Confirmed					Phosphorous deficiency	Allen et al. (2005); Bari et al. (2006); Pant et al. (2008); Jia et al. (2009)

+ : induced; − : repressed

In addition, spatio-temporal accumulation of miRNAs at specific developmental stages and in certain tissues/cell types also causes striking developmental abnormalities in several cases (Baker et al., 2005; Guo et al., 2005; Wu and Poethig, 2006; Gandikota et al., 2007; Kutter et al., 2007). An additional layer of regulation in miRNA action uncovered by bioinformatics revealed that most of the miRNA targets encode transcriptional families that control major events in plant development (Rhoades et al., 2002; Juarez et al., 2004; Mallory et al., 2004a; Kim et al., 2005; Williams et al., 2005; Byrne, 2006). These transcription factor families include HD-ZIP III (Chitwood et al., 2007; Zhou et al., 2007; Boualem et al., 2008), SCARECROW-like (SCL; Llave et al., 2002b), APETALA2 (AP2) and AP2-like (Aukerman and Sakai, 2003; Chen, 2004; Zhao et al., 2007), the auxin response factor (ARF) family (Mallory et al., 2005; Sorin et al., 2005; Yang et al., 2006), the squamosa promoter binding protein (SBP) family, and CUP-SHAPED COTYLEDON (CUC1) and CUC2 (Laufs et al., 2004; Baker et al., 2005; Guo et al., 2005).

Furthermore, several miRNAs have been found to modulate key components of hormone-signaling pathways, hormone homeostasis, and plant responses to hormones as evidenced by the presence of many plant miRNAs (miR159, miR160, miR164, miR167, miR393, miR397b, miR402, miR413) in tissues treated by ABA, JA, SA, and other phytohormones (Sunkar and Zhu, 2004; Zhang et al., 2005; Reyes and Chau, 2007; Schommer et al., 2008; Liu et al., 2009a). Although several miRNAs have been shown to regulate hormone-signaling network, the overall regulatory scheme is still elusive.

9.2 Mechanisms of miRNA biogenesis and function

Plant miRNAs are the transcriptional products of MIR genes located mainly in the intergenic regions of genomes (Chen, 2004; Kim, 2005; Ying et al., 2008). In some cases, miRNAs are produced from introns, non-protein coding genes, or exons of long non-protein coding transcripts (Rodriguez et al., 2010). MIR genes are generally transcribed by RNA polymerase II with some exceptions, which are transcribed by RNA polymerase III into a long miRNA precursor termed a primary transcript (pri-miRNA). pri-miRNA is then processed into a stem-loop structure with a two-nucleotide overhang at its 3′ end (Park et al., 2002; Papp et al., 2003), which is called precursor miRNA (pre-miRNA). pre-miRNA is further processed to a mature miRNA duplex (miRNA: miRNA*) by a Dicer homologue, called Dicer-like 1 (DCL1) with the assistance of two proteins, HYPONASTIC LEAVES1 (HYL1) and SERRATE (SE) (Bartel, 2004; Kurihara and Watanabe 2004; Gasciolli et al., 2005; Kurihara et al., 2006). In fact, scientific evidence indicates that plant pri-miRNAs are longer than the sequence needed to form a stem-loop structure (Aukerman and Sakai, 2003; Jones-Rhoades and Bartel, 2004; Xie et al., 2005). The next crucial step in miRNA biogenesis is methylation. The last nucleotide of the 3′ end is methylated by the small RNA-specific methyl transferase, HUA ENHANCER1 (HEN1) (Yu et al., 2005). HEN1 adds methyl group to the ribose in either strand of the duplex. Addition of methyl group is thought to protect the duplex against degradation by endonucleases (Li et al., 2005) and further prevent the miRNAs acting as primers for the RNA-dependent RNA polymerase (RNRP) from generating dsRNA using the target mRNA as template (Yu et al., 2005). In *Arabidopsis*, *hen1* defective in methylation, small RNAs have an additional unidine nucleotide on their 3′ ends indicating that methylation functions to protect miRNAs from the poly-uridylation (Yang et al., 2006). All this processing is believed to occur in nuclear dicing bodies (D-bodies) within the nucleus of the plant cell (Fang and Spector, 2007). Following methylation, the mature miRNA duplexes are then transported to the cytoplasm by a nuclear exporting protein HASTY (HST), a plant homologue of animal Exportin 5 protein (Bollman et al., 2003; Park et al., 2005). After being exported to the cytoplasm, individual mature miRNA is incorporated into RNA-induced silencing complex (RISC) with the help of Argonaute (AGO) proteins (Baumberger and Baulcombe, 2005). Generally, only one strand, called the guide strand, is incorporated into RISC and the other strand, called the passenger strand denoted with an asterisk (*), is degraded due to its lower level of stability. Strand selection and incorporation into RISC depends on asymmetry exhibited by the 5′ ends of the duplex (Khvorova et al., 2003; Schwarz et al., 2003). In some cases, both strands can be viable and functional (Okamura et al., 2008). Research suggests that plant miRNA* carries important regulatory functions (German et al., 2008; Mi et al., 2008; Takeda et al., 2008).

miRNA-loaded RISC regulates the target gene expression by binding to the target mRNA in a sequence-specific manner and then cleaving the target (Llave et al., 2002b; Baumberger and Baulcombe, 2005), or preventing the translation (Aukerman and Sakai, 2003; Chen, 2004; Arteaga-Vazquez et al., 2006; Gandikota et al., 2007; Brodersen et al., 2008; Vazquez et al., 2010). In addition, specific miRNAs can silence genes at the transcriptional level by chromatin methylation (Bao et al., 2004; Mallory et al., 2004b; Qi et al., 2006). Increasing molecular evidence favors the idea that miRNAs are assigned to different AGO proteins based on the 5′ nucleotide of the guide strand, which results in different RISCs (Mi et al., 2008). In plants, high complementarity of miRNAs to their target mRNAs supports cleavage of target mRNAs by the miRNA-loaded RISC (Jones-Rhoades et al., 2006). In addition to this, resulting mRNA cleavage products are then further degraded by other mechanisms (Shen and Goodman, 2004; Souret et al., 2004). In fact, it is no longer true that mRNA cleavage is widespread in plants because increasing evidence has shown translation repression as a predominant mechanism (Brodersen and Voinnet, 2009). miRNA regulatory activity has immediate effects on normal growth and development (Llave, 2004).

9.3 miRNA-mediated functions in plants

Currently, the miRNA database shows 24,521 mature miRNA sequences (Release 20.0, June 2013; http://www.mirbase.org/index.shtml), which include 6843 miRNAs from 64 plant species. The majority of these miRNAs were identified from model plants like *Arabidopsis thaliana*, *Oryza sativa*, *Physcomitrella patens*, *Populus trichocarpa*, and *Vitis vinifera*, owing to the fact that complete genomes of these plants have already been sequenced and are readily available. However, the rate of miRNA identification in crop plants has increased rapidly due to the availability of high-throughput sequencing methods and improved computational and experimental protocols. Examples include alfalfa, wheat, soybean, peanut, barley, etc. (Yao et al., 2007; Subramanian et al., 2008; Jagadeeswaran et al., 2009; Lelandais-Briere et al., 2009; Joshi et al., 2010; Zhao et al., 2010; Schreiber et al., 2011; Kim et al., 2012; Li et al., 2012; Wang et al., 2012b; Zhang et al., 2012; Liang et al., 2013; Lin and Lai, 2013; Mantri et al., 2013).

The plethora of plant miRNA targets a large number of genes involved in biological processes including plant growth and development as well as responses to environmental stresses and pathogen invasion (Chen, 2004, 2005; Willmann and Poethig, 2005; Jones-Rhoades et al., 2006; Mallory and Vaucheret, 2006; Nogueira et al., 2006; Sunkar et al., 2006; Laporte et al., 2007; Wang and Li, 2007; Zhang et al., 2007; Navarro et al., 2008; Berger et al., 2009; Chitwood et al., 2009; Chuck et al., 2009; Husbands et al., 2009; Liu and Chen 2009; Meng et al., 2009; Rodriguez et al., 2010). However, the functional mechanisms of miRNAs remain to be elucidated (Liang et al., 2013).

Loss- and gain-of-function mutants are important genetic resources for identification of functions of genes. The same strategies can be used to identify the functions of miRNAs. However, because an miRNA family contains many members and each member can target multiple genes or the same sets of genes (Sieber et al., 2007), loss of function of an miRNA member may not generate phenotypes or generated phenotypes are complex, which makes it difficult to confirm miRNA function. Therefore, so far, only few miRNA mutants have successfully been applied for the identification of miRNA functions in plants (Aukerman and Sakai, 2003; Palatnik et al., 2003; Baker et al., 2005; Guo et al., 2005; Kim et al., 2005; Williams et al., 2005).

In *Arabidopsis*, over 184 miRNAs have been identified, which are predicted to regulate more than 600 genes including 225 known targets (Griffiths-Jones et al., 2008; Alves et al., 2009). Some of these miRNAs have been analyzed at the molecular level for their roles in the regulation of target genes (Llave et al., 2002a; Reinhart et al., 2002; Chen, 2004; Laufs et al., 2004; Duan et al., 2006). The regulatory roles of miRNAs in plants have been established primarily through overexpression or by generating plants that express miRNA-resistant versions (Chen, 2004; Park et al., 2005; Schwab et al., 2005; Gandikota et al., 2007; Li et al., 2010a; Khan et al., 2011; Bustos-Sanmamed et al., 2013; Turner et al., 2013).

9.4 Genome-wide miRNA profiling under abiotic stresses

With a combination of different computational and experimental approaches, several groups have demonstrated the regulatory function of many miRNAs in various plant growth, developmental, and reproductive processes (Llave et al., 2002a; Aukerman and Sakai, 2003; Palatnik et al., 2003; Achard et al., 2004; Chen, 2004; Juarez et al., 2004; Laufs et al., 2004; Mallory et al., 2004a; Guo et al., 2005; Kim et al., 2005; Lauter et al., 2005; Jones-Rhoades et al., 2006; Nikovics et al., 2006; German et al., 2008; Zhang et al., 2008; Lelandais-Briere et al., 2009; Xie et al., 2010, 2011; Yu et al., 2011; Dong et al., 2012; Kim et al., 2012; Wang et al., 2007, 2012a; Dehury et al., 2013). The discovery of the regulatory role of miRNAs in plant growth and development diverted the attention of scientists towards identification of stress-related miRNAs (Reinhart et al., 2002; Jones-Rhoades and Bartel, 2004; Sunkar and Zhu, 2004; Fujii et al., 2005; Lu et al., 2005; Aung et al., 2006; Chiou et al., 2006). To discover stress-related miRNAs, high-throughput sequencing methods are currently widely adopted. The methods involve the construction of small RNA libraries from different tissues of plants treated under stress conditions such as drought, salinity, cold, and ABA (Sunkar and Zhu, 2004).

In a pioneering study, genome-wide profiling and analysis of miRNAs under drought stress in rice were performed using microarray (Zhao et al., 2007) and miRNA sequences annotated in miRBase (Griffiths-Jones, 2004). Although several miRNAs showed differential expression on microarray, significant up-regulation of miR169g under drought stress came under intense research.

As a matter of fact, the presence of two drought-responsive elements (DRE) in the promoter region of miR169g suggested that miR169g may be regulated directly by rice CBF/DREB transcription factors (Dobouzet et al., 2003; Zhao et al., 2007).

Using a similar microarray approach, a study based on systematic expression analysis of miRNAs under abiotic stresses in 2-week old *Arabidopsis* seedlings detected 10, 4, and 10 miRNAs regulated by high salinity, drought, and cold, respectively (Liu et al., 2008). Furthermore, computational analysis of stress-regulated miRNAs showed the presence of sequence motifs (AREs-anaerobic induction elements, ABREs-ABA-response elements) among the promoters of the miRNAs, suggesting that these miRNAs are involved in various stress-response processes (Liu et al., 2008).

Microarray analysis in *Populus tremula* showed expression dynamics of miR398 under salt stress with high to low and final accumulation of the miR398 level during different stages of stress (Jia et al., 2009). However, this expression pattern was absent in *Arabidopsis* where miR398 expression was steadily suppressed. Similarly, Ding et al. (2009) identified 98 miRNAs from 27 families revealing significant changes in expression under salt stress in salt-tolerant and salt-sensitive maize lines. miR168 showed a salt stress-responsive pattern similar to that in *Arabidopsis*.

Zhou et al. (2010) used microarray for genome-wide identification and analysis of miRNAs responsive to drought in rice, across a wide range of development stages from tillering to inflorescence formation. Significantly more miRNA sequences were included in the microarray chip because more miRNAs were available in miRBase in 2010. Overall, the results showed that 14 miRNAs were up-regulated, while 16 miRNAs were down-regulated. This study also analyzed the promoters of 18 miRNAs in order to reveal a relationship between a drought-induced miRNA, target gene, and the relevant promoter. miR169 and miR171 were found to have five kinds of *cis*-elements in upstream regions, which are consistent with aforementioned studies. Interestingly, miR854 was found to be induced by drought stress in rice. However, its mechanism of action remains elusive (Zhou et al., 2010). Similarly, several other studies used other high-throughput techniques for exploring the miRNA profiles in response to salinity, cold, drought, auxins, mechanical and submergence stresses (Rajagopalan et al., 2006; Fahlgren et al., 2007; Lu et al., 2008; Sunkar et al., 2008; Ding et al., 2009; Jia et al., 2009; Liu et al., 2009b; Jian et al., 2010; Song et al., 2011; Wang et al., 2011; Barakat et al., 2012; Eldem et al., 2012; Li et al., 2013; Ozhuner et al., 2013; Shuai et al., 2013; Yanik et al., 2013).

Although a large number of plant miRNAs have been identified and their corresponding targets were *in silico* predicted, the roles of some of these miRNAs in diverse physiological processes have not been elucidated yet.

9.5 Involvement of miRNAs in plant stresses
9.5.1 Abiotic stresses

High-throughput gene expression analysis revealed a modulated expression of several hundred plant genes under different abiotic stresses (Seki et al., 2002; Bartels and Sunkar, 2005; Shinozaki and Yamaguchi-Shinozaki, 2007). In addition to developmental regulation, it is speculated that miRNAs might be involved in specific physiological responses to different stresses (Sunkar and Zhu, 2004). There has been strong evidence leading to the understanding that miRNAs are hypersensitive to both abiotic and biotic stresses as well as diverse physiological processes (Sunkar and Zhu, 2004; Lu et al., 2005). Functional analysis reveals that several plant miRNAs play vital roles in plant tolerance to abiotic stresses (Fujii et al., 2005; Chiou et al., 2006; Sunkar et al., 2006; Chiou, 2007; Chen et al., 2012a,b; Shen et al., 2013; Xu et al., 2013). The first indication of miRNA's involvement in abiotic stresses came from a bioinformatics approach (Jones-Rhoades and Bartel, 2004; Zhang et al., 2005), followed by miRNA cloning from stressed *Arabidopsis*, digging out new miRNAs that had not been previously cloned from plants grown under normal conditions (Sunkar and Zhu, 2004; Fujii et al., 2005; Chiou et al., 2006; Chiou, 2007).

Several miRNAs demonstrated regulatory roles in different plant species under drought, salinity, and extreme temperature such as miR156, miR159, miR164, miR165, miR167, miR168, miR169; miR171; miR319; miR393; miR395, miR396; miR397; miR398; miR399; miR402, miR408 (Sunkar and Zhu, 2004; Liu et al., 2008; Arenas-Huertero et al., 2009; Barrera-Figueroa et al., 2011; Frazier et al., 2011; Kantar et al., 2011; Kulcheski et al., 2011; Sunkar et al., 2012). High expression of miR395 upon sulfate starvation provided the clue that miRNAs can be induced by environmental factors (Jones-Rhoades and Bartel, 2004). miR395 targets ATP sulfurylas (APS) and sulfate transporter. APS is involved in sulfate assimilation into cysteine amino acid and 5'-adenylyl-sulfate. A later compound is further converted to sulfate esters by cytosolic APS activity (Kawashima et al., 2009). On the other hand, miR395 also targets low-affinity sulfate transporter SULTR2;1 mRNA and regulates distribution of sulfate. miR395 activity and sulfate availability are negatively correlated. Therefore, under high sulfate condition, miR395 activity is down-regulated and vice versa (Chiou, 2007; Sunkar et al., 2007). This further highlights one important miRNA function, that miRNA regulation is not only dependent on the structural features of the target genes (Liu et al., 2009c) but is also related to the biological regulation of the target genes.

Similarly, miR399 expression was highly induced whereas mRNA encoding a ubiquitin-conjugating enzyme (UBC) was reduced by low phosphate stress. Further study showed that miR399 targets UBC mRNA, thereby reducing the accumulation of UBC mRNA. As a result of this interaction, induction of phosphate (Pi) transporter gene (AtPT1) is stopped, which further restricted the attenuation of primary root elongation (Fujii et al., 2005; Chiou et al., 2006). Transgenic plants constitutively expressing miR399 showed suppressed UBC mRNA accumulation even under high Pi. In fact, miR399 targets the 5' UTR region of UBC mRNA and this regulation is important for conferring tolerance to plants under phosphate starvation. In a different set of experiments, transgenic *Arabidopsis* plants with constitutive expression of miR399 accumulated more Pi, exhibited no inhibition of primary root growth (miRNA deregulation), and reduced induction of a Pi transporter gene by low Pi stress. Interestingly, in grafting experiments, miR399 was shown to be a phloem-mobile long-distance signal for the regulation of Pi homeostasis, suggesting a role of miRNAs in systemic nutrient homeostasis (Pant et al., 2008).

Oxidative stress adaptation is manifested by the activation of superoxide dismutase (SOD) genes in plants. Interestingly, the up-regulation of the two CSD (CSD1 and CSD2) genes was found to be dependent on changes in miR398 levels. Under normal conditions, miR398 cleaves the mRNAs accumulated by the transcription of these two genes. This miRNA is transcriptionally down-regulated to release its suppression of CSD1 and CSD2 genes under oxidative stress (Sunkar et al., 2006). This is an example of positive gene regulation by miRNA. Additionally, transgenic *Arabidopsis* plants overexpressing an miR398-resistant form of CSD2 showed more CSD2 mRNA accumulation compared to transgenic plants overexpressing wild-type CSD2 and in continuation showed much improved tolerance to high light, heavy metals, and other oxidative stresses (Sunkar et al., 2006).

In plants, copper is an essential micronutrient, involved in a variety of physiological functions including photosynthetic and respiratory electron transport, oxidative stress protection, cell wall metabolism, and ethylene perception (Marschner, 1995). With respect to copper nutrition, Yamasaki et al. (2007) proposed that the miR398-mediated down-regulation of Cu/Zn SOD may allow plants to save copper for other essential functions such as plastocyanin, which is active in green photosynthetic tissues. Consistent with the bioinformatics approach (Jones-Rhoades and Bartel, 2004), miR398 targets mRNAs that encode Cu proteins CSD1, CSD2, and COX5b-1. Therefore, as a regulator of abundant Cu proteins, miR398 is a key factor in the control of Cu homeostasis in the plant cell. The situation becomes more complicated with the identification of three miRNAs, miR397, miR408, and miR857, in several independent studies. These miRNAs are predicted to target the transcripts for other copper enzymes in *Arabidopsis*, rice and poplar (Bonnet et al., 2004; Jones-Rhoades and Bartel, 2004; Sunkar and Zhu, 2004; Lu et al., 2005; Sunkar et al., 2006; Fahlgren et al., 2007). Recently, Abdel-Ghany and Pilon (2008) provided evidence that miR397, miR408, and miR857 are also involved in copper metabolism and these are accumulated when copper is deficient and disappear when copper is sufficient. Based on data, these authors proposed that copper-regulated miRNAs have systemic roles in plants to regulate the abundance of a number of copper enzymes in response to fluctuations in copper availability.

9.5.2 Biotic stresses

In addition to abiotic stresses, biotic stresses such as bacteria, fungi, viruses, nematodes, and insects also significantly affect plant growth and productivity. Plants being sessile in nature respond to these biotic stresses via their anatomical, physiological, and molecular mechanisms. Many studies have revealed the role of miRNAs in plant disease resistance responses (Kasschau et al., 2004; Navarro et al., 2006; Fahlgren et al., 2007; Lu et al., 2007; He et al., 2008; Jin, 2008; Bazzini et al., 2009; Katiyar-Agawall and Jin, 2010; Li et al., 2010b; Blevins et al., 2011; Du et al., 2011; Hu et al., 2011; Lang et al., 2011; Zhang et al., 2011b). Perception of flagellin is an important step in plant resistance to *Pseudomonas syringae* bacterium (Gómez-Gómez and Boller, 2002) and miR393 was the first miRNA to be demonstrated to play an important role in plant antibacterial mechanism. miR393 regulates auxin signaling by binding to auxin receptor TIR1 (transport inhibitor response 1) and ABF2&ABF3 (auxin signaling F-box protein 2&3). Although there is no direct evidence of miR393 involvement in antibacterial activity in plants, auxin signaling by miR393 seems to be important for resistance. The model described by Navarro et al. (2006) predicts that control of auxin signaling by miR393 restricts *P. syringae* growth in *Arabidopsis*.

Fahlgren et al. (2010) performed a small RNA expression profiling on *Arabidopsis* leaves inoculated with *pst DC3000 hrcC* at 1 and 3 hours post-inoculation. Three miRNAs (miR160, miR167, and miR393) were found to be highly induced, while one miRNA (miR825) was down-regulated after infection. In addition to these results, strong evidence from *Arabidopsis* miRNA-deficit mutants *dcl1* and *hen1* showing enhanced growth of *pst DC3000 hrcC* bacterium supports the role of miRNAs in plant defense against plant pathogens.

Zhang et al. (2011a) studied the expression of 12 target genes (including ARF8, ARF10, ARF16, ARF17, TIR1, ABF2, ABF3, MYB33, and MYB65) of miR160, miR167, miR393, and miR159 in *Arabidopsis* using *Pseudomonas* infection. The expression profiles of these genes clearly showed that the expression of these target genes was negatively correlated with the accumulation of their corresponding miRNAs. This illustrates that bacteria-regulated miRNAs that target auxin perception and signaling are up-regulated during infection, suggesting that plant miRNAs play a vital role in plant defense signaling by regulating and fine tuning multiple plant hormone pathways.

Similarly, the role of miRNAs in fungal infection was also recently revealed. Xin et al. (2010) studied the expression of miRNAs in powdery mildew-infected wheat and identified 125 putative wheat stress-responsive long non-protein coding (npc) RNAs. Similarly, some of the conserved miRNAs exhibited differential expression in response to powdery mildew infection, e.g., miR393, miR444, and miR827 were up-regulated while miR156, miR159, miR164, miR171, and miR396 were down-regulated. Besides this, negative correlation exists between miRNAs and their targets in roots infected by the plant parasitic nematode *H. Schachtii* (Bazzini et al., 2007). These results suggest that miRNAs are involved in plant—nematode interaction.

Various miRNA families showed differential expression in response to infection by viruses in many plant species including *Arabidopsis* (Bazzini et al., 2009; Blevins et al., 2011; Hu et al., 2011), rice (Du et al., 2011), *Brassica rapa* (He et al., 2008), and tomato (Lang et al., 2011). So far, no evidence of naturally occurring plant miRNAs with antiviral activity is available (Perez-Quintero et al., 2010). Therefore, plants use the established RNA silencing machinery either to degrade viral RNA or to target viral DNAs for methylation (Hohn and Vazquez, 2011; Pantaleo, 2011).

9.6 Overexpression of miRNAs to resolve their functions in abiotic stresses in plants

Plants face a variety of abiotic stresses; however, drought, salt, cold, and heat are among the most common abiotic stresses responsible for limiting agricultural productivity worldwide (Chen et al., 2012a; Sun, 2012). Several studies in various plant species have shown that miRNAs play important regulatory roles in plant response to environmental stresses by either targeting mRNAs for mRNA cleavage or translational repression (Jones-Rhoades and Bartel, 2004; Zhang et al., 2007). Defining specific functions for an miRNA in plants is tedious because, as described previously, many miRNAs are present in gene families. A simple way to overcome this problem is to overexpress an miRNA gene. However, this could lead to down-regulation of many targets at the same time. Various functional studies using either overexpression of miRNAs or the target genes revealed roles of miRNAs in plant adaptation to abiotic stresses.

In *Arabidopsis*, miR169 is down-regulated, whereas its target NFYA5 is up-regulated in response to drought (Li et al., 2008). Moreover, molecular analyses have shown that plants overexpressing NFYA5 are drought tolerant compared to plants overexpressing miR169, which are drought sensitive. These analyses clearly demonstrate the importance of miR169-mediated NFYA5 regulation for drought stress tolerance in *Arabidopsis*. In contrast to this, up-regulation of miR169 for drought tolerance has been documented in tomato (Zhang et al., 2011b). In this case, up-regulation of miR169 in response to drought stress results in down-regulation of expression of three nuclear factor Y subunit genes (SINF-YA1/2/3). Indeed, functional analyses have confirmed that transgenic plants overexpressing miR169c showed better drought tolerance probably due to reduction in stomatal opening and a decreased transpiration rate, which helps in preventing water loss from leaf (Zhang et al., 2011b). In *Prunus persica*, *Panicum virgatum*, and *Medicago truncatula*, miR169 is also down-regulated (Li et al., 2008), while in rice, *Glycine max*, and *Populus euphratica* it is also up-regulated under drought stress (Li et al., 2011; Qin et al., 2011; Zhou et al., 2010). In rice, miR169g reached the highest level in roots after 6 hours of drought stress while shoots showed high levels of this miRNA after 24 hours of stress treatment, suggesting that miR169 is induced more prominently in the roots than in the shoots (Zhao et al., 2007). This result indicates that miRNAs are differentially expressed under drought stress between different tissues or developmental stages (Reinhart et al., 2002). As discussed earlier, the existence of the two DREs in the promoter of MIR169g results in miR169g accumulation under stress and supports the role of miR169 in drought stress response (Zhao et al., 2007).

Another interest is that miR169 is also involved in the regulation of nodule development in legumes (Combier et al., 2006). In *Medicago truncatula*, *MtHAP2-1* is targeted by miR169 and expression of miRNA is specifically regulated during symbiotic interactions. Biological nitrogen fixation by nodules plays a significant role in plants to withstand stress conditions. Nodule development is blocked in transgenic plants overexpressing miR169, suggesting that miR169 negatively regulates an MtHAP2-1 mRNA level, which is essential for the differentiation of nodule cells (Combier et al., 2006). This study revealed that the regulatory role of miRNA and its target can be divergent under different stresses in different plants (Castolano et al., 2007; Jia et al., 2009).

In another example, transgenic *Arabidopsis* plants overexpressing miR396 exhibited drought tolerance compared to wild-type plants (Liu et al., 2009a). Previously, miR396 is predicted to regulate seven GRF genes in *Arabidopsis*. To resolve this, authors examined the expression of drought stress-regulated genes like COR15, COR47, and RD22, and found no significant difference in expression of these genes in both transgenic and wild-type plants (Liu et al., 2009a,b,c). However, transgenic *Arabidopsis* leaves showed low stomatal densities compared to wild-type leaves, which could be the possible reason for enhanced drought tolerance. In addition, these findings further suggested that transgenic plants showed enhanced drought tolerance due to combined effects of the decreased expression level of *AtGRF* genes and the AN3 gene. The AN3 gene is supposed to be responsible for lower leaf stomatal density (Horiguchi et al., 2005). Consistent with this, Wang et al. (2010) suggested that miR396 negatively regulates cell proliferation in leaves of transgenic *Arabidopsis* by controlling the entry into the mitotic cell cycle, which ultimately arrests cell division.

In an attempt to overexpress stress-responsive miRNAs, Jung and Kang (2007) demonstrated the functional role of miR417 in *Arabidopsis* under various abiotic stresses. Transgenic *Arabidopsis* showed retarded seed germination compared to wild-type plants in the presence of high salt or ABA, which suggests that miR417 may act as a negative regulator of seed germination in *Arabidopsis* under high salt stress.

In *Arabidopsis*, miR398 is predicted to be a key regulator for copper homeostasis by cleaving target genes such as CSD1, CSD2, and COX5b-1, suggesting a control of specific target expression by miR398 under abiotic stress (Yamasaki et al., 2007; Brodersen and Voinnet, 2009; Jagadeeswaran et al., 2009). On the other hand, Jia et al., (2009) noticed a dynamic regulation of miR398 expression in *P. tremula* during salt stress. Dynamic regulation refers to the phenomenon that miR398 expression was initially induced upon 3−4 hours of ABA or salt stress, then declined after 48 hours and finally accumulated again over a prolonged stress (72 hours). However, an opposite trend in miR398 expression was noticed in *Arabidopsis* under salt stress in previous studies. These results revealed that different plant species exhibit differential expression of miRNAs under different/same abiotic stresses (Jia et al., 2009). However, this report did not show effects of this differential/dynamic miRNA regulation at a protein level.

Feng et al. (2010) described the presence and overexpression of *Arabidopsis* miR398b in tobacco. The functional characterization of miR398b was performed in two tobacco lines showing accumulation of miR398b at different levels. Molecular analysis of transgenic tobacco plants showed that overproduction of miR398b resulted in down-regulation of antioxidant enzyme activities, probably due to post-transcriptional regulation of its target genes. In addition, transgenic plants showed retarded seed germination and seedling growth under copper and salt stresses, suggesting that miR398b exists in tobacco and acts in a conserved way similar to *Arabidopsis* (Feng et al., 2010).

Xia et al. (2012) revealed the overexpression of miR393 in rice, which resulted in two new phenotypes (increased tillers and early flowering) and two previously observed phenotypes (reduced tolerance to drought and salt and hyposensitivity to auxin). Increased tillers and early flowering time are desirable characters but impaired stress tolerance posed a dilemma regarding its use in rice molecular breeding (Gao et al., 2011; Xia et al., 2012).

Alteration in different miRNA's expression seems crucial for the attenuation of plant growth and development under stress. However, these contrasting/conflicting findings highlight the importance of detailed characterization of stress-regulated miRNAs in plants.

9.7 Innovative approaches for elucidating gene function

The publication of the genome sequence of *Arabidopsis* in 2000 (The Arabidopsis Information Resource, 2000) and the subsequent development of genomic tools have allowed the determination of the biological function of a gene, accelerating research on different aspects of plant biology. Historically, large-scale T-DNA and transposon insertion mutagenesis have proven to be exceptionally useful genetic tools for different model species such as rice, *Arabidopsis* and alfalfa (Weigel et al., 2000). Similarly, genome-wide mapping of mutations can only be performed in plants having a well-defined genetic map like wheat, rice, barley, etc. (Harushima et al., 1998; Sun et al., 2005; Wenzl et al., 2006; Peleg et al., 2008; Wang et al., 2008). Although gene identification in non-model plant species is mostly a difficult task due to technical reasons, information from *Arabidopsis* can be easily validated particularly in crop plants. Recent advances in genomic techniques, for example, microarray and availability of robust sequencing technologies, have made possible the identification of genetic polymorphism and mutation mapping in plants with large and complex genomes (Shendure and Ji, 2008; Ansorge, 2009; Lister et al., 2009).

Several lines of research indicate that plants have no antibody-based immune system as found in animals (Waterhouse and Helliwell, 2003). Gene silencing was first used to build resistance to viral diseases in plants. Although the mechanism was not well understood, resistance to virus was achieved by transforming plants with virus genes or segments and was referred to as pathogen-derived resistance (PDR) (Goldbach et al., 2003). Therefore, transgene-induced gene silencing was used in the past for demonstration to acquire resistance against several viruses (Lindbo et al., 1993;

Dougherty et al., 1994; Baulcombe, 1996; Wesley et al., 2001; Goldbach et al., 2003; Prins et al., 2008). This resulted in concerted efforts focusing on this mechanism that is just in its infancy. Later on, antisense suppression, virus-induced gene silencing (VIGS), TGS, and RNAi were used either individually or cooperatively to effectively tackle virus problem in plants.

9.7.1 Artificial miRNA technology

In recent years, our knowledge of repertoire of RNA-mediated functions has been greatly increased, with the discovery of small non-coding RNAs, which play a central part in a process called RNA silencing. Silencing is a commonly used reverse genetics tool for knocking down gene expression to elucidate or manipulate biological function of novel or agriculturally important genes (Park et al., 2009). RNA silencing induced by double-stranded RNA (dsRNA) molecules such as short hairpins, short interfering RNAs (siRNAs), and long dsRNAs has developed into a standard tool in gene function studies (gene knockdown). The dsRNA silencing technique involves ectopic expression of long dsRNAs derived from inverted repeat hairpin (hp) RNA precursor (Watson et al., 2005). Ma et al. (2004) clearly demonstrated that transgenic plants expressing the hairpin RNA (hpRNA) constructed from a segment of *rice dwarf virus* (RDV) displayed high resistance or attenuated viral symptoms. In addition, a similar strategy was used against several other viruses to develop virus-resistant plants (García and Simón-Mateo, 2006; Simón-Mateo and García, 2006; Niu et al., 2006). However, this process can generate a large number of siRNAs with varying sizes (21−24 nucleotides), which may cause unintended nonspecific gene suppression (Jackson et al., 2003; Jackson and Linsley, 2003; Xu et al., 2004). To overcome this potential problem, artificial miRNA (amiRNA) technology was developed based on the RNAi mechanism (Zeng et al., 2002; Parizotto et al., 2004; Alvarez et al., 2006; Niu et al., 2006; Schwab et al., 2006; Tang et al., 2007; Ossowski et al., 2008; Brew-Appiah et al., 2010). This technique lies in utilizing the endogenous miRNA precursors to generate new target-specific miRNAs for gene silencing in both animals and plants (Alvarez et al., 2006; Schwab et al., 2006; Qu et al., 2007; Duan et al., 2008; Khraiwesh et al., 2008; Ossowski et al., 2008; Warthmann et al., 2008; Eamens et al., 2011; Molnar et al., 2009; Zhao et al., 2009). Computational analysis and genome-wide expression profiling demonstrated that amiRNA results in more accurate gene silencing compared to other methods (Schwab et al., 2005; Duan et al., 2008; Park et al., 2009; Tang et al., 2010). Alterations in the sequence of both members of the duplex without changing structural features such as mismatches or bulges often leads to high-level accumulation of an miRNA of desired sequence (Vaucheret et al., 2004). This makes it possible to modify existing natural miRNA sequences and to create amiRNA to target any gene of interest through post-transcriptional gene silencing of the corresponding transcript (Zeng et al., 2002; Parizotto et al., 2004; Alvarez et al., 2006; Niu et al., 2006; Schwab et al., 2006; Warthmann et al., 2008). So far, no evidence of naturally occurring plant miRNAs with antiviral activity is available. However, genetically modified plants and viruses provided evidence that complementarity between a plant miRNA and virus genome is enough for antiviral activity. Based on this, several plants displaying resistance to viruses have been generated. Niu et al. (2006) first used this approach to target turnip yellow mosaic virus (TYMV) and turnip mosaic virus (TuMV), and transgenic plants exhibited resistance against both viruses. Consistent with above, transgenic tobacco plants demonstrated resistance against cucumber mosaic virus (CMV), when expressing amiRNAs derived against genes/segments in the virus genome (Chen et al., 2004; Qu et al., 2007).

Subsequently, transgenic plants also exhibited high resistance to both subgroup strains (Duan et al., 2008). Ai et al., (2011) demonstrated that multiple virus-specific resistances in transgenic plants can be developed by co-expression of several amiRNAs targeting multiple viruses. They showed that transgenic tobacco plants carrying amiRNAs targeting sequences encoding silencing suppressors HC-Pro from PVY and p25 from PVX, conferred highly specific resistance against PVY and PVX infection. Further, transgenic plants maintained resistance even under high viral pressure. In addition, these findings are consistent with previous reports (Niu et al., 2006; Qu and Fang, 2007; Duan et al., 2008; Park et al., 2009). However, keeping high virus mutation rate in view, choice of target is very important and highly conserved sequences should be selected as targets to minimize the risks.

amiRNAs were used in *Arabidopsis* where they effectively interfere with reporter gene expression (Parizotto et al., 2004). Later, efficient amiRNA-mediated gene silencing has been demonstrated for endogenous genes. Genome-wide expression analyses have shown that amiRNAs have high specificity comparable to endogenous miRNAs (Schwab et al., 2005, 2006). The amiRNA sequence can be easily optimized to quantitatively silence one or several target transcripts without affecting the expression of other transcripts. Similarly, gene silencing efficiency can be increased by the use of strong promoters and it seems that there are a few, if any, non-autonomous effects (Alvarez et al., 2006; Schwab et al., 2006; Ossowski et al., 2008). Several studies have demonstrated the usefulness of amiRNA vectors based on their specificity and versatility not only in *Arabidopsis*, tomato, and tobacco (Parizotto et al., 2004; Alvarez et al., 2006; Niu et al., 2006; Schwab et al., 2006; Qu and Fang, 2007; Eamens et al., 2010, 2011; Park et al., 2009) but also in *Physcomitrella patens* and *Chamydomonas reinhardtii* (Khraiwesh et al., 2008; Molnar et al., 2009; Zhao et al., 2009b). amiRNA vectors have also been recognized as second-generation RNAi vectors (Tang et al., 2007). In addition, Web MicroRNA Designer (http://wmd2.weigelworld.org) further facilitated the application of the amiRNA technique by providing computational incorporation of nucleotide mismatches so that they complement imperfectly the target sequences (Park et al., 2009).

In 2008, Warthmann et al. demonstrated the classical utility of amiRNA technology by simultaneously silencing three genes in both japonica and indica rice varieties. Their results clearly showed that amiRNAs can efficiently trigger gene silencing in monocot crop and in the meantime amiRNA can also be effectively utilized for modulation of characters of interest. Interestingly, the transgenes were not only stably inherited but also remained active in the progeny. This technology offers a way for time-efficient modifications of the expression of any gene(s), even for genes essentially inaccessible to other reverse genetics techniques. This technology is especially important in the analysis of relative gene contribution in quantitative characters like abiotic stresses.

Most plant miRNA families contain multiple members with individual members often expressed in a spatial and/or temporal manner, which further complicates the problem of knocking down a particular gene function. Moreover, high-throughput sequencing and computational analysis have predicted numerous putative miRNAs with unknown functions (Zhang et al., 2006; Sunkar and Jagadeeswaran, 2008). In addition, many miRNAs belong to multigene families with substantial functional redundancy, and functions of the large majority of miRNAs remain to be elucidated. Therefore, there is a dire need to further develop amiRNA technology for miRNA-mediated gene silencing and for functional analyses of miRNAs in plants (Liang et al., 2010; Tang et al., 2010; Wang et al., 2010).

Eamens et al. (2010) addressed this problem by efficiently silencing endogenous miRNAs in *Arabidopsis* by using amiRNA technology. The authors elegantly demonstrated that an amiRNA designed to target the mature miRNA sequence of individual miRNA family members efficiently knocks down all family members, whereas amiRNA designed to target the stem—loop sequence of an individual miRNA precursor transcript silences the individually targeted family member. Overall, amiRNA-mediated silencing represents a potentially alternative approach for the validation of predicted function of individual members of miRNA families in plants (Devers et al., 2013).

9.7.2 Target mimicry

Target mimicry is a phenomenon conceptually similar to miRNA sponge in animals, used to reduce the activity of endogenous miRNA. miRNA sponges are basically long sequences carrying multiple miRNA binding sites. These transcripts compete with endogenous target mRNA, subsequently reducing the efficiency of the corresponding miRNA (Ebert et al., 2007; Ebert and Sharp, 2010). Franco-Zorrilla et al. (2007) were the first to demonstrate that a non-coding transcript in *Arabidopsis* regulates the miRNA activity involved in phosphate homeostasis by mimicking its target site. In fact, Pi starvation induces the expression of both miR399, which cleaves *PHO2* (*PHOSPHATE 2*), and *IPS1* (*INDUCED BY PHOSPHATE STARVATION1*) transcripts. *IPS1* encodes a non-coding RNA which contains a region of complementarity with miR399 (Burleigh and Harrison, 1999; Martín et al., 2000; Fujii et al., 2005; Chiou et al., 2006; Shin et al., 2006). In contrast to regular miRNA target sites, the *IPS1*-miR399 pairing is interrupted by an insertion of a three-nucleotide mismatch exactly at the miRNA cleavage site (Franco-Zorrilla et al., 2007). As a result of this mismatch, *IPS1* RNA is not cleaved and instead it sequestrates miR399, thereby leading to reduced miR399 activity.

After this report, it is speculated that target mimicry can be exploited to study the effects of reducing the function of entire miRNA families (Franco-Zorrilla et al., 2007). So far, in two miRNA families, all members have been inactivated and characterized (Allen et al., 2007; Sieber et al., 2007). Similarly, Todesco et al. (2010) reported the generation of several transgenic plants expressing artificial target mimics for most of the known miRNA families. All miRNAs used in this study belong to more abundant and widely conserved families, playing significant roles in plant growth and development. Taken together, using target mimics to interfere with normal miRNA pathway led to several morphological abnormalities. However, these findings are in agreement with experimental results obtained using major approaches to study the functions of miRNAs (miRNA overexpression, expression of mutated target gene, miRNA knockout; William and Poethig, 2007). In addition, target mimicry might be useful for exploring the roles of novel miRNAs with yet unknown functions in the plant kingdom.

9.8 Conclusion and future prospects

The past few years have witnessed an explosive increase in miRNA research in plants. miRNAs have been demonstrated to be involved in gene silencing called RNA interference (RNAi), cosuppression and quelling (Hamilton and Baulcombe, 1999). Similarly, the regulatory roles of miRNAs in plants are quite diverse, and versatility of these molecules is highlighted by their involvement in nearly every cellular process. However, miRNA regulatory networks are still under exploration, and the availability of genome sequences and technological improvement to dissect the genomes will shed light on the fine tuning of miRNA regulatory processes. Similarly, one of the future challenges in miRNA research is to find novel miRNAs in different plant species, which will provide clues as to the dynamics of miRNA evolution.

The estimated overall success rate of artificial microRNA (amiRNA)-based gene silencing is approximately 75% (Ossowski et al., 2008). This provided evidence that amiRNAs have the capacity to replace the existing methods and provide a potentially powerful alternate approach for the validation of predicted miRNA functions in plants. In addition, the emergence of amiRNA as a promising approach has the potential to revolutionize the current plant improvement strategies by modulating agronomically important traits, which contributes significantly to crop yield. On the other hand, amiRNA-directed silencing has the power to provide a potentially powerful alternate approach for function validation of predicted miRNAs in plants (Devers et al., 2013). However, it is empirical to put more effort into the refinement of this technique, which has the potential to be a molecular weapon for crop improvement in the future.

This chapter, however, provides just a snapshot of the many pathways affected by miRNAs. The breadth of regulation by miRNAs as predicted is that miRNA may be utilized as a promising tool for plant improvement with respect to yield, quality, or resistance to various biotic and abiotic stresses. Although our knowledge is expanding about the roles of miRNA regulation in plants, it is only part of the whole iceberg. It is anticipated that many other novel processes will be found under the control of these molecules, as more miRNAs from different organisms become available.

References

Abdel-Ghany, S.E., Pilon, M., 2008. MicroRNA-mediated systemic down-regulation of copper protein expression in response to low copper availability in Arabidopsis. J. Biol. Chem. 283, 15932—15945.

Achard, P., Herr, A., Baulcombe, D.C., Harberd, N.P., 2004. Modulation of floral development by a gibberellins-regulated microRNA. Development 13, 3357—3365.

Ai, T., Zhang, L., Gao, Z., Zhu, C.X., Guo, X., 2011. Highly efficient virus resistance mediated by artificial microRNAs that target the suppressor of PVX and PVY in plants. Plant Biol. 13, 304—316.

Allen, E., Xie, Z., Gustafson, A.M., Carrington, J.C., 2005. MicroRNA-directed phasing during trans-acting siRNA biogenesis in plants. Cell 121, 207—221.

Allen, R.S., Li, J., Stahle, M.I., Dubroue, A., Gubler, F., Miller, A.A., 2007. Genetic analysis reveals functional redundancy and the major target genes of the Arabidopsis miR159 family. PNAS 104, 16371—16376.

Alvarez, J.P., Pekker, I., Goldshmidt, A., Blum, E., Amsellem, Z., Eshed, Y., 2006. Endogenous and synthetic microRNAs stimulate simultaneous, efficient, and localized regulation of multiple targets in diverse species. Plant Cell 18, 1134—1151.

Alves Jr, L., Niemeier, S., Hauenschild, A., Rehsmeier, M., Merkle, T., 2009. Comprehensive prediction of novel microRNA targets in Arabidopsis thaliana. Nucleic Acids Res. 37, 4010—4021.

Ansorge, W.J., 2009. Next-generation DNA sequencing techniques. Nat. Biotech. 25, 195—203.

Arabidopsis Information Resource, 2000. The Arabidopsis Information Resource (TAIR). TAIR, Stanford, California.

Arenas-Huertero, C., Pérez, B., Rabanal, F., Blanco-Melo, D., De la Rosa, C., Estrada-Navarrete, G., et al., 2009. Conserved and novel miRNAs in the legume *Phaseolus vulgaris* in response to stress. Plant Mol. Biol. 70, 385—401.

Arteaga-Vazquez, Caballero-Perez J, Viella-Calzada, J.P., 2006. A family of microRNA present in plants and animals. Plant Cell 18, 3355−3369.

Aukerman, M.J., Sakai, H., 2003. Regulation of flowering time and floral organ identity by a microRNA and its APETALA2-like target genes. Plant Cell 15, 2730−2741.

Aung, K., Lin, S.I., Wu, C.C., Huang, Y.T., Su, C.L., Chiou, T.J., 2006. pho2, a phosphate over accumulator, is caused by a nonsense mutation in a microRNA399 target gene. Plant Physiol. 141, 1000−1011.

Baker, C.C., Sieber, P., Wellmer, F., Meyerowitz, E.M., 2005. The early extra petals1 mutant uncovers a role for microRNA miR164c in regulating petal number in *Arabidopsis*. Curr. Biol. 15, 303−315.

Bao, N., Lye, K.W., Barton, M.K., 2004. MicroRNA binding sites in Arabidopsis class III HDZIP mRNAs are required for methylation of the template chromosome. Develop. Cell 7, 653−662.

Barakat, A., Siram, A., Park, J., Zhebentyayeva, T., Main, D., Abbott, A., 2012. Genome-wide identification of chilling responsive microRNAs in *Prunus persica*. BMC Genomics 13, 481.

Bari, R., Pant, B., Stitt, M., Scheible, W.R., 2006. MicroRNA399 and PHR1 define a phosphate-signalling pathway in plants. Plant Physiol. 141, 988−999.

Barrera-Figueroa, B.E., Gao, L., Diop, N.N., Wu, Z., Ehlers, J.D., Roberts, P.A., et al., 2011. Identification and comparative analysis of drought-associated microRNAs in two cowpea genotypes. BMC Plant Biol. 11, 127.

Bartel, D.P., 2004. MicroRNAs: genomics, biogenesis, mechanism, and function. Cell 116, 281−297.

Bartels, D., Sunkar, R., 2005. Drought and salt tolerance in plants. Cri. Rev. Plant Sci. 24, 23−58.

Baulcombe, D.C., 1996. Mechanisms of pathogen-derived resistance to viruses in transgenic plants. Plant Cell 8, 1833−1844.

Baumberger, N., Baulcombe, D.C., 2005. Arabidopsis ARGONAUTE1 is an RNA slicer that selectively recruits microRNAs and short interfering RNAs. Proc. Nat. Acad. Sci. 102, 3691−3696.

Bäurle, I., Dean, C., 2006. The timing of developmental transitions in plants. Cell 125, 655−664.

Bazzini, A.A., Hopp, H.E., Beachy, R.N., Asurmendi, S., 2007. Infection and co-accumulation of tobacco mosaic virus proteins alter microRNA levels, correlating with symptom and plant development. Proc. Natl. Acad. Sci. 104, 12157−12162.

Bazzini, A.A., Almasia, N.I., Manacorda, C.A., Mongelli, V.C., Conti, G., Maroniche, G.A., et al., 2009. Virus infection elevates transcriptional activity of miR164a promoter in plants. BMC Plant Biol. 9, 152.

Berger, Y., Harpaz-Saad, S., Brand, A., Melnik, H., Sirding, N., Alvarez, J.P., et al., 2009. The NAC-domain transcription factor GOBLET specifies leaflet boundaries in compound tomato leaves. Development 136, 823−832.

Blevins, T., Rajeswaran, R., Aregger, M., Borah, B.K., Schepetilnikov, M., Baerlocher, L., et al., 2011. Massive production of small RNAs from a non-coding region of cauliflower mosaic virus in plant defense and viral counter-defense. Nucleic Acids Res. 39, 5003−5014.

Bollman, K.M., Aukerman, M.J., Park, M.Y., Hunter, C., Berardini, T.Z., Poethig, R.S., 2003. HASTY, the *Arabidopsis* ortholog of exportin 5/MSN5, regulates phase change and morphogenesis. Development 130, 1493−1504.

Bonnet, E., Wuyts, J., Rouzé, P., Van de Peer, Y., 2004. Detection of 91 potential conserved plant microRNAs in Arabidopsis thaliana and Oryza sativa identify important target genes. Proc. Natl. Acad. Sci. 101, 11511−11516.

Boualem, A., Laporte, P., Jovanovic, M., Laffont, C., Plet, J., Combier, J.P., et al., 2008. MicroRNA166 controls root and nodule development in Medicago truncatula. Plant J. 54, 876−887.

Brew-Appiah R.A.T., Rustgi S., Claar M., Langen G., Kogel K.H., Weigel D. (2010) Artificial microRNAs for silencing wheat proteins causing celiac disease. Plant Biology Abs # P08017. http://abstracts.aspb.org/pb2010/public/P08/P08017.html.

Britten, J., Davidson, E.H., 1969. Gene regulation for higher cells: a theory. Science 165, 349−357.

Brodersen, P., Voinnet, O., 2009. Revisiting the principles of microRNA target recognition and mode of action. Nat. Rev. Mol. Cell. Biol. 10, 141−148.

Brodersen, P., Achard, L.S., Rasmussen, M.B., Dunoyer, P., Yamamoto, Y.Y., Sieburth, L., et al., 2008. Widespread translational inhibition by plant miRNAs and siRNAs. Science 320, 1185−1190.

Burleigh, S.H., Harrison, M.J., 1999. The down-regulation of Mt4-like genes by phosphate fertilization occurs systematically and involves phosphate translocation to the shoot. Plant Physiol. 119, 241−248.

Bustos-Sanmamed, P., Mao, G., Deng, Y., Elouet, M., Khan, G.A., Bazin, J., et al., 2013. Overexression of miR160 affects root growth and nitrogen-fixing nodule number in medicago truncatula. Func. Plant Biol. 40, 1208−1220.

Byrne, M.E., 2006. Shoot meristem function and leaf polarity: the role of class III HD-ZIP genes. PLoS Genet. 2, e89.

Carrington, J.C., Ambros, V., 2003. Role of microRNAs in plant and animal development. Science 301, 336−338.

Castolano, M., Castillo, R., Efremova, N., Kuckenberg, M., Zethof, J., Gerats, T., et al., 2007. A conserved microRNA module exerts homeotic control over *Petunia hybrida* and *Antirrhinum majus* floral organ identity. Nat. Genet. 39, 901−905.

Chen, H., Li, Z., Xiong, L., 2012a. A plant microRNA regulates the adaptation of roots to drought stress. FEBS Lett. 586, 1742−1747.

Chen, L., Wang, T., Zhao, M., Tian, Q., Zhang, W.H., 2012b. Identification of aluminium-responsive microRNAs in *Medicago truncatula* by genome-wide high throughput sequencing. Planta 235, 375−386.

Chen, X., 2004. A microRNA as a translational repressor of APETALA2 in *Arabidopsis* flower development. Science 303, 2022−2025.

Chen, X., 2005. MicroRNA biogenesis and function in plants. FEBS Lett. 579, 5923−5931.

Chen, Y., Lohuis, D., Goldbach, R., Prins, M., 2004. High frequency induction of RNA-mediated resistance against cucumber mosaic virus using inverted repeat constructs. Mol. Breed 14, 215−226.

Chiou, T.J., 2007. The role of microRNAs in sensing nutrient stress. Plant Cell Environ. 30, 323−332.

Chiou, T.J., Aung, K., Lin, S.I., Wu, C.C., Chiang, S.F., Su, C.L., 2006. Regulation of phosphate homeostasis by microRNA in Arabidopsis. Plant Cell 18, 412−421.

Chitwood, D.H., Guo, M., Nogueira, F.T., Timmermanns, M.C., 2007. Establishing leaf polarity: the role of small RNAs and positional signals in the shoot apex. Development 134, 813−823.

Chitwood, D.H., Nogueira, F.T., Howell, M.D., Montgomery, T.A., Carrington, J.C., Timmermanns, M.C., 2009. Pattern formation via small RNA mobility. Genes. Develop. 23, 549−554.

Chuck, G., Candela, H., Hake, S., 2009. Big impact by small RNAs in plant development. Curr. Opin. Plant Biol. 12, 81−86.

Combier, J.P., Frugier, F., De Billy, F., Boualem, A., El-Yahyaoui, F., Moreau, S., et al., 2006. *MtHAP2-1* is a key transcriptional regulator of symbiotic nodule development regulated by microRNA169 in *Medicago truncatula*. Genes Develop. 20, 3084−3088.

Dehury, B., Panda, D., Sahu, J., Sahu, M., Sarma, K., Barooah, M., et al., 2013. In silico identification and characterization of conserved miRNAs and their target genes in sweet potato (*Ipomoea batatas* L.) expressed sequence tags (ESTs). Plant Signal Behav. 8, e26543.

Devers, E.A., Teply, J., Reinert, A., Gaude, N., Krajinski, F., 2013. An endogenous artificial microRNA system for unravelling the function of root endosymbiosis related genes in *Medicago truncatela*. BMC Plant Biol. 13, 82.

Ding, D., Zhang, L., Wang, H., Liu, Z., Zhang, Z., Zheng, Y., 2009. Differential expression of miRNAs in response to salt stress in maize roots. Annals Bot. 103, 29−38.

Dobouzet, J.G., Sakuma, Y., Ito, Y., Kasuga, M., Dubouzet, E.G., Miura, S., et al., 2003. *OsDREB* genes in rice, *Oryza sativa* L, encode transcription activators that function in drought-, high salt- and cold responsive gene expression. Plant J. 33, 751−763.

Dong, Q.H., Han, J., Yu, H.P., Wang, C., Zhao, M.Z., Liu, H., et al., 2012. Computational identification of microRNAs in strawberry expressed sequence tags and validation of their precise sequences by miR-RACE. J. Hered. 103, 268—277.

Dougherty, W.G., Lindbo, J.A., Smith, H.A., Parks, T.D., Swaney, S., Proebsting, W.M., 1994. RNA-mediated virus resistance in transgenic plants: exploitation of a cellular pathway possibly involved in RNA degradation. Mol. Plant Microbe Interact. 7, 544—552.

Du, P., Wu, J., Zhang, J., Zhao, S., Zheng, H., Gao, G., et al., 2011. Viral infection induces expression of novel phased microRNAs from conserved cellular microRNA precursors. PLoS Pathol. 7, e1002176.

Duan, C.G., Wang, C.H., Guo, H.S., 2006. Regulation of microRNA on plant development and viral infection. Chinese Sci. Bull. 51, 269—278.

Duan, C.G., Wang, C.H., Fang, R.X., Guo, H.S., 2008. Artificial microRNAs highly accessible to targets confer efficient virus resistance in plants. J. Virol. 82, 11084—11095.

Eamens, A.L., Smith, N.A., Curtin, S.J., Wang, M.B., Waterhouse, P.M., 2010. The *Arabidopsis thaliana* double-stranded RNA binding protein DRB1 directs guide strand selection from microRNA duplexes. RNA 15, 2219—2235.

Eamens, A.L., Agius, C., Smith, N.A., Waterhouse, P.M., Wang, M.B., 2011. Efficient silencing of endogenous microRNAs using artificial microRNAs in *Arabidopsis thaliana*. Mol. Plant 4, 157—170.

Ebert, M.S., Sharp, P.A., 2010. MicroRNA sponges: progress and possibilities. RNA 16, 2043—2050.

Ebert, M.S., Neilson, J.R., Sharp, P.A., 2007. MicroRNA sponges: competitive inhibitors of small RNAs in mammalian cells. Nat. Methods 4, 721—726.

Eldem, V., Celikkol-Akcay, U., Ozhuner, E., Bakr, Y., Uranbey, S., Unver, T., 2012. Genome-wide identification of miRNAs responsive to drought in peach (*Prunus persica*) by high throughput deep sequencing. PloS One 7, e50298.

Fahlgren, N., Howell, M.D., Kasschau, K.D., Chapman, E.J., Sullivan, C.M., Cumbie, J.S., et al., 2007. High-throughput sequencing of Arabidopsis microRNAs: evidence for frequent birth and death of MIRNA genes. PLoS One 2, e219.

Fahlgren, N., Jogdeo, S., Kasschau, K.D., Sullivan, C.M., Chapman, E.J., Laubinger, S., et al., 2010. MicroRNA gene evolution in Arabidopsis lyrata and Arabidopsis thaliana. Plant Cell 22, 1074—1089.

Fang, Y., Spector, D.L., 2007. Identification of nuclear dicing bodies containing proteins for microRNA biogenesis in living Arabidopsis plants. Curr. Biol. 17, 818—823.

Feng, X.M., You, C.X., Qiao, Y., Mao, K., Hao, Y.J., 2010. Ectopic expression of Arabidopsis AtmiR393a gene changes auxin sensitivity and enhances salt resistance in tobacco. Acta Physiol. Plant 32, 997—1003.

Franco-Zorrilla, J.M., Valli, A., Todesco, M., Mateos, I., Puga, M.I., Rubio-Somoza, I., et al., 2007. Target mimicry provides a new mechanism for regulation of microRNA activity. Nat. Genet. 39, 1033—1037.

Frazier, T.P., Sun, G., Burklew, C.E., Zhang, B., 2011. Salt and drought stresses induce the aberrant expression of microRNA genes in tobacco. Mol. Biotech. 49, 159—165.

Fujii, H., Chiou, T.J., Lin, S.I., Aung, K., Zhu, J.K., 2005. A miRNA involved in phosphate starvation response in Arabidopsis. Curr. Biol. 15, 2038—2043.

Gandikota, M., Birkenbihl, R.P., Hohmann, S., Cardon, G.H., Saedler, H., Huijser, P., 2007. The miRNA156/157 recognition element in the 30 UTR of the *Arabidopsis* SBP box gene SPL3 prevents early flowering by translational inhibition in seedlings. Plant J. 49, 683—693.

Gao, P., Bai, X., Yang, L., Lv, D., Pan, X., Li, Y., et al., 2011. *OSMIR393*; a salinity and alkaline stress-related microRNA gene. Mol. Biol. Report 38, 237—242.

Garcia, D., 2008. A miracle in plant development: role of microRNAs in cell differentiation and patterning. Sem. Cell Develop. Biol. 19, 586—595.

García, J.A., Simón-Mateo, C., 2006. A micropunch against plant viruses: artificial microRNAs show promise for combating viral infections in plants. Nat. Biotech. 24, 1358—1359.

Gasciolli, V., Mallory, A.C., Bartel, D.P., Vaucheret, H., 2005. Partially redundant functions of Arabidopsis DICER-like enzymes and a role for DCL4 in producing trans-acting siRNAs. Curr. Biol. 15, 1494—1500.

German, M.A., Pillay, M., Jeong, D.H., Hetawal, A., Luo, S., Janardhanan, P., et al., 2008. Global identification of microRNA-target RNA pairs by parallel analysis of RNA ends. Nat. Biotech. 26, 941—946.

Goldbach, R., Bucher, E., Prins, M., 2003. Resistance mechanisms to plant viruses: an overview. Virus Res. 92, 207—212.

Gómez-Gómez, L., Boller, T., 2002. Flagellin perception: a paradigm for innate immunity. Trends Plant Sci. 7, 251—256.

Griffiths-Jones, S., 2004. The microRNA Registry. Nucleic Acids Res. 32, D109—D111.

Griffiths-Jones, S., Saini, H.K., van Dongen, S., Enright, A.J., 2008. miRBase: tools for microRNA genomics. Nucleic Acids Res. 36, D154—D158.

Guo, H.S., Xie, Q., Fei, J.F., Chua, N.H., 2005. MicroRNA directs mRNA cleavage of the transcription factor *NAC1* to downregulate auxin signals for Arabidopsis lateral root development. Plant Cell 17, 1376—1386.

Hamilton, A.J., Baulcombe, D.C., 1999. A species of small antisense RNA in posttranscriptional gene silencing in plants. Science 286, 950—952.

Harushima, Y., Yano, M., Shomura, A., Sato, M., Shimano, T., Kuboki, Y., et al., 1998. A high-density rice genetic linkage map with 2275 markers using a single F2 population. Genet 148, 479—494.

He, X.F., Fang, Y.Y., Feng, L., Guo, H.S., 2008. Characterization of conserved and novel microRNAs and their targets, including a TuMV-induced TIR-NBS-LRR class R gene-derived novel miRNA in Brassica. FEBS Lett. 582, 2445—2452.

Hohn, T., Vazquez, F., 2011. RNA silencing pathways of plants: silencing and its suppression by plant DNA viruses. Biochem. Biophys. Acta 1809, 588—600.

Horiguchi, G., Kim, G.T., Tsukaya, H., 2005. The transcription factor AtGRF5 and the transcription coactivator AN3 regulate cell proliferation in leaf primordia of Arabidopsis thaliana. Plant J. 43, 68—78.

Hu, Q., Hollunder, J., Niehl, A., Kørner, C.J., Gereige, D., Windels, D., et al., 2011. Specific impact of tobamovirus infection on the Arabidopsis small RNA profile. PLoS One 6, e19549.

Husbands, A.Y., Chitwood, D.H., Plavskin, Y., Timmermanns, M.C., 2009. Signals and pre-patterns: new insights into organ polarity in plants. Genes Develop. 23, 1986—1997.

Jackson, A.L., Linsley, P.S., 2003. Noise amidst the silence: off-target effects of siRNAs? Trends Genet. 20, 521—524.

Jackson, A.L., Bartz, S.R., Schelter, J., Kobayashi, S.V., Burchard, J., Mao, M., et al., 2003. Expression profiling reveals off-target gene regulation by RNAi. Nat. Biotechnol. 21, 635—637.

Jacob, F., Monod, J., 1961. Genetic regulatory mechanisms in the synthesis of proteins. J. Mol. Biol. 3, 318—356.

Jagadeeswaran, G., Saini, A., Sunkar, R., 2009. Biotic and abiotic stress down-regulate miR398 expression in Arabidopsis. Planta 229, 1009—1014.

Jia, X., Wang, W.X., Ren, L., Chen, Q.J., Mendu, V., 2009. Differential and dynamic regulation of miR398 in response to ABA and salt stress in *Populus tremula* and *Arabidopsis thaliana*. Plant Mol. Biol. 71, 51—59.

Jian, X., Zhang, L., Li, G., Zhang, L., Wang, X., Cao, X., et al., 2010. Identification of novel stress regulated microRNAs from *Oryza sativa* L. Genomics 95, 47—55.

Jin, H., 2008. Endogenous small RNAs and antibacterial immunity in plants. FEBS Lett. 582, 2679—2684.

Jones-Rhoades, M.W., Bartel, D.P., 2004. Computational identification of plant microRNAs and their targets, including a stress-induced miRNA. Mol. Cell 14, 787—799.

Jones-Rhoades, M.W., Bartel, D.P., Bartel, D.P., 2006. MicroRNAs and their regulatory roles in plants. Annu. Rev. Plant Biol. 57, 19—53.

Joshi, T., Yan, Z., Libault, M., Jeong, D.H., Park, S., Green, P.J., et al., 2010. Prediction of novel miRNAs and associated target genes in *Glycine max*. BMC Bioinformatics 1, S14.

Juarez, M.T., Kui, J.S., Thomas, J., Heller, B.A., Timmermanns, M.C., 2004. MicroRNA-mediated repression of rolled leaf1 specifies maize leaf polarity. Nature 428, 84–88.

Jung, H.J., Kang, H., 2007. Expression and functional analyses of microRNA417 in *Arabidopsis thaliana* under stress conditions. Plant Physiol. Biochem. 45, 805–811.

Kantar, M., Lucas, S.J., Budak, H., 2011. miRNA expression patterns of *Triticum dicoccoides* in response to shock drought stress. Planta 233, 471–484.

Kasschau, K.D., Xie, Z., Allen, E., Llave, C., Chapman, E.J., Krizan, K.A., et al., 2004. P1/HC-Pro, a viral suppressor of RNA silencing, interferes with Arabidopsis development and miRNA function. Develop. Cell 4, 205–217.

Katiyar-Agawall, S., Jin, H., 2010. Role of small RNAs in host-microbe interactions. Annu. Rev. Phytopathol. 48, 225–246.

Kawashima, C.G., Yoshimoto, N., Maruyama-Nakashita, A., Tsuchiya, Y.N., Saito, K., Takahashi, H., et al., 2009. Sulphur starvation induces the expression of microRNA-395 and one of its target genes but in different cell types. Plant J. 57, 313–321.

Khan, G.A., Declerck, M., Sorin, C., Hartmann, C., Crespi, M., Leladais-Briere, C., 2011. MicroRNAs as regulators root development and architecture. Plant Mol. Biol. 77, 47–58.

Khraiwesh, B., Ossowski, S., Weigel, D., Reski, R., Frank, W., 2008. Specific gene silencing by artificial microRNAs in *Physcomitrella patens*: an alternative to targeted gene knockouts. Plant Physiol. 148, 684–693.

Khvorova, A., Reynolds, A., Jayasena, S.D., 2003. Functional siRNAs and miRNAs exhibit strand bias. Cell 115, 209–216.

Kim, B., Yu, H.J., Park, S.G., Shin, J.Y., Oh, M., Kim, N., et al., 2012. Identification and profiling of novel microRNAs in the *Brassica rapa* genome based of small RNA deep sequencing. BMC Plant Biol. 12, 218.

Kim, J., 2005. Regulation of short-distance transport of RNA and protein. Curr. Opin. Plant Biol. 8, 45–52.

Kim, J., Jung, J.H., Reyes, J.L., Kim, Y.S., Kim, S.Y., Chung, K.S., et al., 2005. microRNA directed cleavage of ATHB15 mRNA regulates vascular development in Arabidopsis inflorescence stems. Plant J. 42, 84–94.

Kulcheski, F.R., de Oliveira, L.F., Molina, L.G., Almerao, M.P., Rodrigues, F.A., Marcolino, J., et al., 2011. Identification of novel soybean microRNAs involved in abiotic and biotic stresses. BMC Genomics 12, 307.

Kurihara, Y., Watanabe, Y., 2004. Arabidopsis micro-RNA biogenesis through Dicer-like 1 protein functions. Proc. Natl. Acad. Sci. 101, 12753–12758.

Kurihara, Y., Takashi, Y., Watanabe, Y., 2006. The interaction between DCL1 and HYL1 is important for efficient and precise processing of pri-miRNA in plant microRNA biogenesis. RNA 12, 206–212.

Kutter, C., Schob, H., Stadler, M., Meins Jr, F., Si-Ammour, A., 2007. MicroRNA-mediated regulation of stomatal development in Arabidopsis. Plant Cell 19, 2417–2429.

Lai, E.C., Tomancak, P., Williams, R.W., Rubin, G.M., 2003. Computational identification of Drosophila microRNA genes. Genome Biol. 4, R42.

Lang, Q.L., Zhou, X.C., Zhang, X.L., Drabek, R., Zuo, Z.X., Ren, Y.L., et al., 2011. Microarray-based identification of tomato microRNAs and time course analysis of their response to Cucumber mosaic virus infection. J. Zhejiang Univ. Sci.B 12, 116–125.

Laporte, P., Merchan, F., Amor, B.B., Wirth, S., Crespi, M., 2007. Riboregulators in plant development. Biochem. Soc. Trans. 35, 1638–1642.

Laufs, P., Peaucelle, A., Morin, H., Traas, J., 2004. MicroRNA regulation of the CUC genes is required for boundary size control in Arabidopsis meristems. Development 131, 4311–4322.

Lauter, N., Kampani, A., Carlson, S., Goebel, M., Moose, S.P., 2005. MicroRNA172 down regulates glossy15 to promote vegetative phase change in maize. Proc. Natl. Acad. Sci. 102, 9412–9417.

Lee, Y., Kim, M., Han, J., Yeom, K.H., Lee, S., Baek, S.H., et al., 2004. MicroRNA genes are transcribed by RNA polymerase II. EMBO J. 23, 4051−4060.

Lelandais-Briere, C., Naya, L., Sallet, E., Calenge, F., Frugier, F., Hartmann, C., et al., 2009. Genome-wide Medicago truncatula small RNA analysis revealed novel microRNAs and isoforms differentially regulated in roots and nodules. Plant Cell 21, 2780−2796.

Li, B., Duan, H., Li, H., Deng, X.W., Yin, W., Xia, X., 2013. Global identification of miRNAs and targets in *Populus euphratica* under salt stress. Plant Mol. Biol. 81, 525−539.

Li, H., Deng, Y., Wu, T., Subramaniun, S., Yu, O., 2010a. Misexpression of miR482, miR1512, and miR1515 increases soybean nodulation. Plant Physiol. 153, 1759−1770.

Li, H., Dong, Y., Wang, Y., Yang, J., Liu, X., 2011. Characterization of the stress associated microRNAs in Glycine max by deep sequencing. BMC Plant Biol. 11, 170−181.

Li, J., Yang, Z., Yu, B., Liu, J., Chen, X., 2005. Methylation protects miRNAs and siRNAs from a 30-end uri-dylation activity in Arabidopsis. Curr Biol. 15, 1501−1507.

Li, T., Chen, J., Qiu, S., Zhang, Y., Wang, P., Yang, L., et al., 2012. Deep sequencing and microarray hybrid-ization identify conserved and species specific microRNAs during somatic embryogenesis in hybrid yellow poplar. PLoS One 7, e43451.

Li, W.X., Oono, Y., Zhu, J., He, X.J., Wu, J.M., Iida, K., et al., 2008. The Arabidopsis NFYA5 transcription factor is regulated transcriptionally and post-transcriptionally to promote drought resistance. Plant Cell 20, 2238−2251.

Li, Y., Zhang, Q.Q., Zhang, J., Wu, L., Qi, Y., Zhou, J.M., 2010b. Identification of microRNAs involved in pathogen-associated molecular pattern triggered plant innate immunity. Plant Physiol. 152, 2222−2231.

Liang, G., Yang, F., Yu, D., 2010. MicroRNA395 mediates regulation of sulfate accumulation and allocation in Arabidopsis thaliana. Plant J. 62, 1046−1057.

Liang, L., Li, Y., He, H., Wang, F., Yu, D., 2013. Identification of miRNAs and miRNA-mediated regulatory pathways in *Carica papaya*. Planta 238, 739−752.

Lin, Y., Lai, Z.X., 2013. Comparative analysis reveals dynamic changes in miRNA and their targets and expression during somatic embryogenesis in longan (*Dinocarpus longan* Lour.). PLoS One 8, e60337.

Lindbo, J.A., Silva-Rosales, L., Proebsting, W.M., Dougherty, W.G., 1993. Induction of a highly specific anti-viral state in transgenic plants: implications for regulation of gene expression and virus resistance. Plant Cell 5, 1749−1759.

Lister, R., Gregory, B.D., Ecker, J.R., 2009. Next is now: new technologies for sequencing of genomes, tran-scriptomes and beyond. Curr. Opin. Plant Biol. 12, 107−118.

Liu, D., Song, Y., Chen, Z., Yu, D., 2009a. Ectopic expression of miR396 suppresses GRF target gene expres-sion and alters leaf growth in Arabidopsis. Physiol. Plant 136, 223−236.

Liu, H.H., Tian, X., Li, Y.J., Wu, C.A., Zheng, C.C., 2008. Microarray-based analysis of stress-regulated microRNAs in *Arabidopsis thaliana*. RNA 14, 836−843.

Liu, Q., Chen, Y.Q., 2009. Insights into the mechanism of plant development: interactions of miRNAs pathway with phytohormone response. Biochem. Biophysic. Res. Commun. 384, 1−5.

Liu, Q., Zhang, Y.C., Wang, C.Y., Luo, Y.C., Huang, Q.J., Chen, S.Y., et al., 2009b. Expression analysis of phytochrome-regulated microRNAs in rice, implying their regulation roles in plant hormone signalling. FEBS Lett. 583, 723−728.

Liu, T.Y., Chang, C.Y., Chiou, T.J., 2009c. The long-distance signalling of mineral macro-nutrients. Curr. Opin. Plant Biol. 12, 312−319.

Llave, C., 2004. MicroRNAs: more than a role in plant development? Mol. Plant Pathol. 5, 361−366.

Llave, C., Kasschau, K.D., Rector, M.A., Carrington, J.C., 2002a. Endogenous and silencing-associated small RNAs in plants. Plant Cell 14, 1605−1619.

Llave, C., Xie, Z., Kasschau, K.D., Carrington, J.C., 2002b. Cleavage of Scarecrow-like mRNA targets directed by a class of Arabidopsis miRNA. Science 297, 2053–2056.

Lu, S., Sun, Y.H., Amerson, H., Chiang, V.L., 2007. MicroRNAs in loblolly pine (Pinus taeda L.) and their association with fusiform rust gall development. Plant J. 51, 1077–1098.

Lu, S., Sun, Y.H., Chiang, V.L., 2008. Stress-responsive microRNAs in Populus. Plant J. 55, 131–151.

Lu, S.F., Sun, Y.H., Shi, R., Clark, C., Li, L., Chiang, V.L., 2005. Novel and mechanical stress-responsive microRNAs in *Populus trichocarpa* that are absent from Arabidopsis. Plant Cell 17, 2186–2203.

Lu, X.Y., Huang, X.L., 2008. Plant miRNAs and abiotic stress responses. Biochem. Biophys. Res. Commun. 368, 458–462.

Ma, J.B., Ye, K., Patel, D.J., 2004. Structural basis for overhang-specific small interfering RNA recognition by the PAZ domain. Nature 429, 318–322.

Mallory, A.C., Vaucheret, H., 2006. Functions of microRNAs and related small RNAs in plants. Nature Genet. 38, 31–36.

Mallory, A.C., Dugas, D.V., Bartel, D.P., Bartel, B., 2004a. MicroRNA regulation of NAC-domain targets is required for proper formation and separation of adjacent embryonic, vegetative, and floral organs. Curr. Biol. 14, 1035–1046.

Mallory, A.C., Reinhart, B.J., Jones-Rhoades, M.W., Tang, G., Zamore, P.D., Barton, M.K., et al., 2004b. MicroRNA control of PHABULOSA in leaf development: importance of pairing to the microRNA 5' region. EMBO J. 23, 3356–3364.

Mallory, A.C., Bartel, D.P., Bartel, B., 2005. MicroRNA-directed regulation of Arabidopsis AUXIN RESPONSE FACTOR17 is essential for proper development and modulates expression of early auxin response genes. Plant Cell 17, 1360–1375.

Mantri, N., Basker, N., Ford, R., Pang, E., Pardeshi, V., 2013. The role of micro-ribonucleic acids in legumes with a focus on abiotic stress response. Plant Genome 6, 1–14.

Marschner, H., 1995. Mineral Nutrition of Higher Plants. second edition Academic Press, London.

Martín, A.C., del Pozo, J.C., Iglesias, J., Rubio, V., Solano, R., de La Pena, A., et al., 2000. Influence of cytokinins on the expression of phosphate starvation responsive genes in Arabidopsis. Plant J. 24, 559–567.

Mathieu, J., Yant, L.J., Muerdter, F., Kuttner, F., Schmid, M., 2009. Repression of flowering by the miR172 target SMZ. PLoS Biol. 7, e1000148.

Meng, Y., Huang, F., Shi, Q., Cao, J., Chen, D., Zhang, J., et al., 2009. Genome-wide survey of rice microRNAs and microRNA-target pairs in the root of a novel auxin-resistant mutant. Planta 230, 883–898.

Mi, S., Cai, T., Hu, Y., Chen, Y., Hodges, E., Ni, F., et al., 2008. Sorting of small RNAs into Arabidopsis argonaute complexes is directed by the 5' terminal nucleotide. Cell 133, 116–127.

Molnar, A., Bassett, A., Thuenemann, E., Schwach, F., Karkare, S., Ossowski, S., et al., 2009. Highly specific gene silencing by artificial microRNAs in the unicellular alga *Chlamydomonas reinhardtii*. Plant J. 58, 1–10.

Navarro, L., Dunoyer, P., Jay, F., Arnold, B., Dharmasiri, N., Estelle, M., et al., 2006. A plant miRNA contributes to antibacterial resistance by repressing auxin signaling. Science 312, 436–439.

Navarro, L., Jay, F., Nomura, K., He, S.Y., Voinnet, O., 2008. Suppression of the microRNA pathway by bacterial effector proteins. Science 321, 964–967.

Nikovics, K., Blein, T., Peaucelle, A., Ishida, T., Morin, H., Aida, M., et al., 2006. The balance between the *MIR164A* and *CUC2* genes controls leaf margin serration in Arabidopsis. Plant Cell 18, 2929–2945.

Niu, Q.W., Lin, S.S., Reyes, J.L., Chen, K.C., Wu, H.S., Yeh, S.D., Chau, N.H., 2006. Expression of artificial microRNAs in transgenic *Arabidopsis thaliana* confers virus resistance. Nat. Biotech. 24, 1420–1428.

Nogueira, F.T., Sarkar, A.K., Chitwood, D.H., Timmermanns, M.C., 2006. Organ polarity in plants is specified through the opposing activity of two distinct small regulatory RNAs. Cold Spring Harbor Sym. Quan Biol. 71, 157–164.

Nogueira, F.T., Chitwood, D.H., Madi, S., Ohtsu, K., Schnable, P.S., Scanlon, M.J., Timmermans, M.C., 2009. Regulation of small RNA accumulation in the maize shoot apex. PLoS Genet. 5, e1000320.

Okamura, K., Phillips, M.D., Tyler, D.M., Duan, H., Chou, Y.T., Lai, E.C., 2008. The regulatory activity of microRNA* species has substantial influence on microRNA and 3′ UTR evolution. Nat. Struct. Mol. Biol. 15, 354−363.

Ossowski, S., Schwab, R., Weigel, D., 2008. Gene silencing in plants using artificial microRNAs and other small RNAs. Plant J. 53, 674−690.

Ozhuner, E., Eldem, V., Ipek, A., Okay, S., Sakcali, S., Zhang, B., et al., 2013. Boron stress responsive microRNAs and their targets in barley. PLoS One 8, e59543.

Palatnik, J.F., Allen, E., Wu, X., Schommer, C., Schwab, R., Carrington, J.C., Weigel, D., 2003. Control of leaf morphogenesis by microRNAs. Nature 425, 257−263.

Pant, B.D., Buhtz, A., Kehr, J., Scheible, W.R., 2008. MicroRNA399 is a long-distance signal for the regulation of plant phosphate homeostasis. Plant J. 53, 31−38.

Pantaleo, V., 2011. Plant RNA silencing in viral defence. Adv. Exp. Med. Biol. 722, 39−58.

Papp, I., Mette, M.F., Aufsatz, W., Daxinger, L., Schauer, S.E., Ray, A., et al., 2003. Evidence for nuclear processing of plant microRNA and short interfering RNA precursors. Plant Physiol. 132, 1382−1390.

Parizotto, E.A., Dunoyer, P., Rahm, N., Himber, C., Voinnet, O., 2004. In vivo investigation of the transcription, processing, endonucleolytic activity, and functional relevance of the spatial distribution of a plant miRNA. Genes Develop. 18, 2237−2242.

Park, M.Y., Wu, G., Gonzalez-Sulser, A., Vaucheret, H., Poethig, R.S., 2005. Nuclear processing and export of microRNAs in Arabidopsis. Proc. Natl. Acad. Sci. 102, 3691−3696.

Park, W., Li, J., Song, R., Messing, J., Chen, X., 2002. CARPEL FACTORY, a Dicer homolog, and HEN1, a novel protein, act in microRNA metabolism in *Arabidopsis thaliana*. Curr. Biol. 12, 1484−1495.

Park, W., Zhai, J., Lee, J.Y., 2009. Highly efficient gene silencing using perfect complementary artificial miRNA targeting AP1 or heteromeric artificial miRNA targeting AP1 and CAL genes. Plant Cell Rep. 28, 469−480.

Peleg, Z., Saranga, Y., Suprunova, T., Ronin, Y., Roder, M.S., Kilian, A., et al., 2008. High-density genetic map of durum wheat × wild emmer wheat based on SSR and DArT markers. Theor. Appl. Genet. 117, 103−115.

Perez-Quintero, A.L., Neme, R., Zapata, A., Lopez, C., 2010. Plant microRNAs and their roles in defense against viruses: a bioinformatics approach. BMC Plant Biol. 10, 138.

Prins, M., Laimer, M., Noris, E., Schubert, J., Wassenegger, M., Tepfer, M., 2008. Strategies for antiviral resistance in transgenic plants. Mol. Plant Pathol. 9, 73−83.

Qi, Y., He, X., Wang, W.J., Kohany, O., Jurka, J., Hannon, G.J., 2006. Distinct catalytic and non-catalytic roles of ARGONAUTE4 in RNA-directed DNA methylation. Nature 443, 1008−1012.

Qin, F., Shinozaki, K., Yamaguchi-Shinozaki, K., 2011. Achievements and challenges in understanding plant abiotic stress responses and tolerance. Plant Cell Physiol. 52, 1569−1582.

Qu, J., Ye, J., Fang, R., 2007. Artificial microRNA-mediated virus resistance in plants. J. Virol. 81, 6690−6699.

Rajagopalan, R., Vaucheret, H., Trejo, J., Bartel, D.P., 2006. A diverse and evolutionarily fluid set of microRNAs in *Arabidopsis thaliana*. Genes Develop. 20, 3407−3425.

Reinhart, B.J., Weinstein, E.G., Rhoades, M.W., Bartel, B., Bartel, D.P., 2002. MicroRNAs in plants. Genes Develop. 16, 1616−1626.

Reyes, J.L., Chau, N.H., 2007. ABA induction of miR 159 controls transcript levels of two MYB factors during Arabidopsis seed germination. Plant J. 49, 592−606.

Rhoades, M.W., Reinhart, B.J., Lim, L.P., Burge, C.B., Bartel, B., Bartel, D.P., 2002. Prediction of plant microRNA targets. Cell 110, 513−520.

Rodriguez, R.E., Mecchia, M.A., Debernardi, J.M., Schommer, C., Weigel, D., Palatnik, J.F., 2010. Control of cell proliferation in Arabidopsis thaliana by microRNA miR396. Development 137, 103–112.

Ru, P., Xu, L., Ma, H., Huang, H., 2006. Plant fertility defects induced by the enhanced expression of microRNA167. Cell Res. 16, 457–465.

Schommer, C., Palatnik, J.F., Aggarwal, P., Chetelat, A., Cubas, P., Farmer, E.E., et al., 2008. Control of jasmonate biosynthesis and senescence by miR319 targets. PLoS Biol. 6, e230.

Schreiber, A.W., Shi, B.J., Huang, C.Y., Langridge, P., Baumann, U., 2011. Discovery of Barley miRNAs through Deep Sequencing of Short Reads. BMC Genomics 12, 129.

Schwab, R., Palatnik, J.F., Riester, M., Schommer, C., Schmid, M., Weigel, D., 2005. Specific effects of microRNAs on the plant transcriptome. Develop. Cell 8, 517–527.

Schwab, R., Ossowski, S., Riester, M., Warthmann, N., Weigel, D., 2006. Highly specific gene silencing by artificial microRNAs in Arabidopsis. Plant Cell 18, 1121–1133.

Schwarz, D.S., Hutvagner, G., Du, T., Xu, Z., Aronin, N., Zamore, P.D., 2003. Asymmetry in the assembly of the RNAi enzyme complex. Cell 115, 199–208.

Seki, M., Narusaka, M., Ishida, J., Nanjo, T., Fujita, M., Oono, Y., et al., 2002. Monitoring the expression profiles of 7000 Arabidopsis genes under drought, cold and high-salinity stresses using a full length cDNA microarray. Plant J. 31, 279–292.

Shen, B., Goodman, H.M., 2004. Uridine addition after microRNA-directed cleavage. Science 306, 997.

Shen, J., Xing, T., Yuan, H., Liu, Z., Jin, Z., Zhang, L., Pei, Y., 2013. Hydrogen sulfide improves drought tolerance in *Arabidopsis thaliana* by microRNA expression. PLoS One 8, e77047.

Shendure, J., Ji, H., 2008. Next-generation DNA sequencing. Nat. Biotech. 26, 1135–1145.

Shin, H., Shin, H.S., Chen, R., Harrison, M.J., 2006. Loss of At4 function impacts phosphate distribution between the roots and the shoots during phosphate starvation. Plant J. 45, 712–726.

Shinozaki, K., Yamaguchi-Shinozaki, K., 2007. Gene networks involved in drought stress response and tolerance. J. Exp. Bot. 58, 221–227.

Shuai, P., Liang, D., Zhang, Z., Yin, W., Xia, X., 2013. Identification of drought responsive and novel Populus trichocarpa microRNAs by high throughput sequencing and their targets using degradome analysis. BMC Genomics 14, 233.

Sieber, P., Wellmer, F., Gheyselinck, J., Riechmann, J.L., Meyerowitz, E.M., 2007. Redundancy and specialization among plant microRNAs: role of the MIR164 family in developmental robustness. Development 134, 1051–1060.

Simon-Mateo, C., Garcia, J.A., 2006. MicroRNA-guided processing impairs Plum pox virus replication, but the virus readily evolves to escape this silencing mechanism. J. Virol. 80, 2429–2436.

Song, Q.X., Liu, Y.F., Hu, X.Y., Zhang, W.K., Ma, B., Chen, S.Y., Zhang, J.S., 2011. Identification of microRNAs and their target genes in developing soybean seeds by deep sequencing. BMC Plant Biol. 11, 5.

Sorin, C., Bussell, J.D., Camus, I., Ljung, K., Kowalczyk, M., Geiss, G., et al., 2005. Auxin and light control of adventitious rooting in Arabidopsis require ARGONAUTE1. Plant Cell 17, 1343–1359.

Souret, F.F., Kastenmayer, J.P., Green, P.J., 2004. AtXRN4 degrades mRNA in Arabidopsis and its substrates include selected miRNA targets. Mol. Cell 15, 173–183.

Subramanian, S., Fu, Y., Sunkar, R., Barbazuk, W.B., Zhu, J.K., Yu, O., 2008. Novel and nodulation regulated microRNAs in soybean roots. MBC Genomics 9, 160.

Sun, D.J., He, Z.H., Xia, X.C., Zhang, L.P., Morris, C., Appels, R., et al., 2005. A novel STS marker for polyphenol oxidase activities in bread wheat. Mol. Breed 16, 209–218.

Sun, G., 2012. MicroRNAs and their diverse functions in plants. Plant Mol. Biol. 80, 17–36.

Sunkar, R., Jagadeeswaran, G., 2008. In silico identification of conserved microRNAs in large number of diverse plant species. BMC Plant Biol. 8, 37.

Sunkar, R., Zhu, J.K., 2004. Novel and stress-regulated microRNAs and other small RNAs from Arabidopsis. Plant Cell 16, 2001–2019.

Sunkar, R., Kapoor, A., Zhu, J.K., 2006. Posttranscriptional induction of two Cu/Zn superoxide dismutase genes in Arabidopsis is mediated by down-regulation of miR398 and important for oxidative stress tolerance. Plant Cell 18, 2051–2065.

Sunkar, R., Chinnusamy, V., Zhu, J., Zhu, J.K., 2007. Small RNAs as big players in plant abiotic stress responses and nutrient deprivation. Trends Plant Sci. 12, 301–309.

Sunkar, R., Zhou, X., Zheng, Y., Zhang, W., Zhu, J.K., 2008. Identification of novel and candidate miRNAs in rice by high throughput sequencing. BMC Plant Biol. 8, 25.

Sunkar, S., Li, Y.F., Jagadeeswaran, G., 2012. Functions of microRNAs in plant stress responses. Trends Plant Sci. 17, 196–203.

Takeda, A., Iwasaki, S., Watanabe, T., Utsumi, M., Watanabe, Y., 2008. The mechanism selecting the guide strand from small RNA duplexes is different among argonaute proteins. Plant Cell Physiol. 49, 493–500.

Tang, X., Gal, J., Zhuang, X., Wang, W., Zhu, H., Tang, G., 2007. A simple array platform for microRNA analysis and its application in mouse tissues. RNA 13, 1803–1822.

Tang, Y., Wang, F., Zhao, J., Xie, K., Hong, Y., Liu, Y., 2010. Virus based microRNA expression for gene functional analysis in plants. Plant Physiol. 153, 632–641.

Todesco, M., Rubio-Somoza, I., Paz-ares, J., Weigel, D., 2010. A collection of target mimics for comprehensive analysis of microRNA function in *Arabidopsis thaliana*. PLoS Genet. 6, 7.

Turner, M., Nizampatnam, N.R., Baron, M., Coppin, S., Damodaran, S., Adhikari, S., et al., 2013. Ectopic expression of miR160 results in auxin hypersensitivity, cytokinin hyposensitivity, and inhibition of symbiotic nodule development in soybean. Plant Physiol. 162, 2042–2055.

Vaucheret, H., Vazquez, F., Crete, P., Bartel, D.P., 2004. The action of argonaute 1 in the miRNA pathway and its regulation by the miRNA pathway are crucial for plant development. Genes Develop. 18, 1187–1197.

Vazquez, F., Legrand, S., Windels, D., 2010. The biosynthetic pathways and biological scopes of plant small RNAs. Trends Plant Sci. 15, 337–345.

Wang, C., Li, Q.Z., 2007. Identification of differentially expressed miRNAs during the development of Chinese murine mammary gland. J. Genet. Genom 34, 966–973.

Wang, C., Han, J., Kibet, K.N., Kayesh, E., Shangguan, L., Li, X., Fang, J., 2012a. Identification of microRNAs from Amur grapes (Vitis amurensis Rupr.) by deep sequencing and analysis of microRNA variations with bioinformatics. BMC Genomics 13, 122.

Wang, J., Mai, Y.X., Zhang, Y.C., Muo, Q., Yang, H.Q., 2010. MicroRNA171c targeted SCL6-II, SCL6-III and SCL6-IV genes regulate shoot branching I Arabidopsis. Mol. Plant 3, 794–806.

Wang, J., Yang, X., Xu, H., Chi, H., Zhang, M., Hou, X., 2012b. Identification and characterization of microRNAs and their target genes in *Brassica oleracea*. Gene 505, 300–308.

Wang, J.W., Schwab, R., Czech, B., Mica, E., Weigel, D., 2008. Dual effects of miR156-targeted SPL genes and CYP78A5/KLUH on plastochron length and organ size in *Arabidopsis thaliana*. Plant Cell 20, 1231–1243.

Wang, T., Chen, L., Ahao, M., Tian, Q., Zhang, W.H., 2011. Identification of drought responsive microRNAs in *Medicago truncatula* by genome-wide high throughput sequencing. BMC Genomics 12, 367.

Wang, Y., Chen, X., Xiang, C.B., 2007. Stomatal density and bio-water saving. J. Integ. Plant Biol. 57, 477–496.

Warthmann, N., Chen, H., Ossowski, S., Weigel, D., Herv, E.P., 2008. Highly specific gene silencing by artificial miRNAs in rice. PLoS One 3, 1829.

Waterhouse, P.M., Helliwell, C.A., 2003. Exploring plant genomes by RNA-induced gene silencing. Nat. Rev. Genet. 4, 29–38.

Watson, J.M., Fusaro, A.F., Wang, M., Waterhouse, M., 2005. RNA silencing platforms in plants. FEBS Lett. 579, 5982−5987.

Weigel, D., Ahn, J.H., Blazquez, M.A., Borevitz, J.O., Christensen, S.K., Fankhauser, C., et al., 2000. Activation tagging in Arabidopsis. Plant Physiol. 122, 1003−1013.

Wenzl, P., Li, H., Carling, J., Zhou, M., Raman, H., Paul, E., et al., 2006. A high-density consensus map of barley linking DArT markers to SSR, RFLP and STS loci and agricultural traits. BMC Genomics 7, 206.

Wesley, S.V., Helliwell, C.A., Smith, N.A., Wang, M.B., Rouse, D.T., Liu, Q., et al., 2001. Construct design for efficient, effective and high-throughput gene silencing in plants. Plant J. 27, 581−590.

Williams, L., Grigg, S.P., Xie, M., Christensen, S., Fletcher, J.C., 2005. Regulation of Arabidopsis shoot apical meristem and lateral organ formation by microRNA miR166g and its *AtHD-ZIP* target genes. Development 132, 3657−3668.

Willmann, M.R., Poethig, R.S., 2005. Time to grow up: the temporal role of small RNAs in plants. Curr. Opin. Plant Biol. 8, 548−552.

Willmann, M.R., Poethig, R.S., 2007. Conservation and evolution of miRNA regulatory programs in plant development. Curr. Opin. Plant Biol. 10, 503−511.

Wu, G., Poethig, R.S., 2006. Temporal regulation of shoot development in *Arabidopsis thaliana* by miR156 and its target SPL3. Development 133, 3539−3547.

Wu, M.F., Tian, Q., Reed, J.W., 2006. Arabidopsis microRNA167 controls patterns of ARF6 and ARF8 expression, and regulates both female and male reproduction. Development 133, 4211−4218.

Xia, K., Wang, R., Ou, X., Fang, Z., Tian, C., Duan, J., et al., 2012. *OsTIR1* and *OsAFB2* downregulation via *OsmiR393* overexpression leads to more tillers, early flowering and less tolerance to salt and drought in rice. PLoS One 7, e30039.

Xie, F., Frazier, T.P., Zhang, B., 2010. Identification and characterization of microRNAs and their targets in the bioenergy plant switchgrass (*Panicum virgatum*). Planta 232, 417−434.

Xie, F., Frazier, T.P., Zhang, B., 2011. Identification, characterization and expression analysis of microRNAs and their targets in the potato (*Solanum tuberosum*). Gene 473, 8−22.

Xie, Z., Allen, E., Fahlgren, N., Calamar, A., Givan, S.A., Carrington, J.C., 2005. Expression of Arabidopsis MIRNA Genes. Plant Physiol. 138, 2145−2154.

Xin, M., Wang, Y., Yao, Y., Xie, C., Peng, H., Ni, Z., Sun, Q., 2010. Diverse set of microRNAs are responsive to powdery mildew infection and heat stress in wheat (*Triticum aestivum* L.). BMC Plant Biol. 10, 123.

Xu, L., Wang, Y., Zhai, L., Xu, Y., Wang, L., Zhu, X., et al., 2013. Genome-wide identification and character-ization of cadmium-responsive microRNAs and their target genes in radish (*Raphanum sativus* L.) roots. J. Exp. Bot. 64, 4271−4287.

Xu, P., Guo, M., Hay, B.A., 2004. MicroRNAs and the regulation of cell death. Trends Genet. 20, 617−624.

Yamasaki, H., Abdel-Ghany, S.E., Cohu, C.M., Kobayashi, Y., Shikanai, T., Pilon, M., 2007. Regulation of copper homeostasis by micro-RNA in Arabidopsis. J. Biol. Chem. 282, 16369−16378.

Yang, Z., Ebright, Y.W., Yu, B., Chen, X., 2006. HEN1 recognizes 21−24 nt small RNA duplexes and depos-its a methyl group onto the 2′ OH of the 3′ terminal nucleotide. Nucleic Acids Res. 34, 667−675.

Yanik, H., Turktas, M., Dundar, E., Hernendez, P., Dorado, G., Unver, T., 2013. Genome wide identification of alternative bearing-associated microRNAs (miRNAs) in olive (*Olea europaea* L.). BMC Plant Biol. 13, 10.

Yao, Y., Guo, G., Ni, Z., Sunkar, R., Du, J., Zhu, J.K., Sun, Q., 2007. Cloning and characterization of microRNAs from wheat (*Triticum aestivum L.*). Genome Biol. 8, R96.

Ying, S.Y., Chang, D.C., Lin, S.L., 2008. The microRNA (miRNA): overview of the RNA genes that modulate gene function. Mol. Biotechnol. 38, 257−268.

Yu, B., Yang, Z., Li, J., Minakhina, S., Yang, M., Padgett, R.W., et al., 2005. Methylation as a crucial step in plant microRNA biogenesis. Science 307, 932−935.

Yu, H., Song, C., Jia, Q., Wang, C., Li, F., Nicholas, K.K., et al., 2011. Computational identification of microRNAs in apple expressed sequence tags and validation of their precise sequences by miR-RACE. Plant Physiol. 141, 56−70.

Zeng, Y., Wagner, E.J., Cullen, B.R., 2002. Both natural and designed microRNAs can inhibit the expression of cognate mRNAs when expressed in human cells. Mol. Cell 9, 1327−1333.

Zhang, B.H., Pan, X.P., Wang, Q.L., Cobb, G.P., Anderson, T.A., 2005. Identification and characterization of new plant microRNAs using EST analysis. Cell Res. 15, 336−360.

Zhang, B.H., Pan, X., Wang, Q., Cobb, G.P., Anderson, T.A., 2006. Computational identification of microRNAs and their targets. Comp. Biol. Chem. 30, 395−407.

Zhang, B.H., Wang, Q.L., Pan, X.P., 2007. MicroRNAS and their regulatory roles in animals and plants. J. Cell Physiol. 210, 279−289.

Zhang, B.H., Pan, X., Stellwag, E.J., 2008. Identification of soybean microRNAs and their targets. Planta 229, 161−182.

Zhang, J., Zhang, S., Han, S., Wu, T., Li, X., Li, W., Qi, L., 2012. Genome-wide identification of microRNAs in larch and stage-specific modulation of 11 conserved microRNAs and their targets during somatic embryogenesis. Planta 236, 647−657.

Zhang, W., Gao, S., Zhou, X., Chellappan, P., 2011a. Bacteria-responsive microRNAs regulates plant innate immunity by modulating plant hormone networks. Plant Mol. Biol. 75, 93−105.

Zhang, X., Zou, Z., Gong, P., Zhang, J., Ziaf, K., Li, H., et al., 2011b. Over-expression of microRNA169 confers enhanced drought tolerance to tomato. Biotech. Lett. 33, 403−409.

Zhao, B., Liang, R., Ge, L., Li, W., Xiao, H., Lin, H., et al., 2007. Identification of drought-induced microRNAs in rice. Biochem. Biophys. Res. Com. 354, 585−590.

Zhao, B., Ge, L., Liang, R., Li, W., Ruan, K., Lin, H., Jin, Y., 2009. Members of miR-169 family are induced by high salinity and transiently inhibit the NF-YA transcription factor. BMC Mol. Biol. 10, 29.

Zhao, C.Z., Xia, H., Frazier, T.P., Yao, Y.Y., Bi, Y.P., Li, A.Q., et al., 2010. Deep sequencing identifies novel and conserved microRNAs in peanuts (*Arachis hypogaea* L.). BMC Plant Biol. 10, 3.

Zhao, T., Wang, W., Bai, X., Qi, Y., 2009. Gene silencing by artificial microRNAs in Chlamydomonas. Plant J. 58, 157−164.

Zhou, L., Liu, Y., Liu, Z., Kong, D., Duan, M., Luo, L., 2010. Genome-wide identification and analysis of drought-responsive microRNAs in *Oryza sativa*. J. Exp. Bot. 61, 4157−4168.

Zhou, X., Wang, G., Zhang, W., 2007. UV-responsive microRNA genes in Arabidopsis thaliana. Mol. Syst. Biol. 3, 103−109.

Gene Silencing: A Novel Cellular Defense Mechanism Improving Plant Productivity under Environmental Stresses

Renu Bhardwaj, Puja Ohri, Ravinderjit Kaur, Amandeep Rattan, Dhriti Kapoor, Shagun Bali, Parminder Kaur, Anjali Khajuria and Ravinder Singh

10.1 Introduction

Accurate weather and climatic conditions lead to success in agriculture. A slight change in environmental conditions ultimately affects worldwide agricultural production. A number of factors such as soil, insects, pests, pathogens, weeds, etc. are being held responsible for the decline in crop yields. These contributing factors interact with climatic conditions through varied mechanisms such as concentration of CO_2, soil acidification, eutrophication, and survival and dissemination of the pest population in crop breeding programs (Reilly, 1999). In the past, the age-old practices were being employed to manage insect pests and pathogens that affected crop yield. Some of these practices have also been held responsible for the change in crop climate. In addition, with the use of conventional methodologies, these deterrents have developed resistance to these threats. Therefore, scientists across the world are employing more effective and long-lasting techniques to improve crop yields. One such genetic approach is known as gene silencing or RNA interference (RNAi).

RNA interference (or gene silencing) is a novel mechanism for gene down-regulation where the expression of genetic information is interfered with by RNA. It is triggered by endogenous production or by artificial introduction of small interfering double-stranded RNA (siRNA) that have sequences complementary with the targeted gene, thus causing the degradation of its encoded messenger RNA (mRNA) (Bosher and Labouesse, 2000; Agrawal et al., 2003).

The process of RNAi is found in many eukaryotes including animals. It was first discovered in transgenic petunia plants by Napoli et al. (1990) in an attempt to overexpress the chimeric chalcone synthase gene (enhances the purple color of the flowers). But the attempt resulted in the reversible suppression of introduced as well as the homologous genes, called co-suppression. This phenomenon is known as quelling in fungi (Romano and Macino, 1992) and RNA interference in animals (Fire et al., 1998). Later in 2006, Fire and Mello shared the Nobel Prize for Physiology/Medicine for their work on RNA interference in the free-living nematode, *Caenorhabditis elegans*. Earlier, Elbashir et al. (2001a) analyzed the brilliant future of RNAi in therapeutics where the expressed 21 nucleotides of endogenous and heterologous genes were suppressed in different

P. Ahmad (Ed): Emerging Technologies and Management of Crop Stress Tolerance, Volume 1.
DOI: http://dx.doi.org/10.1016/B978-0-12-800876-8.00010-2

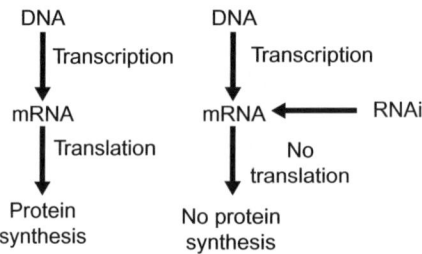

FIGURE 10.1

Simplified process of RNAi.

mammalian cell lines, especially human embryonic kidney and HeLa cells. Later, a number of researchers (Paddison et al., 2002; Paul et al., 2002; Sui et al., 2002) used short hairpin RNAs (shRNAs) to execute sequence-specific silencing in mammalian cells.

RNAi is initiated by short double-stranded RNA molecules (dsRNA) and is controlled by RISC (RNA-induced silencing complex) (Figure 10.1). This limits the level of transcription by either suppressing transcription (transcriptional gene silencing; TGS) or by activating specific RNA degradation (post-transcriptional gene silencing; PGTS) (Agrawal et al., 2003). The central role is played by two types of small RNA molecules, namely, micro-RNA (miRNA) and small interfering RNA (siRNA). RNAi is a phenomenon where a high degree of targeted gene silencing is achieved with less effort. It is highly potent and can be effectively introduced at different developmental stages.

10.2 Elements of RNAi

The basis of RNA silencing can be understood by taking into account both the genetic and biochemical approaches. These screens have been carried out in organisms like fungus (*Neurospora crassa*), algae (*Chlamydomonas reinhardtii*), free-living nematode (*C. elegans*), and plants (*Arabidopsis thaliana*). Analyses carried out in their mutants have led to the description of a number of essential enzymes/factors that are common in these organisms (Figure 10.2). These include: Dicer, siRNA, guide RNAs and RISC, RNA-dependent RNA polymerase, and transmembrane proteins. Some of these elements serve as initiators, while others serve as effectors, amplifiers, or transmitters of the process (Agrawal et al., 2003).

10.2.1 Dicer

The term Dicer was first coined by Bernstein et al. (2001) in *Drosophila*, where siRNAs were produced by implicating an RNAse III-family nuclease in the *Drosophila* extracts. This RNase III-like enzyme produces 22 nucleotide fragments that are similar in size as produced during RNAi (Bernstein et al., 2001). A Dicer is an endoribonuclease in the RNase III family that converts dsRNA into siRNA by cleaving it about 20−25 nucleotides long with a two-base overhang on the 3′ end. It consists of two RNase III domains and one PAZ domain. The distance between these two regions is determined by the length and angle of the connector helix. The nucleases that show specificity for dsRNAs include the

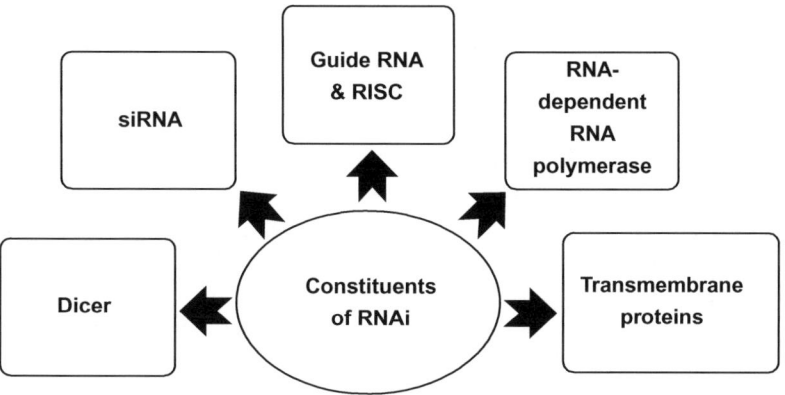

FIGURE 10.2

Different elements of RNAi.

RNase III family members (Nicholson, 1999), which cleave them with 3′ overhangs of two to three nucleotides and 5′-phosphate and 3′-hydroxyl termini (Elbashir et al., 2001b). Presently, many Dicer homologues from many different sources have been identified. Some recombinant Dicers have conjointly been examined *in vitro*. A common ancestry of these proteins has been indicated through the phyletic analysis of the known Dicer-like proteins (Golden et al., 2002).

10.2.2 Small interfering RNA

The most important role in RNAi is played by small RNA (≈22-nt) termed siRNA. siRNAs accumulate as double-stranded RNA molecules with a length of 20−25 base pairs also known as short interfering RNA or silencing RNA. They play a key role in the basic cellular processes such as control of development and the formation of heterochromatin. Small interfering RNA (siRNA) were discovered when RNAi was induced by introducing 500 nucleotide-long exogenous dsRNA in *Drosophila* tissue culture cells (Hammond et al., 2000). These were also observed in *Drosophila* embryo extracts, carrying out RNAi *in vitro* (Zamore et al., 2000) and also in *Drosophila* embryos that were injected with dsRNA (Yang et al., 2000). The role of siRNAs in post-transcriptional gene silencing (PTGS) in plants was first discovered by David Baulcombe's group at the Sainsbury Laboratory in Norwich, England (Hamilton and Baulcombe, 1999).

10.2.3 Guide RNAs and RNA-induced silencing complex

Guide RNAs direct RNA-induced silencing complexes (RISC) to their mRNA targets leading to gene silencing. Earlier, it was proposed that helicases plays a major role in unwinding ds siRNAs. But independent groups like Gregory et al. (2005), Matranga et al. (2005), and Rand et al. (2005) demonstrated that Ago2 is responsible for cleaving the non-incorporated strand of the siRNA duplex and allowing the other strand to assimilate into RISC. Thus, RNAi is a conserved sequence-specific gene regulatory mechanism, which is mediated by RNA-induced silencing complex and is composed of a single-stranded guide RNA and an argonaute protein.

10.2.4 **RNA-dependent RNA polymerase**

RNA-dependent RNA polymerase (RdRP), or RNA replicase, is an enzyme that helps in the replication of RNA from an RNA template. This is similar to DNA-dependent RNA polymerase, which catalyzes the transcription of RNA from a DNA template. RdRP is encoded in the genomes of all RNA-containing viruses (Koonin et al., 1989; Zanotto et al., 1996). The effect of RNAi is both potent and systemic in nature leading to a proposed mechanism in which the silencing effect was triggered and amplified by RNA-dependent RNA polymerases (RdRPs). Accumulation of aberrant transgenic and viral RNAs was seen in transgenic and virus-infected plants that were recognized as templates by RdRP enzymes. These synthesize antisense RNAs form dsRNAs, finally acting as the targets for sequence-specific RNA degradation (Lindbo et al., 1993; Cogoni and Macino, 1997, 1999; Depicker and Montagu, 1997).

10.2.5 **Transmembrane protein (channel or receptor)**

The distribution of gene silencing from one tissue to another has been well characterized in *C. elegans* and plants. To know the mechanism of systemic RNAi, a special transgenic strain of *C. elegans* (HC57) was constructed (Winston et al., 2002). They illustrated that a systemic RNA interference deficient (sid) locus is required to transmit the effects of gene silencing between cells having a green fluorescent protein (GFP) as a marker protein and isolated and characterized *sid1* (a 776 amino acid membrane protein consisting of a signal peptide and 11 putative transmembrane domains) mutants out of 106 sid mutants belonging to three complementation groups (sid1, sid2, and sid3). Thus, the study revealed that *SID1* might act as a channel for the import or export of a systemic RNAi signal or might be necessary for endocytosis of the systemic RNAi signal. These SID homologues are present in humans and mice but absent in *Drosophila* (Fortier and Belote, 2000; Piccin et al., 2001).

10.3 **Mode of action**

Important insights gained till now have elucidated the mechanism of RNAi. Results obtained from the combination of several *in vitro* and *in vivo* investigations have gelled a two-step mechanistic model for RNAi (Figure 10.3). The first phase is the *induction phase* where binding of RNA nucleases to a large dsRNA and its cleavage into discrete RNA fragments (siRNA; ≈ 21 to ≈ 25 nucleotide) takes place. The second phase is the *completion phase* of joining of siRNAs to a multinuclease complex, a RISC that degrades the homologous single-stranded mRNAs. These effector complexes then interfere with gene expression by using a small RNA strand to identify their complementary mRNA and the same is cleaved and degraded (Agrawal et al., 2003).

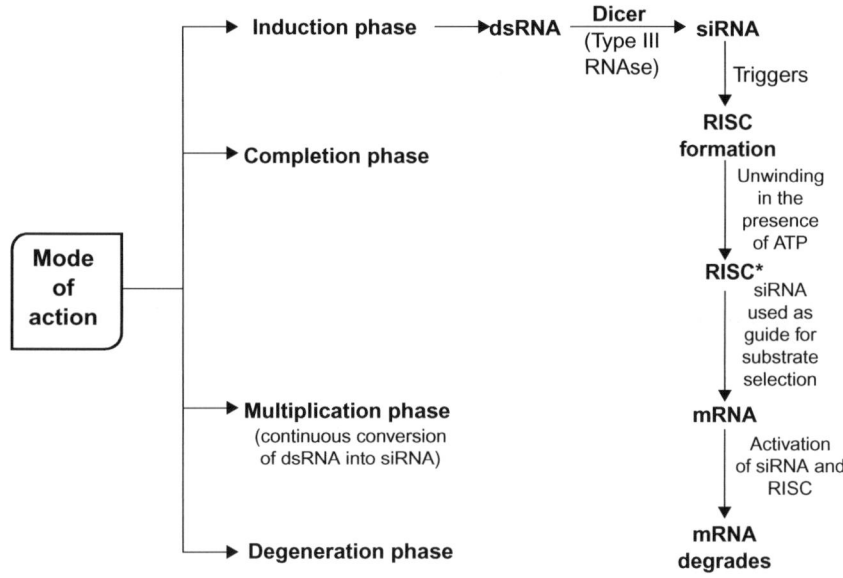

FIGURE 10.3

Mode of action of RNA interference.

10.3.1 Induction phase

In the induction phase, dsRNA is processed into small interfering RNA molecules (siRNA). siRNAs are then incorporated into large ribonucleoprotein complexes. The enzyme that initiates RNAi belongs to the RNase III ribonuclease family displaying specificity for dsRNAs. Based upon the domain structure, RNase III enzymes can be divided into three classes: bacterial RNase III contains a single catalytic domain and a dsRNA binding domain; the Drosha family nucleases contain dual catalytic domains (Filippov et al., 2000); and a third family also contains dual catalytic domains and additional helicase and PAZ motifs (Bernstein et al., 2001). Members of third class of RNAse process dsRNA into siRNAs, thus initiating RNAi (Bernstein et al., 2001).

10.3.2 Completion phase

In the completion phase, siRNAs are incorporated into large ribonucleoprotein complexes. These effector complexes then interfere with the gene expression by using the small RNA strand to identify their complementary mRNA leading to its cleavage and degradation (Agrawal et al., 2003). Dicer-mediated cleavage produces tiny dsRNA intermediates that are triggered by dsRNA precursors that dissociate into "competent" single strands to act as guides for RISCs.

10.3.3 Multiplication of siRNAs

Only few molecules of dsRNA are required to degrade continuously reproducing target mRNA for a longer period of time. However, the conversion of dsRNA into siRNAs resulted in some multiplication, but it is not satisfactory to bring about such sequential mRNA degradation. Convincing biochemical and genetic evidence has been provided by Lipardi et al., (2001) and Sijen et al., (2001) regarding RdRP (RNA-dependent RNA polymerase), which plays a critical role in amplifying RNAi effects. Both single-stranded RNAs (equivalent to target mRNA) and dsRNAs serve as templates for copying by RdRP. Furthermore, new full-length dsRNAs were formed rapidly and cleaved. Studies have revealed the amplification of siRNAs at different stages of the RNAi reaction and have been demonstrated in plants like *C. elegans*, *N. crassa*, and *Dictyostelium discoideum* but not in flies and mammals (Dykxhoorn et al., 2003).

10.3.4 Degeneration of mRNAs

RISC is formed when the double-stranded siRNAs bind with an RNAi-specific protein complex in the effector step and undergoes activation in the presence of ATP to expose and allow the RISC to perform the downstream RNAi reaction. Zamore et al., (2000) found that when activated by ATP, a 250 kDa precursor RISC, found in *Drosophila* embryo extract, was converted into a 100 kDa complex, which then induces cleavage of substrate. Then inactivated RISC, antisense siRNA pairs with cognate mRNAs. This complex then cuts mRNA in the middle of the duplex region.

10.4 RNAi under environmental stresses

RNAi is being used as an alternative and advantageous biotechnological approach. It is a highly specific, dominant, sequence-based gene silencing technology. This tremendous potentiality of RNAi has been effectively exploited for inducing desirable traits in plants under environmental stresses (abiotic/biotic) (Figure 10.4).

10.4.1 RNAi in plants under abiotic stress

A number of environmental factors, especially abiotic including salinity, heat, cold, drought, etc., are the major causes of considerable reduction in crop productivity. The use of functional genomics

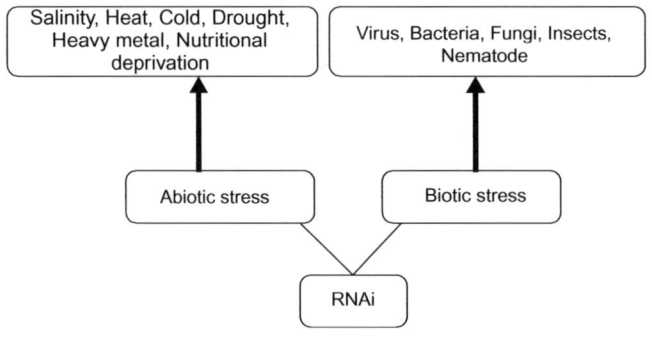

FIGURE 10.4

RNA interference under different environmental stresses.

has provided insight into the effective control strategies to maximize the stress tolerance in plants (Pardo, 2010). The involvement of RNAi (miRNAs and their targets) in abiotic stress responses in plants has been well documented (Khraiwesh et al., 2012; Sun, 2012; Sunkar et al., 2012).

10.4.1.1 Salinity

In worldwide agricultural productivity, salinity is a major constriction. Over the past two decades, the molecular basis of tolerance of plants to these stresses has been investigated profoundly. Excess of salt stress affects approximately 6% of the total arable land (Munns and Tester, 2008). Crop yield was reduced to a large extent due to moderate salt stress, while the survival of crop plants was threatened when exposed to severe salinity (Bartels and Sunkar, 2005). Numerous genes, which are stimulated by this stress, have been identified by various research studies, so that the tolerance to stress can be improved by the overexpression of stress-responsive genes in plants. However, there is incomplete knowledge about the complex genetic interactions underlying plant stress tolerance of transgenic plants. During stress conditions, miRNA expression, in addition to protein-coding genes, is altered in plants (Sunkar and Zhu, 2004; Zhao, 2007; Li et al., 2008; Trindade et al., 2010; Kulcheski, 2011). In numerous plant species, miRNAs have been recognized as stress responsive. Thus, there are major reports regarding the stress-induced alterations in the expression of miRNA. miR169 is one of the major miRNAs having a pivotal role in salt stress responses. Several plant species under salt stress acquire the up-regulation of miR160, miR393, and miR167. Changes in miRNA expression during stress conditions appear to be significant for the promotion of plant growth and development. It was assumed that in response to salt stress, there was differential expression of miRNAs, when comparative analysis was done between two maize (*Zea mays*) varieties that were different in salt sensitivity (salt-tolerant NC286 and salt-sensitive Huangzao4) (Ding et al., 2009). Microarray experiments suggested that members of certain families such as miR162, miR168, miR395, and miR474 were up-regulated, whereas members of the miR156, miR164, miR167, and miR396 families were down-regulated, in salt-shocked maize roots (Ding et al., 2009).

Results have also indicated that in closely related genotypes with complementary stress sensitivities, miRNA profiles are unique. Thus, for the miRNA-guided gene regulation differing in stress-tolerant genotypes, a more inclusive analysis is required, which includes the impact of such regulation on miRNA targets. By improving the stress tolerance in crop plants, certain molecular mechanisms could then be integrated into strategies. miRNA expression of the whole plants was observed in previous studies subjected to stress. There are only a few reports in which miRNA expression profiles of different tissues like roots and shoots were separately analyzed under stress conditions.

The role of miRNAs has also been studied in *Arabidopsis* under salt stress responses. Sequencing of a library of small RNAs isolated from the seedlings of *Arabidopsis* exposed to different stresses led to recognition of several stress-related miRNAs (Sunkar and Zhu, 2004), which play a significant role in the adaptive responses towards abiotic stress and were supported by miRNAs like miR168, miR171, and miR396, which were found to be responsive to high salinity stress in *Arabidopsis* (Liu et al., 2008). Up-regulation of genes like miR156, miR158, miR159, miR165, miR167, miR168, miR169, miR171, miR319, miR393, miR394, miR396, and miR397 in *Arabidopsis* were reported in response to salt stress where miR398 accumulation was found to decrease (Liu et al., 2008), whereas enhanced accumulation of miRS1 and miR159.2 was observed in the case of *Phaseolus vulgaris* under salinity stress (Huertero et al., 2009). It was also found that miR482.2 and miR1450 were up-regulated in *Populus trichocarpa*, while miR171l-n, miR530a, miR1447, miR1445, and miR1446a-e were down-regulated under salt stress (Lu et al., 2008). Certain other members of the miR169 family and miR169g and miR169n were observed to be induced by salt stress (Zhou et al., 2009). In the upstream region of miR169n, the presence of a *cis*-acting ABA-responsive element indicated the regulation of miR169n by ABA.

Similarly, in the case of rice, strong up-regulation was reported in two members (miR169n and miR169g) of the miR169 family (Zhao et al., 2009). In other findings, it was observed that in *Arabidopsis*, expression of miR398 was inhibited and negative correlation was observed in association with its target genes, Cu/Zn superoxide dismutase 1 (CSD1) and CSD2 (Zhao, 2007). A dynamic regulation was displayed by the miR398 expression in *P. tremula*, in which initially miR398 levels were high after 3 h of salt stress, but then were found to reduce after 48 h and reduce further after 72 h of salt stress (Jia et al., 2009). There was negative correlation with CSD1 and CSD2 transcript abundance (Jia et al., 2009). In *P. tremula* and *Arabidopsis*, contrasting regulation for CSD1 and CSD2 mRNAs eventually translates into the activity under salt stress conditions, which was analyzed by the enzymatic activity of CSDs and protein expression levels.

10.4.1.2 Heat stress

Fluctuations in temperature during the day and different seasons are common. Plants have to reprogram their gene expression profiles to adjust to such dramatic shifts in temperature. There are certain contradictory reports having authentic responses which may be recognized to the differential responses of unrelated species of plants. Average growing season temperatures in different parts of the world have increased due to increasing greenhouse gas emissions along with decreased annual precipitation. Plant growth and development, pollen viability, grain filling, and maturation processes are negatively affected by high temperature. Reports have suggested that several miRNAs were induced by heat stress in poplar and wheat. Xin (2011) studied variations in the induction kinetics of miRNAs between two cultivars heat-tolerant and heat-susceptible were shown in the miRNA profiles of wheat seedlings when exposed to heat stress (Xin, 2011). Out of 32 miRNA families of wheat, nine conserved miRNAs were found to be heat responsive. Under heat stress, miR172 was considerably reduced while miR156, miR168, miR159, miR160, miR166, miR393, miR169, and miR827 were up-regulated.

For the protection of reproductive organs, a novel thermotolerance mechanism was discovered in plants. This mechanism involves the down-regulation of its target CSD genes, like CSD1 and CSD2, along with the gene that encodes the copper chaperone for both CSD1 and CSD2 genes by the induction of miR398. These results suggested that CSD1, CSD2, and ccs (copper chaperone for superoxide dismutase) mutants were found to tolerate the heat stress, whereas wild-type plants led to the accumulation of heat stress transcription factors and heat shock proteins, thus causing inhibition in flower damage. Thus, these results led to the fact that miR398 and its targets proved an applicable strategy to enhance the heat tolerance in crops like corn, which suffer due to exposure in long periods of heat stress (Guan et al., 2013).

10.4.1.3 Cold stress

Cold stress is yet another abiotic stress factor that has adverse effects on the growth of plants as it causes tissue damage and stunts growth. In response to cold stress, the expression of miR319 (a conserved miRNA family) was found to change in sugarcane (Thiebaut et al., 2011), rice (Lv et al., 2010), and *Arabidopsis* (Sunkar and Zhu, 2004; Liu et al., 2008). It has been reported that in rice exposed to cold stress conditions, miR319 was down-regulated while its targets, OsPCF5, OsPCF6, OsPCF7, OsPCF8, and OsTCP21, were up-regulated under cold stress conditions (Zhou and Hong, 2013). In the case of *Oryza sativa* L., the MIR319 gene family consists of two members such as Osa-MIR319a and Osa-MIR319b. Yang et al. (2013) observed that in transgenic rice seedlings, overexpression of osa-miR319 led to increase in cold tolerance (4°C) after chilling acclimation (12°C). When sugarcane cultivars (TAMBO FEPAGRO and RB931011) were exposed to 4°C, it has been studied that there was up-regulation of miR319 and down-regulation of the targets in both varieties, but alterations in expression were postponed in the cold-tolerant cultivar (Thiebaut et al., 2011). It has been concluded that recently identified 21−24 nucleotide small RNAs (miRNA) and small-interfering RNAs (siRNAs) regulate the expression of genes at the post-transcriptional level. They contribute to the stress-induced alterations in profiles of mRNAs or proteins and thus the RNAi technique emerged as a powerful tool to increase cold tolerance after chilling acclimation.

10.4.1.4 Drought stress

In the present scenario of advanced technology, there are still a large number of areas where people depend on rain for irrigation of crops. For such areas, drought conditions are one of the common problems. However, scientists are working on the identification of such genes that are effected due to drought stress using various high-throughput technologies like genome-wide gene expression and proteomics (Zhu, 2002; Bartels and Sunkar, 2005; Yamaguchi-Shinozaki and Shinozaki, 2006). Research has also shown that drought stress affects miRNA expression (Li et al., 2008; Zhao et al., 2009), and miR393, miR160, miR169 and miR167 are the most commonly affected genes in plants in response to drought stress (Sunkar, 2010). In *Arabidopsis*, drought stress-responsive miRNAs are miR171, miR167, miR396, and miR168 (Liu et al., 2008). Proteins of nuclear factor Y (NFY) are reported to have an important role in response of plants under stress conditions. It was also observed that the transcription factor for the Nfya5 upon drought stress is regulated by miR169, and overexpression of miR169-resistant *nfya5* transgene leads to closure of stomata (Li et al., 2008). NF transcription factor in its Y subunit is also made a target by miR169 in *O. sativa* (Li et al., 2010a). ABA gathering under drought stress is controlled by the *myb33* and *myb101*- genes, which were regulated by miR159 (Reyes and Chua, 2007). miR398 and miR408 were up-regulated in *Medicago trunculata* and down-regulated in *cox5b* and plantacyanin in water stress (Trindade et al., 2010). *P. vulgaris* and *Populus* have also been studied for drought response where miRS1, miR1514a, miR2119, miR171l-n, miR1445, miR1446a-e, and miR1447 were identified for controlling water-deficient stress (Lu et al., 2008; Arenas-Huertero et al., 2009).

10.4.1.5 Heavy metal stress

Plants need heavy metals for performing their physiological as well as biochemical processes like growth, development, etc. Metals like copper (Cu), iron (Fe), and zinc (Zn) are essential for plants, whereas metals like cadmium (Cd), lead (Pb), mercury (Hg), and aluminum (Al) are non-essential. Non-essential metals are toxic even at low concentrations unlike the essential metals, which are called "micronutrients" for plants at low level, but are toxic at high concentrations (Ding and Zhu, 2009; Huang et al., 2009; Mendoza-Soto et al., 2012). The primary response of plants to heavy metal toxicity is the generation of reactive oxygen species (ROS) by interruption of their enzyme systems (Romero-Puertas et al., 2007). Root length reduction in plants is a primary symptom of heavy metal toxicity of Cu, Cd, Al, Zn, etc. (Mendoza-Soto et al., 2012). Gene expression at transcriptional and post-transcriptional levels is needed for plants to respond to various abiotic stresses including metal stress (Jacoby et al., 2002; Krämer et al., 2007; Farinati et al., 2010).

Metal toxicity response in plants has been studied using different microarrays and deep sequencing of small RNA libraries, and shows the involvement of miRNAs (Jones-Rhoades et al., 2006; Hobert, 2008; Ben Amor et al., 2009). Different plants species were used for studying the involvement of miRNA like *Brassica napus*, *P. vulgaris*, *A. thaliana*, and *O. sativa* (Gielen et al., 2012) after exposure to different heavy metals like Cu, Zn, Mn, As, etc. In *A. thaliana*, miRNAs have an important role in response to heavy metal stress (Jones-Rhoades and Bartel, 2004; Fujii et al., 2005; Sunkar et al., 2006). The plant RNAi identifies miRNA as an RNA-induced silencing complex (RISC) and hence starts degradation of target mRNA(s) (Jones-Rhoades et al., 2006). Several miRNAs, such as miR319, miR390, miR393, and miR398, have important functions in the regulation of metal stress in plants. miR398 is the main regulatory enzyme in regulation of Cu metal stress. Under high copper stress, its suppression is important for the stimulation of CSD1 and CSD2 mRNA levels (Sunkar et al., 2006; Yamasaki et al., 2007; Abdel-Ghany and Pilon, 2008). Cadmium toxicity in plants is controlled by conserved miRNAs, e.g., miR160, miR164, and miR167 in rice plants as observed by Huang et al. in 2009. Analysis showed that during Cd stress miR390, miR168, miR156, miR162, miR166, and miR171 were down-regulated while up-regulation of miR528 was seen (Ding et al., 2011). Cd stress has also been studied in roots of *B. napus*, which showed down-regulation of miR393, miR396, miR171, and miR156 (Xie et al., 2007). Mercury-responsive miRNAs (miR2681) were identified in *M. trunculata*. However, similar studies showed that miR171, miR396, and miR390 were down-regulated and miR164, miR395, miR167, miR172, and miR169 were up-regulated (Zhou et al., 2012). It was reported by Chen et al. (2012) that expression of miR159, miR160, miR319, miR396, miR390, miR159, and miR160 was down-regulated under high Al toxicity using high-throughput genome sequencing using small RNA. *M. truncatula* was also studied for toxicity of Cd and Hg, where up-regulation of miR171, miR319, miR393, and miR529 and down-regulation of miR166 and miR398 were observed (Zhou et al., 2008).

10.4.1.6 Nutritional deprivation

Mineral nutrients are essential to plants and indirectly to all living organisms that are dependent on plants. Of the 16 essential elements for plants except carbon (C) and oxygen (O), all are carried through plant roots from soil. Plants experience frequent fluctuations in soil nutrient status during their life cycle. Diverse factors like water content and pH change the availability of nutrients in the soil. Plants must change their structural design and metabolism to adapt to different nutrient conditions. Nitrogen (N) is an indispensable macronutrient accessible to plants mostly in the form of nitrate in agricultural soils and is a major environmental factor that limits crop productivity globally. Transgenic studies have discovered that miRNAs regulate plant adaptive responses to nutrient deprivation. The expression of *Arabidopsis* miR169 was reported to be down-regulated while its targets NFYA (nuclear factor Y, subunit A) family members were up-regulated by nitrogen starvation. Expression of miR169 precursors was studied and observation was made that miR169a was substantially down-regulated in both roots and shoots under nitrogen deprivation. It has also been observed that transgenic *Arabidopsis* plants overexpressing miRNA169a accumulated lesser amounts of N and were more sensitive to N stress as compared to the wild type (Zhao et al., 2011).

Sulfur (S) is another macronutrient that plays a significant role in plant growth and development (Liang and Yu, 2010). Studies have revealed that the expression of miR395 is drastically up-regulated under sulfate starvation conditions and its induction is controlled by transcription factor SLIM1 (SULFUR LIMITATION 1) (Kawashima et al., 2008, 2011; Liang et al., 2010). ATP sulfurylases (encoded by APS genes) and sulfate transporter 2;1 (SULTR2;1, also called AST68) are two gene families targeted by miR395 and are involved in the sulfate metabolism pathway (Liang et al., 2010). In sulfate-deficient plants, miR395 in cooperation with the transcription factor SLIM1 maintain optimal concentration of ATP sulfurylase transcripts to facilitate increased flux by sulfate assimilation pathway (Kawashima et al., 2008, 2011; Liang et al., 2010). In APS4-RNAi transgenic plants, it was observed that silencing the APS4 gene leads to overaccumulation of sulfate. Although in the shoots of miR395 overexpressing transgenic plants, sulfate was overaccumulated, and, in addition, the distribution of sulfate from older to younger leaves was also impaired (Liang et al., 2010).

Phosphorus (Pi) is a key limiting factor in natural and agricultural soils. It has been reported that at low Pi stress, expression of miR399 was up-regulated, while the target ubiquitin conjugating enzyme (UBC) was down-regulated. Transgenic studies demonstrated that there was more accumulation of Pi in transgenic *Arabidopsis* plants with constitutive expression of miR399 as compared to wild type. It was observed that there was less inhibition in the growth of primary root and less induction of Pi transporter gene in transgenic plants expressing the UBC mRNA without 5′ UTR (miRNA-deregulated) under low Pi conditions in comparison of wild-type plants (Fujii et al., 2005). Further studies demonstrated the overexpression of miR156b precursor in switchgrass. Relatively low levels of miR156 overexpression were sufficient to enhance biomass yield while producing plants with normal flowering time. Moderate levels of miR156 led to better biomass although the plants produced were non-flowering. However, high levels of miR156 resulted in severely stunted growth (Fu et al., 2012). In transgenic switchgrass, the degree of morphological changes depends on miR156 level.

10.4.2 **RNAi in plants under biotic stress**

Biotic stress factors besides abiotic stresses are another restraint to global agricultural potency. These include viruses, bacteria, fungi, insects, and nematodes (Li et al., 2010b, 2012; Zhang et al., 2011; Eckardt, 2012; Romanel et al., 2012). Therefore, the involvement of the RNAi approach has provided new insights for controlling large numbers of plant diseases caused by these deterrents.

10.4.2.1 Viruses

Viruses are known to cause maximum loss to crop productivity. Therefore, there is an urgent need to develop eco-friendly, non-chemical, and publicly acceptable control measures (Jagtap et al., 2011). Various "pathogen-derived resistance" (PDR) approaches, where resistance to a determined pathogen is obtained from its own genetic material, have been used to develop resistant plants (Shepherd et al., 2009). Plant viruses usually have RNA genomes and replicates via dsRNA intermediates and hence serve as potent inducers of RNAi during early replication and as silencing targets in later infections. As RNAi is an antiviral mechanism, this phenomenon was first utilized by Waterhouse and Graham (1998) to develop potato virus-resistant plants that harbor vectors for the simultaneous expression of both the sense and antisense transcripts of the viral helper-component proteinase (HCPro) gene showing complete immunity to potato virus Y (PVY). Then, in 2003, Kasschau et al. (2003) induced miR156 and miR164 in transgenic *Arabidopsis* plants, which express P1/HC-Pro (turnip mosaic virus (TuMV)-encoded RNA silencing suppressor).

Further, He et al. (2008) characterized and verified bra-miR1885 (a novel miRNA) specifically activated by TuMV infection that targeted disease-resistant transcripts TIR-NBS-LRR class for slicing. Later on, a number of researchers used RNAi to develop virus-resistant crops like potato (Missiou et al., 2004), papaya (Kertbundit et al., 2007), *Nicotiana* (Nazmul-Haque et al., 2007), and cassava (Vanderschuren et al., 2009).

10.4.2.2 Bacteria

In bacteria, RNAi has been used effectively by targeting two genes, namely, *iaaM* and *ipt*, to protect plants against phytopathogenic bacteria *Agrobacterium tumefaciens*, responsible for inducing resistance to crown gall disease (Dunoyer et al., 2006). In *Arabidopsis*, positive restriction of bacterial infection (*Pseudomonas syringae*) by repression of auxin signaling was reported by Navarro et al. (2006). This was shown by induction of a bacterial pattern-associated molecular pattern (PAMP) peptide flg22 and negative regulation of the F-box auxin receptors TIR1, AFB2, and AFB3. Further, transgenic *Arabidopsis* plants overexpressing a TIR1 paralogue (partially refractory to miR393) were found to be susceptible to virulent P to DC3000. On the other hand, plants that overexpressed miR393 were more resistant to bacterial infection along with some alterations such as reduction in apical dominance with multiple shoots. In another report, use of high-throughput sequencing technology strongly revealed an up-regulation of miR393 during *P. syringae* pv. tomato (DC3000hrcC) infection of plants (Fahlgren et al., 2007). Additionally, Fahlgren et al. (2007) reported that miR160 and miR167, which target transcription factors in the ARF family, were significantly induced up to five-fold and six-fold, respectively. Thus, indicating that down-regulation of auxin signaling by manipulating miRNAs and their targets could be an effective means to genetically engineer bacterial pathogen resistance varieties of plants. Also, Jagadeeswaran et al. (2009) analyzed down-regulation of miR398, which targets coding for two Cu/Zn superoxide dismutases (CSD1 and CSD2) during bacterial infections. The results demonstrated up-regulation of CSD1 upon bacterial infection, which is in accordance with the down-regulation of miR398 during biotic stress.

10.4.2.3 Fungi

RNAi-mediated resistance against fungal disease in plants was developed by Hernández et al. (2009). In the study, resistance in *N. tabacum* against black shank disease caused by *Phytophthora parasitica* was developed by silencing the glutathione S-transferase gene in sense and anti-sense orientation to an RNAi vector. Xin et al. (2010) reported 24 miRNAs responsive to powdery mildew caused by the obligate biotrophic fungus *Blumeria graminis* f. sp. *tritici* (*Bgt*) in wheat. Some of these were down-regulated, such as miRNA156, miRNA159, miRNA164, miRNA171, and miRNA396, while others were up-regulated, such as miR393, miR444, and miR827, during infection. In rice, involvement of miRNA in response to the blast fungus *Magnaporthe oryzae* was carried out by Campo et al. (2013). The results reported the discovery of a new rice miRNA, osa-miR7695, that negatively regulated the natural resistance-associated macrophage protein 6 (*OsNramp6*). Moreover, transgenic rice plants overexpressing Osa-miR7696 were found to be highly resistant to the infection.

10.4.2.4 Insects

RNAi has also been used to control insect population besides functional genomic research. Baum et al. (2007) reported that transgenic corn plants transformed to express dsRNA for tubulin genes were resistant to western corn rootworm in terms of lesser damage to plants, stunting in larvae, and even mortality. Similar results have also been demonstrated by Mao et al. (2007) in cotton boll-worm (*Helicoverpa armigera*). In the study, reduced larval growth was found in larvae feeding on plant material that expresses the specific cytochrome P450 gene (CYP6AE14) dsRNA. Price and Gatehouse (2008) in another investigation demonstrated that when dsRNA was fed as a component in diet, it efficiently down-regulated target genes in insects. Bt transgenic technology is well established in a number of plant species such as cotton, legumes, cereals, etc. These transgenic Bt crops produce an insecticidal crystalline protein from *Bacillus thuringiensis* (Cry toxins) and has proved an effective means to control insect pests. However, insects have developed resistance to Bt crops (Jagtap et al., 2011). Therefore, the RNAi approach has emerged as an effective biotechnological substitute with multitudinal assets over Bt transgenics to control insects (Bravo and Soberôn, 2008). The RNAi-mediated insect-resistant plants are specific, commanding, sequence based, eco-friendly, and reduce off-target effects that crop up when sequence equivalence concedes novel small RNAs to debase mRNA for genes that were not the expected silencing spots.

10.4.2.5 Nematodes

Nematodes also cause severe damage to crops by draining their nutrients. Bioengineered crops have been developed that express dsRNA targeting root-knot nematode (RKN) parasitism genes to disrupt the parasitic processes which represent a viable and flexible means of developing novel and durable RKN-resistant crops (Huang et al., 2006). Tobacco plants that express dsRNA hairpin structures against root-knot nematodes have also been developed (Fairbairn et al., 2007). In transgenic *A. thaliana*, Sindhu et al. (2009) studied RNAi data for cyst nematodes (*Heterodera schachtii*) related to four parasitism genes. It was analyzed that in plants expressing corresponding RNAi constructs, the targeted genes were specifically reduced in nematodes feeding on them. Further, additive effects of ectopically expressed dsRNA on RKN development have been reported by Charlton et al. (2010). In the study, two genes involved in different aspects of *Meloidogyne incognita* development were targeted, which resulted in $\geq 50\%$ reduction in nematode numbers in roots and retardation of development of the egg-producing females. Similar investigations have also been carried out by Ibrahim et al. (2011), Tan et al. (2013), and Tsygankova et al. (2013) where different genes were targeted to determine their efficacy to reduce nematode development on the host plant. In addition, Khraiwesh et al. (2012) reported down-regulation of miRNAs (miR161, miR164, miR167a, miR172c, miR396c, miR396a, b, and miR398a) during soybean cyst nematode (SCN) *H. schachtii* infection in *Arabidopsis*. Moreover, in soybean, Li et al., (2012) marked 101 miRNAs corresponding to 40 families during comparative analysis amid *H. glycines* pathogenesis, out of which, 20 miRNAs were found to be expressed differentially within SCN-susceptible and -resistant cultivars.

10.5 Conclusion and future prospects

In the past few years, RNAi has become a key research tool that has revolutionized the possibilities by causing "knockdowns" of targeted gene activities. MicroRNAs have emerged as an outstanding class of gene regulatory factors that has the potential to improve various intricate traits and play important functions in agricultural productivity. However, current achievements in this approach have produced speculation for its role in RNA-transcribed resistant crops. It is routinely used by scientists as it offers many advantages over more traditional knockdown technologies. RNAi-based transgenic plants can be used by molecular plant breeders in plant breeding programs to engineer crop resistance. The use of RNAi in relation to molecular resources is in its infancy. Therefore, proper mining of available resources such as manipulating new pathways for RNAi, microarray, laser capture microdissection, next generation sequencing, EST sequences, etc. could be employed for more successful control of pathogens. Moreover, an offshoot of RNAi is required to effectively utilize multiple targets or their promoters to produce new quality traits so as to reduce the burden of pathogens on agricultural production. Also, transfer of signal in plants in response to environmental stress needs to be revised and clearly understood. Finally, envision should be made to design and use the knockdown libraries to obtain resistant crops to challenge multiple infections in the field.

References

Abdel-Ghany, S.E., Pilon, M., 2008. MicroRNA-mediated systemic down-regulation of copper protein expression in response to low copper availability in *Arabidopsis*. J. Biol. Chem. 283, 15932–15945.

Agrawal, N., Dasaradhi, P.V.N., Mohmmed, A., Malhotra, P., Bhatnagar, R.K., Mukherjee, S.K., 2003. RNA interference, biology, mechanism and applications. Microbiol. Mol. Biol. Rev. 67, 657–685.

Arenas-Huertero, C., Pérez, B., Rabanal, F., Blanco-Melo, D., De la Rosa, C., Estrada-Navarrete, G., et al., 2009. Conserved and novel miRNAs in the legume *Phaseolus vulgaris* in response to stress. Plant Mol. Biol. 70, 385–401.

Bartels, D., Sunkar, R., 2005. Drought and salt tolerance in plants. Crit. Rev. Plant Sci. 24, 23–58.

Baum, J.A., Bogaert, T., Clinton, W., Heck, G.R., Feldmann, P., Illagan, O., et al., 2007. Control of coleopteran insect pests through RNA interference. Nat. Biotechnol. 25, 1322–1326.

Ben Amor, B., Wirth, S., Merchan, F., Laporte, P., Aubenton-Carafa, Y., Hirsch, J., et al., 2009. Novel long non-protein-coding RNAs involved in *Arabidopsis* differentiation and stress responses. Genome. Res. 19, 57–69.

Bernstein, E., Caudy, A.A., Hammond, S.M., Hannon, G.J., 2001. Role for a bidentate ribonuclease in the initiation step of RNA interference. Nature 409, 363–366.

Bosher, J.M., Labouesse, M., 2000. RNA interference, genetic wand and genetic watchdog. Nat. Cell. Biol. 2, 31–36.

Bravo, A., Soberôn, M., 2008. How to cope with insect resistance to Bt toxins. Trends Biotechnol. 26, 573–579.

Campo, S., Peris-Peris, C., Sire, C., Moreno, A.B., Donaire, L., Zytnicki, M., et al., 2013. Identification of a novel microRNA (miRNA) from rice that targets an alternatively spliced transcript of the Nramp6 (Natural resistance-associated macrophage protein 6) gene involved in pathogen resistance. New Phytol. 199, 212–227.

Charlton, W.L., Harel, H.Y.M., Bakhetia, M., Hibbard, J.K., Atkinson, H.J., McPherson, M.J., 2010. Additive effects of plant expressed double-stranded RNAs on root-knot nematode development. Int. J. Parasitol. 40, 855–864.

Chen, L., Wang, T., Zhao, M., Tian, Q., Zhang, W.H., 2012. Identification of aluminum-responsive microRNAs in Medicago truncatula by genome-wide high-throughput sequencing. Planta 235, 375–386.

Cogoni, C., Macino, G., 1997. Conservation of transgene-induced posttranscriptional gene silencing in plants and fungi. Trends Plant Sci. 2, 438–443.

Cogoni, C., Macino, G., 1999. Gene silencing in *Neurospora crassa* requires a protein homologous to RNA-dependent RNA polymerase. Nature 399, 166–169.

Depicker, A., Van Montagu, M., 1997. Post-transcriptional gene silencing in plants. Curr. Opin. Cell Biol. 9, 373–382.

Ding, D., Zhang, L., Wang, H., Liu, Z., Zhang, Z., Zheng, Y., 2009. Differential expression of miRNAs in response to salt stress in maize roots. Ann. Bot. 103, 29–38.

Ding, Y., Zhu, C., 2009. The role of microRNAs in copper and cadmium homeostasis. Biochem. Biophys. Res. Commun. 386, 6–10.

Ding, Y., Chen, Z., Zhu, C., 2011. Microarray-based analysis of cadmium-responsive microRNAs in rice (*Oryza sativa*). J. Exp. Bot. 62, 3563–3573.

Dunoyer, P., Himber, C., Voinnet, O., 2006. Induction, suppression and requirement of RNA silencing pathways in virulent *Agrobacterium tumefaciens* infections. Nat. Genet. 38, 258–263.

Dykxhoorn, D.M., Novina, C.D., Sharp, P.A., 2003. Killing the messenger, short RNAs that silence gene expression. Nat. Rev. Mol. Cell Biol. 4, 457–467.

Eckardt, N.A., 2012. A microRNA cascade in plant defense. Plant Cell 24, 840.

Elbashir, S., Harborth, J., Lendeckel, W., Yalcin, A., Weber, K., Tuschl, T., 2001a. Duplexes of 21-nucleotide RNAs mediate RNA interference in cultured mammalian cells. Nature 411, 494–498.

Elbashir, S.M., Lendeckel, W., Tuschl, T., 2001b. RNA interference is mediated by 21- and 22-nucleotide RNAs. Genes. Dev. 15, 188–200.

Fahlgren, N., Howell, M.D., Kasschau, K.D., Chapman, E.J., Sullivan, C.M., Cumbie, J.S., et al., 2007. High-throughput sequencing of *Arabidopsis* microRNAs, evidence for frequent birth and death of MIRNA Genes. PLoS One 2, e219.

Fairbairn, D.J., Cavallaro, A.S., Bernard, M., Mahalinga-Iyer, J., Graham, M.W., Botella, J.R., 2007. Host-delivered RNAi, an effective strategy to silence genes in plant-parasitic nematodes. Planta 226, 1525–1533.

Farinati, S., DalCorso, G., Varotto, S., Furini, A., 2010. The *Brassica juncea* BjCdR15, an ortholog of *Arabidopsis* TGA3, is a regulator of cadmium uptake, transport and accumulation in shoots and confers cadmium tolerance in transgenic plants. New. Phytol. 185, 964–978.

Filippov, V., Solovyev, V., Filippova, M., Gill, S.S., 2000. A novel type of RNase III family proteins in eukaryotes. Genesis 245, 213–221.

Fire, A., Xu, S., Montgomery, M., Kostas, S., Driver, S., Mello, C., 1998. Potent and specific genetic interference by double-stranded RNA in *Caenorhabditis elegans*. Nature 391, 806–811.

Fortier, E., Belote, J.M., 2000. Temperature-dependent gene silencing by an expressed inverted repeat in *Drosophila*. Genesis 26, 240–244.

Fu, C., Sunkar, R., Zhou, C., Shen, H., Zhang, J.Y., Matts, J., et al., 2012. Overexpression of miR156 in switchgrass (*Panicum virgatum* L.) results in various morphological alterations and leads to improved biomass production. Plant Biotechnol. J. 10, 443–452.

Fujii, H., Chiou, T.J., Lin, S.I., Aung, K., Zhu, J.K., 2005. A miRNA involved in phosphate starvation response in *Arabidopsis*. Curr. Biol. 15, 2038–2043.

Gielen, H., Remans, T., Vangronsveld, J., Cuypers, A., 2012. MicroRNAs in metal stress, specific roles or secondary responses? Int. J. Mol. Sci. 13, 5826–15847.

Golden, T.A., Schauer, S.E., Lang, J.D., Pien, S., Mushegian, R., Grossniklaus, A.U., et al., 2002. SHORT INTEGUMENTS1/SUSPENSOR1/CARPEL FACTORY, a Dicer homolog, is a maternal effect gene required for embryo development in *Arabidopsis*. Plant Physiol. 130, 808–822.

Gregory, R.I., Chendrimada, T.P., Cooch, N., Shiekhattar, R., 2005. Human RISC couples microRNA biogenesis and posttranscriptional gene silencing. Cell 123, 631–640.

Guan, Q., Lu, X., Zeng, H., Zhang, Y., Zhu, J., 2013. Heat stress induction of miR398 triggers a regulatory loop that is critical for thermotolerance in *Arabidopsis*. Plant J. 74, 840–851.

Hamilton, A., Baulcombe, D., 1999. A species of small antisense RNA in post-transcriptional gene silencing in plants. Science 286, 950—952.

Hammond, S.M., Berstein, E., Beach, D., Hannon, G.J., 2000. An RNA-directed nuclease mediates post-transcriptional gene silencing in *Drosophila* cells. Nature 404, 293—296.

He, X.F., Fang, Y.Y., Feng, L., Guo, H.S., 2008. Characterization of conserved and novel microRNAs and their targets, including a TuMV-induced TIR—NBS—LRR class R gene-derived novel miRNA in *Brassica*. FEBS Lett. 582, 2445—2452.

Hernández, I., Chacón, O., Rodriguez, R., Portieles, R., Pujol, Y.L.M., Borrás-Hidalgo, O., 2009. Black shank resistant tobacco by silencing of glutathione S-transferase. Bioch. Bioph. Res. Co. 387, 300—304.

Hobert, O., 2008. Gene regulation by transcription factors and microRNAs. Science 319, 1785—1786.

Huang, G., Allen, R., Davis, E.L., Baum, T.J., Hussey, R.S., 2006. Engineering broad root-knot resistance in transgenic plants by RNAi silencing of a conserved and essential root-knot nematode parasitism gene. Proc. Natl. Acad. Sci. 103, 14302—14306.

Huang, S.Q., Peng, J., Qiu, C.X., Yang, Z.M., 2009. Heavy metal-regulated new microRNAs from rice. J. Inorg. Biochem. 103, 282—287.

Huertero, C.A., Pérez, B., Rabanal, F., Blanco-Melo, D., De la Rosa, C., Estrada-Navarrete, G., et al., 2009. Conserved and novelmi RNAs in the legume *Phaseolus vulgaris* in response to stress. Plant Mol. Biol. 70, 385—401.

Ibrahim, H.M.M., Alkharouf, N.W., Meyer, S.L.F., Aly, M.A.M., El-Din, Gamal, Abd El, K.Y., et al., 2011. Post-transcriptional gene silencing of root-knot nematode in transformed soybean roots. Exp. Parasitol. 127, 90—99.

Jacoby, M., Weisshaar, B., Vicente-Carbajosa, J., Tiedemann, J., Kroj, T., Parcy, F., 2002. bZIP transcription factors in *Arabidopsis*. Trends Plant Sci. 7, 106—111.

Jagadeeswaran, G., Saini, A., Sunkar, R., 2009. Biotic and abiotic stress down-regulate miR398 expression in *Arabidopsis*. Planta 229, 1009—1014.

Jagtap, U.B., Gurav, R.G., Bapat, V.A., 2011. Role of RNA interference in plant improvement. Naturwissenchaften 98, 473—492.

Jia, X., Ren, L., Chen, Q.J., Li, R., Tang, G., 2009. UV-B-responsive microRNAs in *Populus tremula*. J. Plant Physiol. 166, 2046—2057.

Jones-Rhoades, M.W., Bartel, D.P., 2004. Computational identification of plant micro-RNAs and their targets, including a stress-induced miRNA. Mol. Cell. 14, 787—799.

Jones-Rhoades, M.W., Bartel, D.P., Bartel, B., 2006. MicroRNAs and their regulatory roles in plants. Annu. Rev. Plant Biol. 57, 19—53.

Kasschau, K.D., Xie, Z., Allen, E., Llave, C., Chapman, E.J., Krizan, K.A., et al., 2003. P1/HC-Pro, a viral suppressor of RNA silencing, interferes with *Arabidopsis* development and miRNA function. Dev. Cell. 4, 205—217.

Kawashima, C.G., Yoshimoto, N., Maruyama-Nakashita, A., Tsuchiya, Y.N., Saito, K., Takahashi, H., et al., 2008. Sulphur starvation induces the expression of microRNA-395 and one of its target genes but in different cell types. Plant J. 57, 313—321.

Kawashima, C.G., Matthewman, C.A., Huang, S., Lee, B.R., Yoshimoto, N., Koprivova, A., et al., 2011. Interplay of SLIM1 and miR395 in the regulation of sulfate assimilation in *Arabidopsis*. Plant J. 66, 863—876.

Kertbundit, S., Pongtanom, N., Ruanjan, P., Chantasingh, D., Tanwanchai, A., Panyim, S., 2007. Resistance of transgenic papaya plants to papaya rings pot virus. Biol. Plantarum 51, 333—339.

Khraiwesh, B., Zhu, J.K., Zhu, J., 2012. Role of miRNAs and siRNAs in biotic and abiotic stress responses of plants. Biochim. Biophys. Acta 1819, 137—148.

Koonin, E.V., Gorbalenya, A.E., Chumakov, K.M., 1989. Tentative identification of RNA-dependent RNA polymerases of dsRNA viruses and their relationship to positive strand RNA viral polymerases. FEBS. Lett. 252, 42−46.

Krämer, U., Talke, I.N., Hanikenne, M., 2007. Transition metal transport. FEBS. Lett. 581, 2263−2272.

Kulcheski, F.R., 2011. Identification of novel soybean microRNAs involved in abiotic and biotic stresses. BMC. Genomics. 12, 307.

Li, W.X., Oono, Y., Zhu, J.H., He, X.J., Wu, J.M., Iida, K., et al., 2008. The *Arabidopsis* NFYA5 transcription factor is regulated transcriptionally and posttranscriptionally to promote drought resistance. Plant Cell 20, 2238−2251.

Li, X., Wang, X., Zhang, S., Liu, D., Duan, Y., Dong, W., 2012. Identification of soybean microRNAs involved in soybean cyst nematode infection by deep sequencing. PLoS One 7, e39650.

Li, Y., Zhang, Q.Q., Zhang, J., Wu, L., Qi, Y., Zhou, J.M., 2010b. Identification of microRNAs involved in pathogen-associated molecular pattern-triggered plant innate immunity. Plant Physiol. 152, 2222−2231.

Li, Y.F., Zheng, Y., Addo-Quaye, C., Zhang, L., Saini, A., Jagadeeswaran, G., et al., 2010a. Transcriptome-wide identification of microRNA targets in rice. Plant J. 62, 742−759.

Liang, G., Yu, D., 2010. Reciprocal regulation among miR395, APS and SULTR2; 1 in *Arabidopsis thaliana*. Plant Signal. Behav. 5, 1257−1259.

Liang, G., Yang, F., Yu, D., 2010. MicroRNA395 mediates regulation of sulfate accumulation and allocation in *Arabidopsis thaliana*. Plant J. 62, 1046−1057.

Lindbo, J.A., Silva-Rosales, L., Proebsting, W.M., Dougherty, W.G., 1993. Induction of a highly specific anti-viral state in transgenic plants, implications for regulation of gene expression and virus resistance. Plant Cell. 5, 1749−1759.

Lipardi, C., Wei, Q., Paterson, B.M., 2001. RNAi as random degradation PCR, siRNA primers convert mRNA into dsRNA that are degraded to generate new siRNAs. Cell 101, 297−307.

Liu, H.H., Tian, X., Li, Y.J., Wu, C.A., Zheng, C.C., 2008. Microarray-based analysis of stress-regulated microRNAs in *Arabidopsis thaliana*. RNA 14, 836−843.

Lu, S.F., Sun, Y.H., Chiang, V.L., 2008. Stress-responsive microRNAs in *Populus*. Plant J. 55, 131−151.

Lv, D.K., Bai, X., Li, Y., Ding, X.D., Ge, Y., Cai, H., et al., 2010. Profiling of cold-stress-responsive miRNAs in rice by microarrays. Genesis 459, 39−47.

Mao, Y.B., Cai, W.J., Wang, J.W., Hong, G.J., Tao, X.Y., Wang, L.J., et al., 2007. Silencing a cotton boll-worm P450 monooxygenase gene by plant-mediated RNAi impairs larval tolerance of gossypol. Nat. Biotechnol. 25, 1307−1313.

Matranga, C., Tomari, Y., Shin, C., Bartel, D.P., Zamore, P.D., 2005. Passenger-strand cleavage facilitates assembly of siRNA into Ago2-containing RNAi enzyme complexes. Cell 123, 607−620.

Mendoza-Soto, A.B., Sanchez, F., Hernández, G., 2012. MicroRNAs as regulators in plant metal toxicity response. Front Plant Sci. 3, 105.

Missiou, A., Kalantidis, K., Boutla, A., Tzortzakaki, S., Tabler, M., Tsagris, M., 2004. Generation of transgenic potato plants highly resistant to potato virus Y (PVY) through RNA silencing. Mol. Breed. 14, 185−197.

Munns, R., Tester, M., 2008. Mechanisms of salinity tolerance. Annu. Rev. Plant Biol. 59, 651−681.

Napoli, C., Lemieux, C., Jorgensen, R., 1990. Introduction of a chimeric chalcone synthase gene into petunia results in reversible co-suppression of homologous genes in transgenic plant. Plant Cell. 2, 279−289.

Navarro, L., Dunoyer, P., Jay, F., Arnold, B., Dharmasiri, N., Estelle, M., et al., 2006. A plant miRNA contributes to antibacterial resistance by repressing auxin signalling. Science. 312, 436−439.

Nazmul-Haque, A.K.M., Tanaka, Y., Sonoda, S., Nishiguchi, M., 2007. Analysis of transitive RNA silencing after grafting in transgenic plants with the coat protein gene of sweet potato feathery mottle virus. Plant Mol. Biol. 6, 35−47.

Nicholson, A.W., 1999. Function, mechanism and regulation of bacterial ribonucleases. FEMS. Microbiol. Rev. 23, 371–390.

Paddison, P.J., Caudy, A.A., Bernstein, E., Hannon, G.J., Conklin, D.S., 2002. Short hairpin RNAs (shRNAs) induce sequence-specific silencing in mammalian cells. Genes. Dev. 16, 948–958.

Pardo, J.M., 2010. Biotechnology of water and salinity stress tolerance. Curr. Opin. Biotechnol. 21, 1–12.

Paul, C.P., Good, P.D., Winer, I., Engelke, D.R., 2002. Effective expression of small interfering RNA in human cells. Nat. Biotechnol. 20, 505–508.

Piccin, A., Salameh, A., Benna, C., Sandrelli, F., Mazzotta, G., Zordan, M., et al., 2001. Efficient and heritable knockout of an adult phenotype in *Drosophila* with a GAL-4 driven hairpin RNA incorporating a heterologous spacer. Nucleic. Acids. Res. 29, 55–60.

Price, D.R.G., Gatehouse, J.A., 2008. RNAi-mediated crop protection against insects. Trends Biotechnol. 26, 393–400.

Rand, T.A., Petersen, S., Du, F., Wang, X., 2005. Argonaute2 cleaves the anti-guide strand of siRNA during RISC activation. Cell 123, 621–629.

Reilly, J.M., 1999. Climate change and agriculture, the state of the scientific knowledge. Climate Change 43, 645–650.

Reyes, J.L., Chua, N.H., 2007. ABA induction of miR159 controls transcript levels of two MYB factors during *Arabidopsis* seed germination. Plant J. 49, 592–606.

Romanel, E., Silva, T., Corrêa, R., Farinelli, L., Hawkins, J., Schrago, C.G., et al., 2012. Global alteration of microRNAs and transposon-derived small RNAs in cotton (*Gossypium hirsutum*) during Cotton leafroll dwarf polerovirus (CLRDV) infection. Plant Mol. Biol. 80, 443–460.

Romano, N., Macino, G., 1992. Quelling transient inactivation of gene expression in *Neurospora crassa* by transformation with homologous sequences. Mol. Microbiol. 6, 3343–3353.

Romero-Puertas, M.C., Corpas, F.J., Rodriguez-Serrano, M., Gomez, M., del Rio, L.A., Sandalio, L.M., 2007. Differential expression and regulation of antioxidative enzymes by cadmium in peaplants. J. Plant Physiol. 164, 1346–1357.

Shepherd, D.N., Martin, D.P., Thomson, J.A., 2009. Transgenic strategies for developing crops resistant to geminivirus. Plant Sci. 176, 1–11.

Sijen, T., Fleenor, J., Simmer, F., Thijssen, K.L., Parrish, S., Timmons, L., et al., 2001. On the role of RNA amplification in dsRNA-triggered gene silencing. Cell 107, 465–476.

Sindhu, A.S., Maier, T.R., Mitchum, M.G., Hussey, R.S., Davis, E.L., Baum, T.J., 2009. Effective and specific in planta RNAi in cyst nematodes, expression interference of four parasitism genes reduces parasitic success. J. Exp. Bot. 60, 315–324.

Sui, G., Soohoo, C., Affar el, B., Gay, F., Shi, Y., et al., 2002. A DNA vector-based RNAi technology to suppress gene expression in mammalian cells. Proc. Natl. Acad. Sci. USA 99, 5515–5520.

Sun, G., 2012. MicroRNAs and their diverse functions in plants. Plant Mol. Biol. 80, 17–36.

Sunkar, R., 2010. MicroRNAs with macro-effects on plant stress responses. Semin. Cell. Dev. Biol. 21, 805.

Sunkar, R., Zhu, J.K., 2004. Novel and stress-regulated micro RNAs and other small RNAs from *Arabidopsis*. Plant Cell. 16, 2001–2019.

Sunkar, R., Kapoor, A., Zhu, J.K., 2006. Post-transcriptional induction of two Cu/Zn superoxide dismutase genes in *Arabidopsis* is mediated by down-regulation of miR398 and important for oxidative stress tolerance. Plant Cell. 18, 2051–2065.

Sunkar, R., Li, Y.F., Jagadeeswaran, G., 2012. Functions of microRNAs in plant stress responses. Trends Plant Sci. 17, 196–203.

Tan, J.A., Jones, M.G., Fosu-Nyarko, J., 2013. Gene silencing in root lesion nematodes (*Pratylenchus* spp.) significantly reduces reproduction in a plant host. Exp. Parasitol. 133, 166−178.

Thiebaut, F., Rojas, C.A., Almeida, K.L., Grativol, C., Domiciano, G.C., Lamb, C.R.C., et al., 2011. Regulation of miR319 during cold stress in sugarcane. Plant Cell. Environ. 35, 502−512.

Trindade, I., Capitão, C., Dalmay, T., Fevereiro, M.P., Santos, D.M., 2010. miR398 and miR408 are up-regulated in response to water deficit in *Medicago trunculata*. Planta 231, 705−716.

Tsygankova, V.A., Yemets, A.I., Iutinska, H.O., Beljavska, L.O., Galkin, A.P., Blume, Y.B., 2013. Increasing the resistance of rape plants to the parasitic nematode *Heterodera schachtii* using RNAi technology. Cytol. Genet. 47, 222−230.

Vanderschuren, H., Alder, A., Zhang, P., Gruissem, W., 2009. Dose dependent RNAi-mediated geminivirus resistance in the tropical root crop cassava. Plant Mol. Biol. 64, 549−557.

Waterhouse, P.M., Graham, W.M.B., 1998. Virus resistance and gene silencing in plants can be induced by simultaneous expression of sense and antisense RNA. Proc. Natl. Acad. Sci. USA 95, 13959−13964.

Winston, W.M., Molodowitch, C., Hunter, C.P., 2002. Systemic RNAi in *C. elegans* requires the putative transmembrane protein SID-1. Science 295, 2456−2459.

Xie, F.L., Huang, S.Q., Guo, K., Xiang, A.L., Zhu, Y.Y., Nie, L., et al., 2007. Computational identification of novel microRNAs and targets in *Brassica napus*. FEBS. Lett. 581, 1464−1474.

Xin, M., 2011. Identification and characterization of wheat long non-protein coding RNAs responsive to powdery mildew infection and heat stress by using microarray analysis and SBS sequencing. BMC. Plant Biol. 11, 61.

Xin, M., Wang, Y., Yao, Y., Xie, C., Peng, H., Ni, Z., et al., 2010. Diverse set of microRNAs are responsive to powdery mildew infection and heat stress in wheat (*Triticum aestivum* L.). BMC. Plant Biol. 10, 123.

Yamaguchi-Shinozaki, K., Shinozaki, K., 2006. Transcriptional regulatory networks in cellular responses and tolerance to dehydration and cold stresses. Annu. Rev. Plant. Biol. 57, 781−803.

Yamasaki, H., Abdel-Ghany, S.E., Cohu, C.M., Kobayashi, Y., Shikanai, T., Pilon, T.M., 2007. Regulation of copper homeostasis by micro-RNA in *Arabidopsis*. J. Biol. Chem. 282, 16369−16378.

Yang, C., Li, D., Mao, D., Liu, X., Li, C., Li, X., et al., 2013. Over expression of microRNA319 impacts leaf morphogenesis and leads to enhanced cold tolerance in rice (*Oryza sativa* L.). Plant Cell. Environ. 36, 2207−2218.

Yang, D., Lu, H., Erichson, J.W., 2000. Evidence that processed small dsRNA may mediate sequence specific mRNA degradation during RNAi in *Drosophila* embryos. Curr. Biol. 10, 1191−1200.

Zamore, P.D., Tuschl, T., Sharp, P.A., Bartel, D.P., 2000. RNAi, double-stranded RNA directs the ATP-dependent cleavage of mRNA at 21- to 23-nucleotide intervals. Cell 101, 25−33.

Zanotto, P.M., Gibbs, M.J., Gould, E.A., Holmes, E.C., 1996. A re-evaluation of the higher taxonomy of viruses based on RNA polymerases. J. Virol. 70, 6083−6096.

Zhang, W., Gao, S., Zhou, X., Chellappan, P., Chen, Z., Zhou, X., et al., 2011. Bacteria responsive microRNAs regulate plant innate immunity by modulating plant hormone networks. Plant Mol. Biol. 75, 93−105.

Zhao, B., 2007. Identification of drought-induced microRNAs in rice. Biochem. Biophys. Res. Commun. 354, 585−590.

Zhao, B.T., Ge, L.F., Liang, R.Q., Li, W., Ruan, K.C., Lin, H.X., 2009. Members of miR-169 family are induced by high salinity and transiently inhibit the NF-YA transcription factor. BMC. Mol. Biol. 10, 29.

Zhao, M., Ding, H., Zhu, J.K., Zhang, F., Li, W.X., 2011. Involvement of miR169 in the nitrogen-starvation responses in *Arabidopsis*. New. Phytol. 190, 906−915.

Zhou, M., Hong, L., 2013. MicroRNA-mediated gene regulation, potential applications for plant genetic engineering. Plant Mol. Biol. 83, 59−75.

Zhou, X., Sunkar, R., Jin, H., Zhu, J.K., Zhang, W., 2009. Genome-wide identification and analysis of small RNAs originated from natural antisense transcripts in *Oryza sativa*. Genome. Res. 19, 70—78.

Zhou, Z.S., Huang, S.Q., Yang, Z.M., 2008. Bioinformatic identifcation and expression analysis of new microRNAs from *Medicago trunculata*. Biochem. Biophys. Res. Commun. 374, 538—542.

Zhou, Z.S., Zeng, H.Q., Liu, Z.P., Yang, Z.M., 2012. Genome-wide identification of *Medicago trunculata* microRNAs and their targets reveals their different regulation by heavy metal. Plant Cell. Environ. 35, 86—99.

Zhu, J.K., 2002. Salt and drought stress signal transduction in plants. Annu. Rev. Plant Biol. 53, 247—273.

The Role of Carbohydrates in Plant Resistance to Abiotic Stresses

Marina S. Krasavina, Natalia A. Burmistrova and Galina N. Raldugina

11.1 Introduction

The world's population is growing, but the land suitable for intensive agriculture is constantly decreasing due to the deterioration of their quality and changes in the climate. This creates conditions for the increasing lack of food for humans and animals. Numerous environmental stresses adversely affect plant growth and productivity and human health. Most of Earth's plants are suffering from low or high temperature, drought, salinity, and contamination with toxic substances. Drought and salinity affect more than 10–20% of arable land, with a harvest reduction of main cereals up to 50–70% (Boyer, 1982; Bray et al., 2000; Wood, 2005; Alcázar et al., 2006). Unfortunately, the situation is aggravated each year as stresses become more intense and frequent. Unfavorable environmental factors induce morphological, metabolic, and physiological disruptions leading to plant development inhibition, yield reduction, and, in extreme cases, plant death. Active irrigation of arid lands results in increasing soil salinity, which aggravates the problem of fresh water availability for plants, animals, and humans. Industrial manufacturing pollutes the soil and air with toxic substances, including heavy metals. Violations of the biological balance in ecosystems activate the development of pathogenic organisms—fungi, mixomycetes, viruses, and bacteria—causing great harm to all living organisms. The frequency of biotic and abiotic stresses highlights the critical issue of agrobiology—an improvement of plant resistance to environmental disturbances. Breeding and transgenesis require an understanding of plant mechanisms in a natural environment to overcome harmful factor action. Plants are living in an ever-changing environment—during the day and throughout the development cycle, large increases in temperature, humidity, illumination, and many other parameters occur. In order to survive, plants need to monitor these changes and to readjust their own metabolic and physiological processes at different levels: the whole plant, organ, tissue, cell, and molecule. The situation becomes more complex as plants are subjected, as a rule, to combined factors. Therefore, the understanding of their interaction and the formation of complex plant responses to the combination of unfavorable environmental conditions are needed. However, the behaviors of stress signal perception and spreading, and the specific plant responses towards them, are poorly studied. Even harder to understand is the interaction of different signaling pathways.

P. Ahmad (Ed): Emerging Technologies and Management of Crop Stress Tolerance, Volume 1.
DOI: http://dx.doi.org/10.1016/B978-0-12-800876-8.00011-4

In addition to specific plant responses to various environmental factors, there are general mechanisms of plant resistance. One of the earliest effects of practically all adverse factors is a disturbance in osmoregulation, redox, and acid—base balance in the cells. The recovery of these parameters contributes to plant stability under various stressful conditions. The maintenance of photosynthesis efficiency, carbohydrate and nitrogen metabolism, and the integrity of membranes and intracellular organelles are also required to overcome the toxic action of various environmental factors. In addition, the synthesis of specific protective proteins plays an important role in plant resistance.

In recent years, a large number of articles, reviews, and monographs on plant responses to various stresses have appeared. To consider all of them is impossible. In this chapter, we will only briefly discuss the changes in carbohydrate metabolism under stressful conditions, and the effects of various osmotically active substances on plant resistance.

11.2 Osmotic balance during the action of unfavorable environmental factors

Osmotic disbalance induces deep changes in plant metabolism. Generally, three abiotic stressors affect plants: drought, high or low temperature, and salinity. These factors disturb water relations, resulting in cell dehydration. The accumulation of osmotically active compounds in the cells is the main way to resist water deficit. The accumulation of osmolytes reduces the cell osmotic potential, which increases the water-accumulating capacity and favors the cell water upkeep (Thomashow, 1999; Xin and Browse, 2000). In these cases, osmolyte accumulation increases resistance to all stresses: reducing cell water loss under water deficit, preventing the uptake of toxic ions (Na^+ and Cl^-) under salt stress, and decreasing the extracellular ice formation at low temperatures (Mao et al., 2010).

Compounds accumulated under stress belong to compatible solutes. Among them are amino acids and their derivatives (proline, glycine, betaine, beta-alanine betaine, and proline betaine), polyamines, quaternary compounds, amines and polyols (glycerols, mannitol, and sorbitol), trehalose, fructans, and methylated inositol. Compatible solutes may accumulate in large amounts without disturbing cell metabolism. This may be determined by the presence of common properties from all compatible solutes: high solubility, non-toxicity, compatibility with intracellular compounds, and absence of effect on their metabolic activity.

11.2.1 Plant resistance and accumulation of soluble sugars

By now, the opinion that the accumulation of metabolites is the most common and reliable way to avoid the toxic effects of stresses is firmly established. For many plants growing in a cold climate, on saline soils, or subjected to short-term chilling and water deficit, the accumulation of compatible solutes was demonstrated. Along with soluble proteins and amino acids (proline primarily), predominant accumulation of soluble sugars occurs (Guy, 1990; Patton et al., 2007; Lee et al., 2012). Carbohydrates—hexoses (glucose and fructose), disaccharides (sucrose, trehalose), and oligosaccharides (raffinose, stachyose)—are important compatible osmolytes (Gusta et al., 1996; Kozlowski and Pallardy, 2002; Choudhary et al., 2005; Shao et al., 2006; Mokhamed et al., 2006;

Morsy et al., 2007; Patton et al., 2007; Guy et al., 2008; Welling and Palva, 2008; Hussain et al., 2011). There are many reports about sugar accumulation under low temperature (Guy, 1990; Castonguay et al., 1995; Guy et al., 2008), drought, or water deficit (for example, Whittaker et al., 2001), osmotic stress (Wang et al., 2000), and salinity (Balibrea et al., 1997; Jouve et al., 2004; cited in Wang et al., 2000). A comparison of non-dormant alfalfa roots and those in deep winter dormancy showed that the latter contained more sugars (Cunningham et al., 2003). In another work performed with tissues of two potato clones differing in their tolerance to cold, salinity, and osmotic stress (Evers et al., 2007), it was found that salinity and osmotic stress induced sucrose accumulation in both clones. At the same time, glucose accumulated under salinity only in the salt-tolerant clone. In the less salt-tolerant clone sensitive to cold, the glucose content even decreased under salt conditions. These data suggest a specific role of glucose in plant resistance.

It should be noted that different carbohydrates accumulate depending on the stress type. For example, chilling of rice plants resulted in the accumulation of galactose and raffinose, whereas salinity and water deficit induced the accumulation of glucose, trehalose, and mannitol (Morsy et al., 2007). In potato, drought induced the accumulation of sugars, polyols, oligosaccharides (raffinose predominantly), proline, and glycine betaine. Such accumulation was transient: rewatering reduced the content of these compounds (Taji et al., 2002; Valliyodan and Nguyen, 2006; Legay et al., 2011).

11.2.2 Regulation of the low molecular sugar (glucose, fructose, and sucrose) contents in plants

Processes resulting in sugar accumulation in stressed plant cells are poorly studied. After carbon fixation in leaves, synthesized carbohydrates are distributed in all cells, tissues, and organs. The organism viability and its capability of counteracting the medium challenges depend on an adequate supply of nutrients to all sink organs.

Photosynthetic carbon fixation depends on many environmental factors. It was shown long ago that plant transfer from room temperature to cold (4°C) or salt conditions inhibits photosynthesis (Krapp and Stitt, 1995; Strand et al., 1997). Salinity suppressed photosynthesis and reduced the chlorophyll contents in olive leaves (Ben Ahmed et al., 2010). Such inhibition is a consequence of a stomatal conductance decrease. This first and especially important response allows a rapid adaptation to available water shortage and disturbance in the leaf water balance (Cornic et al., 1989; Galmés et al., 2007; Chaves et al., 2009). Stomatal closure suppresses CO_2 entry into the cells, its transport through mesophyll, and thus retards photosynthetic CO_2 fixation by leaves, which will be reflected in the sugar contents. It might be a disturbance in the carbon balance that induces rapid changes in the expression of photosynthetic genes, which enhances physiological disturbances in leaves (Krapp and Stitt, 1995; Strand et al., 1997; Evers et al., 2010; Ben Ahmed et al., 2010; Liu et al., 2012). Under unfavorable conditions, a decrease in the leaf area is also frequently observed, and this also reduces the amount of carbon fixed by leaves. A decrease in the rate of CO_2 assimilation leads to the inhibition of $NADP^+$ regeneration in the Calvin cycle (Cruz de Carvalho, 2008) and to reactive oxygen species (ROS) accumulation. The inhibition of carboxylation by Rubisco occurring at the lowered CO_2 concentration in leaves also results in ROS synthesis (Cruz de Carvalho, 2008; Legay et al., 2011).

Under stress conditions, genes encoding proteins functioning at all stages of photosynthesis may be inhibited. Among them are genes of light reactions (PSI, PSII, ATP synthase, and electron carriers), Calvin cycle, photorespiration, and chlorophyll binding proteins. In various plant species, the absence of expression of these genes occurred under the influence of cold, drought, and salinity (Hannah et al., 2005; Svensson et al., 2006). Under stressful conditions, the leaf blade area is often reduced. This also reduces the amount of photosynthetically fixed carbon.

In some cases, the relationship between photosynthesis inhibition by stressors and plant resistance can be followed. Thus, cold stress inhibited photosynthesis in tomato; most of the related genes were inhibited and to a higher degree in cold-sensitive than cold-tolerant plants. Thus, the expression of genes encoding ferredoxin $NADP(^+)$-oxidoreductase and glycolate oxidase was more inhibited in cold-sensitive plants than in cold-tolerant ones. The authors concluded that photosynthesis, especially photosystem II, of tolerant plants is less sensitive to stress (Liu et al., 2012).

11.2.2.1 Phloem transport

The ways of carbohydrate distribution over the plant and their entry into cells should be considered first. It is known that trioses are formed during fixation of atmospheric CO_2 in chloroplasts of mature leaf mesophyll; they are transported into the cytosol where sucrose is synthesized. Sucrose is transported through mesophyll mainly via plasmodesmata, i.e., without exit from the cell, although the movement of sucrose leaking from mesophyll cells along the apoplasts with its subsequent uptake by the sink cells is not excluded. However, such a path is slower and energetically costly (Ayre, 2011; Chen et al., 2012; Liesche and Schulz, 2012).

Reaching the vascular bundles of the donor leaf, sucrose accumulates in the phloem-conducting complexes consisting of sieve elements and companion cells. These cells are connected by numerous plasmodesmata with increased permeability. Therefore, the complex of sieve element and companion cell is a functional unit. Sucrose entering into the conducting complex (phloem loading) occurs in the terminals of small vascular bundles. The ways of phloem loading differ in different plant species and even in the leaves of a single species at different developmental stages (Liesche et al., 2011; Davidson et al., 2011). In most heat-loving plants originating from tropical regions, phloem is loaded symplastically, e.g., via plasmodesmata connecting mesophyll parenchymal cells and bundle sheath with the complex of sieve elements and companion cells (Gamalei, 1989, 1991; van Bel and Gamalei, 1992). In plants with symplastic loading, sucrose is the main carbohydrate loaded into the phloem, which is transported from the sites of its synthesis (mesophyll cells) to small veins along the gradient of its concentration (Reidel et al., 2009; Rennie and Turgeon, 2009).

In other plants, carbohydrate accumulation in the loading sites occurs in specialized companion cells, so-called intermediary cells. In these cells, a "polymer trapping" mechanism is functioning. In this case, sucrose also diffuses symplastically from mesophyll cells to intermediary cells where it converts partly into oligosaccharides (raffinose and stachyose) in some plants and into polyols (mannitol and sorbitol) in others. Sucrose concentration reduces in these cells, which favors its diffusion from the mesophyll. Molecular weights of oligosaccharides and polyols are higher than that of sucrose, and this hampers their reverse exit into the mesophyll cells. Synthesized compounds diffuse only into the sieve element through plasmodesmata with the high size exclusion limit. In such a way, they are trapped in the phloem and are transported along with sucrose to sink organs (long-distance transport) (Zhang and Turgeon, 2009; Turgeon and Wolf, 2009).

Phloem of herbaceous plants of temperate climates is loaded mainly apoplastically. Plasmodesmata are absent or scarce at the boundary between the conductive complex sieve element/companion cell and neighboring cells. Therefore, before entering the small veins of the phloem, sucrose exits into the apoplast and is then absorbed by the conducting complex of the phloem by transporters actively pumping sucrose against a concentration gradient (Dinant and Lemoine, 2010; Turgeon, 2010). As a result of active transport, the high concentrations of osmotically active substances, mainly sucrose, accumulate in the phloem of the thin terminals of leaf vascular bundles, leading to water entrance and to increased hydrostatic pressure therein. In tissues of sink organs, carbohydrates are metabolized, their concentrations decrease, and water exits from the cells along the water potential gradient. The hydrostatic gradient arises between phloem sieve elements of source and sink organs; it is a driving force of carbohydrate mass (bulk) flow along the phloem (Knoblauch and Peters, 2010; Liesche and Schulz, 2012).

The rates of phloem loading and unloading, the long-distance transport, and the distribution between cells determine the sugar import by cells and the potential of their accumulation in them. The rates of photosynthesis, intercellular transport over leaf, the intensity of phloem terminal loading, and the efflux from the leaf along the phloem determine the sucrose content in the donor leaf cells. The rates of phloem transport and phloem unloading determine the sucrose content in sink tissues.

11.2.2.2 Sucrose metabolization

Along with participation in osmotic regulation, sugars, mainly mono- and disaccharides, affect plant survival due to their intracellular metabolization. Soluble carbohydrates play a crucial role in the plant cell cycle: source of energy, structural components, and intermediates for the synthesis of other organic molecules and carbon-storing compounds.

Before entering in metabolic reactions, sucrose is subjected to cleavage with invertase or sucrose synthase (Koch, 1996; Sturm and Jang, 1999). As a result, two instead of one molecule appears, and this increases the turgor pressure in cells. Thus, both enzymes are involved in the control of the cell osmotic potential. The concentration of sucrose and the products of its cleavage can influence the photosynthesis rate via the reaction product regulation or via direct action on photosynthetic gene expression (McCormick et al., 2008). Moreover, functioning of invertase and sucrose synthase reduces cytosolic sucrose content in the conductive complex of donor leaves, which may adversely affect the transport of assimilates.

Invertase (β-fructofuranosidase) plays an important role in sugar distribution and in the control of source—sink relations (Roitsch, 1999; Sturm, 1999; Sturm and Jang, 1999). The enzyme hydrolyzes sucrose into two hexoses, glucose and fructose. Both hexoses serve substrates for further metabolization, being involved in glycolysis. The products of glycolysis are used for diverse anabolic processes, including the synthesis of fatty acids, lipids, amino acids, and proteins. In addition, phosphorylated hexoses are metabolized in the pentose-phosphate cycle, producing substrates (riboses) for the synthesis of nucleotides, nucleic acids, and reducing equivalents for numerous redox reactions. Entering into the citric acid cycle, the glycolysis products form a set of organic acids serving as substrates for amino acid synthesis.

When sucrose is cleaved by sucrose synthase, fructose is formed, in the same way as with invertase, and it is involved in the respiratory metabolism. However, in addition to fructose 2,6-bisphosphate, uridine diphosphate glucose is produced and is involved in the processes of polysaccharide synthesis, including cell wall polysaccharides and starch (Ruan et al., 2003, 2010).

Processes of oxidative and glycolytic phosphorylation generate the energetic potential directly used for metabolic reactions or stored in the cell under ATP. Finally, starch, the main plant storage compound, polysaccharides, cell wall structural components, and numerous secondary substances are synthesized on the basis of sucrose and products from its metabolization. Thus, sugars provide cells with the building material and energy for cell component synthesis; therefore, they are needed during all stages of the plant life cycle, from germination and vegetative growth to the reproductive stage providing for grain and fruit yield. Therefore, a special sugar importance for overcoming stress-induced disturbances seems obvious. An increase in the sugar content activates metabolic reactions, including the protective compound synthesis, whereas an increase in the cell energetic status is necessary for the running of these reactions. All this is required for the struggle against stresses. In this connection, it can be concluded that sugars play a crucial role in plant cell metabolism and plant defense against harmful action of unfavorable environmental factors.

Changes in activities of enzymes metabolizing sugars may be one of the factors inducing their accumulation in the cells. The important role of carbohydrate metabolism in plant resistance underlines the fact that some genes of the corresponding proteins are activated by stress (Theerawitaya et al., 2012; Kumar et al., 2013; Wang et al., 2013). For example, osmotic stress induced by the high concentrations of sorbitol in medium activates sucrose phosphate synthase, the enzyme synthesizing sucrose, in the cell culture of sweet potato (*Ipomoea batatas*) and wheat. As a result, the content of sucrose in the cells increased. Sucrose consumption for the synthesis of starch and oligosaccharides, in contrast, slightly reduced the content of sucrose. Enzymes cleaving sucrose interfered in this balance: cytoplasmic invertase and sucrose synthase (functioning in the direction of the sucrose decay) were activated (Wang et al., 2000). Sucrose synthase activation and sucrose accumulation in cereal stems were observed when water stress began during grain filling. In this case, as a result of early aging, the remobilization of fructans stored in the stem was activated and compounds were transported to the grain. This accelerated the process of grain filling (Yang et al., 2004; Yang and Zhang, 2006). A need in sucrose redistribution under drought conditions was demonstrated in the work with potato clones differing in resistance (Legay et al., 2011). More sucrose was accumulated in the tolerant clone, despite the fact that the main enzyme of its synthesis, sucrose phosphate synthase, was less active in these experiments. Since drought induces starch-degradation enzymes, the authors suggested that sucrose accumulation occurred due to the usage of reserve carbohydrates (Schafleitner et al., 2007; Legay et al., 2011).

Under stresses, activities of sucrose-degrading and sucrose-synthesizing enzymes increase, i.e., the futile cycle in the cytoplasm is activated, multiplying metabolic signals. Small changes in each of the one-way paths will lead to large changes in the net rate of carbon storage (Geigerberger et al., 1997). Thus, the accelerated turnover of sucrose increases the sensitivity of its metabolic pathways to stress (Wang et al., 2000).

11.2.3 Involvement of di- and oligosaccharides in plant resistance

Trehalose (α,α-1,1-diglucose), as distinct from glucose, fructose, and sucrose, is a chemically weakly reactive, relatively stable disaccharide. Nevertheless, the plants contain many genes of trehalose metabolism, which implies a significance of this disaccharide as a regulator of basic and stress-induced metabolism. Trehalose functions as a storage carbohydrate, a stabilizer of proteins and membranes. This disaccharide is a widespread osmolyte. At millimolar (or even trace) amounts, it was detected only in some plants, for example, in resurrection species, where it evidently functions as a protectant against dehydration. This function of trehalose is related to its hydrophilicity, which allows it to participate in the regulation of the water balance in plant cells (Iturriaga et al., 2009). The maintenance of the cell water content is required under all stresses destroying water relations: low temperatures, salinity, and drought. Like other osmolytes, trehalose accumulates under stresses, thus contributing to the improvement of plant resistance. In fact, potato transformed with the yeast gene of trehalose phosphate synthase 1 (*TPS1*) was more drought tolerant. Under water deficit in leaves of such plants, the stomatal conductance and photosynthesis rate were higher and the leaves detached from these plants maintained turgor longer (Stiller et al., 2008; Kondrák et al., 2011, 2012).

Fructans are water-soluble polymers of fructose. They are involved in the regulation of the osmotic pressure during flower opening, induce a rapid grass regrowth, and stabilize membranes (Valluru and van den Ende, 2008; Keunen et al., 2013; Peshev and van den Ende, 2013). Transgenic plants accumulating fructans are more tolerant to abiotic stresses, for example, to the frosts (Li et al., 2007) and low above-zero temperatures (Kawakami et al., 2008). Transgenic *Lollium* plants with the *FT* gene of wheat accumulated fructans and were also tolerant to freezing (Hisano et al., 2004). Rice plants, which could not accumulate fructans, were very sensitive to chilling (Kawakami et al., 2008). Fructans as reserve carbohydrates may be even more efficient than starch because they are less inhibited in the cold and, being water soluble, are mobilized more rapidly under stresses (Keunen et al., 2013).

Raffinose family oligosaccharides (RFO) are an important component of defense responses. Among these compounds, raffinose, its precursor galactinol, galactose, and somewhat less widespread stachyose have a special significance. For example, among potato osmolytes accumulated under unfavorable conditions, aside sucrose and trehalose, are raffinose, pinitol, proline, and polyamines (Cunningham et al., 2003; Schafleitner et al., 2007). The accumulation of these compounds under stresses was observed in many plant species (Nishizawa et al., 2008; Nishizawa-Yokoi et al., 2008; Peters and Keller, 2009). Grant et al. (2009) presented the list of plant species accumulating (RFO) during cold acclimation. During cold acclimation, alfalfa roots accumulated stachyose, raffinose, and sucrose, whereas the contents of glucose, fructose, and starch were reduced (Castonguay et al., 1995). Under different conditions and in different plant species, various osmolytes accumulate. For example, in *Arabidopsis*, drought, salinity, and cold induced the accumulation of a large amount of raffinose and galactinol, but not stachyose. Raffinose accumulated during cold acclimation in autumn, attaining the highest values in the coldest months in the middle of winter. As temperature increased with spring approaching, the raffinose content decreased (Cox and Stushnoff, 2001; Grant et al., 2009).

Many researchers believe that the content of raffinose-type sugars is correlated with plant resistance to a greater extent than in the case of other soluble sugars (Santarius and Milde, 1977; Castonguay et al., 1995; Taji et al., 2002). In particular, RFO accumulation increased resistance to cold (Pennycooke et al., 2003). When plant resistance to freezing was studied, the frequent conclusion is that, among soluble sugars, only raffinose content is correlated with plant resistance (Rohde et al., 2004; Korn et al., 2008, 2010). The capability to survive during winter frosts was maintained by the accumulation of raffinose and stachyose, but not sucrose. In rice plants subjected to drought or high or low temperature, raffinose accumulated, especially strongly in resistant genotypes; in sensitive cultivars, the content of raffinose could be even reduced (Morsy et al., 2007). Strong changes in the carbohydrate content were observed also in drought-tolerant potato clones: under drought, the accumulation of galactose, galactinol, inositol, proline, and proline analogues in such plants was higher than in sensitive clones (Evers et al., 2010; Legay et al., 2011).

The importance of RFO accumulation in plant stress resistance was supported by other experiments as well. For example, in some woody plants, only the contents of raffinose and stachyose were correlated with cold tolerance, although at lower temperatures not only RFO but also sucrose and glucose accumulated (Stushnoff et al., 1993). A comparison of *Medicago falcate* plants differing in their cold tolerance showed a positive correlation between the content of RFO and tolerance. A similar conclusion was made by Grant et al. (2009) on grape variety leaves differing in cold tolerance. At normal temperature, all varieties contained similar amounts of raffinose, but during cold acclimation more tolerant varieties accumulated this disaccharide much more actively. Hamman et al. (1996) also found that the resistance of grape varieties to freezing was correlated with the content of raffinose, but not sucrose. Under the influence of low temperature, the content of raffinose sharply decreased in sensitive rice cultivars but increased in resistant ones (Morsy et al., 2007). Raffinose accumulation was observed in the early stages of acclimation; this time, shoot growth was retarded, which may be also considered as an adaptive phenomenon. In resistant cultivars, sucrose was also accumulated, whereas the content of hexoses remained unchanged or even reduced.

The correlation was also found between the content of RFO and frost resistance (Rohde et al., 2004). Changes occurring under frost influence, like in the aforementioned examples, are related to cytosol dehydration. At low below-zero temperatures, a decrease in the cell water content occurs due to the extracellular ice formation. The role of raffinose in dehydration is similar to that of other osmolytes, e.g., protection of membrane structure (Hincha et al., 2006; Knaupp et al., 2011). In addition to the osmoprotective role, improved stress resistance is determined by RFO involvement in the maintenance of macromolecule structure, antioxidant, and signaling functions (Nishizawa et al., 2008; Schneider and Keller, 2009; Knaupp et al., 2011).

Unexpectedly, Zuther et al. (2004) found no difference in cold tolerance of *Arabidopsis* lines differing in their capability of raffinose accumulation. In addition, mutation in the gene encoding raffinose synthase did not reduce frost resistance. One of the reasons for such a phenomenon, differing from most effects described in the literature, may be pronounced changes in the content of other sugars and metabolites affecting frost resistance stronger than raffinose. Galactinol might be such a metabolite as it accumulated during cold acclimation of the raffinose synthase-deficient mutant. Galactinol, a precursor of raffinose, accumulates in plants along with raffinose under stress conditions.

RFO biosynthesis occurs by the transfer of galactosyl units to sucrose by specific transferases. Galactose is transferred to sucrose with the formation of raffinose and to raffinose with the formation of stachyose (Saravitz et al., 1987). The first stage of RFO synthesis is catalyzed by galactinol synthase, a key enzyme of biosynthesis of raffinose-type oligosaccharides (Keller and Pharr, 1996). It is well known that this enzyme is activated during some abiotic stresses: water loss, salinity, and high and low temperatures (Zhuo et al., 2013). Different stresses induce various isoforms of the enzyme. Thus, in coffee plants, CaGolS1 isoform was constitutively expressed under normal conditions; it was the main isoform under various stressful conditions as well. CaGolS2 was synthesized only under severe water deficit and salinity, whereas CaGolS3 was synthesized under severe drought (dos Santos et al., 2011). In *Arabidopsis*, *AtGolS1* and *AtGolS2* genes were expressed under drought and salinity conditions, whereas *AtGolS3* expression was induced only by chilling (Taji et al., 2002). During stresses, the activation of these enzyme isoforms increased the content of raffinose and stachyose in plant tissues, as in cucumber leaves and soybean seeds (Handley et al., 1983; Saravitz et al., 1987). A negative correlation between this enzyme activity and sucrose content (Handley et al., 1983) was evidently related to the sucrose consumption in raffinose synthesis.

Transgenic plants with the gene of galactynol synthase (*GolS*) contained more galactinol and raffinose and highlighted the higher resistance to drought and low temperature (Taji et al., 2002). Expression of this gene in *Arabidopsis* was also induced by high temperatures and could play a defense role under heat conditions. In soybean seeds, *GolS* expression was activated immediately before the start of raffinose-type saccharide accumulation. Transgenic T-DNA knockout mutants could not accumulate raffinose (Panikulangara et al., 2004). Due to the increased contents of galactinol, raffinose, and stachyose, overexpression of *MfGolS1* improved tobacco tolerance to chilling and freezing and also to drought and salinity (Zhuo et al., 2013). *Arabidopsis* transformation with the *AtGol2* gene increased the content of galactinol and raffinose as well; it suppressed transpiration and thus improved drought resistance (Taji et al., 2002).

Legay et al. (2011) emphasized that UDP-4-glucose epimerase and raffinose synthase were involved in RFO synthesis along with galactinol synthase. The expression of genes encoding these enzymes was also activated by stresses. Overexpression of these enzymes improved tolerance to drought, for example (Taji et al., 2002; Liu et al., 2007). As a result of their overexpression, tolerant potato clones accumulated more galactinol than sensitive ones. Under the influence of stresses, carbon flux partially deviates from use in glycolysis or synthetic processes (e.g., in the synthesis of starch or cellulose), and is directed to the synthesis of RFO. In this case, the accumulation of separate compounds changed: raffinose accumulated predominantly in the tolerant clone, whereas inositol and galactose accumulated in sensitive one; galactinol accumulated in both clones (Evers et al., 2010; Legay et al., 2011). Redirection of carbon fluxes can be partially explained by the decrease in the content of monosaccharides and sometimes sucrose observed in some experiments.

The role of RFO in the distribution of photosynthetically fixed carbon in the plant is invaluable. It should be emphasized once more that oligosaccharides are the most widespread, after sucrose, transport carbon forms. It is evident from the distribution of the enzyme of their synthesis. In melon plants transporting large amounts of RFO, the *CmGAS1* gene was expressed only in mature source leaves and seeds, whereas the *CmGAS2* gene was expressed only in mature leaves. Mature leaves are the starting point of carbohydrate transport over the plant, and oligosaccharides are synthesized just in them. Expression in leaves is important for the synthesis of the substance forms transporting along the phloem; the synthesis in seeds is evidently required for membrane protection during dehydration (Yazdi-Samadi et al., 1977; Volk et al., 2003).

11.2.4 Proline and polyamines as compatible solutes
11.2.4.1 Proline
Proline is the second osmoticum after carbohydrates regarding prevalence and content in tissues. Numerous reports have appeared about a correlation between the content of proline and plant resistance to unfavorable environmental factors. Dramatic proline accumulation in cells protects them against many harmful effects: excessive insolation, UV radiation, salinity, low and high

temperatures, water and osmotic stresses, and H_2O_2 stress (Bohnert et al., 1995; Hare and Cress, 1997; Kuznetsov and Shevyakova, 1997; Kishor et al., 1995, 2005; Hong et al., 2000; Kumar et al., 2010; Aggarwal et al., 2011; Kaushal et al., 2011; Kondrák et al., 2012; Hayat et al., 2012; Ali et al., 2013). Especially important is proline accumulation in plant tolerance to salinity and drought (Kondrák et al., 2012). Proline functions as an osmolyte, ROS scavenger, redox balancer, cytosolic pH buffer, molecular chaperon, and a stabilizer of protein structure (Szabados and Savoure, 2010). Proline treatment of *Olea europaea* L. leaves neutralized salinity effects on water relations, chlorophyll contents, and photosynthetic activity; it also activated antioxidant enzymes (Ben Ahmed et al., 2010). Treatment with proline improved drought tolerance; in this case, the content of hydrogen peroxide accumulated during stress was reduced.

Proline accumulation can occur due to the stimulation of its synthesis from glutamic acid, the suppression of proline oxidation, or changes in its incorporation in protein synthesis. Under various stresses, proline is synthesized in the cytosol and chloroplasts. The main enzyme catalyzing proline synthesis is Δ1-pyrroline-carboxylate reductase, P5CR (Lehmann et al., 2010). Proline degradation during recovery after stress occurs in mitochondria with the involvement of proline dehydrogenase, PDH (Evers et al., 2010).

A correlation between the content of proline and resistance becomes evident at the comparison of plants resistant and sensitive to stresses. A comparison of more resistant to osmotic stress (0.2 mM mannitol or sorbitol) halophyte *Arabis stelleri* and less resistant *Arabidopsis thaliana* showed that the content of proline was higher in the more resistant *A. stelleri* (Jung et al., 2010). Similarly, cold-tolerant tomato plants contained more proline than sensitive genotype (Liu et al., 2012).

Many studies supported the correlation between proline synthesis and improved resistance. This was shown, for example, for tobacco and wheat (Kishor et al., 1995; Hong et al., 2000), citrus plants, sugarcane, arabidopsis (Nanjo et al., 1999), and other plants. Overexpression in transgenic plants of potato, rice, and cow pea of genes encoding enzymes involved in proline accumulation, e.g., a key enzyme of its biosynthesis, Δ1-pyrroline-carboxylate reductase (P5CR), increases markedly plant resistance to abiotic stresses (Dixon and Paiva, 1995; Kishor et al., 1995, 2005; Hong et al., 2000; Dobra et al., 2010; Pospisilova et al., 2011; Cvikrová et al., 2012). Such transgenic plants initially contained an increased amount of proline; it increased still more in 2 days of plant exposure to stress factors (Xin and Browse, 2000). On the other hand, antisense expression in *Arabidopsis* of the gene of proline degradation, ProDH, also caused strong proline accumulation and higher resistance, e.g., to freezing and salinity (Nanjo et al., 1999; Xin and Browse, 2000). A disturbance in *P5CS1* expression in *Arabidopsis* retarded stress-induced proline synthesis, which was accompanied by plant hypersensitivity to salinity.

Mechanisms of proline action are not completely elucidated. Proline accumulation exerted a positive effect on the water potential and also stimulated the formation of the protective pigments of the xanthophyll cycle (Dobra et al., 2010). Its synthesis activation helps to mitigate the acidification of the cytosol observed under various stresses. In addition, in the presence of proline, the ratio $NADP^+/NADPH$ is maintained (Hare and Cress, 1997). Proline also plays an important role in the activity of the pentose-phosphate pathway producing reducing equivalents needed for the maintenance of many antioxidants in their reduced state. For example, a correlation between the proline content and the reduced glutathione redox state was demonstrated. It is supposed that proline is not only an important participant of redox signaling but can also extinguish free radicals produced under various unfavorable conditions (Hong et al., 2000), including OH^- (Smirnoff and Cumbes, 1989), thus preventing cell death (Chen and Dickman, 2005). By affecting SOD, CAT, and POD activities (Hoque et al., 2007), proline is involved in the maintenance of the redox homeostasis (Hare and Cress, 1997).

Despite numerous indications of the correlation between proline accumulation and resistance of many plant species to various stresses (Bartels and Sunkar, 2005), such correlation is not always evident in all cases (Legay et al., 2011). Thus, drought did not induce the key enzyme of proline synthesis (P5CS) in the tolerant potato clone and only slightly activated it in the sensitive clone. In the earlier studies also performed on potato, expression of *P5CS* was more active in sensitive than tolerant plants (Schafleitner et al., 2005; Vasquez-Robinet et al., 2008). These data imply that pro-line synthesis is not always necessarily required for plant drought resistance. Legay et al. (2011) concludes that the activation of proline synthesis does not characterize plant resistance but is rather related to their response to stress. In much earlier works, researchers, who studied barley stress resistance, also have concluded that proline accumulation is rather the marker of severe stress in the leaves than the component of resistance to drought (Hanson et al., 1977). Thus, in some cases, proline can accumulate more in sensitive plants, which respond to stress stronger. In the sensitive cassava cultivar, under water deficit-, proline content was increased by 25 times compared to the control, whereas in tolerant cultivar proline content increased only nine-fold (Sundaresan and Sudhakaran, 1995). Deyholos (2010) suggested that strong activation of *P5CS* gene expression and proline accumulation observed by various researchers under stress was related to the fact that they performed experiments with sensitive cultivars.

In many ways, proline functions like carbohydrates. Soluble sugars maintain the osmotic potential as well; they are involved in redox reactions and in the maintenance of the structures of macro-molecules and membranes. In this connection, it is important that both accumulate simultaneously. Thus, it was shown that a mutant overexpressing proline accumulated soluble sugars as well (predominantly glucose and fructose), and soluble proteins. The accumulation of these osmotically active compounds reduced markedly the osmotic potential of mutant cells at low temperatures, as compared with the wild type. Such change in the osmotic potential was correlated with plant resis-tance to chilling and freezing (Liu et al., 2012; Wang et al., 2013).

11.2.4.2 Polyamines

Along with carbohydrates and proline, polyamines (putrescine, spermidine, and spermine) can accumulate under stresses (Schafleitner et al., 2007). The multifaceted function of polyamines, their capability of interfering with diverse processes of plant growth and development, and maintaining plant productivity under changing environmental conditions allow some researchers to designate this compound group as a new class of growth substances (Gill and Tuteja, 2010; Alcázar et al., 2006, 2010; Hussain et al., 2011). Aliphatic properties of polyamines determine their involvement in the cell ionic balance and their special role under salinity stress. The polyamine protective role at oxidative stress, drought, salinity, chilling, and metal toxicity is assessed by its antioxidant properties, a capability of membrane and cell wall stabilization. Plants transformed with genes encoding the enzymes of polyamine synthesis—arginine decarboxylase, ornithine decarboxylase, S-adenosylmethionine decarboxylase—are more resistant to abiotic stresses. More drought-resistant potato clones expressed genes of the key enzymes of spermidine metabolism, evidently involved in ROS scavenging.

11.2.5 Effect of main stressors on the osmolyte content

Low, above-zero temperatures induce the accumulation of many compounds (Leshem and Kuiper, 1996; Kaplan et al., 2004). Guy et al. (2008) showed that sugars, proline, and polyamines predominated in *Arabidopsis* and other plant species. The content of metabolites in the *Arabidopsis* leaves may change uniformly during cold action (Kaplan et al., 2004) or demonstrate a maximum in the middle of this treatment (Gray and Heath, 2005). Although many researchers presented different data about the quantitative ratio between different metabolites in cold conditions, all united in the approval of the key role of carbohydrate (glucose, fructose, sucrose, and raffinose) accumulation in plant resistance. The accumulation of these sugars was commonly correlated with cold and frost resistance (Zuther et al., 2004; Hannah et al., 2006). More cold-tolerant plant cultivars accumulated more soluble sugars (Lee et al., 2012).

Cold resistance of different plants may be related to the accumulation of different sugars. For example, an improved tolerance of *Osteospermum ecklonis* plants overexpressing the *Osmyb4* gene was correlated with changes in the contents of soluble sugars, mainly sucrose, and proline (Laura et al., 2010). As distinct from transgenic *O. ecklonis* plants, in blueberry cultivars differing in their tolerance to cold, cold-resistant cultivars accumulated sucrose and stachyose. Similar data were obtained for some other plant species, including oak (Morin et al., 2007) and loquat fruit (Cao et al., 2013) and quaking aspen (Cox and Stushnoff, 2001). Sucrose accumulation improved resistance of red raspberry (Palonen, 1999), but not honeyberry. The content of glucose and fructose was correlated with cold resistance of oak (Morin et al., 2007), whereas that of stachyose and raffinose was correlated with cold resistance of forsythia and quaking aspen (Flinn and Ashworth, 1995; Cox and Stushnoff, 2001).

The formation of extracellular ice withdraws water from the cytosol; thus, the response to freezing is similar to the response to drought: in both cases, cytosol dehydration occurs (Uemura and Steponkus, 2003). It was shown decades ago that the content of soluble carbohydrates is critical for plant survival during winter and subsequent spring development. Overwintered alfalfa plants accumulated more carbohydrates in their tissues. Plants not entering dormancy accumulated less sugar as compared with rewintering and more resistant plants (Cunningham et al., 2003).

The adaptation period (cold acclimation) has great importance for the development of plant resistance to freezing, when plants are subjected to the action of low, above-zero temperatures before frosts. Plants growing in temperate climates are well adapted to low temperatures, whereas tropical plants cannot adapt and are damaged at below-zero temperatures. A comparison of cold hardened and non-hardened rapeseed plants (*Brássica nápus*) showed that plant resistance to freezing depended strictly on the leaf osmotic potential, which was lower in hardened plants (Gusta et al., 2004). Cook et al. (2004) studied the accumulation of metabolites in two *Arabidopsis* ecotypes differing in their capacity in cold acclimation. Although both ecotypes accumulated osmolytes, their amount was higher in the ecotype manifesting the higher capability in cold acclimation. During cold acclimation, sucrose, stachyose, raffinose, amino acids, organic acids, and polyamines were accumulated (Castonguay et al., 1995; Guy et al., 2008) because of the activation of corresponding biosynthetic enzymes (Fowler and Thomashow, 2002; Renaut et al., 2009). These compounds function as compatible osmolytes, suppressing dehydration, stabilizing membranes, and neutralizing toxins; they behave as chelating agents and energy sources (Guy et al., 2008). Uemura and Steponkus (2003) showed that adverse effects of frost on *Arabidopsis* mutants sensitive to freezing occurred because of their low sugar content. Potato tuber tolerance to frost was also correlated with their capability in sugar accumulation due to storage starch mobilization (Rawyler et al., 2002). Simultaneously with sugar accumulation, genes of dehydrins and antifreeze proteins, chaperons, and enzymes of detoxication are activated (Evers et al., 2010). Despite many indications that the acclimatization ability depends on the accumulation of various metabolites, e.g., the regulation of frost resistance by sugars (Tabaei-Aghdaei et al., 2003), other researchers have found no simple correlation between the osmoprotectant accumulation and a capability of cold acclimation (Hannah et al., 2006).

11.2.5.1 Comparison of tolerance to low and high temperatures

High temperatures also induced the accumulation of osmotically active compounds. However, according to some data, the dynamics of this process differed from that observed at low temperatures. Thus, during cold stress, the accumulation of osmotically active substances occurs within the whole period of the low temperature actions. At the same time, most of the heat-induced accumulation occurred during the first 30 minutes of treatment (Kaplan et al., 2004; Gray and Heath, 2005).

Under the influence of high temperatures, several pyruvate- and oxaloacetate-derived amino acids, fumarate and malate (oxaloacetate precursors), some alanine, GABA, and putrescine, and also maltose, sucrose, raffinose, galactinol, and myo-inositol accumulate simultaneously. Cold treatment induces deeper changes in low molecular metabolites. When most of the compounds accumulated under high temperature were accumulated in the cold as well (i.e., accumulation is not specific), most metabolites found at the cold shock are characteristic only for this treatment. Pyruvate- and oxaloacetate-derived amino acids, polyamines, and several carbohydrates, including fructose, sucrose, myo-inositol phosphate, galactinol, and raffinose, were accumulated under both high and low temperature treatments (Kaplan et al., 2004).

Drought, like cold, induces diverse physiological and metabolic changes on various levels of organization: whole plant, organs, tissues, cells, and molecules. Water deficit retards growth, reduces the leaf blade area, decreases the photosynthesis rate, and induces stomata closure, cuticulization, and the appearance of other signs of succulence. Such changes reduce yield under well watering but counteract water loss under drought (Legay et al., 2011).

Under drought, cell dehydration occurs, and all signs of osmotic stress are observed. Plants maintain the osmotic balance by the accumulation of osmotica, which reduce the osmotic potential and maintain the cell turgor needed for growth. The main osmotica accumulated under drought are proline, glycine betaine, sugars, sugar alcohols, and also oligosaccharides of the raffinose type (Taji et al., 2002). The accumulation of these compounds is reversible; they degrade after rewatering (Valliyodan and Nguyen, 2006). The key components of drought tolerance and yield protection are efficient water usage, stomatal control, photosynthesis maintenance, and sucrose export to sink organs.

As distinct from other stressors, an important role in the maintenance of osmotic equilibrium under drought is played by salt accumulation in vacuoles. Drought stimulates abscisic acid (ABA) synthesis and activates ABA-dependent and ABA-independent signaling pathways (Yamaguchi-Shinozaki and Shinozaki, 2006). The accumulation of some metabolites is regulated by ABA-dependent pathways (some amino acids, ethanolamine, glucose, and fructose), but the accumulation of other osmotica does not depend on ABA (raffinose, galactinol, components of the organic acid cycle, and gamma aminobutyric acid (GABA)). Some osmotica are under the control of both signaling pathways (proline, lysine, methionine, and phenylalanine) (Semel et al., 2007; Urano et al., 2009).

Under drought, the content of starch in potato leaves decreased, but the sucrose content was not markedly changed. Evidently, it was of importance to maintain the sucrose level, in spite of a decrease in the photosynthesis rate evidently due to the reduction in the starch synthesis and direction of fixed carbon toward sucrose synthesis. The low starch level was correlated with the high content of inositol; therefore, it was suggested that inositol may act as a signal for the suppression of starch synthesis (Kondrák et al., 2012).

Salt stress both disturbs ionic balance and induces osmotic stress. One of aspects of salinity action is a mass exit from cells of potassium, the main intracellular osmoticum. Treatment with compatible osmolytes retards potassium leakage, maintaining potassium homeostasis, which is important for cell functioning (Chen et al., 2007; Sun et al., 2009; Bose et al., 2013). Like temperature variations and drought, an increase in the salt concentration in medium disturbs osmoregulation in cells, reducing water availability because of a decrease in the water potential. To overcome this stress under salinity, similar defense mechanisms are activated as under other stresses, i.e., osmotica accumulate. Depending on stress duration, different compounds may be accumulated. Thus, in the *Arabidopsis* suspension culture, Gong et al. (2005) and Kim et al. (2007) observed first the accumulation of S-adenosyl-L-methionine (SAM), ethanolamine, cysteine, and aromatic amino acids. After 1 day the content of these compounds decreased, and glycerol, inositol, and S-adenosyl-L-homocysteine accumulated. Later, the accumulation of sucrose, proline, and lactate occurred. ABA can participate in the reorganization of carbohydrate metabolism under salinity stress (Kempa et al., 2008). Both salinity and ABA treatment reduce the content of starch, which implies ABA involvement in starch mobilization.

The set of metabolites accumulated by different plant species differ even in closely related species belonging to the same family. For example, a comparison of *Arabidopsis thaliana* and *Thellungiella halophila* (Brassicaceae) showed that, in the presence of 150 mM NaCl, *Thellungiella* accumulated more proline, glutamic acid, malic acid, and succinic acid, whereas *Arabidopsis* accumulated more fumaric acid and mannitol (Gong et al., 2005). Lugan et al. (2010) believe that both plant species use similar metabolic pathways during salt acclimation but differ quantitatively. *Thellungiella* is more resistant to stresses and contains the higher concentrations of osmotica. Under stress, this plant can lose more water than *Arabidopsis* without a loss of viability. Another example of a variability of stress-accumulated osmotica may be the ratio of K^+ to proline under salinity. Under favorable growth conditions, K^+ is responsible for more than half of the cell osmotic potential. Salinity results in K^+ leakage, and other osmotica, proline in particular, replace it in the maintenance of the osmotic potential (Chen et al., 2007). However, proline cannot completely compensate potassium leakage because of its low intracellular concentration, and sugars become the main osmotica.

11.2.5.2 Comparison of responses to different stresses

Extreme temperature, drought, and salinity induce common physiological disturbances: the retardation of cell division and expansion and plant growth, the inhibition of photosynthesis, disturbances in water relations, osmoregulation, redox state, etc. (Sanchez et al., 2008). Common features were observed in plant responses to temperature stress and drought; mutations were described, and transgenic plants resistant to both stresses were obtained (Bouchabke-Coussa et al., 2008; Kasuga et al., 2006, 2007a; Mattana et al., 2005; Yamaguchi-Shinozaki and Shinozaki, 2006). The accumulation of osmotically active compounds is one of the common defense responses. However, adaptation mechanisms functioning at different stresses somewhat differ. Some examples described above show that, although the accumulation of osmotica occurred under all treatments, the responses differed qualitatively and quantitatively. One more example is experiments performed on *Vitis vinifera* (Cramer et al., 2007). In spite of the maintenance of similar water potential, salinity affected the accumulation of soluble low molecular osmotica weaker than drought. Plants treated with salts accumulated less glucose, proline, and malate, and, as distinct from drought-treated plants, did not accumulate sucrose.

Rice genotypes differing in cold resistance responded differently to cold, drought, and salinity. They manifested different sensitivity to salinity and drought, e.g., the most cold-resistant genotype was more sensitive to water deficit and especially to salt treatment. It might be that these differences may be related to the accumulation of different compounds. Thus, in the cold, the low temperature-resistant ecotype accumulated mainly galactose and raffinose; however, in the cold-sensitive ecotype, their contents even decreased. Under drought, trehalose prevailed among osmoprotectants, whereas mannitol prevailed under salinity. At the same time, the cold-resistant genotype accumulated only small amounts of these compounds (Morsy et al., 2007). In other experiments with rice plants, despite the similar tendency of different genotypes to accumulate glucose, glutamine, and glutamate under osmotic and salt stresses, increased contents of sucrose, threonine, valine, and lactate were observed only under salinity. Thus, under different stresses, the metabolism of sugars and glutamine is controlled differently (Fumagalli et al., 2009).

Even when similar metabolites accumulate under different stresses, the time course of their accumulation may be different. For example, *Arabidopsis* plants subjected to chilling or drought accumulated sugars, proline, and amino acids in their tissues, but in the cold, the accumulation became noticeable as soon as in 8 hours, whereas drought effects were expressed only in several days (Mattana et al., 2005).

One more important difference between the action of cold, drought, and salinity is related to the toxic effects of ions accumulated at high salt concentrations. For neutralization of such ion action (sodium predominantly), the activation of transport mechanisms sequestering them in vacuoles or extruding them from the cytosol into medium is required.

The common responses to different stresses may be related to the fact that many drought-induced genes are also induced by cold and salinity. These genes encode proteins involved in the control of resistance: chaperons, LEAs (late embryogenesis abundant proteins), osmotin, antifreeze proteins, mRNA binding proteins, key enzymes of osmolyte synthesis, proteins of water channels, transporters of sugars and proline, detoxification enzymes, and various proteases. In addition, the above stresses activate genes encoding protein factors involved in the further regulation of signaling and expression of stress-responsive genes: protein kinases, transcription factors, enzymes of phospholipid metabolism. For example, genes regulated by the ESK1 protein are rather connected with genes-controlled osmotic and salt stresses and ABA than with the cold-sensitive CBF/DREB pathway (Xin et al., 2007). However, the *eskimo1* (*esk1*) mutant turned out to be resistant to cold; it accumulated a large amount of proline (Xin and Browse, 2000). This mutant evidently changes the response to the cold but not drought: it accumulates fructose, raffinose, proline, and galactinol, accumulated under all stresses, but not sucrose, trehalose, and glutamine involved in the response to cold stress. It is supposed that the main role of ESK1 is in water saving under its shortage, and the resistance to low temperatures is the consequence of this effect (Bouchabke-Coussa et al., 2008).

11.2.6 Ratio between compounds accumulated under stresses

For a long time, plant resistance was related to the accumulation of all non-structural carbohydrates (TNC, total non-structural carbohydrates, the sum of sugar and starch concentrations). However, total soluble sugars are evidently not the only factor limiting plant survival, in particular under low temperatures. For example, Li et al. (2008a,b) made such a conclusion when studying the reasons for limiting the altitudinal and latitudinal distribution of trees in the mountains. In their experiments, plant survival under more severe conditions depended largely on the ratio of sugar and starch contents in their tissues. The higher was this ratio, the farther north the plants could grow. This observation is in agreement with known data about the inhibitory action of the high starch concentrations on photosynthesis and the reduction of resistance to low temperatures under these conditions (Koch, 1996; Patton et al., 2007).

During plant acclimation to low, above-zero temperatures, a sharp decrease in the starch content was observed simultaneously with the induction of the gene encoding β-amylase (Kozlowski and Pallardy, 2002, Naik et al., 2007; Lee et al., 2012). There is evidently the system of sugar and starch interconversion, which adjusts their ratio to environmental conditions. Such system functions in grasses as well. Thus, Strand et al. (1997) showed that in *Arabidopsis* leaves the sugar/starch ratio at normal temperature (23°C) was much lower than in the cold (5°C). Patton et al. (2007) also concluded that there is a correlation between the sugar/starch ratio and plant cold resistance. The hexose/sucrose ratio is also of importance (Weber et al., 1995; Xiang et al., 2011).

The relative content of hexoses, sucrose, and oligosaccharides is also of great importance. Some of the examples presented above have a greater influence of RFO on the resistance of some plants in comparison with sucrose. The different significance of these sugar groups was clearly observed for rice seedlings tolerant and sensitive to low temperature, salinity, and water deficit. In sensitive seedlings, sucrose, fructose, and glucose accumulated markedly, whereas the content of galactose changed insignificantly. At the same time, in tolerant seedlings, in contrast, sucrose and hexose accumulation was expressed much weaker and under saline conditions was not observed at all, whereas the content of RFO increased substantially. Although the literature did not yield any reports about the relative role of sucrose and RFO in plants with different ratios of carbon transport forms, a connection between RFO accumulation and plant resistance should be expressed especially clearly in plants transporting large amounts of RFO along the phloem. This class of compounds is probably energetically more favorable because it can carry more carbon without breaking turgor of transporting cells. For the same reason, RFO-type carbohydrates are more advantageous as storage compounds.

The ratio of sucrose to raffinose affects seed development, in particular their desiccation. During seed development, the contents of glucose, fructose, and sucrose decreased (Saravitz et al., 1987), but the contents of stachyose and raffinose increased. Their accumulation started simultaneously with the beginning of seed desiccation and activation of galactinol synthase (Li et al., 2011b). Sucrose was accumulated simultaneously with RFO; its proportion was higher in the genotypes with low raffinose contents. In cucumber seeds, raffinose prevailed; it disappeared rapidly during the first 2 days of germination. In 3 days, galactinol synthase activity was up-regulated and the raffinose content gradually increased again (Handley et al., 1983). The accumulation of raffinose but not sucrose was associated with desiccation. Since during all stages of seed and leaf development, galactinol synthase activity was positively correlated with the proportion of RFO in the total carbohydrates and negatively correlated with the sucrose content, this enzyme activity determines the carbon distribution between sucrose and raffinose (Handley et al., 1983; Saravitz et al., 1987).

11.3 Compatible osmolytes and physiology of resistance

Despite the large number of studies on the role of osmolytes in plant resistance, many problems remain unsolved. It is supposed that osmolytes fulfill many functions related to plant resistance: in addition to osmotic adjustment, they can function as low molecular chaperons and stabilizers of the photosystem II complex; they are involved in the maintenance of the structure of membranes, proteins, and enzymes, in the reduction of potassium loss under NaCl salinity, and in ROS scavenging.

11.3.1 Role of osmolytes in the maintenance of macromolecule and membrane structures

The accumulation of compatible osmolytes restores the osmotic equilibrium between the cell compartments and medium. In the presence of osmolytes, injuries to cell membranes induced by unfavorable conditions are reduced, which is expressed in the retardation of ion and electrolyte leakage. In addition, the formation of malondialdehyde from lipid peroxidation decreases (e.g., Deryabin et al., 2005). The role of osmotically active substances in the protection of macromolecules and membrane lipid bilayers against dehydration is well known (Buitink and Leprince, 2004; Shao et al., 2006; Korn et al., 2008, Jain and Roy, 2009; Krasensky and Jonak, 2012). By promoting the retention of intracellular water, osmolytes alter water properties; they may interfere with the interaction between water and biomolecules. In this case, not only properties of osmotica but also properties of the intracellular solvent and their interaction are of importance, e.g., the possibility of molecule salvation. Under low humidity and high medium osmolarity, water exits from the cells, disturbing hydration of macromolecules, which is needed for the maintenance of protein structure in the active state. Direct interaction of osmolytes and polar groups of membrane phospholipids stabilize membranes. For example, sugar molecules can replace water molecules bound with membrane phospholipid head groups or with proteins and stabilize them due to hydrogen bonds (Bentsink et al., 2000; Hincha et al., 2006; Livingston et al., 2009). Sugars maintain membrane hydration and prevent their fusion, thus keeping the space between phospholipid molecules (Valluru and Van den Ende, 2008). In consequence, they maintain the structure of macromolecules and membrane functioning.

Another way of structure protection is manifested under low, below-zero temperatures. When water is frozen, ice crystals damage the cell structures. Extracellular ice formation is especially dangerous. For this ice formation, not only water from the intercellular spaces but also cell water is used. Therefore, osmotic stress is enhanced. The osmolyte accumulation, carbohydrates in particular, favors plant survival at low, below-zero temperatures (Kozlowski and Pallardy, 2002; Patton et al., 2007; Welling and Palva, 2008; Lee et al., 2012). Osmotica favor water retention within the cell and reduce the amount of water in the intercellular spaces used for ice formation. In addition, by reducing the temperature of nucleation, they serve as cryoprotectants and prevent or reduce the rate of ice formation (Xin and Browse, 2000; Morin et al., 2007; Kasuga et al., 2007b). The size of ice crystals is diminished, and this reduces ice-induced structural damage. A specific property of frozen water, a capability of vitrification, favors the protection of intracellular structures. The presence of sugars possessing simultaneously by hydroxyl, keto, and aldehyde groups favors ice transition in such a state; the viscous, glassy environment of phospholipids is produced. The ice amorphous state prevents membrane fusion and drying. Hydrogen bonds are formed between sugar hydroxyl groups and proteins, thus stabilizing protein structure. In addition, aldehyde and keto groups of sugars favor molecule cyclization, and due to intramolecular H-bonds the structure becomes more rigid (Dashnau et al., 2005; Dashnau and Vanderkooi, 2007). The capability of osmolytes to protect membrane structures decreases in the following order: trehalose, lactose, maltose, cellobiose, sucrose, glucose, fructose, sorbitol, raffinose, myo-inositol, and glycerol (Crowe et al., 1984).

A disturbance of osmotic balance affects protein folding and activity. LEAs and HSPs (heat shock proteins) induced by various stresses function as chaperons, stabilizing protein and membrane conformation (Bohnert and Jensen, 1996; Jouve et al., 2004; Legay et al., 2011). Genes encoding chaperons, dehydrins, and LEA proteins are expressed predominantly and are stronger in resistant plants. Small chloroplastic and mitochondrial heat shock proteins (sHSPs) bind to the proteins, providing for their correct folding and preventing protein aggregation (Waters et al., 2008). The capability of osmolytes to maintain membrane structure reduces compound leakage from the cell. The maintenance of the intracellular potassium concentration and, as a consequence, the membrane electric potential, is especially important for cell functioning (Chen et al., 2007).

11.3.2 Sugars and proline affect the generation of ROS under stressful conditions

The accumulation of only osmotica cannot completely protect plants against stresses because toxic stress effects are not limited by disturbances in the osmotic balance. Low temperature, drought, and salinity induce not only osmotic but also other stresses, including oxidative ones (Nishizawa et al., 2008; Gill and Tuteja, 2010; Keunen et al., 2013). ROS generation is the common consequence of most abiotic stresses. Water deficit reduces stomatal conductance and CO_2 availability for leaf cells. A decrease in the CO_2 concentration suppresses carboxylation by Rubisco. Another consequence of the suppressed CO_2 assimilation might be a decrease in the $NADP^+$ regeneration in the Calvin cycle (Cruz de Carvalho, 2008). As a result, the photosynthetic electron transport chain becomes overreduced and superoxide radicals and singlet oxygen are produced in chloroplasts (Shao et al., 2008a,b).

ROS are generated not only under the influence of stresses. Under favorable conditions, they are by-products of photosynthetic and respiratory plant activities. Among them are organic and inorganic peroxides, superoxide anion, hydroxyl radical, singlet oxygen, and hydrogen peroxide. The OH^- (life span of 10^{-9} s) attacks all neighboring molecules; the least active ROS, H_2O_2, is more stable, but it can spread over the cell and pass through membranes. In addition to chloroplasts and mitochondria, most ROS are generated in peroxisomes as well (Navrot et al., 2007; Gill and Tuteja, 2010). Unfavorable environmental conditions (UV radiation, high insolation, drought, salt and temperature stresses, treatment with herbicides, and infection with pathogens) enhance ROS generation (Nishizawa et al., 2008). The capability of successful ROS scavenging contributes to plant resistance, e.g., heat-loving rice plants, to low temperatures (Morsy et al., 2007).

ROS effects on cell metabolism are determined by their concentrations in the cells. At low concentrations, ROS are involved in plant signaling systems, being second messengers in the pathways of stress signal transduction and activating protective mechanisms against biotic and abiotic stresses (Mittler, 2002, 2006; Mittler et al., 2004). ROS also affect new gene expression and thus control many processes, including plant growth and differentiation. In these cases, ROS accumulation is considered as an adaptive process (Roitsch, 1999). Stress-induced high ROS concentrations exert toxic effects on the cells; they induce the oxidative damage to membranes, proteins, lipids, carbohydrates, and nucleic acids; they enhance the formation of toxic compounds, such as malondialdehyde (Apel and Hirt, 2004), and finally lead to the cell death (Mittler, 2002; Foyer and Noctor, 2005; Møller et al., 2007). ROS accumulation is one of the reasons for the yield loss (Bolouri-Moghaddam et al., 2010).

Oxidative damage is one of the key disturbances induced by stresses (Sato et al., 2011). Under normal conditions, the equilibrium exists between ROS generation and scavenging; stresses disturb this equilibrium. In response to the disturbance in the redox balance, leading to oxidative injuries, plants activate some detoxifying enzymes: superoxide dismutase (SOD), ascorbate peroxidase (APX), peroxiredoxin, glutathione-S-transferase, thioredoxin peroxidase, and catalase (CAT) (Foyer and Noctor, 2005; Cruz de Carvalho, 2008). But not always certain antioxidants can improve plant resistance to stresses. It was reported that SOD and CAT were strongly activated in sensitive but not tolerant barnyard grass plants. This implies that enzyme activation does not lead to the improvement of tolerance. Other behavior is characteristic for another group of enzymes, including peroxidase (POD), APX, and glutathione reductase (GR). In sensitive plants, the activities of these enzymes decreased, which resulted in the oxidative stress in spite of the high SOD activity (Abogadallah et al., 2010). More often, the higher the antioxidant activity, the higher the salt resistance of *Plantago maritima* as compared with salt-sensitive *P. media* (Sekmen et al., 2007).

Many metabolic cell components possess redox properties and function as non-enzymatic antioxidants. Among them are ascorbic acid, glutathione, phenolic compounds, alkaloids, non-protein amino acids, α-tocopherol, carotenoids, flavonoids, and anthocyanins (Stoyanova et al., 2011). With the help of enzymatic and non-enzymatic antioxidants, the cell performs the fine regulation of the redox balance, controlling the ratio between pro- and antioxidant processes and maintaining the low ROS concentration required for normal plant functioning, i.e., about 10^{-8} M (Apel and Hirt, 2004; Gill and Tuteja, 2010; Nishizawa et al., 2008). Different plant tissues possess a specific set of antistress mechanisms.

Most compatible solutes can scavenge ROS (Van den Ende and Valluru, 2009). In particular, sugars and proline can be active antioxidants in the cells (Bläsing et al., 2005; Chen and Dickman, 2005; Nishizawa et al., 2008; Knaupp et al., 2011; Keunen et al., 2013). Photosynthetic activity of source leaves results in a transient sugar accumulation. These sugars can directly or indirectly counteract oxidative stress, restoring homeostasis. Sugar starvation induces oxidative stress as well (Bolouri-Moghaddam et al., 2010). Both sugar shortage and excess can disturb respiratory mechanism and electron transport in mitochondria and chloroplasts and induce ROS accumulation (Xiang et al., 2011). It is believed that sugar signaling and expression of sugar-modulated genes relate to the control of oxidative stress (Couée et al., 2006; Rosa et al., 2009). On the one hand, sugars are involved in the ROS-producing metabolic pathways; on the other hand, they activate NADPH-generating reactions via the pentose phosphate cycle and favor ROS scavenging. The interaction between signaling pathways activating sugar-dependent and ROS-dependent genes allows the maintenance of the balance between the effects of carbohydrates and ROS. For example, high photosynthetic activity stimulates both ROS generation and sugar accumulation.

In contrast, sugar accumulation inhibits photosynthesis (Koch, 1996; Rolland et al., 2002, 2006). Low temperatures stimulate photooxidative damages and also lead to sugar accumulation playing the role of stress protectants. Sugar shortage reduces markedly ADP regeneration, suppresses electron transport along the cytochrome *c* pathway, and enhances ROS generation in mitochondria (Dutilleul et al., 2003a,b). Active lipid mobilization also favors ROS accumulation

(Contento et al., 2004). At high sugar content, fatty acid mobilization is suppressed and ROS accumulation is less active. Via their influence on photosynthesis, mitochondrial respiration, and fatty acid metabolism, sugars take a key position in the stabilization of ROS production.

Products of sucrose cleavage, glucose and fructose, play an especial role in these processes. Thus, *Arabidopsis* treatment with glucose increased the reduction status of the NADP system and the redox activation status of AGPase in its leaves (Kolbe et al., 2005), induced thioredoxins (TRXs) (Balmer et al., 2004; Traverso et al., 2007), and improved resistance to oxidative stress (Nishikawa et al., 2005). In its turn, thioredoxins were involved in the regulation of carbohydrate metabolism in the roots, tubers, and leaves (Hendriks et al., 2003; Geigenberger et al., 2005; de Dios Barajas-López et al., 2012). In addition to the involvement in metabolic reactions, sugars may be the substrates of redox reactions. It has long been known a possibility of hexose autooxidation, when glucose functions as a substrate of non-enzymatic oxidation in the presence of compounds capable to redox reactions, e.g., transition metals. In this case, ketoaldehydes, hydrogen peroxide, and various free radicals are generated. Ketoaldehydes can bind to the lysine groups of proteins, inducing their oxidative damage. Hydroxyl radicals oxidize thiol bonds in proteins and induce their structural changes and even fragmentation. Thus, hexose excess can lead to oxidative stress, even when not participating in metabolic reactions (Wolff and Dean, 1987; Russell et al., 2002; Chetyrkin et al., 2011; Lankin et al., 2012).

Sucrose is an excellent antioxidant. At low concentrations, it can function as a signaling molecule; at high concentrations, it can scavenge ROS (Bolouri-Moghaddam et al., 2010). It is characteristic that sugars may affect the cell redox status differently. Thus, glucose increases the total cell redox state, whereas sucrose does not exert such action. Glucose stimulates the flux of carbon components to the respiratory chain, whereas sucrose reduces this flux (Geigenberger et al., 2005).

The antioxidant role of sucrose and glucose was supported in the series of studies performed on potato plants transformed with the yeast invertase gene or treated with sucrose and therefore enriched in sucrose and glucose (Deryabin et al., 2003, 2005, 2007; Sin'kevich et al., 2008, 2009, 2010). A comparison of the content of two ROS (hydrogen peroxide and superoxide radical) and activities of enzymes scavenging them (CAT, POD, and SOD) in control and transformed plants did not demonstrate a correlation between these characteristics. For example, in the leaves of control plants at 22°C, the lower content of H_2O_2 was found simultaneously with the lower activity of CAT and POD. Under hypothermia ($-7°C$), transformants had more H_2O_2 and enhanced activities of enzymes degrading it. The authors supposed that ROS accumulation was correlated not with the antioxidant enzyme activities but rather with the capability of sugars to neutralize them. This property possibly characterizes cold-tolerant plants, as distinct from heat-loving ones using predominantly antioxidant enzymes for ROS scavenging.

Resistance to ROS was correlated with the contents of fructans, sugar alcohols, mannitol, inositol, and sorbitol (Shen et al., 1997; Stoyanova et al., 2011). Transgenic tobacco with increased content of mannitol in chloroplasts manifested an increased resistance to paraquat treatment. In these experiments, mannitol did not reduce the generation of hydroxyl radical but increased a capability of its scavenging (Shen et al., 1997). At the same time, mannitol did not exert injurious action on plants. As distinct from metabolizable sugars (glucose, fructose, and sucrose), it did not inhibit photosynthesis (Bolouri-Moghaddam et al., 2010). Trehalose affected plants similarly *in vitro* (Stoyanova et al., 2011). Transgenic rice accumulating high levels of trehalose was more tolerant to salt, drought, and cold (Garg et al., 2002). Oligosaccharides also play an important role in scavenging hydroxyl and superoxide radicals (Nishizawa et al., 2008; Sin'kevich et al., 2010; Foyer and Shigeoka, 2011; Stoyanova et al., 2011). Transgenic plants overexpression *GolS* (*GolS1*, *GolS2*, *GolS4*) and *RafS* were better scavengers than wild-type plants (Nishizawa et al., 2008).

To date, a large list of genes regulated by sugars has been compiled (Koch, 1996). This list comprises genes controlled simultaneously by sugars and oxidative stress. However, it is unclear whether the expression of sucrose-dependent genes itself results in the protection against ROS (Price et al., 2004). In some cases, such interrelation was obvious. Redox activation of ADP-glucose pyrophosphorylase by sugars is part of the signaling mechanism relating the rate of starch synthesis and carbon availability for other plant tissues. Redox regulation of plastid enzymes may change carbon fixation, starch metabolism, synthesis of lipids, and ascorbic acid in response to stresses.

ROS generation in response to phytohormones frequently affects metabolic processes controlled by sugars (Price et al., 2004; Loreti et al., 2005; Couée et al., 2006). Sugars can suppress some metabolic processes generating ROS, but they can also stimulate them. Orchestration of gene expression in response to sugars is modified at stresses. Sugar starvation enhanced ROS accumulation and expression of genes encoding antioxidant enzymes, CAT in particular (Contento et al., 2004). Like ROS, sugars activate protective pathways (Apel and Hirt, 2004; Mittler, 2002; Orozco-Cardenas and Ryan, 1999) by affecting the expression of pathogenesis-related genes (Badur et al., 1994; Herbers and Sonnewald, 1998; Thibaud et al., 2004); they play a crucial role in stress responses (Loreti et al., 2005; Nishikawa et al., 2005; Sulmon et al., 2004).

Dugas and Bartel (2008) revealed the new way of sugar influence on the redox potential in cells. In this case, sugars also enhance ROS generation under stresses. The authors showed that one of the plant microRNAs, miR398, from *Arabidopsis* induced silencing of mRNA encoding SOD and cytochrome *c*. Stresses inhibited miR398 (Sunkar and Zhu, 2004; Sunkar et al., 2006); their effects were weakened in the presence of sucrose. Thus, sucrose decreased the content of SOD mRNA and protein via its action on miR398. Plants growing in the presence of sucrose under stress accumulated more miR398, and the content of SOD in them was reduced as compared with plants growing without sucrose in medium. Thus, in addition to known sugar participation in the control of gene activity (Roitsch, 1999), they can directly affect microRNAs and reduce the formation of one of the antioxidant enzymes, SOD. Therefore, miR398 can be considered as a sucrose-dependent SOD regulator (Foyer and Noctor, 2009; Cui et al., 2012).

ROS generation in response to phytohormone frequently affects metabolic processes regulated by sugars (Price et al., 2004; Loreti et al., 2005; Couée et al., 2006). Sugars can suppress some metabolic processes generating ROS, such as fatty acid mobilization and peroxisomal β-oxidation, but they can also stimulate ROS formation.

Like sugars, proline is an important participant of redox signaling (Hong et al., 2000). Proline is an antioxidant as well; it is an effective scavenger of the hydroxyl radical and other free radicals (Smirnoff and Cumbes, 1989). For example, proline can scavenge hydroxyl radicals, thus maintaining the redox homeostasis and preventing the cell death (Smirnoff and Cumbes, 1989; Hare and Cress, 1997; Chen and Dickman, 2005). The correlation between proline content and GSH redox state was established. In addition, in the presence of proline, the ratio $NADP^+/NADPH$ was conserved (Hare and Cress, 1997). In this connection, an important role of proline in the regulation of the pentose-phosphate cycle, producing reducing equivalents required for the maintenance of many antioxidants in their reduced state, is emphasized.

11.3.3 Sugars and proline as signaling molecules

Another important function of sugars is their key metabolic signals (Koch, 1996; Price et al., 2004; Gibson, 2004, 2005; Li et al., 2006). Low molecular sugars (glucose, fructose, sucrose) are regulatory molecules controlling the expression of metabolic pathway genes, stress resistance, growth, and development (Koch, 1996; Price et al., 2004; Li et al., 2006; Rolland et al., 2006; Cho and Yoo, 2011; Li et al., 2011a,b). Sugar is needed for all cell functions; however, its high concentrations inhibit growth and development (Rolland et al., 2006). High concentrations of sugars can be attained mainly at the high activity of photosynthesis, suppression of their outflow, or in the sites of phloem loading and unloading. For optimal growth and survival under stressful conditions, plants should monitor and control their sugar levels. This is provided by corresponding signaling pathways.

Interdependence of factors involved in the overall plant response to stresses is summarized in the scheme presented by Huang et al. (2012). It comprises several stages:

1. Specific interaction of extracellular signaling molecules (ligands) with G-protein-coupled receptors (GPCRs) with the change in their conformation. The regulator of G-protein signaling (RGS), AtRGS1, is involved in sugar signaling during seed germination; it functions in the hexokinase (HXK)-independent Glc signaling pathway (Chen et al., 2006). Signal perception includes membrane sensors, such as Ca-channels, membrane physical state (its fluidity), etc.
2. Signal transduction with the involvement of second messengers (Ca, ROS, inositol phosphates, Ca-dependent protein kinases (CDPKs)). An increase in the Ca concentration in the layers of the cytosol closest to the cell membrane under stress is recognized by Ca sensors (Ca binding proteins, Salt Overly Sensitive3 (SOS3)). Ca sensor proteins interact with the MAP-kinase cascade of phosphorylation.
3. Activation of transcription factors (CBF/DREB, ABF, bZIP, MYC/MYB) controlling expression of stress genes (*Hsp, SP1, LEA, COR*).
4. Activation of hormone formation (ABA, salicylic acid, ethylene) multiplying the initial signal and starting a new cycle of signaling.
5. Physiological response including morphological, biochemical, and physiological changes and determining adaptation and resistance.

Initial stages of the signaling pathway have a special significance. In this period, a number of common for different stresses signaling circuit components are active, they can interact with each other. In such a way, resistance is generated not only to the given acting factor, but also to a number other stressors. Such plant property is designated as cross-tolerance. Its operation requires the recognition of the initial alarm signal and which components are evidently changing the properties of the plasma membrane: the membrane lipid matrix, electric potential, permeability for ions and organic signaling molecules, generation of oxidative components, and a disturbance of the redox equilibrium on the plasma membrane and in the apoplast.

Responses to sugar and development are controlled by one and the same protein factors, which imply the interaction of these processes in response to environmental factors. For example, *PLEIOTROPIC REGULATORY LOCUS1* (*PRL1*) suppresses Glc signaling via interaction with SnRK1 kinase. The *prl1* mutant is hypersensitive under light to high glucose content; it coordinates light and sugar signaling. In contrast, the STIMPY/WOX9 protein is a positive coordinator between sugar signaling and growth (cited in Cui et al., 2012).

There is a coordination between responses to sugar and redox signals; however, it is unclear in which way it occurs (Kawakami et al., 2008; Smeekens et al., 2010; Bolouri-Moghaddam et al., 2010). Glucose and sucrose affect redox signals generated by the photosynthetic and respiratory electron transport chains differently. Thus, glucose enhances the carbon flux toward the respiratory chain, whereas sucrose suppresses it. A dependence of sugars on the NADPH/NADP$^+$ ratio, redox activation, and ADP-glucose pyrophosphorylase (AGPase), a key enzyme of starch synthesis, is also different. Glucose but not sucrose enhances redox activation of AGPase and overall redox state. Therefore, the activity of the signal transduction chain depends on the ratio of the hexose and sucrose content (Weber et al., 1995; Xiang et al., 2011). Sugar effect on post-translational redox activation of AGPase is a component of the signaling mechanism connecting the rate of starch synthesis and carbohydrate availability (Geigenberger et al., 2005).

The transcription factor ABI4 (ABA-INSENSITIVE4) plays an important role in the coordination of pathways of redox, hormonal, and sugar signaling. ABI4 transcription factor is a crucial component in the transmission of information about the redox state. It mediates the interaction between the hormonal signaling pathway and the redox system needed for the regulation of plant growth and resistance. In particular, ABI4 is required for ascorbate-dependent growth regulation including the activation of salicylic acid (SA) and the inhibition of jasmonic acid (JA) signaling pathways. It was mentioned above that the occurrence of glucose in medium inhibits plant growth. As distinct from the wild-type seeds, the transgenic seeds with antisense expression of ABI4 (*abi4*) can germinate on medium with the high glucose concentration. Insensitivity to glucose arising at the disturbance of ABI4 signaling is suppressed at the low redox-buffering capacity. The redox center responsible for the oxidative signaling arising under many stresses adjusts plant responses to hormonal and metabolic regulators (Foyer et al., 2012).

Key regulators of root growth and development are SHORT-ROOT (SHR) and SCARECROW (SCR). SCR and ABI4 affect the response to sugars. Since the high content of ABI4 inhibits root growth on glucose, SCR can stimulate growth via the repression of ABI4-mediated sugar response in the root apical meristem. ABI4 is involved in the regulation of not only growth processes but also stress responses. Such facts allow a suggestion that ABI4 is an important component determining the coordination of hexose signaling, redox status, growth, and responses to hormones and stresses (Rook et al., 2001; Li et al., 2006; Dekkers et al., 2008; Cui et al., 2012).

Glucose is the main sugar signaling metabolite controlling gene expression (Gibson, 2005; Rolland et al., 2006; Smeekens et al., 2010; Bolouri-Moghaddam et al., 2010; Cho and Yoo, 2011; Li et al., 2011a,b; Matiolli et al., 2011; Cui et al., 2012). Glucose sensors are hexokinase 1 (HXK1) and SNF1-RELATED PROTEIN KINASE1 (SnRK1). They are the main participants of sugar signaling (Moore et al., 2003) controlling plant energy homeostasis, resistance, and survival (Moore et al., 2003; Baena-González et al., 2007, Baena-González and Sheen, 2008; Smith and Stitt, 2007). Protein phosphorylation is an important element of perception and transduction of signals with information about troubles in the environment (Coello et al., 2011). Stresses activate various protein kinases, in particular mitogen-activated protein kinase, MAPK, calcium-dependent protein kinase, CDPK, and SNF1-related protein kinase, SnRK. Three subfamilies of SnRKs (SnRK1,

SnRK2, and SnRK3) were found in plants (Hrabak et al., 2003). The members of SnRK1 family (KIN10 and KIN11) are involved in the regulation of energetic metabolism, the best usage of energetic resources at energy shortage (Baena-González et al., 2007; Usadel et al., 2008). Specific for plants SnRK2 and SnRK3 are also involved in the responses to stresses; some of SnRK3 are required for the maintenance of K^+/Na^+ homeostasis (Gong et al., 2002). Other SnRK3 regulate responses to Ca and ABA signaling. SnRK2 overexpression makes plants resistant to hyperosmotic stress (drought, salinity, and plant freezing stresses) (Mao et al., 2010).

One of the ways of SnRK action on stress signaling is through the ABA response elements binding proteins (AREBPs), plant-specific members of bZIP transcription factors, which regulate expression of ABA-dependent genes. AREBPs comprise conserved SnRK-binding target sites, which makes them a good substrate for phosphorylation by SnRKs (Kobayashi et al., 2004; Furihata et al., 2006; Zhang et al., 2011). It is supposed that AREBPs may be a center of the signaling network, where many pathways converge (Halford and Hey, 2009). The mutual influence of the carbohydrate signaling pathway and that induced by stress occurs via SnRKs. There are some data that SnRKs can directly phosphorylate and inhibit sucrose phosphate synthase, thus reducing the content of sucrose and its inhibitory action of photosynthesis. A similar effect can be obtained by mutation of the sites of phosphorylation in the molecule of SnRK1, switching off the SnRK1 regulation of sucrose phosphate synthase. SnRKs can activate starch content by changing the redox state and expression of ADP-glucose pyrophosphorylase and α-amylase. A deviation of carbon flux from sucrose to starch results in the photosynthesis activation and the plant yield increase (Mao et al., 2010). The involvement of SnRKs in plant resistance is indicated, e.g., by the role of SnRK2 in response to hyperosmotic stress and ABA treatment. Overexpression of corresponding genes improved drought tolerance (Umezawa et al., 2004; Shin et al., 2007), salt tolerance (Diedhiou et al., 2008), and frost resistance of arabidopsis (Mao et al., 2010).

Additional difficulties arise because of the integration of redox signals and other signals, in particular from sugars in the responses to phytohormones (Couée et al., 2006). For example, the regulation of starch synthesis includes thioredoxin-dependent modulation of the ADP-glucose pyrophosphorylase activity, the enzyme responsible not only for the trehalose metabolism but also for the SnRK cascade, the main regulator of stress and energetic signaling (Baena-González et al., 2007; Cui et al., 2012).

Even so a superficial touch to the regulatory systems of the plant cell gives an indication of the direct carbohydrate involvement in a complex chain of interactions of different signals (Ho et al., 2001; Gibson, 2005; Hartig and Beck, 2006; Koornneef and Pieterse, 2008). The intracellular signaling path does not require such high concentrations of signal metabolites, which are detected under the influence of stressors. The fundamental role in primary metabolism also suggests a rather continuous flow of carbohydrates (from synthesis to metabolization). It is believed that two different functions of carbohydrates—osmotic and regulatory—are separated in space. Maybe sugar is subjected to the general laws of accumulation, which determines the osmotic potential, occurs mainly in the vacuoles and some other structural components of the cell, and the cytoplasmic concentration is not so high (as expected for glucose). The answer to the question of how can metabolites implement fine regulation on the background of their high concentration is awaiting the emergence of methodological approaches to the study of the intracellular localization of sugars.

11.4 **Conclusion and future prospects**

The action of the three most common environmental factors causing tremendous damage to agriculture, abnormal temperature, drought, and salinity, have a common consequence to plants—dehydration. This determines a common strategy of survival, e.g., the accumulation of osmotically active compounds. The central place of carbohydrates in the cell metabolism and their quantitative prevalence over other organic osmotica determine their important role in plant defense. In many studies performed on various plant species, the accumulation of carbohydrates under unfavorable conditions was observed. It is not surprising that accumulated carbohydrates differ in their nature, the time course of accumulation, and different ratios with the accumulation of other metabolites, primarily proline. These parameters reflect the specificity of metabolism in each plant species. However, such accumulation not always favors resistance development. The reason may be that little attention is paid to the ratio between processes that determine resistance of cells, tissues, and the whole plant.

In the analysis of mechanisms of metabolite accumulation, basic attention is usually paid to the processes of their synthesis and degradation. Thus, sucrose accumulation may be explained by the activation under stressful conditions of sucrose phosphate synthase, a main enzyme of sucrose biosynthesis, and amylase degrading starch. Correspondingly, an increase in the content of raffinose-like carbohydrates may be related to the activation of galactinol synthase and raffinose synthase. Synthetic processes require energy consumption, and it is unclear how economically plants can distribute their energy resource, reduced under stress, among various defense responses, especially occurring in their different organs. A comparison of energy costs is too difficult and as yet is not well analyzed.

The significance of intracellular carbohydrate distribution for plant resistance is not essentially considered. In addition, the application of systemic mathematic analysis allowed the comparison of changes in the content of separate carbohydrates in cell compartments of *Arabidopsis* plants differing in resistance. Sucrose synthesized in the cytosol may be transported in the vacuole or plastids, to be used for RFO synthesis, or be cleaved into hexoses. Sucrose transport to the vacuole or plastids allows the accumulation of its larger amounts. Such intracellular sugar redistribution under the influence of cold is activated in resistant but not in cold-sensitive plants. When sucrose can accumulate in the cytosol, the accumulation of glucose is toxic for metabolism. Therefore, hexose phosphorylation by hexokinases, hexose metabolization, or effluxes into plastids or vacuole acquires an importance for plant viability (Nägele et al., 2011, 2012, Nägele and Heyer, 2013).

Another reason for compound accumulation in cells and tissues may be specific changes in their transport between plant cells, tissues, and organs. The main energetically favorable intercellular transport occurs through plasmodesmata. Permeability of these intercellular bridges depends not only on intracellular metabolic processes but also on environmental conditions. Each stress signal is first recognized by the plasma membrane. It may be supposed that a disturbance in the water content in the plasma membrane and generation of water potential gradients on it activate mechano-sensitive ionic channels in the plasma membrane, Ca channels in particular. Ca entrance into the cell, the creation of the Ca gradient, and depolarization of the plasma membrane serve as the first signals of trouble. An increase in the free Ca concentration near the plasma membrane will activate Ca-dependent enzymes localized in the membrane, including callose synthase. Callose deposition around plasmodesmata disturbs their transport activity. As a result, transport of sugars synthesized in the mesophyll cells in the sink cells of the leaf and in the cells of phloem conducting complex is hampered. In this case, sugars accumulate in the mesophyll, but other tissues suffer of their shortage. Thus, resistance of mesophyll cells to dehydration may contradict the overall resistance of the organism.

Callose is deposited not only around plasmodesmata but also on the sieve plates; this interferes with the phloem transport of assimilates and deprives sink organs of carbohydrates, hormones, proteins, and signaling molecules necessary for them to counteract environmental attacks. In fact, we found that, due to the suppression of assimilate transport along the phloem under the influence of low, above-zero temperatures, sugars accumulated not only in the rapeseed leaf blades but also in the petioles (Krasavina et al., unpublished). Limited assimilate transport to sink organs reduced *Arabidopsis* plant cold resistance in the experiments of Nägele et al. (2011). Photosynthetic carbon fixation and sucrose phosphate synthase activity not only produced the sucrose pool for the efflux along the phloem, but specifically activated the synthesis of sugar high-affinity transporter AtSUC1 homologous to AtSUC2 transporter involved in phloem loading (Gottwald et al., 2000; Lundmark et al., 2006). A comparison of *Arabidopsis* plants originated from different climatic zones and thus differing in cold resistance showed that carbon fixation and efflux to sinks increased and were high during the maximal development of resistance in cold-resistant plants, whereas they gradually decreased during long-term cold treatment of sensitive plants (Nägele and Heyer, 2013).

The importance of transport processes for plant viability was clearly demonstrated in studies of pollen sterility induced by low temperatures in the early period of microspore development in the male hametophyte (Oliver et al., 2005, 2007). In contrast to most presented facts, sugars accumulated in anthers in the period of the highest cold sensitivity. In this time, tapetum is symplastically isolated, sugars move apoplastically, and cell wall invertase and hexose transporter should be active. Tapetum, the boundary layer between sporophyte and hametophyte, serves a secretory tissue for microspore nutrition. Cell wall invertase cleaves sucrose coming to the anthers into hexoses, which should be absorbed by the tapetum cells from the apoplast with the help of hexose transporters. However, under cold conditions, activities of both invertase and transporters are inhibited. Therefore, the accumulation of sugars in anthers occurs at the lower invertase activity in the tapetum and pollen grains and a suppressed expression of its gene *OSINV4*. Sugars accumulate in the anther walls, and transport of nutrient compounds to developing microspores is disturbed. In spite of suppressed activity of cell wall invertase, not only sucrose but also hexoses accumulate. It should be emphasized that in this case the content of starch, a source of nutrition for germinating pollen and pollen tube growth, in pollen grains reduced. Unexpectedly, in resistant rice cultivars, sugars did not accumulate in anthers in the cold, invertase activity was not suppressed, and starch did not disappear from pollen grains (Oliver et al., 2005, 2007). A similar situation was observed under water deficit (Koonjul et al., 2005).

These examples emphasize that, in spite of the fact that processes of osmolyte accumulation take a central place in the development of resistance to abnormal temperatures, drought, or salinity, they cannot be automatically considered as indices characterizing plant resistance. Accumulation may be due to unbalancing many interrelated metabolic and signaling processes that are difficult to analyze. A special place in violation of interactions between so many processes belongs to a disruption in the metabolite transport within the cells, between tissues, and between organs.

Despite the importance of the problem and the long period of its study, the influence of unfavorable environmental conditions on various metabolic processes in plants is still poorly understood. It is clear that carbohydrate metabolism uses various ways of influence, contributing to the development of stress resistance. However, the role of carbohydrates in the plant is so diverse and resistance development is so complex that it is difficult to expect a direct correlation between carbohydrate metabolism and plant resistance. One of the approaches to identifying the place of carbohydrate metabolism in the plant response to stress may be to identify the dynamics of changes in various parameters. Such studies are rare. In one study, it was shown that the response of carbohydrate metabolism to water deficit had secondary indirect character. One of the earliest physiological responses to drought is growth retardation and thus a decrease in assimilate consumption. Therefore, during the first 20−40 days of drought, assimilates accumulate. When water deficit becomes more severe, photosynthesis is inhibited and the content of assimilates is sharply reduced (McDowell, 2011). The dynamics of changes in physiological and molecular responses of cells and tissues under short-term exposure to stress factors awaits study.

Different plant species respond to stresses differently. In addition, the response of tolerant and sensitive genotypes to the same stressor differs as well. For example, the accumulation of carbohydrates in tissues of rice plants differing in their tolerance was attained by different routes. In tolerant plants, sugars accumulated due to the activation of photosynthesis and starch degradation, whereas in sensitive plants this was determined only by starch degradation. Different genes are activated in genotypes differing in resistance. The difference in stress severity affects sugar accumulation in tolerant and sensitive plants differently: moderate salinity increases markedly the sugar content in sensitive plants but does not affect it in tolerant plants, whereas severe salt stress increases the content of sucrose in tissues of both genotypes (Theerawitaya et al., 2012). A similar line of research needs a large-scale extension.

These examples emphasize the individuality of plant responses to stresses. Fine mechanisms of the interaction between plant metabolism and stress responses can only be understood by comparing the simultaneous changes of many parameters characterizing the responses of plants differing in their resistance to a certain stress, the responses to several stresses, and the responses of different plant species. Such a task is too complex and labor-consuming to hope for a quick response. Perhaps, it may be useful to use new techniques of computer simulation.

It is impossible to construct an adequate model without an understanding of the interactions between various processes in the plant under stress, e.g., photosynthesis, transpiration, respiration, gas exchange, water relations, synthesis and degradation of soluble carbohydrates, and exchange of metabolites between different organs. Recent work has emphasized the need to consider changes in assimilate transport and distribution over the plant to understand the role of assimilates in plant resistance. The need to accommodate the change in the conductivity of phloem and xylem vessels and the interaction between flows of water and assimilates is emphasized (McDowell, 2011; Frost et al., 2012; Sevanto et al., 2013). The study of the relationship between different transport flows is in its infancy, but it is critical to understand the ways of generating stability of the whole plant.

In recent years, the molecular engineering strategy has great success (Knight and Knight, 2012). Only in the last year have many researchers observed an increased resistance of various transgenic plants expressing foreign genes encoding enzymes, regulatory molecules, and transcription factors involved in responses associated with plant resistance: genes of CBF factors (Soltész et al., 2013), of dehydration-induced factor NAC3 (Liu et al., 2013) involved in the osmoregulation of the gene encoding osmotin, the AREB1 gene related to redox reactions (Li et al., 2013), the gene of CIPK14 protein kinase (Deng et al., 2013), DESD-box helicase (Gill et al., 2013), and many others. However, only a few studies have used transgenic plants overexpressing the genes of carbohydrate metabolism (e.g., Deryabin et al., 2005; Zhai et al., 2013).

The central role of carbohydrate metabolism in the cell life and numerous instances of its dependence on environmental conditions suggests the importance and potential of studying complex and fine mechanisms of carbohydrate participation in the creation of resistant plants.

References

Abogadallah, G.M., Serag, M.M., Quick, W.P., 2010. Fine and coarse regulation of reactive oxygen species in the salt tolerant mutants of barnyard grass and their wild-type parents under salt stress. Physiol. Plant 138, 60–73.

Aggarwal, M., Sharma, S., Kaur, N., Pathania, D., Bhandhari, K., Kaushal, N., et al., 2011. Exogenous proline application reduces phytotoxic effects of selenium by minimising oxidative stress and improves growth in bean (*Phaseolus vulgaris* L.) seedlings. Biol. Trace Elem. Res. 140, 354–367.

Alcázar, R., Marco, F., Cuevas, J.C., Patron, M., Ferrando, A., Carrasco, P., et al., 2006. Involvement of polyamines in plant response to abiotic stress. Biotechnol. Lett. 28, 1867–1876.

Alcázar, R., Altabella, T., Marco, F., Bortolotti, C., Reymond, M., Knocz, C., et al., 2010. Polyamines: molecules with regulatory functions in plant abiotic stress tolerance. Planta. 231, 1237–1249.

Ali, Q., Anwar, F., Ashraf, M., Saari, N., Perveen, R., 2013. Ameliorating effects of exogenously applied proline on seed composition, seed oil quality and oil antioxidant activity of maize (*Zea mays* L.) under drought stress. Int. J. Mol. Sci. 14, 818–835.

Apel, K., Hirt, H., 2004. Reactive oxygen species: metabolism, oxidative stress, and signal transduction. Annu. Rev. Plant Biol. 55, 373–399.

Ayre, B.G., 2011. Membrane transport systems for sucrose in relation to whole-plant carbon partitioning. Mol. Plant 4, 377–394.

Badur, R., Herbers, K., Monke, G., Ludewig, F., Sonnewald, U., 1994. Induction of pathogenesis-related proteins in sugar accumulating tobacco leaves. Photosynthetica 30, 575–582.

Baena-González, E., Sheen, J., 2008. Convergent energy and stress signaling. Trends. Plant Sci. 13, 474–482.

Baena-González, E., Rolland, F., Thevelein, J.M., Sheen, J., 2007. A central integrator of transcription networks in plant stress and energy signaling. Nature 448, 938–942.

Balibrea, M.E., Rus-Alvarez, A.M., Bolarin, M.C., Perez-Alfocea, F., 1997. Fast changes in soluble carbohydrates and proline contents in tomato seedlings in response to ionic and non-ionic isosmotic stresses. J. Plant Physiol. 151, 221–226.

Balmer, Y., Vensel, W.H., Tanaka, C.K., Hurkman, W.J., Gelhaye, E., Rouhier, N., et al., 2004. Thioredoxin links redox to the regulation of fundamental processes of plant mitochondria. Proc. Natl. Acad. Sci. USA 101, 2642–2647.

Bartels, D., Sunkar, R., 2005. Drought and salt tolerance in plants. Crit. Rev. Plant Sci. 24, 23–58.

van Bel, A.J.E., Gamalei, Y.V., 1992. Ecophysiology of phloem loading in source leaves. Plant Cell Environ. 15, 265–270.

Ben Ahmed, C., Ben Rouina, B., Sensoy, S., Boukhriss, M., Ben Abdullah, F., 2010. Exogenous proline effects on photosynthetic performance and antioxidant defense system of young olive tree. J. Agric. Food Chem. 58, 4216–4222.

Bentsink, L., Alonso-Blanco, C., Vreugdenhil, D., Tesnier, K., Groot, S.P.C., Koornneef, M., 2000. Genetic analysis of seed soluble oligosaccharides in relation to seed storability of *Arabidopsis*. Plant Physiol. 124, 1595–1604.

Bläsing, O.E., Gibon, Y., Günther, M., Höhne, M., Morcuende, R., Osuna, D., et al., 2005. Sugars and circadian regulation make major contributions to the global regulation of diurnal gene expression in Arabidopsis. Plant Cell 17, 3257–3281.

Bohnert, H.J., Jensen, R.G., 1996. Metabolic engineering for increased salt tolerance—the next step: comment. Aust. J. Plant Physiol. 23, 661–666.

Bohnert, H.J., Nelson, D.E., Jensen, R.G., 1995. Adaptations to environmental stresses. Plant Cell 7, 1099–1111.

Bolouri-Moghaddam, M.R., Roy, K.L., Li, X.L., Rolland, F., van den Ende, W., 2010. Sugar signaling and antioxidant network connections in plant cells. FEBS J. 277, 2022–2037.

Bose, J., Xie, Y., Shen, W., Shabala, S., 2013. Haem oxygenase modifies salinity tolerance in Arabidopsis by controlling K + retention via regulation of the plasma membrane H + -ATPase and by altering SOS1 transcript levels in roots. J. Exp. Bot. 64, 471–481.

Bouchabke-Coussa, O., Quashie, M.L., Seoane-Redondo, J., Fortabat, M.N., Gery, C., Yu, A., et al., 2008. ESKIMO1 is a key gene involved in water economy as well as cold acclimation and salt tolerance. BMC Plant Biol. 8, 125.

Boyer, J.S., 1982. Plant productivity and the environment. Science 218, 443–448.

Bray, E.A., Bailey-Serres, J., Weretilnyk, E., 2000. Responses to abiotic stresses. In: Buchanan, B.B., Gruissem, W., Jones, R.L. (Eds.), Biochemistry and Molecular Biology of Plants. American Society of Plant Physiologists, Rockville, pp. 1158–1203.

Buitink, J., Leprince, O., 2004. Glass formation in plant anhydrobiotes: survival in the dry state. Cryobiology 48, 215–228.

Cao, S., Yang, Z., Zheng, Y., 2013. Sugar metabolism in relation to chilling tolerance of loquat fruit. Food Chem. 136, 139–143.

Castonguay, Y., Nadeau, P., Lechasseur, P., Chouinard, L., 1995. Differential accumulation of carbohydrates in alfalfa cultivars of contrasting winter hardiness. Crop Sci. 35, 509–516.

Chaves, M.M., Flexas, J., Pinheiro, C., 2009. Photosynthesis under drought and salt stress: regulation mechanisms from whole plant to cell. Ann. Bot. 103, 551–560.

Chen, C., Dickman, M.B., 2005. Proline suppresses apoptosis in the fungal pathogen *Colletotrichum trifolii*. Proc. Natl. Acad. Sci. USA 102, 3459–3464.

Chen, L.Q., Qu, X.Q., Hou, B.H., Sosso, D., Osorio, S., Fernie, A.R., et al., 2012. Sucrose efflux mediated by SWEET proteins as a key step for phloem transport. Science 335, 207–211.

Chen, Y., Ji, F., Xie, H., Liang, J., Zhang, J., 2006. The regulator of G-protein signaling proteins involved in sugar and abscisic acid signaling in *Arabidopsis* seed germination. Plant Physiol. 140, 302–310.

Chen, Z., Cuin, T.A., Zhou, M., Twomey, A., Naidu, B.P., Shabala, S., 2007. Compatible solute accumulation and stress-mitigating effects in barley genotypes contrasting in their salt tolerance. J. Exp. Bot. 58, 4245–4255.

Chetyrkin, S., Mathis, M., Pedchenko, V., Sanchez, O.A., McDonald, W.H., Hachey, D.L., et al., 2011. Glucose autoxidation induces functional damage to proteins via modification of critical arginine residues. Biochemistry 50, 6102–6112.

Cho, Y.-H., Yoo, S.-D., 2011. Signaling role of fructose mediated by FINS1/FBP in *Arabidopsis thaliana*. PLoS. Genet. 7, e1001263.

Choudhary, N.L., Sairam, R.K., Tyagi, A., 2005. Expression of delta1-pyrroline-5-carboxylate synthetase gene during drought in rice (*Oryza sativa* L.). J Biochem. Biophys. 42, 366–370.

Coello, P., Hey, S.J., Halford, N.G., 2011. The sucrose non-fermenting-1-related (SnRK) family of protein kinases: potential for manipulation to improve stress tolerance and increase yield. J. Exp. Bot. 62, 883−893.

Contento, A.L., Kim, S.J., Bassham, D.C., 2004. Transcriptome profiling of the response of *Arabidopsis* suspension culture cells to Suc starvation. Plant Physiol. 135, 2330−2347.

Cook, D., Fowler, S., Fiehn, O., Thomashow, M.F., 2004. A prominent role for the CBF cold response pathway in configuring the low-temperature metabolome of Arabidopsis. Proc. Natl. Acad. Sci. USA 101, 15243−15248.

Cornic, G., Le Gouallec, J.L., Briantais, J.M., Hodges, M., 1989. Effect of dehydration and high light on photosynthesis of two C3 plants: *Phaseolus vulgaris* L. and *Elastostema repens* (Lour.). Hall F. Planta. 177, 84−90.

Couée, I., Sulmon, C., Gouesbet, G., El Amrani, A., 2006. Involvement of soluble sugars in reactive oxygen species balance and responses to oxidative stress in plants. J. Exp. Bot. 57, 449−459.

Cox, S.E., Stushnoff, C., 2001. Temperature related shifts in soluble carbohydrate content during dormancy and cold acclimation in *Populus tremuloides*. Can. J. Res. 31, 730−737.

Cramer, G.R., Ergül, A., Grimplet, J., Tillett, R.L., Tattersall, E.A., Bohlman, M.C., et al., 2007. Water and salinity stress in grapevines: early and late changes in transcript and metabolite profiles. Funct. Integr. Genomics 7, 111−134.

Crowe, L.M., Mouradian, R., Crowe, J.H., Jackson, S.A., Womersley, C., 1984. Effects of carbohydrates on membrane stability at low water activities. Biochim. Biophys. Acta. 769, 141−150.

Cruz de Carvalho, M.H., 2008. Drought stress and reactive oxygen species: production, scavenging and signaling. Plant Signal Behav. 3, 156−165.

Cui, H., Hao, Y., Kong, D., 2012. SCARECROW has a SHORT-ROOT-independent role in modulating the sugar response. Plant Physiol. 158, 1769−1778.

Cunningham, S.M., Nadeau, P., Castonguay, Y., Laberge, S., Volenec, J.J., 2003. Raffinose and stachyose accumulation, galactinol synthase expression, and winter injury of contrasting alfalfa germplasms. Crop Sci. 43, 562−570.

Cvikrová, M., Gemperlová, L., Dobrá, J., Martincová, O., Prásil, I.T., Gubis, J., et al., 2012. Effect of heat stress on polyamine metabolism in proline-over-producing tobacco plants. Plant Sci. 182, 49−58.

Dashnau, J.L., Vanderkooi, J.M., 2007. Computational approaches to investigate how biological macromolecules can be protected in extreme conditions. J. Food Sci. 72, R4−R10.

Dashnau, J.L., Sharp, K.A., Vanderkooi, J.M., 2005. Carbohydrate intramolecular hydrogen bonding cooperativity and its effect on water structure. J. Phys. Chem. B 109, 24152−24159.

Davidson, A., Keller, F., Turgeon, R., 2011. Phloem loading, plant growth form, and climate. Protoplasma 248, 153−163.

Dekkers, B.J.W., Schuurmans, J.A.M.J., Smeekens, S.C.M., 2008. Interaction between sugar and abscisic acid signaling during early seedling development in *Arabidopsis*. Plant Mol. Biol. 67, 151−167.

Deng, X., Zhou, S., Hu, W., Feng, J., Zhang, F., Chen, L., et al., 2013. Ectopic expression of wheat TaCIPK14, encoding a calcineurin B-like protein-interacting protein kinase, confers salinity and cold tolerance in tobacco. Physiol. Plant 149, 367−377.

Deryabin, A.N., Trunova, T.I., Dubinina, I.M., Burakhanova, E.A., Sabelnikova, E.P., Krylova, E.M., et al., 2003. Chilling tolerance of potato plants transformed with a yeast-derived invertase gene under the control of the B33 patatin promoter. Russ. J. Plant Physiol. 50, 449−454.

Deryabin, A.N., Dubinina, I.M., Burakhanova, E.A., Astakhova, N.V., Sabelnikova, E.P., Trunova, T., 2005. Influence of yeast-derived invertase gene expression in potato plants on membrane lipid peroxidation at low temperature. J. Therm. Biol. 30, 73−77.

Deryabin, A.N., Sin'kevich, M.S., Dubinina, I.M., Burakhanova, E.A., Trunova, T.I., 2007. Effect of sugars on the development of oxidative stress induced by hypothermia in potato plants expressing yeast invertase gene. Russ. J. Plant Physiol. 54, 32−38.

Deyholos, M.K., 2010. Making the most of drought and salinity transcriptomics. Plant Cell Environ. 33, 648–654.

Diedhiou, C.J., Popova, O.V., Dietz, K.J., Golldack, D., 2008. The SNF1-type serine–threonine protein kinase SAPK4 regulates stressresponsive gene expression in rice. BMC. Plant Biol. 8, 49.

Dinant, S., Lemoine, R., 2010. The phloem pathway: new issues and old debates. C. R. Biol. 333, 307–319.

de Dios Barajas-López, J., Tezycka, J., Travaglia, C.N., Serrato, A.J., Chueca, A., Thormählen, I., et al., 2012. Expression of the chloroplast thioredoxins f and m is linked to short-term changes in the sugar and thiol status in leaves of *Pisum sativum*. J. Exp. Bot. 63, 4887–4900.

Dixon, N., Paiva, L., 1995. Stress-induced phenylpropanoid metabolism. Plant Cell 7, 1085–1097.

Dobra, J., Motyka, V., Dobrev, P., Malbeck, J., Prasil, I.T., Haisel, D., et al., 2010. Comparison of hormonal responses to heat, drought and combined stress in tobacco plants with elevated proline content. J. Plant Physiol. 167, 1360–1370.

Dos Santos, T.B., Budzinski, I.G., Marur, C.J., Petkowicz, C.L., Pereira, L.F., Vieira, L.G., 2011. Expression of three galactinol synthase isoforms in *Coffea arabica* L. and accumulation of raffinose and stachyose in response to abiotic stresses. Plant Physiol. Biochem. 49, 441–448.

Dugas, D.V., Bartel, B., 2008. Sucrose induction of Arabidopsis miR398 represses two Cu/Zn superoxide dismutases. Plant Mol. Biol. 67, 403–417.

Dutilleul, C., Driscoll, S., Cornic, G., De Paepe, R., Foyer, C.H., Noctor, G., 2003a. Functional mitochondrial complex I is required by tobacco leaves for optimal photosynthetic performance in photorespiratory conditions and during transients. Plant Physiol. 313, 264–275.

Dutilleul, C., Garmier, M., Noctor, G., Mathieu, C., Chétrit, P., Foyer, C.H., et al., 2003b. Leaf mitochondria modulate redox regulation in plants whole cell redox homeostasis, set antioxidant capacity and determine stress resistance through altered signaling and diurnal regulation. Plant Cell 15, 1212–1226.

Evers, D., Bonnechère, S., Hoffmann, L., Hausman, J.-F., 2007. Physiological aspects of abiotic stress response in potato. Belg. J. Bot. 140, 236–245.

Evers, D., Lefèvre, I., Legay, S., Lamoureux, D., Hausman, J.F., Rosales, R.O., et al., 2010. Identification of drought responsive compounds in potato through a combined transcriptomic and targeted metabolite approach. J. Exp. Bot. 61, 2327–2343.

Flinn, C.L., Ashworth, E.N., 1995. The relationship between carbohydrates and flower bud hardiness among three *Forsythia* taxa. J. Am. Soc. Hort. Sci. 120, 607–613.

Fowler, S., Thomashow, M.F., 2002. Arabidopsis transcriptome profiling indicates that multiple regulatory pathways are activated during cold acclimation in addition to the CBF cold response pathway. Plant Cell 14, 1675–1690.

Foyer, C.H., Noctor, G., 2005. Redox homeostasis and antioxidant signaling: a metabolic interface between stress perception and physiological responses. Plant Cell 17, 1866–1875.

Foyer, C.H., Noctor, G., 2009. Redox regulation in photosynthetic organisms: signaling, acclimation, and practical implications. Antioxid. Redox. Signal 11, 861–905.

Foyer, C.H., Shigeoka, S., 2011. Understanding oxidative stress and antioxidant functions to enhance photosynthesis. Plant Physiol. 155, 93–100.

Foyer, C.H., Kerchev, P.I., Hancock, R.D., 2012. The ABA-INSENSITIVE-4 (ABI4) transcription factor links redox, hormone and sugar signaling pathways. Plant Signal Behav. 7, 276–281.

Frost, C.J., Nyamdari, B., Tsai, C-J., Harding, S.A., 2012. The tonoplast-localized sucrose transporter in Populus (PtaSUT4) regulates whole-plant water relations, responses to water stress and photosynthesis. PLOS One 7, e44467.

Fumagalli, E., Baldoni, E., Abbruscato, P., Piffanelli, P., Genga, A., Lamanna, R., et al., 2009. NMR techniques coupled with multivariate statistical analysis: tools to R. analyse *Oryza sativa* metabolic content under stress conditions. J. Agron. Crop Sci. 195, 77–88.

Furihata, T., Maruyama, K., Fujita, Y., Umezawa, T., Yoshida, R., Shinozaki, K., et al., 2006. Abscisic acid-dependent multisite phosphorylation regulates the activity of a transcription activator AREB1. Proc. Natl. Acad. Sci. USA 103, 1988—1993.

Galmés, J., Medrano, H., Flexas, J., 2007. Photosynthetic limitations in response to water stress and recovery in Mediterranean plants with different growth forms. New Phytol. 175, 81—93.

Gamalei, Y., 1989. Structure and function of leaf minor veins in trees and herbs. Trees 3, 96—110.

Gamalei, Y., 1991. Phloem loading and its development related to plant evolution from trees to herbs. Trees 5, 50—64.

Garg, A.K., Kim, J.K., Owens, T.G., Ranwala, A.P., Choi, Y.D., Kochian, L.V., et al., 2002. Trehalose accumulation in rice plants confers high tolerance levels to different abiotic stresses. Proc. Natl. Acad. Sci. USA 99, 15898—15903.

Geigenberger, P., Kolbe, A., Tiessen, A., 2005. Redox-regulation of carbon storage and partitioning in response to light and sugars. J. Exp. Bot. 56, 1469—1479.

Geigerberger, P., Reimholz, R., Geiger, M., Merlo, L., Canale, V., Stitt, M., 1997. Regulation of sucrose and starch metabolism in potato tubers in response to shortterm water deficit. Planta 201, 502—518.

Gibson, S.I., 2005. Control of plant development and gene expression by sugar signaling. Curr. Opin. Plant Biol. 8, 93—102.

Gill, S.S., Tuteja, N., 2010. Polyamines and abiotic stress tolerance in plants. Plant Signal Behav. 5, 26—33.

Gill, S.S., Tajrishi, M., Madan, M., Tuteja, N., 2013. A DESD-box helicase functions in salinity stress tolerance by improving photosynthesis and antioxidant machinery in rice (*Oryza sativa* L. cv. PB1). Plant Mol. Biol. 82, 1—22.

Gong, D.M., Guo, Y., Jagendorf, A.T., Zhu, J.K., 2002. Biochemical characterization of the Arabidopsis protein kinase SOS2 that functions in salt tolerance. Plant Physiol. 130, 256—264.

Gong, Q., Li, P., Ma, S., Rupassara, S.I., Bohnert, H.J., 2005. Salinity stress adaptation competence in the extremophile *Thellungiella halophila* in comparison with its relative *Arabidopsis thaliana*. Plant J 44, 826—839.

Gottwald, J.R., Krysan, P.J., Young, J.C., Evert, R.F., Sussman, M.R., 2000. Genetic evidence for the in planta role of phloem-specific plasma membrane sucrose transporters. Proc. Natl. Acad. Sci. USA 97, 13979—13984.

Grant, T.N., Dami, I.E., Ji, T., Scurlock, D., Streeter, J., 2009. Variation in leaf and bud soluble sugar concentration among *Vitis* genotypes grown under two temperature regimes. Can J. Plant Sci. 89, 961—968.

Gray, G.R., Heath, D., 2005. A global reorganization of the metabolome in Arabidopsis during cold acclimation is revealed by metabolic fingerprinting. Physiol. Plant 124, 236—248.

Gusta, L.V., Wilen, R.W., Fu, P., 1996. Low temperature stress tolerance: the role of abscisic acid, sugars, and heat-stable proteins. Hort Sci. 31, 39—46.

Gusta, L.V., Wisniewski, M., Nesbitt, N.T., Gusta, M.L., 2004. The effect of water, sugars, and proteins on the pattern of ice nucleation and propagation in acclimated and nonacclimated canola leaves. Plant Physiol. 135, 1642—1653.

Guy, C., 1990. Cold acclimation and freezing stress tolerance: role of protein metabolism. Annu. Rev. Plant Physiol. Plant Mol. Biol. 41, 187—223.

Guy, C., Kaplan, F., Kopka, J., Selbig, J., Hincha, D.K., 2008. Metabolomics of temperature stress. Physiol. Plant 132, 220—235.

Halford, N.G., Hey, S.J., 2009. SNF1-related protein kinases (SnRKs) act within an intricate network that links metabolic and stress signalling in plants. Biochem. J. 419, 247—259.

Hamman, R.A., Dami, I.E., Walsh, T.M., Stushnoff, C., 1996. Seasonal carbohydrate changes and cold hardiness of Chardonnay and Riesling grapevines. Am. Enol. Vitic. 47, 31—36.

Handley, L.W., Pharr, D.M., McFeeters, R.F., 1983. Relationship between galactinol synthase activity and sugar composition of leaves and seeds of several crop species. J. Am. Soc. Hort. Sci. 108, 600—605.

Hannah, M.A., Heyer, A.G., Hincha, D.K., 2005. A global survey of gene regulation during cold acclimation in *Arabidopsis thaliana*. PLoS. Genet. 1 (e26), 0179–0196.

Hannah, M.A., Wiese, D., Freund, S., Fiehn, O., Heyer, A.G., Hincha, D.K., 2006. Natural genetic variation of freezing tolerance in *Arabidopsis*. Plant Physiol. 142, 98–112.

Hanson, A., Nelsen, C., Everson, E., 1977. Evaluation of free proline accumulation as an index of drought resistance using two contrasting barley cultivars. Crop Sci. 17, 720–726.

Hare, P.D., Cress, W.A., 1997. Metabolic implications of stressinduced proline accumulation in plants. Plant Growth Regul. 21, 79–102.

Hartig, K., Beck, E., 2006. Crosstalk between auxin, cytokinins, and sugars in the plant cell cycle. Plant Biol. 8, 389–396.

Hayat, S., Hayat, Q., Alyemeni, M.N., Wani, A.S., Pichtel, J., Ahmad, A., 2012. Role of proline under changing environment: a review. Plant Signal Behav. 7, 1456–1466.

Hendriks, J.H.M., Kolbe, A., Gibon, Y., Stitt, M., Geigenberger, P., 2003. ADP-glucose pyrophosphorylase is activated by posttranslational redox-modification in response to light and to sugars in leaves of Arabidopsis and other plant species. Plant Physiol. 133, 838–849.

Herbers, K., Sonnewald, U., 1998. Altered gene expression brought about by inter- and intracellularly formed hexoses and its possible implications for plant-pathogen interactions. J. Plant Res. 111, 323–328.

Hincha, D.K., Zuther, E., Hundertmark, M., Heyer, A.G., 2006. The role of compatible solutes in plant freezing tolerance: a case study on raffinose. In: Chen, T.H.H., Uemura, M., Fujikawa, S. (Eds.), Cold Hardiness in Plants: Molecular Genetics, Cell Biology and Physiology. Oxford University Press, New York, pp. 203–218.

Hisano, H., Kanazawa, A., Kawakami, A., Yoshida, M., Shimamoto, Y., Yamada, T., 2004. Transgenic perennial ryegrass plants expressing wheat fructosyltransferase genes accumulate increased amounts of fructan and acquire increased tolerance on a cellular level to freezing. Plant Sci. 167, 861–868.

Ho, S., Chao, Y., Tong, W., Yu, S., 2001. Sugar coordinately and differentially regulates growth- and stress-related gene expression via a complex signal transduction network and multiple control mechanisms. Plant Physiol. 125, 877–890.

Hong, Z.L., Lakkineni, K., Zhang, Z.M., Verma, D.P.S., 2000. Removal of feedback inhibition of Δ(1)-pyrroline-5-carboxylate synthetase results in increased proline accumulation and protection of plants from osmotic stress. Plant Physiol. 22, 1129–1136.

Hoque, M.A., Okuma, E., Banu, M.N., Nakamura, Y., Shimoishi, Y., Murata, Y., 2007. Exogenous proline mitigates the detrimental effects of salt stress more than exogenous betaine by increasing antioxidant enzyme activities. J. Plant Physiol. 164, 553–561.

Hrabak, E.M., Chan, C.W., Gribskov, M., et al., 2003. The Arabidopsis CDPK-SnRK superfamily of protein kinases. Plant Physiol. 132, 666–680.

Huang, G.-T., Ma, S.-L., Bai, L.-P., Zhang, L., Ma, H., Jia, P., et al., 2012. Signal transduction during cold, salt, and drought stresses in plants. Mol. Biol. Rep. 39, 969–987.

Hussain, S.S., Ali, M., Ahmad, M., Siddique, K.H.M., 2011. Polyamines: natural and engineered abiotic and biotic stress tolerance in plants. Biotechnol. Adv. 29, 300–311.

Iturriaga, G., Suárez, R., Nova-Franco, B., 2009. Trehalose metabolism: from osmoprotection to signaling. Int. J. Mol. Sci. 10, 3793–3810.

Jain, N.K., Roy, I., 2009. Effect of trehalose on protein structure. Protein. Sci. 18, 24–36.

Jouve, L., Hoffmann, L., Hausman, J.F., 2004. Polyamine, carbohydrate, and proline content changes during salt stress exposure of Aspen (*Populus tremula* L.): involvement of oxidation and osmoregulation metabolism. Plant Biol. 6, 74–80.

Jung, Y., Park, J., Choi, Y., Yang, J.G., Kim, D., Kim, B.G., et al., 2010. Expression analysis of proline metabolism-related genes from halophyte *Arabis stelleri* under osmotic stress conditions. J. Integr. Plant Biol. 52, 891–903.

Kaplan, F., Kopka, J., Haskell, D.W., Zhao, W., Schiller, K.C., Gatzke, N., et al., 2004. Exploring the temperature-stress metabolome of Arabidopsis. Plant Physiol. 136, 4159–4168.

Kasuga, J., Mizuno, K., Miyaji, N., Arakawa, K., Fujikawa, S., 2006. Role of intracellular contents to facilitate supercooling capability in beech (*Fagus crenata*) xylem parenchyma cells. CryoLetters 27, 305–310.

Kasuga, J., Arakawa, K., Fujikawa, S., 2007a. High accumulation of soluble sugars in deep supercooling Japanese white birch xylem parenchyma cells. New Phytol. 174, 569–579.

Kasuga, J., Mizuno, K., Arakawa, K., Fujikawa, S., 2007b. Anti-ice nucleation activity in xylem extracts from trees that contain deep supercooling xylem parenchyma cells. Cryobiology 55, 305–314.

Kaushal, N., Gupta, K., Bhandhari, K., Kumar, S., Thakur, P., Nayyar, H., 2011. Proline induces heat tolerance in chickpea (*Cicer arietinum* L.) plants by protecting vital enzymes of carbon and antioxidative metabolism. Physiol. Mol. Biol. Plants 17, 203–213.

Kawakami, A., Sato, Y., Yoshida, M., 2008. Genetic engineering of rice capable of synthesizing fructans and enhancing chilling tolerance. J. Exp. Bot. 59, 793–802.

Keller, F., Pharr, D.M., 1996. Metabolism of carbohydrates in sinks and sources: galactosyl-sucrose oligosaccharides. In: Zamski, E., Schaffer, A.A. (Eds.), Photoassimilate Distribution in Plants and Crops: Source-Sink Relationships. Marcel Dekker, New York, pp. 157–183.

Kempa, S., Krasensky, J., Dal Santo, S., Kopka, J., Jonak, C., 2008. A central role of abscisic acid in stress-regulated carbohydrate metabolism. PLoS One 3 (12), e3935.

Keunen, E., Peshev, D., Vangronsveld, J., Van den Ende, W., Cuypers, A., 2013. Plant sugars are crucial players in the oxidative challenge during abiotic stress: extending the traditional concept. Plant Cell Environ. 10.1111/pce.12061.

Kim, J.K., Bamba, T., Harada, K., Fukusaki, E., Kobayashi, A., 2007. Time-course metabolic profiling in Arabidopsis thaliana cell cultures after salt stress treatment. J. Exp. Bot. 58, 415–424.

Kishor, P.B.K., Hong, Z., Miao, G.-H., Hu, C.A., Verma, D.P.S., 1995. Overexpression of Δ^1-pyrroline-5-carboxylase synthetase increases proline production and confers osmotolerance in transgenic plants. Plant Physiol. 108, 1387–1394.

Kishor, P.B.K., Sangam, S., Amrutha, R.N., Laxmi, P.S., Naidu, K., Rao, K.R.S.S., 2005. Regulation of proline biosynthesis, degradation, uptake and transport in higher plants: its implications in plant growth and abiotic stress tolerance. Curr. Sci. 88, 424–438.

Knaupp, M., Mishra, K.B., Nedbal, L., Heyer, A.G., 2011. Evidence for a role of raffinose in stabilizing photosystem II during freeze–thaw cycles. Planta 234, 477–486.

Knight, M.R., Knight, H., 2012. Low-temperature perception leading to gene expression and cold tolerance in higher plants. New Phytol. 195, 737–751.

Knoblauch, M., Peters, W.S., 2010. Münch, morphology, microfluidics—our structural problem with the phloem. Plant Cell Environ. 33, 1439–1452.

Kobayashi, Y., Yamamoto, S., Minami, H., Kagaya, Y., Hattori, T., 2004. Differential activation of the rice sucrose nonfermenting1-related protein kinase2 family by hyperosmotic stress and abscisic acid. Plant Cell 16, 1163–1177.

Koch, K.E., 1996. Carbohydrates modulate gene expression in plants. Annu. Rev. Plant Physiol. Plant Mol. Biol. 47, 509–540.

Kolbe, A., Tiessen, A., Schluepmann, H., Paul, M., Ulrich, S., Geigenberger, P., 2005. Trehalose 6-phosphate regulates starch synthesis via posttranslational redox activation of ADPglucose pyrophosphorylase. Proc. Natl. Acad. Sci. USA 102, 11118–11123.

Kondrák, M., Marincs, F., Kalapos, B., Juhász, Z., Bánfalvi, Z., 2011. Transcriptome analysis of potato leaves expressing the trehalose-6-phosphate synthase gene of yeast. PLoS One 6, e23466.

Kondrák, M., Marincs, F., Antal, F., Juhász, Z., Bánfalvi, Z., 2012. Effects of yeast trehalose-6-phosphate synthase on gene expression and carbohydrate contents of potato leaves under drought stress conditions. BMC. Plant Biol. 12, 74—86.

Koonjul, P.K., Minhas, J.S., Nunes, C., Sheoran, I.S., Saini, H.S., 2005. Selective transcriptional down-regulation of anther invertases precedes the failure of pollen development in water-stressed wheat. J. Exp. Bot. 56, 179—190.

Koornneef, A., Pieterse, C.M.J., 2008. Cross talk in defence signaling. Plant Physiol. 146, 839—844.

Korn, M., Peterek, S., Mock, H.-P., Heyer, A.G., Hincha, D.K., 2008. Heterosis in the freezing tolerance, and sugar and flavonoid contents of crosses between Arabidopsis thaliana accessions of widely varying freezing tolerance. Plant Cell Environ. 31, 813—827.

Korn, M., Gärtner, T., Erban, A., Kopka, J., Selbig, J., Hincha, D.K., 2010. Predicting Arabidopsis freezing tolerance and heterosis in freezing tolerance from metabolite composition. Mol. Plant 3 (224), 235.

Kozlowski, T.T., Pallardy, S.G., 2002. Acclimation and adaptive responses of woody plants to environmental stresses. Bot. Rev. 68, 270—334.

Krapp, A., Stitt, M., 1995. An evaluation of direct and indirect mechanisms for the "sink- regulation" of photosynthesis in spinach: changes in gas exchange, carbohydrates, metabolites, enzyme activities and steady-state transcript levels after cold-girdling source leaves. Planta 195, 313—323.

Krasensky, J., Jonak, C., 2012. Drought, salt, and temperature stress-induced metabolic rearrangements and regulatory networks. J. Exp. Bot. 63, 1593—1608.

Kumar, R.R., Sharma, S.K., Goswami, S., Singh, G.P., Singh, R., Singh, K., et al., 2013. Characterization of differentially expressed stress-associated proteins in starch granule development under heat stress in wheat (*Triticum aestivum* L.). Indian. J. Biochem. Biophys. 501, 126—138.

Kumar, V., Shriram, V., Kishor, P.B.K., Jawali, N., Shitole, M.G., 2010. Enhanced proline accumulation and salt stress tolerance of transgenic *indica* rice by over-expressing P5CSF gene. Plant Biotechnol. Rep. 4, 37—48.

Kuznetsov, V.V., Shevyakova, N.I., 1997. Stress responses of tobacco cells to high temperature and salinity: proline accumulation and phosphorylation of polypeptides. Physiol. Plant 100, 320—326.

Lankin, V.Z., Konovalova, G.G., Tikhaze, A.K., Nedosugova, L.V., 2012. The influence of glucose on the free radical peroxidation of low density lipoproteins in vitro and in vivo. Biomed. Khim. 58, 339—352, in Russian.

Laura, M., Consonni, R., Locatelli, F., Fumagalli, E., Allavena, A., Coraggio, I., et al., 2010. Metabolic response to cold and freezing of *Osteospermum ecklonis* overexpressing *Osmyb4*. Plant Physiol. Biochem. 48, 764—771.

Lee, J.H., Yu, D.J., Kim, S.J., Choi, D., Lee, H.J., 2012. Intraspecies differences in cold hardiness, carbohydrate content and β-amylase gene expression of *Vaccinium corymbosum* during cold acclimation and deacclimation. Tree. Physiol. 32, 1533—1540.

Legay, S., Lefèvre, I., Lamoureux, D., Barreda, C., Luz, R.T., Gutierrez, R., et al., 2011. Carbohydrate metabolism and cell protection mechanisms differentiate drought tolerance and sensitivity in advanced potato clones (*Solanum tuberosum* L). Funct. Integr. Genomics 11, 275—291.

Lehmann, S., Funck, D., Szabados, L., Rentsch, D., 2010. Proline metabolism and transport in plant development. Amino Acids 39, 949—962.

Leshem, Y.Y., Kuiper, P.J.C., 1996. Is there a GAS (general adaptation syndrome) response to various types of environmental stress? Biol. Plant 38, 1—18.

Li, H.J., Yang, A.F., Zhang, X.C., Gao, F., Zhang, J.R., 2007. Improving freezing tolerance of transgenic tobacco expressing sucrose: sucrose 1-fructosyltransferase gene from *Lactuca sativa*. Plant Cell Tiss. Organ Cult. 89, 37—48.

Li, M.-H., Xiao, W.-F., Wang, S.-G., Cheng, G.-W., Cherubini, P., Cai, X.-H., et al., 2008a. Mobile carbohydrates in Himalayan treeline trees I. Evidence for carbon gain limitation but not for growth limitation. Tree. Physiol. 28, 1287—1296.

Li, M.H., Xiao, W.F., Shi, P., Wang, S.G., Zhong, Y.D., Liu, X.L., et al., 2008b. Nitrogen and carbon source–sink relationships in trees at the Himalayan treelines compared with lower elevations. Plant Cell Environ. 31, 1377–1387.

Li, P., Wind, J.J., Shi, X., Zhang, H., Hanson, J., Smeekens, S.C., et al., 2011a. Fructose sensitivity is suppressed in Arabidopsis by the transcription factor ANAC089 lacking the membrane-bound domain. Proc. Natl. Acad. Sci. 108, 3436–3441.

Li, X., Zhuo, J., Jing, Y., Liu, X., Wang, X., 2011b. Expression of a GALACTINOL SYNTHASE gene is positively associated with desiccation tolerance of *Brassica napus* seeds during development. J. Plant Physiol. 168, 1761–1770.

Li, X.Y., Liu, X., Yao, Y., Li, Y.H., Liu, S., He, C.Y., et al., 2013. Overexpression of Arachis hypogaea AREB1 gene enhances drought tolerance by modulating ROS scavenging and maintaining endogenous ABA content. Int. J. Mol. Sci. 14, 12827–12842.

Li, Y., Lee, K.K., Walsh, S., Smith, C., Hadingham, S., Sorefan, K., et al., 2006. Establishing glucose- and ABA-regulated transcription networks in Arabidopsis by microarray analysis and promoter classification using a relevance vector machine. Genome Res. 16, 414–427.

Liesche, J., Schulz, A., 2012. *In vivo* quantification of cell coupling in plants with different phloem-loading strategies. Plant Physiol. 159, 355–365.

Liesche, J., Martens, H.J., Schulz, A., 2011. Symplasmic transport and phloem loading in gymnosperm leaves. Protoplasma 248, 181–190.

Liu, H., Dai, X.Y., Xu, Y.Y., Chong, K., 2007. Over-expression of OsUGE-1 altered raffinose level and tolerance to abiotic stress but not morphology in Arabidopsis. J. Plant Physiol. 164, 1384–1390.

Liu, H., Ouyang, B., Zhang, J., Wang, T., Li, H., Zhang, Y., et al., 2012. Differential modulation of photosynthesis, signaling, and transcriptional regulation between tolerant and sensitive tomato genotypes under cold stress. PloS One 7, e50785.

Liu, X., Liu, S., Wu, J., Zhang, B., Li, X., Yan, Y., et al., 2013. Overexpression of Arachis hypogaea NAC3 in tobacco enhances dehydration and drought tolerance by increasing superoxide scavenging. Plant Physiol. Biochem. 70C, 354–359.

Livingston, D.P., Hincha, D.K., Heyer, A.G., 2009. Fructan and its relationship to abiotic stress tolerance in plants. Cell. Mol. Life. Sci. 66, 2007–2023.

Loreti, E., Poggi, A., Novi, G., Alpi, A., Perata, P., 2005. A genome-wide analysis of the effects of sucrose on gene expression in *Arabidopsis* seedlings under anoxia. Plant Physiol. 137, 1130–1138.

Lugan, R., Niogret, M.F., Leport, L., Guégan, J.P., Larher, F.R., Savouré, A., et al., 2010. Metabolome and water homeostasis analysis of Thellungiella salsuginea suggests that dehydration tolerance is a key response to osmotic stress in this halophyte. Plant J. 64, 215–229.

Lundmark, M., Cavaco, A.M., Trevanion, S., Hurry, V., 2006. Carbon partitioning and export in transgenic *Arabidopsis thaliana* with altered capacity for sucrose synthesis grown at low temperature: a role for metabolite transporters. Plant Cell Environ. 29, 1703–1714.

Mao, X., Zhang, H., Tian, S., Chang, X., Jing, R., 2010. TaSnRK2.4, an SNF1-type serine/threonine protein kinase of wheat (*Triticum aestivum* L.) confers enhanced multistress tolerance in *Arabidopsis*. J. Exp. Bot. 61, 683–696.

Matiolli, C.C., Tomaz, J.P., Duarte, G.T., Prado, F.M., Del Bem, L.E.V., Silveira, A.B., et al., 2011. The Arabidopsis bZIP gene AtbZIP63 is a sensitive integrator of transient abscisic acid and glucose signals. Plant Physiol. 157, 692–705.

Mattana, M., Biazzi, E., Consonni, R., Locatelli, F., Vannini, C., Provera, S., et al., 2005. Overexpression of Osmyb4 enhances compatible solute accumulation and increases stress tolerance of Arabidopsis thaliana. Physiol. Plant 125, 212–223.

McCormick, A.J., Cramer, M.D., Watt, D.A., 2008. Regulation of photosynthesis by sugars in sugarcane leaves. J. Plant Physiol. 165, 1817−1829.

McDowell, N.G., 2011. Mechanisms linking drought, hydraulics, carbon metabolism, and vegetation mortality. Plant Physiol. 155, 1051−1059.

Mittler, R., 2002. Oxidative stress, antioxidants and stress tolerance. Trends. Plant Sci. 7, 405−410.

Mittler, R., 2006. Abiotic stress, the field environment and stress combination. Trends. Plant Sci. 11, 15−19.

Mittler, R., Vanderauwera, S., Gollery, M., Van Breusegem, F., 2004. The reactive oxygen gene network of plants. Trends Plant Sci. 9, 490−498.

Mokhamed, A.M., Kuznetsov, V.L.V., Raldugina, G.N., Kholodova, V.P., 2006. Osmolyte accumulation in different rape genotypes under sodium chloride salinity. Russ. J. Plant Physiol. 53, 649−655.

Moore, B., Zhou, L., Rolland, F., Hall, Q., Cheng, W.-H., Liu, Y.-X., et al., 2003. Role of the Arabidopsis glucose sensor HXK1 in nutrient, light and hormonal signaling. Science 3000, 332−336.

Morin, X., Améglio, T., Ahas, R., Kurz-Besson, C., Lanta, V., Lebourgeois, F., et al., 2007. Variation in cold hardiness and carbohydrate concentration from dormancy induction to bud burst among provenances of three European oak species. Tree. Physiol. 27, 817−825.

Morsy, M.R., Jouve, L., Hausman, J.F., Hoffmann, L., Stewart, J.M., 2007. Alteration of oxidative and carbohydrate metabolism under abiotic stress in two rice (*Oryza sativa* L.) genotypes contrasting in chilling tolerance. J. Plant Physiol. 164, 157−167.

Møller, I.M., Jensen, P.E., Hansson, A., 2007. Oxidative modifications to cellular components in plants. Annu. Rev. Plant Biol. 58, 459−481.

Nägele, T., Heyer, A.G., 2013. Approximating subcellular organisation of carbohydrate metabolism during cold acclimation in different natural accessions of *Arabidopsis thaliana*. New Phytol. 198, 777−787.

Nägele, T., Kandel, B.A., Frana, S., Meissner, M., Heyer, A.G., 2011. A systems biology approach for the analysis of carbohydrate dynamics during acclimation to low temperature in *Arabidopsis thaliana*. FEBS J. 278, 506−518.

Nägele, T., Stutz, S., Hörmiller, I.I., Heyer, A.G., 2012. Identification of a metabolic bottleneck for cold acclimation in *Arabidopsis thaliana*. Plant J. 72, 102−114.

Naik, D., Dhanaraj, A.L., Arora, R., Rowland, L.J., 2007. Identification of genes associated with cold acclimation in blueberry (*Vaccinium corymbosum* L.) using a subtractive hybridization approach. Plant Sci. 173, 213−222.

Nanjo, T., Kobayashi, M., Yoshiba, Y., Kakubari, Y., Yamaguchi Shinozaki, K., Shinozaki, K., 1999. Antisense suppression of proline degradation improves tolerance to freezing and salinity in *Arabidopsis thaliana*. FEBS Lett. 461, 205−210.

Navrot, N., Rouhier, N., Gelhaye, E., Jacquot, J.P., 2007. Reactive oxygen species generation and antioxidant systems in plant mitochondria. Physiol. Plant 129, 185−195.

Nishikawa, F., Kato, M., Hyodo, H., Ikoma, Y., Sugiura, M., Yano, M., 2005. Effect of sucrose on ascorbate level and expression of genes involved in the ascorbate biosynthesis and recycling pathway in harvested broccoli florets. J. Exp. Bot. 56, 65−72.

Nishizawa, A., Yabuta, Y., Shigeoka, S., 2008. Galactinol and raffinose constitute a novel function to protect plants from oxidative damage. Plant Physiol. 147, 1251−1263.

Nishizawa-Yokoi, A., Yabuta, Y., Shigeoka, S., 2008. The contribution of carbohydrates including raffinose family oligosaccharides and sugar alcohols to protection of plant cells from oxidative damage. Plant Signal Behav. 3, 1016−1018.

Oliver, S.N., Van Dongen, J.T., Alfred, S.C., Mamun, E., Zhao, X., Saini, A.S., et al., 2005. Cold-induced repression of the rice anther-specific cell wall invertase gene *OSINV4* is correlated with sucrose accumulation and pollen sterility. Plant Cell Environ. 28, 1534−1551.

Oliver, S.N., Dennis, E.S., Dolferus, R., 2007. ABA regulates apoplastic sugar transport and is a potential signal for cold-induced pollen sterility in rice. Plant Cell Physiol. 48, 1319–1330.

Orozco-Cardenas, M., Ryan, C.A., 1999. Hydrogen peroxide is generated systemically in plant leaves by wounding and systemin via the octadecanoid pathway. Proc. Natl. Acad. Sci. USA 96, 6553–6557.

Palonen, P., 1999. Relationship of seasonal changes in carbohydrates and cold hardiness in canes and buds of three red raspberry cultivars. J. Am. Soc. Hortic. Sci. 124, 507–513.

Panikulangara, T.J., Eggers-Schumacher, G., Wunderlich, M., Stransky, H., Schöffl, F., 2004. Galactinol synthase1. A novel heat shock factor target gene responsible for heat-induced synthesis of raffinose family oligosaccharides in *Arabidopsis*. Plant Physiol. 136, 3148–3158.

Patton, A.J., Cunningham, S.M., Volenec, J.J., Reicher, Z.J., 2007. Differences in freeze tolerance of Zoysia grasses: II. Carbohydrate and proline accumulation. Crop Sci. 47, 2170–2181.

Pennycooke, J.C., Jones, M.L., Stushnoff, C., 2003. Down-regulating α-galactosidase enhances freezing tolerance in transgenic petunia. Plant Physiol. 133, 901–909.

Peshev, D., van den Ende, W., 2013. Sugars as antioxidants in plants. In: Tuteja, N., Gill, S.S. (Eds.), Crop Improvement Under Adverse Conditions. Springer-Verlag, Berlin, Heidelberg, p. 295.

Peters, S., Keller, F., 2009. Frost tolerance in excised leaves of the common bugle (*Ajuga reptans* L.) correlates positively with the concentrations of raffinose family oligosaccharides (RFOs). Plant Cell Environ. 32, 1099–1107.

Pospisilova, J., Haisel, D., Vankova, R., 2011. Responses of transgenic tobacco plants with increased proline content to drought and/or heat stress. Am. J. Plant Sci. 2, 318–324.

Price, J., Laxmi, A., St Martin, S.K., Jang, J.C., 2004. Global transcription profiling reveals multiple sugar signal transduction mechanisms in Arabidopsis. Plant Cell 16, 2128–2150.

Rawyler, A., Arpagaus, S., Braendle, R., 2002. Impact of oxygen stress and energy availability on membrane stability of plant cell. Ann. Bot. 90, 499–507.

Reidel, E.J., Rennie, E.A., Amiard, V., Cheng, L., Turgeon, R., 2009. Phloem loading strategies in three plant species that transport sugar alcohols. Plant Physiol. 149, 1601–1608.

Renaut, J., Planchon, S., Oufir, M., Hausman, J.-F., Hoffmann, L., Evers, D., 2009. Identification of proteins from potato leaves submitted to chilling temperature. In: Gusta, L.V., Wisniewski, M.E., Tanino, K. (Eds.), Plant Cold Hardiness. From Laboratory to the Field. CAB International, Wallingford, Oxfordshire, pp. 279–292.

Rennie, E.A., Turgeon, R., 2009. A comprehensive picture of phloem loading strategies. Proc. Natl. Acad. Sci. USA 106, 14162–14167.

Rohde, P., Hincha, D.K., Heyer, A.G., 2004. Heterosis in the freezing tolerance of crosses between two *Arabidopsis thaliana* accessions (Columbia-0 and C24) that show differences in non-acclimated and acclimated freezing tolerance. Plant J. 38, 790–799.

Roitsch, T., 1999. Source-sink regulation by sugar and stress. Curr. Opin. Plant Biol. 2, 198–206.

Rolland, F., Moore, B., Sheen, J., 2002. Sugar sensing and signaling in plants. Plant Cell 14 (Suppl), S185–S205.

Rolland, F., Baena-Gonzalez, E., Sheen, J., 2006. Sugar sensing and signaling in plants: conserved and novel mechanisms. Annu. Rev. Plant Biol. 57, 675–709.

Rook, F., Corke, F., Card, R., Munz, G., Smith, C., Bevan, M.W., 2001. Impaired sucrose-induction mutants reveal the modulation of sugar-induced starch biosynthetic gene expression by abscisic acid signalling. Plant J. 26, 421–433.

Rosa, M., Prado, C., Podazza, G., Interdonato, R., González, J.A., Hilal, M., et al., 2009. Soluble sugars—metabolism, sensing and abiotic stress: a complex network in the life of plants. Plant Signal Behav. 4, 388–393.

Ruan, Y.-L., Llewellyn, D.J., Furbank, R.T., 2003. Suppression of sucrose synthase expression represses cotton fibre cell initiation, elongation and seed development. Plant Cell 15, 952−964.

Ruan, Y.-L., Jin, Y., Yang, Y.-J., Li, G.-J., Boyer, J.S., 2010. Sugar input, metabolism, and signaling mediated by invertase: roles in development, yield potential, and response to drought and heat. Mol. Plant 3, 942−955.

Russell, J.W., Golovoy, D., Vincent, A.M., Mahendru, P., Olzmann, J.A., Mentzer, A., et al., 2002. High glucose-induced oxidative stress and mitochondrial dysfunction in neurons. FASEB J. 16, 1738−1748.

Sanchez, D.H., Siahpoosh, M.R., Roessner, U., Udvardi, M., Kopka, J., 2008. Plant metabolomics reveals conserved and divergent metabolic responses to salinity. Physiol. Plant 132, 209−219.

Santarius, K.A., Milde, H., 1977. Sugar compartmentation in frost-hardy and partially dehardened cabbage leaf cells. Planta 136, 163−166.

Saravitz, D.M., Pharr, D.M., Carter, T.E., 1987. Galactinol synthase activity and soluble sugars in developing seeds of four soybean genotypes. Plant Physiol. 83, 185−189.

Sato, Y., Masuta, Y., Saito, K., Murayama, S., Ozawa, K., 2011. Enhanced chilling tolerance at the booting stage in rice by transgenic overexpression of the ascorbate peroxidase gene, *OsAPXa*. Plant Cell Rep. 30, 399−406.

Schafleitner, R., Gaudin, A., Rosales, R.O.G., Aliaga, C.A.A., Bonierbale, M., 2005. Proline accumulation and real-time PCR expression analysis of genes encoding enzymes of proline metabolism in relation to drought tolerance in Andean potato. Acta. Physiol. Plant 29, 19−26.

Schafleitner, R., Rosales, R.O.G., Gaudin, A., Aliaga, C.A.A., Martinez, G.N., Marca, L.R.T., et al., 2007. Capturing candidate drought tolerance traits in two native Andean potato clones by transcription profiling of field grown plants under water stress. Plant Physiol. Biochem. 45, 673−690.

Schneider, T., Keller, F., 2009. Raffinose in chloroplasts is synthesized in the cytosol and transported across the chloroplast envelope. Plant Cell Physiol. 50, 2174−2182.

Sekmen, A.H., Türkan, I., Takio, S., 2007. Differential responses of antioxidative enzymes and lipid peroxidation to salt stress in salt-tolerant *Plantago maritima* and salt-sensitive *Plantago media*. Physiol. Plant 131, 399−411.

Semel, Y., Schauer, N., Roessner, U., Zamir, D., Fernie, A.R., 2007. Metabolite analysis for the comparison of irrigated and non-irrigated field grown tomato of varying genotype. Metabolomics 3, 289−295.

Sevanto, S., McDowell, N.G., Dickman, L.T., Pangle, R., Pockman, W.T., 2013. How do trees die? A test of the hydraulic failure and carbon starvation hypotheses. Plant Cell Environ. 37, 153−161.

Shao, H.-B., Chu, L.-Y., Lu, Z.-H., Kang, C.-M., 2008a. Primary antioxidant free radical scavenging and redox signaling pathways in higher plant cells. Int. J. Biol. Sci. 4, 8−14.

Shao, H.B., Shao, M.A., Liang, Z.S., 2006. Osmotic adjustment comparison of 10 wheat (*Triticum aestivum* L.) genotypes at soil water deficits. Colloids. Surf. B. Biointerfaces. 47, 132−139.

Shao, H.B., Chu, L.Y., Shao, M.A., Jaleel, C.A., Mi, H.M., 2008b. Higher plant antioxidants and redox signaling under environmental stresses. C. R. Biol. 331, 433−441.

Shen, B., Jensen, R.G., Bohnert, H.J., 1997. Increased resistance to oxidative stress in transgenic plants by targeting mannitol biosynthesis to chloroplasts. Plant Physiol. 113, 1177−1183.

Shin, R., Alvarez, S., Burch, A.Y., Jez, J.M., Schachtman, D.P., 2007. Phosphoproteomic identification of targets of the Arabidopsis sucrose nonfermenting-like kinase SnRK2.8 reveals a connection to metabolic processes. Proc. Natl. Acad. Sci. USA 104, 6460−6465.

Sin'kevich, M.S., Sabelnikova, E.P., Deryabin, A.N., Astakhova, N.V., Dubinina, I.M., Burachanova, E.A., et al., 2008. The changes in invertase activity and the content of sugars in the course of adaptation of potato plants to hypothermia. Russ. J. Plant Physiol. 55, 449−454.

Sin'kevich, M.S., Deryabin, A.N., Trunova, T.I., 2009. Characteristics of oxidative stress in potato plants with modified carbohydrate metabolism. Russ. J. Plant Physiol. 56, 168−174.

Sin'kevich, M.S., Naraykina, N.V., Trunova, T.I., 2010. Involvement of sugars in the antioxidant defense against paraquat-induced oxidative stress in potato transformed with yeast invertase gene. Dokl. Biol. Sci. 434, 338–340.

Smeekens, S., Ma, J.K., Hanson, J., Rolland, F., 2010. Sugar signals and molecular networks controlling plant growth. Curr. Opin. Plant Biol. 13, 274–279.

Smirnoff, N., Cumbes, Q.J., 1989. Hydroxyl radical scavenging activity of compatible solutes. Phytochemistry 37, 1057–1060.

Smith, A.M., Stitt, M., 2007. Coordination of carbon supply and plant growth. Plant Cell Environ. 30, 1126–1149.

Soltész, A., Smedley, M., Vashegyi, I., Galiba, G., Harwood, W., Vágújfalvi, A., 2013. Transgenic barley lines prove the involvement of TaCBF14 and TaCBF15 in the cold acclimation process and in frost tolerance. J. Exp. Bot. 64, 1849–1862.

Stiller, I., Dulai, S., Kondrák, M., Tarnai, R., Szabó, L., Toldi, O., et al., 2008. Effects of drought on water content and photosynthetic parameters in potato plants expressing the trehalose-6-phosphate synthase gene of *Saccharomyces cerevisiae*. Planta 227, 299–308.

Stoyanova, S., Geuns, J., Hideg, E., Van Den Ende, W., 2011. The food additives inulin and stevioside counteract oxidative stress. Int. J. Food Sci. Nutr. 62, 207–214.

Strand, Å., Hurry, V., Gustafsson, P., Gardeström, P., 1997. Development of *Arabidopsis thaliana* leaves at low temperatures releases the suppression of photosynthesis and photosynthetic gene expression despite the accumulation of soluble carbohydrates. Plant J. 12, 605–614.

Sturm, A., 1999. Invertases: primary structures, functions and roles in plant development and sucrose partitioning. Plant Physiol. 121, 1–7.

Sturm, A., Jang, G.-Q., 1999. The sucrose-cleaving enzymes of plants are crucial for development, growth and carbon partitioning. Trends. Plant Sci. 4, 401–407.

Stushnoff, C., Remmele, R.L., Essensee, V., McNeil, M., 1993. Low temperature induced biochemical mechanisms: implications for cold acclimation and de-acclimation. In: Jackson, M.B., Black, C.R. (Eds.), Interacting Stresses on Plants in a Changing Climate, 116. NATO ASI series, pp. 647–657.

Sulmon, C., Gouesbet, G., Couee, I., El Amrani, A., 2004. Sugar-induced tolerance to atrazine in Arabidopsis seedlings: interacting effects of atrazine and soluble sugars on psbA mRNA and D1 protein levels. Plant Sci. 167, 913–923.

Sun, J., Chen, S.-L., Dai, S.-X., Wang, R.-G., Li, N.-Y., Shen, X., et al., 2009. Ion flux profiles and plant ion homeostasis control under salt stress ion flux profiles and plant ion homeostasis control under salt stress. Plant Signal Behav. 4, 261–264.

Sundaresan, S., Sudhakaran, P.R., 1995. Water stress-induced alterations in the proline metabolism of drought-susceptible and drought tolerant cassava (*Manihot esculenta*) cultivars. Physiol. Plant 94, 635–642.

Sunkar, R., Zhu, J.K., 2004. Novel and stress-regulated microRNAs and other small RNAs from Arabidopsis. Plant Cell 16, 2001–2019.

Sunkar, R., Kapoor, A., Zhu, J.K., 2006. Posttranscriptional induction of two Cu/Zn superoxide dismutase genes in Arabidopsis is mediated by downregulation of miR398 and important for oxidative stress tolerance. Plant Cell 18, 2051–2065.

Svensson, J.T., Crosatti, C., Campoli, C., Bassi, R., Stanca, A.M., Close, T.J., et al., 2006. Transcriptome analysis of cold acclimation in barley albina and xantha mutants. Plant Physiol. 141, 257–270.

Szabados, L., Savoure, A., 2010. Proline: a multifunctional amino acid. Trends. Plant Sci. 15, 89–97.

Tabaei-Aghdaei, R.S., Pearce, R.S., Harrison, P., 2003. Sugars regulate cold-induced gene expression and freezing-tolerance in barley cell cultures. J. Exp. Bot. 54, 1565–1575.

Taji, T., Ohsumi, C., Iuchi, S., Seki, M., Kasuga, M., Kobayashi, M., et al., 2002. Important roles of drought- and cold-inducible genes for galactinol synthase in stress tolerance in *Arabidopsis thaliana*. Plant J. 29, 417–426.

Theerawitaya, C., Boriboonkaset, T., Cha-Um, S., Supaibulwatana, K., Kirdmanee, C., 2012. Transcriptional regulations of the genes of starch metabolism and physiological changes in response to salt stress rice (Oryza sativa L.) seedlings. Physiol. Mol. Biol. Plants 18, 197–208.

Thibaud, M.C., Gineste, S., Nussaume, L., Robaglia, C., 2004. Sucrose increases pathogenesis-related PR-2 gene expression in *Arabidopsis thaliana* through an SA-dependent but NPR1-independent signaling pathway. Plant Physiol. Biochem. 42, 81–88.

Thomashow, M.F., 1999. Plant cold acclimation, freezing tolerance genes and regulatory mechanisms. Annu. Rev. Plant Physiol. Plant Mol. Biol. 50, 571–599.

Traverso, J.A., Vignols, F., Cazalis, R., Pulido, A., Sahrawy, M., Cejudo, F.J., et al., 2007. PsTRX*h*1 and PsTRX*h*2 are both pea *h*-type thioredoxins with antagonistic behavior in redox imbalances. Plant Physiol. 143, 300–311.

Turgeon, R., 2010. The role of phloem loading reconsidered. Plant Physiol. 152, 1817–1823.

Turgeon, R., Wolf, S., 2009. Phloem transport: cellular pathways and molecular trafficking. Annu. Rev. Plant Biol. 60, 207–221.

Uemura, M., Steponkus, P.L., 2003. Modification of the intracellular sugar content alters the incidence of freeze-induced membrane lesions of protoplasts isolated from *Arabidopsis thaliana* leaves. Plant Cell Environ. 26, 1083–1096.

Umezawa, T., Yoshida, R., Maruyama, K., Yamaguchi-Shinozaki, K., Shinozaki, K., 2004. SRK2C, a SNF1-related protein kinase 2, improves drought tolerance by controlling stress-responsive gene expression in Arabidopsis thaliana. Proc. Natl. Acad. Sci. USA 101, 17306–17311.

Urano, K., Maruyama, K., Ogata, Y., Morishita, Y., Takeda, M., Sakurai, N., et al., 2009. Characterization of the ABA-regulated global responses to dehydration in Arabidopsis by metabolomics. Plant J. 57, 1065–1078.

Usadel, B., Blasing, O.E., Gibon, Y., Retzlaff, K., Hohne, M., Gunther, M., et al., 2008. Global transcript levels respond to small changes of the carbon status during progressive exhaustion of carbohydrates in Arabidopsis rosettes. Plant Physiol. 146, 1834–1861.

Valliyodan, B., Nguyen, H.T., 2006. Understanding regulatory networks and engineering for enhanced drought tolerance in plants. Curr. Opin. Plant Biol. 9, 189–195.

Valluru, R., van den Ende, W., 2008. Plant fructans in stress environments: emerging concepts and future prospects. J. Exp. Bot. 59, 2905–2916.

Van den Ende, W., Valluru, R., 2009. Sucrose, sucrosyl oligosaccharides, and oxidative stress: scavenging and salvaging? J. Exp. Bot. 60, 9–18.

Vasquez-Robinet, C., Mane, S.P., Ulanov, A.V., Watkinson, J.I., Stromberg, V.K., De Koeyer, D., et al., 2008. Physiological and molecular adaptations to drought in Andean potato genotypes. J. Exp. Bot. 59, 2109–2123.

Volk, G.M., Haritatos, E.E., Turgeon, R., 2003. Galactinol synthase gene expression in melon. J. Am. Soc. Hort. Sci. 128, 8–15.

Wang, H.-L., Lee, P.-D., Chen, W.-L., Huang, D.-J., Su, J.-C., 2000. Osmotic stress-induced changes of metabolism in cultured sweet potato cells. J. Exp. Bot. 51, 1991–1999.

Wang, X., Chang, L., Wang, B., Wang, D., Li, P., Wang, L., et al., 2013. Comparative proteomics of Thellungiella halophila leaves under different salinity revealed chloroplast starch and soluble sugar accumulation played important roles in halophyte salt tolerance. Mol. Cell. Proteomics 12, 2174–2195.

Waters, E.R., Aevermann, B.D., Sanders-Reed, Z., 2008. Comparative analysis of the small heat shock proteins in three angiosperm genomes identifies new subfamilies and reveals diverse evolutionary patterns. Cell Stress Chaperones 13, 127–142.

Weber, H., Borisjuk, L., Heim, U., Buchner, P., Wobus, U., 1995. Seed coat-associated invertases of fava bean control both unloading and storage functions: cloning of cDNAs and cell type-specific expression. Plant Cell 7, 1835–1846.

Welling, A., Palva, E.T., 2008. Molecular control of cold acclimation in trees. Physiol. Plant 127, 167−181.

Whittaker, A., Bochicchio, A., Vazzana, C., Lindsey, G., Farrant, J., 2001. Changes in leaf hexokinase activity and metabolite levels in response to drying in the desiccation-tolerant species *Sporobolus staphianus* and *Xerophyta viscosa*. J. Exp. Bot. 52, 961−969.

Wolff, S.P., Dean, R.T., 1987. Glucose autoxidation and protein modification: the potential role of autoxidative glycosylation in diabetes. Biochem J. 245, 243−250.

Wood, A.J., 2005. Eco-physiological adaptations to limited water environments. In: Jenks, M.A., Hasegawa, P.M. (Eds.), Plant Abiotic Stress. Blackwell Publishing, pp. 1−13.

Xiang, L., Le Roy, K.L., Bolouri-Moghaddam, M.-R., Vanhaecke, M., Lammens, W., Rolland, F., et al., 2011. Exploring the neutral invertase-oxidative stress defence connection in *Arabidopsis thaliana*. J. Exp. Bot. 62, 3849−3862.

Xin, Z., Browse, J., 2000. Cold comfort farm: the acclimation of plants to freezing temperatures. Plant Cell Environ. 23, 893−902.

Xin, Z., Mandaokar, A., Chen, J., Last, R.L., Browse, J., 2007. Arabidopsis ESK1 encodes a novel regulator of freezing tolerance. Plant J. 49, 786−799.

Yamaguchi-Shinozaki, K., Shinozaki, K., 2006. Transcriptional regulatory networks in cellular responses and tolerance to dehydration and cold stresses. Annu. Rev. Plant Biol. 57, 781−803.

Yang, J., Zhang, J., 2006. Grain filling of cereals under soil drying. New Phytol. 169, 223−236.

Yang, J., Zhang, J., Wang, Z., Zhu, Q., Liu, L., 2004. Activities of fructan- and sucrose-metabolizing enzymes in wheat stems subjected to water stress during grain filling. Planta 220, 331−343.

Yazdi-Samadi, B., Rinne, R.W., Seif, R.D., 1977. Components of developing soybean seeds: oil, protein, sugars, starch, organic acids, and amino acids. Agron. J. 69, 481−486.

Zhai, Y., Wang, Y., Li, Y., Lei, T., Yan, F., Su, L., et al., 2013. Isolation and molecular characterization of GmERF7, a soybean ethylene-response factor that increases salt stress tolerance in tobacco. Gene 513, 174−183.

Zhang, C.K., Turgeon, R., 2009. Downregulating the sucrose transporter VpSUT1 in *Verbascum phoeniceum* does not inhibit phloem loading. Proc. Natl. Acad. Sci. USA 106, 18849−18854.

Zhang, H., Mao, X., Jing, R., 2011. SnRK2 acts within an intricate network that links sucrose metabolic and stress signaling in wheat. Plant Signal Behav. 6, 652−654.

Zhuo, C., Wang, T., Lu, S., Zhao, Y., Li, X., Guo, Z., 2013. A cold responsive galactinol synthase gene from Medicago falcata (MfGolS1) is induced by myo-inositol and confers multiple tolerances to abiotic stresses. Physiol Plant. 149 (1), 67−78.

Zuther, E., Büchel, K., Hundertmark, M., Stitt, M., Hincha, D.K., Heyer, A.G., 2004. The role of raffinose in the cold acclimation response of Arabidopsis thaliana. FEBS Lett. 576, 169−173.

Role of Glucosinolates in Plant Stress Tolerance

P.S. Variyar, A. Banerjee, Jincy J. Akkarakaran and P. Suprasanna

12.1 Introduction

Plants elicit multiple responses when exposed to a complex array of biotic (e.g., pathogen infection and herbivore feeding) and abiotic (e.g., nutrient levels and light conditions) stress factors. These stress factors induce signaling cascades that activate ion channels, kinases, production of reactive oxygen species (ROS), and accumulation of hormones such as salicylic acid (SA), ethylene (ET), jasmonic acid (JA) and abscisic acid (ABA) (Mittler, 2006; Jain, 2013). These signals eventually induce expression of specific subsets of defense genes that produce an overall defense response (Mittler et al., 2004; Mantri et al., 2011; Arbona et al., 2013). Activation of defense systems affects both the primary and secondary metabolism resulting in a substantial and significant variation in plant metabolome within and between species. Chemical defenses form a part of the plant's inherent immune system. Plants produce a wide diversity of secondary metabolites that play a prominent role in defense against herbivores and pathogens. Some of them also act as defense against abiotic stress and in communication with its own species and with other organisms. Plant secondary chemistry is phenotypically plastic and varies in response to both biotic and abiotic factors. The major classes of secondary metabolites include the terpenoids and phenolics as well as the nitrogen (N)- and sulfur (S)-containing compounds synthesized primarily from amino acids.

Brassicaceae plants that include cruciferous vegetables such as cabbage, broccoli, cauliflower, kale, etc. are some of the most popular vegetables consumed the world over and considered to be a good source of bioactive phytochemicals. Sulfur-containing glucosides — glucosinolates (GSLs) — are one of the most important phytochemicals of *Brassica* vegetables responsible for their characteristic flavor and odor (Fahey et al., 2001; Martínez-Ballesta et al., 2013). Structurally, they are anions made up of thiohydroximates containing an S-linked β-glucopyranosyl residue and an O-linked sulfate residue with a variable amino acid-derived side chain. Unlike the other major classes of natural plant products, glucosinolates comprise a relatively small but diverse group of secondary metabolites that are generally limited to the species of the order Brassicales. These compounds represent a large chemical family that includes over 130 different compounds with varying structural subgroups (Clarke, 2010). They are largely responsible for the nutraceutical and pharmacological value of *Brassica* vegetables and have been implicated in defense against insects and pathogens and thus possess a bio-protective role. Efforts to improve specific quality attributes of plant foods, for example GSL, through breeding for quantitative food processing traits, are both promising and challenging (Hennig et al., 2013; Banerjee et al., 2014).

P. Ahmad (Ed): *Emerging Technologies and Management of Crop Stress Tolerance, Volume 1.*
DOI: http://dx.doi.org/10.1016/B978-0-12-800876-8.00012-6

The level of glucosinolate metabolites has been shown to be induced upon root colonization by *Trichoderma* (Brotman et al., 2013). Following tissue damage, GSLs are hydrolyzed *in vivo* by endogenous enzymes, myrosinase, to unstable aglycones that further rearrange to a variety of products including isothiocyanates, thiocyanates, and nitriles, the nature of which depends upon the condition of hydrolysis and the structure of the GSLs (Halkier and Gershenzon, 2006). GSLs and their hydrolytic products have been documented to have significant antimicrobial and insecticidal activities (Aires et al., 2009). These compounds are known to accumulate in *Brassica* tissues after infestation by various pathogens restricting either the spread of fungal infection or inhibiting subsequent infections. Profound (and species-specific) temporal changes in GSLs have also been reported during insect herbivory (Yan and Chen, 2007). Glucosinolate structure and levels have been shown to influence host plant suitability for generalist and specialist herbivore and their levels were altered in response to herbivory (Halkier and Gershenzon, 2006).

The concentration and type of GSLs and their hydrolysis in plants has been shown to be regulated by genetic fluctuation, by environmental factors as well as by developmental cues (Martínez-Ballesta et al., 2013). Changes in total as well as different GSL subgroups have been reported in broccoli as a result of changes in salinity suggesting a role for these compounds in the leaf water response (Martínez-Ballesta et al., 2013). Induction of GSLs in broccoli mediated by ultraviolet-B (UV-B) was found to be associated with up-regulation of genes responsive to fungal and bacterial pathogens, thus demonstrating their role as stress alleviators (Mewis et al., 2012a). These compounds are induced in response to plant signaling molecules such as SA, JA, and methyl jasmonate (MeJA), the nature and extent of elicitation being dependent on the type of elicitors (Yan and Chen, 2007). Levels of these compounds are reported to be effected under temperature and heavy metal stress and also by post-harvest storage conditions (Yan and Chen, 2007). Thus, while selenium was found to affect the content of glucosinolates in a concentration-dependent manner, cadmium stress produced no change in glucosinolate production in *B. rapa* (Kim and Juvic, 2011; Jakovljević et al., 2013). Glucosinolate concentration also increased as a result of temperature stress showing seasonal variation in *Brassica* plants (Martínez-Ballesta et al., 2013).

Van Dam et al. (2009) summarized information on the GSL levels in both root and shoot in the same plant. The authors analyzed constitutive root and shoot glucosinolates of 29 plant species, which showed that roots have higher concentrations and a greater diversity of glucosinolates than shoots. Roots have significantly higher levels of the aromatic 2-phenylethyl glucosinolate, possibly related to the greater effectiveness and toxicity of its hydrolysis products in soil. It was also seen that in shoots, the most dominant indole glucosinolate is indol-3-ylglucosinolate, whereas in roots, its methoxy derivatives are dominated. The regulation of GSL metabolism at different levels and the diverse physiological function of their metabolites indicate a complex metabolic network. Studies at the genetic level have shown that GSL metabolism interacts with cellular signaling and metabolic pathways and is regulated at different levels. The extent of GSL hydrolysis that effects interaction with microbes and herbivores is also known to be controlled at the genetic level (Sønderby et al., 2010). There is, however, a lack of understanding at the molecular level on the functional aspects such as signaling transduction pathways, control at transcriptional, translational and post-translational levels, subcellular compartmentation, and interaction with many other metabolic pathways. Further studies are needed to understand the sophisticated signaling network that connects environmental factors with GSL metabolism. Knowledge on these and related aspects can aid in metabolic engineering of *Brassica* crops for better quality, nutrition, and disease resistance.

Considerable interest in optimizing GSL content and composition for plant protection and human health has made GSLs a dynamic area in plant metabolomic research.

12.2 Glucosinolate structure, isolation, and analysis

The first general structure of GSL was proposed in 1897 by Gadamer (Fahey et al., 2001), which proposed that the side chain is linked to the nitrogen rather than to the carbon atom (Figure 12.1). Several approaches have been proposed for classification of GSLs into subgroups. The most common approach is based on the nature of the biosynthetic precursor amino acids. Another approach classifies them into aliphatic, aromatic, and indolic, or aliphatic, benzenic, and indolic derivatives. They are also grouped based on their tendency for forming specific breakdown products. Specific nomenclature of individual GSLs is generally based on naming the entire anionic structure (the central carbon (C) as well as the connected, substituted S and N) as glucosinolate and adding the systematic name of the side chain as a radical.

Glucosinolates are known to be regulated both developmentally and environmentally in various organs and tissues depending on the type of biotic and abiotic stresses. Depending on the developmental stage, tissue, and photoperiod, the distribution pattern of GSLs differs between species and ecotypes as well as between and within individual plants (Table 12.1; Martínez-Ballesta et al., 2013). Tissue-level glucosinolate accumulation has a major influence on its hydrolysis to bioactive products. Vascular tissue has been found to be the site of glucosinolate biosynthesis with endoplasmic reticulum as the subcellular location of GSL biosynthetic enzymes. Plants containing GSLs always possess a thioglucosidase called myrosinase that catalyzes the degradation of GSL substrates when plant tissue is disrupted as a result of wounding or insect and pathogen attack (Sønderby et al., 2010). The majority of the myrosinase enzymes act on multiple GSL substrates, although some of them have high specificity. Myrosinases are localized in specific cells named myrosin cells or myrosin idioblasts that contain protein-rich vacuolar-type structures termed myrosin grains. Glucosinolates, however, have been reported to be localized in vacuoles in non-specific cells together with ascorbic acid, which modulates myrosinase activity.

Substantial degradation of GSLs by myrosinase enzymes when extracted with cold organic solvent necessitates the use of hot aqueous alcohols such as methanol:water (70:30) for their isolation from plant materials (Clarke, 2010). This process denatures the enzyme and prevents hydrolysis of these compounds thereby facilitating their quantitative isolation. A prior separation into groups normally precedes their identification and quantification by instrumental methods. The presence of

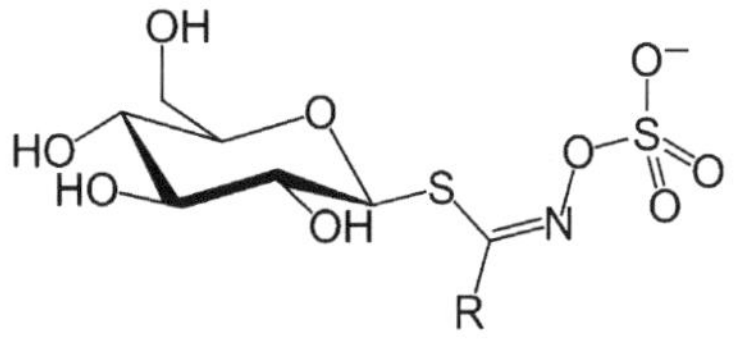

FIGURE 12.1

Structure of glucosinolate.

Table 12.1 Glucosinolates Found in Different Food Sources

Trivial Name	R Side Chain	Food Source
Glucocapparin	Methyl	Capers
Glucolepidin	Ethyl	Radish
—	Propyl	Cabbage
Glucoputranjivin	Isopropyl	Radish
Sinigrin	2-Propenyl	Cabbage
Glucoiberin	3-Methylsulfinylpropyl	Cabbage
Glucoibervirin	3-Methylthiopropyl	Cabbage
Glucocheirolin	3-Methylsulfonylpropyl	Cow's milk
Glucocapparisflexuosain	Butyl	Cabbage
Gluconapin	3-Butenyl	Cabbage
Progoitrin	(2R)-2-Hydroxy-3-butenyl	Cabbage
Epiprogoitrin	(2S)-2-Hydroxy-3-butenyl	Sea kale
Glucoerucin	4-Methylthiobutyl	Cabbage
Glucoraphanin	4-Methylsulfinylbutyl	Broccoli
Glucoerysolin	4-Methylsulfonylbutyl	Cabbage
Dehydroerucin	4-Methylthiobut-3-enyl	Daikon's radish
Glucoraphenin	4-Methylsulfinylbut-3-enyl	Radish
Glucobrassicanapin	4-Pentenyl	Chinese cabbage
Glucoberteroin	5-Methylthiopentyl	Cabbage
Glucoalyssin	5-Methylsulfinylpentyl	Rocket
Gluconapoleiferin	2-Hydroxy-pent-4-enyl	Swede
Glucosiberin	7-Methylsulfinylheptyl	Watercress
Glucohirsutin	8-Methylsulfinyloctyl	Watercress
4-Hydroxyglucobrassicin	4-Hydroxy-3-indolylmethyl	Cabbage
Glucobrassicin	3-Indolylmethyl	Cabbage
4-Methoxyglucobrassicin	4-Methoxy-3-indolylmethyl	Cabbage
Neoglucobrassicin	N-Methoxy-3-indolylmethyl	Cabbage
Glucotropaeolin	Benzyl	Cabbage
Glucosinalbin	p-Hydroxybenzyl	Mustard
Gluconasturtiin	2-Phenylethyl	Cabbage
Glucobarbarin	(2S)-2-Hydroxy-2-phenylethyl	Land cress
Glucosibarin	(2R)-2-Hydroxy-2-phenylethyl	White mustard

sulfate groups facilitates binding of these compounds to an anion exchange column and thus allows separation of either the intact GSLs or "desulfo" derivatives after enzymatic desulfation (Clarke, 2010). Direct analysis of volatile isothiocyanates and nitriles produced from GSLs by gas chromatography-mass spectrometry (GC-MS) can also provide proof of the presence of corresponding GSL in intact plants. Use of high-pressure liquid chromatography-mass spectrometry (HPLC-MSn) for detection of intact GSLs in crude extracts is currently a powerful routine method comparable in specificity with the classical methods of identification (Clarke, 2010). Capillary electrophoresis for simultaneous quantification of GSLs and their hydrolysis products has also been reported (Clarke, 2010). Use of modern MSn equipment with ion traps allows for highly sophisticated analysis of side chain structures and validation of elucidated GSL structures. Even with highly sophisticated MS-based detection methods, comparison of chromatographic retention time with authentic standard and one additional characteristic property such as retention time in a different chromatographic system, a characteristic UV spectrum, a mass spectrum, or nuclear magnetic resonance (NMR) data is a must to suggest a tentative identification of a given GSL.

12.3 **Biosynthesis of glucosinolates**

Biosynthesis of GSLs involves three independent stages, namely: (1) chain elongation of selected precursor amino acids (mainly methionine) by addition of methylene groups; (2) formation of core glucosinolate structure by reconfiguration of the amino acid moiety; and (3) secondary modification of the amino acid side chain by hydroxylations, methylations, oxidations, or desaturations. While the construction of core anionic structure from amino acids involves a number of common steps, a number of diverse steps are involved in formation of side chain and other diversifications. Aliphatic GSLs are derived from alanine, leucine, isoleucine, valine, and methionine, while benzenic GSLs are formed from phenylalanine and tryptophan and indolic GSLs from tryptophan (Sønderby et al., 2010).

Synthesis of the core GSL structure is achieved in five steps (Figure 12.2). The first step involves oxidation of precursor amino acids to aldoximes by side chain-specific cytochrome P450 monooxygenase of the CYP79 family. Further oxidation by cytochrome P450 of the CYP83 family leads to aci-nitro compounds or nitrile oxides. The nitro compounds formed are strong

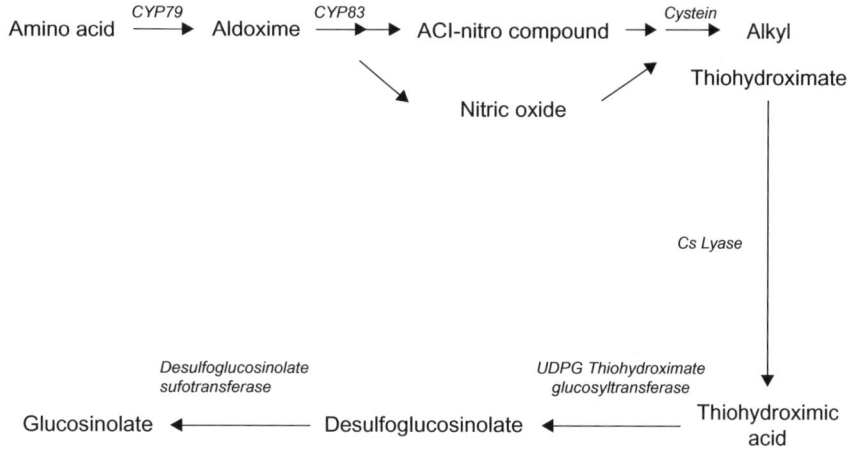

Biosynthesis of GLS core structure

FIGURE 12.2

Biosynthesis of glucosinolate core structure.

electrophiles that react spontaneously with thiols to form S-alkylthiohydroximate conjugates that further undergo cleavage into unstable thiohydroximates, pyruvate, and ammonia by the action of a C−S lyase. Glucosyl transferase catalyses thiohydroximate-specific S-glycosylation. The final step is the 3′-phosphoadenosine 5′-phosphosulfate-dependent sulfation of desulfoglucosinolates (Sønderby et al., 2010).

The amino acid elongation is similar to the valine-to-leucine conversion and involves five steps which include initial and final transamination, acetyl-CoA condensation, isomerization, and oxidative decarboxylation. Methylthioalkylmalate (MAM) synthases that catalyze the condensation reaction have been characterized in *Arabidopsis* and *Eruca sativa*. Methionine side chain elongation occurs in the chloroplast and elongated α-keto acid can either be transaminated and enter the core GSL pathway or undergo additional elongation steps with insertion of up to nine methylene units. The variation in side chain length of methionine-derived GSL is controlled by three partially redundant MAM genes (Sønderby et al., 2010).

Secondary modification of the side chains involving various types of oxidations, eliminations, alkylations, and esterifications is generally considered as the final stage in GSL synthesis. An extensive natural variation of aliphatic glucosinolates has been noted in *Arabidopsis* with two α-ketogluterate-dependent dioxygenases controlling the production of alkenyl and hydroxyalkyl GSLs (Kliebenstein et al., 2001).

An interdependent metabolic control of aliphatic and indolyl GSL branches has been proposed indicating a homeostatic control of GSL synthesis. This is achieved by a reciprocal negative feedback regulation between both the branches using intermediates or end products of glucosinolate biosynthesis as inhibitors. Limited NADPH supply has also been proposed for the interdependence of the two pathways wherein inhibition of one branch would lead to increased NADPH availability for the other. In addition, side chain elongation can lead to extra yield of NADH that can be converted to NADPH via the malate dehydrogenase and maleic enzyme reactions. Thus, side chain elongations can provide NADPH independently of the pentose-phosphate pathway thereby increasing GSL production. Considerable variation is thus noted in the total as well as individual GSL content of methionine-derived and indolyl GSL in leaves and seeds, respectively (Grubb and Abel, 2006).

12.4 Role of glucosinolates in stress alleviation

Loss of cellular integrity as a consequence of stress induced by wounding, insect, or pathogen attack leads to hydrolysis of GSLs by the enzyme myrosinase. GSLs and their hydrolytic products are frequently investigated for their role as a plant defense system against insects, herbivores, and certain microbial pathogens. It has been shown that infection with fungal pathogen can induce local synthesis of myrosinase and the possibility of such a mechanism under other stress response is also proposed. Environmental factors influence secondary metabolism as plants under stress produce more secondary metabolites, more so as the growth is often limited more than in photosynthesis, and carbon fixation is predominantly invested to secondary metabolite production (Endara and Coley, 2011). It has been very well reported that environmental factors, such as light (Engelen-Eigles et al., 2006), temperature (Velasco et al., 2007), salinity (Qasim et al., 2003; López-Berenguer et al., 2009), water (Champolivier and Merrien, 1996; Rask et al., 2000), CO_2 (Schonhof et al., 2007a), and drought (Radovich et al., 2005) may affect glucosinolate levels (Table 12.2).

Table 12.2 Impact of Abiotic Stress on Glucosinolate Accumulation in Different *Brassica* Plant Species

Plant Species	Glucosinolate Content	Stress	Treatment Condition
Brassica oleracea L. var. *italic*	Increase	Salinity	NaCl (40, 80 mM), during 2 weeks
Brassica rapa L.	Increase		NaCl (20, 40, 60 mM), during 5 days
Brassica campestris L. ssp. *chinensis* var. *communis*			NaCl (50 and 100 mM for 2 weeks)
Brassica oleracea L. var. *capitata*	Increase	Drought	Severe stress 2 weeks
Brassica oleracea L. var. *italica*	Increase		Severe stress 2 weeks
Brassica napus L.	Increase		Severe stress more than 1 week
Brassica rapa ssp. *rapifera* L.	Increase		Mild stress—25% of available water
Brassica carinata L.	Increase/no effect		Mild and severe stress (40, 23, 17 and 15% of available water)
Brassica oleracea L. var. *gemmifera*	No effect		Mild stress (30% of available water)
Brassica napus L.	No effect		Mild stress
Brassica oleracea L.	Decrease		Mild and severe stress (40—45% of available water)
Arabidopsis thaliana L.	Decrease		Severe stress
Arabidopsis thaliana (L.)	Decrease		Mild stress (50% of available water)
Arabidopsis thaliana (L.)	Decrease		Water logging (200% of available water)
Brassica rapa L.	Increase	Temperature	Elevated temperature (21—34°C)
Brassica rapa L.	Decrease		Low—medium temperature (15—27°C)
Brassica oleracea L.	Increase		Elevated temperature (32°C)
Brassica oleracea L.	Decrease during day/ increase during night	Light cycling	14 h/10 h day/night*
Arabidopsis thaliana L.	Increase upon light/ decrease upon darkness		16 h/8 h d/n or continuous darkness
Brassica oleracea L. var. *italica*	Increase upon light		16 h/8 h d/n or continuous darkness
Arabidopsis thaliana	Slight increase	UV-B radiation	1.55 Wm^{-2}
Brassica oleracea L. var. *italica*	Increase		Up to 0.9 kJm^{-2} d^{-1}
Brassica oleracea L. var. *italica*	Increase	Nutrient availability	N-limitation (1 gr N pot^{-1})

(Continued)

Table 12.2 (Continued)

Plant Species	Glucosinolate Content	Stress	Treatment Condition
Brassica rapa ssp. *rapifera* L	Increase		S-supply (60 kg S ha^{-1})
Brassica oleracea L. var. *italica*	No effect		S-supply (150 kg/ha)
Brassica oleracea L. *capitata*	Increase		S-supply (110 kg S ha^{-1})
Brassica napus	Increase		S-supply (100 kg S ha^{-1})
Tropaeolum majus	Increase		S-supply (8.3 mM SO$_4$$^{2-}$)
Brassica oleracea L. var. *italica*	No effect		S-limitation (15 kg/ha)
Arabidopsis thaliana L.	Increase		K-deficiency (lack KNO$_3$ for 2 weeks)
Brassica rapa L.	Decrease		K-deficiency (lack of nutrient solution for 5 days)
Brassica oleracea L. var. *italica*	Increase		Se-supply (5.2 mM Na$_2$ SeO$_4$)
Brassica oleracea L. var. *italica*			B-deficiency (9−12 µg gr DW^{-1})
Cabbage and kale	Increase	Cadmium	Cd (5 and 10 mg Cd kg^{-1} soil)
Thlaspi caerulescens	Increase		

Source: Modified after Martínez-Ballesta et al., 2013.

As *Brassica* crops contain high amounts of sulfur-containing amino acids and glucosinolates, glucosinolate metabolism and the effects of sulfur and nitrogen nutrition have been studied (Schnug et al., 1993; Krumbein et al., 2002; Salac et al., 2006; Schonhof et al., 2007b). It is evident that when broccoli plants were supplied with low sulfur or nitrogen, a decrease in glucosinolates was noted, whereas total glucosinolate levels were elevated at sufficient nitrogen supply or high sulfur levels, and were lower at low sulfur supply with an optimal nitrogen supply (Aires et al., 2006; Schonhof et al., 2007a). Similarly, glucosinolate levels in turnip were found to be strongly regulated by nitrogen and sulfur application (Kim et al., 2002). In field experiments, nitrogen and sulfur supply showed a clear influence on individual glucosinolates as it may favor the hydroxylation step converting but-3-enyl glucosinolate to 2-hydroxybut-3-enyl glucosinolate. Compared to indole glucosinolates, aliphatic glucosinolates show a greater sensitivity to sulfur deficiency probably because they are synthesized from methionine (Zhao et al., 1994). Some *B. napus* cultivars with reduced contents of aliphatic glucosinolates were more sensitive to sulfur deficiency (Schnug, 1990), which suggests a role of aliphatic glucosinolates in the survival strategy to mineral stress. Sulfur fertilisation leads to increases in glucosinolate content in most cases. Increases of over 10-fold have sometimes been reported. For example, the benzyl glucosinolate content of *Tropaeolum majus* was increased over 50-fold by fertilising a particular cultivar with 8.3mM sulfate (Matallana et al., 2006).

12.4.1 **Biotic stress**

During their lifetime, plants have to deal with a variety of environmental stresses including biotic stresses such as those from microbial pathogens and herbivores. As plants are not in a position to move from their unfavorable environment, they have evolved a broad range of defense mechanisms. The role of GSLs in combating biotic stress has been well recognized. GSLs exhibit growth inhibition or feeding deterrence to a wide range of general herbivores such as birds, slugs, and generalist insects (Rask et al., 2000; Barth and Jander, 2006). Plants respond to herbivore or insect damage by accumulating higher GSL levels and thus increase their resistance to such biotic stresses. Glucosinolates, the characteristic secondary compounds of Brassicaceae, as well as proteinase inhibitors, remained unaffected by UV in all plants, demonstrating independent regulation pathways for different metabolites (Kuhlmann and Müller, 2009a,b). Mewis et al. (2012b), however, demonstrated an increase in aliphatic GSLs in *Arabidopsis thaliana* when fed by phloem-feeding aphids, the green peach aphid (*Myzus persicae*), cabbage aphid (*Brevicoryne brassicae*), and generalist caterpillar species *Spodoptera exigua*. Interestingly, the content of indole GSLs were found to be unchanged. GSL levels have been demonstrated to reduce damage by generalist herbivores. Volatiles produced by GSLs can also provide indirect protection to plants by attracting natural enemies of herbivores such as parasitoids. Several reports exist on the toxicity of GSL hydrolysis products to bacteria and fungi (Mayton et al., 1996; Brader et al., 2001). Pedras and Sorensen (1998) demonstrated an inhibitory action by various isothiocyanates derived from GSLs on germination and growth of a fungal pathogen *Leptosphaeria maculans*. Aromatic isothiocyanates were found to be more toxic than aliphatic isothiocyanates and the fungal toxicity of the latter decreased with increase in side chain length. In a study on the antimicrobial effect of crude extracts from *Arabidopsis*, Tierens et al. (2001) identified 4-methylsulfonyl butyl isothiocyanate as the major active compound with a broad spectrum of antimicrobial activity. Thus, the possible protective role of GSL-derived isothiocyanate against pathogens was demonstrated. Investigation of the level of GSLs in different *Brassica* cultivars by several workers indicated changes in GSL pattern when inoculated by fungal pathogens. These changes were mostly due to increase of indole and aromatic GSLs, although increase of aliphatic GSLs was also noted.

12.4.2 **Abiotic stress**

All abiotic stresses are important environmental factors that reduce plant growth and yield. Plants respond and adapt to these stresses in order to survive. Signaling pathways are induced in response to environmental stresses. Several signaling molecules have been identified in plant defense responses. These include JA, SA, and ET, which have been demonstrated to operate independently and/or synergistically in different signal transduction pathways. JA and SA have been shown to be involved in the induction of different GSLs (Yan and Chen, 2007). Different signal transduction pathways activate specific biosynthetic and secondary modifying enzymes, leading to altered levels of specific GSLs. The induction of GSLs by several defense pathways strongly indicates that these compounds play a role in plant defense.

Salt stress is a major abiotic stress reducing the productivity of crops in many areas of the world. Salinity affects the water balance resulting in osmotic damage. Osmotic adjustment is a plant adaptation mechanism used to maintain water balance in plants. In their studies on the effect of salinity stress on GSL content, Keling and Zhujun (2010) found a considerable influence of NaCl stress on the GSL content and composition in pakchoi (*Brassica campestris* L. ssp. *chinensis* var. *communis*) shoots. At 50 mM NaCl, the contents of total GSLs as well as aliphatic and indole GSL significantly increased. A significant increase in indole GSLs and a decrease in aromatic GSL (gluconasturtiin) were, however, noted at 100 mM NaCl. Glucoalyssin, gluconapin, glucobrassicin, and neglucobrassicin were significantly enhanced at 50 mM NaCl, while only the content of glucomapin and glucobrassicin increased at 100 mM NaCl.

Drought stress resulted in considerably elevated leaf GSL content of *Brassica carinata* varieties with the magnitude of GSL concentration varying with the stage of development and intensity of the drought stress (Schreiner et al., 2009a). Increase in leaf GSL concentrations correlated with relative water content with reduced water content leading to higher leaf GSL concentration. *Brassica oleracea* L., plants grown for two weeks under drought stress showed decreased levels of indolyl GS when compared to well-watered plants, while water logging conditions resulted in slight increases within the GS contents (Khan et al., 2011). Imbalance in sulfur to nitrogen ratio may result in the alteration of nutrient uptake due to water deficit resulting in the accumulation of GSLs as sulfur sink. Further, stresses such as low water availability change the hormonal distribution of plants leading to a cascade of signal transduction pathways that result in the expression of stress-responsive genes. Particularly, stress hormones like ABA, JA, ethylene, and SA that play a very important role in biotic and abiotic stress resistance are known to increase the concentrations of GSLs (Yan and Chen, 2007).

UV-B radiation acts as an environmental stress and triggers various responses in plants. These include changes in growth, development, morphology, and physiological aspects. In recent years, some researchers have reported the effect of UV-B on GSL metabolism. Microarray data have shown that the genes related to the biosynthesis of flavonoids, glucosinolates, and terpenoids were differently expressed after UV-B radiation. A study on the effect of UV radiation on *Tropaeolum majus* demonstrated that appropriate UV-B dosage could increase the glucotropaeolin concentration (Schreiner et al., 2009b). Wang et al. (2011) showed that UV-B radiation induced the production of GSLs. Continuous UV-B exposure for 12 h, however, inhibited the expression of glucosinolate metabolism-related genes resulting in a significant decline in glucosinolate content, particularly that of indolic glucosinolates. In another study on UV-B-mediated induction of GSLs, Mewis et al. (2012c) reported the induction of of 4-methylsulfinyl butyl GSL and 4-methoxy-indol-3-ylmethyl GSL in sprouts of *Brassica oleracea* var. *italica* (broccoli). Accumulation of defensive GSLs was accompanied by increased expression of genes associated with salicylate and JA signaling defense pathways and up-regulation of genes responsive to fungal and bacterial pathogens. Enhanced GSL formation had a negative effect on the growth of aphid *Myzuz persicae* and attack by caterpillar *Pieris brassicae*. Levels of these compounds are also reported to be effected under temperature stress. The TU8 mutant of *Arabidopsis* deficient in glucosinolate metabolism and pathogen-induced auxin accumulation showed less tolerance to elevated temperatures than wild-type plants (Ludwig-Müller et al., 2000). Seasonal variation in the concentration of aliphatic, aromatic, and indole GSLs was noted in different varieties of *Brassica oleraceae* (Cartea et al., 2008). Similar effects with increase in aliphatic glucosinolates (particularly glucoraphanin) were observed in broccoli kept at daily mean temperatures between 7 and 13°C (mean radiation of $10-13$ mol m^{-2} day^{-1}) (Schonhof et al., 2007c).

In the authors' laboratory (Banerjee et al., 2014), the cabbage leaves subjected to gamma radiation stress were found to have an enhanced sinigrin content. No effect of myrosinase activity was, however, noted, thus providing high retention of glucosinolates and facilitating improved release of these nutraceutically significant compounds during mastication of the vegetable. Thus, exposure to such abiotic stress was demonstrated to provide improved benefit in terms of enhancing intake of potentially important health protective and promoting compounds in *Brassica* vegetables (Banerjee et al., 2014).

Heavy metal stress also can lead to changes in GSL content. While selenium was found to affect the content of GSLs in a concentration-dependent manner, cadmium stress produced no change in GSL production in *B. rapa* (Kim and Juvic, 2011; Jakovljević et al., 2013). GSL concentration also increased as a result of temperature stress showing seasonal variation in *Brassica* plants. In *Thlaspi caerulescens*, a metal hyperaccumulator with a high requirement of zinc, GSL levels (particularly sinalbin) increased in roots but decreased in leaves and shoots. Zinc had a clearly distinctive effect on the specific group of indolyl GSLs in *T. caerulescens* with a drastic reduction in both roots. Post-harvest storage conditions of *Brassica* vegetables are also known to influence GSL and related isothiocyanate content. Content of these compounds was found to decrease in vegetables such as broccoli, brussel sprouts, cauliflower, and green cabbage when stored in a domestic refrigerator (4−8°C) for 7 days unlike when stored at ambient temperature (Song and Thornalley, 2007). Storage of vegetables at very low temperature (−85°C) can result in freeze−thaw fracture of plant cells leading to significant loss of GSLs as a consequence of their conversion to isothiocyanates during thawing (Song and Thornalley, 2007). Tamara et al. (2013) found that GSLs in leaves and root could be more involved in ameliorating S deficiency rather than plant defense in the short term in cadmium (Cd) stress; however, total GSL levels in the stem during the long term could serve as a GSL storage organ implying possible roles of GSL in Cd stress.

12.5 Genes involved in glucosinolate biosynthesis

The main genetic pathway of glucosinolate biosynthesis has been identified in *Arabidopsis* using genetic and biochemical approaches. Several enzymes and transcription factors involved in the GSL biosynthesis have been studied in the model plant, *Arabidopsis*, and in a few other *Brassica* crop species (Baskar et al., 2012). Figure 12.3 presents the genetic machinery involved in different aspects of GSL sysnthesis. Six MYB factors, namely, MYB28, MYB29, MYB76, MYB34, MYB51, and MYB122, have been found to be transcriptional regulators in the biosynthesis of glucosinolate in *Arabidopsis*. While MYB28, MYB29, and MYB76 specifically transactivate genes related to aliphatic glucosinolate biosynthetic pathway (MAM3, CYP79F1, and CYP83A1) (Gigolashvili et al., 2007b, 2008), MYB34, MYB51, and MYB122 are regulators of the indolic glucosinolate biosynthetic pathway (TSB1, CYP79B2, and CYP79B3) (Celenza et al., 2005; Gigolashvili et al., 2007a). Wang et al. (2011) used the comparative genomic analysis method of *Arabidopsis thaliana* and *Brassica rapa* and identified 102 putative genes in *B. rapa* as the orthologues of 52 *Arabidopsis* glucosinolate genes. The glucosinolate genes in *B. rapa* and *A. thaliana* shared 59−91% nucleotide sequence identity. Microarray experiments have also shown that CYP79B2, an important gene involved in the biosynthesis of indolic glucosinolates (Chen and Andreasson, 2001), is downregulated by brassinosteroids (Goda et al., 2002). Both MYB34 and MYB51, which encode transcriptional factors of indolic glucosinolate biosynthesis, contain a BZR1 binding site in their promoters (Sun et al., 2010). Further, Guo et al. (2012) investigated the role of

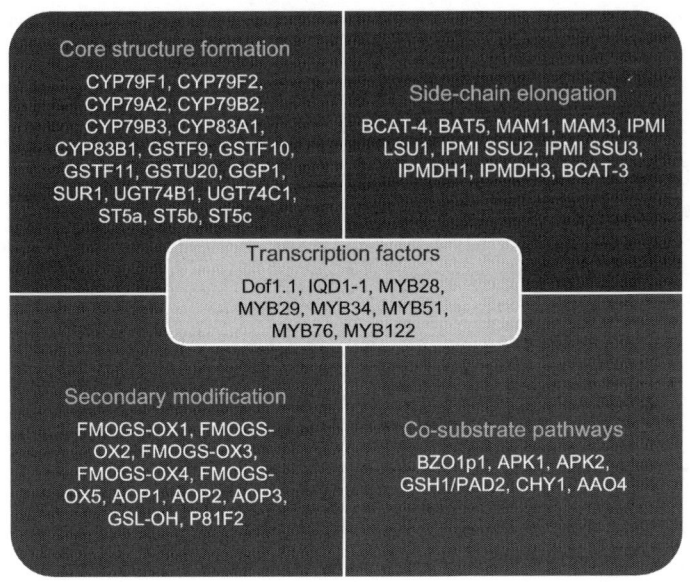

FIGURE 12.3

Genes involved in different stages of glucosinolate biosynthetic pathway.

brassinosteroids in glucosinolate biosynthesis in *Arabidopsis* using mutants and transgenic plants. Zang et al. (2009) identified glucosinolate synthesis genes in *Brassica rapa* on the basis of cDNA and BAC libraries with about 21.5% of the genes as partial CDS sequences and many BrGS genes anchored only on the BAC, rather than on chromosomes. The authors also identified glucosinolate biosynthetic genes by comparative genomic analysis between *B. rapa* and *A. thaliana*. Augustine et al. (2013) analyzed four MYB28 genes that are differentially expressed and regulated in both a tissue- and cell-specific manner in controlling aliphatic glucosinolate biosynthesis in *B. juncea*.

Several myrosinase genes from *Sinapis alba*, *Brassica napus*, and *Arabidopsis thaliana* have been isolated and characterized indicating that myrosinases are encoded by a multigene family consisting of three subgroups (Xu et al., 2004). Myrosinase in the *Brassica* family is encoded by a gene family, which consists of three subfamilies, namely, MA (Myr1), MB (Myr2), and MC (Baskar et al., 2012). Several myrosinase-associated proteins, such as epithiospecifier modifier 1 (ESM1), ESP, and MVP1, have been identified in *Arabidopsis*, which are mainly involved in the generation of diversified GSL metabolic products (Baskar et al., 2012).

12.6 Gene expression profiling in response to environmental cues

Plant glucosinolate metabolism is responsive to many environmental factors. Generally, glucosinolate degradation products serve as defense compounds against pathogens and generalist herbivores, and as attractants to glucosinolate-adapted specialists (Rask et al., 2000; Barth and Jander, 2006).

Several glucosinolate hydrolysis products have been reported to display toxicity to fungi and bacteria (Mayton et al., 1996; Brader et al., 2001). Glucosinolate levels in oilseed rape were positively correlated with resistance to pathogens (Li et al., 1999) with some exceptions (Giamoustaris and Mithen, 1997). The best *in vivo* evidence for the defense role of glucosinolates came from an *MAM1* mutant study, where decreased glucosinolate levels in *Arabidopsis* caused susceptibility to *Fusarium oxysporum* (Tierens et al., 2001). Pathogen infection can also change glucosinolate profiles. When *Brassica* plants were infected by *Leptosphaeria maculans*, glucosinolate levels were induced, but myrosinase levels were not affected (Siemens and Mitchell-Olds, 1998). JA and SA signaling pathways may be involved in the regulation of glucosinolate levels (Li et al., 2006). Currently, there is more literature on plant interactions with insect herbivores. When glucosinolate levels increased in *B. napus* and *Sinapis alba*, feeding by generalist insects decreased significantly, while feeding by specialist insects (e.g., *Pieris rapae*) greatly increased and caused severe damage (Giamoustaris and Mithen, 1995). The damage led to a systemic increase in indole glucosinolate and often in total glucosinolate levels. For example, when seedlings of oilseed rape and mustard were attacked by *Xea* beetles, there was a tremendous increase in the concentration of indole glucosinolates, but no significant changes in aliphatic glucosinolates (Bodnaryk, 1992; Bartlet et al., 1999). In one case when feeding with turnip root fly, the concentrations of aliphatic glucosinolates actually dropped (Hopkins et al., 1998).

Mewis et al. (2006) analyzed glucosinolate accumulation levels and gene expression of glucosinolate biosynthetic genes in response to feeding by four herbivores in *Arabidopsis thaliana* (L.) wild-type (Columbia) and mutant lines that were affected in defense signaling. Herbivory on wild-type plants led to increased aliphatic glucosinolate content for three of four herbivores tested, namely, *Myzus perscae* (Sulzer), *Brevicoryne brassicae* (L.), and *Spodoptera exigua* Hübner. The lepidopteran *Pieris rapae* L. did not affect the levels of aliphatic glucosinolate in the wild type, except for an increase in indole glucosinolates. Increased expression of genes of aliphatic glucosinolate biosynthesis was observed after feeding by all species, while mutations in jasmonate (*coi1*), salicylate (*npr1*), and ethylene signaling process (*etr1*) showed varied gene expression, glucosinolate profile, and insect performance compared to wild type. As against in wild type, the gene transcripts of aliphatic glucosinolate biosynthesis did not generally increase in the mutants. Both glucosinolate content and gene expression data indicated that salicylate and ethylene signaling repress some jasmonate-mediated responses to herbivory. This possibly indicates that all three modes of signaling processes are involved in responses to herbivores.

Plant interactions with the environment influences glucosinolate metabolism and is constantly regulated by different environmental factors including UV-B radiation. Wang et al. (2011) studied the glucosinolate content and expression of glucosinolate metabolism-related genes in response to enhanced UV-B radiation ($1.55 \text{ W} \cdot \text{m}^{-2}$) and the succeeding dark recovery process in *Arabidopsis thaliana* rosette leaves. Induction of glucosinolates was observed in the 1 h of enhanced UV-B radiation exposure, whereas, after continuous exposure for 12 h, the expression of glucosinolate metabolism-related genes was significantly inhibited and the glucosinolate content was declined, especially that of indolic glucosinolates. Upon exposure to darkness for 12 h for partially recovery, both glucosinolate gene expression and the content returned to the control levels. The results of Wang et al. (2011) showed up-regulation of some genes (MYB51, OBP2, MYB76, SOT16, and TGG1) in the initial 1 h of UV-B exposure. These genes were also induced by JA and wounding (Skirycz et al., 2006; Dombrecht et al., 2007; Staswick, 2008). The transcription factors (MYB51,

OBP2, and MYB76) can positively mediate glucosinolate biosynthesis (Skirycz et al., 2006; Dombrecht et al., 2007; Gigolashvili et al., 2008). MYB51 and MYB76 specifically activate indolic and aliphatic glucosinolate biosynthesis, respectively, which might be responsible for the temporary and initiative increase of 4MSOB and I3M. The 3 h of treatment led to down-regulation of MYB28 and MYB29 (the genes encoding the transcription factors that regulate aliphatic glucosinolate biosynthesis) and MYB34 (regulating indolic glucosinolate biosynthesis), and structural genes (CYP79F1, CYP83B1, MAM1, CYP79B2, and CYP79B3) encoding enzymes catalyze the synthesis of aliphatic glucosinolates. Subsequent exposure led to decline in the expression of the majority of genes and the glucosinolate contents. Microarray analysis experiments with UV-B exposed *Tropaeolum majus* also indicated differential expression of genes related to the biosynthesis of flavonoids, glucosinolates, and terpenoids (Hectors et al., 2007).

Schweizer et al. (2013) showed that a triple mutant for *MYC2*, *MYC3*, and *MYC4*, three basic helix—loop—helix transcription factors that are known to additively control jasmonate-related defense responses, was shown to have a highly reduced expression of GSL biosynthesis genes. The myc2 myc3 myc4 (myc234) triple mutant was almost completely devoid of GS and was extremely susceptible to the generalist herbivore *Spodoptera littoralis*.

12.7 Signaling networks

The biotic and abiotic factors such as pathogen challenge, herbivore damage, mechanical wounding, or altered mineral nutrition are known to regulate glucosinolate profiles (Wittstock and Halkier 2002; Agrawal and Kurashige 2003; Mewis et al., 2005). Sulfur limitation is shown to repress most glucosinolate pathway genes (Hirai et al., 2005). The bioinformatics approach has enabled identification of core pathway genes and predicted additional enzymes with roles in glucosinolate biosynthesis. In addition to this modulation, plant hormones such as jasmonates, SA, and ET associated with specific and broad-spectrum defense responses can also affect glucosinolate content (Brader et al., 2001; Kliebenstein et al., 2002; Mikkelsen et al., 2003a; Mewis et al., 2005). Jasmonates known to be involved in responses to insect attack and necrotrophic pathogens have shown increased indolyl and specific aliphatic glucosinolates (Brader et al., 2001; Mikkelsen and Halkier, 2003b), possibly via multiple signaling pathways (Kliebenstein et al., 2002). These studies have demonstrated the utility of mutants defective in hormone synthesis or signaling through the regulation of specific indole glucosinolate production by SA as well as interactions of jasmonate and salicylate signaling.

NPR1 (*nonexpresser of PR genes 1*), ETR1 (*ethylene receptor 1*), and COI1 (*coronatine insensitive 1*) are important for SA, ET, and JA signaling, respectively. It is also evident that insect feeding-induced glucosinolate biosynthesis requires the functions of regulatory proteins NPR1 and ETR1, but not COI1 (Mewis et al., 2005). As the SA and JA pathways seem to be mutually antagonistic (Glazebrook et al., 2003), blocking of JA signaling or increased SA signaling caused reduction of glucosinolate levels and vice versa (Mikkelsen et al., 2003a,b; Mewis et al., 2005; Li et al., 2006). While ET signaling may influence other resistance traits than glucosinolates, NPR1 may be a point of crosstalk of multiple signaling pathways (Glazebrook et al., 2003; Mewis et al., 2005). Mewis et al. (2005) found that exogenous application of JA did not mimic plant responses to insect feeding in terms of glucosinolate metabolism. Potassium starvation was shown to induce the expression of

JA biosynthetic genes and glucosinolate metabolic genes. Resupply of potassium down-regulated the transcription of these genes (Armengaud et al., 2004). Defense responses are not only activated at the site of wounding but also distantly in other remote plant parts. Chen et al. (2013) suggested that higher contents of indole glucosinolates in systemic leaves might arise from the induction of a long-distance signal produced in local leaves as well as from JA synthesized in systemic leaves.

Studies have demonstrated that wound, pathogen, and hormones like JA and ET induced expression of transcriptional factor genes (Schenk et al., 2000) and that nuclear proteins regulate glucosinolate metabolism (Yan and Chen, 2007). Among the transcription factor genes, ATR1, a Myb transcription factor, regulates a number of genes of tryptophan biosynthesis and indole glucosinolate biosynthesis. Specifically, ATR1 participates in the JA-mediated induction of indole glucosinolate biosynthesis (Celenza et al., 2005). Transcription factors AtDof1 and Myb51 also control indole glucosinolates. While AtDof1 regulates CYP83B1 and is inducible by generalists and JA (Skirycz et al., 2006), both AtDof1 and Myb51 activate ATR1 and myrosinase binding proteins (Skirycz et al., 2006; Gigolashvili et al., 2007b). It has also been shown that overexpression of AtDof1 caused changes in aliphatic glucosinolate levels, similar to overexpression of a novel calmodulin binding nuclear protein IQD1, which led to high levels of both indole and aliphatic glucosinolates (Levy et al., 2005). The study suggested that regulation of aliphatic glucosinolate metabolism may not occur at transcriptional level and crosstalk may in fact operate between regulatory pathways of indole glucosinolates and aliphatic glucosinolates. Hirai et al. (2007) investigated Myb28 and Myb29 as master transcription factors; while Myb28 regulates the pathway from methionine to aliphatic glucosinolates, and is essential for the basal-level control of aliphatic glucosinolate biosynthesis, Myb29 has a role in JA-mediated aliphatic glucosinolate biosynthesis (Hirai et al., 2007).

The regulation of cellular processes in a cell are fine-tuned by post-translational modifications, especially protein phosphorylation and redox regulation (Paget and Buttner, 2003; Chen and Harmon, 2006). Several protein kinases, including receptor-like protein kinases, protein phosphatases, and MAP kinases, are shown to respond to pathogens, JA, and hydrogen peroxide treatments (Desikan et al., 1999; Schenk et al., 2000). Desikan et al. (2001) found that protein kinases, phosphatases, and calmodulin proteins as well as myrosinases and myrosinase binding proteins were affected by oxidative stress often impacting the accumulation of hydrogen peroxide and ROS (Apel and Hirt, 2004). Abiotic and biotic stresses cause changes in glucosinolate metabolism and hence it can be assumed that redox modifications may also play an important role in regulating glucosinolate metabolism.

12.8 Metabolic engineering of glucosinolates

Metabolic engineering of glucosinolates can be achieved by targeting either the biosynthetic or the transcription factors of the GSL biosynthetic pathway. More than 20 genes with potential regulatory function in GSL metabolism and several transcription factors have been identified in *Arabidopsis* and other plant species. Manipulation of these transcription factors appears to be more effective for the control of metabolic pathways than that of genes encoding single enzyme in plants (Capell and Christou, 2004). Further efforts in this direction will certainly provide the required insights to facilitate the modification of the complex GSL biosynthesis of plants in the near future. Baskar et al. (2012) described different strategies including overexpression of *CYP79A1*, *CYP71E1*, MAM1, *CYP79F1, CYP83A1*, etc., to produce genetically engineered plants with altered GSL profiles.

In addition to plants, microbial source has also been attempted as a great potential for large-scale production of desirable GSLs for the benefit of human health (Mikkelsen et al., 2012) through the stable expression of multigene pathways from *Arabidopsis* to yeast.

12.9 Conclusion and future prospects

Glucosinolates are a diverse group of secondary metabolites largely responsible for the nutraceutical and pharmacological value of *Brassica* vegetables and have been implicated in defense against insects and pathogens and thus possess a bio-protective role. Recent years have witnessed great progress in the understanding of glucosinolate biosynthesis in model plants like *Arabidopsis thaliana* using different biochemical and reverse genetics approaches. Cooperation between researchers involved in plant breeding and food technology could lead to using food technological parameters as breeding traits to identify genetic loci associated with food processing to breed vegetables with higher retention of glucosinolates (Hennig et al., 2013).

Abiotic stresses, such as salinity, drought, extreme temperatures, light and nutrient deprivation, modulate the glucosinolate profiles through different mechanisms through the involvement of hormones or signaling molecules. The accumulation of glucosinolates is also defined by the magnitude and duration of the stress impact, for example plant–pathogen interactions where the plant water availability and herbivore feeding or pathogen attack come into the picture. It has also been seen that exogenous glucosinolate hydrolysis products (isothiocyanates) alleviate the impact of drought or elevated temperatures. It is yet to be investigated how molecular mechanisms operate in this alleviation process, and also allocation and/or redistribution of glucosinolates in response to environmental changes. As the isothiocyanates produced from the Val- and isoleucine-derived glucosinolates are volatile, metabolically engineered plants producing these glucosinolates have novel properties with great potential for improvement of resistance to herbivorous insects and for biofumigation. Regulation and control of glucosinolate metabolism also needs to be investigated at different levels of signaling and metabolic network that control this pathway. It is also desirable to identify genes/loci in different *Brassica* species so that they might be used to manipulate aliphatic glucosinolates towards favorable forms (Li et al., 2008; Baskar et al., 2012). Understanding the dynamics of the glucosinolate biosynthesis network will not only advance our basic knowledge of this bioactive molecule complex but also augment research efforts towards metabolic engineering.

References

Agrawal, A., Kurashige, N.S., 2003. A role for isothiocyanates in plant resistance against the specialist herbivore *Pieris rapae*. J. Chem. Ecol. 29, 1403–1415.

Aires, A., Rosa, E., Carvalho, R., 2006. Effect of nitrogen and sulfur fertilization on glucosinolates in the leaves and roots of broccoli sprouts (*Brassica oleracea* var. italica). J. Sci. Food Agricul. 86, 1512–1516.

Aires, A., Mota, V.R., Saavedra, M.J., Rosa, E.A., Bennett, R.N., 2009. The antimicrobial effects of glucosinolates and their respective enzymatic hydrolysis products on bacteria isolated from the human intestinal tract. J. Appl. Microbiol. 106, 2086–2095.

Apel, K., Hirt, H., 2004. Reactive oxygen species: metabolism, oxidative stress, and signal transduction. Annu. Rev. Plant Biol. 55, 373–399.

Arbona, V., Manzi, M., Cd, Ollas, Gómez-Cadenas, A., 2013. Metabolomics as a tool to investigate abiotic stress tolerance in plants. Int. J. Mol. Sci. 14, 4885—4911.

Armengaud, P., Breitling, R., Amtmann, A., 2004. The potassium-dependent transcriptome of Arabidopsis reveals a prominent role of jasmonic acid in nutrient signaling. Plant Physiol. 136, 2556—2576.

Augustine, R., Majee, M., Gershenzon, J., Bisht, N.C., 2013. Four genes encoding MYB28, a major transcriptional regulator of the aliphatic glucosinolate pathway, are differentially expressed in the allopolyploid *Brassica juncea*. J. Exp. Bot. 64, 4907—4921.

Banerjee, A., Variyar, P.S., Chatterjee, S., Sharma, A., 2014. Effect of post harvest radiation processing and storage on the volatile oil composition and glucosinolate profile of cabbage. Food Chem. 151, 22—30.

Barth, C., Jander, G., 2006. *Arabidopsis* myrosinases TGG1 and TGG2 have redundant function in glucosinolate breakdown and insect defense. Plant J. 46, 549—562.

Bartlet, E., Kiddle, G., Williams, I., Wallsgrove, R., 1999. Wound-induced increases in the glucosinolate content of oilseed rape and their effect on subsequent herbivory by a crucifer specialist. Entomol. Exp. Appl. 91, 163—167.

Baskar, V., Gururani, M.A., Yu, J.W., Park, S.W., 2012. Engineering glucosinolates in plants: current knowledge and potential uses. Appl. Biochem. Biotechnol. 168, 1694—1717.

Bodnaryk, R.P., 1992. Effects of wounding on glucosinolates in the cotyledons of oilseed rape and mustard. Phytochem. 31, 2671—2677.

Brader, G., Tas, E., Palva, E.T., 2001. Jasmonate-dependent induction of indole glucosinolates in Arabidopsis by culture filtrates of the nonspecific pathogen Erwinia carotovora. Plant Physiol. 126, 849—860.

Brotman, Y., Landau, U., Cuadros-Inostroza, Á., Takayuki, T., Fernie, A.R., et al., 2013. Trichoderma-plant root colonization: escaping early plant defense responses and activation of the antioxidant machinery for saline stress tolerance. PLoS Pathog 9, e1003221.

Capell, T., Christou, P., 2004. Progress in plant metabolic engineering. Curr. Opin. Biotechnol. 15, 148—154.

Cartea, M.E., Velasco, P., Obregon, S., Padilla, G., De Haro, A., 2008. Seasonal variation in glucosinolate content in *Brassica oleracea* crops grown in northwestern Spain. Phytochemistry 69, 403—410.

Celenza, J.L., Quiel, J.A., Smolen, G.A., Merrikh, H., Silvestro, A.R., Normanly, J., et al., 2005. The Arabidopsis ATR1 Myb transcription factor controls indolic glucosinolate homeostasis. Plant Physiol. 137, 253—262.

Champolivier, L., Merrien, A., 1996. Effects of water stress applied at different growth stages to *Brassica napus* L. var. oleifera on yield, yield components and seed quality. Eur. J. Agron. 5, 153—160.

Chen, S., Andreasson, E., 2001. Update on glucosinolate metabolism and transport. Plant Physiol. Biochem. 39, 743—758.

Chen, S., Harmon, A.C., 2006. Advances in plant proteomics. Proteomics 6, 5504—5516.

Chen, Y., Feiab, M., Wangab, Y., Chenc, S., Yana, X., 2013. Proteomic investigation of glucosinolate systematically changes in Arabidopsis Rosette leaves to exogenous methyl jasmonate. Plant Biosyst. 10.1080/11263504.2013.819044.

Clarke, D.B., 2010. Glucosinolates, structures and analysis in food. Anal. Methods 2, 310—325.

van Dam, N.M., Tytgat, T.O.G., Kirkegaard, J.A., 2009. Root and shoot glucosinolates: a comparison of their diversity, function and interactions in natural and managed ecosystems. Phytochem. Rev. 8, 171—186.

Desikan, R., Clarke, A., Hancock, J.T., Neill, S.J., 1999. H_2O_2 activates a MAP kinase-like enzyme in *Arabidopsis thaliana* suspension cultures. J. Exp. Bot. 50, 1863—1866.

Desikan, R., Mackerness, S.A.H., Hancock, J.T., Neill, S.J., 2001. Regulation of the Arabidopsis transcriptome by oxidative stress. Plant Physiol. 127, 159—172.

Dombrecht, B., Xue, G.P., Sprague, S.J., Kirkegaard, J.A., Ross, J.J., Reid, J.B., et al., 2007. MYC2 differentially modulates diverse jasmonate-dependent functions in Arabidopsis. Plant Cell. 19, 2225—2245.

Endara, M.J., Coley, P.D., 2011. The resource availability hypothesis revisited: a meta-analysis. Funct. Ecol. 25, 389—398.

Engelen-Eigles, G., Holden, G., Cohen, J.D., Gardner, G., 2006. The effect of temperature, photoperiod, and light quality on gluconasturtiin concentration in watercress (*Nasturtium offcinale* R. Br.). J. Agric. Food Chem. 54, 328−334.

Fahey, J., Zalcmann, A., Talalay, P., 2001. The chemical diversity and distribution of glucosinolates and iso-thiocyanates among plants. Phytochem. 56, 5−51.

Giamoustaris, A., Mithen, R., 1995. The effect of modifying the glucosinolate content on leaves of oilseed rape (Brassica napus spp. Oleifera) on its interaction with specialist and generalist pests. Ann. Appl. Biol. 126, 347−363.

Giamoustaris, A., Mithen, R., 1997. Glucosinolates and disease resistance in oilseed rape (*Brassica napus* ssp. oleifera). Plant Pathol. 46, 271−275.

Gigolashvili, T., Berger, B., Mock, H., Müller, C., Weisshaar, B., Flügge, U., 2007a. The transcription factor HIG1/MYB51 regulates indolic glucosinolate biosynthesis in *Arabidopsis thaliana*. Plant J. 50, 886−901.

Gigolashvili, T., Engqvist, M., Yatusevich, R., Berger, B., Müller, C., Flügge, U., 2007b. The R2R3-MYB transcription factor HAG1/MYB28 is a regulator of methionine-derived glucosinolate biosynthesis in *Arabidopsis thaliana*. Plant J. 51, 247−261.

Gigolashvili, T., Engqvist, M., Yatusevich, R., Müller, C., Flügge, U.I., 2008. HAG2/MYB76, HAG3/MYB29 exert a specific and coordinated control on the regulation of aliphatic glucosinolate biosynthesis in *Arabidopsis thaliana*. New Phytol. 177, 627−642.

Glazebrook, J., Chen, W.J., Estes, B., Chang, H.S., Nawrath, C., Metraux, J.P., et al., 2003. Topology of the network integrating salicylate and jasmonate signal transduction derived from global expression phenotyping. Plant J. 34, 217−228.

Goda, H., Shimada, Y., Asami, T., Fujioka, S., Yoshida, S., 2002. Microarray analysis of brassinosteroid-regulated genes in Arabidopsis. Plant Physiol. 130, 1319−1334.

Grubb, C.D., Abel, S., 2006. Glucosinolate metabolism and its control. Trends Plant Sci. 11, 89−100.

Guo, R., Qian, H., Shen, W., Liu, L., Zhang, M., Cai, C., et al., 2012. BZR1 and BES1 participate in regulation of glucosinolate biosynthesis by brassinosteroids in Arabidopsis. J. Exp. Bot. 64, 2401−2412.

Halkier, B.A., Gershenzon, J., 2006. Biology and biochemistry of glucosinolates. Annu. Rev. Plant Biol. 57, 303−333.

Hectors, K., Prinsen, E., De Coen, W., Jansen, M.A., Guisez, Y., 2007. *Arabidopsis thaliana* plants acclimated to low dose rates of ultraviolet B radiation show specific changes in morphology and gene expression in the absence of stress symptoms. New Phytol. 175, 255−270.

Hennig, K., Verkerk, R., van Boekel, M.A.J.S., Dekker, M, Bonnema, G, 2013. Food science meets plant science: a case study on improved nutritional quality by breeding for glucosinolate retention during food processing. Trends Food Sci. Tech. . 10.1016/j.tifs.2013.10.006.

Hirai, M.Y., Klein, M., Fujikawa, Y., Yano, M., Goodenowe, D.B., Yamazaki, Y., et al., 2005. Elucidation of gene-to-gene and metabolite to gene networks in Arabidopsis by integration of metabolomics and transcriptomics. J. Biol. Chem. 280, 25590−25595.

Hirai, M.Y., Sugiyama, K., Sawada, Y., Tohge, T., Obayashi, T., Suzuki, A., et al., 2007. Omics-based identification of Arabidopsis Myb transcription factors regulating aliphatic glucosinolate biosynthesis. Proc. Natl. Acad. Sci. USA 104, 6478−6483.

Hopkins, R.J., Griffiths, D.W., Birch, A.N.E., McKinlay, R.G., 1998. Influence of increasing herbivore pressure on modification of glucosinolate content of Swedes (*Brassica napus* ssp rapifera). J. Chem. Ecol. 24, 2003−2019.

Jain, M., 2013. Emerging role of metabolic pathways in abiotic stress tolerance. J. Plant Biochem. Physiol. 1, 108. 10.4172/jpbp.1000108.

Jakovljević, T., Cvjetko, M., Sedak, M., Đokić, M., Bilandžić, N., Vorkapić-Furač, J., Redovniković, I.R., 2013. Balance of glucosinolates content under Cd stress in two Brassica species. Plant Physiol. Biochem. 63, 99−106.

Keling, H., Zhujun, Z., 2010. Effects of different concentrations of sodium chloride on plant growth and glucosinolate content and composition in pakchoi. African J. Biotechnol. 9, 4428−4433.

Khan, M.A.M., Ulrichs, C.H., Mewis, I., 2011. Drought stress—impact on glucosinolate profile and performance of phloem feeding cruciferous insects. Acta Hort 917, 111−117.

Kim, H.S., Juvic, J.A., 2011. Effect of selenium fertilization and methy jasmonate treatment on glucosinolate accumulation in broccoli florets. J. Am. Soc. Hort. Sci. 136, 239−246.

Kim, S.J., Matsuo, T., Watannabe, M., Watannabe, Y., 2002. Effect of nitrogen and sulphur application on the glucosinolate concentration in vegetable turnip rape (*Brassica rapa* L.). Soil. Sci. Plant Nutr. 48, 43−49.

Kliebenstein, D.J., Kroymann, J., Brown, P., Figuth, A., Pedersen, D., Gershenzon, J., Mitchell-Olds, T., 2001. Genetic control of natural variation in Arabidopsis glucosinolate accumulation. Plant Physiol. 126, 811−825.

Kliebenstein, D.J., Pedersen, D., Barker, B., Mitchell-Olds, T., 2002. Comparative analysis of quantitative trait loci controlling glucosinolates, myrosinase and insect resistance in *Arabidopsis thaliana*. Genetics 161, 325−332.

Krumbein, A., Schonhof, I., Rühlmann, J., Widell, S., 2002. Influence of sulphur and nitrogen supply on flavour and health-affecting compounds in *Brassicaceae*. Plant Nutrition Developments in Plant and Soil Sci. 92, 294−295.

Kuhlmann, F., Müller, C., 2009a. Development-dependent effects of UV radiation exposure on broccoli plants and interactions with herbivorous insects. Environ. Exp. Bot. 66, 61−68.

Kuhlmann, F., Müller, C., 2009b. Independent responses to ultraviolet radiation and herbivore attack in broccoli. J. Exp. Bot. 60, 3467−3475.

Levy, M., Wang, Q.M., Kaspi, R., Parrella, M.P., Abel, S., 2005. Arabidopsis IQD1, a novel calmodulin-binding nuclear protein, stimulates glucosinolate accumulation and plant defense. Plant J. 43, 79−96.

Li, J., Brader, G., Kariola, T., Palva, E.T., 2006. WRKY70 modulates the selection of signaling pathways in plant defense. Plant J. 46, 477−491.

Li, J., Hansen, B.G., Ober, J.A., Kliebenstein, D.J., Halkier, B.A., 2008. Subclade of flavin-monooxygenases involved in aliphatic glucosinolate biosynthesis. Plant Physiol. 148, 1721−1733.

Li, Y., Kiddle, G.A., Bennett, R.N., Wallsgrove, R.M., 1999. Local and systemic changes in glucosinolates in Chinese and European cultivars of oilseed rape (*Brassica napus*) after inoculation with Sclerotinia sclerotiorum (stem rot). Ann. Appl. Biol. 134, 45−58.

López-Berenguer, C., Martínez-Ballesta, M.C., Moreno, D.A., Carvajal, M., García-Viguera, C., 2009. Growing hardier crops for better health: salinity tolerance and the nutritional value of broccoli. J. Agric. Food Chem. 57, 572−578.

Ludwig-Müller, J., Krishna, P., Forreiter, C., 2000. A glucosinolate mutant of Arabidopsis is thermosensitive and defective in cytosolic Hsp90 expression after heat stress. Plant Physiol. 123, 949−958.

Mantri, N., Patade, V., Suprasanna, P., Ford, Rebecca, Pang, Edwin, 2011. Abiotic stress responses in plants—present and future. In: Ahmad, Parvaiz, Prasad, M.N.V. (Eds.), Environmental Adaptations to Changing Climate: Metabolism, productivity and Sustainability. Springer, pp. 1−20.

Martínez-Ballesta, M.C., Moreno, D.A., Carvaja, M., 2013. The physiological importance of glucosinolates on plant response to abiotic stress in *Brassica*. Int. J. Mol. Sci. 14, 11607−11625.

Matallana, L., Kleinwaechter, M., Selmar, D., 2006. Sulfur is limiting the glucosinolate accumulation in nasturtium in vitro plants (*Tropaeolum majus* L.). J. Applied Bot. Food Qual. 80, 1−5.

Mayton, H.S., Oliver, C., Vaughn, S.F., Loria, R., 1996. Correlation of fungicidal activity of Brassica species with allyl isothiocyanate production in macerated leaf tissue. Phytopathology 86, 267−271.

Mewis, I., Appel, H.M., Hom, A., Raina, R., Schultz, J.C., 2005. Major signaling pathways modulate (*Arabidopsis thaliana* L.) glucosinolate accumulation and response to both phloem feeding and chewing insects. Plant Physiol. 138, 1149−1162.

Mewis, I., Tokuhisa, J.G., Schultz, J.C., Appel, H.M., Christian, U., Jonathan, G., 2006. Gene expression and glucosinolate accumulation in *Arabidopsis thaliana* in response to generalist and specialist herbivores of different feeding guilds and the role of defense signaling pathways. Phytochemistry 67, 2450−2462.

Mewis, I., Mohammed, A., Khan, M., Glawischnig, E., Schreiner, M., Ulrichs, C., 2012a. Water stress and aphid feeding differentially influence metabolite composition in (*Arabidopsis thaliana* L.). PLoS One 7, 1−15.

Mewis, I., Khan, M.A.M., Glawischnig, E., Schreiner, M., Ulrichs, C.H., 2012b. Water stress and aphid feeding differentially influence metabolite composition in (*Arabidopsis thaliana* L.). PLoS One 70, 11.

Mewis, I., Schreiner, M., Nguyen, C.N., Krumbein, A., Ulrichs, C., Lohse, M., Zrenner, R., 2012c. UV-B irradiation changes specifically the secondary metabolite profile in broccoli sprouts: induced signaling overlaps with defense response to biotic stressors. Plant Cell. Physiol. 53, 1546−1560.

Mikkelsen, M.D., Petersen, B.L., Glawischnig, E., Jensen, A.B., Andreasson, E., Halkier, B.A., 2003a. Modulation of CYP79 genes and glucosinolate profiles in Arabidopsis by defense signaling pathways. Plant Physiol. 131, 298−308.

Mikkelsen, M.D., Halkier, B.A., 2003b. Metabolic engineering of valine- and isoleucine-derived glucosinolates in Arabidopsis expressing CYP79D2 from cassava. Plant Physiol. 131, 773−779.

Mikkelsen, M.D., Buron, L.D., Salomonsen, B., Olsen, C.E., Hansen, B.G., Mortenson, U.H., Halkier, B.A., 2012. Microbial production of indolyl glucosinolate through engineering of a multi-gene pathway in a versatile yeast expression platform. Metab. Eng. 1, 104−111.

Mittler, R., 2006. Abiotic stress, the field environment and stress combination. Trends Plant Sci. 11, 15−19.

Mittler, R., Vanderauwerab, S., Gollerya, M., Breusegemb, F.V., 2004. Reactive oxygen gene network of plants. Trends Plant Sci. 9, 490−498.

Paget, M.S., Buttner, M.J., 2003. Thiol-based regulatory switches. Annu. Rev. Genet. 37, 91−121.

Pedras, M.S.C., Sorensen, J.L., 1998. Phytoalexin accumulation and production of antifungal compounds by the crucifer wasabi. Phytochem. 49, 1959−1965.

Qasim, M., Ashraf, M., Ashraf, M.Y., Rehman, S.U., Rha, E.S., 2003. Salt induced changes in two canola cultivars differing in salt tolerance. Biol. Plantarum. 46, 629−632.

Radovich, T.J.K., Kleinhenz, M.D., Streeter, J.G., 2005. Irrigation timing relative to head development influences yield components, sugar levels and glucosinolate concentrations in cabbage. J. Am. Soc. Hortic. Sci. 130, 943−949.

Rask, L., Andreasson, E., Ekbom, B., Eriksson, S., Pontoppidan, B., Meijer, J., 2000. Myrosinase: gene family evolution and herbivore defense in Brassicaceae. Plant Mol. Biol. 42, 93−113.

Salac, I., Haneklaus, S., Bloem, E., Booth, E., Sutherland, K., Walker, K., Schnug, E., 2006. Influence of sulfur fertilization on sulfur metabolites, disease incidence and severity of fungal pathogens in oil-seed rape in Scotland. Landbauforschung Völkenrode 56, 1−4.

Schenk, P.M., Kazan, K., Wilson, I., Anderson, J.P., Richmond, T., Somerville, S.C., Manners, J.M., 2000. Coordinated plant defense responses in Arabidopsis revealed by microarray analysis. Proc. Natl. Acad. Sci. USA. 97, 11655−11660.

Schnug, E., 1990. Sulphur nutrition and quality of vegetables. Sulfur. Agric. 14, 3−7.

Schnug, E., Haneklaus, S., Murphy, D., 1993. Impact of sulfur fertilization on fertilizer nitrogen efficiency. Sulfur. Agric. 17, 8−12.

Schonhof, I., Klaring, H.P., Krumbein, A., Schreiner, M., 2007a. Interaction between atmospheric CO_2 and glucosinolates in broccoli. J. Chem. Ecol. 33, 105−114.

Schonhof, I., Blankenburg, D., Muller, S., Krumbein, A., 2007b. Sulfur and nitrogen supply influence growth, product appearance and glucosinolate concentration of broccoli. J. Plant Nutr. Soil Sci. 170, 65−72.

Schonhof, I., Klaring, H.P., Krumbein, Claussen W, Schreiner, M., 2007c. Effect of temperature increase under low radiation conditions on phytochemicals and ascorbic acid in greenhouse grown broccoli. Agric. Ecosyst. Environ. 19, 103−111.

Schreiner, M., Beyene, B., Krumbein, A., Stutzel, H., 2009a. Ontogenetic changes of 2-propenyl and 3-indolylmethyl glucosinolates in *Brassica carinata* leaves as affected by water supply. J. Agric. Food Chem. 57, 7259−7263.

Schreiner, M., Krumbeina, A., Mewis, I., Ulrichs, C., Huyskens-Keil, S., 2009b. Short-term and moderate UV-B radiation effects on secondary plant metabolism in different organs of nasturtium (*Tropaeolum majus* L.). Innov. Food Sci. Emerg. Technol. 10, 93−96.

Schweizer, F., Fernández-Calvo, P., Zander, M., Diez-Diaz, M., Fonseca, S., Glauser, G., et al., 2013. Arabidopsis basic helix-loop-helix transcription factors MYC2, MYC3, and MYC4 regulate glucosinolate biosynthesis, insect performance, and feeding behavior. Plant Cell. 25, 3117−3132.

Siemens, D.H., Mitchell-Olds, T., 1998. Evolution of pest-induced defenses in Brassica plants: tests of theory. Ecology 79, 632−646.

Skirycz, A., Reichelt, M., Burow, M., Birkemeyer, C., Rolcik, J., Kopka, J., et al., 2006. DOF transcription factor AtDof1.1 (OBP2) is part of a regulatory network controlling glucosinolate biosynthesis in Arabidopsis. Plant J. 47, 10−24.

Song, L., Thornalley, P.J., 2007. Effect of storage, processing and cooking on glucosinolate content of *Brassica* vegetables. Food Chem. Toxicol. 45, 216−224.

Staswick, P.E., 2008. JAZing up jasmonate signaling. Trends Plant Sci. 13, 66−71.

Steinbrenner, A.D., Agerbirk, N., Orians, C.M., Chew, F.S., 2012. Transient abiotic stresses lead to latent defense and reproductive responses over the *Brassica rapa* life cycle. Chemoecology 22, 239−250.

Sun, Y., Fan, X.Y., Cao, D.M., He, K., Tang, W., Zhu, J.Y., et al., 2010. Integration of brassinosteroid signal transduction with the transcription network for plant growth regulation in *Arabidopsis*. Developmental Cell 19, 765−777.

Sønderby, I.E., Fernando, G., Halkier, B.A., 2010. Biosynthesis of glucosinolates—gene discovery and beyond. Trends Plant Sci. 15, 283−290.

Tamara, J., Marina, C., Marija, S., Maja, Đ., Bilandzic, N., Vorkapić-Furac, J., Redovniković, I.R., 2013. Balance of glucosinolates content under Cd stress in two Brassica species. Plant Physiol. Biochem. 63, 99−106.

Tierens, K., Thomma, B.P.H., Brouwer, M., Schmidt, J., Kistner, K., Porzel, A., et al., 2001. Study of the role of antimicrobial glucosinolate-derived isothiocyanates in resistance of Arabidopsis to microbial pathogens. Plant Physiol. 125, 1688−1699.

Velasco, P., Cartea, M.E., Gonzalez, C., Vilar, M., Ordas, A., 2007. Factors affecting the glucosinolate content of kale (*Brassica oleracea* acephala group). J. Agric. Food Chem. 55, 955−962.

Wallbank, B.E., Wheatley, G.A., 1976. Volatile constituents from cauliflower and other crucifers. Phytochemistry 15, 763−766.

Wang, Y., Xu, W., Yan, X., Wang, Y., 2011. Glucosinolate content and related gene expression in response to enhanced UV-B radiation in Arabidopsis. African J. Biotechnol. 10, 6481−6491.

Wittstock, U., Halkier, B.A., 2002. Glucosinolate research in the Arabidopsis era. Trends Plant Sci. 7, 263−270.

Xu, Z., Escamilla-Treviño, L., Zeng, L., Lalgondar, M., Bevan, D., Winkel, B., et al., 2004. Functional genomic analysis of Arabidopsis thaliana glycoside hydrolase family 1. Plant Mol. Biol. 55, 343−367.

Yan, X., Chen, S., 2007. Regulation of plant glucosinolate metabolism. Planta 226, 1343−1352.

Zang, Y., Kim, H., Kim, J., Lim, M., Jin, M., Lee, S., et al., 2009. Genome-wide identification of glucosinolate synthesis genes in *Brassica rapa*. FEBS J. 276, 3559−3574.

Zhao, F., Evans, E., Bilsborrow, P.E., Syers, J.K., 1994. Influence of nitrogen and sulphur on the glucosinolate profile of rapeseed (*Brassica napus* L.). J. Sci. Food Agric. 64, 295−304.

Plant Responses to Iron, Manganese, and Zinc Deficiency Stress

13

Theocharis Chatzistathis

13.1 Introduction

Micronutrient content in soils depends greatly on parent material. According to Robinson et al. (2009), trace element loading in soils is a function of parent material plus subsequent atmospheric or waterborne deposition. High nickel concentration in serpentine soils is a characteristic example of the influence of parent material on trace element content. Parent material influences total soil micronutrient content, while soil type and properties affect micronutrient availability for plants. For example, Mn concentrations in soils were found to vary from traces (in some Podzol, highly leached, soils of Poland) to more than 10,000 ppm (in unleached alkali soils of Chad, Africa) (Aubert and Pinta, 1977). There are many soil factors influencing micronutrient solubility and availability, such as pH, cation exchange capacity (CEC), organic matter, $CaCO_3$ content, soil texture and moisture.

Iron, manganese, and zinc are among the most important and most studied micronutrients. Iron is involved in chlorophyll formation and in the photosynthetic electron transport (Molassiotis et al., 2006); Mn is a constituent of the complex of oxygen liberation of photosystem II (PSII) and of the isoenzyme of superoxide dismutase with Mn (MnSOD), while it also participates in the water splitting and O_2-evolving system of photosynthesis and plays a crucial role in the photosynthetic electron transport in PSII. In addition, Mn is involved in the metabolism of carbohydrates (reactions of glycolysis), in the reactions of the Krebs cycle, as well as in the biosynthetic pathway of shikimic acid (the intermediate products of this biosynthetic pathway—such as some phenols—are related with increased resistance of plants to fungi and insect attack) (Huber and Wilhelm, 1988; Marschner, 1995). Zinc is necessary for the synthesis of tryptophan (former substance of auxin-IAA), while it is also closely involved in the nitrogen (N) metabolism of plants, as well as in protein (under Zn starvation protein levels are markedly reduced and amino acids and amides are accumulated) and RNA synthesis (Bashir et al., 2012). Under adverse soil conditions plants adapt different strategies to withstand the stress. The most important of these mechanisms and adaptation strategies include exudations from their root system, antioxidant defense mechanisms, adaptation of root morphology to soil conditions in order to exploit greater soil volumes, phytosiderophore release, acidification of the rhizosphere, symbiosis with mycorrhiza, etc.

The purpose of this chapter is to throw light on adaptation strategies developed by plants in order to survive and sufficiently produce yield under Fe, Zn, and Mn deficiency conditions.

P. Ahmad (Ed): Emerging Technologies and Management of Crop Stress Tolerance, Volume 1.
DOI: http://dx.doi.org/10.1016/B978-0-12-800876-8.00013-8

13.2 Iron deficiency in soils

Making up 5% of Earth's crust, Fe is the fourth most abundant element in the geosphere. The Fe content of soils varies from 1 to 20% (on average 3.2%), but its normal concentration in plants is only 0.005% (Graham and Welch, 2000). The main reason is that Fe in soils exists mainly in the forms of hydrogen oxide, phosphate, and other deposited compounds (Grusak et al., 1999), which are insoluble forms. It is known that the chemical composition of parent material has a direct effect on chemical properties of soil (Alifragis, 2008). High Fe content of soils does not necessarily guarantee a high uptake by plants, since Fe solubility is influenced by many soil and other factors. Among the most important Fe-containing minerals are those of hematite, magnetite, biotite, siderite, goethite, granodiorite, pyrite, vivianite, chromite, cubanite, germanite, etc. (Arrieta and Grez, 1971).

The critical Fe level in soils by using DTPA solution is 4.5 mg/kg dw (Lindsay and Norvell, 1978). Irmak et al. (2008) found that Fe concentrations in calcareous soils of the Cukurova region in Turkey varied from 2.60 to 6.0 mg/kg dw in 2005 and from 6.96 to 12.70 mg/kg dw in 2006.

13.3 Soil factors influencing Fe availability and uptake

There are many soil factors influencing Fe solubility and uptake by plants. The most important of these factors are pH, $CaCO_3$ content, organic matter, cation exchange capacity (CEC), soil moisture and oxygen, salinity, the phosphoric ion content, the availability of other nutrients, the presence of some bacteria such as *Thiobacillus* and *Ferrobacillus*, the content of Fe oxides and hydroxides, etc. The influence of each of these factors on Fe availability is briefly described below:

1. **pH.** Iron belongs to the group of micronutrients that are mostly influenced by pH (Alifragis, 2008). Iron chlorosis is a major nutritional disorder in crops growing in calcareous soils. According to Fodor et al. (2012), as pH increased from 4.5 to 7.5, the root ferric chelate reductase (FCR) activity of cucumber plants significantly decreased.
2. **Cation exchange capacity (CEC).** Cation exchange capacity is one of the most important soil properties governing the cycling and availability of trace elements (included Fe) in soils (Kabata and Pendias, 2001).
3. **Organic matter.** Iron exhibits a great affinity to form mobile organic complexes and chelates. These complexes are important for the supply of Fe to plant roots (Kabata and Pendias, 2001), especially under alkaline/calcareous soil conditions.
4. **$CaCO_3$ content.** Iron deficiency is the most common micronutrient constraint provoked by high bicarbonate content. Under such soil conditions, Fe chlorosis occurs. In calcareous soils, the total Fe content in soils may be high, but the available fraction for plants is insufficient (Alcantara et al., 2003).
5. **Soil moisture.** Under flooded (anaerobic) conditions, Fe^{3+} is converted to Fe^{2+} due to the lack of soil oxygen. In these cases, Fe availability for plants is significantly increased; according to Fan et al. (2012), with increasing water shortages in China, rice cultivation is gradually shifting away from continuously flooded conditions to partly or even completely aerobic ones.
6. **The availability of other nutrients.** The interaction between Mn and Fe is the most studied one. Generally, under conditions of Mn toxicity, Fe uptake is significantly reduced (El-Jaoual and Cox, 1998; Chatzistathis, 2008). This is ascribed to the antagonism between Mn^{2+} and

Fe^{2+} for the same absorption sites in the root system of plants. Nevertheless, in some cases the influence of Mn^{2+} on Fe^{2+} uptake may be synergistic. Many times, different genotypes of the same plant species may exert different (synergistic or antagonistic) effects. Chatzistathis (2008) found that under excess Mn conditions Fe uptake was significantly suppressed in the olive cultivars "Picual," "Koroneiki," and "FS-17," while it was enhanced in the cultivar "Manaki." Apart from Mn, other nutrients that have been found to influence the availability and uptake of Fe are Ca, Mg, Zn, P, Cu, etc.

7. **The quantity of Fe oxides-hydroxides and the presence of Fe-oxidizing or reducing bacteria.** There is an equilibrium between Fe^{2+} and Fe^{3+}, greatly depending on soil moisture and oxygen. In wet soils, Fe^{3+} (that of Fe oxides-hydroxides) is converted into Fe^{2+} (accessible to plants). This transformation may be realized by some bacteria genera, such as *Sulfobacillus thermosulfidooxidans*, *Sulfobacillus acidophilus*, and *Acidimicrobium ferrooxidans*, which have been found to be capable of reducing ferric (Fe^{3+}) to ferrous (Fe^{2+}) iron when they were grown under oxygen limitation conditions (Bridge and Johnson, 1998).

13.4 Physiological roles and symptoms of Fe deficiency in plants

Iron is involved in chlorophyll formation and in the photosynthetic electron transport. It was found that Fe starvation resulted in the reduction of the total chlorophyll content of peach rootstocks "GF-677" and "Cadaman," as well as in the decrease of the photosynthetic rate, stomatal conductance, and maximum quantum yield of PSII (F_v/F_m) (Molassiotis et al., 2006). In addition, several enzymes belonging to both respiratory chain and to the tricarboxylic acid cycle are Fe-containing proteins. Iron deficiency also affects N metabolism in cucumber (*Cucumis sativus* L.) plants, since Fe plays a crucial role, being a metal cofactor of enzymes of the reductive assimilatory pathway; particularly, it has been found that under Fe deficiency nitrate reductase activity decreased both at leaf and root level (Borlotti et al., 2012). The morphology and ultrastructure of the mitochondria are also affected by Fe deficiency (Vigani, 2012). The possible roles of mitochondria and chloroplasts, as cellular Fe sensing and signaling sites, offering a new perspective on the integrated regulation of Fe homeostasis in plants, are fully analyzed (Vigani et al., 2013).

When plants suffer from Fe deficiency, the most characteristic macroscopic symptom in their leaves is Fe chlorosis (Figures 13.1–13.3). Iron chlorosis is the result of reduced chlorophyll concentration.

13.4.1 Strategy I plants

In strategy I plant species (developed by all species, with the exception of *Poaceae*), the root system of tolerant plants exposed to Fe deficiency has an enhanced ability to induce H^+ extrusion. It was found that the tolerant to Fe starvation peach rootstock "GF-677," exposed to ($-$ Fe) treatment, induced a strong H^+ extrusion, compared to the sensitive rootstock "Cadaman" (Molassiotis et al., 2006). Iron-efficient dicotyledonous are able to improve Fe uptake by enhancing root ferric chelate reductase and ATPase enzyme activities and also by increasing the release of reductants into the rhizosphere (Romheld and Marschner, 1986). Wu et al. (2012) found that Fe deprivation in a portion of the root system of *Malus xiaojinensis* induced a dramatic increase in Fe(III) reductase activity and proton extrusion in the Fe-supplied portion, suggesting that Fe deficiency responses were mediated by systemic auxin (IAA) signaling.

FIGURE 13.1

Typical Fe chlorosis.

Source: http://keys.lucidcentral.org/keys/sweetpotato/key/
Sweetpotato%20Diagnotes/Media/Html/TheProblems/
MineralDeficiencies/IronDeficiency/Iron%20deficiency.html.

FIGURE 13.2

Iron deficiency (lime-induced chlorosis).

Source: http://www.wwgreenhouses.com/more-info/
plant-problems.html.

13.4.2 Strategy II plants

In strategy II plant species suffering from Fe starvation, the synthesis and secretion of phytosiderophores (PS) are enhanced in the environment of the rhizosphere. Genotypic differences in Fe deficiency stress tolerance are correlated with the quantities of phytosiderophores (PS) that are released by roots in order to mobilize Fe (plants may uptake the complexes Fe—PS) under conditions of Fe deficiency (Takagi et al., 1984).

13.5 Physiological mechanisms and adaptation strategies of plants under Fe deficiency conditions

1. **Root exudations.** In some strategy I species (in strategy I plants, developed by dicotyledonous species, the action of proton-extruding H^+-ATPases are responsible for the solubilization of Fe through rhizosphere acidification) (Mlodzinska, 2012) the release of organic compounds, such as phenolics, flavins, sugars, and organic acids could help in the solubilization of Fe-containing compounds (Lopez-Millan et al., 2009). Increased exudation of organic substances under conditions of Fe starvation may be an adaptation mechanism of some strategy I plants. In strategy II species, there is an increase in the synthesis and secretion of phytosiderophores (PS) to the rhizospheric environment (Lopez-Millan et al., 2012); afterwards, the Fe—PS complexes are easily taken up by plants. In many grasses, PS are secreted only during the morning after being accumulated during the rest of the day and in the night (Walter et al., 1995). In roots of *Hordeum vulgare* L., five phytosiderophores (deoxymugineic acid, mugineic acid,

FIGURE 13.3

Iron deficiency.

Source: http://www.yates.co.nz/problem-solver/problems/iron-deficiency/.

epihydroxymugineic acid, avenic acid, and hydroxyavenic acid) were identified under Fe deficiency conditions (Tsednee et al., 2012). According to Molassiotis et al. (2005a), Fe(III) chelate reductase activity was higher in the (− Fe) treated roots than in the (+ Fe) treated ones of the peach rootstock GF-677 (*Prunus amygdalus* x *Prunus Persica*). Beyond the classical responses of strategy I plants, *Parietaria diffusa* showed a greater production and exudation of phenolics and organic acids, when grown in the presence of high bicarbonate content (limited Fe availability) (Donnini et al., 2012).

2. **The reducing capacity and acidification of the rhizosphere.** The reducing capacity of the root system consists of liberating H^+ to the rhizospheric environment; after the liberation of H^+ pH is decreased and Fe solubility is increased. Romera et al. (1991) studied the characterization of the tolerance to Fe chlorosis in different peach rootstocks and found that, when plants grown without Fe, "Nemaguard" was the most chlorotic rootstock, developed the least reducing capacity, and mobilized less Fe from the root system to young leaves, compared to the other genotypes. According to Slatni et al. (2012), induction of root Fe(III) reductase activity is necessary for Fe uptake and can be coupled to the rhizosphere acidification capacity linked to the H^+-ATPase activity.

3. **Root morphology adaptations.** Under low Fe concentrations, the increase of root hair length allows the effective uptake of Fe; this is a strategy adopted by many plant species in order to survive under conditions of low Fe availability. According to Schmidt et al. (2003), the increase of root hair length and number of transfer cells are positively correlated with the amount of detectable H^+-ATPases. These morphological and physiological responses in roots under Fe starvation might be regulated and induced by ethylene and/or auxin signaling (Schmidt et al., 2003).

4. **Physiological and biochemical adaptations.** It was found that ethylene (and possibly salicylic acid) is involved in the regulation of Fe reduction in explants of peach trees (Molassiotis et al., 2005a). Another possible explanation for the role of ethylene in the regulation of Fe reduction in peach trees under Fe deficiency is that it may directly affect transcription or translation of the *FRO2* gene, responsible for the synthesis of the Fe(III) chelate reductase (Molassiotis et al., 2005a). According to Romera et al. (2011), auxin, ethylene, and NO increase under Fe deficiency, which would be necessary for the up-regulation of Fe acquisition genes in strategy I plants. In the study of Kabir et al. (2012), it was found that the natural variation efficiency

between the Fe-inefficient *Pisum sativum* L. genotype "Parafield" and the Fe-efficient "Santi" was associated with up-regulation of strategy I mechanisms, as well as with enhanced citrate and ethylene synthesis. As in the study of Molassiotis et al. (2005a), the Fe-chelate reductase gene *FRO1* and the Fe-transporter RIT-1 were up-regulated in the Fe-deficient roots of "Santi" (Fe-efficient genotype) (Kabir et al., 2012). Slatni et al. (2012) mention that Fe uptake is related to the expression of a Fe^{2+} transporter (IRT1). Under Fe deficiency an overaccumulation of IRT-1 proteins was observed, especially around the cortex cells of nodules. According to Zamboni et al. (2012), 97 differentially expressed transcripts, by comparing roots of Fe-deficient and Fe-sufficient tomato plants, were related to the physiological responses of tomato roots to the nutrient stress, resulting in an improved Fe uptake, including regulatory aspects, translocation, root morphological modification, and adaptation in primary metabolic pathways, such as glycolysis and TCA cycle. At molecular level, the study of Vigani et al. (2012) revealed a set of new genes overexpressed under Fe deficiency, such as those coding for calmodulin (calmodulin protein accumulated in Fe-deficient root apexes), SNAP, TIM23, and V-PPase in the root system of *Cucumis sativus* plants (Vigani et al., 2012).

5. **Antioxidant defense mechanism of plant tolerance.** A common consequence of most abiotic and biotic stresses (like Fe deficiency) is the increased production of reactive oxygen species (ROS), i.e., superoxide (O_2^-), hydrogen hyperoxide (H_2O_2), hydroxyl radicals (OH^-), etc. The production of ROS leads to peroxidation of membrane lipids, base mutation, breakage of DNA strands, and destruction of proteins (Mittler, 2002). Under stress conditions (like under Fe deficiency), plant cells have an antioxidant defense mechanism against ROS consisting of: (1) non-enzymatic low-molecular-weight antioxidants, such as ascorbate, α-tocopherol, glutathione, and carotenoids and (2) enzymatic antioxidant system, according to the following reactions:

$$O_2 + e^- \rightarrow O_2^-$$

$$O_2^- + O_2^- + 2H^+ \rightarrow H_2O_2 + O_2 \text{ (with the aid of SOD)}$$

$$2H_2O_2 \rightarrow 2H_2O + O_2$$

(with the aid of peroxidase and catalase) (Marschner, 1995)

Great differences among genotypes exist concerning tolerance mechanisms to Fe starvation. Molassiotis et al. (2005b), who studied the antioxidant mechanisms in five *Prunus* rootstocks ("Peach seedling," "Barrier," "Cadaman," "Saint Julien 655/2," and "GF-677") differing in resistance to Fe deficiency, found that the tolerance of these genotypes was associated with induction of an antioxidant defense mechanism (superoxide dismutase, peroxidase, catalase, isoforms of SOD, non-enzymatic antioxidants, etc.).

13.6 Manganese deficiency in soils

The Mn concentration of different rocks varies a lot; generally, the greatest Mn contents have been found in basic eruptive rocks (basalt, gabbro), varying between 1000 and 2000 ppm. Manganese contents also vary widely in acid eruptive (granite, rhyolite), metamorphic (schists), as well as in certain sedimentary rocks, varying between 200 and 1200 ppm. Average contents are found in limestones (400–600 ppm), while in sands Mn contents are relatively low (20–500 ppm). Commonly, the manganese range in rocks varies from 350 to 2000 ppm (Kabata and Pendias, 2001), while Lindsay (1979) reports that the mean Mn concentration of the lithosphere is about 900 ppm.

The basic Mn forms in soils are the following: (1) soluble Mn (that of soil solution, i.e., Mn^{2+}), (2) exchangeable Mn, (3) organically bound soluble Mn (that which can be taken up by plants as an organic complex), (4) easily reducible Mn, (5) inert and insoluble Mn (that of the lattice of minerals, as well as that contained in Mn oxides-hydroxides), and (6) insoluble complexes of Mn with organic matter (Smith and Paterson, 1990). From the above-mentioned Mn forms, only the soluble (Mn^{2+}) and the exchangeable Mn are the forms that can be directly taken up by plants. The basic Mn sources in soils are different minerals, by which, through the processes of weathering, hydrolysis, and dilution, Mn^{2+} is "liberated" in soil solution (then it can be absorbed by plants).

There is an equilibrium between soluble Mn (Mn^{2+}) and insoluble (inert, in the form of Mn^{4+}) in soils, according to the following reaction:

$$MnO_2 + 4H^+ + 2e^- \rightarrow Mn^{2+} + 2H_2O \qquad (13.1)$$

Many factors influence the transformation of Mn^{4+} to soluble Mn^{2+}, a form that can be absorbed by plants. Conditions of good soil aeration and alkalinity are the most important factors that favor the prevalence of insoluble Mn (Mn^{4+}). In contrast, when soils are humid/or waterlogged and soil oxygen is poor, Mn is reduced from Mn^{4+} to Mn^{2+} according to the Eq. 13.1.

13.7 **Soil factors influencing Mn availability**

There are many soil factors influencing the concentration of soluble Mn (that of Mn^{2+}) in soils. The most important factors are pH, the organic matter content, the content in $CaCO_3$, soil moisture, redox potential, soil texture, soil microorganisms, the phosphoric ion content, the content of soils in Fe and Al oxides-hydroxides, the interaction with other nutrients, etc. The influence of each one of these factors is briefly described below:

1. **pH and $CaCO_3$ content.** The solubility of Mn is increased when soil pH is decreased. Manganese uptake is closely related to soil pH, more than that of other micronutrients (Marschner, 1988). Generally, the highest Mn availability occurs at pH 5–5.5 (Alifragis, 2008). In the same way, when the content of $CaCO_3$ increases, Mn solubility is decreased.
2. **Soil moisture and redox potential.** In wet or waterlogged soils, Mn^{4+} is converted into Mn^{2+} (a form that can be absorbed by plants) due to the lack of oxygen. In dry soils, Mn is prevalent in insoluble, unavailable for plants form (Mn^{4+}). Soil moisture (thus, Mn reduction) is closely associated with redox potential in soils. According to Grass et al. (1973), the reduction of Mn takes place when the redox potential is lower than 300 mV.
3. **The organic matter content.** Manganese forms complexes with organic matter. Usually these complexes are insoluble; thus, Mn availability for plants decreases. However, there are cases when organic Mn complexes constitute about 80–95% of the total soluble Mn forms; in addition, their stability is usually increased with the increase of soil pH. So, the importance of

organic Mn complexes for mineral nutrition of plants is important under alkaline soil conditions (Alifragis, 2008). Wei et al. (2006) found that the decomposition of organic matter would have provided protons to the soil solution and resulted in the dissolution and reduction of Mn, increasing its availability in some soils after 18 years of cropping and fertilization.

4. **The phosphoric ion content of soils.** High phosphoric ion content in soils may lead to the formation of insoluble Mn−phosphate substances (Alifragis and Papamichos, 1994). In that case, Mn availability is significantly decreased and plants may suffer from Mn deficiency.

5. **Soil texture.** Soil texture influences Mn availability. More specifically, in sandy soils Mn is leached easier than in clayey ones (Alifragis and Papamichos, 1994).

6. **Soil microorganisms.** Manganese oxidation may take place through biological processes (basically through aerobic bacteria). These bacteria are not very sensitive to soil acidity, but they are very active in pH ranging from 6 to 7.5 (Smith and Paterson, 1990). According to Douka (1973), bacteria of the genus *Pseudomonas* and *Citrobacter*, found in western Peloponnesus, Greece, are capable of oxidizing (through enzymatic or non-enzymatic processes) Mn^{2+} into Mn oxides (Mn^{4+}). Other kinds of microorganisms that may oxidize Mn^{2+} into Mn oxides are the bacteria of genus *Arthrobacter* sp., which are particularly active in pH ranging from 5.7 to 7.5 (Marschner, 1988).

13.8 Physiological roles and symptoms of Mn deficiency in plants

Manganese plays a crucial role in many biochemical as well as physiological functions of plants. Concerning biochemical functions, Mn is a constituent of three enzyme complexes; it is constituent of the complex of oxygen liberation of PSII, of the isoenzyme of superoxide dismutase with Mn (MnSOD), and of the acid phosphatases complex (Burnell, 1988). The most important and unique physiological role of Mn appears to be its participation in the water splitting and O_2-evolving system of photosynthesis; it also plays a basic role in the photosynthetic electron transport system. Apparently, the Mn fraction that is loosely bound in chloroplasts is associated with O_2 evolution, whereas the firmly bound Mn fraction is involved in the electron pathway in photosynthesis (Kabata and Pendias, 2001). Superoxide dismutase (SOD) is a group of metalloenzymes, widely distributed in biological systems, which catalyze the transformation of ROS into free oxygen and H_2O_2. ROS, which derive from the union of one electron with free oxygen (O_2), may cause lipid peroxidation, as well as damage to the cell membranes, amino acids and nucleic acids, and ultimately leads to cell death. Finally, acid phosphatases are metalloenzymes found to contain Mn in their molecule. These enzymes, which catalyze the hydrolysis of phosphoric monoesters under acid conditions, have been isolated from a wide variety of vegetal species. The most studied and famous phosphatase is that found in *Ipomoea batatas* plants, with molecular weight 110 kDa, containing 1−2 atoms of Mn in its molecule (Burnell, 1988). Manganese is also an activator of about 35 enzymes, catalyzing reactions of oxidation−reduction, decarboxylations, reactions of hydrolysis, etc. In Box 13.1 are presented some of the most important enzymes, which are activated by Mn. Generally, Mn is involved in the metabolism of carbohydrates (reactions of glycolysis) and of N, in the reactions of the Krebs cycle, in photosynthesis, in the biosynthetic pathway of shikimic acid, as well as in the activities of the enzymes IAA oxidase, polyphenol oxidase, allantoate amylohydrolase, etc.

BOX 13.1 SOME ENZYMES THAT ARE ACTIVATED BY MN

- NAD-malic enzyme
- NADP-malic enzyme
- NAD-isocitrate dehydrogenase
- NADP-isocitrate dehydrogenase
- Hydroxylamine reductase
- Hexokinase
- Phosphoglucokinase
- Pyruvate kinase
- UDP-glucose pyrophosphorylase
- NAD kinase
- Adenosine kinase
- Arginine kinase
- Phosphoglycomutase
- Arginase
- Alkaline phosphatase
- Acid phosphatase
- PEP carboxykinase
- PEP carboxylase
- Enolase, glutamine synthetase, pyruvate carboxylase, peroxidase, Mn-superoxide dismutase (MnSOD)

Source: *Modified from Burnell, 1988.*

Apart from the above-mentioned roles of Mn in basic physiological and biochemical functions, Mn is also involved as cofactor in many other key reactions, which lead to the biosynthesis of secondary metabolites of high importance for plants. The total of these key reactions constitutes the metabolic pathway of shikimic acid. A significant number of phenols, which are related to plant resistance to fungi and insect attack, are derived from intermediate metabolites of the biosynthetic pathway of shikimic acid. Caffeic, chlorogenic, and ferrulic acid are among the most important phenols related with plant resistance to fungi and insect attack (Burnell, 1988; Huber and Wilhelm, 1988; Marschner, 1995).

The critical Mn deficiency level for most plant species varies from 15 to 25 ppm (Kabata and Pendias, 2001). Below these concentrations, crops suffer from Mn starvation. In that case, macroscopic symptoms, such as small yellow spots on leaves and interveinal chlorosis, are usual. This kind of chlorosis may sometimes be confused with that of Fe deficiency; however, in the cases of Fe starvation the whole leaf remains chlorotic (Figure 13.4).

13.9 Physiological mechanisms and adaptation strategies of plants under Mn deficiency conditions

There are many physiological mechanisms and adapted strategies of plants under conditions of Mn starvation. Some of the most important mechanisms include the acidification of the rhizosphere, the root exudation of low-molecular-weight organic compounds, the development of microbial

FIGURE 13.4

Contrast between Fe (left) and Mn (right) deficiency in *Vitis vinifera* L. plants. The chlorosis caused by iron deficiency is much more uniform as compared to the mosaic pattern of manganese deficiency, and with iron deficiency the veins change color as well.

Source: http://www.omafra.gov.on.ca/IPM/english/grapes/plant-nutrition/iron.html.

associations in the rhizospheric environment, as well as the development of internal (enhanced transport from root system to shoots and/or enhanced photosynthetic efficiency) and external (uptake efficiency) Mn utilization efficiency mechanisms. More specifically:

1. **Root exudation.** Gherardi and Rengel (2004) found that low-molecular-weight carboxylates, commonly found in plant root exudates, have the potential to increase the availability of Mn in the rhizosphere. According to them, release of various compounds into the rhizosphere by plant roots may be a mechanism by which certain species and genotypes are able to tolerate conditions of low Mn availability better than less tolerant ones. More specifically, different rates of exudation of carboxylates may play a role in differential genotypic tolerance to Mn deficiency in Lucerne. Among the most important carboxylates exudated by the roots are malonate, citrate, and succinate (they were the majority of the total identifiable carboxylate exudation) (Gherardi and Rengel, 2004).

2. **The development of microbial populations (Mn reducers) in the rhizosphere.** Microbial populations in the rhizosphere of wheat genotypes was found to be correlated with the concentration of DTPA-extractable Mn in the rhizosphere, as well as with the shoot dry matter and total shoot Mn uptake (Marschner et al., 2003). Microorganisms that oxidize Mn from Mn^{2+} to Mn^{4+} (as some bacteria genus mentioned above) decrease their availability to plants. Under conditions of Mn deficiency, the number of Mn reducing microorganisms was higher in the rhizosphere of tolerant to Mn deficiency wheat genotypes than in that of sensitive ones; in contrast, there were no significant differences among tolerant and sensitive genotypes to sufficient Mn supply (Rengel et al., 1996). Rengel (1997) also found that the rhizosphere of wheat plants contained an increased proportion of Mn reducers under Mn deficiency than under Mn sufficiency.

3. **The acidification of the rhizosphere.** Some vegetal species are able to acidify their rhizosphere under alkaline conditions (reduction of Mn^{4+} to Mn^{2+}, which can be absorbed by plants) (Reuter et al., 1988). According to Rengel (2001), Mn use-efficient genotypes are capable of increasing soil Mn available pools through changing chemical (acidification) and microbiological properties of their rhizosphere.

4. **Manganese use efficiency mechanisms.** According to Jiang (2008), more than one mechanism has arisen in wheat to confer tolerance to Mn deficiency. More specifically, the ability of the wheat genotype "C8MM" to accumulate higher amounts of Mn is the basis for the enhanced Mn efficiency of that cultivar, compared to that of "Paragon." In contrast, "Maris Butler" was found to have a high internal use efficiency for Mn, rather than a high Mn uptake mechanism. In many cases, it seems that the internal use efficiency mechanisms of Mn (like improved photosynthetic efficiency) are more important than the external (high uptake) mechanisms. High internal use efficiency of Mn for tree species may be related to enhanced transport of Mn from root system to the shoots. Indeed, Mn-efficient olive cultivars are related to enhanced transport of Mn from the root system to shoots, rather than to high Mn uptake. In contrast, Mn inefficient olive genotypes were found to present low Mn transport from root to shoots (low internal Mn use efficiency), despite their high Mn uptake rates (Chatzistathis et al., 2009).

5. **Root system colonization by arbuscular mucorrhizal fungi (AMF).** It has been found that AMF colonization enhances Mn uptake under alkaline and/or Mn-deficient soil conditions (Wu et al., 2011).

13.10 Zinc deficiency in soils

Generally, as soil pH increases, Zn solubility is significantly decreased, e.g., under calcareous/alkaline soil conditions Zn deficiency is often observed. Generally, DTPA extractable Zn concentrations below 0.5 mg/kg soil are considered deficient (Lindsay and Norvell, 1978). Phattarakul et al. (2012) studied the soil and rice Zn status in China, India, Lao PDR, Thailand, and Turkey and found that in a pH ranging from 4.8 to 8.8, the DTPA extractable Zn concentrations varied from 0.5 (corresponding to the highest pH values) to 6.5 mg/kg soil (corresponding to the most acidic soil conditions). Critical Zn soil levels for the expression of deficiency symptoms depend on the extractant solution used for the determination of Zn, as well as on the kind of species (crops). Critical soil Zn levels for sorghum (cvs. "PARC-SS-1" and "Potohar 4-8") were 3.1−3.4 mg/kg, according to the DTPA test extractant and 3.5−3.7 mg/kg, according to the AB-DTPA extractant solution. These critical quantities were much higher (7.2−8.0 mg/dm^3) when Mehlich-3 was used for extraction (Rashid et al., 1997). Critical soil Zn deficiency values in rice crops were found to vary from 0.3 to 4.2 mg/kg soil, depending on the "plant available" extractant method (DTPA-TEA, AB-DTPA, EDTA, AB-EDTA, 0.05 M HCl, 0.1 M HCl, Mehlich-3) (Impa and Johnson-Beebout, 2012).

13.11 Soil factors influencing Zn availability

There are many soil factors influencing Zn solubility and availability for plants. Among these factors, the most important are pH, calcium carbonate, organic matter content, parent material, soil moisture, soil texture, the quantity of Fe oxides and hydroxides, the concentrations of other nutrients, which may limit or enhance Zn solubility, etc. The influence of each one of these factors on Zn solubility and availability for plants is briefly described below:

1. **pH and CaCO₃ content.** As pH and $CaCO_3$ increased, Zn availability is significantly decreased. According to Mengel and Kirkby (2001), adsorption and occlusion of Zn by carbonates are the major causes of poor Zn availability and appearance of Zn deficiency symptoms on calcareous soils.

2. **Organic matter.** Zinc interacts with soil organic matter, and both soluble and insoluble Zn organic complexes are formed. On average, about 60% of the soluble Zn in soils occurs as soluble Zn organic complexes. The contribution of these complexes to plant Zn mineral nutrition is valuable under calcareous/alkaline soil conditions. According to Kabata and Pendias (2001), Zn organic complexes may account for the availability of this trace element at higher pH values.

3. **Salinity.** High pH is the main problem of alkaline soils. Under adverse soil conditions, the solubility of Zn is significantly decreased and the crops may suffer from Zn deficiency symptoms.

4. **Soil moisture.** Soil moisture may influence Zn availability in two ways: under water shortage conditions Zn solubility is considerably decreased and it is not available for plants. Rego et al. (2007) found that dry lands in the semi-arid regions of India, suffering from water shortage, are more susceptible to widespread deficiency of Zn than other soil types with sufficient soil water content. On the other hand, in flooded soils, where leaching occurs (especially in light, sandy soils), Zn availability may be also reduced.

5. **Parent material.** The level of Zn in soils is very much related to the parent material and to the quantity of Zn-containing minerals. For example, soils originating from basic igneous rocks are high in Zn. Zinc is contained in many oxides, such as (ZnO), sulfides (ZnS), sphalerite-(ZnFe) S, carbonate salts ($ZnCO_3$), etc. The two Zn silicates $ZnSiO_3$ and Zn_2SiO_4 (willemite) also occur in some soils (Kabata and Pendias, 2001; Mengel and Kirkby, 2001; Alifragis, 2008).

6. **The kind of clay minerals and the quantity of Fe oxides and hydroxides.** Montmorillonite is the clay mineral that is basically present in soil; Zn may enter its layer lattice structures and become very immobile, thus it is not available for plants (Kabata and Pendias, 2001). So, in montmorillonitic soils it is possible to face Zn deficiency than in other soils, like caolinitic ones. According to Abd-Elfattah and Wada (1981), the highest selective adsorption of Zn was found by Fe oxides, halloysite, and allophane.

7. **Interactions with other nutrients.** In soils receiving rich P fertilization, insoluble Zn phosphate substances are formed, so Zn availability is reduced. From that point of view, excess P fertilizations should be avoided. Zinc availability and uptake by plants may be also reduced due to competition with other cations (e.g., Ca, Mg, Na) at the root surface (Marschner, 1995). The negative influence of excess Ca on Zn uptake may be direct (due to competition for cation uptake in the root surface, i.e., Ca^{2+}/Zn^{2+}), or indirect, via the elevation of soil pH and decrease of Zn solubility.

13.12 Physiological roles and symptoms of Zn deficiency in plants

Zinc is completely necessary for the synthesis of tryptophan (former substance of auxin-IAA), while it is also a constituent of other enzymes, such as metalloenzymes and dehydrogenases (dehydrogenase of glutamic acid, dehydrogenase of L-galactic acid, etc.) (Therios, 1996). Furthermore, Zn is very closely involved in the N metabolism of plants, protein synthesis (under Zn starvation protein levels are markedly reduced and amino acids and amides are accumulated), and RNA synthesis (RNA polymerase contains Zn and when there is lack of Zn this enzyme is inactivated and RNA synthesis is impaired), as well as in carbohydrate and lipid metabolism (Bashir et al., 2012).

According to Kabata and Pendias (2001), the deficiency of Zn in plants has been determined from 10 to 20 ppm and it also depends on plant species and genotype. General symptoms of Zn deficiency for fruit trees are the formation of small leaves and small fruit size. Unevenly distributed clusters, or rosettes of small stiff leaves, are formed at the ends of the young shoots. Frequently, the die-off and leaves fall prematurely. In apple trees, the disease occurs in the early part of the year and is known as rosette or little-leaf (Therios, 1996). In *Citrus* trees the main veins remain green, as if the leaves were recovering from Fe deficiency (Figure 13.5). A variety of visual typical Zn deficiency symptoms has been noted for different plant species, such as whitish-brown necrotic patches on leaf blades of wheat plants (Cakmak et al., 1997), interveinal chlorosis in *Juglans regia* (Therios, 1996), and chlorotic bands forming on either side of the midrib of the leaf in the monocots (Mengel and Kirkby, 2001).

13.13 Physiological mechanisms and adaptation strategies of plants under Zn deficiency conditions

There are some tolerance mechanisms and adaptation strategies adopted by plants in order to overcome Zn starvation. Some of the most important strategies include the release of phytosiderophore, the differential antioxidant capacity of genotypes, the adjustment of root system to adverse soil conditions in order to increase Zn uptake, the formation of arbuscular mycorrhizal fungus (AMF), the enhanced mobilization and translocation of Zn among tissues, etc.

1. **Root exudations and phytosiderophore release.** Wheat genotypes tolerant to Zn deficiency when grown under conditions of Zn starvation, released greater amounts of the phytosiderophore 2-deoxymugineic acid than the sensitive genotypes (Rengel, 1997). In soils with limited Zn availability, Zn uptake may be enhanced by the exudation of low-molecular-weight compounds, like malate and mugineic acid family phytosiderophores (Suzuki et al., 2006; Peng et al., 2009). Under high salinity conditions phytosiderophore release by plants is enhanced in order to increase Zn solubility and uptake by plants. According to Daneshbakhsh et al. (2013), the highest amount of phytosiderophores by the roots of the wheat genotype "Kavir," grown hydroponically, were released at the highest salinity level (120 mM NaCl). Furthermore, a good relationship was found between root exudation of phytosiderophores and differential tolerance to Zn deficiency, when durum (sensitive to Zn starvation) and bread wheat (tolerant to Zn deficiency) were tested (Cakmak et al., 1994, 1996).

2. **Differential genotypic antioxidant ability.** The production of ROS under any kind of stress is a very serious problem and a concern regarding the functional permeability of membranes. The effect of Zn stress on antioxidant ability has been thoroughly studied by Daneshbakhsh et al. (2012) and Khoshgoftarmanesh et al. (2006). The differential tolerance to salt stress among

FIGURE 13.5

Zinc deficiency symptoms in Citrus trees.

Source: http://www.vaniperen.com/Products/Trace-elements/Non-chelated/Oligo-Zinc-Sulphate-11-Liquid.aspx.

wheat genotypes was highly related to their tolerance to Zn deficiency. Elevated tolerance to high salinity-induced Zn deficiency was probably associated with significant differences in antioxidant defense capacity among them (significantly greater root activity of catalase and superoxide dismutase) (Khoshgoftarmanesh et al., 2006; Daneshbakhsh et al., 2012).

3. **AMF colonization.** In soils with low Zn availability, the root system may be colonized by mycorrhiza in order to increase Zn uptake (Ortas et al., 2002; Cavagnaro et al., 2010; Han et al., 2012; Scagel and Lee, 2012). In the study of Balakrishnan and Subramanian (2012), it was found that in mycorrhizal maize plants (inoculated with *Glomus intraradices*) siderophore production was higher than in the non-mycorrhizal ones, irrespective of Zn levels. In addition, Zn grain contents in mycorrhizal plants were 15% greater than in the non-mycorrhizal ones (Balakrishnan and Subramanian, 2012). Apart from the increase in the siderophore release, another possible explanation for the increased Zn uptake rates of AMF colonized plants may be the decrease in pH of the rhizospheric environment. Mohammad et al. (2005) also reported that the decrease in pH in the rhizosphere of wheat plants inoculated with AMF could be the possible mechanism for the increased Zn uptake.

4. **The adjustment of root system morphology.** Plants tend to alter the morphology of their root system under nutrient deficient conditions for efficient nutrient acquisition. Enhanced root growth under Zn deficiency, both in length and number of roots, has been associated with tolerance to Zn deficiency in lowland rice genotypes (Impa and Johnson-Beebout, 2012). The formation of fine roots may be an adaptation mechanism of plants in order to sufficiently cover their Zn nutritional needs (Suzuki et al., 2006).

5. **Enhanced mobilization and translocation mechanisms of Zn among plant tissues.** Zinc may be bound to soluble low-molecular-weight proteins or light organic compounds in xylem fluids (Kabata and Pendias, 2001). In barley, for example, the uptake of Zn—deoxymugineic acid (DMA) complexes is preferred to Zn^{2+} through roots (Suzuki et al., 2006). In contrast, rice plants take up less Zn—DMA complexes compared to Zn^{2+} (Suzuki et al., 2008). Despite the

fact that the uptake of Zn^{2+} is preferred over Zn—DMA complexes by rice plants, translocation of Zn from root system to shoot is greater as Zn—DMA complexes than as Zn^{2+} (Ishimaru et al., 2011). The internal allocation processes include translocation of Zn from root to shoots, from older to younger leaves, from leaves to grain, as well as from root to stems and to grain (Impa and Johnson-Beebout, 2012). Translocation and remobilization of Zn was found to be critical for grain Zn accumulation when its availability is restricted during grain filling (Kutman et al., 2012). Palmgren et al. (2008) suggested that in wheat and barley Zn remobilization contributes to grain Zn accumulation more than uptake. Wu et al. (2010) reported that genotypes having a pronounced ability to mobilize and translocate Zn from source tissues to grain are highly Zn efficient as compared to those having limited Zn mobilization ability, thus the first category should be selected and preferred for cultivation in Zn-deficient soils.

13.14 Conclusion and future perspectives

In conclusion, Fe, Mn, and Zn availability has been found to be influenced by many soil factors, such as pH, $CaCO_3$, CEC, salinity, organic matter, soil microorganisms, the quantity of Fe oxides and hydroxides, etc. Adverse soil conditions limit their solubility and availability to plants. Plants develop important adaptation strategies and physiological mechanisms of tolerance under stress. Some of these strategies include the acidification of the rhizosphere, the production and release of root exudates and phytosiderophores from their root system, the adjustment of root morphology and the production of root absorptive hairs in order to increase micronutrient uptake, the development of suitable antioxidant mechanism system, AMF colonization, the enhanced remobilization and translocation mechanisms from plant tissues having greater Fe, Mn, and Zn concentrations to other ones containing insufficient micronutrient levels, etc. Since some of the physiological mechanisms have not yet been fully explained and understood more efforts are required in the near future to provide a thorough insight to these mechanisms. To this direction, it is absolutely necessary to understand the adaptation strategies by plants at the molecular level. The production of transgenic plants having the ability to overproduce root exudates and phytosiderophores, may be a good solution for reclaming Fe, Mn, and Zn-deficient soils. Furthermore, the production of genotypes with enhanced antioxidant mechanisms under stress conditions could be another milestone in understanding plant adaptability under micronutrient stress.

References

Abd-Elfattah, A., Wada, K., 1981. Adsorption of lead, copper, zinc, cobalt and cadmium by soils that differ in cation-exchange materials. J. Soil Sci. 32, 271.

Alcantara, E., Cordeiro, A.M., Barranco, D., 2003. Selection of olive varieties for tolerance to iron chlorosis. J. Plant Physiol. 160, 1467—1472.

Alifragis, D., 2008. Soil: Genesis, Properties and Classification, vol. I. Aivazi Publications, Thessaloniki, Greece, pp. 487—492. In Greek

Alifragis, D., Papamichos, N., 1994. Fertility of Forest Soils. Dedousis Publications, Thessaloniki, Greece, In Greek.

Arrieta, L., Grez, R., 1971. Solubilization of iron-containing minerals by soil microorganisms. Appl. Environ. Microbiol. 22, 487–490.

Aubert, H., Pinta, M., 1977. Trace Elements in Soils. Elsevier Scientific Publishing Company, Amsterdam, pp. 43–53.

Balakrishnan, N., Subramanian, K.S., 2012. Mycorrhizal symbiosis and bioavailability of micronutrients in maize grain. Maydica 57, 129–138.

Bashir, K., Ishimaru, Y., Nishizawa, N.K., 2012. Molecular mechanisms of zinc uptake and translocation in rice. Plant Soil 361, 189–201.

Borlotti, A., Vigani, G., Zocchi, G., 2012. Iron deficiency affects N metabolism in cucumber (*Cucumis sativus* L.) plants. BMC Plant Biol. 12, 189.

Bridge, T., Johnson, D., 1998. Reduction of soluble iron and reductive dissolution of ferric iron-containing minerals by moderately thermophilic iron-oxidizing bacteria. Appl. Environ. Microbiol. 64, 2181–2186.

Burnell, J.M., 1988. The biochemistry of manganese in plants. In: Graham, R.D., Hannam, R.J., Uren, N.C. (Eds.), Manganese in Soils and Plants. Proceedings of the International symposium on "Manganese in soils and plants". Kluwer Academic Publishers, Dordrecht, The Netherlands, pp. 125–137.

Cakmak, I., Gulut, K.Y., Marschner, H., Graham, R.D., 1994. Effect of zinc and iron deficiency on phytosiderophore release in wheat genotypes differing in zinc deficiency. J. Plant Nutr. 17, 1–17.

Cakmak, I., Sari, N., Marschner, H., Ekiz, H., Kalayci, M., Yilmaz, A., et al., 1996. Phytosiderophore release in bread and durum wheat genotypes differing in zinc efficiency. Plant Soil 180, 183–189.

Cakmak, I., Ozturk, L., Eker, S., Torun, B., Kalfa, H.I., Yilmaz, A., 1997. Concentration of zinc and activity of copper/zinc superoxide dismutase in leaves of rye and wheat cultivars differing in sensitivity to zinc deficiency. J. Plant Physiol. 151, 91–95.

Cavagnaro, T.R., Dickson, S., Smith, F.A., 2010. Arbuscular mycorrhizas modify plant responses to soil zinc addition. Plant Soil 329, 307–313.

Chatzistathis T. (2008) Investigation of the role of Mn on olive trees' mineral nutrition. Doctoral dissertation 2008. Aristotle University of Thessaloniki, Greece. In Greek.

Chatzistathis, T., Therios, I., Alifragis, D., 2009. Differential uptake, distribution within tissues, and utilization efficiency of Mn, Fe and Zn by olive cultivars "Kothreiki" and "Koroneiki.". Hort Sci. 44, 1994–1999.

Daneshbakhsh, B., Khoshgoftarmanesh, A.H., Shariatmadari, H., Cakmak, I., 2012. Effect of zinc nutrition on salinity-induced oxidative damages in wheat genotypes differing in zinc deficiency tolerance. Acta Physiol. Plant 170, 41–46.

Daneshbakhsh, B., Khoshgoftarmanesh, A.H., Sariatmadari, H., Cakmak, I., 2013. Phytosiderophore release by wheat genotypes differing in zinc deficiency tolerance grown with Zn-free nutrient solution as affected by salinity. J. Plant Physiol. 170, 41–46.

Donnini, S., De Nisi, P., Gabotti, D., Tato, L., Zocchi, G., 2012. Adaptive strategies of *Parietaria diffusa* to calcareous habitat with limited Fe availability. Plant Cell Environ. 35, 1171–1184.

Douka A.E. (1973) Oxidation of Mn^{2+} through the action of two Mn-oxidizing bacteria. Taxonomy and study of their morphology and physiology. Doctoral thesis, National and Kapodestrian University of Athens, Greece. In Greek.

El-Jaoual, T., Cox, D.A., 1998. Manganese toxicity in plants. J. Plant Nutr. 21, 353–386.

Fan, X., Karim, R., Chen, X., Zhang, Y., Gao, X., Zhang, F., Zou, C., 2012. Growth and iron uptake of lowland and aerobic rice genotypes under flooded and aerobic cultivation. Com. Soil Sci. Plant Anal. 43, 1811–1822.

Fodor, F., Kovacs, K., Czech, V., Solti, A., Toth, B., Levai, L., et al., 2012. Effect of short term iron citrate treatments at different pH values. J. Plant Physiol. 169, 1615–1622.

Gherardi, M.J., Rengel, Z., 2004. The effect of Mn supply on exudation of carboxylates by roots of Lucerne (*Medicago sativa* L.). Plant Soil 260, 271–282.

Graham, R.D., Welch, R.M., 2000. Plant food micronutrient composition and human nutrition. Com. Soil Sci. Plant Anal. 31, 1627−1640.

Grass, L.B., MacKenzie, A.J., Meek, B.D., Spencer, W.F., 1973. Manganese and iron solubility changes as a factor in tile drain clogging: I. Observations during flooding and drying. Soil Sci. Soc. Am. Proc. 37, 14−17.

Grusak, M.A., Pearson, J.N., Marentes, E., 1999. The physiology of micronutrient homeostasis in field crops. Field Crops. Res. 60, 41−56.

Han, B., Guo, S.R., He, C.X., Yan, Y., Yu, X.C., 2012. Effects of arbuscular mycorrhiza fungi (AMF) on the plant growth, fruit yield and fruit quality of cucumber under salt stress. Chin. J. Appl. Ecol. 23, 154−158.

Huber, D.M., Wilhelm, N.S., 1988. The role of manganese in resistance to plant diseases. In: Graham, R.D., Hannam, R.J., Uren, N.C. (Eds.), Manganese in soils and plants, Proceedings of the International Symposium on "Manganese in Soils and Plants". Kluwer Academic Publishers, Dordrecht, The Netherlands, pp. 155−173.

Impa, S.M., Johnson-Beebout, S.E., 2012. Mitigating zinc deficiency and achieving high grain zinc in rice through integration of soil chemistry and plant physiology research. Plant Soil 361, 3−41.

Irmak, S., Surucu, A.K., Aydin, S., 2008. The effects of iron content of soils on the iron content of plants in the Cukurova region of Turkey. Intern. J. Soil Sci. 3, 109−118.

Ishimaru, Y., Bashir, K., Nishizawa, N.K., 2011. Zn uptake and translocation in rice plants. Rice 4, 21−27.

Jiang, W.Z., 2008. Comparison of responses to Mn deficiency between the UK wheat genotypes Maris Batler, Paragon and the Australian wheat genotype C8MM. J. Integr. Plant Biol. 50, 457−465.

Kabata, A., Pendias, H., 2001. Trace elements in soils and plants. third ed. CRC Press, USA.

Kabir, A.H., Paltridge, N.G., Able, A.J., Paull, J.G., Stangoulis, J.C.R., 2012. Natural variation for Fe-efficiency is associated with up-regulation of strategy I mechanisms and enhanced citrate and ethylene synthesis in *Pisum sativum* L. Planta 235, 1409−1419.

Khoshgoftarmanesh, A.H., Shariatmadari, H., Karimian, N., 2006. Responses of wheat genotypes to zinc fertilization under saline soil conditions. J. Plant Nutr. 29, 1543−1556.

Kutman, B.U., Kutman, B.Y., Ceylan, Y., Ova, E.A., Calemak, I., 2006. Contributions of root uptake and remobilization to grain Zn accumulation in wheat depending on post-anthesis Zn availability and nitrogen nutrition. Plant and Soil 361, 177−187.

Lindsay, W.L., 1979. Chemical Equilibria in Soils. John Wiley and Sons, Inc, New York, pp. 151−160

Lindsay, W.L., Norvell, W.A., 1978. Development of a DTPA soil test for zinc, iron, manganese and copper. Soil Sci. Soc. Am. J. 42, 6421−6428.

Lopez-Millan, A.F., Morales, F., Gogorcena, Y., Abadia, A., Abadia, J., 2009. Metabolic responses in iron deficient tomato plants. J. Plant Physiol. 166, 375−384.

Lopez-Millan, A.F., Grusak, M.A., Abadia, J., 2012. Carboxylate metabolism changes induced by Fe deficiency in barley, a strategy II plant species. J. Plant Physiol. 169, 1121−1124.

Marschner, H., 1988. Mechanisms of manganese acquisition by roots from soils. In: Graham, R.D., Hannam, R.J., Uren, N.C. (Eds.), Manganese in Soils and Plants. Proceedings of the International Symposium on "Manganese in Soils and Plants". Kluwer Academic Publishers, Dordrecht, The Netherlands, pp. 191−204.

Marschner, H., 1995. Mineral nutrition of higher plants. second ed. Academic Press, London, pp. 324−333

Marschner, P., Fu, Q., Rengel, Z., 2003. Manganese availability and microbial populations in the rhizosphere of wheat genotypes differing in tolerance to Mn deficiency. J. Plant Nutr. Soil Sci. 166, 712−718.

Mengel, K., Kirkby, E., 2001. Zinc. In: Mengel, K., Kirkby, E., Kosegarten, H., Appel, T. (Eds.), Principles of Plant Nutrition, fifth ed. Kluwer Academic Publishers, Dordrecht, The Netherlands, pp. 585−597.

Mittler, R., 2002. Oxidative stress, antioxidants and stress tolerance. Trends Plant Sci. 7, 405−410.

Mlodzinska, E., 2012. Alteration of plasma membrane H^+-ATPase in cucumber roots under different iron nutrition. Acta Physiol. Plant 34, 2125−2133.

Mohammad, M.J., Pan, W.L., Kennedy, A.C., 2005. Chemical alteration of the rhizosphere of the mycorrhizal-colonized wheat root. Mycorrhiza 15, 259–266.

Molassiotis, A., Therios, I., Dimassi, K., Diamantidis, G., Chatzissavvidis, C., 2005a. Induction of Fe(III) chelate reductase activity by ethylene and salicylic acid in iron deficient peach rootstock explants. J. Plant Nutr. 28, 669–682.

Molassiotis, A., Tanou, G., Diamantidis, G., Patakas, A., Therios, I., 2006. Effects of 4-month Fe deficiency exposure on Fe reduction, mechanism, photosynthetic gas exchange, chlorophyll fluorescence and antioxidant defence in two peach rootstocks differing in Fe deficiency tolerance. J. Plant Physiol. 163, 176–185.

Molassiotis, A.N., Diamantidis, G.C., Therios, I.N., Tsirakoglou, V., Dimassi, K.N., 2005b. Oxidative stress, antioxidant activity and Fe(III)-chelate reductase activity of five *Prunus* rootstocks explants in response to Fe deficiency. Plant Growth Regul. 46, 69–78.

Ortas, I., Ortakei, D., Kaya, Z., Cinar, A., Onelge, N., 2002. Mycorrhizal dependency of sour orange in relation to phosphorus and zinc nutrition. J. Plant Nutr. 26, 1263–1279.

Palmgren, M.G., Clemens, S., Williams, L.E., Kramer, U., Borg, S., Schjorring, J.K., Sanders, D., 2008. Zinc biofortification of cereals: problems and solutions. Trends Plant Sci. 13, 464–473.

Peng, G.X., Fusuo, Z., Hoffland, E., 2009. Malate exudation by six aerobic rice genotypes varying in zinc uptake efficiency. J. Environ. Qual. 38, 2315–2321.

Phattarakul, N., Rerkasem, B., Li, L.J., Wu, L.H., Zou, C.Q., Ram, H., et al., 2012. Biofortification of rice grain with zinc through zinc fertilization in different countries. Plant Soil 361, 131–141.

Rashid, A., Rafique, E., Bughio, N., Yasin, N., 1997. Micronutrient deficiencies in rainfed calcareous soils of Pakistan. IV. Zinc nutrition of sorghum. Com. Soil Sci. Plant Anal. 28, 455–467.

Rego, T.J., Sahrawat, K.L., Wani, S.P., Pardhasaradhi, G., 2007. Widespread deficiencies of sulphur, boron and zinc in Indian semi-arid tropical soils: On farm crop responses. J. Plant Nutr. 30, 1569–1583.

Rengel, Z., 1997. Root exudation and microflora populations in rhizosphere of crop genotypes differing in tolerance to micronutrient deficiency. Plant Soil 196, 255–260.

Rengel, Z., 2001. Genotypic differences in micronutrient use efficiency in crops. Com. Soil Sci. Plant Anal. 32, 1163–1886.

Rengel, Z., Gutteridge, R., Hirsch, P., Homby, D., 1996. Plant genotype, micronutrient fertilization and take-all colonization influence bacterial populations in the rhizosphere of wheat. Plant Soil 183, 268–277.

Reuter, D.J., Alston, A.M., McFarlane, J.D., 1988. Occurrence and correction of manganese deficiency in plants. In: Graham, R.D., Hannam, R.J., Uren, N.C. (Eds.), Manganese in soils and plants, Proceedings of the International Symposium on "Manganese in Soils and Plants". Kluwer Academic Publishers, pp. 205–224.

Robinson, B.H., Banuelos, G., Conesa, H.M., Evangelou, M.W., Schulin, R., 2009. The phytomanagement of trace elements in soil. Crit. Rev. Plant Sci. 28, 240–266.

Romera, F.J., Alcantara, E., De la Guardia, M.D., 1991. Characterization of the tolerance to Fe chlorosis in different peach rootstocks grown in nutrient solution-II. Iron stress response mechanisms. Plant Soil 130, 121–125.

Romera, F.J., Garcia, M.J., Alcantara, E., Perez-Vicente, R., 2011. Latest findings about the interplay of auxin, ethylene and nitric oxide in the regulation of Fe deficiency responses by strategy I plants. Plant Signal Behav. 6, 167–170.

Romheld, V., Marschner, H., 1986. Mobilization of iron in the rhizosphere of different plant species. Adv. Plant Nutr. 2, 155–204.

Scagel, C.F., Lee, J., 2012. Phenolic composition of basil plants is differentially altered by plant nutrient status and inoculation with mycorrhizal fungi. Hort Sci. 47, 660–671.

Schmidt, W., Michalke, W., Schikora, A., 2003. Proton pumping by tomato roots. Effect of Fe deficiency and hormones on the activity and distribution of plasma membrane H^+-ATPase in rhizodermal cells. Plant Cell Env. 26, 361−370.

Slatni, T., Dell'Orto, M., Ben Salah, I., Vigani, G., Smaoui, A., Gouia, H., et al., 2012. Immunolocalization of H^+-ATPase and IRT1 enzymes in N_2-fixing common bean nodules subjected to Fe deficiency. J. Plant Physiol. 169, 242−248.

Smith, K.A., Paterson, J.E., 1990. Manganese and cobalt. In: Alloway, B.J. (Ed.), Heavy metals in soils, second ed. Blackie Academic and Professional, pp. 225−243.

Suzuki, M., Takahashi, M., Tsukamoto, T., Watanabe, S., Matsuhashi, S., Yazaki, J., et al., 2006. Biosynthesis and secretion of mugineic acid family phytosiderophores in Zn-deficient barley. Plant J. 48, 85−97.

Suzuki, M., Tsukamoto, T., Innoue, H., Watanabe, S., Matsuhasi, S., et al., 2008. Deoxymugineic acid increases Zn translocation in Zn-deficient rice plants. Plant Mol. Biol. 66, 609−617.

Takagi, S., Nomoto, K., Takemoto, S., 1984. Physiological aspect of mugineic acid, a possible phytosiderophore of gramineous plants. J. Plant. Nutr. 7, 469−477.

Therios, I., 1996. Mineral Nutrition and Fertilizers. Dedousis Publications, Thessaloniki, Greece, pp. 174−177. In Greek

Tsednee, M., Mak, Y.W., Chen, Y.R., Yeh, K.C., 2012. A sensitive LC-ESI-Q-TOF-MS method reveals novel phytosiderophores and phytosiderophore-iron complexes in barley. New Phytol. 195, 951−961.

Vigani, G., 2012. Discovering the role of mitochondria in the iron deficiency-induced metabolic responses of plants. J. Plant Physiol. 169, 1−11.

Vigani, G., Chitto, A., De Nisi, P., Zocchi, G., 2012. cDNA-AFLP analysis reveals a set of new genes differentially expressed in cucumber root apexes in response to Fe deficiency. Biol. Plant 56, 502−508.

Vigani, G., Zocchi, G., Bashir, K., Philippar, K., Briat, J.F., 2013. Signals from chloroplasts and mitochondria for Fe homeostasis regulation. Trends Plant Sci. 18, 305−311.

Walter, A., Pich, A., Scholz, G., Marschner, H., Romheld, V., 1995. Effects of iron nutritional status and time of day on concentrations of phytosiderophores and nicotianamine in different root and shoot zones of barley. J. Plant Nutr. 18, 1577−1593.

Wei, X., Hao, M., Shao, M., Gale, W.J., 2006. Changes in soil properties and the availability of soil micronutrients after 18 years of cropping and fertilization. Soil Til Res. 91, 120−130.

Wu, C.Y., Lu, L.L., Yang, X.E., Feng, Y., Wei, Y., Hao, H.L., et al., 2010. Uptake, translocation and remobilization of zinc absorbed at different growth stages by rice genotypes of different Zn densities. J. Agric. Food Chem. 58, 6767−6773.

Wu, Q.S., Li, G.H., Zou, Y.N., 2011. Roles of arbuscular mycorrhizal fungi on growth and nutrient acquisition of peach (Prunus persica L. Batsch) seedlings. J. Anim. Plant Sci. 21, 746−750.

Wu, T., Zhang, H.T., Wang, Y., Jia, W.S., Xu, X.F., Zhang, X.Z., Han, Z.H., 2012. Induction of root Fe(III) reductase activity and proton extrusion by iron deficiency is mediated by auxin-based systemic signalling in *Malus xiaojinensis*. J. Exp. Bot. 63, 859−870.

Zamboni, A., Zanin, L., Tomasi, N., Pezzotti, M., Pinton, R., Varanini, Z., Cesco, S., 2012. Genome-wide microarray analysis of tomato roots showed defined responses to iron deficiency. BMC Genomics 13, 101.

Role of Trace Elements in Alleviating Environmental Stress

Ghader Habibi

14.1 Introduction

Plant metabolism and productivity is influenced adversely by environmental stresses such as salinity, drought, high and low temperature, heavy metals, herbicides, and pathogens (Mittler and Blumwald, 2010). During environmental stress, the rate of reactive oxygen species (ROS) generation in organelles such as chloroplasts, mitochondria, and peroxisomes is dramatically elevated because of an imbalance between ROS production and ROS scavenging (Mittler et al., 2004). The ROS includes superoxide radical ($O_2^{\bullet-}$), hydroxyl radical ($OH^{\bullet}$), hydroperoxyl radical ($HO_2^{\bullet}$), peroxy radical ($ROO^{\bullet}$), singlet oxygen ($^1O_2$), and hydrogen peroxide (H_2O_2), all of which are toxic molecules capable of causing oxidative damage to proteins, DNA, and lipids (Apel and Hirt, 2004; Ahmad and Sharma, 2008; Ding et al., 2010; Miller et al., 2010; Ahmad et al., 2010a,b, 2011, 2012; Ahmad and Umar, 2011; Ahmad and Prasad, 2012a,b).

To protect against these toxic oxygen intermediates, plants possess very efficient enzymatic and non-enzymatic antioxidant defense systems (Miller et al., 2010). The plant cells respond to elevation in ROS levels by enzymatic (such as superoxide dismutase (SOD), ascorbate peroxidase (APX), catalase (CAT), glutathione peroxidase (GPX)) and non-enzymatic antioxidants (such as ascorbic acid (AsA) and glutathione (GSH)), which work in concert to maintain redox homeostasis (Apel and Hirt, 2004; Dietz et al., 2006).

Micronutrients zinc (Zn), copper (Cu), iron (Fe), manganese (Mn), boron (B), molybdenum (Mo), chlorine (Cl), and nickel (Ni) are important constituents for better plant growth and development (Waraich et al., 2011), and also for stress tolerance. It has been well documented that beneficial elements such as aluminum (Al), cobalt (Co), sodium (Na), selenium (Se), and silicon (Si) positively affect plant growth and stress resistance (Pilon-Smits et al., 2009). These plant nutrients are involved in increasing plant resistance to both biotic and abiotic stress factors to maintain productivity and ensure survival (Marschner, 1995; Waraich et al., 2011).

The possible mechanisms of the micronutrient-enhanced tolerance to plants against environmental stresses have not been fully explained. In fact, the ability of trace elements to mitigate various environmental stresses can be attributed to several different mechanisms. Some of these beneficial trace elements have a functional role with respect to plant growth and productivity under variable environmental conditions. In the present chapter, we attempt to highlight and summarize the current knowledge regarding the roles of these trace elements in alleviating environmental stresses.

P. Ahmad (Ed): Emerging Technologies and Management of Crop Stress Tolerance, Volume 1.
DOI: http://dx.doi.org/10.1016/B978-0-12-800876-8.00014-X

14.2 Plant responses to environmental stress

Abiotic stresses such as drought, salinity, and extreme temperature are major causes of crop destructions worldwide. Abiotic stresses trigger common reactions in plants and lead to cellular damages mediated by ROS (Mano, 2002). In chloroplasts, the photosynthetic fixation of CO_2 can regulate the generation of ROS. Under strong light, limited CO_2 fixation decreases adenosine triphosphate (ATP) and NADPH consumption and declines the level of nicotinamide adenine dinucleotide phosphate ($NADP^+$) (Miller et al., 2010). The reduction of $NADP^+$ causes the transport of electrons from photosystem I (PSI) to molecular oxygen resulting in the production of ROS. Over-reduction of the complex I and complex III in the mitochondrial electron transport chain (mtETC) is also a major source of ROS generation during stress (Miller et al., 2010). In plant peroxisomes, glycolate oxidase (GO) produces glycolate and uses O_2 as an electron acceptor to produce H_2O_2 (Mittler et al., 2004; Mhamdi et al., 2012).

Under stress conditions, accumulation of ROS is capable of inducing damage to almost all cellular processes including growth rate, stem elongation, leaf expansion, and stomatal movements (Miller et al., 2010). Enhanced levels of ROS lead to damage to biomolecules such as lipids, proteins, and DNA and can alter intrinsic membrane properties like fluidity. Malondialdehyde (MDA) is one of the final products of peroxidation of unsaturated fatty acids in phospholipids and causes cell membrane damage (Moller et al., 2007). ROS can cause base deletion, strand breakage, and base alkylation and oxidation (Tuteja et al., 2001; Tuteja, 2010).

To keep this damage to a minimum, plants possess enzymatic and non-enzymatic antioxidative defense systems (Creissen and Mullineaux, 2002). Enzymes such as SOD, APX, and CAT play an important role against oxidative stress (Apel and Hirt, 2004). Under stress conditions, increased activity of antioxidant enzymes is important for plants stress tolerance (Ahmad et al., 2010a,b, 2011; Ahmad and Umar, 2011).

It has also been reported that proper plant nutrition is one of the important strategies in mitigating the stress-induced damage in plants. Trace elements are being used in plants to enhance resistance to biotic stresses such as pathogens and also to abiotic stresses.

14.3 Alleviation of environmental stress by trace elements

Plant metabolism is influenced adversely by multiple environmental stresses, and it is important to increase the ability of plants to survive multiple abiotic and biotic stresses (Mittler and Blumwald, 2010). In response to multiple abiotic and biotic stresses, plants possess a wide range of adaptive mechanisms to maintain growth and development and ensure survival. Increasing evidence suggests that suitable plant nutrition is not only required for better plant production, but also beneficial to alleviate different kinds of biotic and abiotic stresses.

The essential plant nutrients are divided into two classes: macronutrients and micronutrients. Previously published reports suggest that micronutrients are components of antioxidant defense enzymes, and modulation in the activity of these enzymes in micronutrient-deficient plants is well substantiated (Pilon-Smits et al., 2009). However, the role of micronutrients in abiotic stress alleviation is not well determined. It has also been reported that there is a good correlation between the accumulation of metals and the improvement of defense against herbivores or pathogens (Poschenrieder et al., 2006; Boyd, 2007; Vesk and Reichman, 2009). The role of micronutrients in stress alleviation is described below.

14.3.1 Zinc

Zinc is required for the activity of a large number of enzymes such as hydrogenase, carbonic anhydrase, Cu/Zn superoxide dismutase, and for the stabilization of ribosomal fractions and synthesis of cytochrome (Tisdale et al., 1984; Cakmak, 2000). It can act as an important trace element involved in the controlling of many different physiological processes.

Zinc nutrition appears to be involved in resistance to many diseases and in maintenance of the gene expression required for the tolerance of environmental stresses in plants (Cakmak, 2000). Zinc has been reported to alleviate abiotic stresses such as high light or temperature intensity, drought, heat, and salt stress (Disante et al., 2010; Peck and McDonald, 2010; Tavallali et al., 2010). There are reports of significant damage to the ultrastructure of chloroplasts subjected to inadequate Zn supply (Chen et al., 2007) because of its interaction with phospholipids and sulfhydryl groups of membrane proteins (Dang et al., 2010; Disante et al., 2010).

14.3.1.1 Activation of antioxidant

Zinc is involved in maintaining the balance of ROS production and scavenging in plants, as Zn is a constituent of SOD, an essential enzyme involved in protection against oxidative stress (Marschner, 1995). SOD activity was suggested to be an indicator of Zn nutritional status of plants (Gill and Tuteja, 2010). Furthermore, it has been well documented that Cu/Zn-SOD is closely related to resistance to stress (Guo et al., 2005), and significant increase in the activities of Cu/Zn-SOD under abiotic stress was observed in different plants. It has been reported that there is a good correlation between the overexpression of Cu/Zn-SOD in transgenic tobacco plants and the tolerance of multiple stresses. In addition, the combined expression of Cu/Zn-SOD and APX in transgenic *Festuca arundinacea* plants leads to elevated tolerance to heavy metal toxicity (Gill and Tuteja, 2010), and the expression of Cu/Zn-SOD in transgenic *Oryza sativa* and *Nicotiana tabacum* leads to enhanced tolerance to salt and water stresses (Badawi et al., 2004).

14.3.1.2 Zinc and plant responses to drought

The soil in dry regions with high calcium carbonate content is often poor in plant-available Zn. In drought conditions, a reduction of the net photosynthetic rate is mediated by a decrease in carbon fixation per unit leaf area or damage of the photosynthetic apparatus (Lal and Edwards, 1996; Castrillo et al., 2001; Bruce et al., 2002). A decrease in stomatal conductance (gs), intercellular CO_2 concentration, and carbonic anhydrase activity has been proposed to reduce photosynthetic rate under Zn deficiency (Cakmak and Engels, 1999; Hacisalihoglu et al., 2003). It was shown that the sensitivity to Zn deficiency stress became more pronounced during drought stress conditions. In addition, Zn-sufficient plants compared with Zn-deficient plants exhibit greater dry matter production and better tolerance to drought stress (Hajiboland and Amirazad, 2010). It has been well documented that applying Zn increased chickpea grain yields under water stress in drought-resistant genotypes (Waraich et al., 2011). It has been reported that zinc sulfate plays a more important role in stomata regulation and ion balance in plant systems to reduce the tensions of drought (Baybordi, 2006; Babaeian et al., 2010). Ajouri et al. (2004) documented that seed priming substantially improved water use efficiency in drought-stressed barley plants.

The possible mechanisms of Zn-enhanced resistance and tolerance of plants to drought stress have been fully explained by Waraich et al. (2011). They reported that Zn is a co-enzyme for production of tryptophane, a precursor to the formation of auxin (Bennett and Skoog, 2002; Waraich et al., 2011). It has been well documented that Zn application enhances the auxin levels in plants. Increase in auxin levels increases the root growth, which in turn improves drought tolerance. In another mechanism, Zn application reduces the generation of ROS. This decline may be related to a reduction in the activity of membrane-bound NADPH oxidase. Under water stress, the activities of antioxidative enzymes such as SOD and CAT are enhanced indicating that Zn decreases the ROS generation and protects cells against oxidative damage (Waraich et al., 2011). It is reported that the uptake and movement of Zn to the xylem in roots of *Arabidopsis thaliana* interacts with membrane transport processes and root anatomy (Claus et al., 2013). As a result, Zn has been suggested to play an important role in the integrity of membranes in plants because of its interaction with sulfhydryl groups of membrane proteins and with membrane-bound NADPH oxidase activity.

14.3.1.3 *Effects of zinc on plant resistance to high light*

High light stress induces damage to the ultrastructure of chloroplasts and leads to leaf chlorosis under Zn deficiency, but not at adequate Zn supply. Zn-enhanced carotenoid content in leaves has been suggested to play a significant role in preventing the photo-inhibitory damage to photosynthetic apparatus during exposure of plants to high light conditions (Munekage et al., 2002; Takahashi et al., 2009). In another mechanism, Zn application reduces the generation of ROS, which might be due to antioxidant enzyme activities (Millaleo et al., 2013).

14.3.1.4 *Zinc-enhanced tolerance to salinity*

Salinity decreases the availability of atmospheric CO_2 because stomatal closure is increased and consuming of NADPH by the Calvin cycle is reduced, which initiates chain reactions that produce more harmful oxygen radicals. Accumulation of ROS is capable of inducing damage to almost all cellular macromolecules including DNA, lipids, proteins, and carbohydrates (Miller et al., 2010; Ding et al., 2010). Suitable plant nutrition is one of the strategies to avoid oxidative damage to cells. Zinc seems to affect the capacity for water uptake and transport in plants that leads to enhanced tolerance to salt stress (Disante et al., 2010). Treatment with Zn promotes the activity of SOD and CAT. Zn facilitates the detoxification of $O_2^{\bullet-}$ to O_2 and H_2O_2 via enhancement of ascorbic acid content. It has been reported that Zn effectively protected *Pistacia vera* L. from salt-induced oxidative stress by inhibiting the peroxidation of membrane lipids by facilitating the proper functioning of membrane proteins. In another mechanism, Zn application repressed Na^+ transport in rice (Aslam et al., 2000) and barley (Abou Hossein et al., 2002) grown in salinized solutions.

Arabidopsis thaliana oxidation-related zinc finger 1 (AtOZF1) is a plasma membrane-localized zinc finger (ZF) protein. Two AtOZF genes (*AtOZF1* and *AtOZF2*) have been reported in the *A. thaliana* genome. AtOZF1 may be involved in abscisic acid (ABA) signaling pathway and the stress response, because expression of AtOZF1 is up-regulated in response to ABA and salinity resulting in abiotic stress tolerance.

14.3.1.5 Alleviation of metal toxicity

It is suggested that the physiological damage induced by Cd toxicity may be prevented by the application of Zn (Wu and Zhang, 2002; Krämer et al., 2007). Some studies have revealed that Zn suppresses the bioaccumulation of Cd by reducing its uptake (Han et al., 2010). It was reported that Zn probably modulates the protection of DNA from Cd-induced damage in several ways: (1) Zn inhibits metal-catalyzed oxidative damage through OH• attack on the DNA or protein by SH group protection, or (2) some transporters might be involved in uptake and transport of Zn both outside the plasma membrane and into the vacuole as well as Cd (Wong and Cobbett, 2009).

14.3.1.6 Resistance against pathogens

Zinc has been reported to alleviate biotic stresses and is involved in resistance to many diseases. The mechanisms of Zn-enhanced resistance of plants to biotic stress have been fully explained by Duffy (2007). He reported that Zn acts as a cofactor for numerous enzymes, and also stimulates root growth. Zinc application to soils decreases attack by root pathogens of tomato, including *Fusarium solani*, *Rhizoctonia solani*, and *Macrophoma phaseoli*, and also *Rhizoctonia* root rots of wheat, cowpea, and medicago (Kalim et al., 2003; Duffy, 2007). Zinc may also be involved in resistance to phytophthora diseases, since low Zn levels in soils and leaf tissues were connected with a high incidence of phytophthora black pod of cocoa.

14.3.2 Manganese

14.3.2.1 Activation of antioxidant by manganese

Manganese is a cofactor for various enzymes that are involved in redox reactions such as Mn-superoxide dismutase in plants (Bowler et al., 1994). All forms of SOD have a key role in the survival of plants under environmental stresses (Gill and Tuteja, 2010; Ahmad et al., 2011; Ahmad and Umar, 2011; Ahmad and Prasad, 2012a). Plants generally contain MnSOD in mitochondria (Bowler et al., 1994). It has been reported that $O_2^{•-}$ is the primary ROS that is formed by monovalent reduction in the electron transport chain (ETC) and it is converted into H_2O_2 by the MnSOD (mitochondrial form of SOD).

In the *Arabidopsis thaliana* genome, one MnSOD gene (*MSD1*) has been detected (Cooke et al., 2003). The variable responses of MnSOD isozyme activity have been observed under various environmental stresses. The activity of MnSOD isozyme is increased by salt stress in *Lycopersicon esculentum* (Gapinska et al., 2008) and *Cicer arietinum*. Furthermore, significant increase in the activity of MnSOD under Cd treatment was observed in *Hordeum vulgare* (Guo et al., 2004), *A. thaliana* (Skorzynska-Polit et al., 2004), and *Triticum aestivum* (Hasan et al., 2008). Some studies have revealed that the transgenic plants with MnSOD overexpression show less oxidative damage under photooxidative stress (Melchiorre et al., 2009), and higher protection of photosynthetic apparatus under drought stress (Wang et al., 2004).

In photosynthetic organisms, Mn plays a critical role in forming an important multiprotein pigment complex embedded in the thylakoid membranes, which participates in catalyzing the water-splitting reaction. The Mn cluster bonds with calcium and chloride ions and with residues of the D1 and CP43 proteins (Barber, 2008), which is required to oxidize water and reduce PSII (Salomon and Keren, 2011). It was shown that the addition of Mn reconstructs oxygen evolution rates in a light-dependent process named photoactivation. This process involves the sequential oxidation and coordinative bonding of the four Mn atoms, calcium, and chloride ions to the C terminus of the mature D1 protein, and the CP43 subunit of PSII (Dasgupta et al., 2008).

14.3.2.2 Resistance to pathogens

Supplying Mn has been shown to alleviate various root and foliar diseases in plants (Huber and Graham, 1999; Heckman et al., 2003). It has been reported that Mn fertilization plays an important role in controlling pathogenic diseases such as powdery mildew, downy mildew, take-all, tan spot, and several others (Huber and Graham, 1999; Heckman et al., 2003; Simoglou and Dordas, 2006).

The possible mechanisms of Mn-enhanced resistance and tolerance of plants to biotic stress have been fully explained. Mn has an important role in lignin and suberin biosynthesis (Marschner, 1995) through activation of several enzymes of the shikimic acid and phenylpropanoid pathways (Marschner, 1995). Both lignin and suberin are important biochemical barriers to fungal pathogen invasion (Vidhyasekaran, 2004). Lignin and suberin have been shown to alleviate powdery mildew in wheat (Krauss, 1999) as they are phenolic polymers resistant to enzymatic degradation (Agrios, 2005). In another mechanism, Mn inhibits the action of aminopeptidase, an enzyme that provides essential amino acids for fungal growth and pectin methylesterase, a fungal enzyme.

14.3.2.3 Alleviation of metal toxicity

Manganese fertilization causes amelioration of Cd-induced root growth inhibition in maize seedlings, which is mediated by a parallel reduction in Cd uptake (Palóve-Balang et al., 2006). Availability of Mn to plants is decreased by the presence of Cd in soil. Supplementation of $MnSO_4$ to Cd-treated lupine plants plays a protective role in photosynthetic tissues against Cd stress (Zornoza et al., 2010). Mn application reduces the generation of ROS. This decline may be related to increase in the antioxidant capacity of root together with high leaf Mn concentration (Peng et al., 2008).

14.3.2.4 Alleviation of temperature stress by manganese

Manganese plays an important role in photosynthesis and nitrogen metabolism. A decline in the photochemical activity of photosystem II (PSII), lower oxygen evolution rates, lower maximal photosynthetic yield of PSII values, and faster QA reoxidation rates are the characteristics of Mn deficiency (Salomon and Keren, 2011). It can reduce the adverse effects of temperature stress (high and low) indirectly by enhancing the photosynthetic rate and nitrogen metabolism in plants. Manganese plays a role in detoxification of ROS through increase in antioxidative compounds and enzymatic activities (Aktas et al., 2005; Aloni et al., 2008) under temperature stress.

14.3.3 Iron

14.3.3.1 Iron and plant stress responses

Iron plays an essential role in vegetative and reproductive plant growth (Briat et al., 2007; Chu et al., 2010a,b; Roschzttardtz et al., 2011). Fe can act as a prooxidant through the Fenton reaction (Halliwell and Gutteridge, 1984). Iron is either a constituent or a cofactor of many antioxidant enzymes such as catalase, non-specific peroxidases, APX, and Fe superoxide dismutase. Three FeSOD genes (*FSD1*, *FSD2*, and *FSD3*) have been reported in the *A. thaliana* genome. During early chloroplast development, heteromeric *FSD2* and *FSD3* work as H_2O_2 scavengers via protecting the chloroplast nucleoids from H_2O_2 (Myouga et al., 2008). POD isoforms are differently affected in plants grown under Fe deficiency (Ranieri et al., 2001). Molassiotis et al. (2006) showed that an increased level of H_2O_2 may be the result of insufficient activation of heme-containing POD and CAT, due to low Fe availability in sunflower and peach. Root cell wall lignification is affected in *Pyrus communis* grown under Fe deficiency, which provides structural and durability to plant tissues (Srivastava et al., 2007; Kim et al., 2008). Cell wall lignification has been reported to be involved in plant tolerance to salt and drought stress (Lee et al., 2007). Production of ROS requires tight control of iron transport and homeostasis (Kosman, 2010).

14.3.3.2 Fe and plant pathogen defenses
It is also found that silicon increases the synthesis of fungal antibiotics by soil bacteria. Foliar application of silicon increases the resistance against pathogens in wheat and banana leaves (Graham and Webb, 1991). Iron has recently been identified as a central factor regulating plant pathogen defense (Brissot et al., 2011). Si and Fe are able to interact and many Fe molecules can be condensed in silica. Fe^{2+}- and Fe^{3+}-chelating siderophores can bind directly to silica (Fleck et al., 2011).

14.3.3.3 Alleviation of metal toxicity
Iron can alleviate the negative effects of Cd by retaining both quantity and quality of chloroplast. It has been well documented that Fe limits Cd uptake and root-to-shoot transport and enhances photosynthetic pigment accumulation. Increase in the availability of Fe to the growth medium modulates the cation binding capacity of the cell wall, which in turn improves the heavy metal toxicity tolerance. Ferredoxins are necessary for the light-induced process and the decrease in Fe results in the decrease in ferredoxin content in Cd-treated almond seedlings (Nada et al., 2007). It has been demonstrated that the supplementation of Indian mustard with 40 μM Fe alleviates oxidative damages and helps in stabilizing thylakoid membrane under Cd stress. Phytosiderophores can chelate other heavy metals besides Fe. Iron acquisition by phytosiderophores was suggested to prevent Cd stress in graminaceous plants.

14.3.3.4 Iron and circadian oscillations
There is a correlation between the circadian rhythm and Fe homeostasis (Harmer, 2009; Imaizumi, 2010). A disordered circadian rhythm affects Fe homeostasis. Fe is involved in maintaining the period length of circadian rhythm. Circadian clock regulators such as TIME FOR COFFEE (TIC) modulate the expression of the AtFer1 (ferritin gene) (Duc et al., 2009) and this expression of AtFer1 was up-regulated by excess Fe. Since Fe is a signal to the central oscillator to modulate circadian period length for plant growth and development, Fe deficiency may induce disorganized chloroplast maturation in *Arabidopsis* (Haydon et al., 2012; Chen et al., 2013).

14.3.4 Copper
14.3.4.1 Copper and plant stress responses
Under environmental stresses, elevated ROS remove electrons from the lipids of cell membranes and cause lipid peroxidation, thereby leading to cell death. MDA is one of the final products of lipid peroxidation. Intracellular Cu is utilized to escape toxicity caused by ROS such as superoxide

and H_2O_2 (Brewer, 2010). Accumulation of free amino acids and proline has been found under Cu^{2+} exposure in plants (Al-Hakimi and Hamada, 2011; Azooz et al., 2012).

Three isoforms of Cu/Zn-SOD, cytosolic (CSD1), chloroplastic (CSD2), and peroxisomal (CSD3) forms, have been recognized in *Arabidopsis* (Chu et al., 2005). It has been well documented that the overproduction of three isoforms of Cu/Zn-SOD causes enhanced tolerance to oxidative stress (Faize et al., 2011).

14.3.4.2 Copper-enhanced resistance to drought

Copper application to crops has been reported to alleviate the adverse effects of water stress by improving the nitrogen metabolism and greater accumulation of soluble phenolic compounds and lignin (Waraich et al., 2011). Copper is also an important micronutrient essential for lignin synthesis, which is needed for cell wall durability and avoidance of wilting. Drought stress adversely affects all these parameters. It is known that the Cu-supplemented *Capsicum annuum* and *Raphanus sativus* show higher levels of shikimate dehydrogenase and POD activity and higher levels of lignin (Diaz et al., 2001; Chen et al., 2002). After excess Cu addition, the increased activation of glucose-6-phosphate dehydrogenase, shikimate dehydrogenase, PAL, CAT, caffeic acid-POD, polyphenol oxidase, laccase, and β-glucosidase were also observed (Akgül et al., 2007). Later studies have revealed that the simultaneous overexpression of Cu/Zn-SOD in the cytosol of transgenic *Nicotiana tabacum* plants alleviates water stress damage because of higher water use efficiency, better photosynthetic rate, lowered lipid peroxidation, and reduced H_2O_2 levels in transgenic *N. tabacum* (Faize et al., 2011).

14.3.4.3 Copper-enhanced resistance to pathogens

In nature, heavy metals not only cause the stress in plants but exposure to heavy metals may lead to protection against pathogens (Poschenrieder et al., 2006). Cu application may induce disease resistance in several ways: (1) proper Cu nutrition increases the activation of peroxidase and laccase that leads to the higher levels of lignifications (Diaz et al., 2001; Passardi et al., 2005), (2) excess Cu is toxic to pathogens, therefore Cu accumulation by the plant may suppress pathogen infection, and (3) Cu can act as elicitors of plant defense mechanisms (Maksymiec, 2007). Plants may also defend themselves against pathogens via secondary metabolites such as phytoanticipins and phytoalexins.

Ethylene is released from the plant in response to both biotic and abiotic stresses (Abeles et al., 1992). Since excess Cu stimulates the biosynthesis of ethylene (Arteca and Arteca, 2007), plant defense reaction induced by copper could also be regulated by ethylene.

14.3.5 Boron

Boron is an essential metal involved in widespread physiological activity including photosynthesis, carbohydrate and nitrogen metabolism, the ascorbate/gluthathione system, and auxin transit (Ruuhola et al., 2011).

14.3.5.1 Boron and plant stress responses

Boron significantly influences the ability of plants to produce variable levels of ascorbate in different carrot and tomato genotypes grown in the nutrient solution (Gunes et al., 2006; Eraslan et al., 2008;

Singh et al., 2012). These plants show a common stimulation of ascorbate after high B supply. When the plants were concurrently treated with B and Na_2SO_4, the concentration of ascorbate significantly increased. In contrast, in citrus (Han et al., 2009) high B concentration reduced ascorbate concentration. Tripeptide glutathione (GSH) is one of the crucial metabolites against ROS because of its ability to regenerate ascorbate via the AsA-GSH cycle (Foyer and Noctor, 2011). It has been reported that B reduces Al toxicity via stimulating GSH accumulation in the leaves and the increase in its concentration in the roots (Ruiz et al., 2006). Reports have revealed that the high boron supply increases the status of GSH pools in pear leaves (Wang et al., 2011). It has been well documented that the high B supply increases antioxidative enzyme activities (Gunes et al., 2006; Eraslan et al., 2008; Ardic et al., 2009). It was shown that B supply increases SOD activity in tomato (Kaya et al., 2009), citrus leaves (Han et al., 2009), and in two chickpea cultivars (Ardic et al., 2009).

14.3.5.2 Mitigation of chilling stress by boron

Low temperature stress causes inhibition of growth and development (Xu et al., 2008), which leads to significant enhancement in the generation of ROS. B deficiency induces sensitivity to photooxidative damage in leaf cells and leads to the lower chilling tolerance of plants (Räisänen et al., 2007; Lehto et al., 2010). There are several explanations for this effect. First, B deficiency may lead to the accumulation of carbohydrates (Han et al., 2009); therefore, carbohydrate accumulation provides the precursors for the synthesis of phenolic compounds via the shikimate pathway (Ruuhola et al., 2011). Second, B nutrition reduces the production of ROS species and enhances the photosynthetic rate under low temperatures (Pennycooke et al., 2005; Waraich et al., 2011). Accumulation of phenolics in the cell walls is thought to be involved in resistance to mechanical stress caused by extracellular ice formation (Solecka and Kacperska, 2003).

14.3.5.3 Boron and plant responses to drought

Drought stress induces a range of physiological and biochemical responses in plants and reduces sugar transport, flower retention, pollen formation, seed germination, and grain production (Shinozaki et al., 2003; Bartels and Sunkar, 2005). B and Ca foliar sprays are most commonly used to correct detrimental effects of drought. It was shown that B may induce drought tolerance in several ways. Proper B supply increases seed and grain production via improving sugar transport, flower retention, and pollen formation. It was shown that the application of B increases the tolerance of plants to drought-induced oxidative damage by enhancing the activities of antioxidants such as CAT and POD (Li et al., 2003).

14.3.5.4 Boron-enhanced tolerance to salinity

It has been reported that B application reduces salinity effects on plant. There are several reasons for this effect: (1) absorption of both B and Cl are reduced because of antagonism effect between these two elements, (2) B reduces root uptake and shoot accumulation of Cl (Yermiyahu et al., 2008), and (3) significant decrease in the growth inhibition under combined B toxicity and salt stress was observed (Yermiyahu et al., 2008). It has been demonstrated that the combined effect of both Ca^{2+} and B can mitigate the harmful effects caused by salinity.

14.3.5.5 Induction of plant resistance to disease

The function that B has in reducing disease susceptibility could be because of (1) the function of B in cell wall structure, (2) the function of B in cell membrane permeability, stability, or function, or (3) its role in metabolism of phenolics or lignin (Figure 14.1). B is involved in the stability and integrity of the plasma membrane and cell wall (Marschner, 1995; Dordas and Brown, 2000; Brown et al., 2002). B can control or reduce the severity of several diseases such as tobacco mosaic virus in bean, and tomato yellow leaf curl virus in tomato (Graham and Webb, 1991). The beneficial role of B in reducing crucifer plants' susceptibility to *Plasmodiophora brassicae* has also been shown. It has been well documented that the B supply reduces biotic stress induced by *Fusarium solani* in soybeans, *Verticillium alboatrum* in tomato and cotton, and *Blumeria graminis* in wheat (Marschner, 1995).

14.3.6 Molybdenum

In bacteroids of legume nodules, the nitrogen-fixing enzyme nitrogenase and xanthine dehydrogenase/oxidase involved in ureide biosynthesis are molybdoenzymes. Other molybdoenzymes have also been identified in plants including aldehyde oxidase (AO) that is involved in ABA biosynthesis, and sulfite oxidase that can convert sulfite to sulfate (Mendel and Haensch, 2002). ABA is a key component of the signaling system that is involved in plant adaptive responses to environmental stress. In addition, a mitochondrial Mo^{2+} carrier protein has been reported in *Arabidopsis* (Tan et al., 2010).

14.3.6.1 Molybdenum-induced tolerance to drought

The AO enzyme is involved in oxidation of abscisic aldehyde to ABA (Seo et al., 2004). Enhanced expression of AO activity leads to ABA accumulation and increased drought tolerance in maize plants. In *Arabidopsis*, the molybdenum cofactor sulfurase gene (LOS5) encodes the sulfurylated form of a molybdenum cofactor (MoCo) involved in regulation of AO activity (Bittner et al., 2001).

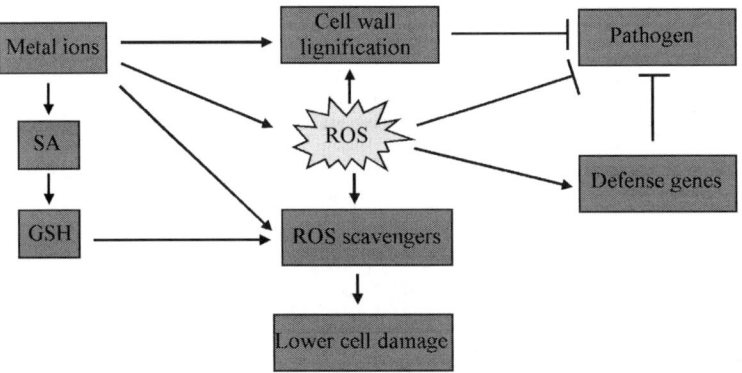

FIGURE 14.1

The possible mechanisms of excess metal-enhanced resistance and tolerance of plants to abiotic stresses. Abbreviations: GSH: glutathione; ROS: reactive oxidative species; SA: salicylic acid.

The function that LOS5 overexpression has in reducing drought stress could be because of (1) the function of LOS5 overexpression in overproducing ABA leading to proline accumulation and (2) the function of LOS5 overexpression in mobilizing ROS-scavenging enzymes and activating signaling molecules that regulate ROS-scavenging genes leading to enhanced drought stress tolerance. In conclusion, overexpression of LOS5 induces ABA accumulation in maize subjected to drought stress (Lu et al., 2013). Accumulation of ABA is thought to be involved in reduction of water loss, activation of stress-regulated genes, and alleviation of drought stress in maize.

14.3.6.2 Molybdenum and low temperature stress
It has been reported that application of Mo improves the cold resistance of winter wheat (Wang et al., 1995), which is possibly due to increase in the activities of Mo-containing enzymes (Vunkova-Radeva et al., 2003) and antioxidative enzymes (Sun et al., 2006), increased nitrogen-containing compounds (Hu et al., 2002) and chlorophyll biosynthesis (Sun et al., 2006). A number of cold-responsive (COR) genes that are regulated by both ABA-dependent and ABA-independent pathways have been identified in plants (Sharma et al., 2005).

14.3.7 Chlorine
14.3.7.1 Effects of chlorine on plant resistance to drought
Chlorine is an essential cofactor for the activation of polypeptides associated with the water-splitting complex of PSII (Marschner, 1995). It has been demonstrated that the accumulation of Cl in plant cells increases tissue hydration and turgor pressure. Drought is one of the main environmental factors that limit plant productivity. In coconut plant, Cl is involved in the stomatal regulation, particularly during the dry season. Cl has been shown to help plants in counteracting the negative effects of drought stress by maintaining a relatively high level of leaf gas exchanges (Hajiboland, 2012).

14.3.7.2 Controlling plant diseases
Chlorine has been reported to alleviate biotic stresses (Mann et al., 2004) and is involved in protection of corn plants from stalk rot, wheat from stripe rust, take-all and septoria (Graham and Webb, 1991; Mann et al., 2004; Heckman, 2007). Cl was proven to alleviate salinity stress by suppression of nitrification and increase the availability of Mn leading to host resistance to diseases in grain crops (Huber, 1996). Higher concentrations of Cl in plant tissues can also enhance water retention and turgor when roots have been attacked by pathogens. Cl supply in plant tissues decreases the amount of malate from roots. This action takes away pathogens of an organic substrate (Huber, 1996).

14.3.8 Nickel and plant stress responses
Nickel is thought to be involved in the activation of several enzymes such as NiFe-hydrogenase, superoxide dismutase, and Ni-dependent glyoxylase (Mulrooney and Hausinger, 2003). In *Oryza sativa*, Ni^{2+} influences the lipid composition and H-ATPase activity of the plasma membrane. Other symptoms observed in Ni^{2+}-treated plants were related with changes in water balance. High uptake of Ni^{2+} induces a decline in water content of dicot and monocot plant species (Nagajyoti et al., 2010). Evidence proposes that Ni exposure may induce an elevation in GSH (Zengin and

Munzuroglu, 2005; Meng et al., 2007). This effect is thought to be connected with the participation of peroxidase and proline (Nadgórska-Socha et al., 2013).

14.4 Effects of beneficial elements on plant stress responses

Aluminum (Al), cobalt (Co), sodium (Na), selenium (Se), and silicon (Si) are considered beneficial elements for plants. All have positive effects on plant growth and stress resistance, and thus may be applied as fertilizer.

It has been well documented that all of these elements promote growth favorably and affect plant stress resistance (Pilon-Smits et al., 2009; Shen et al., 2010). While none of these elements is essential for all plants, Na and Si are essential for certain plant taxa and some hyperaccumulator species; these beneficial elements have been suggested to be essential, while Co is essential only for the microbial partners of some plants. These beneficial elements alleviate biotic stresses such as pathogens and herbivory, and mitigate the salt- and drought-induced damage in plants (Ahmad et al., 2010a; Hayat and Ahmad, 2011; Tahir et al., 2012; Yusuf et al., 2012).

14.4.1 Sodium

Negative effects of Na at excess levels (salt stress) have been studied more than as a beneficial element. It has been well documented that Na^+ is closely related to resistance to salt stress in plants (Pilon-Smits et al., 2009), and significant increase in the activities of specific Na^+ transporters under salt stress was observed. These transporters export Na^+ out of the root, into the vacuole, or into the shoot phloem for transport to the root.

14.4.1.1 Salt and drought tolerance

In tonoplast, Na^+/H^+ antiporters (NHX proteins) sequestrate excess Na^+ into the vacuole; they are important salt-tolerant determinants. Two major types of plant NHX have been identified (Rodríguez-Rosales et al., 2009). In the plasma membrane, the SOS1-like (salt overly sensitive 1 type) NHX protein may play an important role in long-distance Na^+ transport in plants (Olias et al., 2009). The *Arabidopsis AtNHX1* gene catalyzes Na^+ accumulation in vacuoles and mitigates the toxic effects of Na^+ on cellular metabolism. NHX proteins also play crucial roles in pH status (Yamaguchi et al., 2001; Yoshida et al., 2009) and K^+ homeostasis (Munns and Tester, 2008; Rodríguez-Rosales et al., 2009; Leidi et al., 2010). Overexpression of *AtNHX5* gene in paper mulberry plants showed high drought tolerance because of increased leaf water content and leaf chlorophyll contents, accumulated more proline and soluble sugars, and less membrane damage under water deficiency and salt stress (Rodriguez-Rosales et al., 2009; Leidi et al., 2010). It is suggested that the *AtNHX5* gene could enhance the resistance of plants to abiotic stresses by encouraging the accumulation of osmolytes to mitigate the osmotic stress caused by multiple environmental stresses (Li et al., 2011).

14.4.1.2 Sodium and induction of crassulacean acid metabolism

Crassulacean acid metabolism (CAM) is a metabolic strategy to maintain photosynthesis under stress conditions. High salinity (400 μM NaCl) induces a shift from C_3 photosynthesis to CAM in

Mesembryanthemum crystallinum plants (Niewiadomska et al., 2011). In previous work (Habibi and Hajiboland, 2012), we showed that a strong diurnal rhythm in the activity of some antioxidative enzymes has been observed in C_3/CAM intermediate plants. In plants that use C_4 or CAM photosynthetic pathways, Na is an essential element (Ohnishi et al., 1990). It was reported that the C_4/CAM plants utilize phosphoenolpyruvate (PEP) to fix CO_2 for photosynthesis, and Na is needed for the regeneration of PEP from pyruvate.

14.4.1.3 Alleviation of K^+ deficiency

Na^+ can replace K^+ as a cofactor for certain enzymes, and as osmoregulator for stomatal movement and cell expansion. This is particularly beneficial when K^+ levels are limiting. In plant species that can replace K^+ with Na^+ (natrophilic vs. natrophobic species), Na^+ shows favorable effects on plant growth, and thus may be applied as fertilizer (Marschner, 1995). Because Na^+ affects stomatal movement differently than K^+, addition of Na to plants can have a positive effect on leaf water status and water use efficiency. Some halophytic plant species accumulate high levels of Na^+ as a salt resistance mechanism in their natural habitat. In natrophilic species, low levels of Na^+ can have a beneficial effect on growth under K^+ deficient conditions or moderate drought stress.

14.4.2 Silicon

14.4.2.1 Silicon and plant responses to drought

Silicon application to crops has been reported to enhance their tolerance of water stress (Ma, 2004; Hattori et al., 2007). Mechanisms of Si-mediated alleviation of damage caused by drought stress are poorly understood (Chen et al., 2011). Several authors suggested that Si causes an improvement of water use efficiency and stimulation of the antioxidative defense system (Liang et al., 2007; Cooke and Leishman, 2011). Increase in production of antioxidants and decline of ROS generation mediated by Si causes reduction of photo-oxidative damage, maintenance of chloroplast membrane integrity, and thus enhancement of plant drought tolerance (Waraich et al., 2011). Si fertilization leads to increased volume and weight of roots (Ahmed et al., 2011; Sonobe et al., 2011; Ahmed et al., 2013). Si decreases water loss because of silicification of leaf surfaces (Cooke and Leishman, 2011), which decreases cuticular transpiration as well as reduces the diameter of stomatal pores (Snyder et al., 2007). All these factors can ultimately enhance drought resistance of Si-supplemented plants. In previous work (Habibi and Hajiboland, 2013), we concluded that supplementation of water-deficient pistachio plants with Si alleviates the adverse effects of drought due to its enhancement of photochemical efficiency and photosynthetic gas exchange, as well as an activation of the antioxidant defense capacity in this species (Figure 14.2).

14.4.2.2 Silicon-enhanced tolerance to salinity

It has been well documented that the Si-supplemented plant species exhibits beneficial effects on growth and yield (Richmond and Sussman, 2003; Pilon-Smits et al., 2009). It has been reported that Si effectively protects *O. sativa* (Lekklar and Chaidee, 2011), *T. aestivum* (Tahir et al., 2012), *Z. mays* (Moussa, 2006), and *B. napus* (Hashemi et al., 2010) from salt-induced oxidative stress by enhancing Na^+ exclusion and inhibiting the peroxidation of membrane lipids through stimulation of enzymatic and non-enzymatic antioxidants (Hasanuzzaman and Fujita, 2011). In particular, Si was proven to alleviate salinity stress by enhanced photosynthetic activity, stimulation of ROS

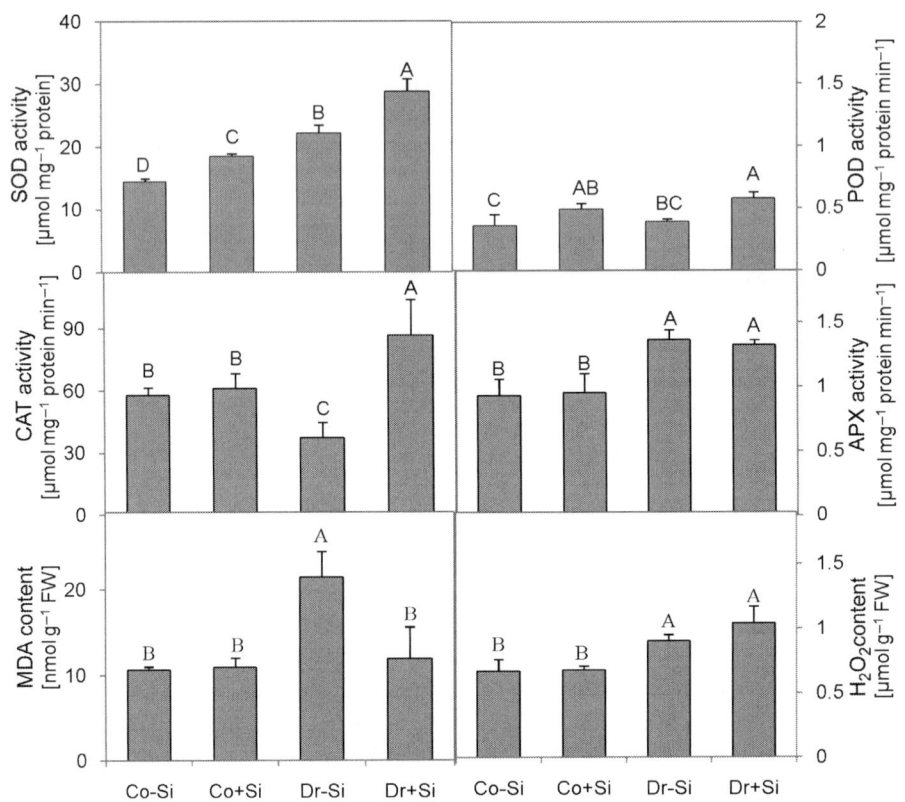

FIGURE 14.2

The specific activity of SOD (μmol mg^{-1} protein), POD (μmol mg^{-1} protein min^{-1}), CAT (μmol mg^{-1} protein min^{-1}), APX (μmol mg^{-1} protein min^{-1}), and concentration of MDA and H$_2$O$_2$ (μmol g^{-1} FW) in pistachio plants grown for 45 days under control and drought conditions with or without Si application. Each value is the mean $\pm$ SD of four replicates. Bars indicated with the same letter are not significantly different ($p < 0.05$) (adapted from Habibi and Hajiboland, 2013). Abbreviations: APX: ascorbate peroxidase; CAT: catalase; H$_2$O$_2$: hydrogen peroxide; MDA: malondialdehyde; POD: peroxidase; SOD: superoxide dismutase.

dismutation, immobilization of toxic Na$^+$, enhanced K$^+$ uptake, and higher K$^+$:Na$^+$ selectivity (Ahmed et al., 2013). The possible mechanisms of Si-enhanced resistance and tolerance of plants to salt stress have also been explained by Wang et al. (2011). They reported that Si-supplemented *Medicago sativa* exhibited increase in activity of APX, CAT, and POD. The improvement of plant water status, chlorophyll content, K$^+$ concentrations in shoots, and membrane permeability is maintained by Si application under salt stress conditions in wheat (Tahir et al., 2012).

14.4.2.3 Induction of plant resistance to disease by silicon

Foliar applications of silicon increase the resistance against pathogens in plant species (Bowen et al., 1992). Much of the research on Si-enhanced resistance against biotic stresses has been focused on rice (Kamenidou et al., 2009). The beneficial effects of Si in reducing rose plants' susceptibility to *Podosphaera pannosa* have been shown. It is also found that silicon increases the accumulation of fungi toxic phenolic compounds and silica depositions at the site of infection and the formation of papillae and deposition of callose and H_2O_2 (Shetty et al., 2012; Bockhaven et al., 2013).

Recent evidence proposes that silicon application may induce disease resistance in several ways: (1) polymerized Si can delay infections via inhibiting fungal germ tube penetration of the epidermis, (2) Si nutrition affects plant hormone homeostasis (Robert-Seilaniantz et al., 2011), (3) Si induces the accumulation of photorespiratory enzymes in Si-treated plants under stress, and finally (4) Si may act as a signal in triggering natural defense responses by stimulating the activity of chitinases, peroxidases, polyphenol oxidases (Dallagnol et al., 2011), and may also interact either directly or indirectly with various signal transduction components in both dicots and monocots, resulting in enhanced signaling capacity in Si-treated plants.

14.4.2.4 UV screening of grasses by plant silica layer

It has been well documented that the Si supply reduces stress induced by UV radiation in soybeans and wheat (Shen et al., 2010; Yao et al., 2011a). Increase in silica depositions in or near the epidermis of the leaves decreases the penetration of UV through the sclerenchyma and even the mesophyll (Goto et al., 2003; Schaller et al., 2012b). Schaller et al. (2012a, 2013) showed that both silica depositions and phenolic substances protect grass plants from UV radiation. In ecosystems with Si-accumulating species, large amounts of Si were absorbed each year via plant root (Cornelis et al., 2010; Melzer et al., 2010).

14.4.2.5 Alleviation of metal toxicity

Silicon can also alleviate heavy metal toxicity symptoms. It has been well documented that Si enhances metal uptake and root-to-shoot transport. Increase in the availability of metal to the growth medium by Si application modulates the cation binding capacity of the cell wall, which in turn improves the heavy metal toxicity tolerance (Pilon-Smits et al., 2009). In another mechanism, Si application stimulates the generation of antioxidants. It is reported that the uptake and movement of Si to the cytoplasm induces complexation of Si with toxic metal ions and leads to production of Zn, Cd, and Al silicates. In this case, Si was shown to bind to the toxic metal ions and thus promotes the sequestration of the metals in the vacuoles (Pilon-Smits et al., 2009).

Silicon has been shown to help plants in counteracting the negative effects of Al toxicity in most strongly acid soils by decreasing the toxic Al^{3+} concentration in solution via forming Al−Si complexes (Roy et al., 1998). It has been reported that Si effectively protected conifers (Ryder et al., 2003), *Hordeum vulgare* (Hammond et al., 1995), and *Zea mays* (Barcelo et al., 1993) from Al toxicity-induced damages.

It has been demonstrated that Si application reduces the toxicity of high Mn. This decline may be related to an apoplastic activity of PODs and phenols (Führs et al., 2009). Si may modify the Mn binding capacity of the cell wall in cucumber. Some studies have revealed that the Si-supplemented cowpea (Führs et al., 2009), pumpkin (Iwasaki and Matsumura, 1999), and cucumber (Shi et al., 2005) show less oxidative damage under Mn toxicity conditions. Si fertilization causes amelioration of Cd-induced root growth inhibition in rice seedlings, which is mediated by an increase in shoot and root dry weights (Shi et al., 2005).

14.4.3 Selenium

14.4.3.1 Selenium and alleviation of environmental stress

Addition of Se to the plant growth medium can reduce the excess $O_2^{\bullet-}$ and/or H_2O_2 generation in plants subjected to diverse environmental stresses, such as in *Sorghum bicolor* under high temperature (Djanaguiraman et al., 2010), wheat seedlings exposed to UV-B radiation stress (Yao et al., 2011b) and cold stress (Chu et al., 2010b), rapeseed seedlings under salt and drought stress (Hasanuzzaman et al., 2011), *Trifolium repens L.* during water stress (Wang, 2011), and *Phaseolus aureus* under As stress (Malik et al., 2012). Application of Na_2SeO_3 was observed to efficiently ameliorate the Cd toxicity in marine red alga and the Pb toxicity in *Vicia faba* L. by decreasing the accumulation of $O_2^{\bullet-}$ and H_2O_2 (Kumar et al., 2012). Se addition decreases ROS levels in plants subjected to stresses, and this decline may be related to (1) an increase in the dismutation of $O_2^{\bullet-}$ into H_2O_2 (Cartes et al., 2010), (2) together with direct quenching of $O_2^{\bullet-}$ by Se compounds, and (3) the regulation of antioxidative enzymes (Figure 14.3).

In the absence of Se, H_2O_2 was eliminated primarily by the APX, but the H_2O_2 was detoxified mainly by glutathione peroxidase (GPX) following the application of this trace element (Ríos et al., 2009). In Se-treated *Brassica napus* seedlings subjected to excess Cd, salt, and drought stress, GPX played a unique role in counteracting oxidative damages (Hasanuzzaman and Fujita, 2011). In different plant cell organelles including cytosol, chloroplasts, mitochondria, peroxisome, and apoplast, a large family of diverse GPX isozymes catalyze the reduction of H_2O_2 (Anjum et al., 2011).

Se addition has been shown to reduce stress-induced damages caused by Al toxicity in ryegrass (Cartes et al., 2010), UV-B stress in wheat seedlings (Yao et al., 2010), water-deficient stress in *Trifolium repens* L. (Wang, 2011), Cd toxicity in marine red algae (Kumar et al., 2012), As toxicity in mungbean (Malik et al., 2012), and high temperature stress in grain sorghum (Djanaguiraman et al., 2010). Several authors have also observed enhanced SOD activity after Se application. In the Se accumulator *Pteris vittata*, later studies have revealed that SOD activity is inhibited at low Se levels but is enhanced at a high level of Se (Feng and Wei, 2012). After excess Se addition, the increased activation of APX, GSH, CAT, and GR was also shown (Djanaguiraman et al., 2010; Yao et al., 2010; Feng and Wei, 2012; Malik et al., 2012).

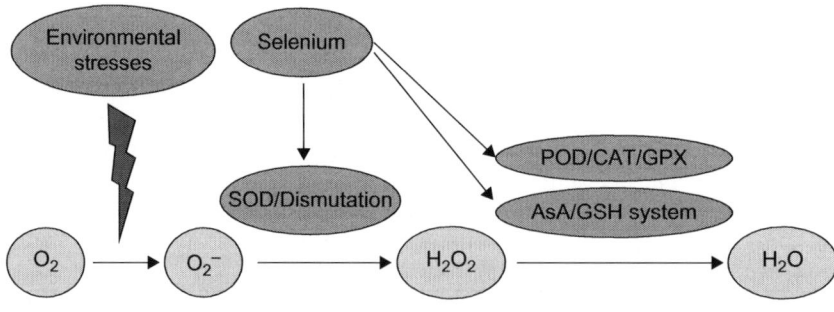

FIGURE 14.3

The roles of selenium in quenching of reactive oxidative species (ROS) under stress conditions. Abbreviations: APX: ascorbate peroxidase; AsA: ascorbic acid; CAT: catalase; GPX: glutathione peroxidase; GSH: glutathione; H_2O_2: hydrogen peroxide; POD: peroxidase; SOD: superoxide dismutase; O_2^-: superoxide radical.

14.4.3.2 Selenium and plant response to drought

Hasanuzzaman and Fujita (2011) reported that Se-pretreated rapeseed seedlings exhibited elevated ascorbate and GSH levels and significantly enhanced APX, DHAR, MDAR, GR, GPX, and CAT activity under drought stress. It was shown that the Se-supplemented water-deficit buckwheat exhibits significantly higher stomatal conductance (g_s). A significantly higher photochemical efficiency of PSII was obtained in Se- and water-deficit plants, which was possibly due to improvement of plant water management during treatment. In Habibi (2013), the barley plants treated with selenate exhibited increased levels of enzymes that detoxify H_2O_2, especially ascorbate peroxidase (APX) and GPX, thereby improving stress resistance (Ríos et al., 2009).

14.4.3.3 Selenium-enhanced tolerance to salinity

Selenium seems to affect antioxidant enzyme activities and proline content in plants and leads to enhanced tolerance to salt stress (Terry et al., 2000; Kong et al., 2005). Additionally, Se enhanced the salt tolerance of *Glycin max* seedlings by protecting the cell membrane against lipid peroxidation (Djanaguiraman et al., 2005). It has been reported that Se effectively protects *Pistacia vera* L. from salt-induced oxidative stress by inhibiting the peroxidation of membrane lipids and the percentage of electrolyte leakage (Walaa et al., 2010). As a result, se-supplemented, water-deficit plants exhibit better protection from salt-stressed damage because of higher CAT, APX, MDHAR, DHAR, GR, GST, and GPX activities and lower levels of H_2O_2 and MDA as compared to salt stress alone.

14.4.3.4 Resistance to pathogens

Selenium has been reported to alleviate biotic stresses and is involved in protecting plants from a wide variety of herbivores as well as from fungal and pest infections (Quinn et al., 2007). In particular, Se was proven to alleviate pathogen stress by up-regulation of jasmonic acid and ethylene, sulfate/selenate assimilation, and the production of defense-related proteins (Tamaoki et al., 2008); stimulation of volatile Se (dimethylselenide) production by plants deters herbivores and pathogens. In Se hyperaccumulators, such as *Astragalus bisulcatus* and *Brassica oleracea*, Se is accumulated as relatively non-toxic methyl-selenocysteine (Lyi et al., 2005).

14.4.3.5 Alleviation of temperature stress

Temperature stress (high and low) can promote ROS accumulation in the chloroplasts, decrease antioxidant activity, which leads to membrane damage, and decrease photosynthetic capacity (Djanaguiraman et al., 2009). Studies have shown that Se can protect plants from high and low temperature stresses (Valadabadi et al., 2010) because of its structural role in the synthesis of glutathione peroxidase enzyme. In *Stanleya pinnata*, Freeman et al. (2010) reported that Se can delay leaf senescence and increase the carbon supply to developing grain under high temperature stress because of elevated expression of genes involved in sulfur assimilation, antioxidant activities, and defense genes of the jasmonic acid and salicylic acid pathway.

14.4.3.6 Alleviation of metal toxicity

Selenium fertilization causes amelioration of heavy metal-induced growth inhibition. Se has been shown to protect plants against damage caused by As, Pb, Cd, Zn, Cu, and chromium (Cr) (Feng et al., 2011; Kumar et al., 2012; Malik et al., 2012). Se can alleviate the negative effects of heavy metals by inhibition of uptake and translocation of heavy metals from the root to shoot. It has been well documented that Se limits heavy metal uptake and induces the accumulation of heavy metals, such as As in *Thunbergia alata* (Bluemlein et al., 2009) and Al in ryegrass (Cartes et al., 2010). It has been demonstrated that the supplementation of *Brassica napus* with a proper dose of Se alleviates oxidative damage under Cd stress. Se-enhanced ascorbate content has been suggested to play a significant role in reducing the MDA and H_2O_2 levels during exposure of plants to Cd stress conditions (Hasanuzzaman et al., 2012). Later studies have revealed that the exogenous application of Se increases the tolerance of plants to Cd-induced oxidative damage by enhancing the activities of APX, GR, GPX, and CAT (Feng et al., 2013).

14.4.4 Cobalt and plant stress responses

In leguminous plants such as *Pisum sativum*, the application of Co increases growth and nodule number and weight. Co is required for the structure of cobalamin (vitamin B_{12}), which is needed for the activity of several nitrogen-fixation enzymes in *Rhizobium* (Palit et al., 1994). It can act as an essential metal involved in controlling of symbiotic *Rhizobia* that live in the nodules of leguminous plants. In leguminous plants, nodules are major sources of ROS production because of high O_2 consumption. To keep this damage to a minimum, nodules possess enzymatic and non-enzymatic antioxidative defense systems (Marschner, 1995). Addition of Co causes growth improvement, increases the nodulation and leghemoglobin concentration and CAT activity in legume plants, and results in stress tolerance. Co-sufficient plants compared with Co-deficient plants exhibit greater enhancement of drought tolerance in seeds. It has been well documented that applying Co decreases leaf senescence via inhibition of ethylene biosynthesis. Co also improves biotic stress resistance indirectly in medicinal plants by accumulation of alkaloids (Palit et al., 1994). In hyperaccumulators of Co, the efficacy of cobalt in protection from herbivory or pathogens could be related to the high tissue Co levels, as was shown for other hyperaccumulated elements (Pilon-Smits et al., 2009).

14.4.5 Aluminum and plant stress responses

Aluminum is one of the main factors responsible for inhibiting root growth in acidic tropical soils through stimulation of lignin deposition in the cell wall. Furthermore, a significant increase in the root growth inhibition under combined Al and drought stress was observed in *Phaseolus coccineus* genotypes (Butare et al., 2011). However, Al partially alleviates the negative effects of water stress in this plant. Al supplementation has been reported to alleviate B deficiency and is involved in fractionation of B in the young and old leaves and roots. Al may modify the B binding capacity of the cell wall in tea leaves. Later studies have revealed that the Al-supplemented tea plants showed a high CO_2 assimilation rate, greater B root—shoot transport and less oxidative damage under B deficiency stress (Hajiboland, 2012). Al may also be involved in resistance to black root rot pathogen (*Thielaviopsis basicola*) and mycelial growth and sporangial germination of potato late blight pathogen (*Phytophthora infestans*).

14.5 Conclusion and future prospects

Micronutrients play an important role in stress alleviation by changing the activity of enzymes, producing certain metabolites involved in plant stress responses, and by modulating the signal transduction pathways within the plant body. In addition, trace metal hyperaccumulation provides a defense against herbivores and pathogens. Since the most extensively investigated mechanisms for micronutrient-mediated stress alleviation have involved the antioxidative stress response, the relationship between antioxidants (GPX, SOD, CAT, APX, and GSH) and micronutrient elements should be taken into account in future studies. Furthermore, excess Mo, Zn, and Cu stimulate the biosynthesis of hormones such as ABA, auxin, and ethylene, which are involved in mediating expression of stress-responsive genes. Plant defense reactions induced by these elements could be regulated by hormones. However, future research is needed to solve the nature of this regulation. Thus, it is also important to solve the correlations between trace element oscillations and hormonal signaling pathways. Plants growing in natural conditions are concurrently affected by a number of stress factors (drought, salinity, metal toxicity, UV radiation, etc.). Understanding the roles of beneficial elements in plant ecological processes under combined stress conditions may be another potential area of further studies. It is interesting to note that other potentially beneficial elements such as silver, chromium, and iodine were studied in different species of terrestrial plants, and could be considered beneficial elements to counteract the harmful effects of environmental stress.

References

Abeles, F.B., Morgan, P.W., Salveit, M.E.J., 1992. Ethylene in Plant Biology. Academic Press, San Diego, USA, pp. 56−119.

Abou Hossein, E.A., Shehata, M.M., El-Sherif, M.A., 2002. Phosphorus nutrition of barley plant as affected by zinc, manganese and organic matter application to saline soils. Egypt. J. Soil Sci. 42, 331−345.

Agrios, N.G., 2005. Plant Pathology. fifth ed. Elsevier, Amsterdam, p. 635.

Ahmad, P., Prasad, M.N.V., 2012a. Abiotic Stress Responses in Plants: Metabolism, Productivity and Sustainability. Springer, New York.

Ahmad, P., Prasad, M.N.V., 2012b. Environmental Adaptations and Stress Tolerance in Plants in the Era of Climate Change. Springer Science + Business Media, New York.

Ahmad, P., Sharma, S., 2008. Salt stress and phyto-biochemical responses of plants. Plant Soil Environ. 54, 89−99.

Ahmad, P., Umar, S., 2011. Oxidative Stress: Role of Antioxidants in Plants. Studium Press, New Delhi.

Ahmad, P., Jaleel, C.A., Salem, M.A., Nabi, G., Sharma, S., 2010a. Roles of enzymatic and non-enzymatic antioxidants in plants during abiotic stress. Crit. Rev. Biotechnol. 30, 161−175.

Ahmad, P., Jaleel, C.A., Sharma, S., 2010b. Antioxidative defence system, lipid peroxidation, proline metabolizing enzymes and biochemical activity in two genotypes of *Morus alba* subjected to NaCl stress. Russ. J. Plant Physiol. 57, 509−517.

Ahmad, P., Nabi, G., Ashraf, M., 2011. Cadmium-induced oxidative damage in mustard [*Brassica juncea* (L.). Czern. & Coss.] plants can be alleviated by salicylic acid. South Afr. J. Bot. 77, 36−44.

Ahmad, P., Hakeem, K.R., Kumar, A., Ashraf, M., Akram, N.A., 2012. Salt-induced changes in photosynthetic activity and oxidative defense system of three cultivars of mustard (*Brassica juncea* L.). Afr. J. Biotechnol. 11, 2694−2703.

Ahmed, M., Hassen, F.U., Qadeer, U., Aslam, M.A., 2011. Silicon application and drought tolerance mechanism of sorghum. Afr. J. Agric. Res. 6, 594−607.

Ahmed, M., Kamran, A., Asif, M., Qadeer, U., Iqbal Ahmed, Z., Goyal, A., 2013. Silicon priming: a potential source to impart abiotic stress tolerance in wheat: a review. AJCS 7, 484−491.

Ajouri, A., Asgedom, H., Becker, M, 2004. Seed priming enhances germination and seedling growth of barley under conditions of P and Zn deficiency. J. Plant Nutr. Soil Sci. 167, 630–636.

Akgül, M., Çöpür, Y., Temiz, S., 2007. A comparison of kraft and kraftsodium borohydrate brutia pine pulps. Build. Environ. 42, 2586–2590.

Aktas, H., Karni, L., Chang, D.C., Turhan, E., Bar-Tal, A., Aloni, B., 2005. The suppression of salinity associated oxygen radicals production in pepper (*Capsicum annuum* L.) fruit, by manganese, zinc and calcium in relation to its sensitivity to blossom end rot. Physiol. Plant 123, 67–74.

Al-Hakimi, A.M., Hamada, A.M., 2011. Ascorbic acid, thiamine or salicylic acid induced changes in some physiological parameters in wheat grown under copper stress. Plant Protec. Sci. 47, 92–108.

Aloni, B., Karni, L., Deventurero, G., Turhan, E., Aktas, H., 2008. Changes in ascorbic acid concentration, ascorbate oxidase activity, and apoplastic pH in relation to fruit development in pepper (*Capsicum annuum* L.) and the occurrence of blossom end rot. J. Hortic. Sci. Biotechnol. 83, 100–105.

Anjum, N.A., Umar, S., Iqbal, M., Khan, N.A., 2011. Cadmium causes oxidative stress in mungbean [*Vigna radiata* (L.). Wilczek] by affecting antioxidant enzyme systems and ascorbate-glutathione cycle metabolism. Russ. J. Plant Physiol. 58, 92–99.

Apel, K., Hirt, H., 2004. Reactive oxygen species: metabolism, oxidative stress, and signal transduction. Annu. Rev. Plant Biol. 55, 373–399.

Ardic, M., Sekmen, A.H., Tokur, S., Ozdemir, F., Turkan, I., 2009. Antioxidant response of chickpea plants subjected to boron toxicity. Plant Biol. 11, 328–338.

Arteca, R.N., Arteca, J.M., 2007. Heavy-metal-induced ethylene production in *Arabidopsis thaliana*. J. Plant Physiol. 164, 1480–1488.

Aslam, M., Ranjha, A.M., Akhtar, J., 2000. Salinity tolerance of rice affected by zinc application. Pak. J. Biol. Sci. 3, 2055–2057.

Azooz, M.M., Abou-Elhamd, M.F., Al-Fredan, M.A., 2012. Biphasic effect of copper on growth, proline, lipid peroxidation and antioxidant enzyme activities of wheat (*Triticum aestivum* cv. Hasaawi) at early growing stage. AJCS 6, 688–694.

Babaeian, M., Heidari, M., Ghanbari, A., 2010. Effect of water stress and foliar micronutrient application on physiological characteristics and nutrient uptake in sunflower (*Helianthus annuus L.*). Iran. J. Crop Sci. 12, 311–391.

Badawi, G.H., Yamauchi, Y., Shimada, E., Sasaki, R., Kawano, N., Tanaka, K., et al., 2004. Enhanced tolerance to salt stress and water deficit by overexpressing superoxide dismutase in tobacco (*Nicotiana tabacum*) chloroplasts. Plant Sci. 166, 919–928.

Barber, J., 2008. Crystal structure of the oxygen-evolving complex of photosystem II. Inorg. Chem. 47, 1700–1710.

Barcelo, J., Guevara, P., Poschenrieder, C., 1993. Effect of silicon on amelioration of aluminium toxicity in teosinte (*Zea mays* L. ssp. Mexicana). Plant Soil 154, 249–255.

Bartels, D., Sunkar, R., 2005. Drought and salt tolerance in plants. Crit. Rev. Plant Sci. 24, 23–58.

Baybordi, A., 2006. Effect of Fe, Mn, Zn and Cu on the quality and quantity of wheat under salinity stress. J. Water Soil Sci. 17, 140–150.

Bennett, J.P., Skoog, F., 2002. Preliminary experiments on the relation of growth-promoting substances to the rest period in fruit trees. Plant Physiol. 13, 219–225.

Bittner, F., Oreb, M., Mendel, R.R., 2001. ABA3 is a molybdenum cofactor sulfurase required for activation of aldehyde oxidase and xanthine dehydrogenase in *Arabidopsis thaliana*. J. Biol. Chem. 276, 40381–40384.

Bluemlein, K., Klimm, E., Raab, A., Feldmann, J., 2009. Selenite enhances arsenate toxicity in *Thunbergia alata*. Environ. Chem. 6, 486–494.

Bockhaven, J.V., De Vleesschauwer, D., Hófte, M., 2013. Towards establishing broad-spectrum disease resistance in plants: silicon leads the way. J. Exp. Bot. 64, 1281−1293.

Bowen, P., Menzies, J., Ehret, D., Samuels, L., Glass, A.D.M., 1992. Soluble silicon sprays inhibit powdery mildew development on grape leaves. J. Am. Soci. Hortic. Sci. 117, 906−912.

Bowler, C., VanCamp, W., VanMontagu, M., Inze, D., 1994. Superoxide dismutase in plants. Crit. Rev. Plant Sci. 13, 199−218.

Boyd, R.S., 2007. The defense hypothesis of elemental hyperaccumulation: status, challenges and new directions. Plant Soil 293, 153−176.

Brewer, G.J., 2010. Copper toxicity in the general population. Clin. Neurophysiol. 121, 459−460.

Briat, J.F., Curie, C., Gaymard, F., 2007. Iron utilization and metabolism in plants. Curr. Opin. Plant Biol. 10, 276−282.

Brissot, P., Bardou-Jacquet, E., Jouanolle, A.M., Loreal, O., 2011. Iron disorders of genetic origin: a changing world. Trends Mol. Med. 17, 707−713.

Brown, P.H., Bellaloui, N., Sah, R.N., Bassil, E., Hu, 2002. Uptake and transport of boron. In: Goldbach, H.E., Rerkasem, B., Wimmer, M., Brown, P.H., Thellier, M., Bell, R.W. (Eds.), Boron in Plant and Animal Nutrition. Kluwer Academic, New York, pp. 87−102.

Bruce, W.B., Edmeades, G.O., Barker, T.C., 2002. Molecular and physiological approaches to maize improvement for drought tolerance. J. Exp. Bot. 53, 13−25.

Butare, L., Rao, I., Lepoivre, P., Polania, J., Cajiao, C., Cuasquer, J., et al., 2011. New genetic sources of resistance in the genus *Phaseolus* to individual and combined stress factors of aluminium toxicity and progressive soil drying. Euphytica 185, 385−404.

Cakmak, I., 2000. Role of zinc in protecting plant cells from reactive oxygen species. New Phytol. 146, 185−205.

Cakmak, I., Engels, C., 1999. Role of mineral nutrients in photosynthesis and yield formation. In: Rengel, Z. (Ed.), Mineral Nutrition of Crops: Fundamental Mechanisms and Implications. The Haworth Press, New York, USA, pp. 141−168.

Cartes, P., Jara, A.A., Pinilla, L., Rosas, A., Mora, M.L., 2010. Selenium improves the antioxidant ability against aluminium-induced oxidative stress in ryegrass roots. Ann. Appl. Biol. 156, 297−307.

Castrillo, M., Fernandez, D., Calcagno, A.M., Trujillo, I., Guenni, L., 2001. Responses of ribulose-1,5-bisphosphate carboxylase, protein content, and stomatal conductance to water deficit in maize, tomato, and bean. Photosynthetica 39, 221−226.

Chen, E.L., Chen, Y.A., Chen, L.M., Liu, Z.H., 2002. Effect of copper on peroxidase activity and lignin content in *Raphanus sativus*. Plant Physiol. Biochem. 40, 439−444.

Chen, W., Yang, X., He, Z., Feng, Y., Hu, F., 2007. Differential changes in photosynthetic capacity, 77K chlorophyll fluorescence and chloroplast ultrastructure between Zn-efficient and Zn-inefficient rice genotypes (*Oryza sativa* L.) under low Zn stress. Plant Physiol. 132, 89−101.

Chen, W., Yao, X., Cai, K., Chen, J., 2011. Silicon alleviates drought stress of rice plants by improving plant water status, photosynthesis and mineral nutrient absorption. Biol. Trace Elem. Res. 142, 67−76.

Chen, Y., Wang, Y., Shin, L.J., Wu, J.F., Shanmugam, V., Tsednee, M., et al., 2013. Iron is involved in the maintenance of circadian period length in *Arabidopsis*. Plant Physiol. 161, 1409−1420.

Chu, C.C., Lee, W.C., Guo, W.Y., Pan, S.M., Chen, L.J., Li, H.M., et al., 2005. A copper chaperone for superoxide dismutase that confers three types of copper/zinc superoxide dismutase activity in *Arabidopsis*. Plant Physiol. 139, 425−436.

Chu, H.H., Chiecko, J., Punshon, T., Lanzirotti, A., Lahner, B., Salt, D.E., et al., 2010a. Successful reproduction requires the function of *Arabidopsis* Yellow Stripe-Like1 and Yellow Stripe-Like3 metal-nicotianamine transporters in both vegetative and reproductive structures. Plant Physiol. 154, 197−210.

Chu, J.Z., Yao, X.Q., Zhang, Z.N., 2010b. Responses of wheat seedlings to exogenous selenium supply under cold stress. Biol. Trace Elem. Res. 136, 355–363.

Claus, J., Bohmann, A., Chavarría-Krauser, A., 2013. Zinc uptake and radial transport in roots of *Arabidopsis thaliana*: a modelling approach to understand accumulation. Annal. Bot. 112, 369–380.

Cooke, J., Leishman, M.R., 2011. Is plant ecology more siliceous than we realise? Trends Plant Sci. 16, 61–68.

Cooke, M.S., Evans, M.D., Dizdaroglu, M., Lunec, J., 2003. Oxidative DNA damage: mechanisms, mutation, and disease. FASEB J. 17, 1195–1214.

Cornelis, J.T., Delvaux, B., Cardinal, D., Andre, L., Ranger, J., Opfergelt, S., 2010. Tracing mechanisms controlling the release of dissolved silicon in forest soil solutions using Si isotopes and Ge/Si ratios. Geochim. et Cosmochim. Acta 74, 3913–3924.

Creissen, G.P., Mullineaux, P.M., 2002. Molecular biology of the ascorbate-glutathione cycle in higher plants. In: Inzé, D., Montganu, M.V. (Eds.), Oxidative Stress in Plants. Taylor & Francis, UK, pp. 247–270.

Dallagnol, L.J., Rodrigues, F.A., DaMatta, F.M., Mielli, M.V.B., Pereira, S.C., 2011. Deficiency in silicon uptake affects cytological, physiological, and biochemical events in the rice–*Bipolaris oryzae* interaction. Phytopathology 101, 92–104.

Dang, H.R., Li, Y., Sun, X., Zhang, Y., 2010. Absorption, accumulation and distribution of zinc in highly-yielding winter wheat. Agric. Sci. China 9, 965–973.

Dasgupta, J., Ananyev, G.M., Dismukes, G.C., 2008. Photoassembly of the water oxidizing complex in photosystem II. Coord. Chem. Rev. 252, 347–360.

Diaz, J., Bernal, M.A., Pomar, F., Merino, F., 2001. Induction of shikimate dehydrogenase and peroxidase in pepper (*Capsicum annuum* L.) seedlings in response to copper stress and its relation to lignifications. Plant Sci. 161, 179–188.

Dietz, K.J., Jacob, S., Oelze, M.L., Laxa, M., Tognetti, V., de Miranda, S.M., et al., 2006. The function of peroxiredoxins in plant organelle redox metabolism. J. Exp. Bot. 57, 1697–1709.

Ding, M., Hou, P., Shen, X., 2010. Salt-induced expression of genes related to Na^+/K^+ and ROS homeostasis in leaves of salt-resistant and salt-sensitive poplar species. Plant Molecul. Biol. 73, 251–269.

Disante, K.B., Fuentes, D., Cortina, J., 2010. Response to drought of Zn-stressed (*Quercus suber* L.) seedlings. Environ. Exp. Bot. 70, 96–103.

Djanaguiraman, M., Devi, D.D., Shanker, A.K., Sheeba, A., Bangarusamy, U., 2005. Selenium—an antioxidative protectant in soybean during senescence. Plant Soil 272, 77–86.

Djanaguiraman, M., Sheeba, J.A., Devi, D.D., Bangarusamy, U., 2009. Cotton leaf senescence can be delayed by nitrophenolate spray through enhanced antioxidant defence system. J. Agron. Crop Sci. 195, 213–224.

Djanaguiraman, M., Prasad, P.V.V., Seppänen, M., 2010. Selenium protects sorghum leaves from oxidative damage under high temperature stress by enhancing antioxidant defense system. Plant Physiol. Biochem. 48, 999–1007.

Dordas, C., Brown, P.H., 2000. Permeability of boric acid across lipid bilayers and factors affecting it. J. Mem. Biol. 175, 95–105.

Duc, C., Cellier, F., Lobréaux, S., Briat, J.F., Gaymard, F., 2009. Regulation of iron homeostasis in *Arabidopsis thaliana* by the clock regulator time for coffee. J. Biol. Chem. 284, 36271–36281.

Duffy, B., 2007. Zinc and plant disease. In: Datnoff, L.E., Elmer, W.H., Huber, D.M. (Eds.), Mineral Nutrition and Plant Disease. American Phytopathological Society, Paul, Minnesota, pp. 155–178.

Eraslan, F., Inal, A., Pilbeam, D.J., Gunes, A., 2008. Interactive effects of salicylic acid and silicon on oxidative damage and antioxidant activity in spinach (*Spinacia oleracea* L. CV. Matador) grown under boron toxicity and salinity. Plant Growth Regul. 55, 207–219.

Faize, M., Burgos, L., Faize, L., Piqueras, A., Nicolas, E., Barba-Espin, G., et al., 2011. Involvement of cytosolic ascorbate peroxidase and Cu/Zn-superoxide dismutase for improved tolerance against drought stress. J. Exp. Bot. 62, 2599–2613.

Feng, R.W., Wei, C.Y., 2012. Antioxidative mechanisms on selenium accumulation in *Pteris vittata* L., a potential selenium phytoremediation plant. Plant Soil Environ. 58, 105−110.

Feng, R., Weic, C., Tu, S., 2013. The roles of selenium in protecting plants against abiotic stresses. Environ. Exp. Bot. 87, 58−68.

Feng, R.W., Wei, C.Y., Tu, S.X., Tang, S.R., Wu, F.C., 2011. Detoxification of antimony by selenium and their interaction in paddy rice under hydroponic conditions. Microchem. J. 97, 57−61.

Fleck, A.T., Nye, T., Repenning, C., Stahl, F., Zahn, M., Schenk, M.K., 2011. Silicon enhances suberization and lignification in roots of rice (*Oryza sativa*). J. Exp. Bot. 62, 2001−2011.

Foyer, C.H., Noctor, G., 2011. Ascorbate and glutathione: the heart of the redox hub. Plant Physiol. 155, 2−18.

Freeman, J.L., Tamaoki, M., Stushnoff, C., Quinn, C.F., Cappa, J.J., Devonshire, J., et al., 2010. Molecular mechanisms of selenium tolerance and hyperaccumulation in *Stanleya pinnata*. Plant Physiol. 153, 1630−1652.

Führs, H., Götze, S., Specht, A., Erban, A., Gallien, S., Heintz, D., et al., 2009. Characterization of leaf apoplastic peroxidases and metabolites in *Vigna unguiculata* in response to toxic manganese supply and silicon. J. Exp. Bot. 60, 1663−1678.

Gapinska, M., Skiodowska, M., Gabara, B., 2008. Effect of short- and long-term salinity on the activities of antioxidative enzymes and lipid peroxidation in tomato roots. Acta Physiol. Plant 30, 11−18.

Gill, S.S., Tuteja, N., 2010. Reactive oxygen species and antioxidant machinery in abiotic stress tolerance in crop plants. Plant Physiol. Biochem. 48, 909−930.

Goto, M., Ehara, H., Karita, S., Takabe, K., Ogawa, N., Yamada, Y., et al., 2003. Protective effect of silicon on phenolic biosynthesis and ultraviolet spectral stress in rice crop. Plant Sci. 164, 349−356.

Graham, D.R., Webb, M.J., 1991. Micronutrients and disease resistance and tolerance in plants. In: Mortvedt, J.J., Cox, F.R., Shuman, L.M., Welch, R.M. (Eds.), Micronutrients in Agriculture. Soil Science Society of America Inc, Madison, Wisconsin, USA, pp. 329−370.

Gunes, A., Soylemezoglu, G., Inal, A., Bagci, E.G., Coban, S., 2006. Antioxidant and stomatal responses of grapevine (*Vitis vinifera* L.) to boron toxicity. Sci. Hortic. 110, 279−284.

Guo, L.H., Wu, X.L., Gong, M., 2005. Roles of glutathione reductase and superoxide dismutase in heat-shock-induced cross adaptation in maize seedlings. Plant Physiol. Commun. 41, 429−432.

Guo, T., Zhang, G., Zhou, M., Wu, F., Chen, J., 2004. Effects of aluminum and cadmium toxicity on growth and antioxidant enzyme activities of two barley genotypes with different Al resistance. Plant Soil 258, 241−248.

Habibi, G., 2013. Effect of drought stress and selenium spraying on photosynthesis and antioxidant activity of spring barley. Acta Agric. Slov. 101, 167−177.

Habibi, G., Hajiboland, R., 2012. Comparison of photosynthesis and antioxidative protection in *Sedum album* and *Sedum stoloniferum* (Crassulaceae) under water stress. Photosynthetica 50, 508−518.

Habibi, G., Hajiboland, R., 2013. Alleviation of drought stress by silicon supplementation in pistachio (*Pistacia vera* L.). Folia Hort. 25, 21−29.

Hacisalihoglu, G., Hart, J.J., Wang, Y., Cakmak, I., Kochian, L.V., 2003. Zinc efficiency is correlated with enhanced expression and activity of Cu/Zn superoxide dismutase and carbonic anhydrase in wheat. Plant Physiol. 131, 595−602.

Hajiboland, R., 2012. Effect of micronutrient deficiencies on plants' stress responses. In: Ahmad, P., Prasad, M.N.V. (Eds.), Abiotic Stress Responses in Plants: Metabolism, Productivity and Sustainability. Springer, New York, pp. 283−331.

Hajiboland, R., Amirazad, F., 2010. Drought tolerance in Zn-deficient red cabbage (*Brassica oleracea* L. var. *capitata*) plants. Hort Sci. 37, 88−98.

Halliwell, B., Gutteridge, J., 1984. Oxygen toxicity, oxygen radicals, transition metals and disease. Biochem. J. 219, 1−14.

Hammond, K.E., Evans, D.E., Hodson, M.J., 1995. Aluminium/Silicon interactions in barley (*Hordeum vulgare* L.) seedlings. Plant Soil 173, 89−95.

Han, S., Tang, N., Jiang, H.X., Yang, L.T., Lee, Y., 2009. CO_2 assimilation, photosystem II photochemistry, carbohydrate metabolism and antioxidant system of citrus leaves in response to boron stress. Plant Sci. 176, 143−153.

Han, S.H., Kim, D.H., Lee, J.C., 2010. Cadmium and zinc interaction and phytoremediation potential of seven *Salix caprea* clones. J. Ecol. Field Biol. 33, 245−251.

Harmer, S.L., 2009. The circadian system in higher plants. Annu. Rev. Plant Biol. 60, 357−377.

Hasan, S.A., Hayat, S., Ali, B., Ahmad, A., 2008. 28-Homobrassinolide protects chickpea (*Cicer arietinum*) from cadmium toxicity by stimulating antioxidants. Environ. Pollut. 151, 60−66.

Hasanuzzaman, M., Fujita, M., 2011. Selenium pretreatment upregulates the antioxidant defense and methylglyoxal detoxification system and confers enhanced tolerance to drought stress in rapeseed seedlings. Biol. Trace Elem. Res. 143, 1758−1776.

Hasanuzzaman, M., Hossain, M.A., Fujita, M., 2011. Selenium-induced up-regulation of the antioxidant defense and methylglyoxal detoxification system reduces salinity-induced damage in rapeseed seedlings. Biol. Trace Elem. Res. 143, 1704−1721.

Hasanuzzaman, M., Hossain, M.A., Fujita, M., 2012. Exogenous selenium pretreatment protects rapeseed seedlings from cadmium-induced oxidative stress by upregulating antioxidant defense and methylglyoxal detoxification systems. Biol. Trace Elem. Res. 149, 248−261.

Hashemi, A., Abdolzadeh, A., Sadeghipour, H.R., 2010. Beneficial effects of silicon nutrition in alleviating salinity stress in hydroponically grown canola, *Brassica napus* L. plants. Soil Sci. Plant Nutr. 56, 244−253.

Hattori, T., Sonobe, K., Inanaga, S., An, P., Tsuji, W., Araki, H., et al., 2007. Short term stomatal responses to light intensity changes and osmotic stress in sorghum seedlings raised with and without silicon. Environ. Exp. Bot. 60, 177−182.

Hayat, S., Ahmad, A., 2011. Brassinosteroids: a Class of Plant Hormone. Springer, Dordrecht.

Haydon, M.J., Kawachi, M., Wirtz, M., Hillmer, S., Hell, R., Krümer, U., 2012. Vacuolar nicotianamine has critical and distinct roles under iron deficiency and for zinc sequestration in *Arabidopsis*. Plant Cell 24, 724−737.

Heckman, J.R., 2007. Chlorine. In: Barker, A.V., Pilbeam, D.J. (Eds.), Handbook of plant nutrition. CRC Press, Taylor and Francis Group, Boca Raton, FL, pp. 279−292.

Heckman, J.R., Clarke, B.B., Murphy, J.A., 2003. Optimizing manganese fertilization for the suppression of take-all patch disease on creeping bentgrass. Crop Sci. 43, 1395−1398.

Hu, C., Wang, Y., Wei, W., 2002. Effect of molybdenum applications on concentrations of free amino acids in winter wheat at different growth stages. J. Plant Nutri. 25, 1487−1499.

Huber, D.M., Graham, R.D., 1999. The role of nutrition in crop resistance and tolerance to disease. In: Rengel, Z. (Ed.), Mineral Nutrition of Crops: Fundamental Mechanisms and Implications. Food Product Press, New York, pp. 205−226.

Huber, M.D., 1996. The role of nutrition in the take-all disease of wheat and other small grains. In: Engelhard, W.A. (Ed.), Management of Diseases with Macro- and Microelements. APS, Minneapolis, USA, pp. 46−74.

Imaizumi, T., 2010. *Arabidopsis* circadian clock and photoperiodism: time to think about location. Curr. Opin. Plant Biol. 13, 83−89.

Iwasaki, K., Matsumura, A., 1999. Effect of silicon on alleviation of manganese toxicity in pumpkin (Cucurbita moschata Duch cv. Shintosa). Soil Sci. Plant Nutri. 45, 909−920.

Kalim, S.Y.P., Luthra, S.K., Gandhi, I., 2003. Role of zinc and manganese in resistance of cowpea. J. Plant Dis. Protect. 110, 235−243.

Kamenidou, S., Cavins, T.J., Marek, S., 2009. Evaluation of silicon as a nutritional supplement for greenhouse zinnia production. Sci. Hortic. 119, 297–301.

Kaya, C., Levent, A., Dikilitas, M., Ashraf, M., Koskeroglu, S., 2009. Supplementary phosphorus can alleviate boron toxicity in tomato. Sci. Hortic. 121, 284–288.

Kim, Y.H., Kim, C.Y., Song, W.K., Park, D.S., Kwon, S.Y., Lee, H.S., et al., 2008. Overexpression of sweet potato *swpa4* peroxidase results in increased hydrogen peroxide production and enhances stress tolerance in tobacco. Planta 227, 867–881.

Kong, L., Wang, M., Bi, D., 2005. Selenium modulates the activities of antioxidant enzymes, osmotic homeostasis and promotes the growth of sorrel seedlings under salt stress. Plant Growth Regul. 45, 155–163.

Kosman, D.J., 2010. Redox cycling in iron uptake, efflux, and trafficking. J. Biol. Chem. 285, 26729–26735.

Krämer, U., Talke, I.N., Hanikenne, M., 2007. Transition metal transport. FEBS Lett. 581, 2263–2272.

Krauss, A., 1999. Balanced Nutrition and Biotic Stress. IFA agricultural conference on managing plant nutrition, Barcelona, Spain.

Kumar, M., Bijo, A.J., Baghel, R.S., Reddy, C.R.K., Jha, B., 2012. Selenium and spermine alleviates cadmium induced toxicity in the red seaweed *Gracilaria dura* by regulating antioxidant system and DNA methylation. Plant Physiol. Biochem. 51, 129–138.

Lal, A., Edwards, G.E., 1996. Analysis of inhibition of photosynthesis under water stress in the C4 species *Amaranthus cruentus* and *Zea mays*: electron transport, CO_2 fixation and carboxylation capacity. Aust. J. Plant Physiol. 23, 403–412.

Lee, B.R., Kim, K.Y., Jung, W.J., Avice, J.C., Ourry, A., Kim, T.H., 2007. Peroxidases and lignification in relation to the intensity of water-deficit stress in white clover (*Trifolium repens* L.). J. Exp. Bot. 58, 1271–1279.

Lehto, T., Ruuhola, T., Dell, B., 2010. Boron in forest trees and ecosystems. For. Ecol. Manag. 260, 20563–22069.

Leidi, E.O., Barragán, V.N., Rubio, L., et al., 2010. The AtNHX1 exchanger mediates potassium compartmentation in vacuoles of transgenic tomato. Plant J. 61, 495–506.

Lekklar, C., Chaidee, A., 2011. Roles of silicon on growth of Thai jasmine rice (*Oryza sativa* L. cv. KDML105) under salt stress. Agric. Sci. J. 42, 45–48.

Li, M., Wang, G.X., Lin, J.S., 2003. Application of external calcium in improving the PEG-induced water stress tolerance in liquorice cells. Bot. Bull. Acad. Sin. 44, 275–284.

Li, M., Li, Y., Li, H., Wu, G., 2011. Overexpression of *AtNHX5* improves tolerance to both salt and drought stress in (*Broussonetia papyrifera* L.) Vent. Tree Physiol. 31, 349–357.

Liang, Y., Sun, W., Zhu, Y.G., Christie, P., 2007. Mechanisms of silicon-mediated alleviation of abiotic stresses in higher plants: a review. Environ. Pollut. 147, 422–428.

Lu, Y., Li, Y., Zhang, J., Xiao, Y., Yue, Y., Duan, L., et al., 2013. Overexpression of *Arabidopsis* molybdenum cofactor sulfurase gene confers drought tolerance in maize (*Zea mays* L.). PLoS ONE 8, 521–526.

Lyi, S.M., Heller, L.I., Rutzke, M., Welch, R.M., Kochian, L.V., Li, L., 2005. Molecular and biochemical characterization of the selenocysteine Se-methyltransferase gene and Se-methylselenocysteine synthesis in broccoli. Plant Physiol. 138, 409–420.

Ma, J.F., 2004. Role of silicon in enhancing the resistance of plants to biotic and abiotic stresses. Soil Sci. Plant Nutr. 50, 11–18.

Maksymiec, W., 2007. Signaling responses in plant to heavy metal stress. Acta Physiol. Plant 29, 177–187.

Malik, J.A., Goel, S., Kaur, N., Sharma, S., Singh, I., Nayyar, H., 2012. Selenium antagonises the toxic effects of arsenic on mungbean (*Phaseolus aureus* Roxb.) plants by restricting its uptake and enhancing the antioxidative and detoxification mechanisms. Environ. Exp. Bot. 77, 242–248.

Mann, R.L., Kettlewell, P.S., Jenkinson, P., 2004. Effect of foliar applied potassium chloride on septoria leaf blotch of winter wheat. Plant Pathol. 53, 653–659.

Mano, J., 2002. Early events in environmental stresses in plants' induction mechanisms of oxidative stress. In: Inzé, D., Montganu, M.V. (Eds.), Oxidative Stress in Plants. Taylor & Francis, UK, pp. 217−246.

Marschner, H., 1995. Mineral Nutrition of Higher Plants. Academic Press, London, UK.

Melchiorre, M., Robert, G., Trippi, V., Racca, R., Lascano, H.R., 2009. Superoxide dismutase and glutathione reductase overexpression in wheat protoplast: photooxidative stress tolerance and changes in cellular redox state. Plant Growth. Regul. 57, 57−68.

Melzer, S.E., Knapp, A.K., Kirkman, K.P., Smith, M.D., Blair, J.M., Kelly, E.F., 2010. Fire and grazing impacts on silica production and storage in grass dominated ecosystems. Biogeochemistry 97, 263−278.

Mendel, R.R., Haensch, R., 2002. Molybdoenzymes and molybdenum cofactor in plants. J. Exp. Bot. 53, 1689−1698.

Meng, Q., Zou, J., Jiang, W., Liu, D., 2007. Effect of Cu^{2+} concentration on growth, antioxidant enzyme activity and malondialdehyde content in garlic (*Allium sativum* L.). Acta Biol. Crac. 49, 95−101.

Mhamdi, A., Noctor, G., Baker, A., 2012. Plant catalases: peroxisomal redox guardians. Arch. Biochem. Biophys. 525, 181−194.

Millaleo, R., Alberdi, M., Ivanov, A.G., Krol, M., Hüner, N.P.A., Huang, P., et al., 2013. Excess manganese differentially inhibits photosystem I versus II in *Arabidopsis thaliana*. J. Exp. Bot. 64, 343−354.

Miller, G., Suzuki, N., Ciftci-Yilmaz, S., Mittler, R., 2010. Reactive oxygen species homeostasis and signaling during drought and salinity stresses. Plant Cell Environ. 33, 453−467.

Mittler, R., Blumwald, E., 2010. Genetic engineering for modern agriculture: challenges and perspectives. Annu. Rev. Plant Biol. 61, 443−462.

Mittler, R., Vanderauwera, S., Gollery, M., Van Breusegem, F., 2004. Reactive oxygen gene network of plants. Trends Plant Sci. 9, 490−498.

Molassiotis, A., Tanou, G., Diamantidis, G., Patakas, A., Therios, I., 2006. Effects of 4-month Fe deficiency exposure on Fe reduction mechanism, photosynthetic gas exchange, chlorophyll fluorescence and antioxidant defense in two peach rootstocks differing in Fe deficiency tolerance. J. Plant Physiol. 163, 176−185.

Moller, I.M., Jensen, P.E., Hansson, A., 2007. Oxidative modifications to cellular components in plants. Annu. Rev. Plant Biol. 58, 459−481.

Moussa, H.R., 2006. Influence of exogenous application of silicon on physiological response of salt-stressed maize (*Zea mays* L.). Int. J. Agri. Biol. 8, 293−297.

Mulrooney, S.B., Hausinger, R.P., 2003. Nickel uptake and utilization by microorganisms. FEMS Microbiol. Rev. 27, 239−261.

Munekage, Y., Hojo, M., Meurer, J., Endo, T., Tasaka, M., Shikanai, T., 2002. *PGR5* is involved in cyclic electron flow around photosystem I and is essential for photoprotection in *Arabidopsis*. Cell 110, 361−371.

Munns, R., Tester, M., 2008. Mechanisms of salinity tolerance. Annu. Rev. Plant Biol. 59, 651−681.

Myouga, F., Hosoda, C., Umezawa, T., Iizumi, H., Kuromori, T., Motohashi, R., et al., 2008. Heterocomplex of iron superoxide dismutases defends chloroplast nucleoids against oxidative stress and is essential for chloroplast development in *Arabidopsis*. Plant Cell 20, 3148−3162.

Nada, E., Ferjani, B.A., Ali, R., Imed, B.R.B.M., Makki, B., 2007. Cadmium-induced growth inhibition and alteration of biochemical parameters in almond seedlings grown in solution culture. Acta Physiol. Plant 29, 57−62.

Nadgórska-Socha, A., Kafel, A., Kandziora-Ciupa, M., Gospodarek, J., Zawisza-Raszka, A., 2013. Accumulation of heavy metals and antioxidant responses in *Vicia faba* plants grown on monometallic contaminated soil. Environ. Sci. Pollut. Res. 20, 1124−1134.

Nagajyoti, P.C., Lee, K.D., Sreekanth, T.V.M., 2010. Heavy metals, occurrence and toxicity for plants: a review. Environ. Chem. Lett. 8, 199−216.

Niewiadomska, E., Bilger, W., Gruca, M., Mulisch, M., Miszalski, Z., Krupinska, K., 2011. CAM-related changes in chloroplastic metabolism of *Mesembryanthemum crystallinum* L. Planta 233, 275−285.

Ohnishi, J.I., Flugge, U.I., Heldt, H.W., Kanai, R., 1990. Involvement of Na^+ in active uptake of pyruvate in mesophyll chloroplasts of some C_4 plants. Plant Physiol. 94, 950−959.

Olias, R., Eljakaoui, Z., Li, J., Alvarez De Morales, P., Marin-Manzano, M.C., Pardo, J.M., et al., 2009. The plasma membrane Na^+/H^+ antiporter SOS1 is essential for salt tolerance in tomato and affects the partitioning of Na^+ between plant organs. Plant Cell Environ. 32, 904−916.

Palit, S., Sharma, A., Talukder, G., 1994. Effect of cobalt on plants. Bot. Rev. 60, 149−181.

Palóve-Balang, P., Kisová, A., Pavlovkin, J., Mistrík, I., 2006. Effect of manganese on cadmium toxicity in maize seedlings. Plant Soil Environ. 52, 143−149.

Passardi, F., Cosio, C., Pend, C., Dumand, C., 2005. Peroxidases have more functions than a Swiss army knife. Plant Cell Rep. 24, 255−265.

Peck, A.W., McDonald, G.K., 2010. Adequate zinc nutrition alleviates the adverse effects of heat stress in bread wheat. Plant Soil 337, 355−374.

Peng, K., Luo, C., You, W., Lian, C.H., Li, X., Shen, Z., 2008. Manganese uptake and interactions with cadmium in the hyperaccumulator *Phytolacca americana*. J. Hazard. Mat. 154, 674−681.

Pennycooke, J.C., Cox, S., Stushnoff, C., 2005. Relationship of cold acclimation, total phenolic content and antioxidant capacity with chilling tolerance in petunia (*Petunia* × *hybrida*). Environ. Exp. Bot. 53, 225−232.

Pilon-Smits, E.A.H., Quinn, C.F., Tapken, W., Malagoli, M., Schiavon, M., 2009. Physiological functions of beneficial elements. Curr. Opin. Plant Biol. 12, 267−274.

Poschenrieder, C., Tolrà, R., Barceló, J., 2006. Can metals defend plants against biotic stress? Trends Plant Sci. 11, 288−295.

Quinn, C.F., Galeas, M.L., Freeman, J.L., Pilon-Smits, E.A.H., 2007. Selenium: deterrence, toxicity, and adaptation. Integr. Environ. Assess. Manag. 3, 460−462.

Räisänen, M., Repo, T., Lehto, T., 2007. Cold acclimation was partially impaired in boron deficient Norway spruce seedlings. Plant Soil 292, 271−282.

Ranieri, A., Castagna, A., Baldan, B., Soldatini, G.F., 2001. Iron deficiency differently affects peroxidase isoforms in sunflower. J. Exp. Bot. 52, 25−35.

Richmond, K.E., Sussman, M., 2003. Got silicon? The non-essential beneficial plant nutrient. Curr. Opin. Plant Biol. 6, 268−272.

Ríos, J.J., Blasco, B., Cervilla, L.M., Rosales, M.A., Sanchez-Rodriguez, E., Romero, L., et al., 2009. Production and detoxification of H_2O_2 in lettuce plants exposed to selenium. Ann. Appl. Biol. 154, 107−116.

Robert-Seilaniantz, A., Grant, M., Jones, J.D.G., 2011. Hormone crosstalk in plant disease and defense: more than just jasmonate salicylate antagonism. Annu. Rev. Phytopath. 49, 317−343.

Rodríguez-Rosales, M.P., Gálvez, F.J., Huertas, R., Aranda, M.N., Baghour, M., Cagnac, O., et al., 2009. Plant NHX cation/proton antiporters. Plant Signal Behav. 4, 265−276.

Roschzttardtz, H., Séguéla-Arnaud, M., Briat, J.F., Vert, G., Curie, C., 2011. The FRD3 citrate effluxer promotes iron nutrition between symplastically disconnected tissues throughout *Arabidopsis* development. Plant Cell 23, 2725−2737.

Roy, A.K., Sharma, A., Talukder, G., 1998. Some aspects of aluminum toxicity. Bot. Rev. 54, 145−178.

Ruiz, J.M., Rivero, R.M., Romero, L., 2006. Boron increases synthesis of glutathione in sunflower plants subjected to aluminium stress. Plant Soil 279, 25−30.

Ruuhola, T., Keinänen, M., Keski-Saari, S., Lehto, T., 2011. Boron nutrition affects the carbon metabolism of silver birch seedlings. Tree Physiol. 31, 1251−1261.

Ryder, M., Gerard, F., Evans, D.E., Hodson, M.J., 2003. The use of root growth and modeling data to investigate amelioration of aluminium toxicity by silicon in *Picea abies* seedlings. J. Inorg. Biochem. 97, 52−58.

Salomon, E., Keren, N., 2011. Manganese limitation induces changes in the activity and in the organization of photosynthetic complexes in the *Cyanobacterium synechocystis*. Plant Physiol. 155, 571−579.

Schaller, J., Brackhage, C., Dudel, E., 2012a. Silicon availability changes structural carbon ratio and phenol content of grasses. Environ. Exp. Bot. 77, 283−287.

Schaller, J., Brackhage, C., Gessner, M.O., Bäuker, E., Gert Dudel, E., 2012b. Silicon supply modifies C:N:P stoichiometry and growth of Phragmites australis. Plant Biol. 14, 392−396.

Schaller, J., Brackhage, C., Paasch, S., Brunner, E., Bäucker, E., Dudel, E.G., 2013. Silica uptake from nanoparticles and silica condensation state in different tissues of *Phragmites australis*. Sci. Total Environ. 442, 6−9.

Seo, M., Aoki, H., Koiwai, H., Kamiya, Y., Nambara, E., et al., 2004. Comparative studies on the Arabidopsis aldehyde oxidase (AAO) gene family revealed a major role of AAO3 in ABA biosynthesis in seeds. Plant Cell Physiol. 45, 1694−1703.

Sharma, P., Sharma, N., Deswal, R., 2005. The molecular biology of the low temperature response in plants. Bio. Essays 27, 1048−1059.

Shen, X., Zhou, Y., Duan, L., Li, Z., Eneji, A.E., Li, J., 2010. Silicon effects on photosynthesis and antioxidant parameters of soybean seedlings under drought and ultraviolet-B radiation. J. Plant Physiol. 167, 1248−1252.

Shetty, R., Jensen, B., Shetty, N.P., Hansen, M., Hansen, C.W., Starkey, K.R., et al., 2012. Silicon induced resistance against powdery mildew of roses caused by *Podosphaera pannosa*. Plant Pathol. 61, 120−131.

Shi, Q., Bao, Y., Zhu, Y., He, Y., Qian, Q., Yu, J., 2005. Silicon-mediated alleviation of Mn toxicity in *Cucumis sativus* in relation to activities of superoxide dismutase and ascorbate peroxidase. Phytochemistry 66, 1551−1559.

Shinozaki, K., Yamaguchi-Shinozaki, K., Seki, M., 2003. Regulatory network of gene expression in the drought and cold stress responses. Curr. Opin. Plant Biol. 6, 410−417.

Simoglou, K., Dordas, C., 2006. Effect of foliar applied boron, manganese and zinc on tan spot in winter durum wheat. Crop Prot. 25, 657−663.

Singh, D.P., Belay, J., McInerney, J.K., Day, L., 2012. Impact of boron, calcium and genetic factors on vitamin C, carotenoids, phenolic acids, anthocyanins and antioxidant capacity of carrots (*Daucus carota*). Food Chem. 132, 1161−1170.

Skorzynska-Polit, E., Drazkiewicz, M., Krupa, Z., 2004. The activity of the antioxidative system in cadmium-treated *Arabidopsis thaliana*. Biol. Plant 47, 71−78.

Snyder, G.H., Matichenkov, V.V., Datnoff, L.E., 2007. Silicon. In: Barker, A.V., Pilbeam, D.J. (Eds.), Handbook of Plant Nutrition. CRC Press, Taylor and Francis Group, Boca Raton, FL, pp. 551−568.

Solecka, D., Kacperska, A., 2003. Phenylpropanoid deficiency affects the course of plant acclimation to cold. Physiol. Plant 119, 253−262.

Sonobe, K., Hattori, T., An, P., Tsuji, W., Eneji, A.E., Kobayashi, S., et al., 2011. Effect of silicon application on sorghum root responses to water stress. J. Plant Nutr. 34, 71−82.

Srivastav, V., Schinkel, H., Witzell, J., Hertzberg, M., Torp, M., Srivastava, M.K., et al., 2007. Downregulation of high-isoelectric-point extracellular superoxide dismutase mediates alterations in the metabolism of reactive oxygen species and developmental disturbances in hybrid aspen. Plant J. 49, 135−148.

Sun, X.C., Hu, C.X., Tan, Q.L., 2006. Effects of molybdenum on antioxidative defense system and membrane lipid peroxidation in winter wheat under low temperature stress. J. Plant Physiol. Mol. Biol. 32, 175−182.

Tahir, M.A., Aziz, T., Farooq, M., Sarwar, G., 2012. Silicon-induced changes in growth, ionic composition, water relations, chlorophyll contents and membrane permeability in two salt-stressed wheat genotypes. Arch. Agron. Soil. Sci. 58, 247−256.

Takahashi, S., Milward, S.E., Fan, D.Y., Chow, W.S., Badger, M.R., 2009. How does cyclic electron flow alleviate photoinhibition in *Arabidopsis*? Plant Physiol. 149, 1560−1567.

Tamaoki, M., Freeman, J.L., Pilon-Smits, E.A.H., 2008. Cooperative ethylene and jasmonic acid signaling regulates selenite resistance in *Arabidopsis thaliana*. Plant Physiol. 146, 1219–1230.

Tan, Y.F., Toole, N., Taylor, N.L., Millar, A.H., 2010. Divalent metal ions in plant mitochondria and their role in interactions with proteins and oxidative stress-induced damage to respiratory function. Plant Physiol. 152, 747–761.

Tavallali, V., Rahemi, M., Eshghi, S., Kholdebarin, B., Ramezanian, A., 2010. Zinc alleviates salt stress and increases antioxidant enzyme activity in the leaves of pistachio (*Pistacia vera* L.) seedlings. Turk. J. Agr. Forest. 34, 349–359.

Terry, N., Zayed, A.M., de Souza, M.P., Tarun, A.S., 2000. Selenium in higher plants. Annu. Rev. Plant Physiol. Plant Mol. Biol. 51, 401–432.

Tisdale, S.L., Nelson, W.L., Beaten, J.D., 1984. Zinc in Soil Fertility and Fertilizers. fourth ed. Macmillan Publishing Company, New York, pp. 382–391.

Tuteja, N., 2010. Cold, salt and drought stress. Plant Stress Biology: From Genomics Towards System Biology. Wiley-Blackwell, Weinheim, Germany, pp. 137–159.

Tuteja, N., Singh, M.B., Misra, M.K., Bhalla, P.L., Tuteja, R., 2001. Molecular mechanisms of DNA damage and repair: progress in plants. Crit. Rev. Biochem. Mol. Biol. 36, 337–397.

Valadabadi, S.A., Shiranirad, A.H., Farahani, H.A., 2010. Ecophysiological influences of zeolite and selenium on water deficit stress tolerance in different rapeseed cultivars. J. Ecol. Nat. Environ. 2, 154–159.

Vesk, P.A., Reichman, S., 2009. Hyperaccumulators and herbivores—a Bayesian meta-analysis of feeding choice trials. J. Chem. Ecol. 35, 289–296.

Vidhyasekaran, P., 2004. Concise Encyclopaedia of Plant Pathology. Food Products Press, The Haworth Reference Press, p. 619.

Vunkova-Radeva, R., Yaneva, I., Strumin, P., 2003. Mo-containing enzymes responses to low temperature stress of winter wheat grown on acid soil. Bulg. J. Plant Physiol. 14, 382–383.

Walaa, A.E., Shatlah, M.A., Atteia, M.H., Sror, H.A.M., 2010. Selenium induces antioxidant defensive enzymes and promotes tolerance against salinity stress in cucumber seedlings (*Cucumis sativus*). Arab. Univ. J. Agric. Sci. 18, 65–76.

Wang, C.Q., 2011. Water-stress mitigation by selenium in *Trifolium repens* L. J. Plant Nutri. Soil Sci. 174, 276–282.

Wang, J.Z., Tao, S.T., Qi, K.J., Wu, J., Wu, H.Q., 2011. Changes in photosynthetic properties and antioxidative system of pear leaves to boron toxicity. Afr. J. Biotechnol. 10, 19693–19700.

Wang, Y., Ying, Y., Chen, J., Wang, X.C., 2004. Transgenic *Arabidopsis* overexpressing Mn-SOD enhanced salt-tolerance. Plant Sci. 167, 671–677.

Wang, Y.H., Wei, W.X., Tan, Q.L., 1995. A study on molybdenum deficiency and molybdenum application of winter wheat in yellow-brown soil of Hubei province. Soil Fertil. 8, 24–28.

Waraich, E.A., Amad, R., Ashraf, M.Y., Saifullah, A.M., 2011. Improving agricultural water use efficiency by nutrient management. Acta Agric. Scand—Soil. Plant Sci. 61, 291–304.

Wong, C.K.E., Cobbett, C.S., 2009. HMA P-type ATPases are the major mechanism for root-to-shoot Cd translocation in *Arabidopsis thaliana*. New Phytol. 181, 71–78.

Wu, F.B., Zhang, G.P., 2002. Alleviation of cadmium-toxicity by application of zinc and ascorbic acid in barley. J. Plant Nutr. 25, 2745–2761.

Xu, P.L., Guo, Y.K., Bai, J.G., Shang, L., Wang, X.J., 2008. Effects of long-term chilling on ultrastructure and antioxidant activity in leaves of two cucumber cultivars under low light. Physiol. Plant 132, 467–478.

Yamaguchi, T.S., Fukuda-Tanaka, Y., Inagaki, N., Saito, K., Yonekura-Sakakibara, Y., Tanaka, T., et al., 2001. Genes encoding the vacuolar Na^+/H^+ exchanger and flower coloration. Plant Cell Physiol. 42, 451–461.

Yao, X.Q., Chu, J.Z., Ba, C.J., 2010. Antioxidant responses of wheat seedlings to exogenous selenium supply under enhanced ultraviolet-B. Biol. Trace Elem. Res. 136, 96–105.

Yao, X.Q., Chu, J., He, X., Ba, C., 2011a. Protective role of selenium in wheat seedlings subjected to enhanced UV-B radiation. Russ. J. Plant Physiol. 58, 283–289.

Yao, X.Q., Chu, J.Z., Cai, K.Z., Liu, L., Shi, J.D., Geng, W.Y., 2011b. Silicon improves the tolerance of wheat seedlings to ultraviolet-B stress. Biol. Trace Elem. Res. 143, 507–517.

Yermiyahu, U., Ben-Gal, A., Keren, R., Redi, R.J., 2008. Combined effect of salinity and boron on plant growth and yield. Plant Soil 304, 73–87.

Yoshida, K., Miki, N., Momonoi, K., Kawachi, M., Katou, K., Okazaki, Y., et al., 2009. Synchrony between flower opening and petal-color change from red to blue in morning glory, *Ipomoea tricolor* cv. Heavenly Blue. Proc. Jpn. Acad. Ser. B. Phys. Biol. Sci. 85, 187–197.

Yusuf, M., Fariduddin, Q., Varshney, P., Ahmad, A., 2012. Salicylic acid minimizes nickel and/or salinity-induced toxicity in Indian mustard (*Brassica juncea*) through an improved antioxidant system. Environ. Sci. Pollut. Res. 19, 8–18.

Zengin, F., Munzuroglu, O., 2005. Effect of some heavy metals on content of chlorophyll, proline and some antioxidant chemicals in bean (*Phaseolus vulgaris* L.) seedlings. Acta Biol. Crac. 47, 157–164.

Zornoza, P., Sánchez-Pardo, B., Carpena, R.R.O., 2010. Interaction and accumulation of manganese and cadmium in the manganese accumulator *Lupinus albus*. J. Plant Physiol. 167, 1027–1032.

Nutritional Stress in Dystrophic Savanna Soils of the Orinoco Basin: Biological Responses to Low Nitrogen and Phosphorus Availabilities

Danilo López-Hernández, Rosa Mary Hernández-Hernández,
Ismael Hernández-Valencia and Marcia Toro

15.1 Introduction

Savannas represent one of the largest extensions of land in the world. In South America, savannas occupy an area of more than 269×10^6 ha extended in Brazil (204×10^6 ha), Colombia (23×10^6 ha), Venezuela (25×10^6 ha), Guyana (4×10^6 ha), and Bolivia (13×10^6 ha) (Rippstein et al., 2001), thus giving to those biomes a high potential for agricultural expansion. In its natural form (with little human intervention), savannas are characterized by associations of herbaceous vegetation with the presence of scattered trees and with patterns in the seasonality of water availability defined by a marked dry season.

Within the areas occupied by savannas in South America, the most important ones are the Brazilian Cerrado and the savannas between Colombia and Venezuela, locally known as the Orinocos llanos. The llanos, with an extension of about 500,000 km², is the largest region of savannas located to the north of South America (López-Hernández and Hernández-Valencia, 2008). In Venezuelan and Colombian llanos (Figure 15.1), we can find well-drained and flooded, eutrophic and oligotrophic savannas (Huber, 2007); however, the most extended are well-drained and dystrophic savannas also known as *Trachypogon* savannas due to the dominance of species of this genus (Ramia, 1967).

It is a well-known fact that not all the surface occupied by llanos is suitable for cropping, as soils suffer from physical and chemical limitations, which together with the marked climatic seasonality reduces the spectrum of possibilities for agricultural production (López-Hernández et al., 2005). Moreover, recently implanted crops in neotropical savannas, as in other regions of the tropics, are also subjected immediately to biological stress (pests) induced by fungi, bacteria, and endemic insects, which cohabit in the environment and which also compromise agricultural production (López-Hernández et al., 2005). Notwithstanding the above, the large tracts of savannas in South America are the main alternative to expansion into tropical areas of greater ecological fragility, e.g., slopes of mountains and the Amazon rainforest (López-Hernández, 1995).

P. Ahmad (Ed): Emerging Technologies and Management of Crop Stress Tolerance, Volume 1.
DOI: http://dx.doi.org/10.1016/B978-0-12-800876-8.00015-1

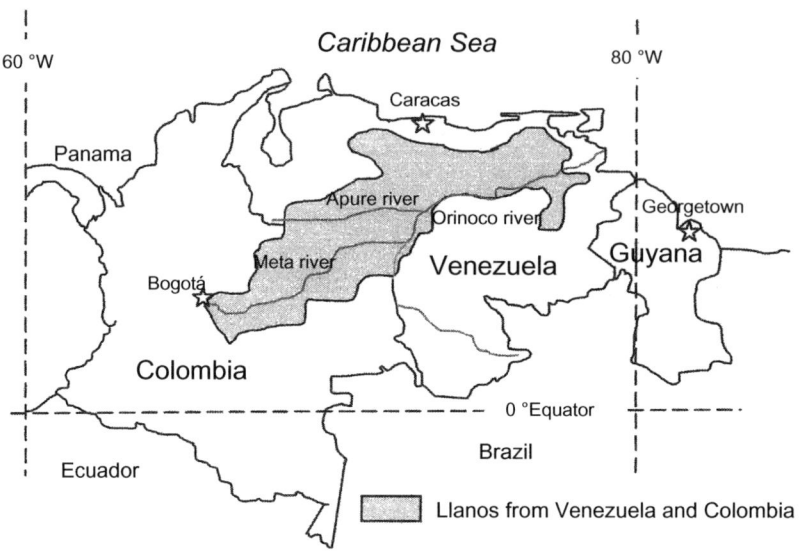

FIGURE 15.1

Location of the llanos of the Orinoco basin.

Human presence in the savannas of South America is relatively recent, between 7000 and 9000 years before present (Berrio et al., 2002). After the encounter of cultures and for centuries, the activities of the sparse human populations settled in the savannas of the neotropics did not have a very marked effect on the landscape (López-Hernández, 1995). It has been pointed out that indigenous communities practice controlled burning; so, their knowledge is important to pursue strategies of prescriptive burning to control fuel loads (Rodriguez et al., 2009; Sleto and Rodríguez, 2013) and reduce carbon (C) emissions to the atmosphere. However, natural fires existed long before the emergence of humans and they are inextricably linked to the presence of savannas, as it has been recently reviewed by Beerling and Osborne (2006).

In the case of natural savannas (e.g., without significant cattle raising activity), particularly referred in this chapter, the regular burning of vegetation is a management tool. In this case, although fire frequency can be very variable (every 1 to 5 years) according to the type of savannas and to human uses, its influence on C and also on nutrient cycling (particularly nitrogen (N) cycling) can be remarkable indeed (López-Hernández and Hernández-Valencia, 2008). When savanna burns, a significant decline in the effective input of organic matter and nutrients to the soil is induced, as a result, N can be limited to plant growing (Hernández-Valencia and López-Hernández, 2002). In tropical savannas, phosphorus (P) also has been recognized as one of the most limited nutrients for plant production because of soil acidity and the high P adsorption induced by the reactivity of phosphate with iron (Fe) and aluminum (Al) sesquioxides, which are the final products of deep weathering (López-Hernández, 1977). Although important amounts of fertilizer have been added to some savanna soils as a management practice to overcome P adsorption, most of the added fertilizer is irreversibly fixed, so from an economical point of view the P fertilizer practice in the savannas is in general highly inefficient (Oberson et al., 1999). Thus, N and P cycling in savanna ecosystems is especially sensitive to frequent fires resulting in the loss of N and P.

For this reason, much research with emphasis in nutritional aspects has been carried out in savannas to characterize their ecological features and to assess the impact of different land uses. In this review, we would like to present information related to: (1) the environment in which the well-weathered soils of Orinoco savannas were developed, (2) the main inputs/outputs of N and P in well-drained savannas, and (3) how some biological processes in savanna ecosystems can ameliorate N and P availabilities for native plants.

15.2 Main environmental features of the savannas of the Orinoco basin

15.2.1 Climate in the Orinoco basin

Venezuela is under the influence of the Intertropical Convergence Zone (ITCZ) and consequently most of the climatological records for the meteorological stations located in the llanos region show a seasonal pattern for precipitation and temperature. Monthly mean temperature is isohyperthermic and ranges between 24°C and 29°C, with minimum values from December to February. According to the amount of precipitation, two seasons are recognized: (1) a dry season of 5−7 months (November to May) and (2) a rainy season (May to November) that concentrates at least 65% of annual total precipitation (Walter and Medina, 1971). The total annual precipitation ranges from 900 to 2041 mm with a tendency to increase toward the southern part of the Venezuelan territory, the Andean piedmont and the western plains (Duno de Stefano and Huber, 2007). The potential evapotranspiration is usually higher than the precipitation, with the predominance of semi-arid conditions, which favors the presence of xeromorphic species.

15.2.1.1 General characteristic of the soils of the Orinoco basin

Neotropical savannas are usually related to acid soils with high P-fixation capacity and consequently low P availability, low organic matter, nitrogen and exchangeable base contents (exchangeable bases <5 cmol kg^{-1} soil), but high Al saturation (López-Hernández and Hernández-Valencia, 2008). An analysis of the main chemical and physical characteristics of different savanna soils in Venezuela confirms this assumption (Table 15.1). Soil pH ranged between 4.8 and 6.3 units, whereas the highest values recorded for nutrients were 5.3 mg P kg^{-1} for available P, 1800 mg N kg^{-1} for total N, 1.7% for organic C, and 1.2 cmol exchangeable bases kg^{-1}. The dystrophic nature of soils is a consequence of parent materials with low nutrient contents and/or strong weathering, which results in accumulations of resistant primary minerals (e.g., quartz), clays with low exchange activities, and the presence of amorphous and crystalline Fe and Al sesquioxides with strong P-fixation capacities.

Soil texture in *Trachypogon* savanna is usually dominated by sands (43−96%) (Table 15.1) and coarse gravels but an increase in clay content could be found in the subsurface due to the accumulation of clays and the formation of argilic, kandic, oxic, or cambic horizons. On the other hand, a hardpan of laterite or plinthite could be present in soils with alternating periods of flooding and drought. Precipitation is seasonal and has an influence on soil water content with periods of excessive wet and eventually anaerobic conditions, alternated with periods where the soil becomes dry and below the permanent wilting point. Anaerobic conditions are usually found in subsuperficial

Table 15.1 Soil Chemical and Physical Characteristics (0–30 cm) of Selected *Trachypogon* Savannas from Venezuela

Location	pH	Available P (mg kg⁻¹)	Total N (mg kg⁻¹)	Total C (%)	Σ Exchangeable Bases (cmol kg⁻¹)	Exchangeable Aluminum (cmol kg⁻¹)	Sand (%)	Silt (%)	Clay (%)
Calabozo	5.8	2.9	503	0.8	0.3	0.8	62.1	12.6	25.3
Puerto Ayacucho	5.5	4.8	393	0.6	1.2	–	93.0	2.8	4.2
Uverito	4.5	0.9	227	0.3	0.4	0.4	94.8	2.0	3.2
La Iguana	5.2	4.0	320	0.7	1.1	0.9	93.3	0.1	6.6
Cabruta	5.6	2.0	40	0.1	0.3	0.3	96.0	6.4	3.5
Anzoátegui-A	5.2	2.4	1100	0.8	0.5	0.5	61.5	15.0	23.5
Anzoátegui-B	5.3	5.3	1800	0.8	0.4	0.2	73.0	12.5	14.5
Anzoátegui-C	5.6	4.2	1100	0.8	0.3	0.1	71.5	15.0	13.5
Anzoátegui-D	5.4	3.2	1200	0.6	0.4	0.1	77.5	12.5	10.0
Anzoátegui-E	5.9	3.5	600	0.2	0.5	0.1	75.0	2.5	22.5
Bolívar	6.3	2.9	700	0.6	0.5	0.1	72.5	5.0	22.5
Cojedes-A	5.2	4.8	1100	1.2	0.6	0.5	56.8	10.5	32.7
Cojedes-B	5.1	2.9	1300	0.9	0.3	0.9	43.0	13.2	43.7
Cojedes-C	5.5	5.0	1400	1.7	0.6	0.7	55.8	11.5	30.5

Source: Calabozo (López-Hernández and Hernández-Valencia, 2008), Puerto Ayacucho (Sánchez et al., 1985), La Iguana (1988), Uverito (Hernández-Valencia and Bautis, 2005), Cabruta (Susach, 1984), Anzoátegui, Bolívar, Cojedes (Medina and Bilbao, 1991).

layers where water drainage is poor due to high clay and silt content, or the presence of a lateritic layer or concretions (López-Hernández and Hernández-Valencia, 2011).

According to the Soil Taxonomy classification system (USDA, 2010), the soils of *Trachypogon* savannas are mainly identified as oxisols, ultisols, in those with argilic, oxic, kandic, and cambic subsurface horizons, with or without a lateritic hardpan, while entisols are related to sandy soils from alluvial sediments or eolian deposits (e.g., dunes). For the World Reference Base for Soil Resource (IUSS, 2007), soil types are mainly identified as ferralsols, acrisols, and lixisols.

Physical and chemical features of savanna soils have an influence on plant physiognomy and, consequently, on the use of water, nutrients, and, finally, on primary production. Savannas are dominated by grasses. Most herbaceous species have an intensive superficial root system, whereas trees have extensive roots able to exploit water and nutrients from deeper soil layers (Medina, 1993). February et al. (2013) found for a broadleaved woody savanna in Australia that regardless of the type of grass, both tree and grass roots are concentrated in the top 20 cm of the soil. While trees have greater root production and contribute with a finer root biomass, grass roots contribute with a disproportional amount of N and C to the soil relative to total root biomass. They postulate that grasses maintain soil nutrient pools and provide biomass for regular fires that prevent forest trees from establishing, while savanna trees are important for increasing soil N content, cycling, and mineralization rates.

Open savannas are related to superficial hardpans, which interfere with root tree development, or with flooded soils. On the other hand, tree density increases when the hardpan is absent or deeper and/or water availability increases in the tree root zone, but flooding is not present (Medina, 1993). A similar relationship between physiognomy and soil fertility has also been observed in Brazilian cerrado where open savannas are associated with less fertile soils, especially those with low P and calcium (Ca) availabilities and high Al percent saturation (Lopes and Cox, 1977). Both physical (fast drainage, presence of hardpan) and chemical (low nutrient availability and acidity) features are important constraints for agricultural production, and must be overcome by using machinery and great amounts of fertilizers and limestone (Lozano et al., 2012).

15.2.1.2 *Vegetation of savannas in the Orinoco basin*

The species of plants of the savannas, whether herbaceous or tree components, are adapted to oligotrophic conditions. The floristic variation in the ecosystem depends in good part on the higher or lower contents of clay, which can affect the water conditions of the soils and the biological transformation of organic matter. Productivity may vary depending on small changes in the physical, chemical, and biological characteristics affecting soil fertility. The general low productivity in the Orinoco basin, at the same time, contributes to the low levels of organic matter found in its soils, with values around 1 to 2% or less (Hernández-Hernández and López-Hernández, 2002).

Grass species are the most important floristic component of tropical well-drained savannas of the Orinoco basin. Legumes and sedges also coexist in the herbaceous layer; however, their coverage is lower than 10% in comparison with grasses (Velasquez, 1965; Susach, 1984). San José et al. (1985) studied seven sites of *Trachypogon* savannas from low to moderate fertility and found that the above-ground biomass of grasses ranged from 80 to 100%, while legumes reached a maximum of 3.8% in the location with the highest pH (5.5) and base saturation (81%). The lower biomass of legumes in neotropical savannas has been related to the high content of Al and, in turn, low P availability which hinders the formation of effective N-fixing symbiosis (Medina, 1987).

In *Trachypogon* savannas, species richness is about 285 species of angiosperms belonging to 55 families, but dominance is restricted to only a few like Poaceae, Papilionaceae, Cyperaceae, and Asteraceae, with Poaceae being represented by *Trachypogon* sp., the genus with the higher coverage (Riina et al., 2007). Other species of grasses are *Axonopus canescens, A. anceps, Andropogon selloanus, Leptocoryphium lanatum, Paspalum carinatum, Sporobolus indicus, S. cubensis*, and several species of the genus *Aristida*; legumes of the genera *Mimosa, Cassia, Desmodium, Eriosema, Galactia, Indigofera, Phaseolus, Stylosanthes, Tephrosia*, and *Zornia*; and sedges of the genera *Rhynchospora* and *Bulbostylis*. As previously described, tree presence is variable from absent in open savannas to dense in savanna parkland where crowns are closer, but with minimal overlapping. Trees can also be found scattered or in groups, and most of them are pyrophyte species like *Byrsonima crassifolia, Curatella americana*, and *Bodwichia virgilioides* (López-Hernández and Hernández-Valencia, 2008).

15.2.1.3 Effects of fire on nutrient cycling in savannas

Fire is considered one of the most important factors when explaining the origin and maintenance of savannas, and acts as a selective force which favors the dominance of fire-tolerant species (Sarmiento, 1984). Vegetation is burned during the dry season in order to eliminate senescent leaves, as a plague control, and to stimulate the production of new and more palatable pastures for stock breeding. Loss of nutrients by fires is related to burning efficiency, which in turn depends on the amount and the heterogeneity of the accumulated above-ground biomass, its moisture content, the wind speed, and the oxygen diffusion to the combustion zone (Bilbao and Medina, 1996). Beside burning efficiency, the long-term loss of nutrients by burning depends on fire frequency, which is usually annual, biennial, or triennial (Coutinho, 1988). The accumulation of high amounts of fuel load increases burning efficiency, but when fires have been burning in the same year, recent burn scars are very effective at stopping fires (Bilbao et al., 2010).

Fires produce strong mineralization and loss of nutrients. In the case of gaseous elements, like N and S, losses occur by volatilization and the spread of ashes, whereas in the case of sedimentary elements, like P and bases (potassium (K), sodium (Na), calcium (Ca), and magnesium (Mg)), losses by fire are due just to the spread of ashes. Hernández-Valencia and López-Hernández (2002) assessed the nutrient losses in the herbaceous layer of *Trachypogon* savanna, comparing the amount of N, P, Ca, Mg, and K, before the burning of the above-ground vegetation and after the combustion in the remnant tissues and ashes. Results indicated that fires expelled to the atmosphere as gases and ashes as follows: $1.7 \, kg \, P \, ha^{-1}$, $5.9 \, kg \, N \, ha^{-1}$, $3.7 \, kg \, Ca \, ha^{-1}$, $2.8 \, kg \, Mg \, ha^{-1}$ and $5.0 \, kg \, K \, ha^{-1}$, which represent more of 90% of the nutrient contents in the above-ground biomass, including litter. The high burning efficiency was related to the accumulation of dry foliages, its homogeneity, and the rapid fire spread enhanced by winds. In relation to their biogeochemical cycles, sedimentary nutrients like P, Ca, Mg, and K showed higher concentration in the ashes compared to the above-ground biomass; however, N was more diluted in the ashes due to its volatilization at high temperatures. Ash depositions returned to the soil between 21 and 34% of the sedimentary nutrients (P, Ca, Mg, and K), but only 1% N, and lowered the net losses to $1.1 \, kg \, P \, ha^{-1}$, $5.8 \, kg \, N \, ha^{-1}$, $2.4 \, kg \, Ca \, ha^{-1}$, $1.9 \, kg \, Mg \, ha^{-1}$, and $3.9 \, kg \, K \, ha^{-1}$; however, those measures have uncertainties because of the suspension of ashes by winds and the high variability in ash depositions.

15.3 **Nutritional stresses in well-drained savannas—nitrogen as a limiting element**

The soil characteristics, the climate, the geomorphology and net primary productivity in the savannas determine the low levels of N reported for soils of the Colombian and Venezuelan llanos. Neotropical savanna ecosystems include values of N, ranging from the extremely low of 0.2 g kg^{-1} in the first 10 cm of soils of *Trachypogon* Eastern Llanos, Venezuela, presenting more than 80% of sand composition (Gómez, 2004), to values of 1.2 g N kg^{-1} on savannas of higher clay contents in the Central Llanos, Venezuela (Hernández-Hernández et al., 2000). Hétier et al. (1989) showed N ranges between 0.6 g N kg^{-1} and 0.2 g N kg^{-1} for savannas of the Western Llanos, Venezuela, with up to 18% of clay at the surface and 40% of clay at 1 m deep.

The main form of total soil N is the organic, representing in many cases 96 to 98% of the soil (Sprent, 1999). In general, water-soluble (readily available) in well-drained *Trachypogon* savannas may reach around 30 to 40 mg N kg^{-1} soil, and within the mineral forms (López-Hernández, 2013), ammonium (NH_4) is more abundant than nitrate (NO_3). Although, NH_4 and NO_3 content changes during the rainy and dry seasons in savannas, in an ultisol, NH_4 (20 mg NH_4 kg^{-1}) duplicates NO_3 (10 mg NO_3 kg^{-1}) and this proportion is kept until 30 cm of depth (Lozano et al., 2011).

After the burning of savanna ecosystems, the fast combustion of the plant material and other organic debris together with the ash dispersion and volatilization of the element with gaseous cycles (e.g., C, N, and S) produce a drastic decline in the effective input of organic matter and nutrient to the soil (Frost and Robertson, 1985; Sanhueza and Crutzen, 1998; Hernández-Valencia and López-Hernández, 2002). Because many gaseous forms of N are formed under burning effects, N cycling is especially sensitive to frequent fires resulting in the net loss of N in savannas, unless losses are compensated by other mechanisms including biological fixation. Below we will examine the inputs and outputs of N in savannas and information related to some biological processes involved in savanna ecosystems which can ameliorate N deficiencies.

15.3.1 **Main nitrogen inputs**

15.3.1.1 *Atmospheric depositions (wet and dry deposition)*

In savannas, ash produced by fires *in situ* or in nearby areas can be deposited as dry deposition or washed out from the atmosphere and dissolved in the rain. In this way, precipitation can return nutrients lost by fires and other volatilization processes to the soil. Inorganic inputs of N from bulk precipitation are relatively low in savannas, unless the savanna is located near an industrial or polluted area (López-Hernández et al., 2013). Mineral N inputs ($NO_3 + NH_4$) by precipitation in Venezuelan savannas ranged from 2.2 kg N ha^{-1} yr^{-1} in savannas of Calabozo (Montes and San José, 1989) to 6.2 kg N ha^{-1} yr^{-1} in a place more distant from urban activities at Estación Experimental, La Iguana (Table 15.2) according to Chacón (1988). Sanhueza and Crutzen (1998) also reported wet deposition of mineral N from 1.12 to 4.6 ha^{-1} yr^{-1} for the Orinoco region. Those precipitation inputs in inorganic N are very similar to the value presented for savannas of *Loudetia* located at comparable geographical latitude at Lamto, Ivory Coast, Africa (1.3—2.3 kg NO_3 ha^{-1} yr^{-1} and 3.0 kg NH_4 ha^{-1} yr^{-1}) according to Villecourt and Roose (1978), and to the value presented by Bustamante et al. (2006) in a cerrado area protected from fire in Central Brazil. Detailed

Table 15.2 N Inputs and Outputs (kg N ha^{-1} yr^{-1}) in Savannas of La Iguana (Venezuela)

Location	Unburned Savanna	Burned Savanna
Wet deposition	6.2 (mineral)	6.2 (mineral)
Dry deposition	10.5*	10.5*
N$_2$ fixation		
Soil−plant system	7.8	13.7
Microbial crusts	4.0	2.4
Legumes	?	?
Σ inputs	28.5	32.8
Fire	0	8.2
Leaching	2.1	2.1
Denitrification, NO and N$_2$O*	7.5*	7.5*
Σ outputs	9.6	17.8

Source: *Sanhueza and Crutzen, 1998 and Cárdenas et al., 1993.

information on organic N sources in *Trachypogon* savannas located in Central Venezuela is almost non-existent, although Pacheco et al. (2004) reported that soluble organic N inputs may be considered an important contribution, making up 75% of total N in rainfall; however, more work needs to be done to confirm this information. Nonetheless, in Lamto savannas (Ivory Coast), soluble organic N inputs have been reported to be as much as 14.5 kg ha^{-1} yr^{-1} (Villecourt and Roose, 1978; Abbadie, 2006). In general, typical mineral N inputs in Africa are reported to be very low (Giller et al., 1997).

As to dry deposition, Sanhueza and Crutzen (1998) presented information concerning the Venezuelan savannas during the wet and dry seasons with values ranging from 4.5 to 10.5 kg N ha^{-1} yr^{-1}) (Table 15.2), most of them NH$_4^+$ + NH$_3$ forms. In the savannas of Lamto, Abbadie (2006) reported a dry deposition of 3.5 and 2.1 kg N ha^{-1} yr^{-1} for the ammonium and nitrate form, respectively.

15.3.1.2 Biological nitrogen fixation

In the savannas of the Orinoco basin, fires volatize every year up to 19−30% of the N required for the net primary production of savanna herbaceous vegetation (Chacón et al., 1992; López-Hernández et al., 2006). N volatilized by fires and N losses due to leaching and erosion are not compensated for by the scarce inorganic N inputs from the precipitation mentioned above. This situation could cause a progressive reduction in the potential productive capacity of savannas in the absence of additional mechanisms of N input, apart from precipitation (Abbadie, 1983; Chacón, 1988; Chacón et al., 1991; Cook, 1994). Consequently, N budgets in dystrophic savannas must be balanced by biological N fixation.

In general, N fixation can occur in savannas from three different sources: (1) rhizobium-legume symbiosis; (2) organisms located in the rhizosphere of grasses; and (3) microbial crusts on the soil surface formed by cyanobacteria.

15.3.1.2.1 Nitrogen fixation by rhizobium Symbiosis

Neotropical savannas are characterized by their great diversity of herbaceous and woody legumi-nous species, although their relative importance in the plant canopy of well-drained savannas is small compared with the herbaceous component (Chacón et al., 1992). Previous studies in *Trachypogon* savannas indicated that, although most of the native leguminous species formed nodules, they seemed inactive or with reduced N fixation compared to cultivated legumes due to the soil acidity and poor base contents (Chacón et al., 1991; Medina and Bilbao, 1991). Few studies have been done to document N fixation by native legumes under natural conditions in Orinoco's savannas, although evidence through natural abundance of ^{15}N and relative abundance of ureids suggest N fixation for a few species (Medina and Bilbao, 1991; Izaguirre-Mayoral et al., 1992). Therefore, this mechanism of N fixation seems to be less crucial for the N economy of savannas than the other N-fixing mechanisms mentioned above; however, Bustamante et al. (2006) have emphasized that to have a reliable assessment of legumes to the N budget of savannas it is neces-sary to have information related to legume density and seasonal N fixation.

15.3.1.2.2 Nitrogen fixation by organisms located in the rhizosphere of savanna grasses

Nitrogen fixation by non-symbiotic organisms associated with the rhizosphere, rhizoplane, and endorhizosphere could be an important source of available N in nutrient-poor soils of savannas. Estimates of N fixation by the acetylene reduction method (nitrogenase activity, NA) in the soil—plant system in savannas of *Trachypogon plumosus* and *Paspalum carinatum* located at Estación Experimental La Iguana, Central Venezuela (Table 15.2) reported values of 13.7 and 7.8 kg ha^{-1} yr^{-1} for burned and protected plots (López-Hernández et al., 2006), respectively. During the dry season, the low water availability in savannas could be a major limiting factor for microbial and NA activity. As in the case of protected plot (unburned), the increase in NA in the dry season indi-cates the effect of rhizosphere conditions (moisture content, exudates, pH) on associated microorganisms.

Similar values for NA have been reported by Balandreau and Villemin (1973) and Balandreau (1975) in *Loudetia* and *Andropogon* savannas (12 kg ha^{-1} yr^{-1} and 9 kg ha^{-1} yr^{-1}, respectively) in the Ivory Coast (Africa). Non-symbiotic N fixation has also been reported to have a significant importance to N budget in West African savannas (Robertson and Rosswall, 1986) with a contribu-tion of Ca (approximately 12 kg ha^{-1} yr^{-1}). In general, rhizospheric N fixation becomes significant in areas affected by limited N availability, a situation very common in well-weathered tropical savan-nas. These reports make clear that N$_2$ fixation mediated by leguminous symbionts seems to be rela-tively low as compared to the activity of free-living organisms associated with the grass roots.

15.3.1.2.3 Nitrogen fixation by microbial crust system

The well-drained savannas present pedoclimatic features, which can lead to extreme conditions of water stress during certain months of the year (dry season), contributing to the formations of bio-crusts, typical of arid and semi-arid ecosystems on the surface of the soil (Chacín et al., 2011). According to Schlesinger et al. (1996) and Housman et al. (2007), their presence produces islands of fertility in the soil that can be modulated by climatic and edaphic factors as well as management. Normally, in Orinoco savannas, the soil crusts are located between patches of *Trachypogon* tillers and other grasses, contributing to the fertility of the soil and affecting decomposition and mineralization of biological processes and biogeochemical cycles. Soil crusts function as bridges of transfers of resources between plants, through their roots in connection with mycorrhizal fungi that form networks, and carry on in a bidirectional way C and N between plants and crusts. Symbiotic associations are established according to the needs of the fungi and the primary producers, which use the C and transport and transform nutrients such as N (Allen, 2007). Their presence implies clear effects on the N cycling, either through biological fixation by algae and cyanobacteria that form the crust (López-Hernández et al., 2006), or via decomposition of organic compounds from metabolites of biocrust organisms and the accumulation of organic material that are concentrated in crusts (Hernández-Hernández et al., 2013). The soils of the Venezuelan savannas, with varying degrees of index of aridity, have crusts dominated by cyanobacteria, with N values between 0.6 and 0.9 g kg^{-1} compared with 0.2 and 0.3 g N kg^{-1} for the bare soil (Hernández-Hernández et al., 2013), whereas the values for available P were not much different in the soil with crusts (0.54−0.94 mg kg^{-1}) in relation to the bare soils (0.59−0.62 mg kg^{-1}). Concerning the inorganic N forms, ammonium forms were always higher compared to nitrate forms (30 mg NH_4 kg^{-1} vs. 5 mg NO_3 kg^{-1}) in both crusted soil and bare soil (16 mg NH_4 kg^{-1} vs. 9 mg NO_3 kg^{-1}), respectively (Chacín et al., 2011).

These values are closely related to a more active cycling of labile compounds of C in the soil of savannas in the presence of crusts. There is a greater biological activity in soils of savannas with crusts up to six times greater than in a soil without crusts (Chacín et al., 2011). Also, the crusts give physical protection to the impact of rain drops in the wet season and improve the structure of soils (Chamizo et al., 2012).

Taking into account the microbial crust cover and NA measurements, López-Hernández et al. (2006) have estimated N_2 fixation rates by microbial crusts of 2.4 and 4.0 kg ha^{-1} yr^{-1} in burned and protected *Trachypogon* savannas, respectively (Table 15.2). In any case, fixation should be less than 13.5 kg ha^{-1} yr^{-1}, a value which would be met if the maximum NA recorded was maintained during the 214 days that cyanobacteria remained active. Moisture content in the microbial crusts strongly influences NA activity, which is limited to the rainy season. During the dry season no NA was recorded in *Trachypogon* savannas. This could indicate that cyanobacteria are very active and responsible for N fixation during the wet season, whereas between rainy seasons they are dormant due to the severe drying of the soil. The larger plant cover in the protected plot maintains, at a microclimatic level, a relative higher crust moisture content creating more favorable conditions for NA during the rainy season (Stewart et al., 1977).

15.3.2 Main nitrogen outputs

15.3.2.1 Losses by fires

Burning of vegetation during the dry season is the rule in savannas (Vareschi, 1962; Vuattoux, 1976). This factor is supposed to be pivotal for the origin and maintenance of savannas (Beerling and Osborne, 2006) and behaves like a selective force which maintains the dominance of fire-tolerant species.

After the fire's passage, more than 90−95% of the herbaceous aerial cover is lost, and most of the organic elements are transformed into CO, CO_2, NO, and NO_2 (Sanhueza and Crutzen, 1998). In unfertile neotropical *Trachypogon* savannas located at La Iguana Experimental Station, Venezuela, losses of 8.2 kg N ha^{-1} yr^{-1} have been measured (Table 15.2), whereas in the more fertile soils of Lamto Experimental Station in Ivory Coast, Africa, losses of 17 and 24 kg N ha^{-1} yr^{-1} have been reported for *Loudetia* and *Andropogon* savannas, respectively (Abbadie, 1983, 2006). The higher N losses from Lamto savannas are in good agreement with their higher aboveground primary production. Losses of N by biomass burning at La Iguana savannas are in accordance with more detailed information of gas emissions (NH_3, N_2, NOx, RCN) in other savannas of the Orinoco basin presented by Sanhueza and Crutzen (1998) and Bustamante et al. (2006), ranging from 8.8 to 26.5 kg N ha^{-1} yr^{-1}.

15.3.2.2 Nitrogen losses by leaching

The loss of N by internal drainage is relatively small in well-drained savannas, which is understandable since the water-soluble levels of N forms are always scarce, except perhaps at the beginning of the rainy season when substantial organic matter decomposition occurs. The losses of N in the savannas of Estación Experimental La Iguana in Guarico Central were estimated at 2.1 kg N ha^{-1} yr^{-1} (Chacón et al., 1991; Table 15.2), whereas for a *Loudetia—Andropogon* savanna at Lamto, Ivory Coast, 5.6 kg N ha^{-1} yr^{-1} was reported to leave the system through internal drainage, mostly in organic form (Villecourt and Roose, 1978; Abbadie, 2006) coming from microorganisms or recent dead plant material. In savanna ecosystems, information on nutrient losses by leaching is scarce and perhaps related to methodological inconveniences for designing and installing an effective system to catch leachates (López-Hernández and Hernández-Valencia, 2011). As a consequence, more accurate measures are still needed.

15.3.2.3 NH₃ losses

The mineralization of the soil organic matter in native soils might produce ammonium and at high pH some of the NH_4^+ exists as NH_3. However, in acidic, well-weathered savanna ecosystems, soil emissions of NH_3 are considered negligible (Abbadie, 1983; Sanhueza and Crutzen, 1998; López-Hernández, 2013).

NH_3 emissions in the budget presented in Table 15.2 are not included since animal waste and fertilizer applications are negligible, particularly in the case of the experimental site located at La Iguana Experimental Station.

15.3.2.4 Denitrification and emissions of NO and N₂O

Losses by denitrification in savannas of the Orinoco basin should be insignificant due to the low levels of nitrate and organic matter and the acidic pH of the soils. Similar information was presented by Abbadie (2006) in Lamto savannas, who reported that denitrification is highly variable in the space between soils. NO_3 pools in the soils of the tropical savannas of South America are low, which means that the rates of nitrification are also negligible; nonetheless, the levels of the intermediate NH_4 are, in general, significant.

In the savannas of the Orinoco region, Sanhueza and Crutzen (1998) have estimated an annual production of 0.18—0.63 kg ha^{-1} and 0.36—6.3 kg ha^{-1} for N_2O and N_2, respectively. However, in Orinoco savannas, soil emissions of NO are largely uncertain, ranging from 0.3 to 3.0 kg ha^{-1} yr^{-1} depending on the amount and distribution of rainfall (Cárdenas et al., 1993); all together N_2O, N_2, and NO losses in *Trachypogon* savannas account for about 7.5 kg ha^{-1} yr^{-1} (Table 15.2).

15.4 Nutritional stresses in well-drained savannas—phosphorus as a limiting element

The total content of P in soils of the savannas of the Orinoco basin is in general low and corresponds to well-weathered soil according with the model of Walker and Syers (1976) for ecosystem development. Values can be as low as 50 mg kg^{-1} in sandy Quartzipsamments and above 600 mg kg^{-1} in some clay soils.

Soil P can be found in different forms of organic (Po) and inorganic phosphorus (Pi), which can be discriminated as: (1) active or labile forms represented by Pi soluble or sorbed on soil surfaces and labile Po (easily mineralizable Po) associated to the soil microorganisms and organic matter, (2) Pi associated with amorphous and some crystalline Al and Fe phosphates and Po in humic and fulvic acids with lower availability than labile forms and involved in long-term transformations, (3) calcium-bound Pi which predominates in primary minerals like apatite-type minerals or alkaline soils, and (4) recalcitrant P, composed by the more chemically stable organic P and insoluble inorganic P and represented by P in particulate organic matter, occluded or strongly fixed in the mineral matrix (López-Contreras et al., 2007). Plants absorb soluble inorganic P ion forms (HPO_4^-, HPO_4^{2-}) from the labile fraction, which is usually termed as available P. The amount of available P is highly dependent on pH and a maximum is reached at neutral values (6−7 pH units), but despite this favorable condition, available P rarely surpasses 15% of the total P content (Brady and Weil, 2008).

Figure 15.2 shows the distribution of the soil P fractions mentioned above in four different *Trachypogon* savannas in Venezuela, expressed as a percentage of total P. Despite the differences in textural conditions and climatic regimes, the picture resembles a similar pattern, with a dominance of recalcitrant and very low contents of associated P to primary minerals or calcium bound P, as usually found in highly weathered soils. Available plant P, which is included in the labile

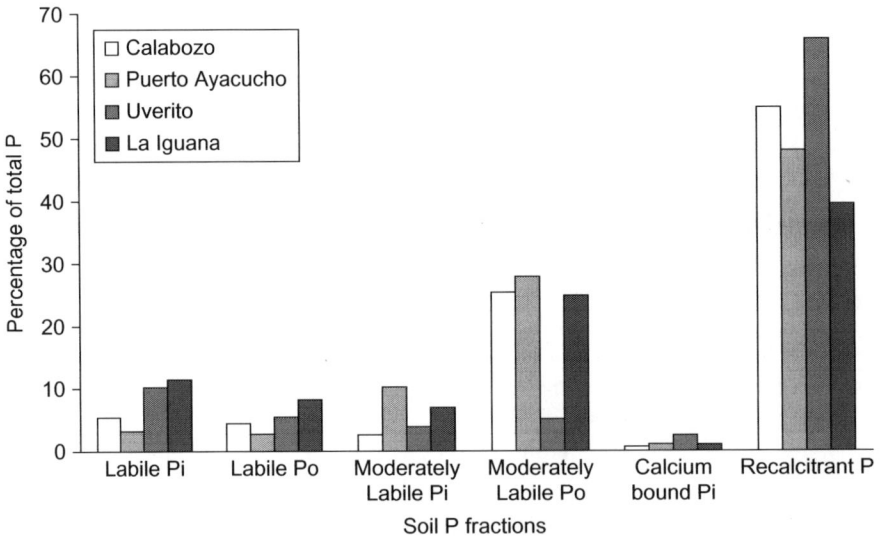

FIGURE 15.2

Distribution of soil P fractions in four *Trachypogon* savannas of Venezuela.

inorganic P pool, accounted for less than 13%. Their absolute values are below 5 mg kg^{-1} soil and could be considered low for agricultural purposes. However, labile organic P accounted for 2 to 9% of total P, whereas moderately labile organic P accounted for 5 to 30% of total P. Both forms can perform as P suppliers in the short (labile) to long (moderately labile) term.

As presented above, the periodical burning of savanna ecosystems produces a drastic decline in the effective input of organic matter and nutrient to the soil. In the case of elements with sedimentary biogeochemical cycles (e.g., P and nutrient bases K, Ca, and Mg), the gaseous losses are irrelevant or non-existent; however, part of the *in situ* produced ashes can be lost by wind dispersion and lateral movement with precipitation waters (Hernández-Valencia and López-Hernández, 2002), therefore inducing a potential P limitation to plants. Moreover, the scarcely available P forms deposited in ashes can be easily adsorbed to insoluble P forms for the soil matrix containing Al and Fe sesquioxides, which easily adsorb or precipitate orthophosphate (López-Hernández, 1977). Now we will examine the inputs and outputs of P in savannas, and information related to some biological processes involved in savanna ecosystems, which can ameliorate P deficiencies.

15.4.1 Main phosphorus inputs

In burned savannas, ash produced by fires and dust are settled down or are washed out from the atmosphere by rain and dissolved or suspended in the droplets. Inorganic inputs of P from bulk precipitation are relatively low in the well-drained savannas of Venezuela, because most of them are located far from industrial or polluted areas. This was demonstrated for Estación Biológica de los Llanos, where input values ranged from 0.31 to 0.52 kg ha^{-1} yr^{-1} (Table 15.3) and water P concentration from 0.03 to 0.10 mg L^{-1} (Montes and San José, 1989; Hernández-Valencia, 1996). The pattern observed in P inputs via precipitation has two peaks, one at the beginning of the rainy season and the other during the period of heavy rainfall.

At a global scale, it has been suggested that substantial amounts of P can be transported over extremely long distances in the form of dust (Swap et al., 1992; Okin et al., 2004). In the case of tropical forest ecosystems, it has been registered that dust inputs originating in the Sahara/Sahel region in Africa can incorporate 1 to 4 kg P ha^{-1} yr^{-1} in the northeastern Amazon basin, concluding that the

Table 15.3 Annual P Budget for a Well-drained Savanna in Calabozo (Venezuela)

Transferences	kg P ha^{-1}
Inputs	
— Precipitation	0.5
Outputs	
— Fire	1.1
— Cattle extraction	0.1
— Leaching	0.3
Balance	−1.0

Source: Hernández-Valencia, 1996.

high productivity of the Amazon forest is fueled, at least partially, from this transported dust. Is a similar situation occurring in the relatively nearby savannas of the Orinoco basin? We do not have enough information to corroborate this assertion; however, any P input in savannas of the Orinoco is important since the extremely weathered environments of some regions have a concomitant serious P depletion.

P released from parent material by weathering has not been assessed in well-drained savannas; however, we consider it negligible, because the sediments are mainly composed of P forms very resistant to weathering (Hernández-Valencia and López-Hernández, 1999).

15.4.2 Main phosphorus outputs

15.4.2.1 Losses by fires

In tropical savannas, burning of vegetation produces a strong mineralization pulse and loss of nutrients once ashes are spread outside the ecosystem. Studies in Calabozo savannas in Estación Biológica de los Llanos, comparing the P content in the above-ground biomass before and after the burning of the vegetation, indicated that about 1.7 kg P ha^{-1} was expelled to the atmosphere (Table 15.3). This amount represents an important part of the above-ground biomass, including litter (Hernández-Valencia and López-Hernández, 2002). Although temperature was not recorded during the experimental period, the high burning efficiency (more than 90%) seems to be related to the accumulation of dry foliage and the rapid fire spread that was enhanced by winds. Ash deposition returned about 21% of the P transferred to the atmosphere by fires to the soil and lowered the net losses to 1.1 kg P ha^{-1}. However, those measurements could have significant uncertainties, because of the suspension of ash by winds and the high variability in the measured ash depositions. These results demonstrate the role of fire as net exporter of P and its contribution to the global balance of other elements in the atmosphere (i.e., C, S, and N). Also the need to manage fire to reduce carbon and other nutrient emissions from savanna ecosystems is emphasized. Reducing fire frequency could be an easy way to address this situation; if fires were produced on a biennial or triennial basis, the associated N and P losses could be compensated by inputs through precipitation, as was shown for the Brazilian Cerrado (Coutinho, 1988).

15.4.2.2 Losses by leaching

Nutrient losses from soil by leaching are likely in well-drained savannas due to high permeability of the soil. Leaching is usually restricted to the wet season and when an excess of water drains out of the rooting zone (Hernández-Valencia, 1996). For well-drained savannas the information on P losses by leaching is scant. At Estación Biológica de los Llanos, P concentrations were below 0.1 µg L^{-1} in leachates sampled at 1 m depth; these values are typical for tropical weathered soils with concomitant high P-sorbing capacities (López-Hernández et al., 1981). Drainage out of the rooting zone ($>$ 1m) was recorded in three of the six rainy-season months and P concentration did not present any seasonal pattern. The amount of P that is lost by leaching ranged between 0.23 and 0.29 kg P ha^{-1} yr^{-1} (Table 15.3).

15.5 Strategies that are used by native savanna plants to enhance nitrogen and phosphorus conservation and uptake

The information presented suggests that the plants of the native savannas under frequent fires and strong seasonality are limited by N and P or both, yet the rates of many biological processes in this biome continues to be significant; a similar situation occurs in the case of the tropical rain forests with strong P limitation (Reed et al., 2011). The answer to the apparent paradox lies in the existence of the adaptations that the organisms of this biome have evolved to effectively overcome such low N and P availabilities. These strategies may be of two categories: (1) enhancing N and P conservation and efficiency and (2) enhancing N fixation and P incorporation and uptake. Biological N fixation was thoroughly discussed above.

15.5.1 Enhancing nitrogen and phosphorus conservation and efficiency

15.5.1.1 Nutrient uptake, translocation, and biomass production

The mechanisms involved in the adaptation of savanna plants to soils of low nutrient availability have scientific and practical importance (Medina and Bilbao, 1991). Natural selection favors species with competitive ability to overcome oligotrophic conditions. Most of them usually have low nutrient demands, low nutrient contents in their tissues, and, in turn, a low organic matter production. Since savannas are areas of stock breeding, intensive animal production is restricted by the poor production and palatability of forages (López-Hernández and Hernández-Valencia, 2008).

A review of 15 studied sites corroborates the mean low productivity values for *Trachypogon* savannas (López-Hernández and Hernández-Valencia, 2008), where above-ground production (3.78 mg ha^{-1} yr^{-1}) surpassed mean values for below-ground production (2.64 mg ha^{-1} yr^{-1}) and finally the overall primary production reached 6.74 mg ha^{-1} yr^{-1}, values that are within the lowest in the world production of savanna ecosystems (Sarmiento 1984; López-Hernández et al., 2012).

Organic matter production by the herbaceous layer shows a seasonal trend correlated with precipitation and soil water availability (Hernández-Valencia and López-Hernández, 1997). The burning of the vegetation during the dry season stimulates plant growth and the demand for nutrients increases. As a consequence, due to their higher metabolic activity the youngest tissues reach the highest nutrient concentrations at that growth period. As long as the rainy season proceeds, biomass and nutrient storage increases and nutrient concentration decreases due to dilution or translocations from old to young leaves. Thus, the translocation to young leaves and roots can operate as efficient mechanisms for nutrient conservation. In the dry season, organic matter productivity and nutrient uptake decrease, therefore the nutrient contents in foliage reach their lowest concentrations. This behavior has been found for N, P, and S, while Ca and Mg show the opposite behavior (Medina et al., 1978; Hernández-Valencia and López-Hernández, 2000). In general, low levels of N, P, Ca, and S are commonly found in the foliage of grasses in well-drained savannas in South America. The maximum amounts recorded after fires (dry season) ranged between $6.2-8.9$ mg N g^{-1} and $1.8-3.1$ mg P g^{-1} while the maximum at the standing crop (wet season) ranged between $0.9-8.1$ mg N g^{-1} and $0.9-2.0$ mg P g^{-1} (Medina et al., 1978; Sánchez et al., 1985; Hernández-Valencia and López-Hernández, 2000).

Information about N and P uptake for primary production is scarce when compared to productivity data. Comprehensive studies on N cycles estimated values from 27.0 to 48.0 kg N ha^{-1} yr^{-1} and from 29.6 to 30.3 kg N ha^{-1} yr^{-1} for the unburned and burned savannas, respectively (López-Hernández and Hernández-Valencia, 2008), while in the case of P, reported phosphorus uptake was about 5.9 to 8.8 kg P ha^{-1} yr^{-1} in burned savanna (Hernández-Valencia and López-Hernández, 2000). According to Medina (1987), the ability of savanna grasses to extract nutrients from poor soils is probably more related to a dense root system for exploring the soil volume and, therefore, a

Table 15.4 Root/shoot Ratios for Different *Trachypogon* Savannas from Venezuela at the Maximum Standing Crop

	Location	Root/Shoot Ratio
Burned savannas	Calabozo	0.22–0.94
	Cabruta	1.75
	La Iguana	0.95–1.32
Unburned savannas	Calabozo	0.23–0.43
	Cabruta	0.57
	La Iguana	0.17–0.34

Source: Calabozo (López-Hernández and Hernández-Valencia, 2008), La Iguana (1988), Cabruta (Susach, 1984).

Table 15.5 Total N Contents and Available P in Soils of Well-drained Savannas of the Orinoco Basin under the Canopies of Dominant Tree Species and *Trachypogon plumosus*

Specie	Total N (g kg^{-1})	Available P (mg kg^{-1})
Byrsonima crassifolia	0.63 ± 0.01	4.07 ± 0.09
Curatella americana	0.69 ± 0.01	2.74 ± 0.08
Bowdichia virgilioides	1.14 ± 0.01	7.01 ± 0.08
Trachypogon plumosus	0.65 ± 0.11	2.70 ± 0.08

higher root/shoot ratio than more fertile soils. Data from seasonal savannas of Venezuela indicate that the root/shoot at maximum standing crop ranged from 0.22 to 1.75 (Table 15.4). Unburned savannas showed lower root/shoot ratios due to the accumulation of above-ground biomass.

There is no information about nutrient uptake of savannas trees in the Orinoco basin; however, several patterns have been distinguished among grasses and trees (Medina, 1993). Grasses and woody plants from semi-arid savannas in general have higher nutrient contents than those from dystrophic humid tropical savannas; however, trees demand more nutrients than xeromorphic C_4 species, and thereby are associated with soils with better nutritional status. Evergreen trees are more frequent in dystrophic savannas, especially if fire is frequent, and produce leaves with lower nutrient contents than deciduous trees. Leguminous trees have usually higher N and P content and consequently higher N and P demands than non-legumes, which explains the low coverage of legumes observed in most P-depleted savannas. It has also been demonstrated that trees improve nutrient contents in the soil beneath them in comparison with the soil associated with the herbaceous layer of a treeless savanna (Table 15.5; Hernández-Hernández, 2008). The input of litter with a richer nutrient content has been argued as a mechanism to explain this finding (Kellman, 1979), but also it must be considered that trees can concentrate nutrients beneath the soil by pumping them from the subsurface to the topsoil, by acting as traps of dust and ashes suspended in the air and by using perching bird droppings.

15.5.1.2 Role of microbial biomass

The microbial biomass (MB) mediates the transformation (mineralization–immobilization) of biogenic nutrients (C, N, P, and S) between the inorganic and organic forms; therefore, MB plays an important role in nutrient cycling. It also acts as an important sink for plant nutrient conservation and utilization in tropical savannas. In the soils of the Orinoco savannas, in general, the contents of soil microbial biomass carbon (C-MB) are low ($100-200$ mg kg^{-1}), which is in agreement with the low soil organic matter content of the soils (López-Hernández and Hernández-Valencia, 2008).

15.5.1.2.1 Microbial forms of nitrogen

In savannas, it has been reported that very low values of microbial N respond to a typical temporal variability of the ecosystem, which tends to increase significantly during rainy season. In Venezuelan eastern savannas, Gómez (2004) has reported microbial biomass N (N-MB), in the dry season, ranging from 3 to 6 mg kg^{-1} in a sandy soil, those values being lower than in savannas with higher silt–clay components, where the contents of microbial N are approximately 35 mg kg^{-1} (Hernández-Hernández et al., 2012). In general, it can be emphasized that in the Venezuelan savannas the N-MB followed a similar pattern to the C-MB, e.g., low values (around $10-30$ mg kg^{-1}) associated with the low levels of organic matter. Nutrients in microbial biomass have been shown to provide pulses of nutrient release in seasonally dry ecosystems (Hernández-Hernández and López-Hernández, 1998); thus, this source of N for primary production could indicate the potential provision and conservation role in the N economy of the savanna.

15.5.1.2.2 Microbial forms of phosphorus

In the case of P, microorganisms release available P and readily mineralizable organic labile P after cell lysis. Microorganisms also produce enzymes like phosphatases that catalyze the conversion of organic P to inorganic available P for plant requirements. Values of microbial P for *Trachypogon* savannas range from 2.6 to 12.4 mg kg^{-1} (Hernández-Valencia and López-Hernández, 1999; López-Contreras et al., 2007), and account for about 6 to 10% of total soil P content. These values are higher than available P pools and show the importance of microorganisms in the transformation, storage, and supply of P for plants in these P-depleted soils. At Estación Biológica de los Llanos, microbial P ranged between 3.7 and 13.7 mg kg^{-1} and no seasonal trends were observed (Hernández-Valencia, 1996). However, Singh et al. (1991) found for Indian savannas that microbial C, N, and P decreased in the rainy season and then increased in the dry season. According to those authors, the higher soil moisture during the rainy season favors cell lyses and nutrient release through microbial activity.

15.5.1.3 Role of pedo-fauna in the generation of biogenic "hotspots" in Trachypogon *savannas*

In savanna ecosystems, termites and earthworms are among the most conspicuous components of soil macro-fauna and play an important role in the decomposition processes of organic matter and nutrient cycling (Lavelle et al., 1994; López-Hernández, 2001; Chapuis-Lardy et al., 2011).

Invertebrates belonging to macro-fauna are key species of soil functioning. They participate in litter decomposition, mix organic and mineral matter, create and maintain soil structure by digging burrows and modifying aggregation, and regulate microbial diversity and activity (Lavelle et al., 1994; Renard et al., 2013). The gut and biogenic structures (burrows, casts, nests) of earthworms and termites are specific habitats where soil microbial activities are either stimulated or attenuated

(Martin, 1991). Earthworms are important actors in the regeneration of compacted soils by reduced tillage systems (Capowiez et al., 2012) and freshly egested earthworm casts, i.e., the by-products of gut passage, are generally characterized by an intense mineralization of organic matter and the release of nutrients for plants (Lavelle et al., 1994; López-Hernández et al., 1993; Ngo et al., 2012). Conversely, the mineralization in old casts is reduced, which allows for carbon and nutrient storage in soil in the long term (Martin, 1991; Lavelle et al., 1994). Thus, biogenic structures produced by earthworms (casts), ants, and termites have been considered as "hotspots" for nutrients and places to potentially start a successional development from savannas to a "micro-forest" (López-Hernández et al., 2012; Erpenbach et al., 2013).

15.5.1.3.1 Nitrogen contents in biogenic structures of earthworms (casts) and termite mounds and adjacent soils

In *Trachypogon* savannas using a conservative estimate of earthworm population density, it has been estimated that 3 and 34 kg ha^{-1} yr^{-1} of inorganic N in the savanna and introduced pasture, respectively, may be released in fresh casts (Decaëns et al., 1999). If underground and above-ground cast productions are taken into consideration (e.g., 14 and 114 Mg ha^{-1} for the natural and the introduced-pasture savanna, respectively) then a significant contribution to the N budget (internal recycling) in those ecosystems derives from deposition of casts, particularly in the man-affected savannas.

The role of termite activity in the N economy in savannas is more difficult to assess. Termites are recognized as the most important decomposers in tropical forest and savannas, strongly affecting soil organic matter and nutrient dynamics. In general, an increase in N content has been reported in mounds (which act as hotspots) compared with the associated soils, and it may affect ecosystem processes on certain spatial and temporal scales (Decaëns et al., 1999; López-Hernández, 2001).

15.5.1.3.2 Phosphorus contents in biogenic structures of earthworms (casts) and termite mounds and adjacent soils

Earthworm activity creates a series of geochemical and biological effects that are especially important for the cycling of P in soils (Chapuis-Lardy et al., 2011). The availability of P is strongly influenced by the physical adsorption or fixation of P in soil, which is much related to soil type (Figure 15.3). López-Hernández et al. (1993), using isotopic techniques, reported higher P availability (P in solution) and E_1 (P isotopically exchanged in 1 min) in fresh casts of *P. corethrurus* for two soils, one from the savanna of Lamto, Ivory Coast, with a moderate to low P-sorption capacity, and the other from Laguna Verde, Veracruz, Mexico, with a high P-sorption capacity. These changes in P availability were ascribed to (1) the greater pH of the gut content, (2) changes in sorption complexes induced by competition for sorbing sites between orthophosphates and carboxyl groups of a mucus glycoprotein produced by the earthworm in its gut, and (3) an increase in microbial activity during digestion (López-Hernández et al., 2012).

Earthworm casts usually contain larger amounts of organic P (e.g., for *P. corethrurus*; Chapuis-Lardy et al., 1998), probably derived from a selective ingestion of soil particles (López-Hernández et al., 1993; Chapuis-Lardy et al., 1998). Several studies have reported increased enzymatic activity and the stimulation of microbial activity in earthworm casts (Sharpley and Syers, 1976; López-Hernández et al., 1993; Brossard et al., 1996; Le Bayon and Binet, 2006). Devliegher and Verstraete (1996) showed that the increase of inorganic P in soils in the presence of earthworms

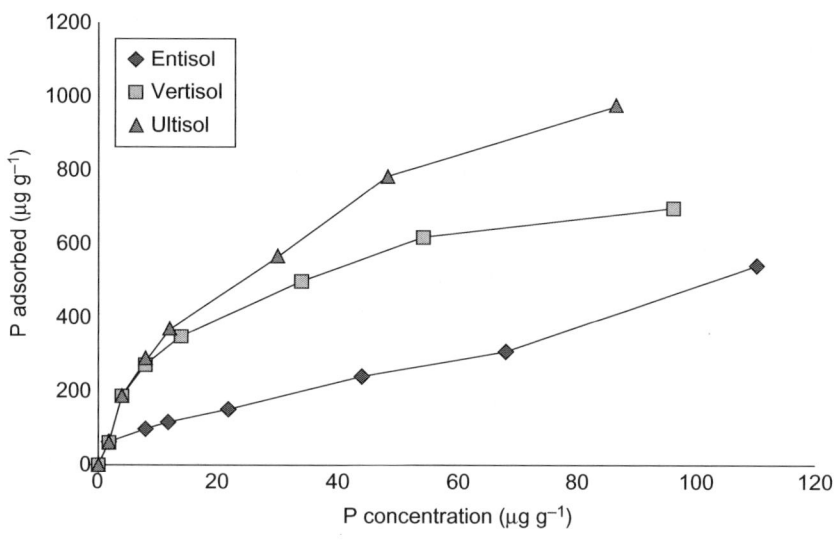

FIGURE 15.3

Phosphate adsorption isotherms in selected soils of Estación Experimental La Iguana.

was due to the production of alkaline phosphatases by earthworms (Satchell and Martin, 1984; Park et al., 1990) and a stimulation of microbial acid phosphatase production by earthworm activity (Satchell and Martin, 1984).

A similar role in the decomposition processes of organic matter for tropical and subtropical environments has been ascribed to termites in Africa and Australia; however, in neotropical savannas the importance of termites is considered secondary (Chapuis-Lardy et al., 2011). In a comprehensive P fractionation study done in soil and associated termite mounds of *Nasutitermes ephratae*, a common plant debris-feeding termite from savannas of the Orinoco Llanos (López-Hernández, 2001), it was found that all P concentrations were significantly higher in the *Nasutitermes* mounds than in the adjacent soils. Evidence suggests that nutrients accumulate in the mound from the nearby areas; therefore, the abundant termite populations could play an important role in controlling nutrient cycling in tropical savannas, in which available nutrients can often be a limiting factor for plant growth and development.

15.5.2 Enhancing nitrogen fixation and phosphorus acquisition and uptake

Under increasing acidity, soil exchangeable aluminum (Al^{+3}) tends to increase to toxic levels with a concomitant deficiency in available forms of P (López-Hernández, 1977). Savanna plants need to cope with the minuscule amounts of P existing in the soil solution through a series of adaptations that in many cases involve the abundant microorganism population living in the rhizosphere zone. The diversity and abundance of microorganisms influence the function of the ecosystems through soil nutrient cycling (N, P, and C) (Pineda et al., 2013). Many of those microorganisms cohabit in the rhizosphere soil zone where roots develop and the microorganisms grow actively (Richardson et al., 2009; Bhattacharyya and Jha, 2012). In natural conditions, most of the tropical plants are adapted to different ecological niches associated with soil microorganisms such as arbuscular mycorrhiza (AM) (Janos, 1985; Mora et al., 2013), phosphate solubilizing bacteria (PSB) (Toro et al., 1996, 1997; Mora et al., 2013), and symbiotic N-fixing bacteria as *Rhizobium* (Izaguirre-Mayoral et al., 1992; Mora et al., 2013), which can play a key role in nutrient cycling and in the protection of the plant to environmental stress (Olivares et al., 2013).

These microorganisms have great potential to contribute to improving local fertility problems and consequently they might have a potential use to generate biofertilizers (Richardson et al., 2009; Zhang et al., 2010; Olivares et al., 2013). Next, we analyze the presence and functionality of some of these microbial groups in the savanna soils, and their relevance to P and N nutrition.

15.5.2.1 Mycorrhizal association

Mycorrhizal association is a symbiosis established since ancient times between fungi of the phylum Glomeromycota (Schüßler and Walker, 2010) and the roots of plants; 80% of the land plants possess such association (Smith and Read, 2008; Brundrett, 2009). These fungi are symbionts that are forced and require a source of carbon from the host to complete their life cycle; the resulting structure in the root is known as arbuscular mycorrhiza (AM), frequently present in terrestrial ecosystems. Only some families have lost the ability to form this symbiosis (10%), e.g., Chenopodiaceae and Brassiceae, among others (Smith and Read, 2008; Brundrett, 2009), consequently, AM presence is widely extended in nature.

The AM exert a beneficial effect on plant growth and development, because the fungi develop an extensive network of mycelium out of theroots, able to take elements like P, usually depleted in the rhizosphere due to slow diffusion in soil solution and fast root uptake (Smith and Read, 2008; Brundrett, 2009). Thus, a mycorrhized plant has an additional mechanism of nutrient incorporation than non mycorrhized plant. For this reason, some authors consider this symbiotic association as a biofertilizer and recognize its usefullness in farming systems, remediating contaminated soils, among other applications (Barea et al., 2002a, 2013).

Symbiotic AM can contribute greatly to plant P nutrition and may be induced by P deficiency and the presence of organic matter in soil (López-Gutiérrez et al., 2004 a,b). AM contribution may be indirect due to AM hyphae exploring a larger soil volume and inducing physiological changes that favor the establishment of P solubilizing and mineralizing microorganisms in the mycorrhizosphere (Pineda et al., 2013). Alternatively, it may be directed as a consequence of the production of extracellular phosphatases such as acidic phosphatase activity (APA) and the access to distant P sources otherwise not available to the host plant (Li et al., 1991; Tarafdar and Marshner, 1994; Joner and Johansen, 2000; Shen et al., 2011).

It seems possible that some specific mechanisms must be operating in the plant rhizosphere to allow plant growth in savannas. Thus, AM, soil microorganisms, and APA in the rhizosphere are considered as crucial mechanisms for P availability and plant P uptake. López-Gutiérrez et al. (2004a,b) focus on the biochemical processes occurring in the rhizosphere of native plant species (*Trachypogon plumosus*) of P-deficient savanna soils. Therefore, microbial activity, as measured by viable plate counts and dehydrogenase activity (DHA), mineralization activity (APA), and AM dynamics was studied in the rhizosphere of *T. plumosus* growing on three seasonal savanna soils differing in P content and soil order (Table 15.6).

15.5.2.1.1 Mycorrhizal status in the studied savanna soils

AM colonization, expressed as a percentage of AM-colonized root length (%CRL) in *T. plumosus*, was higher during the rainy season in entisol and vertisol but not significantly different among seasons in ultisol (López-Gutiérrez et al., 2004b; Figure 15.4). The higher AM colonization as well as the increase in AM spore number during the rainy season (data not presented) concurs with the higher plant P demand. Nevertheless, seasonality, intrinsic characteristics of the plant—fungus combination and also soil properties can alter these parameters (Smith and Read, 2008). Moreover, AM spore number is relatively high in these soils since 3.6 spores g^{-1} dry soil is considered to be a high density (López-Gutiérrez et al., 2004b). Therefore, *T. plumosus* shows a particularly high mycorrhizal status for a grass in terms of AM infective potential spore density and root colonization when compared with reports from savannas of the Venezuelan Amazonia, which include values of 54.7% of CRL for *Panicum pilosum* (Poaceae) and only 29.8% for *Andropogon bicornis* (Poaceae) (St. John and Uhl, 1983). It has been reported that grasses with fine roots, abundant root hairs, and rapid growth are less sensitive to AM colonization (St. John, 1980).

The fact that AM colonization is, in general, high and increases during the rainy season (Figure 15.4), when P demand by plants also increases, suggests the importance of AM symbiosis as a mechanism for P uptake of *T. plumosus* in savanna ecosystems. This is emphasized when considering that one of the main strategies of grass species for P uptake in acid soils is a preferential distribution of resources to the roots and *T. plumosus* possesses a dense and fibrous radical system that allows an intensive exploration of the soil substrate (Medina and Bilbao, 1991).

Table 15.6 General Characterization of Savanna Soils in Estación Experimental La Iguana

Soil order	Entisol	Vertisol	Ultisol
pH	5.4	5.7	5.4
Sand (%)	94.4	38.8	73.2
Silt (%)	1.6	29.2	13.2
Clay (%)	4.0	32.0	13.6
Texture	Sand	Clay loam	Sandy loam
P-Total (mg kg^{-1})	90.1	137.3	102.1
P-NaHCO$_3$ (mg kg^{-1})	9.2	1.6	2.6
N (%)	0.08	0.11	0.11
C (%)	0.54	5.1	2.36
Fe (mg kg^{-1})	25.4	75.6	109.6
Al (meq 100 g^{-1})	0.85	0.20	0.70
H (meq 100 g^{-1})	0.20	0.15	0.20
CEC	2.42	15.66	8.28

Source: Adapted from López-Gutiérrez et al., 2004b.

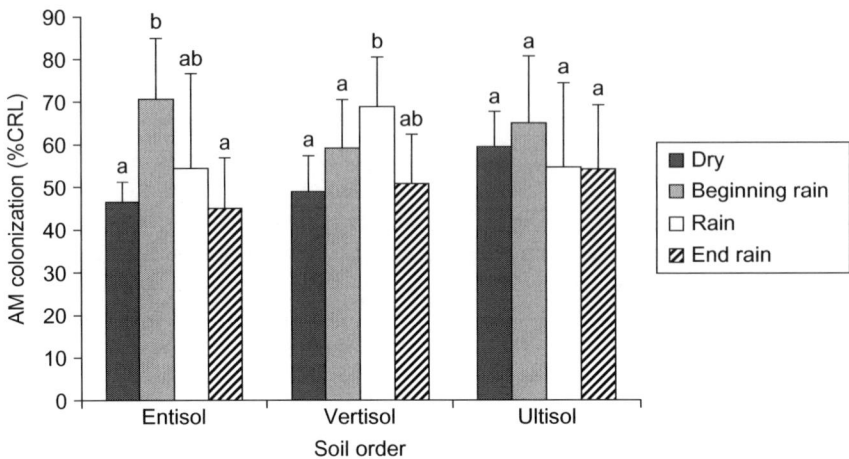

FIGURE 15.4

Mean of the percentage of AM-colonized root length in the rhizosphere of *T. plumosus* during different seasons in Estación Experimental La Iguana. Values followed by the same letter do not differ significantly according to LSD test ($n = 6$, $P < 0.05$). Error bars represent SEM. Different seasons are compared in each soil order.

Taken from López-Gutiérrez et al., 2004b.

Regardless of the presence or not of seasonal changes in the different mycorrhizal parameters studied, the sustained high level of AM colonization throughout the different seasons indicates that *T. plumosus* requires AM symbiosis to achieve an adequate P nutrition in savanna ecosystems.

15.5.2.1.2 Enzymatic activities in the studied savanna soils

Studies on the enzymatic activities of savanna soils in two contrasting seasons indicated that APA significantly increased during the rainy season in all soil orders examined (López-Gutiérrez et al., 2004a). This increase, however, was proportionally higher for entisol (Figure 15.5). These results suggest that P mineralization, as expressed by APA, increases during the rainy season, when plant nutrient uptake and nutrient leaching are higher.

Dehydrogenases are not expected to be free in soil; therefore, DHA expresses living cell activity and that is why it is an indicator of biological activity (López-Gutiérrez et al., 2004a,b). DHA significantly decreased during the rainy season in all soil orders (Figure 15.6). Even though several soil factors such as soil texture affect DHA, this enzymatic activity can be used to detect soil seasonal changes. Furthermore, DHA can correlate with changes in the microbial biomass due to long-term soil amendment. In *Trachypogon* savannas, DHA activities (Figure 15.6) appear to be well related to organic matter contents, which, in turn, control biological activities, thus DHA in entisol with poor soil carbon content (0.54%, Table 15.6) values was very low; on the contrary values increased in vertisol and ultisol with a higher soil carbon content (5.1 and 2.36%, respectively). In other studies done in ultisol and oxisol, extremely poor and acid soils of Uverito savannas

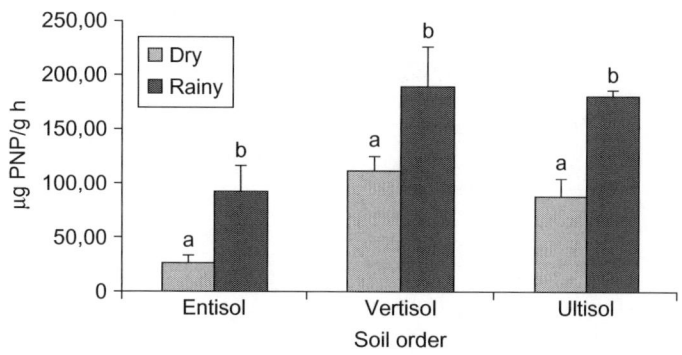

FIGURE 15.5

Seasonal changes in acid phosphatase activity in the rhizosphere of *T. plumosus* in three acid savanna soils (means + SD). Values followed by the same letter do not differ significantly. Duncan's test ($P = 0.05$).

Taken from López-Gutiérrez et al., 2004a.

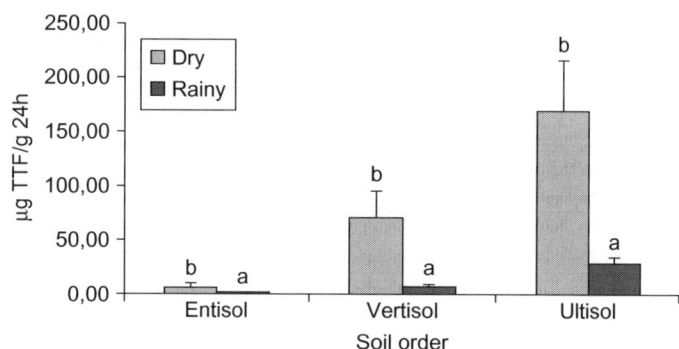

FIGURE 15.6

Seasonal changes in dehydrogenase activity in the rhizosphere of *T. plumosus* in three acid savanna soils (means + SD). Values followed by the same letter do not differ significantly. Duncan's test ($P = 0.05$).

Taken from López-Gutiérrez et al., 2004a.

evidenced an increase of APA, β-glucosidase and protease B-AA activity during the rainy season (Gómez and Paolini, 2011).

Furthermore, microorganisms in the rhizosphere may contribute to P nutrition through the synthesis and release of phosphatases when P is not available (Hayat et al., 2010; Shen et al., 2011). The decrease in DHA and bacterial counts during the rainy season, however, suggests that the increase in APA during that period is not mediated by rhizospheric microorganisms, i.e., soil extracellular phosphatases may have been already present in the soil or have plant root origin (Shen et al., 2011). López-Gutiérrez et al. (2004a) noted that in savanna ecosystems, as well as other tropical ecosystems affected by climatic seasonality, the APA appears to be an important mechanism of adaptation for the acquisition of the P. During the rainy season, the APA increased, but the

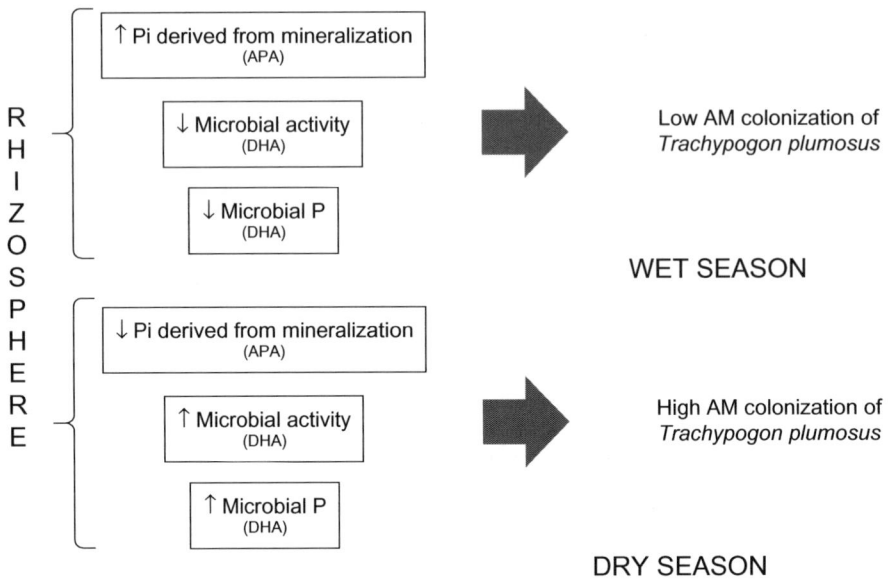

FIGURE 15.7

Interrelationships among AM colonization, microbial biomass, and mineralization (APA production) in *Trachypogon plumosus* rhizosphere during wet and dry seasons.

rhizosphere microorganisms do not appear to be involved in the mineralization activity, but seem to contribute to the cycling of P by the immobilization and release P during the dry and rainy seasons, respectively. Native plants such as *T. plumosus* were highly colonized by mycorrhiza, suggesting that this may constitute an important adaptive mechanism of native species to soils deficient in P. A summary of the interrelationships among AM colonization, microbial biomass, and mineralization (APA production) in *Trachypogon plumosus* rhizosphere during wet and dry seasons is presented in Figure 15.7.

15.5.2.2 Releasing of organic acids and competition for phosphorus sorbing places in the rhizosphere

Many soil and rhizosphere microorganisms have the ability to solubilize phosphate (Richardson, 2001; Hayat et al., 2010). As early as the 1950s, Sperber (1958) found that organic acids, such as lactic, glycolic, citric, and oxalic, produced by microbial activities in soils were able to solubilize insoluble sources of phosphates. Moreover, it was noted that there is a relationship between the amount of phosphate solubilized and the soil production of organic acids (citric and oxalic) produced by the fungi *Aspergillus niger*, *A. terreus*, and *Sclerotium rolfsu*. Microorganisms can lead to the dissolution of insoluble phosphates by the production of inorganic or organic acids and/or the pH decrease, thus freeing available phosphate. In addition, these organic acids excreted by the microorganisms can act as chelating agents of metals present in insoluble sources as phosphate rock, also releasing soluble phosphates. The interaction between organic acids and sorbing places in soil materials has been very well studied by using the model of the adsorption isotherms (López-Hernández, 1977). Particular attention has been given to the competition between orthophosphate and organic anions in sesquioxides and clay materials (Nagarajah et al., 1968; Shen et al., 2011) and in soils (López-Hernández et al., 1986).

The rock phosphates are natural phosphate sources that are a more economical option for agriculture use, particularly in countries where the availability of raw material for the production of soluble phosphate is complicated (Toro et al., 1997). Leon et al. (1986) showed an increase in the number of soil microbial solubilizers after fertilizing with rock phosphate. Toro et al. (2008) also obtained a similar effect by applying phosphate rocks in a system managed with conservation techniques in savanna soils. These results suggest that the use of rock phosphate as fertilizer is favored by the proliferation of microbiota with an ability to solubilize P-insoluble sources. The interaction of phosphate microbial solubilizers and the arbuscular mycorrhizal fungi is positive and helps plant nutrition; when the source used is the rock phosphate, both microorganisms act synergistically (Toro et al., 1997, 1998) emphasizing the promising use of these organisms for agricultural management (Barea et al., 2002b; Richardson et al., 2009; Bhattacharyya and Jha, 2012).

In a study to characterize the ability of phosphate solubilizing bacteria to dissolve insoluble phosphate compounds and rock phosphate from native acidic soil of the savannas of Guárico State, Venezuela, it was found that these bacteria showed high ability to solubilize phosphate from Fe and Al compounds, and, to a lesser extent, from phosphate rock (Osorio, 2007). The species of bacteria found in the rhizospheres studied were mainly: *Bacillus pumilus*, *Burkholderia cepacia*, *Bacillus circulans*, and *Pseudomonas fluorescens*. By using the technique of high resolution liquid chromatography it was noted that the most active bacteria in the solubilization process generated acetic, tartaric, malic, butyric, and oxalic acids.

15.6 Conclusion and future prospects

Savannas from the Orinoco basin have acid soils with low nutrient content, particularly N and P deficiencies, which together with the marked climatic seasonality reduce the spectrum of possibilities for agricultural production. Despite the environmental constraints, these ecosystems are important centers of agricultural and forestry activities in Venezuela and Colombia and represent the main future alternative to prevent expansion into tropical areas of greater ecological fragility as mountainous landscapes and the Amazonian rainforest. Plants growing on limited N and P soils are usually adapted through two strategies: (1) enhancing N and P conservation and use efficiency and (2) enhancing N fixation and P incorporation and uptake. Vegetation of tropical savannas of the Orinoco basin usually have low nutrient requirement and therefore low nutrient content, but a high use efficiency of N and P. On the other hand, soil biota such as termites, earthworms, arbuscular mycorrhiza, phosphate solubilizing bacteria, and symbiotic N-fixing bacteria as *Rhizobium* can play a key role in nutrient cycling and in the protection of the plant to environmental stress. These microorganisms have great future potential to contribute to improving local fertility problems and consequently they might have a potential use to generate biofertilizers to be used once the savanna is transformed in more intensive agricultural systems.

Savannas from the Orinoco basin are under the influence of different factors, which affect nutrient cycling and should be investigated. Natural herbaceous vegetation with low nutrient requirements has been replaced by exotic and more productive pastures for cattle raising. Also, large-scale plantations of pine (*Pinus caribea*) and eucalyptus (*Eucalyptus* spp.) have been established since the end of the 1960s (Rondón et al., 2006). Vegetal production intensification results in net changes in the amounts of C stored in the soil and the introduction of African grasses have displaced native species, converting relatively diverse and open savanna communities into monospecific grassland stands (Williams and Baruch, 2000). New questions arise from the influence of different fire regimes (Hutley et al., 2013) and changes in land use (San José et al., 2008; Groover et al., 2012) on the role of savannas as a sink or source of C, and how higher temperatures and inputs of C and N through atmospheric fertilization modify ecosystem structure and function (Higgins and Scheiter, 2012; Wake, 2012), particularly on biomass production and nutrient cycling.

Acknowledgments

We thank Fondo Nacional de Ciencia y Tecnología (Proyecto S1-2000000649) and Consejo de Desarrollo Científico y Humanístico-UCV (Proyecto 03.31.4109.98) for partially financing the research involved in this chapter. Thanks to M. Auxi Toro for the English revision of the manuscript. We are grateful to the personnel at the Estación Biológica los Llanos (SVCN) and Estación Experimental La Iguana (USR) who allowed us to use their installations, gave us technical assistance during field sampling, and provided us with information pertinent to the studied subjects.

References

Abbadie, L., 1983. Contribution à l'étude de la production primaire et du cycle de l'azote dans les savanes de Lamto (Côte d'Ivoire). Programme MAB savanes. Travaux des chercheurs de la Station de Lamto (Côte d'Ivoire).

Abbadie, L., 2006. Nitrogen inputs to and outputs from the soil-plant system. Lamto structure, functioning and dynamics of a savanna ecosystem. In: Abbadie, L., Gignoux, J., Le Roux, J., Lepage, M. (Eds.), Ecological Studies. Springer, NY, USA, pp. 255–275.

Allen, M.F., 2007. Mycorrhizal fungi: highways for water and nutrients in arid soils. Vadose Zone J. 6, 291–297.

Balandreau, J., 1975. Activité nitrogénasique dans la rhizosphère de quelques graminées. Thèse de doctorat es sciences naturelles. Université de Nancy I, France.

Balandreau, J., Villemin, G., 1973. Fixation biologique de l'azote moléculaire en savanes de Lamto (basse Côte d'Ivoire), résultats préliminaires. Rev. Ecol. Biol. Sol. 10, 25–33.

Barea, J.M., Azcón, R., Azcón-Aguilar, C., 2002a. Mycorrhizosphere interactions to improve plant fitness and soil quality. Antonie van Leeuwenhoek 81, 343–351.

Barea, J.M., Toro, M., Orozco, M.O., Campos, E., Azcón, R., 2002b. The application of isotopic (^{32}P and ^{15}N) dilution techniques to evaluate the interactive effect of phosphate-solubilizing rhizobacteria, mycorrhizal fungi and *Rhizobium* to improve the agronomic efficiency of rock phosphate for legume crops. Nut. Cyc. Agroeco. 63, 35–42.

Barea, J.M., Pozo, M.J., López-Ráez, J.M., Aroca, R., Ruíz-Lozano, J.M., Ferrol, N., et al., 2013. Arbuscular mycorrhizas and their significance in promoting soil-plant systems sustainability against environmental stresses. In: Rodelas, B., González-López, J. (Eds.), Beneficial Plant-Microbial Interactions: Ecology and Applications. CRC Press, USA, pp. 353–387.

Beerling, D., Osborne, C., 2006. The origin of the savanna biome. Glob. Change Biol. 12, 2023–2031.

Berrio, J.C., Hooghiemstra, H., Behling, H., Botero, P., Van der Borg, K., 2002. Late-quaternary savanna history of the Colombian Ilanos Orientales from Lagunas Chenevo and Mozambique: a transect synthesis. The Holocene 12, 35–48.

Bhattacharyya, P.N., Jha, D.K., 2012. Plant growth-promoting rhizobacteria (PGPR): emergence in agriculture. World J. Microbiol. Biotechnol. 28, 1327–1350.

Bilbao, B., Medina, E., 1996. Types of grassland fires and nitrogen volatilization in tropical savannas of Calabozo, Venezuela. In: Levine, J.S. (Ed.), Biomass Burning and Global Change. The MIT Press, pp. 569–574.

Bilbao, B.A., Leal, A.V., Méndez, C.L., 2010. Indigenous use of fires and forest loss in Canaima National Park, Venezuela. Assessment of and tools for alternatives strategies of fire management in Pemon indigenous land. Hum. Ecol. 38, 663–677.

Brady, N.C., Weil, R.R., 2008. The Nature and Properties of Soils. 14th edition Pearson-Prentice Hall, Upper Saddle River, NJ.

Brossard, M., Lavelle, P., Laurent, J.Y., 1996. Digestion of a vertisol by the endogeic earthworm *Polypheretima elongate*, Megascolecidae, increases soil phosphate extractability. Eur. J. Soil Biol. 32, 107–111.

Brundrett, M.C., 2009. Mycorrhizal associations and other means of nutrition of vascular plants: understanding the global diversity of host plants by resolving conflicting information and developing reliable means of diagnosis. Plant Soil 320, 37–77.

Bustamante, M.M.C., Medina, E., Asner, G.P., Nardoto, G.B., García-Montiel, D.C., 2006. Nitrogen cycling in tropical and temperate savannas. Biogeochemistry 79, 209–237.

Capowiez, Y., Samartino, S., Cadoux, S., Bouchant, P., Richard, G., Boizard, H., 2012. Role of earthworms in regenerating soil structure after compaction in reducing tillage system. Soil Biol. Biochem. 55, 93–103.

Cárdenas, L., Rondón, A., Johansson, C., Sanhueza, E., 1993. Effects of soil moisture, temperature, and inorganic nitric oxide emissions from acidic tropical savannah soils. J. Geo. Phy. Res. 98, 14783–14790.

Chacín E., León W., Hernández-Hernández R.M. (2011) Influencia de las costras biológicas en fracciones de materia orgánica de suelos de sabanas. Memorias IX Congreso Venezolano de Ecología, Porlamar, Venezuela.

Chacón, P., 1988. Dynamique de la matière organique et de l'azote dans une savane à *Trachypogon* du Venezuela. PhD thesis. University of Paris VI, France.

Chacón, P., López-Hernández, D., Lamotte, M., 1991. Le cycle de l'azote dans une savane à *Trachypogon* au centre du Venezuela. Rev. Écol. Biol. Sol. 28, 67–75.

Chacón, P., Lamotte, M., López-Hernández, D., 1992. Dynamique de la matière organique de la strate herbacée dans une savane á *Trachypogon* du Venezuela, Comptes Rendus de l'Academie des Sciences, Paris, 315 (3), 209–212.

Chamizo, S., Cantón, Y., Miralles, I., Domingo, F., 2012. Biological soil crust development affects physicochemical characteristics of soil surface in semi arid ecosystems. Soil Biol. Biochem. 49, 96–105.

Chapuis-Lardy, L., Brossard, M., Lavelle, P., Schouller, E., 1998. Phosphorus transformation in a ferralsol through ingestion of *Pontoscolex corethrurus*, a geophagus earthworm. Eur. J. Soil Biol. 34, 61–67.

Chapuis-Lardy, L., Le Bayon, R.C., Brossard, M., López-Hernández, D., Blanchart, E., 2011. Role of soil macrofauna in P cycling. In: Bünemann, E.K., Oberson, A., Frossard, E. (Eds.), Phosphorus in action—biological processes in soil phosphorus cycling. Springer Soil, Biology Series 26. Springer, NY USA.

Cook, G.D., 1994. The fate of nutrients during fires in a tropical savanna. Austral. Ecol. 19, 359–365.

Coutinho, L.M., 1988. Influencia del fuego en el cerrado del Brazil. Bol. Soc. Ven. Cien. Nat. 145, 61–83.

Decaëns, T., Rangel, A.F., Asakawa, N., Thomas, R.J., 1999. Carbon and nitrogen dynamics in ageing earthworm casts in grassland of the eastern plains of Colombia. Biol. Fert. Soil 30, 20–28.

Devliegher, W., Verstraete, W., 1996. *Lumbricus terrestris* in a soil core experiment: effects of nutrient-enrichment processes (NEP) and gut-associated processes (GAP) on the availability of plant nutrients and heavy metals. Soil Biol. Biochem. 28, 489–496.

Duno de Stefano, R., Huber, O., 2007. Clima. In: Duno de Stefano, R., Huber, O., Aymard, G. (Eds.), Catálogo ilustrado de la flora vascular de los llanos venezolanos. Fudena, Fundación Polar, FIBV, Caracas, pp. 43–46.

Erpenbach, A., Bernhardt-Römermann, M., Wittig, R., Thiombiano, A., Hanh, K., 2013. The influence of termite induced heterogeneity on savanna vegetation along a climatic gradient in West Africa. J. Trop. Ecol. 29, 11–23.

February, E.C., Cook, G.D., Richards, A.E., 2013. Root dynamics influence tree-grass coexistence in an Australian savanna. Aust. Ecol. 38, 66–75.

Frost, P.G.H., Robertson, F., 1985. The ecological effects of fire in savannas. In: Walker, B.H. (Ed.), Determinants of tropical savannas. IVBS # 3. IRL Press, pp. 93–140.

Giller, K.E., Cadish, G., Ehaliotis, C., Adams, E., Sakala, W., Mafongoya, P., 1997. Building soil nitrogen capital in Africa. Replenishing Soil Fertility in Africa. SSSA Special Publication No 51, pp. 151–192.

Gómez, I., 2004. Parámetros microbiológicos y bioquímicos en suelos de los Llanos Orientales de Venezuela bajo diferentes usos de la tierra y prácticas de manejo, PhSc thesis, Instituto Venezolano de Investigaciones Científicas, Caracas, Venezuela.

Gómez, I., Paolini, J., 2011. Conversion effects of native savanna to pine plantations on soil enzyme activities. Suelos Ecuatoriales 41, 98–104.

Groover, S.P.P., Livesley, S.J., Hutley, L.B., Jamali, H., Fest, B., Beringer, J., et al., 2012. Land use change and the impact on greenhouse gas exchange in north Australian savanna soils. Biogeosciences 9, 423–437.

Hayat, R., Ali, S., Amara, U., Khalid, R., Ahmed, I., 2010. Soil beneficial bacteria and their role in plant growth promotion: a review. Ann. Microbiol. 60, 579–598.

Hernández-Hernández, R.M., 2008. Dinámica y manejo de la materia orgánica en suelos de sabanas bien drenadas. Acta Cient. Venez. 28, 69–84.

Hernández, R.M., López-Hernández, D., 1998. Efecto de la intensidad de la labranza sobre diversas fracciones de la materia organica y la estabilidad estructural de un suelo de sabana. Ecotrópicos 2, 69–82.

Hernández-Hernández, R.M., López-Hernández, D., 2002. Microbial biomass, mineral nitrogen and carbon content in savanna soil aggregates under conventional and no-tillage. Soil Biol. Biochem. 34, 1563–1570.

Hernández-Hernández, R.M., Florentino, A., López-Hernández, D., 2000. Efecto de la siembra directa y la labranza convencional sobre la estabilidad estructural y otras propiedades fisicas de un suelo de sabana. Agron. Trop. 50, 9–29.

Hernández-Hernández, R.M., Lozano, Z., Bravo, C., Moralea, J., Toro, M., Ramírez, E., et al., 2012. Manejo agroecológico de suelos de sabanas bien drenadas con unidades de producción cereal-ganado. Mimeographed. CLCS, Buenos Aires, Argentina.

Hernández-Hernández, R.M., González, I., Chacín, E., León, W., 2013. Influence of the biological crusts on microbial populations, biological activities and labile organic matters fractions in savanna soils. Proceeding II International Workshop on Biological Soil Crusts: Biological Soil Crusts in a Changing World, Madrid, Spain.

Hernández-Valencia, I., 1996. Dinámica del fósforo en una sabana de *Trachypogon* de los Llanos Altos Centrales. PhD thesis. Universidad Central de Venezuela, Caracas, Venezuela.

Hernández-Valencia, I., Bautis, M., 2005. Cambios en el contenido de fósforo en el suelo superficial por la conversión de sabanas en pinares. Bioagro. 17, 69–78.

Hernández-Valencia, I., López-Hernández, D., 1997. Flujo de materia orgánica en el estrato herbáceo de una sabana de *Trachypogon* sometida a quema y pastoreo. Acta Biol. Ven. 17, 1–15.

Hernández-Valencia, I., López-Hernández, D., 1999. Allocation of phosphorus in a tropical savanna. Chemosphere 39, 199–207.

Hernández-Valencia, I., López-Hernández, D., 2000. Dinámica del fósforo en el estrato herbáceo de una sabana de *Trachypogon* sometida a quema y pastoreo. Acta Biol. Ven. 20, 49–62.

Hernández-Valencia, I., López-Hernández, D., 2002. Pérdida de nutrimentos por quema de vegetación en una sabana de *Trachypogon*. Rev. Biol. Trop. 50, 1013−1019.

Hétier, J.M., Sarmiento, G., Aldana, T., Zuvia, M., Acevedo, D., Thiery, J.M., 1989. The fate of nitrogen under maize and pasture cultivated on an alfisol in the Western Llanos Venezuela. Plant Soil 114, 295−302.

Higgins, S.I., Scheiter, S., 2012. Atmospheric CO_2 forces abrupt vegetation shifts locally, but not globally. Nature 408, 209−212.

Housman, D.C., Yeager, C.M., Darby, B.J., Sanford Jr, R.L., Kuske, C.H.R., Neher, D.A., Belnap, J., 2007. Heterogeneity of soil nutrients and sub surface biota in a dryland ecosystem. Soil Biol. Biochem. 39, 2138−2149.

Huber, O., 2007. Sabanas de los llanos venezolanos. In: Duno de Stefano, R., Aymard, G., Huber, O. (Eds.), Flora vascular de los llanos de Venezuela. FIBV, Caracas, Venezuela, pp. 73−86.

Hutley, L., Beringer, J., Cook, G., Razon, E., 2013. Savanna carbon turnover: quantifying source and sink dynamics from frequent and infrequent disturbance events in north Australian savannas. Geophys. Res. Abs. 15.

IUSS. 2007. World reference base for soil resources. World Soil Resources Reports No 103. FAO, Rome, Italy.

Izaguirre-Mayoral, M.L., Carballo, O., Flores, S., Mallorca, M., Oropeza, T., 1992. Quantitative analysis of the symbiotic N2-fixation, non-structural carbohydrates and chlorophyll content in sixteen native legume species collected in different savanna sites. Symbiosis 12, 293−312.

Janos, D., 1985. VA mycorrhizas in humid tropical ecosystems. In: Safir, G. (Ed.), Ecophysiology of VA Mycorrhizal Plants. CRC Press, Boca Raton, Florida, pp. 107−133.

Joner, E.J., Johansen, A., 2000. Phosphatase activity of external hyphae of two arbuscular mycorrhizal fungi. Mycol. Res. 104, 81−86.

Kellman, M., 1979. Soil enrichment by neotropical trees. J. Ecol. 565−577.

Lavelle, P., Dangerfield, M., Fragoso, C., Eschenbrenner, V., López-Hernández, D., Pashanashi, B., Brussard, L., 1994. The relationship between soil macrofauna and tropical soil fertility. In: Woomer, O.L., Swift, M.J. (Eds.), The Biology Management of Tropical Soil Fertility. TSBF: A Wiley-Sayce Publication, pp. 137−169.

Le Bayon, R.C., Binet, F., 2006. Earthworms change the distribution and availability of phosphorous in organic substrates. Soil Biol. Biochem. 38, 235−246.

Leon, L.A., Fenster, W.E., Hammond, L.L., 1986. Agronomic potential of eleven phosphate rocks from Brazil, Colombia, Peru and Venezuela. Soil Sci. Soc. Am. J. 50, 798−802.

Li, X.L., George, E., Marshner, H., 1991. Extension of the phosphorus depletion zone in VA-mycorrhizal white clover in a calcareous soil. Plant Soil 136, 41−48.

Lopes, A.S., Cox, F.R., 1977. A survey of the fertility status of surface soils under cerrado vegetation in Brazil. Soil Sci. Soc. Am. J. 41, 742−747.

López-Contreras, A.Y., Hernández-Valencia, I., López-Hernández, D., 2007. Fractionation of soil phosphorus in organic amended farms located on sandy soils of Venezuelan Amazonian. Biol. Fert. Soil 43, 771−777.

López-Gutiérrez, J.C., Toro, M., López-Hernández, D., 2004a. Seasonality of organic phosphorus mineralisation in the rhizosphere of the native savanna grass, *Trachypogon plumosus*. Soil Biol. Biochem. 36, 1675−1684.

López-Gutiérrez, J.C., Toro, M., López-Hernández, D., 2004b. Arbuscular mycorrhyza and enzymatic activities in the rhizosphere of *Trachypogon plumosus* in three acid savanna soils. Soil Agric. Environ. 103, 405−411.

López-Hernández, D., 1977. La química del fósforo en suelos ácidos. Ediciones de la Biblioteca. Universidad Central de Venezuela, Caracas.

López-Hernández, D., 1995. Balance de elementos en una sabana inundada. Mantecal, Edo. Apure. Venezuela. Acta Biol. Ven. 15, 55–88.

López-Hernández, D., 2001. Nutrient dynamics (C, N and P) in termite mounds of *Nasutitermes ephratae* from savannas of the Orinoco llanos (Venezuela). Soil Biol. Biochem. 33, 747–753.

López-Hernández, D., 2012. Earthworm populations in savannas of the Orinoco basin. A review of studies in long-term agricultural-managed and protected ecosystems. Agriculture 2, 87–108.

López-Hernández, D., 2013. N biogeochemistry and cycling in two well drained savannas: a comparison between the Orinoco Basin (Llanos-Venezuela) and Ivory Coast (Western-Africa). Chem. Ecol. 29, 280–295.

López-Hernández, D., Hernández-Valencia, I., 2008. Nutritional aspects in Trachypogon savannas as related to nitrogen and phosphorus cycling. International Commission on Tropical Biology and Natural Resources. In: Del Claro, Kleber , et al., (Eds.), Encyclopedia of Life Support Systems EOLSS), Developed under the Auspices of the UNESCO. Eolss Publishers, Oxford, UK.

López-Hernández, D., Hernández-Valencia, I., 2011. P cycling and biogeochemistry in well drained and flooded Venezuelan savannas. In: Verees, B., Szigethy, J. (Eds.), Horizons in Earth Science Research, vol 4. Nova Science Publishers Inc, pp. 99–125.

López-Hernández, D., Coronel, I., Álvarez, L., 1981. Uso de isotermas de adsorción para evaluar requerimientos de fósforo. I: Isotermas de adsorción de los suelos. Turrialba 31, 169–180.

López-Hernández, D., Siegert, G., Rodríguez, J.V., 1986. Competitive adsorption of phosphate with malate and oxalate by tropical soils. Soil Sci. Soc. Amer. J. 50, 1460–1462.

López-Hernández, D., Lavelle, P., Fardeau, J.C., Niño, M., 1993. Phosphorus transformations in two P-sorption contrasting tropical soils during transit of *Pontoscolex corethrurus* (Glossoscolesidae: Oligochaeta). Soil Biol. Biochem. 25, 789–792.

López-Hernández, D., Hernández-Hernández, R.M., Brossard, M., 2005. Historia del uso reciente de tierras de las sabanas de América del Sur. Estudios de casos de las sabanas del Orinoco. Interciencia 30, 623–660.

López-Hernández, D., Santaella, S., Chacón, P., 2006. Contribution of free-living organisms to N-budget in *Trachypogon* savannas. Eur. J. Soil Biol. 42, 43–50.

López-Hernández, D., Brossard, M., Fournier, A., 2012. Savanna biomass production, N biogeochemistry, and cycling: a comparison between Western Africa (Ivory Coast and Burkina Faso) and the Venezuelan Llanos. Recent Res Research Singpost Kerala. India Soil Sci. 3, 1–33.

López-Hernández, D., Sequera, D., Vallejo, O., Infante, C., 2013. Sugar cane nutrient requirements and the role of atmospheric deposition supplying supplementary fertilisation in a Venezuelan sugar cane plantation. Atmosfera 26, 337–348.

Lozano, Z., Hernández-Hernández, R.M., Bravo, C., Delgado, M., 2011. Cultivos de cobertura y fertilización fosfórica y su efecto sobre algunas propiedades químicas del suelo en un sistema mixto maíz-ganado. Venesuelos 19, 45–54.

Lozano, Z., Hernández-Hernández, R.M., Bravo, C., Rivero, C., Toro, M., Delgado, M., 2012. Disponibilidad de fósforo en un suelo de las sabanas bien drenadas venezolanas, bajo diferentes coberturas y tipos de fertilización. Interciencia 37, 820–827.

Martin, A., 1991. Short- and long-term effects of the endogenic earthworm *Millsonia anomala* (Omodeo) (Megascolecidae, Oligochaeta) of tropical savannas, on soil organic matter. Biol. Fertil. Soil 11, 234–238.

Medina, E., 1987. Nutrients requirement, conservation and cycles in the herbaceous layer. In: Walker, B. (Ed.), Determinant of Savannas. IUBS Monographs Series. IRL Press Ltd, Oxford, UK, pp. 39–65.

Medina, E., 1993. Mineral nutrition: tropical savannas. Prog. Bot. 54, 237–253.

Medina, E., Bilbao, B., 1991. Significance of nutrient relations and symbiosis for the competitive interaction between grasses and legumes in tropical savannas. In: Esser, G., Overdieck, D. (Eds.), Modern ecology: basic and applied aspects. Elsevier Science Publishing, Amsterdam, pp. 295–319.

Medina, E., Mendoza, A., Montes, R., 1978. Nutrient balance and organic matter production in the *Trachypogon* savannas of Venezuela. Trop. Agric. 55, 243–253.

Montes, R., San José, J.J., 1989. Chemical composition and nutrient loading by precipitation in the *Trachypogon* savannas of the Orinoco llanos, Venezuela. Biogeochemistry 7, 241–256.

Mora, E., Toro, M., López-Hernández, D., 2013. A survey of arbuscular mycorrhizae, *Rhizobium* and phosphate solubilizing bacteria in low fertility savanna soils in Central Venezuela (Estación Experimental La Iguana). In: Miransari, M. (Ed.), Soil Microbiology and Biotechnology. Studium Press LLC, Houston, TX, USA, pp. 97–114.

Nagarajah, S., Posner, A.M., Quirk, J.P., 1968. Desorption of phosphate from kaolinite by citrate and bicarbonate. Soil Sci. Soc. Amer. Proc. 32, 507–510.

Ngo, P.T., Rumpel, C., Doanc, T.T., Jouquet, P., 2012. The effect of earthworms on carbon storage and soil organic matter composition in tropical soil amended with compost and vermicompost. Soil Biol. Biochem. 50, 214–220.

Oberson, A., Friesen, D.K., Tiessen, H., Morel, C., Stahel, W., 1999. Phosphorus status and cycling in native savanna and improve pastures on a acid low-P Colombian oxisol. Nut. Cyc. Agroeco. 55, 77–88.

Okin, G.S., Mahowald, N., Chadwick, O.A., Artaxo, P., 2004. Impact of desert dust on the biochemistry of phosphorus in terrestrial ecosystems. Glob. Biochem. Cyc. 18, GB2005.

Olivares, J., Bedmar, E., Sanjuán, J., 2013. Biological nitrogen fixation in the context of global change. Mol. Plant Microbe. Interact. 26, 486–494.

Osorio E. 2007. Caracterización de los ácidos orgánicos producidos por microorganismos del suelo y su aplicación en la solubilización de fosfatos inorgánicos. Trabajo Especial de Grado, Escuela de Química, Facultad de Ciencias, UCV. Caracas, Venezuela.

Pacheco, M., Donoso, L., Sanhueza, E., 2004. Soluble organic nitrogen in Venezuelan rains. Tellus 56, 393–395.

Park, S.C., Smith, T.J., Bisesi, M.S., 1990. Hydrolysis of bis (4-nitro phenyl) phosphate by the earthworm *Lumbricus terrestris*. Soil Biol. Biochem. 22, 729–730.

Pineda, A., Dicke, M., Pieterse, C.M.J., Pozo, M.J., 2013. Beneficial microbes in a changing environment: are they always helping plants to deal with insects? Func. Ecol. 27, 574–586.

Ramia, M., 1967. Tipos de sabanas de los llanos venezolanos. Bol. Soc. Ven. Cien. Nat. 27, 264–288.

Reed, S.C., Towsend, A.R., Taylor, P.G., Cleveland, C.C., 2011. Phosphorus cycling in tropical forests growing on highly weathered soils. In: Bünemann, E.K., Oberson, A., Frossard, E. (Eds.), Phosphorus in Action: Biological Processes in Soil Phosphorus Cycling. Springer Soil, Biology Series 26. Springer, NY, USA, pp. 199–213.

Renard, D., Birk, J.J., Zengerlé, A., Lavelle, P., Glaser, B., Blatrix, R., McKey, D., 2013. Ancient human agricultural practices can promote activities of contemporary non-human soils ecosystem engineers: a case of study in coastal savanna of French Guinea. Soil Biol. Biochem. 62, 46–56.

Richardson, A.E., 2001. Prospects for using soil microorganisms to improve the acquisition of phosphorus by plants. Aust. J. Plant Physiol. 28, 897–906.

Richardson, A.E., Barea, J.M., McNeill, A.M., Prigent-Combaret, C., 2009. Acquisition of phosphorus and nitrogen in the rhizosphere and plant growth promotion by microorganisms. Plant Soil. 321, 305–339.

Riina, R., Duno de Stefano, R., Aymard, G., Fernández, A., Huber, O., 2007. Análisis de la diversidad florística de los llanos de Venezuela. In: Duno de Stefano, R., Huber, O., Aymard, G. (Eds.), Catálogo ilustrado de la flora vascular de los llanos Venezolanos. Fudena. Fundación Polar. FIBV, Caracas, pp. 107–122.

Rippstein, G., Amezquita, E., Escobar, G., Grollier, C., 2001. Condiciones naturales de la sabana. In: Rippstein, G., Escobar, G., Motta, F. (Eds.), Agroecología y biodiversidad de las sabanas en los llanos Orientales de Colombia. CIAT, Colombia, pp. 1–21.

Robertson, G.P., Rosswall, T., 1986. Nitrogen in West Africa: the regional cycle. Ecol. Monog. 56, 45–72.

Rodriguez, I., Leal, A., Sánchez-Rose, I., Vessuri, H., Bilbao, B., 2009. Facing up the challenging of interdisciplinary research in the Gran Sabana (Venezuela). Hum. Ecol. 37, 787–789.

Rondón, M.A., Acevedo, D., Hernández, R.M., Rubiano, Y., Rivera, M., Amezquita, E., et al., 2006. Carbon sequestration potential of the neotropical savannas of Colombia and Venezuela. In: Lal, R., Cerri, C., Bernoux, M., Etchevers, J., Pellegrino, C. (Eds.), Carbon Sequestration in Soils of Latin America. Haworth Press Inc, New York, pp. 213–243.

San José, J., Montes, R., García Miragaya, J., Orihuela, B.E., 1985. Bioproduction of the *Trachypogon* savanna in a latitudinal cross section of the Orinoco Llanos, Venezuela. Acta Oecol/Oecol Plant 6, 24–43.

San José, J., Montes, R., Grace, J., Nikonova, N., 2008. Land-use changes alter radiative energy and water vapor fluxes of a tall grass Andropogon field and a savanna woodland continuum in the Orinoco lowlands. Tree Physiol. 28, 425–435.

Sánchez, P.V., Guinand, M., Gonzalez, V., 1985. Efectos del fuego sobre el balance nutricional de una sabana de *Trachypogon* del Territorio Federal Amazonas, Venezuela. Acta Biol. Ven. 12, 1–8.

Sanhueza, E., Crutzen, P.J., 1998. Budgets of fixed nitrogen in the Orinoco savannah region. Glob. Biogeochem. Cyc. 12, 653–666.

Sarmiento, G., 1984. The Ecology of Neotropical Savannas. Harvard University Press, Cambridge.

Satchell, J.E., Martin, K., 1984. Phosphatase activity in earthworm faeces. Soil Biol. Biochem. 16, 191–194.

Schlesinger, W.H., Raikes, J.A., Hartley, A.E., Cross, A.F., 1996. On the spatial pattern of soil nutrients in desert ecosystems. Ecology 77, 364–374.

Schüßler, A., Walker, C., 2010. The Glomeromycota. A Species list with New Families and New Genera. The Royal Botanic Garden Edinburgh, Scotland.

Sharpley, A., Syers, J., 1976. Potential role of earthworms casts for the phosphorus enrichment of run-off water. Soil Biol. Biochem. 8, 341–346.

Shen, J., Yuan, L., Zhang, J., Li, H., Bai, Z., Chen, X., et al., 2011. Phosphorus dynamics: from soil to plant. Plant Physiol. 156, 997–1005.

Singh, R.S., Srivastava, S.C., Raghubanshi, A.S., Singh, J.S., Singh, J.P., 1991. Microbial C, N and P in a dry tropical savanna effects of burning and grazing. J. Appl. Ecol. 28, 868–879.

Sleto, B., Rodríguez, I., 2013. Burning, fire prevention and landscape productions among the Pemón, Gran Sabana, Venezuela: toward an intercultural approach to wildland fire management in neotropical savannas. J. Environ. Mgm. 115, 155–166.

Smith, S.E., Read, D.J., 2008. Mycorrhizal Symbiosis. Third ed. Elsevier.

Sperber, J.I., 1958. The incidence of apatite solubilizing microorganisms producing organic acids. Aust. J. Agr. Res. 9, 778–781.

Sprent, J., 1999. The Ecology of the Nitrogen Cycle. Cambridge University Press, New York.

St. John, T.V., 1980. Root size, root hairs and mycorrhizal infection: a re-examination of Baylis's hypothesis with tropical tress. New Phytol. 84, 483–487.

St. John, T.V., Uhl, C., 1983. Mycorrhizae in the rain forest of San Carlos de Río Negro, Venezuela. Acta Cient Ven. 34, 233–237.

Stewart, W.D.P., Sampaio, M.J., Isichei, A.O., Silvester-Bradley, R., 1977. Nitrogen fixation by soil algae of temperate and tropical soils. In: Döbereiner, J., Burris, R., Hollaender, A. (Eds.), Limitations and Potentials for Biological Nitrogen Fixation in the Tropics, vol 10. Basic Life Sciences, Plenum Press, New York. USA, pp. 41–63.

Susach, F., 1984. Caracterización ecológica de las sabanas de un sector de los Llanos Centrales Bajos de Venezuela. Tesis Doctoral. Postgrado de Ecología. Facultad de Ciencias, Universidad Central de Venezuela, Caracas, Venezuela.

Swap, R., Garstang, M., Greco, S., Talbot, R., Kallberg, P., 1992. Saharan dust in the Amazonian basin. Tellus. B. Chem. Phys. Meteorol. 44, 133–1149.

Tarafdar, J.C., Marshner, H., 1994. Phosphatase activity in the rhizosphere of VA mycorrhizal wheat supplied with inorganic and organic phosphorus. Soil Biol. Biochem. 26, 387–395.

Toro, M., Azcón, R., Herrera, R., 1996. Effects on yield and nutrition of mycorrhizal and nodulated *Pueraria phaseoloides* exerted by P-solubilizing rhizobateria. Biol. Fert. Soils 21, 23–29.

Toro, M., Azcón, R., Barea, J.M., 1997. Improvement of arbuscular mycorrhiza development by inoculation of soil with phosphate-solubilizing rhizobacteria to improve rock phosphate bioavailability (P32) and nutrient cycling. App. Env. Mic. 63, 4408–4412.

Toro, M., Azcón, R., Barea, J.M., 1998. The use of isotopic dilution techniques to evaluate the interactive effects of *Rhizobium* genotype mycorrhizal fungi, phosphate-solubilizing rhizobacteria and rock phosphate on nitrogen and phosphorus acquisition by *Medicago sativa*. New Phytol. 138, 265–273.

Toro, M., Bazo, I., López, M., 2008. Micorrizas arbusculares y bacterias promotoras de crecimiento vegetal, biofertilizantes nativos de sistemas agrícolas bajo manejo conservacionista. Agron. Trop. 58, 78–83.

USDA, 2010. Keys to Soil Taxonomy, Soil Survey Staff. Natural Resource Conservation Service. Eleventh ed. US Agriculture Department.

Vareschi, V., 1962. La quema como factor ecológico en los llanos. Bol. Soc. Ven. Cien. Nat. 23, 9–26.

Velasquez, J., 1965. Estudio fitosociológico acerca de los pastizales de las sabanas de Calabozo. Estado Guárico. Bol. Soc. Ven. Cien. Nat. 25, 59–101.

Villecourt, P., Roose, E., 1978. Charge en azote et en éléments minéraux divers des eaux de pluie, de pluvio-lessivage et de drainage dans la savane de Lamto (Côte d'Ivoire). Rev. Ecol. Biol. Sol. 16, 9–15.

Vuattoux, R., 1976. Contribution à l'étude de l'évolution des strates arborées et arbustives dans la savane de Lamto (Côte d'Ivoire). Deuxième note. Ann. Univ. Abidjan, série C, tome XII, 35–63.

Wake, B., 2012. Regime change: savannah shift. Nat. Clim. Chang. 2, 571–571.

Walker, T., Syers, J., 1976. The fate of phosphorus during pedogenesis. Geoderma 15, 1–19.

Walter, H., Medina, E., 1971. Caracterización climática de Venezuela sobre la base de climadiagramas de estaciones particulares. Bol. Soc. Ven. Cien. Nat. 29, 211–240.

Williams, D.G., Baruch, Z., 2000. African grass invasion in the Americas: ecosystem consequences and the role of ecophysiology. Biol. Invasions 2, 123–140.

Zhang, F., Shen, J., Zhang, J., Zuo, Y., Li, L., Chen, X., 2010. Rhizosphere processes and management for improving nutrient use efficiency and crop productivity: implications for China. Adv. Agron. 107, 1–32.

Silicon and Selenium: Two Vital Trace Elements that Confer Abiotic Stress Tolerance to Plants

Mirza Hasanuzzaman, Kamrun Nahar and Masayuki Fujita

16.1 Introduction

Doubtless, we all are in some way or other affected by the damage to economic plants caused by environmental stresses; this is particularly alarming in view of the shrinking agricultural land area and the continuously expanding population on Earth. Rapid climate change throughout the world has exposed plants to various environmental adversities that prevent them from reaching their full genetic potential and limit their productivity (Hasanuzzaman et al., 2012a). Abiotic stresses are the greatest restriction for crop production worldwide and account for yield reductions of as much as 50% (Rodríguez et al., 2005; Acquaah, 2007). Crop plants, as sessile organisms, encounter unavoidable abiotic stresses during their life cycles, including salinity, drought, extreme temperatures, heavy metal (HM) toxicity, flooding, UV-B radiation, ozone, etc., which all pose a serious challenge to plant growth, metabolism, and productivity (Ahmad and Prasad, 2012a,b; Hasanuzzaman et al., 2012a). Now is the right time to be strategic: first by understanding the reasons—fundamental to complex—for yield reductions so that precise research planning can be brought about to cope with ever-changing environmental conditions or stresses. With that view, plant scientists are now sacrificing their time searching for ways to make the plants adaptive even under adverse growing conditions. Researchers are trying to understand the effects of environmental stresses on plants so that they can modify the plant's external growing condition by applying different exogenous protectants including trace elements and even by molecular mechanisms.

Many reports have appeared on silicon (Si) research in the last few decades, but this element continues to be an anomaly—an underappreciated element and in most cases one not considered as essential for plant function. However, in certain plant species like rice (*Oryza* spp.), Si is absorbed from soil in amounts that are even higher than those of the essential macronutrients (Datnoff et al., 2001). The beneficial effects of Si on growth have been reported in many crop plants in addition to rice, such as wheat (*Triticum aestivum*), barley (*Hordeum vulgare*), and cucumber (*Cucumis sativus*). However, plant species differ greatly in their ability to accumulate Si (Epstein, 1999; Ma et al., 2002; Richmond and Sussman, 2003). Numerous research reports have provided evidence for the notion that Si may also play a vital role in conferring plants with tolerance to adverse environmental factors. Hence, Si is considered as a beneficial element for plants growing under stressful

P. Ahmad (Ed): Emerging Technologies and Management of Crop Stress Tolerance, Volume 1.
DOI: http://dx.doi.org/10.1016/B978-0-12-800876-8.00016-3

conditions (Sacaa, 2009). Silicon-induced alleviation of abiotic stresses such as drought, salinity, high temperature (HT), chilling, UV radiation, nutrient imbalance, and metal toxicity have been observed by many researchers (Raven, 2001; Ma and Takahashi, 2002; Liang et al., 2007; Ma and Yamaji, 2008; Hasanuzzaman and Fujita, 2011a; Ahmed et al., 2012).

Selenium (Se) is a widely studied trace element in humans and animals due to its role in the antioxidant defense system, which is needed for the maintenance of health and hormone balance. For the past few decades, the beneficial role of Se in plants has been investigated by several groups of researchers but the question is still unresolved whether Se is an *essential* micronutrient for plants (Terry et al., 2000). Although Se has not been defined as essential, its role as a beneficial element in plants has been revealed in many plant studies (Xue et al., 2001; Hasanuzzaman et al., 2010). In many plant species, Se exerts a positive effect on plant growth and productivity at low concentrations (Terry et al., 2000; Xue et al., 2001; Turakainen et al., 2004; Hasanuzzaman et al., 2010). However, the precise mechanisms underlying the beneficial role of Se in plants have not been clearly elucidated yet (Turakainen, 2007). Some plant species supplemented with Se have shown enhanced resistance to certain abiotic stresses like salinity (Djanaguiraman et al., 2005; Hawrylak-Nowak, 2009; Hasanuzzaman et al., 2011), drought (Yao et al., 2009; Hasanuzzaman et al., 2010; Hasanuzzaman and Fujita, 2011b), extreme temperature (Chu et al., 2010; Hawrylak-Nowak et al., 2010; Djanaguiraman et al., 2010), metal toxicity (Hawrylak et al., 2007; Pedrero et al., 2008; Filek et al., 2008; Srivastava et al., 2009; Cartes et al., 2010), and UV radiation (Valkama et al., 2003; Yao et al., 2010a,b). Different plant studies have shown that Se could help in detoxification of reactive oxygen species (ROS) and thus enhanced plant tolerance to oxidative stress (Djanaguiraman et al., 2005; Hasanuzzaman et al., 2010; Hasanuzzaman and Fujita, 2011b).

In this chapter, we review the recent findings on the diverse roles of Si and Se on plant growth and development. Special emphasis is given to their roles in conferring abiotic stress tolerance to plants. We also shed light on the mechanisms of uptake and transport of these elements in plants.

16.2 Silicon uptake and transport in plants

Silicon is the second most abundant element after oxygen (O_2) in soils, with a value as high as 26%. It has strong affinity with O_2 and thus it always exists as silica (SiO_2) or silicate (SiO_4^{4-}). In nature, Si does not occur as an elemental form but is found as a compound within many minerals which form rocks. Minerals containing Si are resistant to weathering processes and decomposition; hence, the amount of Si in the soil solution is low (Brogowski, 2000). The Si content of soils can vary considerably, from <1 to 45% of dry weight (Sommer et al., 2006), and its presence in the form of silicic acid [$Si(OH)_4$] (or its ionized form, $Si(OH)_3O_2$, which predominates at pH 9) allows its uptake by plants. Therefore, by nature, plants have a great scope to uptake Si into their tissues. Although there is extensive literature on the uptake, transport, and distribution of Si in plants, the mechanisms by which these processes occur is still a matter of research because the nature of uptake and transport and the concentration of Si in plant tissues varies greatly between species (i.e., 0.1−10% of shoot dry weight). These differences are mainly due to the differential characteristics and capacities of Si uptake and transport among various species (Epstein, 1994; Liang et al., 2007; Ma et al., 2007).

In their review, Sacaa (2009) summarized three types of plants based on their capacity for Si accumulation (Table 16.1). In general, monocotyledonous plants, especially the plants in the Poaceae family, are able to take up more Si than are other plant families. Among the crop plants of this family, rice has the highest capability to uptake Si (Tamai and Ma, 2003) so that rice contains much higher Si levels than do barley, maize, rye, sorghum, and wheat. Some other plant families (namely, Equisetaceae, Cyperaceae, and Balsaminaceae) also show high Si accumulation. Table 16.2 presents the differences of Si concentration in some plant species (Broadley et al., 2012). At the same levels of soil Si, rice was shown to accumulate 39.1 mg g^{-1} dry weight of Si, whereas chickpea (*Cicer arietinum*) accumulated only 3.0 mg g^{-1} dry weight. According to Takahashi et al. (1990), plants possess three methods of Si uptake: (1) active, (2) passive, and (3) rejective. For instance, rice, which is an Si accumulator, takes up Si through an active process (Ma et al., 2006), while some dicots such as cucumber (*Cucumis sativus*), muskmelon (*Cucumis melo*), strawberry (*Fragaria × Ananassa*), and soybean (*Glycine max*) take up Si passively (Takahashi et al., 1990; Ma et al., 2001a; Mitani and Ma, 2005; Liang et al., 2007). Most of the dicotyledonous plants absorb Si passively and some cannot take up Si into their tissues (Liang et al., 2007). Some crops like tomato (*Solanum lycopersicum*) exclude Si from uptake (Mitani and Ma, 2005). According to Ma and Takahashi (2002), the Si content in the rice tops was 20-fold higher than that in the roots, while the Si content in tomato tops was one-tenth that in the roots. Their explanation was that rice roots can take up large amounts of Si, which could be translocated from the roots to the tops in rice, whereas tomato could not take up and translocate Si. This difference was due to the dissimilar ability of the roots to take up Si (Ma and Takahashi, 2002). Among the crops, rice is

Table 16.1 Types of Plant on the Basis of Si Accumulating Capacity

Types	Accumulation Capacity	Example
Si accumulators	Higher than 1% Si on dry weight	Rice and some wetland grasses
Si non-accumulators	Less than 1% Si on dry weight	Most dicotyledonous plants like chickpea, tomato, etc.
Intermediate	Intermediate (1—3%) Si on dry weight	Rye, oat, etc.

Table 16.2 Variations in Si Concentration in Some Plants Species under Same Levels of Soil Si

Plant Species	Family	Si Concentration (mg g^{-1} dry Weight)
Oryza sativa	Poaceae	39.1
Triticum aestivum	Poaceae	15.4
Cucurbita maxima	Cucurbitaceae	13.4
Cucurbita pepo	Cucurbitaceae	19.8
Cicer arietinum	Fabaceae	3.0
Cucumis sativus	Cucurbitaceae	22.9
Zea mays	Poaceae	21.0

the most potent accumulator of Si, while other cereal crops such as wheat (Casey et al., 2004) and barley (Barber and Shone, 1966) and some cyperaceous plants take up Si actively. However, not all cereal crops take up Si actively; for instance, certain cereals like oats take it up passively (Jones and Handreck, 1965). In their report, Ma et al. (2001b) claimed that the lateral roots of rice play an important role in Si uptake but the root hairs do not. Apart from these species differences, different parts of the same plant may show significant differences in Si accumulation. For instance, in rice, Si content was measured as 0.5, 50, 130, 230, and 350 g kg^{-1} in polished endosperm, bran, straw, hulls, and nodes (joints), respectively (Van Hoest, 2006).

Ma et al. (2002) compared two rice varieties, Nipponbare (*japonica*) and Kasalath (*indica*), under different doses of Si. When the plants were grown in a nutrient solution containing 0.15 mM Si(OH)$_4$, the shoot Si content was higher in Nipponbare than in Kasalath. In contrast, when grown in a solution containing 1.5 mM Si(OH)$_4$, the Si content was nearly the same in both varieties. These variations were due to different mechanisms involved in accumulation of Si in the two varieties (Ma and Takahashi, 2002). In their advanced study, Ma et al. (2007) investigated the genotypic differences in Si uptake and expression of Si transporter genes in these same two rice varieties. Both Si transporter genes (*Lsi1* and *Lsi2*) were more highly expressed in Nipponbare than in Kasalath, so that Nipponbare was a higher Si accumulator. In addition, Nipponbare had a more extensive root system and the genes *Lsi1* and *Lsi2* were localized at the distal and proximal sides, respectively, of both exodermis and endodermis of the roots; hence, this variety was capable of greater Si uptake. These results confirmed that the genotypic differences in the Si uptake were due to the differences in abundance of Si transporters in roots of different rice varieties.

Ma et al. (2003) found great variations in Si content, ranging from 1240 to 3600 mg kg^{-1}, among 400 barley cultivars grown in the same soil. Mitani and Ma (2005) investigated the comparative uptake potential of three different crop species (rice, cucumber, and tomato) and found that shoots of these plants showed variations in Si accumulation. Their explanation was that this was due to the differences in radial transport of Si from the rhizosphere to the root cortical cells and subsequent transport to the xylem. Among the species tested, rice accumulated the largest amount of Si in the root—cell symplasm, at levels three- and five-fold higher than in cucumber and tomato. As a result, the Si transport in the xylem sap in rice was also the highest, and was 20- and 100-fold higher than in cucumber and tomato, respectively. The lower Si accumulation in cucumber and tomato was explained by the lower densities of the Si transporter that transports Si from the external solution to the cortical cells. In addition, the absence or defects in the Si transporter that moves Si from the cortical cells to the xylem was also considered a causal factor (Mitani and Ma, 2005). In a recent review, Ma et al. (2011) described the process of Si transport from roots to panicles in plants and showed that this transport is mediated by different transporters such as *Lsi1*, *Lsi2*, etc. In rice, transport of Si by these transporters is driven by a proton gradient (Figure 16.1). However, the expression patterns and cellular localization of these transporters differs depending on the plant species.

Ma et al. (2011) also reported that the differences in Si uptake were also partly mediated by root anatomy. Rice roots, for example, possess two Casparian strips at the exodermis and endodermis (Figure 16.1; Ma et al., 2011). However, in maize and barley, only one Casparian strip is present at the endodermis. Rice roots also contain well-differentiated anatomical structures, namely, highly developed aerenchyma and destroyed cortical cells between the exodermis and endodermis. Maize and barley, on the other hand, do not possess such structures. These differences in

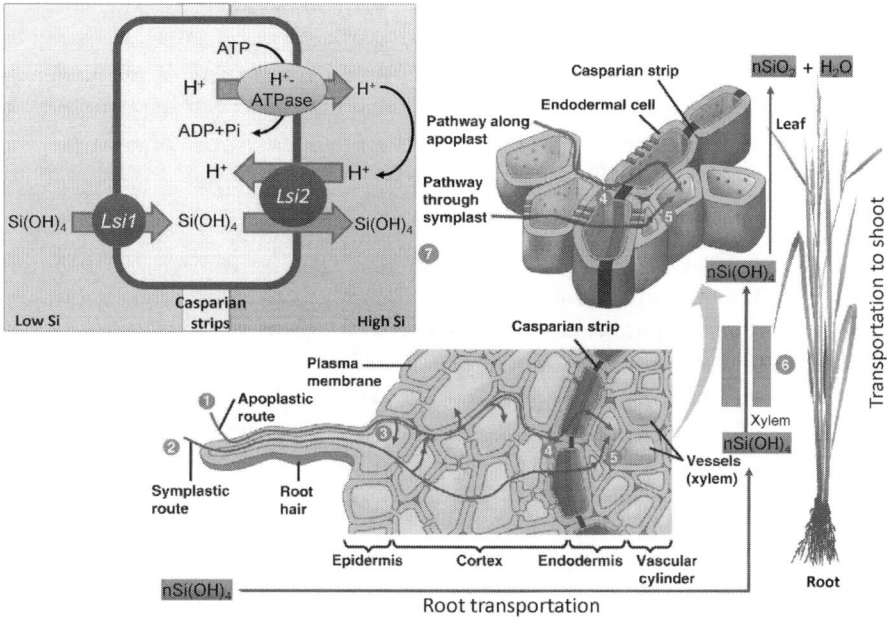

FIGURE 16.1

Transport of Si from root to other parts of the plants. 1. Apoplastic route of Si uptake. 2. Symplastic route of Si uptake. 3. Si transport to adjacent cells. 4. Transport through the Casparian strip. 5. Transport into xylem vessels. 6. In the shoot, Si is polymerized into silica and deposited in the bulliform cells (silica body). 7. Si uptake system in roots exodermis/endodermis cell (Mitani et al., 2009; Ma et al., 2011).

anatomical structures and localizations of transporters are predicted to explain at least part of the variation in Si uptake capacity among different species (Ma et al., 2011).

When Si is uptaken by roots through transporters, it is translocated to the shoots by transpirational volume flow through xylem; more than 90% of the Si uptaken by roots is translocated to the shoots (Ma et al., 2002). Because it follows the transpiration stream, Si is largely accumulated and distributed in the leaves (Dagmar et al., 2003). In most cases, Si uptake occurs in the mature regions of the roots, rather than the root tips (Yamaji and Ma, 2007). The transport of Si from the xylem to the leaf cells involves another transporter *Lsi6* in rice (Yamaji et al., 2008). The transportation of Si sometimes depends on its concentration in the sap. When the concentration of Si $(OH)_4$ exceeds 2 mM, $Si(OH)_4$ was polymerized to form silica gel ($SiO_2 \bullet nH_2O$) (Ma and Yamaji, 2006; Figure 16.1). High concentrations of Si are very common in some grasses. For instance, in rice and wheat, the concentration of Si in the xylem sap can exceed 2 mM and in such cases monomeric silicic acid (H_4SiO_4) was also identified as the major form of Si in the xylem (Figure 16.1; Mitani and Ma, 2005; Mitani et al., 2005; Ma and Yamaji, 2006). Silicon is further concentrated in the shoots through loss of water and polymerization, which converts $Si(OH)_4$ to colloidal $Si(OH)_4$ and finally to $SiO_2 \bullet nH_2O$ with increasing $Si(OH)_4$ concentration (Ma and Takahashi, 2002). In some species, like rice, Si is sometimes deposited as a thin layer in the space immediately beneath the thin cuticle layer, to form a cuticle Si double layer in the leaf blades. This is one of the ways that rice plants are protected from multiple abiotic and biotic stresses (Ma and Yamaji, 2006). Silicon distribution in the aerial parts of plants depends on the intensity of transpiration. The xylem transpiration stream transports $Si(OH)_4$ to the leaves, where it accumulated in older tissues (Sacaa, 2009).

At the cell level, Si is transported through two different processes: (1) passive transport of Si $(OH)_4$ across membranes and (2) active transport of $Si(OH)_4$ across membranes (Figure 16.1). The default condition of Si transport across the cell membranes is that the solute crosses the membrane by diffusion across the lipid components of the membrane. The active transport of $Si(OH)_4$ can be recognized at the cell level by the demonstration of $Si(OH)_4$ movement from a lower to a higher concentration. In some cases, the occurrence of active Si transport is deduced from the occurrence of intercellular SiO_2 deposition in cells growing in a medium with $Si(OH)_4$ at lower than saturated concentrations (Raven, 2001). At the tissue level, the movement of $Si(OH)_4$ through plant tissues can involve apoplasmic and symplasmic pathways. During apoplasmic movement, $Si(OH)_4$ is transported through cell walls and water-filled intercellular spaces, etc., while during symplasmic movement it is transported through the cytosol, including the cytosolic sleeve of the plasmodesmata and the conducting cells of phloem (Raven, 2001; Figure 16.1). Both of these processes involve diffusion and mass flow.

Different physiological processes also affect Si transport among the plant parts. For instance, in rice, transpiration plays a certain role in translocation and accumulation of Si to the tops. The Si concentration is high in leaf blades and husks where the transpiration rate is high. Although rice roots play an important role in active uptake of Si, the Si content in the roots is much lower than that found in the tops. Therefore, Si taken up by the roots is supposedly translocated to the shoot along with the transpiration steam and then concentrated and finally physically gelled in rapidly transpiring organs. Nutrient ions present in the soil also sometimes affect Si uptake. For example, excess nitrogen is reported to decrease the Si content and number of silica bodies in rice (Ishizuka and Tanaka, 1950). Different metabolic inhibitors like pyruvate and acetate were found to disturb Si uptake (Ma and Takahashi, 2002). Some environmental factors like low temperature (LT) sometimes inhibit transportation (Mitani and Ma, 2005).

16.3 Selenium uptake and metabolism in plants

Selenium has a chemical nature similar to sulfur (S). In nature, Se can exist in different oxidative states, namely, in the -2 (selenide, Se^{2-}), 0 (elemental Se), -4 (selenite, SeO_3^{2-}), and -6 (selenate, SeO_4^{2-}) oxidation states (Brown and Shrift, 1982). Selenium may also exist as organic complexes. The forms of Se present in soil mostly depend on pH and redox potential (Eh); the predominant form is SeO_4^{2-} in alkaline and well-oxidized soils (pe + pH >15), SeO_3^{2-} in well-drained mineral soils with pH from acidic to neutral (7.5 $<$ pe + pH <15), and selenide under reduced soil conditions (pe + pH <7.5) (Elrashidi et al., 1987). In soil, Se is found in small amounts (ranging from 0.01 to 2 mg kg^{-1}). However, in seleniferous areas, its content may be as high as 1200 mg kg^{-1} total Se and 38 mg kg^{-1} soluble Se (Mayland et al., 1989; Pezzarossa and Petruzzelli, 2001).

Table 16.3 Se Concentration in Shoots of Different Plant Species Grown in Soils with the Same Se Level ($2-4$ mg Se kg^{-1})

Plant Species	Se Concentration (mg Se kg^{-1})
Astragalus pectinalus	4000
Stanleya pinnnata	330
Gutierrezia fremontii	70
Zea mays	10
Helianthus annuus	2

Plant species differ strongly in Se uptake and accumulation as well as in their tolerance capacity (Table 16.3; Broadley et al., 2012). Their differences in the capacities to accumulate and tolerate Se place plants into different classifications: (1) non-accumulators, (2) indicators, and (3) accumulators (Terry et al., 2000; Dhillon and Dhillon, 2003; White et al., 2004). Shrift (1973) hypothesized that Se is an essential microelement in accumulator plants, based on the following evidence: accumulator plants grow only on seleniferous soils and accumulate higher quantities of Se than do non-accumulator plants; the growth of accumulator plants is stimulated by adding small amounts of Se to the growth solution, whereas the growth of non-accumulator plants is inhibited; the assimilation path of Se in the accumulator plants differs substantially from non-accumulator plants. Apart from these categories, some plant species are called Se hyperaccumulators, which are further divided into two groups: the primary Se accumulators are capable of accumulating thousands of milligrams of Se kg^{-1} (>4000 mg kg^{-1}), and the secondary accumulators hundreds of milligrams Se kg^{-1} (Terry et al., 2000). One of the promising Se hyperaccumulator plant families is the Brassicaceae, which includes mustard (*Brassica juncea* L.), broccoli (*B. oleracea botrytis* L.), and canola (*B. napus* spp. *oleifera* L.), which have been classified as primary accumulators (Whanger, 2002). However, most cultivated crop plants have a low tolerance to high Se levels. Generally, they contain less than 25 µg Se g^{-1} dry weight (DW) and are considered to be non-accumulators. Potato (*Solanum tuberosum*) is classified as an Se non-accumulator (White et al., 2004) and cannot tolerate high concentrations of Se. Plant species greatly vary in their potential to uptake Se into their tissues. The critical tissue Se concentrations that decreased the yield were 105 µg g^{-1} DW in *B. juncea*, 77 µg g^{-1} DW in *Z. mays*, 42 µg g^{-1} DW in *O. sativa*, and 19 µg g^{-1} DW in *T. aestivum*. These levels were attained by Se addition as SeO$_3^{2-}$ at 5 µg g^{-1} soil for *B. juncea* and *Z. mays*, 4 µg g^{-1} soil for *T. aestivum,* and 10 µg g^{-1} soil for *O. sativa* (Rani et al., 2005).

Accumulation of Se also differs greatly among the plant organs in the same plant species. With some exceptions, most of the plants accumulate more Se in upper parts (stem and leaf) than in roots (Zayed et al., 1998). Studies on potato (*Solanum tuberosum*) plants by Turakainen (2007) indicated that, at early stages, Se concentration was higher in young upper leaves. However, at maturity, Se content declined in the aerial parts, but a rigorous accumulation took place in the tubers (Turakainen et al., 2006; Turakainen, 2007). Selenium accumulation was also influenced by the application methods. Foliar spray of SeO$_4^{2-}$ on tea leaves considerably increased Se content in the cells (Hu et al., 2003). Smrkolj et al. (2006) observed a linear relationship between exogenous Se spraying and the accumulation of Se in *Pisum sativum* seeds. Other factors, like presence of ions (Cl^-, SO_4^{2-}, PO_4^{3-}), salinity, and trace elements, also affect the uptake and metabolism of Se in plants (Gupta et al., 1982). The interactions between Se and other ions may be due to chemical reactions either in the soil or in the plant, or to the dilution effect due to increased plant growth.

Selenium is taken up from soil by plant roots mainly as SeO_4^{2-}, SeO_3^{2-}, or organoselenium compounds. The biosynthesis of most Se compounds in plants follows the same pathways that lead to isologous S compounds (Çakır et al., 2012). However, SeO_4^{2-} is taken up faster than SeO_3^{2-} but acquires organoselenium compounds, namely, selenocysteine (SeCys) and selenomethionine (SeMet) (White et al., 2007). Thereafter, it is metabolized (via the S assimilation pathway) into SeCys, SeMet, and other Se analogues of S-based metabolites (Ellis and Salt, 2003). The most abundant bioavailable form of Se in soils is SeO_4^{2-}, which is taken up via sulfate transporters. It then forms SeO_3^{2-} and Se^{2-} through reduction and SeCys through combination with *O*-acetylserine. SeCys is further metabolized to form Se-cystathionine, Se-homocysteine, and SeMet. Glutathione (GHS) plays a role in Se metabolism, serving as a reductant for the reduction of SeO_3^{2-} to Se^{2-} (Sors et al., 2005).

The mechanisms of SeO_3^{2-} uptake by plants are not well understood. Early studies suggested that SeO_3^{2-} may enter root cells passively by diffusion (Terry et al., 2000). However, SeO_3^{2-} uptake is at least partly active (Li et al., 2008). Both SeCys and SeMet can be non-specifically incorporated into proteins, replacing Met and Cys (Figure 16.2; Valdez Barillas et al., 2011). Further, SeCys can also be converted into Se^{2-} and alanine (Ala) through breakdown, or to methyl-SeCys through methylation. However, the methylation reaction is generally observed in Se-hyperaccumulator plants, where it acts as one of the mechanisms of Se tolerance since methyl-SeCys cannot be incorporated into proteins (Figure 16.2; Freeman et al., 2006; Valdez Barillas et al., 2011). Later, this methyl-SeCys is metabolized to a volatile compound, dimethyldiselenide (DMDSe). Non-hyperaccumulators also volatilize Se as DMSe using SeMet as a starter (Figure 16.2; Terry et al., 2000). The cellular metabolism of Se and its sequestration patterns in plant tissues are dissimilar in non-hyperaccumulators and hyperaccumulators. In non-hyperaccumulators, Se is mainly sequestered in the vascular tissues (Freeman et al., 2006), while hyperaccumulators accumulate Se mainly in the leaf hairs and the vacuoles of epidermal cells (Freeman et al., 2006).

Through plants, Se is easily transported from the soil to the human food chain (Figure 16.3). It is mixed with various agricultural products and fodders to a variable extent, depending on the concentration and bioavailability of Se in soil (Koivistoinen, 1980; Yläranta, 1983a, 1985). In fact, the availability of Se in soils is restricted due to reduced weathering status and acidity (Sippola, 1979; Yläranta, 1983b). Quantities of Se in the soil solution are also governed by the solubility of its adsorbed forms and by the biological transformation to other forms (Figure 16.3).

16.4 Involvement of silicon and selenium in plant growth, development, and physiology

Although Si is considered a non-essential element, exogenous application promotes the growth of many plant species (Epstein, 1999). Increases in different growth parameters following Si application is dose and plant specific, and also depends on different edaphic and environmental conditions (Agurie et al., 1992; Hossain et al., 2002; Ali et al., 2009).

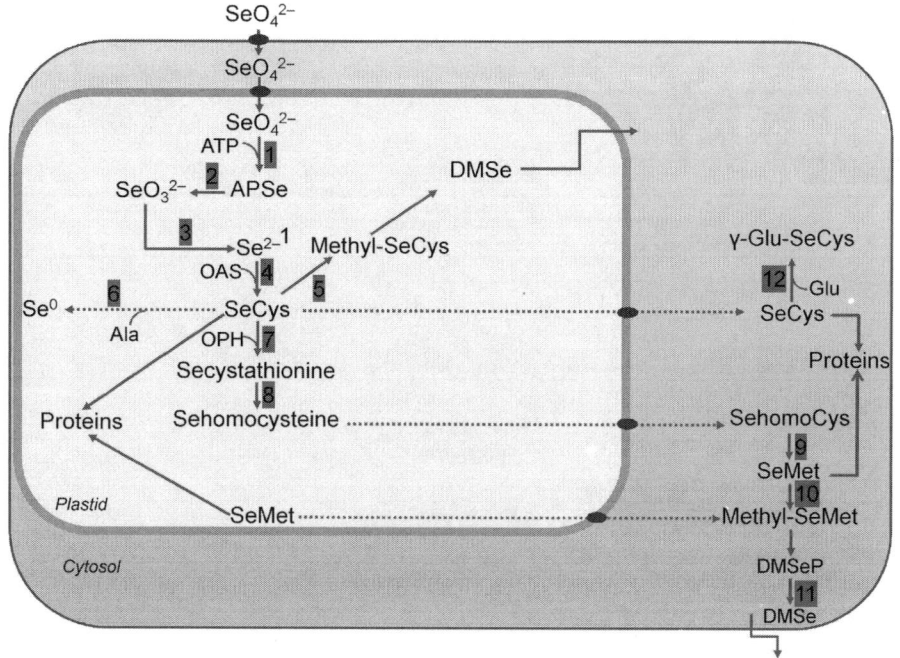

FIGURE 16.2

Schematic overview of Se metabolism in plants. APSe: adenosine phosphoselenate, OAS: O-acetylserine, OPH: O-phosphohomoserine, SeCys: selenocysteine, SeMet: selenomethionine, DMSeP: dimethyl selenoproprionate, DMSe: dimethylselenide. Numbers denote known enzymes. (1) ATP sulfurylase, (2) adenosine phosphosulfate reductase, (3) sulfite reductase (or glutathione), (4) OAS thiol lyase, (5) SeCys methyltransferase, (6) SeCys lyase, (7) cysthathionine-γ-synthase, (8) cysthathionine-β-lyase, (9) methionine synthase, (10) methionine methyltransferase, (11) DMSeP lyase, (12) γ-glutamylcysteine synthetase (Pilon-Smits and Quinn, 2010; Çakır et al., 2012).

From Pilon-Smits and Quinn, 2010.

Application of low concentrations of Si $(50-100\,\mu g\,g^{-1})$ to cowpea (*Vigna unguiculata*) increased the relative yield of root and shoot, nitrogen (N), phosphorus (P), and calcium (Ca) concentrations (Mali and Aery, 2008). It also significantly enhanced the growth, number, and fresh weight of nodules (Mali and Aery, 2008). Silicon is also reported to increase biomass production of wheat plants (Tahir et al., 2012), *Prosopis juliflora* (Bradbury and Ahmad, 1990), and *Zinnia elegans* (Kamenidou et al., 2009). Silicon (12 ml L^{-1} OryMax) application increased the internode diameter in rice by 17%, increased the internode wall thickness by 13%, and resulted in an Si content on dry weight basis of 13.7% in stems and 13.1% in husks (Nhan et al., 2012). The effects of Si $(0.31-80\,mg\,L^{-1})$ on growth and somatic embryogenesis were studied in *Phragmites australis*. Silicon had positive effects in terms of callus production, number of plants regenerated per callus, number of differentiated roots per callus, number of somatic embryos, increase in fresh weight, and the obtained results were clearly Si dose and genotype dependent (Máthé et al., 2012). Silicon

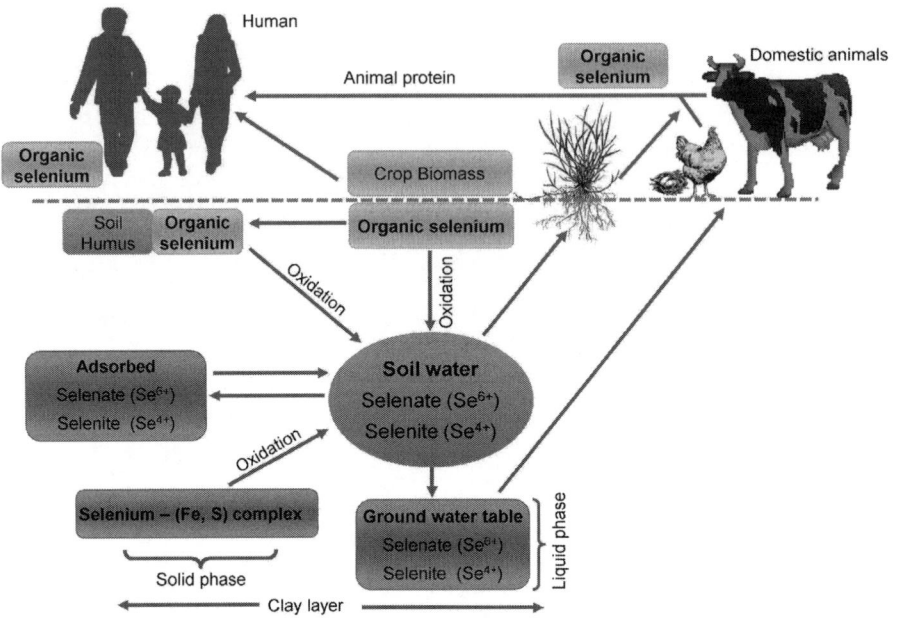

FIGURE 16.3

Selenium in the soil–plant–water consumer system. Plant roots take up SeO_4^{2-} or SeO_3^{2-} forms of Se from the soil water. The Se concentration in the soil solution depends on the solubility of the forms of Se present and the biological transformation of organic forms.

Adapted from Hasanuzzaman et al., 2010.

fertilizer (SiO_2) improved the N, P, Ca, magnesium (Mg), and S levels in soybean (*Glycine max*) plants. It also increased dry matter and number of plants per unit area. The same fertilizer increased white oat dry matter by 18% (Castro and Crusciol, 2013). Fresh weight and diameter of flower of *Paeonia lactiflora* Pall. were increased by Si (Zhao et al., 2013). Again, Si application increased the layer number of thickened sclerenchyma cells, thickness of cell walls, lignin content of inflorescence stems, cortex and xylem Si content, and the strength of stalks, and improved the quality of the cut flowers (Zhao et al., 2013).

The effects of Si on the reproductive development of plants are well recognized. In *G. max*, Si-deficient plants showed anomalies in pod development when compared to Si-supplemented plants (Miyake and Takahashi, 1985). Silicon application ameliorated abnormal flower development in tomato (Miyake and Takahashi, 1985). Silicon influenced inflorescence production in *Gerbera jamesonii* (Savvas et al., 2002). Ma and Yamaji (2008) reported that silicate stimulated the growth and development of flower and inflorescences of Gramineaceous plants. In *O. sativa*, Si influenced panicle development and grain fertility (Ma and Yamaji, 2008). Silicon suppressed anther dehiscence in rice by 85%. It also improved diameter and number of pollen grains in stigma and subsequent fertilization (Li et al., 2005). Silicon also improves photosynthesis and other physiological processes. Silicon-induced increases in water use efficiency were reported in rice (Nwugo and Huerta, 2008) and cucumber (Feng et al., 2010). Addition of Si increased the chlorophyll (chl) content and ribulose bisphosphate carboxylase (Rubisco) activity in cucumber (Adatia and Besford, 1986). Significant increases in growth and photosynthetic parameters, including SPAD value, net photosynthetic rate, cellular CO_2 concentration, stomatal conductance (g_s), transpiration rate, and chl fluorescence efficiency, were recorded in barley following Si (1 and 2 mM) application (Ali et al., 2013). Stomatal movement is variously affected by Si. Besides promoting rapid and elevated stomatal movement under normal growth conditions, under stress conditions, Si is supposedly incorporated into epidermal tissue, where it may reduce the diameter of stomatal pores and thus control stomatal movement, which might be beneficial under certain stress conditions (Efimova and Dokynchan, 1986), possibly by reinforcement of cell walls (Epstein, 1999). Thus, Si treatment has been confirmed to reduce stomatal resistance in some crops like *Vitis vinifera* L. (Soylemezoglu et al., 2009), rice (Yeo et al., 1999), and tomato (Romero-Aranda et al., 2006) under salinity stress. When *Z. mays* plants were treated with nine levels of Si (0, 0.4, 0.8, 1.2, 1.6, 2.0, 2.4, 2.8, and 3.2 mM) improved gas exchange and photosynthetic attributes were noted, along with improved growth relative to the control. Si could improve transpiration, g_s, net CO_2 assimilation rate, and leaf substomatal CO_2 concentration quite remarkably. Silicon levels of 0.8, 1.6, and 2.8 mM showed better results overall (Parveen and Ashraf, 2010). In rice leaves, Si at 12 ml L^{-1} supplied as OryMax increased the chl a content by 43%, chl b content by 47%, and total carotenoid content by 47%; whereas, Si at 6 g L^{-1} supplied as SilySol increased these parameters by 45, 54, and 15%, respectively (Nhan et al., 2012). Si promoted the development of an erect plant and improved photosynthetic efficiency (Pulz et al., 2008). Silicon deposition decreased evapotranspiration water loss (Ma and Yamaji, 2006). Silicon increases a plant's capacity for transpiration, thereby allowing cooling and aiding bulk flow of soluble nutrients from the soil (Soylemezoglu et al., 2009).

16.5 Effect of silicon and selenium in improving yield of crop plants

The improvement in yields of different crop plants due to exogenous application of Si has been reported by many authors (Rafi and Epstein, 1999; Kim et al., 2012; Castro and Crusciol, 2013). Silicon deficiency reduced pod yield in *G. max* (Miyake and Takahashi, 1985), while Si application increased the yield of tomato in terms of fruit number and fruit weight (Miyake and Takahashi, 1985). Effects of SiO_2 fertilizer were studied on different growth and yield attributes of *G. max*, *T. aestivum*, *A. sativa*, and *Z. mays*. When compared to control *G. Max* plants, the pods per plant increased by 8%, grains per pod by 19%, mass of 100 grains by 21%, and total grain yield by 49% in response to Si. In white oat, Si increased the yield attributes and yield as shown by a 15% increase in panicle number m^{-2}, an 11% increase in grain-filled spikelets panicle^{-1}, a 6% increase in the mass of 1000 grains, and a 70% increase in grain yield. The yield attributes of maize were increased by Si, with increases in grains ear^{-1} and grain yield of 23 and 18%, respectively (Castro and Crusciol, 2013). Silicon also increased yields in *T. aestivum* (Rafi and Epstein, 1999), *S. officinirum* (Savant et al., 1999), and *H. vulgare* (Liang, 1999). In rice, Si (12 mL L^{-1} OryMax) increased the number of panicles pot^{-1} (7%), the number of filled grains panicle^{-1} (22%), the filled grain ratio (4%), the weight of 1000 grains, and the yield pot^{-1} (15%) (Nhan et al., 2012). Kim et al. (2012) also found positive effects of Si supplementation in rice; Si application reduced the lodging index (13.7%) and increased plant height (12.2$-\approx$16.7%), pushing resistance

(10.5− ≈ 13.8%), and yield up to 15.1%. Addition of Si to the nutrient solution increased the shelf-life, and improved the quality and quantitative yield of *Valerianella locusta* (Gottardi et al., 2012).

A beneficial role of Se has been observed in plants that are capable of accumulating large amounts of this element (Shanker, 2006). However, the role of Se is mostly dose dependent (Hasanuzzaman et al., 2010, 2012b). Hamilton (2004) reported three levels of biological activity of Se: (1) trace concentrations required for normal growth and development; (2) moderate concentrations that can be stored to maintain homeostatic functions; and (3) elevated concentrations that can result in toxic effects. In ryegrass (*Lolium perenne*) and lettuce (*Lactuca sativa*), Se exerted beneficial effects at low concentrations (0.1 mg kg^{-1} soil), while it showed toxic effects at high concentrations (> 10 mg kg^{-1} and 1.0 mg kg^{-1}, for ryegrass and lettuce, respectively) (Hartikainen et al., 2000; Xue et al., 2001). The positive growth responses of plants to Se added at small concentrations have been attributed to the antioxidative effect of Se, which counteracts oxidative stress (Seppänen et al., 2003). Selenium also has a demonstrated effect on seed germination. Carvalho et al. (2003) reported that elevated concentration of Se (> 29 mg kg^{-1} soil) inhibited the growth and germination of *S. lycopersicum* and *Raphanus sativus* seeds. In contrast, priming of seeds with SeO_3^{2-} increased the germination of *Momordica charantia* seeds (Chen and Sung, 2001). Turakainen (2007) demonstrated that Se supplementation increased the carbohydrate accumulation in the young upper leaves and in stolons of potato plants, although this was not correlated with increased production of photoassimilates as net photosynthesis did not differ among Se treatments (Turakainen, 2007). Bekheta et al. (2008) observed marked increases in growth and synthesis of photosynthetic pigments (chl *a*, *b* and carotenoids) in *Gerbera jamesonii* following Se supplementation (5−20 mg L^{-1}). In our laboratory, we observed better phenotypic appearance of *B. napus* seedlings supplemented with exogenous Se (Figure 16.4).

16.6 Protective roles of silicon and selenium under abiotic stress

Silicon not only exerts beneficial effects on plant growth and productivity, it also plays a vital role in alleviating the damage caused by several abiotic stresses including salinity, drought, HT, chilling, UV radiation, HM toxicity, and nutrient imbalance (Liang et al., 2007; Ma and Yamaji, 2008; Table 16.4). The beneficial effects of Si on crops are depicted in Figure 16.5. These effects are achieved through the deposition of Si as $SiO_2 \bullet nH_2O$ in leaves and stems of plants. Some of the effects of Si are also mediated by the interaction between $Si(OH)_4$ and other elements such as Al. However, unlike other essential plant nutrients, the function of Si in plants might involve mechanical defense rather than physiological changes (Ma et al., 2001a). Like Si, Se has also been reported to play important protective roles in conferring tolerance to certain abiotic stresses like salinity, drought, HT, chilling, HMs, and UV irradiation (Hasanuzzaman et al., 2010, 2011b, 2012b; Table 16.5). One of the roles of Se in exerting beneficial effects on plant growth and stress tolerance is the enhancement of the antioxidant capacity (Hasanuzzaman et al., 2010, 2011, 2012b).

16.6.1 Salinity

The role of silicon in alleviating salt stress in plants has been studied widely in many plant species including *T. aestivum* (Ahmad et al., 1992; Saqib et al., 2008; Hashemi et al., 2010), *Z. mays*

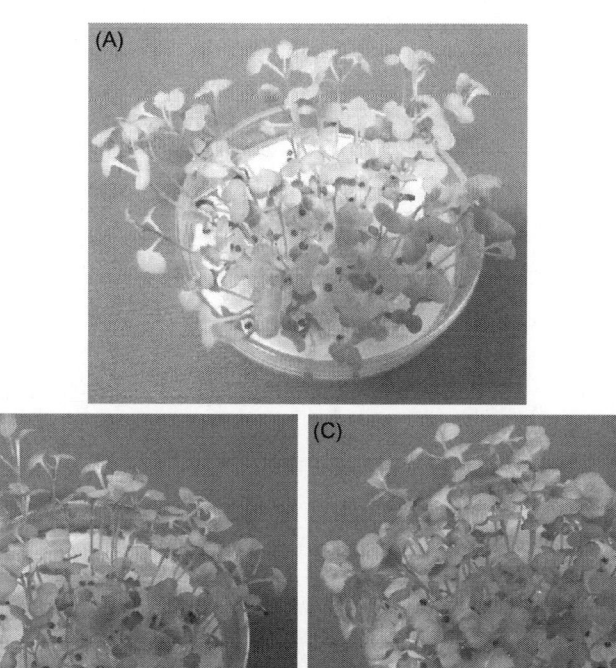

FIGURE 16.4

Phenotypic appearance of *B. napus* seedling supplemented with Se. (A) Control; (B) Pretreated with 50 μM Na₂SeO₄ (24 h); and (C) 100 μM Na₂SeO₄ (24 h). Twelve-day-old seedlings were pretreated with Se and then grown for more than 48 h.

(Liang et al., 2003), *L. esculentum* (Al-Aghabary et al., 2004), *O. sativa* (Gong et al., 2006), *C. sativus* (Zhu et al., 2004), *B. napus* (Hashemi et al., 2010), *S. officinarum* (Ashraf et al., 2010a,b), and *H. vulgare* (Liang et al., 2003). In *Spartina densiflora*, exogenous Si (500 μM) improved growth parameters, relative growth rate, and leaf elongation rate under salinity (680 mM NaCl), which was associated with higher net photosynthetic rate, greater water-use efficiency, and balanced nutrient concentrations (Mateos-Naranjo et al., 2013). Wheat plants supplemented with 0.25 and 0.50 mM Na₂SiO₃ and subsequently exposed to 100 mM NaCl showed amelioration of the negative effects of salinity by improvements in dry matter, chl content, and Pro content (Tuna et al., 2008). Shi et al. (2013) showed that Si (3 mM) could modulate various physiological processes that improved the growth of *O. sativa* under saline conditions (NaCl 50 mM). In canola seedlings (*cv*. Okapi), Si (2 and 4 mM K₂SiO₃) application increased leaf area, leaf fresh weight, and seed yield during salt stress conditions (100, 200, and 300 mM NaCl) (Bybordi, 2012).

Table 16.4 Beneficial Effects of Exogenous Application of Si in Plants Grown under Various Abiotic Stresses

Plant Species	Si Treatment	Stressors	Protective Effects	References
Saccharum officinarum	2 mM Ca_2SiO_4	Salinity, 100 mM NaCl	Reduced tissues Na^+ concentration Improved K^+ uptake, K^+/Na^+ ratios, and Ca^{2+} content Increased shoot and root dry matter	Ashraf et al. (2010a)
Vitis vinifera	4 mM Si ($Na_2Si_3O_7$)	Salinity, 20 mM NaCl	Reduced stomatal resistance Reduced MDA and H_2O_2 contents Increased APX activity	Soylemezoglu et al. (2009)
Brassica napus	2 and 4 mM K_2SiO_3	Salinity, 300 mM NaCl	Increased leaf area, leaf fresh weight, seed yield, and photosynthesis Increased APX and NR activities Increased chl content	Bybordi (2012)
Zea mays	0.4, 0.8, 1.2, 1.6, 2.0, 2.4, 2.8 and 3.2 mM Si $(OH)_4$	Salinity, 150 mM NaCl	Improved growth Increased net CO_2 assimilation rate, g_s, transpiration, and leaf substomatal CO_2 concentration	Parveen and Ashraf (2010)
Spartina densiflora	500 µM Na_2SiO_3	Salinity, 171 and 680 mM NaCl	Improved growth associated with higher net photosynthetic rate, water-use efficiency, and g_s Reduced tissue Na^+ content	Mateos-Naranjo et al. (2013)
Triticum aestivum	Si 50 mg kg^{-1} and 150 mg kg^{-1}	Drought, 50%, 75%, and 100% of field capacity (FC)	Increased plant biomass, plant height, and spike weight Increased tissue Si concentration and uptake	Ahmad et al. (2007)
Zea mays	0.8 mM Na_2SiO_3	Drought, 50% water deficit	Increased dry mass, tissue nutrient content, and water-use efficiency Reduced leaf wilting	Janislampi (2012)
Glycine max	1.70 mM Na_2SiO_3	Drought, −0.5 MPa	Increased root and shoot dry matter and the ratio of root/shoot Increased leaf water potential and growth Increased net photosynthetic rate, g_s, and chl content Decreased free Pro content, lipid peroxidation, and electrolyte leakage	Shen et al. (2010)

(Continued)

Table 16.4 (Continued)

Plant Species	Si Treatment	Stressors	Protective Effects	References
Sorghum bicolor	200 mg L^{-1} Si	Drought, withholding irrigation	Increased net photosynthetic rate Decreased shoot to root ratio by increasing root growth	Ahmed et al. (2011)
Oryza sativa	20, 40, 100 mg L^{-1} SiO$_2$	HT, 42°C	Reduced electrolyte leakage Increased level of polysaccharide in cell wall with an increased ratio of polysaccharide in cell wall to total carbohydrate	Agarie et al. (1998)
Oryza sativa	1.5 mM Na$_2$SiO$_3$	HT, 39°C	Increased number of pollen grains than those with higher diameter Increased anther dehiscence percentage, pollination, and fertilization	Li et al. (2005)
Cucumis sativus	0.1 and 1 mM K$_2$SiO$_3$	Chilling (15/8°C)	Reduced leaves withering Increased activities of SOD, GPX, APX, MDHAR, GR, and GSH Increased AsA content Decreased levels of H$_2$O$_2$, O$_2^{\bullet-}$, and MDA	Liu et al. (2009)
Triticum aestivum	0.1 and 1.0 mM K$_2$SiO$_3$	Freezing, -5°C	Increased leaf water content Improved activities of antioxidant enzymes AsA, GSH, SOD, and CAT Reduced H$_2$O$_2$, free Pro and MDA content	Liang et al. (2008)
Citrus limon	50, 150, and 250 mg L^{-1} K$_2$SiO$_3$	Freezing, 0.5°C for 28 d	Increased phenolics and flavonoids concentration Improved fruit quality Reduced chilling injury	Mditshwa et al. (2013)
Oryza sativa	1.25 and 2.5 mM Na$_2$SiO$_3$	HM, 100 μM K$_2$Cr$_2$O$_7$	Increased seedling height, dry biomass, and soluble protein content Reduced Cr uptake and translocation Improved antioxidant defense	Zeng et al. (2011)
Brassica chinensis	1.5 mM K$_2$SiO$_3$	HM, 0.5, and 5 mg L^{-1} Cd	Increased shoot and root biomass Decreased Cd uptake and root-to-shoot transport Increased SOD, CAT, APX, reduced MDA and H$_2$O$_2$ concentrations	Song et al. (2009)

(Continued)

Table 16.4 (Continued)

Plant Species	Si Treatment	Stressors	Protective Effects	References
Cucumis sativus	1 mM Na_2SiO_3	HM, 100 μM $CdCl_2$	Reversed chlorosis, protected the chloroplast from disorganization	Feng et al. (2010)
			Increased pigment contents, intercellular CO_2 concentration, g_s, and net photosynthetic rate	
			Improved water-use efficiency	
Zea mays	1 mM Si as Si $(OH)_4$	HM, 200, or 500 μM $MnSO_4$	Ameliorated chloroplast damage and photoinhibition	Doncheva et al. (2009)
			Improved detoxification and compartmentation of Mn	
Picea abies	0.2, 0.5, and 1.0 mM Si	HM, 0.2, 0.5, and 1.0 mM Al	Ameliorated adverse effects on cell wall thickening, degree of vacuolation, and the degeneration of mitochondria, Golgi bodies, endoplasmic reticulum, and nucleus	Prabagar et al. (2011)
			Reduced cell death	
Vitis vinifera	4 mM $Na_2Si_3O_7$	HM, 20 mg kg^{-1} H_3BO_3	Reduced tissue B concentration	Soylemezoglu et al. (2009)
			Increased activities of CAT and APX	
			Reduced Pro, H_2O_2, and MDA content	
Oryza sativa	Si fertilizer $(CaSiO_3)$ @ 40 g m^{-2}	UV-B radiation 250 to 350 nm	Improved cell walls of sclerenchyma, vascular bundle sheath and metaxylem vessel cells, cellulose, non-cellulosic polysaccharides, lignin, and phenolic acids	Goto et al. (2003)
Glycine max	1.70 mM Si	UV-B radiation 290–320 nm	Increased root and shoot dry weight and their ratio	Shen et al. (2010)
			Increased net photosynthetic rate and g_s	
			Decreased H_2O_2 content	

Al-Aghabary et al. (2004) suggested that Si enhanced the photochemical efficiency of tomato plants under salinity stress. Silicon could also improve plant defense systems and help to detoxify ROS induced by salt stress; this helps to increase chl and enhances the photochemical efficiency of photosystem II (PSII). At 150 mM NaCl, exogenously applied Si levels (0.8, 1.6, and 2.8 mM) in the rooting medium improved net CO_2 assimilation rate, g_s, transpiration, and leaf substomatal CO_2 in *Z. mays,* which was positively correlated with plant photosynthetic and growth attributes (Parveen and Ashraf, 2010). Barley root tonoplasts maintained integrity, stability, and functions

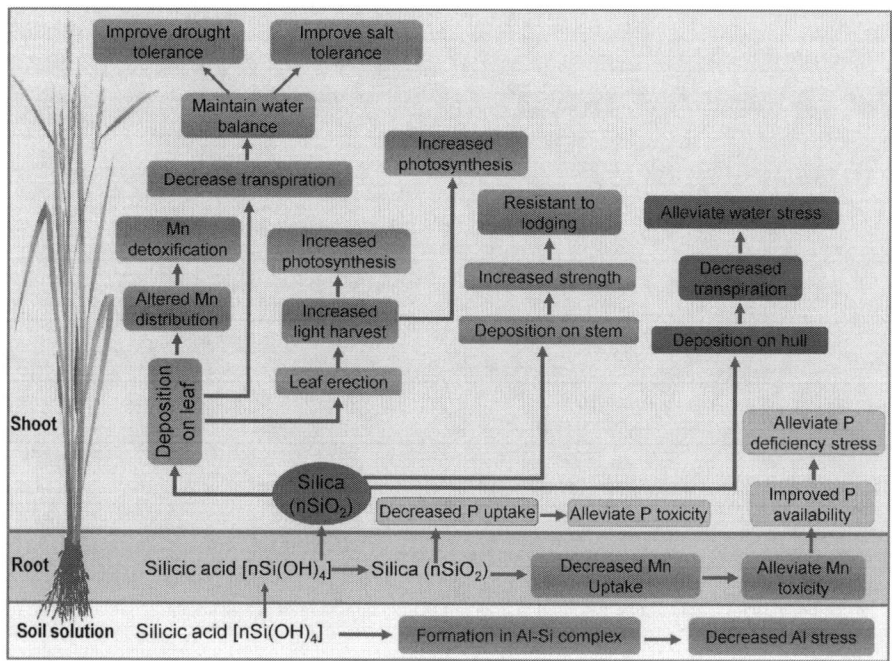

FIGURE 16.5

Silicon-induced improvement in plant processes towards abiotic stress tolerance.

following Si mediation of salt stress (Liang et al., 2005). In *Spartina densiflora*, Si (500 μM) application improved photosynthesis under 680 mM NaCl stress, which was confirmed by higher net photosynthetic rate and greater water-use efficiency. Si also exerted beneficial effects on the photochemical (PSII) apparatus and chl concentrations. Moreover, when compared to salt stress alone, Si supplementation of plants exposed to NaCl resulted in higher g_s that maintained higher intercellular CO_2. Silicon addition alleviated the adverse effects of salinity on quantum efficiency of PSII (Mateos-Naranjo et al., 2013). Addition of 3 mM silicate under 50 mM NaCl stress improved net photosynthetic rate, g_s, and transpiration in rice (Shi et al., 2013).

Motomura et al. (2002) indicated that the undesirable effects of Na during salt stress could be reduced by increasing the K/Na ratio in plants. In salt-stressed *H. vulgare*, Si was reported to activate the root plasma membrane H-ATPase pump, which helped to increase K conductivity and to improve the K/Na ratio (Liang et al., 2006a). Silicon also reduced the concentration of Na and Cl in *H. vulgare* (Inal et al., 2009) and grapevine (Soylemezoglu et al., 2009). One of the ways to maintain optimum cytosolic K^+/Na^+ ratio is via restriction of Na^+ influx into root cells or into the xylem stream (Chinnusamy et al., 2005). Silicate crystals deposited in the epidermal cells create a barrier to ion movement and their balance is managed (Romero-Aranda et al., 2006; Ashraf et al., 2010a,b). In the endodermis and rhizodermis, deposition and polymerization of Si block Na^+ influx through the apoplasmic pathway, as documented in the roots of *O. sativa* (Yeo et al., 1999). Silicate crystals deposited in the epidermal cells create a barrier to water loss through cuticles,

Table 16.5 Beneficial Effects of Exogenous Application of Se in Plants Grown under Various Abiotic Stresses

Plant Species	Se Treatment	Stressors	Protective Effects	References
Brassica napus	25 μM Na_2SeO_4, 48 h	Salinity, 100 and 200 mM NaCl, 48 h	Increased AsA and GSH contents; GSH/GSSG ratio; and activities of APX, MDHAR, DHAR, GR, GST, GPX, and CAT Reduced levels of H_2O_2 and MDA	Hasanuzzaman et al. (2011)
Cucumis sativus	5, 10, or 20 μM Na_2SeO_4, 11 d	Salinity, 50 mM NaCl, 11 d	Decreased content of Cl^- in the shoot tissues Increased Pro accumulation Enhanced antioxidant defense Decreased lipid peroxidation	Hawrylak-Nowak (2009)
Sorrel (*Rumex patientia* × *R. tianshanicus*)	1–5 μM Na_2SeO_4, 43 d	100 mM NaCl, 43 d	Stimulated growth Increased activities of SOD and POD Increased accumulation of water-soluble sugar	Kong et al. (2005)
Trifolium repens	5 μM Na_2SeO_4, 0–5 d	Drought, PEG-6000 (− 1.0 MPa), 0–5 d	Decreased levels of H_2O_2, TBARS, DHA, and GSSG Increased the levels of AsA and GSH and AsA/DHA and GSH/GSSG ratios Improved the activities of MDHAR, DHAR, and GR	Wang et al. (2011)
Brassica napus	15 and 30 g L^{-1} as Na_2SeO_3	Drought, limited irrigation at early stem elongation	Increased plant height, number of pods in plant, number of seeds in pod, seed yield, biological yield, harvest index, and oil yield	Zahedi et al. (2009)
Triticum aestivum	Se (Na_2SeO_3) 0.5 mg kg^{-1}, 20 d	Drought of 30% FC	Increased root activity Increased Pro content Increased activities of POD and CAT	Xiaoqin et al. (2009a)
Triticum aestivum	1.0, 2.0, and 3.0 mg kg^{-1} Na_2SeO_3, 20 d	Water stress of 30% FC, 20 d	Increased root activity	Xiaoqin et al. (2009b)

(Continued)

Table 16.5 (Continued)

Plant Species	Se Treatment	Stressors	Protective Effects	References
			Increased chl, carotenoids, and Pro content	
			Increased activities of POD and CAT	
			Decreased MDA content	
Fagopyrum esculentum	Solution of 1 g m^{-3} Na_2SeO_4 as foliar spray, 7–8 weeks	Reduction of water by 50%, 8 weeks	Improved g_s, potential photochemical efficiency of PSII, respiratory potential	Tadina et al. (2007)
			Increased yield	
Zea mays	20 g ha^{-1} Na_2SeO_4	Drought, withholding water	Enhanced activities of SOD	Sajedi et al. (2011)
			Reduced MDA content	
			Improved grain yield	
Triticum aestivum	Seed soaking in 0, 25, 50, 75, and 100 μM of Na_2SeO_4, 30 or 60 min	Drought, withholding water for 1 week	Increased root length and total biomass	Nawaz et al. (2013)
			Increased stress tolerance index	
			Increased total sugar content and total free amino acids	
Brassica napus	Pretreatment with 25 μM Se (Na_2SeO_4), 48 h	Drought, 10 and 20% PEG-6000, 48 h	Increased activities of APX, DHAR, MDHAR, GR, GST, GPX, and CAT	Hasanuzzaman and Fujita (2011b)
			Decreased GSSG content, H_2O_2	
			Decreased lipid peroxidation	
Sorghum bicolor	Foliar spray of Na_2SeO_4 (75 mg L^{-1}), 7–28 d	HT, 40/30°C, 7–28 d	Increased antioxidant enzyme	Djanaguiraman et al. (2010)
			Decreased ROS levels and membrane damage	
			Increased antioxidant defense	
			Increased grain yield	
Cucumis sativus	2.5, 5, 10, or 20 μM Na_2SeO_4	10°C/5°C for 24 h, for a further 24 h at 20°C/15°C, and then transferred to 25/20°C (re-warming) for 7 d	Improved shoot fresh weight	Hawrylak-Nowak et al. (2010)
			Increase of Pro content	
			Reduced MDA level	

(Continued)

Table 16.5 (Continued)

Plant Species	Se Treatment	Stressors	Protective Effects	References
Triticum aestivum	0.5, 1.0, 2.0, 3.0 mg kg^{-1} Na$_2$SeO$_3$, 72 h	4°C, 72 h	Increased biomass Increased chl, anthocyanins, flavonoids, and phenolic compounds Increased activities of POD and CAT	Chu et al. (2010)
Triticum aestivum	Seeds were soaked in Se (5 mg Se L^{-1}), 5, 10 and 15 h	3°C or 5°C for 14 d and allowed to recover at 22°C for 3 d	Enhanced growth Increased chl, anthocyanin, sugar, and Pro contents Enhanced antioxidant defense system Decreased membrane damage	Akladious (2012)
Sorghum bicolor	Seed soaking with Se (3 and 6 μM L^{-1} Na$_2$SeO$_4$), 6 h	LT, 4°C or 8°C for 7 d	Enhanced growth Increased levels of chl, anthocyanin, sugar, Pro, and AsA Increased enzymatic activities Diminished lipid peroxidation	Abbas (2012)
Brassica napus	Pretreatment with 50 and 100 μM Se (Na$_2$SeO$_4$), 24 h	HM, 0.5 and 1.0 mM CdCl$_2$, 48 h	Increased AsA and GSH contents and GSH/GSSG ratio Increased activities of APX, MDHAR, DHAR, GR, GPX, and CAT Reduced the MDA and H$_2$O$_2$ levels	Hasanuzzaman et al. (2012b)
Brassica napus	2 μM Na$_2$SeO$_4$, 14 d	HM, 400 and 600 μM, CdCl$_2$, 14 d	Reduced oxidative stress by modulating SOD, CAT, APX, and GPX activities Prevented Cd-induced alteration of DNA methylation pattern	Filek et al. (2008)
Pteris vittata	5, 10 μM of Na$_2$SeO$_4$, 5 and 10 d	HM, 150 or 300 μM of Na$_2$HAsO$_4$, 10 d	Improved antioxidant system including thiol and GSH levels Reduced As uptake	Srivastava et al. (2009)

(Continued)

Table 16.5 (Continued)

Plant Species	Se Treatment	Stressors	Protective Effects	References
Brassica oleracea	1 mg L^{-1} Na_2SeO_3, 10 and 40 d	HM, 1 mg L^{-1} $CdCl_2$, 40 d	Increased chl content Improved α-tocopherol level and reduced oxidative damage	Pedrero et al. (2008)
Lolium perenne	1.0, 1.5, 2.0, 5.0, and 10 μM Na_2SeO_3, 20 d	HM, 0.2 mM $AlCl_3$, 20 d	Improved POD activity Reduced $O_2^{\bullet-}$ and lipid peroxidation	Cartes et al. (2010)
Triticum aestivum	1.0 and 2.0 mg Se kg^{-1}, 8 h	UV-B, 40 W, 305 nm, 8 h	Increased root weight and root activity Increased flavonoids and Pro content Increased activities of POD and SOD Reduced MDA and $O_2^{\bullet-}$	Yao et al. (2009)
Euglena gracilis	10^{-7}, 10^{-8}, 10^{-9}, and 10^{-10} M, $Na_2SeO_3 \cdot 5H_2O$, 40 min	UV-A, 320–400 nm, 40 min	Improved light-enhanced dark respiration and photosynthesis	Ekelund and Danilov (2001)

which improves water relations in plant tissue in salt-stressed plants (Romero-Aranda et al., 2006). Supplementation of Si (1.0 mM) to the salt solution (120 mM NaCl) increased plasma membrane H^+-ATPase activity and restored membrane fluidity levels in *H. vulgare* (Liang et al., 2006a). The activity of the plasma membrane H-ATPase was improved by Si, which significantly increased K^+ uptake in barley under salt stress (Liang et al., 2003). Shoot K^+/Na^+ ratios in sugarcane were increased by 150 to 266% following Si treatment of saline growth medium, which boosted its tolerance to a great extent (Ashraf et al., 2010a). According to Mateos-Naranjo et al. (2013), Si may ameliorate nutrient imbalances under salinity stress conditions by maintaining higher concentrations of minerals in the tissues, specifically P. The K/Na ratio of leaves of *Spartina densiflora* was greater in Si-treated plants and these plants also had higher levels of essential nutrients (Si, Al, Cu, Fe, K, and P) in their tissues (Mateos-Naranjo et al., 2013).

Silicon was effective in improving the activity of antioxidative enzymes in different plants under salinity stress. In *B. napus*, Si (2 and 4 mM) increased growth and yield by improving photosynthesis, chl content, and enzyme activities including ascorbate peroxidase (APX) and nitrate reductase (NR) under salt stress (100, 200, and 300 mM NaCl) (Bybordi, 2012). Silicon improved *Vitis vinifera* L. shoot growth and reduced oxidative damage caused by salinity, as indicated by reduced membrane damage and H_2O_2 levels with associated elevation of catalase (CAT) and superoxide dismutase (SOD) activities. Silicon also decreased Pro content (Soylemezoglu et al., 2009). Addition of Si (1.0 mM Si) to the salt treatment (120 mM NaCl) increased the glutathione (GSH) concentration by 20% in a salt-tolerant (Jian 4) *H. vulgare* cultivar compared to 50% in a

Table 16.6 Malondialdehyde (MDA), H_2O_2, Reduced Ascorbate (AsA), Reduced Glutathione (GSH) and Oxidized Glutathione (GSSG) Contents in Rapeseed Seedlings in response to Se under Salt Stress Conditions

Treatment	MDA Content (nmol g^{-1} Fresh Weight)	H_2O_2 Content (nmol g^{-1} Fresh Weight)	AsA Content (nmol g^{-1} Fresh Weight)	GSH Content (nmol g^{-1} Fresh Weight)	GSSG Content (nmol g^{-1} Fresh Weight)
Control	25.58 d	3.70 d	5210.56 a	251.45 e	7.17 d
Na$_1$	43.34 b	6.50 ab	4064.22 b	432.56 c	13.48 b
Na$_2$	58.68 a	7.02 a	3141.51 c	358.68 d	16.95 a
Se	24.22 d	3.95 d	4890.04 a	261.28 e	8.14 cd
Se + Na$_1$	34.71 c	5.25 b	4973.40 a	568.79 a	9.43 c
Se + Na$_2$	43.29 b	6.06 c	3980.97 b	479.86 b	12.06 b

Na$_1$, Na$_2$, Se, Se + Na$_1$, and Se + Na$_2$ indicate 100 mM NaCl, 200 mM NaCl, 25 μM Na$_2$SeO$_4$, 100 mM NaCl + Na$_2$SeO$_4$, 200 mM NaCl + Na$_2$SeO$_4$ treatment, respectively. Values with different letters are significantly different at $P < 0.05$ applying LSD test.

salt-sensitive (Kepin No. 7) cultivar, which supposedly maintained better membrane properties (Liang et al., 2006b). Grapevine (*V. vinifera* rootstocks: 41 B, 1103 P) treated with Si (4 mM Si) significantly reduced MDA: by 20% in 41 B, by 23% in 1103 P, due to increased modulation of CAT, SOD, and APX enzyme activities and Pro content (Soylemezoglu et al., 2009).

Several research results have shown that Se at low concentration provided protection to different plant species against salt stress. Kong et al. (2005) reported that low concentrations (1−5 μM) of Se stimulated growth and enhanced antioxidant enzyme (SOD and POD) activities in leaves of sorrel (*R. patientia* × *R. tianshanicus*) seedlings under salt stress. In contrast, at higher concentrations (10−30 μM), Se showed fewer beneficial effects. In *C. sativus* leaves, Se treatments (5−10 μM) increased the growth, synthesis of photosynthetic pigments, and Pro levels under salt stress (Hawrylak-Nowak, 2009). In our study, we observed beneficial effects of exogenous Se (25 μM Na$_2$SeO$_4$) in *B. napus* seedlings exposed to salt stress (100 and 200 mM NaCl; Hasanuzzaman et al., 2011; Tables 16.6 and 16.7). Selenium treatment had a synergistic effect: in salt-stressed seedlings, it increased the AsA and GSH contents; and the activities of CAT, APX, MDHAR, DHAR, GR, GST, and GPX, which in turn reduced levels of H_2O_2 and MDA when compared to plants exposed to salt stress alone (Tables 16.6 and 16.7). The phenotypic appearance was also better in the Se-treated seedlings (Figure 16.6). These results suggested that exogenous application at low concentration rendered the plants more tolerant to salt stress-induced oxidative damage by enhancing their antioxidant defense systems (Hasanuzzaman et al., 2011).

16.6.2 Drought

For drought tolerance, one of the most important and basic points is conservation of water within the plant body. Silicon could affect this strategy precisely and differently in various cases. Silicon deposition in cuticles or endodermal cells can reduce water loss. In stems and leaves, Si is deposited as hydrated silica (SiO$_2$•nH$_2$O) by following the evapotranspiration path (Sangster et al., 2001);

Table 16.7 Activities of Antioxidant Enzymes in Rapeseed Seedlings in Response to Se under Salt Stress Conditions

Treatment	CAT Activity (μmol min^{-1} mg^{-1} Protein)	APX Activity (μmol min^{-1} mg^{-1} Protein)	MDHAR Activity (nmol min^{-1} mg^{-1} Protein)	DHAR Activity (nmol min^{-1} mg^{-1} Protein)	GR Activity (nmol min^{-1} mg^{-1} Protein)	GST Activity (nmol min^{-1} mg^{-1} Protein)	GPX Activity (nmol min^{-1} mg^{-1} Protein)
Control	28.52 a	0.387 d	47.08 b	153.36 b	26.41 b	23.85 d	0.1628 d
Na$_1$	17.49 cd	0.509 bc	42.29 b	126.68 c	29.86 b	37.28 bc	0.1915 bc
Na$_2$	15.78 d	0.535 b	33.65 c	99.65 d	21.53 c	33.38 c	0.1502 d
Se	25.10 b	0.453 c	46.59 b	139.38 bc	30.04 b	25.35 d	0.1667 cd
Se + Na$_1$	26.79 ab	0.656 a	57.00 a	171.56 a	39.77 a	44.67 a	0.2317 a
Se + Na$_2$	20.75 c	0.628 a	47.69 b	142.72 bc	30.56 b	39.09 b	0.2026 b

Na$_1$, Na$_2$, Se, Se + Na$_1$, and Se + Na$_2$ indicate 100 mM NaCl, 200 mM NaCl, 25 μM Na$_2$SeO$_4$, 100 mM NaCl + Na$_2$SeO$_4$, 200 mM NaCl + Na$_2$SeO$_4$ treatment, respectively. Values with different letters are significantly different at $P < 0.05$ applying LSD test.

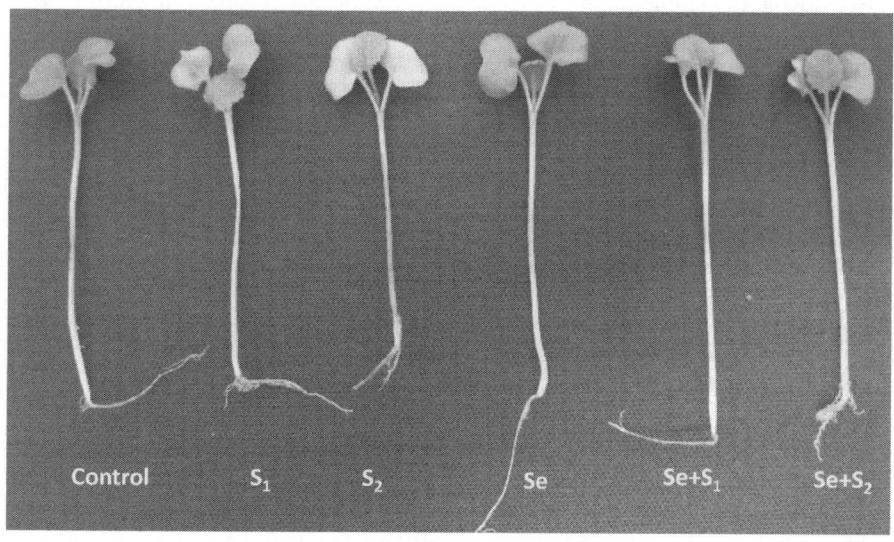

FIGURE 16.6

Phenological appearance of rapeseed seedlings in response to Se under salt stress conditions. S$_1$, S$_2$, Se, Se + S$_1$ and Se + S$_2$ indicate 100 mM NaCl, 200 mM NaCl, Na$_2$SeO$_4$, 100 mM NaCl$^+$ Na$_2$SeO$_4$, 200 mM NaCl$^+$ Na$_2$SeO$_4$ treatment, respectively (Hasanuzzaman et al., 2011).

thus, irrigation with Si reduces evapotranspiration (Gao et al., 2006). In *O. sativa*, Si deposits are 0.1 μM thick (Ma and Takahashi, 2002) and Si could reduce the transpiration rate by 30% (Ma, 2004). Polar monosilicic acid and polymerized $Si(OH)_4$ accumulated mainly in the epidermal cell walls and formed H-bonds between H_2O and $SiO_2 \cdot nH_2O$, which might prevent water loss (Liang et al., 2008). Kaya et al. (2006) found that Si (2 mM Na_2SiO_3) increased leaf relative water content (RWC) by 26.5% in water-stressed corn (50% of FC). In *Z. mays*, Si decreased leaf transpiration under drought, thereby improving leaf water status (Gao et al., 2006). Chen et al. (2011) found that applying 1.5 mM Si to rice under drought-stress conditions increased transpiration rate by 19% in a drought-susceptible line and 53% in a drought-resistant line.

Generally, compared with normal plants, drought-adapted plants have well-developed root systems through which they uptake even the slightest amount of water from deeper soils. In *S. bicolor*, Si fertilization significantly improved the drought tolerance by promoting the formation of a good root system and water uptake (Sonobe et al., 2011). Proline, a key solute for maintaining osmotic adjustment under drought stress, has been increased by Si application (Gunes et al., 2008; Crusciol et al., 2009). Crusciol et al. (2009) report that the Si increased potato plant Pro levels. Kaya et al. (2006) found that 2 mM Na_2SiO_3 decreased 43% of Pro content relative to non-treated drought plants. Pulz et al. (2008) found that the use of calcium and magnesium silicates markedly decreased stem lodging and increased potato plant height and tuber yield in drought conditions (-0.05 MPa water potential). Shen et al. (2010) observed a reduced Pro level in soybean plants subjected to PEG stress (-0.5 MPa). Sorghum plants treated with Si (200 ml L^{-1}) showed reduced drought damage and increases in leaf area index of 65%, specific leaf weight of 12%, leaf dry weight of 164%, shoot dry weight of 80%, root dry weight of 70%, and total dry weight of 75% (Ahmed et al., 2011). Silicon treatment under water deficit conditions resulted in 18, 17, 20, and 30% increases in dry mass in *Z. mays*, *T. aestivum*, *G. max*, and *O. sativa* (Janislampi, 2012).

Besides the positive effects on plant growth and development under drought, Si was effective in improving a number of physiological processes including photosynthesis, g_s, water relations, and nutrient translocation. Si (1.70 mM) addition to drought (-0.5 MPa) affected soybean increased net photosynthetic rate by 21% and g_s by 38%, while reducing transpiration by 29% after a week of drought, when compared to drought stressed plants. Increases in photosynthesis are supposedly associated with increases in activities of photosynthetic enzymes, chl content, and anthocyanin content observed in response to added Si (Shen et al., 2010). Silicon addition to *S. bicolor* increased g_s and transpiration rate and alleviated the photosynthetic reduction caused by drought (Hattori et al., 2005). In *S. bicolor*, drought caused significant damage to different growth parameters of roots, shoots, leaves, and overall growth, but addition of Si (200 ml L^{-1}) alleviated these negative effects by increasing chl contents by 52%, transpiration rate by 80%, and net photosynthetic rate by 62%; together, these changes resulted in a 75% higher total dry weight (Ahmed et al., 2011). Under drought stress, Si upholds the nutrient balance when compared to drought-stressed plants without Si (Gunes et al., 2008). Calcium and magnesium silicate fertilizer application increased P and Mn content and reduced N content in drought-stressed potato (Pulz et al., 2008). Sonobe et al. (2010) found that 1.78 mM Si (SiO_2) increased sorghum root amino acid content and decreased tissue Ca content under drought imposed for 23 days by 15% PEG-6000 (-0.6 MPa) (Janislampi, 2012).

Silicon can enhance the antioxidative machinery through effects on both its enzymatic and nonenzymatic components, thereby reducing drought-induced oxidative stress. Silicon enhanced the stability of lipids and protected the structural and functional integrity of cell membranes in rice plants against drought (Agarie et al., 1998). Supplementation of drought-stressed *Cicer arietinum* with Si significantly increased activity of some antioxidant enzymes (SOD, CAT, APX) in shoots as well as some non-enzymatic antioxidant activity, which reduced lipid peroxidation (MDA), lipoxygenase (LOX) activity, and Pro and H_2O_2 accumulation (Gunes et al., 2007a). Silicon (1.7 mM Si) application significantly reduced the H_2O_2 levels and thus reduced membrane damage as indicated by reduced lipid peroxidation and osmolyte leakage in drought (-0.5 MPa)-affected *G. max* plants. The probable reason for this alleviation of oxidative damage was modulation of SOD, CAT, and peroxidase (POD) activity induced by Si (Shen et al., 2010). Gong et al. (2005) also observed Si (2.11 mM Na_2SiO_3) increased SOD and CAT activity in wheat under drought, which alleviated oxidative damage and improved other indicators of physiological status. According to Gunes et al. (2008), Si applied to the soil reduced sunflower tissue H_2O_2 content.

Several plant studies that focused on the protective effects of Se in under drought stress indicated that the effects of Se are due to its ability to regulate the water status of plants under water deficit. Kuznetsov et al. (2003) reported that the addition of 0.1 or 0.25 mM Se caused a 2–6% increase in leaf water content, thereby increasing the drought resistance. The Se-induced improvement in leaf tissue water status was accompanied by a sharp (two- to four-fold) inhibition of stress-induced accumulation of Pro and a significant inhibition of POX activity (Kuznetsov et al., 2003). These results support the suggestion that Se exerts an antioxidant effect via a decreased concentration of intracellular ROS, which induces the *de novo* biosynthesis of Pro and POX production. The Se-induced effect on the content of intracellular water was also less pronounced at the optimal rate of water supply. In buckwheat (*Fagopyrum esculentum*), Tadina et al. (2007) observed that plants under water deficit exhibited significantly lower g_s, while Se supplementation significantly improved g_s. Selenium also improved the efficiency of PSII, which was attributed to the improvement in plant water status. Yao et al. (2009) suggested that low concentrations of Se (1.0, 2.0, and 3.0 mg Se kg^{-1}) favored the growth of *T. aestivum* seedlings under drought. In addition, Se supplementation increased the Pro, carotenoid, and chl content and the activities of antioxidant enzymes (namely, POD and CAT), which in turn decreased the MDA content. In *B. napus* affected by late season drought stress, foliar application of Se had significant and additive effects on vegetative and reproductive parameters. It increased plant height and improved pollen survival and fertilization, which resulted in higher numbers of pods and seeds; ultimately, the seed yield, biological yield, harvest index, and oil yield improved (Zahedi et al., 2009). While studying *T. aestivum* seedlings under drought conditions, Xiaoqin et al. (2009a) observed 12, 15, and 12% increases in the fresh weight of the shoot, root, and total biomass, respectively, following supplementation with exogenous Se (0.5 mg kg^{-1}). Selenium-induced drought tolerance was also indicated by the decrease in the shoot/root ratio and enhanced activities of POD and CAT (Xiaoqin et al., 2009a). In another study, Xiaoqin et al. (2009b) showed that exogenous Se (1.0 and 2.0 mg kg^{-1}) increased root activity, Pro content, and POD and CAT activities and reduced the MDA content of *T. aestivum* seedlings. Selenium application at 1.0 and 2.0 mg kg^{-1} increased the chl ($a + b$) content by 26 and 32%, carotenoid content by 6 and 9%, and total biomass by 12 and 36%, respectively. Valadabadi et al. (2010) observed that drought stress markedly reduced the physiological and growth indices (namely, dry weight, leaf area index, relative growth rate, and crop growth rate) of *B. napus*, while application of Se (30 g L^{-1}) improved those indices and mitigated the stress-induced damage.

Wang et al. (2011) examined the effect of Se (5 μM Na_2SeO_4) on the AsA–GSH cycle in *Trifolium repens* seedlings subjected to PEG-induced water deficit. They observed that Se

Table 16.8 Malondialdehyde (MDA), H_2O_2 Content, Reduced Ascorbate (AsA), Reduced Glutathione (GSH), and Oxidized Glutathione (GSSG) in Rapeseed Seedlings Induced by Se under Drought Stress Conditions

Treatment	MDA Content (nmol g^{-1} Fresh Weight)	H$_2$O$_2$ Content (nmol g^{-1} Fresh Weight)	AsA Content (nmol g^{-1} Fresh Weight)	GSH Content (nmol g^{-1} Fresh Weight)	GSSG Content (nmol g^{-1} Fresh Weight)
Control	21.16 c	4.44 c	4970.52 d	283.31 d	8.48 d
D$_1$	30.77 b	5.92 b	5923.34 bc	439.93 bc	18.26 c
D$_2$	42.06 a	7.36 a	5203.01 d	414.97 c	32.61 a
Se	19.87 c	4.69 c	5476.88 cd	269.73 d	9.84 d
Se + D$_1$	23.42 c	5.00 c	7006.01 a	514.72 a	17.61 c
Se + D$_2$	29.46 b	5.82 b	6295.85 b	467.83 b	21.52 b

D$_1$, D$_2$, Se, Se + D$_1$, and Se + D$_2$ indicate 10% PEG, 20% PEG, 25 μM Na$_2$SeO$_4$, 10% PEG + Na$_2$SeO$_4$, and 20% PEG + Na$_2$SeO$_4$ treatment, respectively. Values with different letters are significantly different at P < 0.05 applying LSD test.

application decreased the contents of H_2O_2, TBARS, DHA, and GSSG; increased the levels of GSH and AsA; and inhibited the decreases of AsA/DHA and GSH/GSSG ratios when compared to control plants. Selenium supplementation significantly increased the activities of MDHAR, DHAR, and GR. Among the enzymes, GR showed the highest increase in activity compared to DHAR and MDHAR. In our laboratory, we studied the beneficial role of Se pretreatment (25 μM Na$_2$SeO$_4$, 48 h) in *B. napus* seedlings under drought stress (10 and 20% PEG-6000) (Hasanuzzaman and Fujita, 2011b). The seedlings exposed to drought stress showed significant increases in GSH and GSSG content; however, the AsA content increased only under moderate stress (Table 16.8). The MDHAR and GR activity increased only under moderate stress (10% PEG). The activities of DHAR, GST, and GPX significantly increased at all levels of drought, while CAT activity decreased (Table 16.9). Drought stress resulted in a marked increase in the levels of H_2O_2 and MDA (Table 16.8). In contrast, Se-pretreated seedlings exposed to drought stress showed a rise in AsA and GSH content, and up-regulated activities of CAT, APX, DHAR, MDHAR, GR, GST, and GPX when compared with the drought-stressed seedlings without Se. In turn, the Se-treated seedlings showed a considerable decrease in the levels of H_2O_2 and MDA and considerable alleviation of oxidative stress (Tables 16.8 and 16.9; Hasanuzzaman and Fujita, 2011b). Nawaz et al. (2013) found beneficial role of Se priming in conferring drought stress tolerance. In their experiment, seeds of *T. aestivum* were soaked in distilled water or Na$_2$SeO$_4$ solutions (25, 50, 75, and 100 μM) for 30 or 60 min, followed by re-drying and subsequent sowing. Priming with Se significantly increased root length, stress tolerance index, and total biomass of germinated seedlings.

16.6.3 High temperature

Reports on the effects of Si on HT stress are scare. However, some findings have indicated that Si exerts a beneficial effect under heat stress. Agarie et al. (1998) observed that relative to *O. sativa* plants subjected to heat stress alone, plants grown in 100 ppm SiO_2 showed significantly reduced

Table 16.9 Activities of Antioxidant Enzymes in Rapeseed Seedlings Induced by Se under Drought Stress Conditions

Treatment	CAT Activity (μmol min⁻¹ mg⁻¹ Protein)	APX Activity (μmol min⁻¹ mg⁻¹ Protein)	MDHAR Activity (nmol min⁻¹ mg⁻¹ Protein)	DHAR Activity (nmol min⁻¹ mg⁻¹ Protein)	GR Activity (nmol min⁻¹ mg⁻¹ Protein)	GST Activity (nmol min⁻¹ mg⁻¹ Protein)	GPX Activity (nmol min⁻¹ mg⁻¹ Protein)
Control	24.98 a	0.344 bc	37.40 c	201.73 e	33.67 c	45.17 c	0.1177 d
D_1	18.79 b	0.366 b	41.09 b	233.38 cd	41.54 b	56.52 b	0.1390 c
D_2	16.02 b	0.307 c	34.42 c	261.90 bc	39.25 bc	59.24 b	0.1402 c
Se	25.52 a	0.343 bc	37.48 c	212.83 de	36.52 bc	49.98 c	0.1306 c
Se + D_1	25.32 a	0.432 a	45.30 a	270.46 b	54.48 a	67.39 a	0.1748 a
Se + D_2	23.68 a	0.453 a	44.66 a	314.15 a	51.89 a	69.73 a	0.1595 b

D_1, D_2, Se, Se + D_1, and Se + D_2 indicate 10% PEG, 20% PEG, Na_2SeO_4, 10% PEG + Na_2SeO_4, and 20% PEG + Na_2SeO_4 treatment, respectively. Values with different letters are significantly different at $P < 0.05$ applying LSD test.

heat stress symptoms, determined by a 2.5-fold reduction in stress-induced electrolyte leakage. This result suggested that Si might have beneficial effects in maintaining thermal stability of lipids in cell membranes, thereby helping to maintain structural and functional integrity of cell membranes. Besides decreasing the electrolyte leakage, Si addition also increased the levels of polysaccharide in the cell walls, indicated by an increased ratio of cell wall polysaccharide to total carbohydrate. Thus, Si also helped to improve the cell wall structure (Agarie et al., 1998). In *O. sativa*, Si supplementation increased the number and diameter of pollen grains relative to heat treatment (39°C) alone (Li et al., 2005). Silicon application resulted in a dramatic increase in partial dehiscence (135% higher) of the anthers of rice flowers relative to heat stress alone. This consequently developed into a marked increase in fully dehiscence anthers, which was 111% higher in plants in the Si-fertilized treatment when compared to plants exposed to heat stress alone. Anther cracking rate was increased by 130% and the stigma pollination probability was increased by 66%, which enhanced the fertilization of rice flowers (Li et al., 2005). In *Euphorbia pulcherrima* Willd. "Ichiban," Si-treated (50 mg L⁻¹ Si, either as K_2SiO_3, Na_2SiO_3, or $CaSiO_3$) plants were more tolerant to HT (35 ± 1°C) compared to control plants (Son et al., 2011).

Very few reports have also appeared regarding the protective role of Se under HT. Djanaguiraman et al. (2010) investigated the effects of Se foliar spray (75 mg L⁻¹) on leaf photosynthesis, membrane stability, antioxidant enzyme activity, grain yield, and yield components in *S. bicolor* plants grown under HT stress (40°C /30°C). They observed that HT stress decreased chl content, chl *a* fluorescence, photosynthetic rate, and activities of antioxidant enzymes. Heat stress also promoted oxidative stress and membrane damage, which ultimately affected the grain yield negatively. However, Se supplementation prevented membrane damage and improved the antioxidant defense system, which helped in attaining a higher grain yield. Overall, Se application significantly increased photosynthetic rate, g_s, and transpiration rate by 13.2, 12.4, and 8.11%, respectively, compared with the unsprayed control. In addition, foliar spray of Se significantly reduced $O_2^{\bullet-}$ content, H_2O_2 content, MDA level, and membrane injury by 11.5, 35.4, 28.4, and 17.6%, respectively, compared with control plants. Moreover, Se application increased CAT activity in both control and HT stress; however, the maximum increase was observed in HT stress. Across the days of observation, Se application increased CAT and POX enzyme activity by 25.9 and 23.6%, respectively, under HT stress, and 9.2 and 3.3%, respectively, at the control temperature. As a result, an Se spray significantly increased filled seed weight and seed size by 26.3 and 10.7%, respectively, over the untreated controls (Djanaguiraman et al., 2010).

16.6.4 Low temperature

Many reports have indicated a protective role for Si under low temperature (LT) or chilling stress. Silicon at both 0.1 and 1.0 mM improved the shoot dry weight and water status of wheat leaves under a freezing stress of $-5°C$ (Liang et al., 2008). Silicon-treated plants *O. sativa*, *Z. mays*, *C. sativus*, *H. annuus*, and *Benincasa hispida* grown hydroponically showed enhanced tolerance to LT (0−4°C) in terms of root nutrient absorbing capability and prevention of wilting (Liang et al., 2006b). Lemon fruits dipped in 0, 50, 150, and 250 mg L^{-1} solutions of K_2SiO_3 for 30 min showed no chilling injury symptoms when stored at $-0.5°C$ for 28 d. Only 27% of the fruits showed chilling injury when dipped at 50 mg L^{-1} K_2SiO_3, whereas 97% of untreated fruits were injured and showed significant weight loss. Furthermore, treatment with 50 mg L^{-1} reduced the occurrence of chilling injury symptoms (Mditshwa et al., 2013).

The homeostasis of soluble sugar concentration is considered to be one of the osmosis-regulated conditions in plant tissues that confer tolerance in freezing-stressed plants. Compared with the non-Si-amended freezing treatment, the Si-amended freezing treatment decreased the content of soluble sugar significantly in wheat but enhanced freezing tolerance (Zhu et al., 2006; Liang et al., 2008). Silicon also maintained balanced Pro levels for conferring higher chilling tolerance in *T. aestivum* (Zhu et al., 2006; Liang et al., 2008). Under LT, Si improved the photosynthesis and its related parameters. In the wheat leaf, photosynthesis and water-use efficiency were significantly inhibited under freezing stress but these processes were restored by Si (Zhu et al., 2006). Treatment of *T. aestivum* with exogenous Si increased the net photosynthetic rate, g_s, transpiration rate, chl content, and chl *a/b* ratio and decreased the soluble sugar content (Zhu et al., 2006).

Silicon might play a role in maintaining moisture (Gong et al., 2003; Tesfay et al., 2011). Silicon-treated avocado fruit maintained a higher water status compared with non-Si-treated plants during post-harvest storage at 5.5°C. In fruit peel, deposition of Si has been reported to cause impregnation of the intercellular parts, which might cover the fruit stomata with an Si layer and reduce fruit respiration (Hammash and El Assi, 2007); this would concomitantly result in decreased weight loss of avocado under chilling storage of 5.5°C. This Si deposition could also affect the membrane integrity and fruit firmness (Tesfay et al., 2011). Phenolic and flavonoids are correlated with chilling tolerance. Silicon in avocado could also adjust polyphenol oxidase (PPO) activity to regulate total phenolics and increase free polyphenol concentrations in the mesocarp, thereby improving chilling tolerance and thus fruit firmness. A reduced weight loss was observed when compared to untreated fruit following exposure to 5.5°C temperature (Tesfay et al., 2011). Uninjured fruits had higher phenolic and flavonoids contents compared to injured fruit. The chilling susceptible lemon fruit contained lower phenolic and flavonoid concentrations, with higher weight loss and electrolyte leakage, when compared with chilling-resistant lemons (Mditshwa et al., 2013). This electrolyte leakage reduction might be due to Si-induced improvement of antioxidant enzyme activities involved in plant stress defense systems (Crusciol et al., 2009; Keeping and Reynolds, 2009). Low temperatures—both chilling and freezing—induce the production of ROS (Xu et al., 2008), which damages membrane lipids and other cellular and subcellular organelles and leads to cell death (Molassiotis et al., 2006) and exogenous Si application was found beneficial for improving these effects.

Exogenous Si supplementation reduced the percentage of withered cucumber leaves under chilling (15/8°C) stress, which was supposedly correlated with Si deposition in leaves. The other reasons might be improved antioxidant activities including SOD, GPX, APX, MDHAR, GR, GSH, AsA, which would lower the levels of H_2O_2, $O_2^{\bullet-}$, and MDA and reduce the leaf withering percentage (Liu et al., 2009). The addition of Si (1.0 mM) significantly suppressed H_2O_2, free Pro, and MDA under freezing stress ($-5°C$) in *T. aestivum*, which might be due to the observed increases in SOD and CAT activities of 39 and 59%, respectively, and also to the significant augmentation of non-enzymatic antioxidants (namely, AsA and GSH) (Liang et al., 2008). In avocado fruit, K_2SiO_3 application increased the CAT activity, reduced lipid peroxidation, and improved mesocarp electrical conductivity when fruit was stored at 5.5°C for 17 d (Tesfay et al., 2011).

Under LT stress, Se also exerted beneficial effects at low concentration. In *C. sativus*, Se-treated seedlings avoided the membrane damage caused by LT stress (10°C/5°C, 24 h), which was attributed to Se-induced higher Pro levels and higher shoot biomass (Hawrylak-Nowak et al., 2010). In contrast, the seedlings without Se supplementation showed higher membrane damage, as indicated by higher MDA content (Hawrylak-Nowak et al., 2010). Different levels of Se (0.5, 1.0, 2.0, 3.0 mg Se kg^{-1}) were imposed on *T. aestivum* subjected to LT stress (4°C); the lower concentration of Se (1 mg Se kg^{-1}) was found to be more beneficial as it enhanced the content of anthocyanins, flavonoids, and phenolic compounds and activities of antioxidant enzymes (namely, POD and CAT), which reduced oxidative damage and concomitant reduction in the levels of MDA and $O_2^{\bullet-}$ (Chu et al., 2010). Soaking of *T. aestivum* seeds with Se (5 mg L^{-1} for 5, 10, and 15 h) made the seedlings tolerant to LT stress (3 or 5°C). Se improved the shoot and root length, total biomass, and the contents of chl, Pro, and soluble sugars. Low temperature caused membrane damage, electrolyte leakage, and increased lipid peroxidation. Se supplementation, through enhancement of different antioxidant components like anthocyanin and AsA and promotion of enzymatic activities of CAT and PPO, significantly reduced oxidative damage. Furthermore, the Se-induced cold tolerance was confirmed by the appearance of novel protein bands (Akladious 2012). According to Djanaguiraman et al. (2005), Se was able to increase tolerance of *G. max* plants to LT stress by promoting antioxidant capacity and it improved growth and developmental processes of that plant under LT. Abbas (2012) found that SeO_4^{2-} at low concentrations (3 and 6 mg L^{-1}) enhanced growth, levels of chl, anthocyanin, sugar, Pro, and AsA, and enzymatic activities in *S. bicolor* seedlings subjected to LT stress. However, high levels of SeO_4^{2-} (12 mg L^{-1}) resulted in toxic effects. Low SeO_4^{2-} (3 and 6 mg L^{-1}) also diminished lipid peroxidation by enhancing the activities of APX and GPX.

16.6.5 Heavy metals

Among all the abiotic stresses, the effects of Si on heavy metal (HM) stress is probably the most widely studied. Silicon has been shown to reduce HM content within different plants subjected to

HM stresses. Silicon-induced modification of HM binding properties of the cell walls may be an important mechanism for alleviation of HM toxicity (Ye et al., 2012). Proposed mechanisms include coprecipitation of Si and HMs in the cell wall, which would reduce the concentration of biologically active HMs (Neumann and Zur Nieden, 2001); reducing the cell wall porosity by depositing Si in endodermal cell walls; and limiting apoplasmic transport (Shi et al., 2005; da Cunha and do Nascimento, 2009). Silicon addition is documented as reducing the uptake and translocation of HMs (Shi et al., 2005; Nwugo and Huerta, 2008; Kaya et al., 2009).

When Si was applied under Cd stress, it improved the growth parameters in maize including fresh weight, dry weight, primary seminal root, and leaf area, when compared to Cd treatment alone. Silicon might have exerted these positive effects because it increased the cell wall extensibility. Again, by increasing apoplasmic concentration of Cd, Si would considerably decrease the symplasmic content in maize shoots, which might affect other physiological processes related to plant growth (Vaculík et al., 2012). *Cucumis sativus* L. plants treated with Si accumulated higher biomass against Cd stress. The reasons mentioned included the possibility that Si could improve photosynthetic rate by protecting the chloroplast from disorganization and by increasing pigment contents. Application of Si also alleviated the inhibition of photosynthesis and Fv/Fm (ratio of the variable fluorescence to the maximum fluorescence) and actual quantum efficiency of PSII under Cd stress. Moreover, Si alleviated the inhibition of the enzymes of N metabolism including nitrogen reductase (NR), glutamine synthetase (GS), glutamate synthase (GOGAT), and glutamate dehydrogenase (GDH) (Feng et al., 2010). In rice (*O. sativa* L.) under Cd (0, 2, and 4 μM) stress, Si (2 and 4 mM) supplementation increased shoot and root biomass by 125–171% and 100–106%, respectively (Zhang et al., 2008). Silicon can modify the development of root tissues, thereby reducing the uptake and concentration of Cd in other plant parts. Si can also bind toxic Cd ions in the apoplasmic space including the cell wall, or detoxify Cd by sequestering it in vacuoles (Vaculík et al., 2012). Si influenced the development of Casparian bands and suberin lamellae as well as vascular tissues in root, thereby helping to decrease symplasmic and increase apoplasmic concentrations of Cd in maize shoots. These apoplasmic barriers enhanced binding of Cd to the apoplasmic fraction in *Z. mays* shoots (Vaculík et al., 2012).

Silicon also increased growth parameters in *T. aestivum* and *H. vulgare* against boron (B) toxicity (Inal et al., 2009). For B detoxification, some assumptions were confirmed for decreases in B concentrations including the formation B$-$SiO$_4^{4-}$ complexes in the soil and reduction of B translocation from the roots to shoots (Gunes et al., 2007b). Silicon has ameliorative effects as it irreversibly precipitates as amorphous silica (SiO$_2$•nH$_2$O) in the cell walls and lumen, which reduces the HM-induced membrane damage (Epstein, 1999). In *O. sativa*, Si also decreased shoot Cd concentrations by 30–50% and Cd distribution ratio in shoots by 25.3–46%, compared to the treatment without Si supply (Zhang et al., 2008). The mitigation of Cd concentration and its toxicity by Si was also predicted in other plant species including *T. aestivum* (Rizwan et al., 2012), *B. rapa* (Song et al., 2009), *A. hypogaea* (Shi et al., 2010), and *C. sativus* (Feng et al., 2010). Silicon reduced the chromium (Cr) concentration in *O. sativa* shoots by 41% and in roots by 30% in plants grown in Cr (100 μM)-enriched medium; these effects were due to the dramatic decrease in the translocation factor by 16% (Zeng et al., 2011).

Kidd et al. (2001) suggested that the development of aluminum (Al) tolerance in *Z. mays* may occur by an enhanced exudation of phenolic compounds, which leads to complexation of Al. Zinc (Zn) could be detoxified by forming bonds with Si; this was found to be localized in the intercellular space, cytoplasm, nucleus, and vacuolar vesicles of leaf mesophyll cells (Neumann and Zur Nieden, 2001; da Cunha and do Nascimento, 2009). However, Al detoxification was reported to occur by reduction of the Al^{3+} content in symplasm and the formation of hard soluble aluminosilicates or hydroxyaluminosilicates in the apoplasmic space and outer epidermal cell walls (Hodson and Sangster, 1993; Hodson and Evans, 1995; Wang et al., 2004). Prabagar et al. (2011) found that the formation of aluminosilicate complexes in the cell wall and apoplasm of Norway spruce (*Picea abies*) plants create a significant barrier to Al penetration and Si-induced cell damage. Chromium-induced reductions of seedling height, dry biomass, and soluble protein content were alleviated by Si application, which was due mainly to reduced Cr uptake and enhanced antioxidant activity (Zeng et al., 2011). In Norway spruce, the presence of 1.0 mM Si reduced the cell death due to Al stress. Silicon reduced the maximum percentage of cell death by 50, 45, and 35% at Al concentrations of 0.2, 0.5, and 1.0 mM, respectively, after 48 h of exposure (Prabagar et al., 2011). In *H. vulgare*, 2 mM Si application enhanced plant growth relative to the control, and alleviated Cr (100 μM) toxicity, as reflected by a significant increase in growth and photosynthetic parameters, such as SPAD value, net photosynthetic rate, cellular CO_2 concentration, g_s, transpiration rate, and chl fluorescence efficiency (Ali et al., 2013). In *Zea mays*, Si treatment prior to exposure to excess manganese (Mn) caused increased thickness of the leaves and epidermal cells when compared to the control plants without Si pretreatment. Silicon-pretreated maize plants maintained higher concentrations of chl *a* and *b* and carotenoids when compared to plants exposed to Mn stress treatment alone and Si pretreatment also helped to maintain maximum quantum efficiency of PSII photochemistry, the actual efficiency of PSII electron transport, and the rate of photosynthetic O_2 evolution (Doncheva et al., 2009).

Silicon can reduce HM-induced oxidative damage. Silicon improved growth characteristics of *Vitis vinifera* rootstock under B as indicated by decreases in stomatal resistance, lipid peroxidation, membrane damage, and MDA and H_2O_2 levels and increases in CAT and SOD activity (Soylemezoglu et al., 2009). Similarly, LOX (224%) and APX (61%) activity were increased in *H. vulgare* by Si (280 mg kg^{-1}) under sodic-B toxicity conditions, which were correlated with decreased peroxidation of lipid (MDA) by 42%, H_2O_2 by 42%, and membrane permeability by 21% and increased Pro content (53%). Together, these changes could result in increases in fresh (35%) and dry weight (42%) in *H. vulgare* plants. Silicon also decreased the B (18%) and Na (8%) concentration (Gunes et al., 2007a). Silicon alleviates Cr toxicity in rice plants by inhibiting the uptake and translocation of Cr, together with enhanced activities of antioxidant enzymes, SOD, POD, CAT, and APX, which help to reduce TBARS content (Zeng et al., 2011). Alleviation mechanisms against Mn toxicity by Si include Mn adsorption by cell walls, active removal of excess Mn by soluble Si in the apoplasm, and enhanced enzymatic and non-enzymatic antioxidant levels (Horst et al., 1999; Iwasaki et al., 2002a,b; Rogalla and Römhled, 2002; Shi et al., 2005). Song et al. (2011) proved that Si improved antioxidant defense capacity and membrane integrity to prevent Zn toxicity in *O. sativa*. Silicon promoted the expression of free radical metabolizing enzymes, which amplified plant biomass parameters and decreased copper (Cu) uptake and leaf chlorosis (Nowakowski and Nowakowska, 1997; Li et al., 2008; Khandekar and Leisner, 2011).

Selenium has been documented to reduce metal toxicity in several research studies. The modes of action were varied and are still unclear; however, some suggested reasons included improvement of the antioxidant defense system, reduction of metal uptake, formation of non-toxic Se−metal complexes, and phytochelatin activity (Vorobets and Mykiyevich, 2000; Sun et al., 2010).

Moreover, Se is effective at sustaining physiological activities, growth, and developmental processes even in HM toxic environments (Pedrero et al., 2008; Cartes et al., 2010). In *B. oleracea*, Cd phytotoxicity resulted in elevated MDA level and decreased photosynthetic pigment and tocopherol concentrations, but Se treatment effectively alleviated these adverse effects (Pedrero et al., 2008). In *B. napus*, Se (2 µM) conferred tolerance to Cd (400 and 600 µM) stress by reducing lipid unsaturation and peroxidation, modulating the activity of antioxidative enzymes (SOD, CAT, APX, GPX), and preventing Cd-induced changes in the DNA methylation pattern (Filek et al., 2008). Sun et al. (2010) summarized the mechanisms for the positive role of Se on Cd toxicity as: (1) removal of Cd from metabolically active cellular sites, (2) induction of Se to scavenge the Cd-induced ROS generated, and (3) the regulation phytochelatin synthesis-associated enzymes induced by Se. These three actions mitigated the effects of Cd on *A. sativum*. In our recent study, we observed that rapeseed seedlings grown under Cd stress (0.5 and 1.0 mM $CdCl_2$) showed substantial increases in MDA and H_2O_2 levels. The AsA content of the seedlings decreased significantly upon exposure to Cd stress. The amount of GSH increased only at 0.5 mM $CdCl_2$, while GSSG increased at any level of Cd with concomitant decreases in the GSH/GSSG ratio. The activities of antioxidant enzymes also reduced under Cd stress. Importantly, Se-pretreated seedlings exposed to Cd showed increases in the AsA and GSH contents; GSH/GSSG ratio; and the activities of APX, MDHAR, DHAR, GR, GPX, and CAT. However, in most of the cases, pretreatment with 50 µM Se showed better results compared to 100 µM Se. These results indicated that the exogenous application of Se at low concentration increased the tolerance of the plants to Cd-induced oxidative damage by enhancing their antioxidant defenses. The phenotypes of the Se-supplemented seedlings were also superior to the seedlings subjected to Cd stress without Se (Figure 16.7).

The effect of Se (5 or 10 µM of Na_2SeO_4) on *Pteris vittata* plant was studied under arsenic (As, 150 or 300 µM of Na_2HAsO_4) stress. Selenium was effective at increasing the levels of thiols and GSH (increased by 24%) and decreasing lipid peroxidation (by 26−42%) in a dose-dependent manner (Srivastava et al., 2009). In ryegrass, Cartes et al. (2010) proposed two probable mechanisms for Se-induced alleviation of Al toxicity: (1) the improved dismutation of $O_2^{\bullet-}$ to H_2O_2, and (2) the activation of antioxidant enzymes (POD, a H_2O_2 scavenger). In addition, a low concentration of Se (2 µM) also improved root growth and decreased lipid peroxidation in plants grown under Al stress (0.2 mM Al). In contrast, a higher concentration of Se (>2 µM) caused phytotoxicity. Ribeiro et al. (2011) observed that the growth of seedlings of *Stylosanthes humilis* was inhibited by Al^{3+}, while elongation was recovered with Na_2SeO_4 at 1.0 µM. Methyl viologen and H_2O_2, which are ROS-generating compounds, also inhibited seedling elongation and again growth was restored by SeO_4^{2-}. Selenium-induced enhanced oxidizing ability, resistance to oxidation, and reduced membrane lipid peroxidation of rice roots under iron (Fe) stress was reported by Qi et al. (2004) and Peng et al. (2002).

16.6.6 UV radiation

Although studies are scarce, Si has been proved to be beneficial for improving different growth, developmental, physiological, and biochemical parameters in different plants growing under ultraviolet radiation stress. In *G. max* seedling, Si (1.70 mM) application increased photosynthesis by 21.0 and 21.5% under UV-B radiation (UV 290−320 nm) and the combination of UV-B and drought treatment, respectively, with increases seen in shoot dry matter of 23%, root dry matter of 42%,

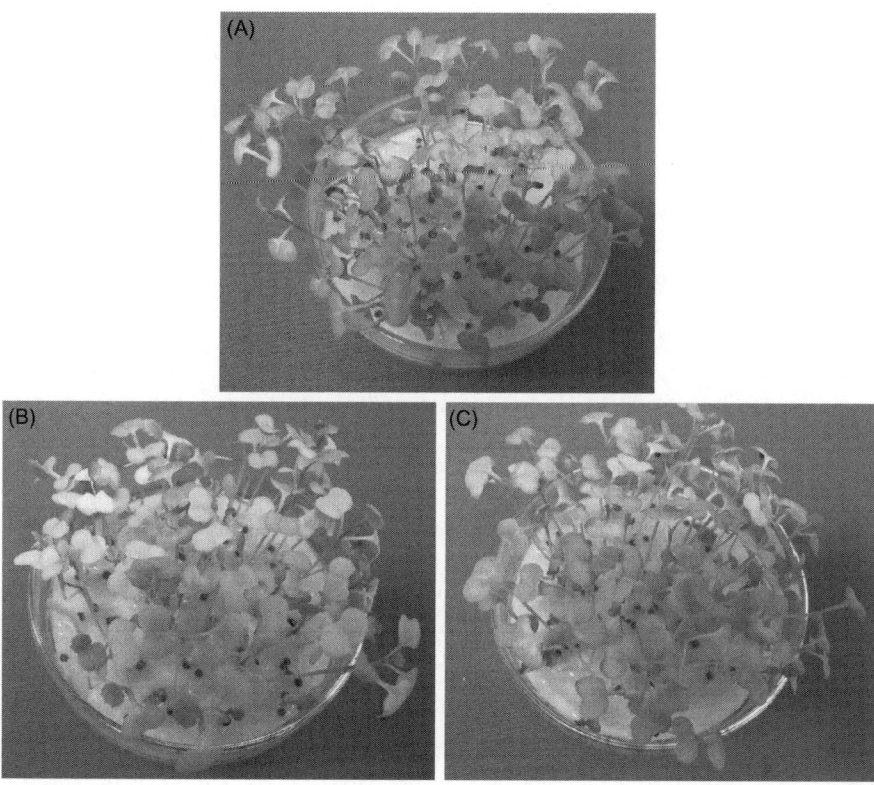

FIGURE 16.7

Phenological appearance of *Brassica napus* leaves induced by Se under Cd stress. (A) Control; (B) 0.5 mM $CdCl_2$, 48 h; (C) Se pretreatment (100 μM Na_2SeO_2, 24 h) followed by treatment with 0.5 mM $CdCl_2$, 48 h.

and root-to-shoot ratio of 16%. Si could also improve the relative water content (RWC) by 19% after a week of Si application in a combined treatment of UV and drought (Shen et al., 2010). In *O. sativa*, Si increased plant height and leaf dry matter relative to treatment without Si. More significant results were obtained for total leaf area and specific leaf area both at the boot and milk ripe stage. The increased values for these two parameters were 208 and 113% for the boot stage and those for the milk stage were 51 and 25%, respectively (Goto et al., 2003). Exogenous Si application could modulate the plant's internal Si and uptake of other nutrients and their contents in soybean plant. Increases in leaf N and P of 9 and 16%, respectively, and decreases in leaf Mg content of 9% and Ca of 24% were seen in rice plants under UV stress. UV-B radiation also increased the allocation of P, potassium (K), and Ca to roots compared with stem and leaves, presumably to ensure sustained nutrient uptake under stress. Silicon application improved the uptake of P and Mg by 11%, which favored the partitioning of dry mass to shoots under UV-B radiation and the allocation of tissue P and Ca to roots (Shen et al., 2009). Gao et al. (2003) also reported that P, Ca, Na, and Si uptake was increased in rice leaves by exogenous Si application under UV stress.

The Si-treated rice plants also showed lower cinnamyl alcohol dehydrogenase (CAD) levels in the cell walls of sclerenchyma, vascular bundle sheath, and metaxylem vessel cells, which might the reason why Si-treated rice plants have a lower UV absorbance of around 280 and 320 nm in the leaf blades compared with plants without Si treatment (Goto et al., 2003). After a week of Si application in UV-B stress conditions, net photosynthetic rate in *G. max* seedlings was significantly increased by 18%, which might be associated with increased g_s (by 15%), and transpiration (by 4%) and significant increases in photosynthetic pigments like chl and anthocyanin in *G. max* seedlings grown under UV-B stress (Shen et al., 2010).

Within the plant, a protective accumulation of anthocyanins and other UV-absorbing compounds like flavonoids and total phenols occurs in response to UV radiation; these compounds may act as solar screens in the leaf (Alexieva et al., 2001; Li et al., 2004; Shen et al., 2010). Phenolics and flavonoids are effective in reducing UV radiation-induced damage as they prevent the accumulation of ROS, as was seen in *Arabidopsis* (Bieza and Lois, 2001). In the rice leaf surface, more specifically in the epidermal cells, phenolic compounds are effective protectors against UV-induced damage as they absorb high amounts of UV radiation. Some insoluble compounds of Si are deposited in the leaf epidermis and act as promoters of constitutive phenols (Goto et al., 2003; Li et al., 2004). UV radiation promoted Si deposition in rice leaves (IRRI, 1991). In the rice leaf epidermis, soluble and insoluble UV-absorbing compounds were increased by 21 and 67%, respectively, due to Si treatment (compared to Si-untreated leaves). Moreover, the cell walls and cell lumina of the epidermis contained more Si-rich insoluble UV-absorbing compounds. This led Li et al. (2004) to conclude that Si increased phenolic compounds in the epidermis, which rendered UV tolerance as evidenced by healthy leaves free from the brown spots and strips of UV damage symptoms present in leaves not treated with Si. At the mature stage, rice glumes consist of about 20% silica on a dry matter basis (Okuda and Takahashi, 1961), which was considered as vital for protecting the kernel from severe damage from UV radiation (Ebata et al., 1989). In *G. max*, UV-B caused substantial membrane damage, as assessed by lipid peroxidation and osmolyte leakage, but Si application significantly reduced the H_2O_2, lipid peroxidation, and membrane damage, and ultimately reduced the leakage of electrolyte. This might have been due to modulation of antioxidant enzymes, phenols, and anthocyanins in relation to other physiological parameters (Shen et al., 2010). Silicate application can increase silica deposits and decrease CAD activity and ferulic and p-coumaric acid contents in rice plants, and these responses are closely related to alterations in the UV defense system (Goto et al., 2003).

Selenium improves plant growth and survival under UV radiation, as reported in several studies. Valkama et al. (2003) investigated the possible ameliorative effects of Se addition (0.1 and 1 mg kg^{-1}) to soil on the detrimental effects of enhanced UV-B radiation in strawberry (*Fragaria* $\times$ *ananassa*) and barley (*H. vulgare*) plants. In contrast, higher concentrations of Se had no protective effects and even increased the sensitivity to UV-B radiation. In buckwheat, Se improved the effective quantum yield of PSII under UV-B radiation; thus, plant height and biomass production were increased (Breznik et al., 2005). Selenium might play positive roles in recovery of the photosynthetic and respiratory machinery, as documented in *Euglena gracilis*. During exposure to different UV radiation (UV-A, 320–400 nm, of 1.02 W m^{-2} plus UV-B, 280–320 nm, of 0.73 W m^{-2}), no effect was seen for Se (10^{-7}, 10^{-8}, 10^{-9}, and 10^{-10} M, Na$_2$SeO$_3$.5H$_2$O) in *E. gracilis*. However, after a 24 h recovery period, Se increased photosynthesis and light-enhanced dark respiration by its protective effects against photodamage and oxidative stress (Ekelund and Danilov, 2001). Se (0.01

and 0.05 mg kg^{-1} soil) improved the antioxidative capacity, protected chloroplast enzymes, and increased shoot yield in *Lactuca sativa* under combined UV-B and UV-C stress (Pennanen et al., 2002). Significant increases in the activities of POD and SOD, together with reduced MDA and $O_2^{\bullet-}$ levels, were documented in *T. aestivum* under UV-B radiation. Selenium also increased root activity and flavonoid and Pro contents in this plant (Yao et al., 2010a,b). Selenium also showed positive effects and increased yield in ryegrass (Hartikainen et al., 2000), lettuce (Xue et al., 2001), potato (Turakainen et al., 2004), and pumpkin (Germ et al., 2005) exposed to UV-B radiation.

16.7 Conclusion and future prospects

The mechanisms underlying the responses and tolerance of plants to different abiotic stresses are still unclear, and require further critical physiological and molecular studies. Exploring the most effective and easiest ways to overcome stressors is one of the emerging tasks for plant scientists. The published literature indicates that Si and Se have vital roles in a wide spectrum of physiological processes in plants. In particular, they have prominent roles in the protection of plants from abiotic stress-induced damage. Although much research has been published on the effects of Si on plants under abiotic stress, the underlying physiological mechanisms by which Si could protect plants from stressful conditions are elusive. The potential of exogenous application of Si has gained significant attention in recent times, and numerous plant studies have revealed its protective roles in oxidative stress tolerance. However, the exact dose of application and methods of application are still under study. Like Si, the facts about Se are also intriguing, enigmatic, and challenging (even capricious) for researchers. Although a number of reports hint at a protective role for Se under abiotic stress conditions, research conducted to date on the physiological role of Se under stressful condition is scarce. Controversy remains regarding the question of whether Se is an essential plant micronutrient. On a cautionary note, the appropriate concentration of exogenous Se is still a matter of research. Complete elucidation of the role of Se as well as detailed protective mechanisms would be helpful for developing stress tolerance in plants. Despite the widespread occurrence of Se deficiency globally, Se toxicity (selenosis) is a problem in some areas. Some soils and mineral deposits are naturally Se rich, and exploitation of these seleniferous soils or fossil fuels can lead to toxic accumulation of Se in the environment. Selenium contamination of sediments, soils, and drainage water is a particular issue in arid seleniferous areas with intensive crop irrigation. Effective enrichment of agricultural crops with Se via soil using Se-enriched fertilizers can be challenging due to varying soil Se concentrations, soil types, soil redox potentials, soil pH, microbiological activity, etc. Furthermore, the high cost of Se fertilizer, in combination with the modest incorporation rate, should be considered. However, the appropriate application of exogenous Se can be an effective protectant for plants for combating abiotic stress. Notably, findings from experimental field research studies are still scarce. Field studies that confirm the ability of exogenous applications of both Si and Se to mitigate abiotic stress are likely to prop up their extended practical application to crop plants.

Acknowledgments

We wish to thank Md. Mahabub Alam, Laboratory of Plant Stress Responses, Faculty of Agriculture, Kagawa University, Japan, for providing several supporting articles and suggestions for improving the manuscript.

We are also highly thankful to Mr. Anisur Rahman and Mr. Md. Hasanuzzaman, Department of Agronomy, Sher-e-Bangla Agricultural University, Dhaka-1207, Bangladesh, for their critical reading of the manuscript draft.

References

Abbas, S.M., 2012. Effects of low temperature and selenium application on growth and the physiological changes in sorghum seedlings. J. Stress Physiol. Biochem. 8, 268—286.

Acquaah, G., 2007. Principles of Plant Genetics and Breeding. Blackwell Publishing, Oxford, UK.

Adatia, M.H., Besford, R.T., 1986. The effects of silicon on cucumber plants grown in recirculating nutrient solution. Ann. Bot. 58, 343—351.

Agarie, S., Hanaoka, N., Ueno, O., Miyazaki, A., Kubota, F., Agata, W., et al., 1998. Effects of silicon on tolerance to water deficit and heat stress in rice plants (*Oryza sativa* L.), monitored by electrolyte leakage. Plant Prod. Sci. 1, 96—103.

Agurie, S., Agata, W., Kubota, F., Kaufman, P.B., 1992. Physiological role of silicon in photosynthesis and dry matter production in rice plants. Crop Sci. 61, 200—206.

Ahmad, F., Rahmatullah, Aziz, T., Maqsood, M.A., Tahir, M.A., Kanwal, S., 2007. Effect of silicon application on wheat (*Triticum aestivum L.*) growth under water deficiency stress. Emir. J. Food Agric. 19, 1—7.

Ahmad, P., Prasad, M.N.V., 2012a. Abiotic Stress Responses In Plants: Metabolism, Productivity and Sustainability. Springer, New York.

Ahmad, P., Prasad, M.N.V., 2012b. Environmental Adaptations and Stress Tolerance of Plants in the Era of Climate Change. Springer, New York.

Ahmad, R., Zaheer, S.H., Ismail, S., 1992. Role of silicon in salt tolerance of wheat (*Triticum aestivum* L.). Plant Sci. 85, 43—50.

Ahmed, M., Asif, M., Goyal, A., 2012. Silicon the non-essential beneficial plant nutrient to enhanced drought tolerance in wheat. In: Goyal, A. (Ed.), Crop Plant. InTech, Rijeka, pp. 31—48.

Ahmed, M., Hassen, F., Qadeer, U., Aslam, M.A., 2011. Silicon application and drought tolerance mechanism of sorghum. Afr. J. Agric. Res. 6, 594—607.

Akladious, S.A., 2012. Influence of different soaking times with selenium on growth, metabolic activities of wheat seedlings under low temperature stress. Afr. J. Biotechnol. 11, 14792—14804.

Al-Aghabary, K., Zhu, Z., Shi, Q., 2004. Influence of silicon supply on chlorophyll content, chlorophyll fluorescence and anti-oxidative enzyme activities in tomato plants under salt stress. J. Plant Nutr. 27, 2101—2115.

Alexieva, V., Sergiev, I., Mapelli, S., Karanov, E., 2001. The effect of drought and ultraviolet radiation on growth and stress markers in pea and wheat. Plant Cell Environ. 24, 1337—1344.

Ali, A., Basra, S.M.A., Ahmad, R., Wahid, A., 2009. Optimizing silicon application to improve salinity tolerance in wheat. Soil Environ. 28, 136—144.

Ali, S., Farooq, M.A., Yasmeen, T., Hussain, S., Arif, M.S., Abbas, F., et al., 2013. The influence of silicon on barley growth, photosynthesis and ultra-structure under chromium stress. Ecotoxicol Environ. Saf. 89, 66—72.

Ashraf, M., Afzal, R.M., Ahmed, R., Mujeeb, F., Sarwar, A., Ali, L., 2010a. Alleviation of detrimental effects of NaCl by silicon nutrition in salt-sensitive and salt-tolerant genotypes of sugarcane (*Saccharum officinarum* L.). Plant Soil 326, 381—391.

Ashraf, M., Rahmatullah, Ahmad, R., Bhatti, S., Afzal, M., Sarwar, A., et al., 2010b. Amelioration of salt stress in sugarcane (*Saccharum officinarum* L.) by supplying potassium and silicon in hydroponics. Pedosphere 20, 153—162.

Barber, D.A., Shone, M.G.T., 1966. The absorption of silica from aqueous solutions by plants. J. Exp. Bot. 17, 569—578.

Bekheta, M.A., Abbas, S., El-Kobisy, O.S., Mahgoub, M.H., 2008. Influence of selenium and paclobutrazole on growth, metabolic activities and anatomical characters of *Gerbera jasmonii* L. Aust. J. Basic. Appl. Sci. 2, 1284—1297.

Bieza, K., Lois, R., 2001. An *Arabidopsis* mutant tolerant to lethal ultraviolet-B levels shows constitutively elevated accumulation of flavonoids and other phenolics. Plant Physiol. 126, 1105—1115.

Bradbury, M., Ahmad, R., 1990. The effect of silicon on the growth of *Prosopis juliflora* growing in saline soil. Plant Soil 125, 71—74.

Breznik, B., Germ, M., Gaberščik, A., Kreft, I., 2005. Combined effects of elevated UV-B radiation and the addition of selenium on common (*Fagopyrum esculentum* Moench) and tartary (*Fagopyrum tataricum* (L.) Gaertn.) buckwheat. Photosynthetica 43, 583—589.

Broadley, M., Brown, P., Cakmak, I., Ma, J.F., Rengel, Z., Zhao, F., 2012. Beneficial elements. In: Marschner, P. (Ed.), Marschner's Mineral Nutrition of Higher Plants, 3rd edition Elsevier, Amsterdam, pp. 249—269.

Brogowski, Z., 2000. Krzen w glebie I rola w żywieniu roślin [Silicon in soil and its role in plant nutrition]. Post Nauk. Rol. 6, 9—16 [in Polish]

Brown, T.A., Shrift, A., 1982. Selenium-toxicity and tolerance in higher plants. Biol. Rev. Camb. Philos. Soc. 57, 59—84.

Bybordi, A., 2012. Effect of ascorbic acid and silicium on photosynthesis, antioxidant enzyme activity, and fatty acid contents in canola exposure to salt stress. J. Integr. Agric. 11, 1610—1620.

Çakır, Ö., Turgut-Kara, N., Arı, Ş., 2012. Selenium metabolism in plants: molecular approaches. In: Montanaro, G., Dichio, B. (Eds.), Advances in Selected Plant Physiology Aspects. Intech, Rijeka, pp. 209—232.

Cartes, P., Jara, A.A., Pinilla, L., Rosas, A., Mora, M.L., 2010. Selenium improves the antioxidant ability against aluminium-induced oxidative stress in ryegrass roots. Ann. Appl. Biol. 156, 297—307.

Carvalho, K.M., Gallardo-Williams, M.T., Benson, R.F., Martin, D.F., 2003. Effects of selenium supplementation on four agricultural crops. J. Agric. Food Chem. 51, 704—709.

Casey, W.H., Kinrade, S.D., Knight, C.T.G., Rains, D.W., Epstein, E., 2004. Aqueous silicate complexes in wheat, *Triticum aestivum* L. Plant Cell Environ. 27, 51—54.

Castro, G.S.A., Crusciol, C.A.C., 2013. Effects of superficial liming and silicate application on soil fertility and crop yield under rotation. Geoderma 195—196, 234—242.

Chen, C.C., Sung, J.M., 2001. Priming bitter gourd seeds with selenium solution enhances germinability and antioxidative responses under sub-optimal temperature. Physiol. Plant 111, 9—16.

Chen, W., Yao, X., Cai, K., Chen, J., 2011. Silicon alleviates drought stress of rice plants by improving plant water status, photosynthesis and mineral nutrient absorption. Biol. Trace Elem. Res. 142, 67—76.

Chinnusamy, V., Jagendorf, A., Zhu, J.K., 2005. Understanding and improving salt tolerance in plants. Crop Sci. 45, 437—448.

Chu, I., Yao, X., Zhang, Z., 2010. Responses of wheat seedlings to exogenous selenium supply under cold stress. Biol. Trace Elem. Res. 136, 355—363.

Crusciol, C.A.C., Pulz, A.L., Lemos, B.L., Soratto, R.P., Lima, G.P.P., 2009. Effects of silicon and drought stress on tuber yield and leaf biochemical characteristics in potato. Crop. Sci. 49, 949—954.

da Cunha, K.P.V., do Nascimento, C.W.A., 2009. Silicon effects on metal tolerance and structural changes in maize (*Zea mays* L.) grown on a cadmium and zinc enriched soil. Water Air Soil Pollut. 197, 323—330.

Dagmar, D., Simone, H., Wolfgang, B., Rüdiger, F., Bäucker, E., Rühle, G., et al., 2003. Silica accumulation in *Triticum aestivum* L., and *Dactylis glomerata* L. Anal. Bioanal. Chem. 376, 399—404.

Datnoff, L.E., Snyder, G.H., Korndörfer, G.H., 2001. Silicon in Agriculture. Elsevier, Amsterdam.

Dhillon, K.S., Dhillon, S.K., 2003. Distribution and management of seleniferous soils. Adv. Agron. 79, 119—184.

Djanaguiraman, M., Devi, D.D., Shanker, A.K., Sheeba, A., Bangarusamy, U., 2005. Selenium—an antioxidative protectant in soybean during senescence. Plant Soil 272, 77—86.

Djanaguiraman, M., Prasad, P.V.V., Seppanen, M., 2010. Selenium protects sorghum leaves from oxidative damage under high temperature stress by enhancing antioxidant defense system. Plant Physiol. Biochem. 48, 999—1007.

Doncheva, S., Poschenrieder, C., Stoyanova, Z., Georgieva, K., Velichkova, M., Barceló, J., 2009. Silicon amelioration of manganese toxicity in Mn-sensitive and Mn-tolerant maize varieties. Environ. Exp. Bot. 65, 189—197.

Ebata, M., Ozeki, T., Inoue, K., Ishikawa, M., Tashiro, T., 1989. Rice flower glumes as an interceptor of ultraviolet rays. Jpn. J. Crop. Sci. 58, 541—548.

Efimova, G.V., Dokynchan, S.A., 1986. An atomo-morphological construction of epidermical tissue of rice leaves and increasing of it protection function under silicon effect. Agric. Biol. 3, 57—61.

Ekelund, N.G.A., Danilov, R.A., 2001. The influence of selenium on photosynthesis and "light-enhanced dark respiration" (LEDR) in the flagellate *Euglena gracilis* after exposure to ultraviolet radiation. Aquat. Sci. 63, 457—465.

Ellis, D.R., Salt, D.E., 2003. Plants, selenium and human health. Curr. Opin. Plant Biol. 6, 273—279.

Elrashidi, M.A., Adriano, D.C., Workman, S.M., Lindsay, W.L., 1987. Chemical-equilibria of selenium in soils—a theoretical development. Soil Sci. 144, 141—152.

Epstein, E., 1994. The anomaly of silicon in plant biology. Proc. Natl. Acad. Sci. USA 91, 11—17.

Epstein, E., 1999. Silicon. Annu. Rev. Plant Physiol. Plant Mol. Biol. 50, 641—664.

Feng, J., Shi, Q., Wang, X., Wei, M., Yang, F., Xu, H., 2010. Silicon supplementation ameliorated the inhibition of photosynthesis and nitrate metabolism by cadmium (Cd) toxicity in *Cucumis sativus* L. Sci. Hortic. 123, 521—530.

Filek, M., Keskinen, R., Hartikainen, H., Szarejko, I., Janiak, A., Miszalski, Z., et al., 2008. The protective role of selenium in rape seedlings subjected to cadmium stress. J. Plant Physiol. 165, 833—844.

Freeman, J.L., Zhang, L.H., Marcus, M.A., Fakra, S., McGrath, S.P., Pilon-Smits, E.A.H., 2006. Spatial imaging, speciation and quantification of selenium in the hyperaccumulator plants *Astragalus bisulcatus* and *Stanleya pinnata*. Plant Physiol. 142, 124—134.

Gao, W., Zheng, Y.F., Slusser, J.R., Heisler, G.M., 2003. Impact of enhanced ultraviolet-B irradiance on cotton growth, development, yield, and qualities under field conditions. Agr. For. Meteorol. 120, 241—248.

Gao, X., Zou, C., Wang, L., Zhang, F., 2006. Silicon decreases transpiration rate and conductance from stomata of maize plants. J. Plant Nutr. 29, 1637—1647.

Germ, M., Kreft, I., Osvald, J., 2005. Influence of UV-B exclusion and selenium treatment on photochemical efficiency of photosystem II, yield and respiratory potential in pumpkins (*Cucurbita pepo* L.). Plant Physiol. Biochem. 43, 445—448.

Gong, H., Chen, K., Chen, G., Wang, S., Zhang, C., 2003. Effects of silicon on growth of wheat under drought. J. Plant Nutr. 26, 1055—1063.

Gong, H., Zhu, X., Chen, K., Wang, S., Zhang, C., 2005. Silicon alleviates oxidative damage of wheat plants in pots under drought. Plant Sci. 169, 313—321.

Gong, H.J., Randall, D.P., Flowers, T.J., 2006. Silicon deposition in the roots reduces sodium uptake in rice (Oryza sativa L.) seedling by reducing bypass flow. Plant Cell Environ. 29, 1970—1979.

Goto, M., Ehara, H., Karita, S., Takabe, K., Ogawa, N., Yamada, Y., et al., 2003. Protective effect of silicon on phenolic biosynthesis and ultraviolet spectral stress in rice crop. Plant Sci. 164, 349—356.

Gottardi, S., Iacuzzo, F., Tomasi, N., Cortella, G., Manzocco, L., Pinton, R., et al., 2012. Beneficial effects of silicon on hydroponically grown corn salad (*Valerianella locusta* (L.) Laterr) plants. Plant Physiol. Biochem. 56, 14—23.

Gunes, A., Pilbeam, D.J., Inal, A., Bagci, E.G., Coban, S., 2007a. Influence of silicon on antioxidant mechanisms and lipid peroxidation in chickpea (*Cicer arietinum* L.) cultivars under drought stress. J. Plant Interact 2, 105−113.

Gunes, A., Inal, A., Bagci, E.G., Coban, S., Sahin, O., 2007b. Silicon increases boron tolerance and reduces oxidative damage of wheat grown in soil with excess boron. Biol. Plant 51, 571−574.

Gunes, A., Pilbeam, D.J., Inal, A., Coban, S., 2008. Influence of silicon on sunflower cultivars under drought stress. I: growth, antioxidant mechanisms, and lipid peroxidation. Commun. Soil Sci. Plant Anal. 39, 1885−1903.

Gupta, U.C., McRae, K.B., Winter, K.A., 1982. Effects of applied selenium on the selenium content of barley and forages and soil selenium depletion rates. Can. J. Soil Sci. 62, 145−154.

Hamilton, S.J., 2004. Review of selenium toxicity in the aquatic food chain. Sci. Total Environ. 326, 1−31.

Hammash, F., El Assi, N., 2007. The Influence of pre-storage waxing and wrapping on quality attributes of stored "Shamouti" oranges. Acta Hortic. 741, 133−140.

Hartikainen, H., Xue, T., Piironen, V., 2000. Selenium as an antioxidant and pro-oxidant in ryegrass. Plant Soil 225, 193−200.

Hasanuzzaman, M., Fujita, M., 2011a. Exogenous silicon treatment alleviates salinity-induced damage in *Brassica* napus L. seedlings by up-regulating the antioxidant defense and methylglyoxal detoxification system. Abstract of Plant Biology 2011, American Society of Plant Biology. <http://abstracts.aspb.org/pb2011/public/P10/P10001.html/> (accessed 12.02.03).

Hasanuzzaman, M., Fujita, M., 2011b. Selenium pretreatment upregulates the antioxidant defense and methylglyoxal detoxification system and confers enhanced tolerance to drought stress in rapeseed seedlings. Biol. Trace Elem. Res. 143, 1758−1776.

Hasanuzzaman, M., Hossain, M.A., Fujita, M., 2010. Selenium in higher plants: physiological role, antioxidant metabolism and abiotic stress tolerance. J. Plant Sci. 5, 354−375.

Hasanuzzaman, M., Hossain, M.A., Fujita, M., 2011. Selenium-induced up-regulation of the antioxidant defense and methylglyoxal detoxification system reduces salinity-induced damage in rapeseed seedlings. Biol. Trace Elem. Res. 143, 1704−1721.

Hasanuzzaman, M., Hossain, M.A., Teixeira da Silva, J.A., Fujita, M., 2012a. Plant response and tolerance to abiotic oxidative stress: antioxidant defense is a key factor. In: Venkateshwarulu, B., Shanker, A.K., Shanker, C., Mandapaka, M. (Eds.), Crop stress and its management: perspectives and strategies. Springer, Berlin, pp. 261−316.

Hasanuzzaman, M., Hossain, M.A., Fujita, M., 2012b. Exogenous selenium pretreatment protects rapeseed seedlings from cadmium-induced oxidative stress by upregulating the antioxidant defense and methylglyoxal detoxification systems. Biol. Trace Elem. Res. 149, 248−261.

Hashemi, A., Abdolzadeh, A., Sadeghipour, H.R., 2010. Beneficial effects of silicon nutrition in alleviating salinity stress in hydroponically grown canola, *Brassica napus* L. plants. Soil Sci. Plant Nutr. 56, 244−253.

Hattori, T., Inanaga, S., Araki, H., An, P., Morita, S., Luxova, M., et al., 2005. Application of silicon enhanced drought tolerance in *Sorghum bicolor*. Physiol. Plant 123, 459−466.

Hawrylak, B., Matraszek, R., Szymańska, M., 2007. Reaction of lettuce (*Lactuca sativa* L.) to selenium in nutrient solution contaminated with nickel. Veg. Crops Res. Bull. 67, 63−70.

Hawrylak-Nowak, B., 2009. Beneficial effects of exogenous selenium in cucumber seedlings subjected to salt stress. Biol. Trace Elem. Res. 132, 259−269.

Hawrylak-Nowak, B., Matraszek, R., Szymańska, M., 2010. Selenium modifies the effect of short-term chilling stress on cucumber plants. Biol. Trace Elem. Res. 138, 307−315.

Hodson, M.J., Evans, D.E., 1995. Aluminium/silicon interactions in higher plants. J. Exp. Bot. 46, 161−171.

Hodson, M.J., Sangster, A.G., 1993. The interaction between silicon and aluminium in *Sorghum bicolor* (L.) Moench: growth analysis and X-ray microanalysis. Ann. Bot. 72, 389−400.

Horst, W.J., Fecht, M., Naumann, A., Wissemeier, A., Maier, P., 1999. Physiology of manganese toxicity and tolerance in *Vigna unguiculata* (L.) Walp. J. Plant Nutr. Soil Sci. 162, 263−274.

Hossain, M.T., Ryuji, M., Soga, K., Wakabayashi, K., Kamisaka, S., Fuji, S., et al., 2002. Growth promotion and increase in cell wall extensibility by silicon in rice and some Poaceae seedlings. J. Plant Res. 115, 23−27.

Hu, Q.H., Xu, J., Pang, G.X., 2003. Effect of selenium on the yield and quality of green tea leaves harvested in early spring. J. Agric. Food Chem. 51, 3379−3381.

Inal, A., Pilbeam, D.J., Gunes, A., 2009. Silicon increases tolerance to boron toxicity and reduces oxidative damage in barley. J. Plant Nutr. 32, 112−128.

IRRI, 1991. Program Report for 1990. International Rice Research Institute, Los Banos Leguna, Philippines.

Ishizuka, Y., Tanaka, A., 1950. Studies on the nitrogen, phosphorus and potassium metabolism of the rice plant. 1. The influence of the nitrogen concentration in the culture solution on the growth of the rice plant, especially on the amount and the form of nitrogen in the plant. J. Sci. Soil Manure 21, 23−28.

Iwasaki, K., Maier, P., Fecht, M., Horst, W., 2002a. Effects of silicon supply on apoplastic manganese concentrations in leaves and their relation to manganese tolerance in cowpea (*Vigna unguiculata* (L.) Walp.). Plant Soil 238, 281−288.

Iwasaki, K., Maier, P., Fecht, M., Horst, W., 2002b. Leaf apoplastic silicon enhances manganese tolerance of cowpea (*Vigna unguiculata*). J. Plant Physiol. 159, 167−173.

Janislampi, K.W., 2012. Effect of silicon on plant growth and drought stress tolerance. MS thesis, Department of Plants, Soils, and Climate, Utah State University. Available from <http://digitalcommons.usu.edu/etd/1360/> (accessed 11.02.13.)

Jones, L.H.P., Handreck, K.A., 1965. Studies of silica in the oat plant. III. Uptake of silica from soils by the plant. Plant Soil 23, 79−96.

Kamenidou, S., Cavins, T.J., Marek, S., 2009. Evaluation of silicon as a nutritional supplement for greenhouse zinnia production. Sci. Hort. 119, 297−301.

Kaya, C., Tuna, A.L., Sonmez, O., Ince, F., Higgs, D., 2009. Mitigation effects of silicon on maize plants grown at high zinc. J. Plant Nutr. 32, 1788−1798.

Kaya, C., Tuna, L., Higgs, D., 2006. Effect of silicon on plant growth and mineral nutrition of maize grown under water-stress conditions. J. Plant Nutr. 29, 1469−1480.

Keeping, M.J., Reynolds, O.L., 2009. Editorial. Silicon in agriculture: new insights, new significance and growing application. Ann. Appl. Biol. 155, 153−154.

Khandekar, S., Leisner, S., 2011. Soluble silicon modulates expression of *Arabidopsis thaliana* genes involved in copper stress. J. Plant Physiol. 168, 699−705.

Kidd, P.S., Llugany, M., Poschenrieder, C., Gunse, B., Barcelo, J., 2001. The role of root exudates in aluminium resistance and silicon-induced amelioration of aluminium toxicity in three varieties of maize (*Zea mays* L.). J. Exp. Bot. 52, 1339−1352.

Kim, Y.H., Khan, A.L., Shinwari, Z.K., Kim, D.H., Waqas, M., Kamran, M., et al., 2012. Silicon treatment to rice (*Oryza sativa* Lcv "Gopumbyeo") plants during different growth periods and its effects on growth and grain yield. Pak. J. Bot. 44, 891−897.

Koivistoinen, P., 1980. Mineral element composition of Finnish foods: N, K, Ca, Mg, P, S, Fe, Cu, Mn, Zn, Mo, Ni, Cr, F, Se, Rb, Al, B, Br, Hg, As, Cd, Pb, and ash. Acta. Agric. Scand. 22, 170.

Kong, L., Wang, M., Bi, D., 2005. Selenium modulates the activities of antioxidant enzymes, osmotic homeostasis and promotes the growth of sorrel seedlings under salt stress. Plant Growth Regul. 45, 155−163.

Kuznetsov, V.V., Kholodova, V.P., Kuznetsov, V.V., Yagodin, B.A., 2003. Selenium regulates the water status of plants exposed to drought. Dokl. Biol. Sci. 390, 266−268.

Li, J., Frantz, J., Leisner, S., 2008. Alleviation of copper toxicity in *Arabidopsis thaliana* by silicon addition to hydroponic solutions. J. Am. Soc. Hortic. Sci. 133, 670−677.

Li, W.B., Shi, X.H., Wang, H., Zhang, F.S., 2004. Effects of silicon on rice leaves resistance to ultraviolet-B. Acta. Bot. Sin. 46, 691−697.

Li, W.B., Wang, H., Zhang, F.S., 2005. Effects of silicon on anther dehiscence and pollen shedding in rice under high temperature stress. Acta. Agron. Sin. 31, 134−136.

Liang, Y., 1999. Effects of silicon on enzyme activity and sodium, potassium and calcium concentration in barley under salt stress. Plant Soil 209, 217−224.

Liang, Y., Chen, Q., Liu, Q., Zhang, W., Ding, R., 2003. Exogenous silicon (Si) increases antioxidant enzyme activity and reduces lipid peroxidation in roots of salt-stressed barley (*Hordeum vulgare* L.). J. Plant Physiol. 160, 1157−1164.

Liang, Y., Zhang, W.Q., Chen, J., Ding, R., 2005. Effect of silicon on H^+-ATPase and H^+-PPase activity, fatty acid composition and fluidity of tonoplast vesicles from roots of salt stressed barley (*Hordeum vulgare* L.). Environ. Exp. Bot. 53, 29−37.

Liang, Y., Zhang, W., Chen, Q., Liu, Y., Ding, R., 2006a. Effect of exogenous silicon (Si) on H^+-ATPase activity, phospholipids and fluidity of plasma membrane in leaves of salt-stressed barley (*Hordeum vulgare* L.). Environ. Exp. Bot. 57, 212−219.

Liang, Y., Hua, H.X., Zhu, Y.G., Zhang, J., Cheng, C.M., Römheld, V., 2006b. Importance of plant species and external silicon concentration to active silicon uptake and transport. New Phytol. 172, 63−72.

Liang, Y., Sun, W., Zhu, Y.G., Christie, P., 2007. Mechanisms of silicon mediated alleviation of abiotic stresses in higher plants: a review. Environ. Pollut. 147, 422−428.

Liang, Y., Zhu, J., Li, Z., Chu, G., Ding, Y., Zhang, J., et al., 2008. Role of silicon in enhancing resistance to freezing stress in two contrasting winter wheat cultivars. Environ. Exp. Bot. 64, 286−294.

Liu, J.J., Lin, S.H., Xu, P.L., Wang, X.J., Bai, J.G., 2009. Effects of exogenous silicon on the activities of antioxidant enzymes and lipid peroxidation in chilling-stressed cucumber leaves. Agr. Sci. China. 8, 1075−1086.

Ma, J.F., 2004. Role of silicon in enhancing the resistance of plants to biotic and abiotic stresses. Soil Sci. Plant Nutr. 50, 11−18.

Ma, J.F., Takahashi, E., 2002. Soil, Fertilizer, and Plant Silicon Research in Japan. Elsevier, Amsterdam.

Ma, J.F., Yamaji, N., 2006. Silicon uptake and accumulation in higher plants. Trends Plant Sci. 11, 392−397.

Ma, J.F., Yamaji, N., 2008. Functions and transport of silicon in plants. Cell Mol. Life Sci. 65, 3049−3057.

Ma, J.F., Miyake, Y., Takahashi, E., 2001a. Silicon as a beneficial element for crop plants. In: Datonoff, L., Snyder, G., Korndorfer, G. (Eds.), Silicon in Agriculture. Elsevier, Amsterdam, pp. 17−39.

Ma, J.F., Goto, S., Tamai, K., Ichii, M., 2001b. Role of root hairs and lateral roots in silicon uptake by rice. Plant Physiol. 127, 1773−1780.

Ma, J.F., Tamai, K., Ichii, M., Wu, K., 2002. A rice mutant defective in active Si uptake. Plant Physiol. 130, 2111−2117.

Ma, J.F., Higashitani, A., Sato, K., Tateda, K., 2003. Genotypic variation in Si content of barley grain. Plant Soil 249, 383−387.

Ma, J.F., Tamai, K., Yamaji, N., Mitani, N., Konishi, S., Katsuhara, M., et al., 2006. A silicon transporter in rice. Nature 440, 688−691.

Ma, J.F., Yamaji, N., Tamai, K., Mitani, N., 2007. Genotypic difference in silicon uptake and expression of silicon transporter genes in rice. Plant Physiol. 145, 919−924.

Ma, J.F., Yamaji, N., Mitani-Ueno, N., 2011. Transport of silicon from roots to panicles in plants. Proc. Jpn. Acad. Ser. B. 87, 377−385.

Mali, M., Aery, N.C., 2008. Silicon effects on nodule growth, dry-matter production, and mineral nutrition of cowpea (*Vigna unguiculata*). J. Plant Nutr. Soil Sci. 171, 835−840.

Mateos-Naranjo, E., Andrades-Moreno, L., Davy, A.J., 2013. Silicon alleviates deleterious effects of high salinity on the halophytic grass *Spartina densiflora*. Plant Physiol. Biochem. 63, 115−121.

Máthé, C., Mosolygó, Á., Surányi, G., Beke, A., Demeter, Z., Tóth, V.R., et al., 2012. Genotype and explant-type dependent morphogenesis and silicon response of common reed (*Phragmites australis*) tissue cultures. Aquat. Bot. 97, 57−63.

Mayland, H.F., James, L.H., Sonderegger, J.L., Panter, K.E., 1989. Selenium in seleniferous environments. In: Jacobs, L.W. (Ed.), Selenium in agriculture and the environment, 23. Soil Sci Soc Amer Spec Pub, Madison, pp. 15−50.

Mditshwa, A., Bower, J.P., Bertling, I., Mathaba, N., Tesfay, S.Z., 2013. The potential of postharvest silicon dips to regulate phenolics in citrus peel as a method to mitigate chilling injury in lemons. Afr. J. Biotechnol. 12, 1482−1489.

Mitani, N., Ma, J.F., 2005. Uptake system of silicon in different plant species. J. Exp. Biol. 56, 1255−1261.

Mitani, N., Ma, J.F., Iwashita, T., 2005. Identification of the silicon form in xylem sap of rice (*Oryza sativa* L.). Plant Cell Physiol. 46, 279−283.

Mitani, N., Yamaji, N., Ma, J.F., 2009. Identification of maize silicon influx transporters. Plant Cell Physiol. 50, 5−12.

Miyake, Y., Takahashi, E., 1985. Effect of silicon on the growth of soybean plants in a solution culture. Soil Sci. Plant Nutr. 31, 625−636.

Molassiotis, A., Sotiropoulos, T., Tanou, G., Diamantidis, G., Therios, I., 2006. Boron-induced oxidative damage and antioxidant and nucleolytic responses in shoot tips culture of the apple rootstock EM9 (*Malus domestica* Borkh). Environ. Exp. Bot. 56, 54−62.

Motomura, H., Mita, N., Suzuki, M., 2002. Silica accumulation in long-lived leaves of *Sasa veitchii*. Ann. Bot. Lond. 90, 149−152.

Nawaz, F., Ashraf, M.Y., Ahmad, R., Waraich, E.A., 2013. Selenium (Se) seed priming induced growth and biochemical changes in wheat under water deficit conditions. Biol. Trace Elem. Res. 151, 284−293.

Neumann, D., Zur Nieden, U., 2001. Silicon and heavy metal tolerance of higher plants. Phytochemistry 56, 685−692.

Nhan, P.P., Dong, N.T., Nhan, H.T., Chi, N.T.M., 2012. Effects of OryMax[SL] and Siliysol[MS] on growth and yield of MTL560 rice. World Appl. Sci. J. 19, 704−709.

Nowakowski, W., Nowakowska, J., 1997. Silicon and copper interaction in the growth of spring wheat seedlings. Biol Plant 39, 463−466.

Nwugo, C.C., Huerta, A.J., 2008. Effects of silicon nutrition on cadmium uptake, growth and photosynthesis of rice plants exposed to low-level cadmium. Plant Soil 311, 73−86.

Okuda, H., Takahashi, E., 1961. Effects of silica on nutrition and physiology of rice plant. J. Sci. Soil Manual 473−479.

Parveen, N., Ashraf, M., 2010. Role of silicon in mitigating the adverse effects of salt stress on growth and photosynthetic attributes of two maize (*Zea mays* L.) cultivars grown hydroponically. Pak. J. Bot. 42, 1675−1684.

Pedrero, Z., Madrid, Y., Hartikainen, H., Cámara, C., 2008. Protective effect of selenium in broccoli (*Brassica oleracea*) plants subjected to cadmium exposure. J. Agric. Food Chem. 56, 266−271.

Peng, X.L., Liu, Y.Y., Luo, S.G., 2002. Effects of selenium on lipid peroxidation and oxidizing ability of rice roots under ferrous stress. J. Northeast Agric. Univ. 19, 9−15.

Pennanen, A., Xue, T., Hartikainen, H., 2002. Protective role of selenium in plant subjected to severe UV irradiation stress. J. Appl. Bot. 76, 66−76.

Pezzarossa, B., Petruzzelli, G., 2001. Selenium contamination in soil: sorption and desorption processes. In: Selim, H.M., Sparks, D.L. (Eds.), Heavy Metals Release in Soils. CRC Press, Boca Raton, pp. 191−206.

Pilon-Smits, E.H.A., Quinn, C.F., 2010. Selenium metabolism in plants. In: Hell, R., Mendel, R.R. (Eds.), Cell biology of metals and nutrients, Plant Cell Monographs. Springer, Berlin, p. 17.

Prabagar, S., Hodson, M.J., Evans, D.E., 2011. Silicon amelioration of aluminium toxicity and cell death in suspension cultures of Norway spruce (*Picea abies* (L.) Karst.). Environ. Exp. Bot. 70, 266−276.

Pulz, A.L., Crusciol, C.A.C., Lemos, L.B., Soratto, R.P., 2008. Influência de silicato e calcário na nutrição, produtividade e qualidade da batata sob deficiência hídrica. Rev. Bras. Ciênc. Solo. 32, 1651–1659.

Qi, X., Liu, Y.Y., Song, T.X., 2004. Effects of selenium on root oxidizing ability and yield of rice under ferrous stress. J. Northeast Agric. Univ. 11, 19–22.

Rafi, M.M., Epstein, E., 1999. Silicon absorption by wheat. Plant Soil 211, 223–230.

Rani, N., Dhillon, K.S., Dhillon, S.K., 2005. Critical levels of selenium in different crops grown in an alkaline silty loam soil treated with selenite-Se. Plant Soil 277, 367–374.

Raven, J.A., 2001. Silicon transport at the cell and tissue level. In: Datnoff, L.E., Snyder, G.H., Korndorfer, G. H. (Eds.), Silicon in Agriculture. Elsevier, Amsterdam, pp. 41–55.

Ribeiro, D.M., Mapeli, A.M., Antunes, W.C., Barros, R.S., 2011. A dual role of selenium in the growth control of seedlings of *Stylosanthes humilis*. Agric. Sci. 2, 78–85.

Richmond, K.E., Sussman, M., 2003. Got silicon? The non-essential beneficial plant nutrient. Curr. Opin. Plant Biol. 6, 268–272.

Rizwan, M., Meunier, J.D., Miche, H., Keller, C., 2012. Effect of silicon on reducing cadmium toxicity in durum wheat (*Triticum turgidum* L. cv. Claudio W.) grown in a soil with aged contamination. J. Hazard. Mater 209–210, 326–334.

Rodríguez, M., Canales, E., Borrás-Hidalgo, O., 2005. Molecular aspects of abiotic stress in plants. Biotechnol. Appl. 22, 1–10.

Rogalla, H., Römhled, V., 2002. Role of apoplast in silicon-mediated manganese tolerance of *Cucumis sativus* L. Plant Cell Environ. 25, 549–555.

Romero-Aranda, M.R., Jurado, O., Cuartero, J., 2006. Silicon alleviates the deleterious salt effect on tomato plant growth by improving plant water status. J. Plant Physiol. 163, 847–855.

Sacaa, E., 2009. Role of silicon in plant resistance to water stress. J. Elementol. 14, 619–630.

Sajedi, N., Madani, H., Naderi, A., 2011. Effect of microelements and selenium on superoxide dismutase enzyme, malondialdehyde activity and grain yield maize (*Zea mays* L.) under water deficit stress. Not. Bot. Horti. Agrobot. 39, 153–159.

Sangster, A.G., Hodson, M.J., Tubb, H.J., 2001. Silicon deposition in higher plants. In: Datnoff, L.E., Snyder, G.H., Korndorfer, G.H. (Eds.), Silicon in Agriculture. Elsevier, Amsterdam, pp. 85–113.

Saqib, M., Zörb, C., Schubert, S., 2008. Silicon-mediated improvement in the salt resistance of wheat (*Triticum aestivum*) results from increased sodium exclusion and resistance to oxidative stress. Funct. Plant Biol. 35, 633–639.

Savant, N.K., Korndorfer, G.H., Datnoff, L.E., Snydre, G.H., 1999. Silicon nutrition and sugarcane production: a review. J. Plant Nutr. 22, 1853–1903.

Savvas, D., Manos, G., Kotsiras, A., Souvalotis, S., 2002. Effects of silicon and nutrient induced salinity on yield, flower quality and nutrient uptake of gerbera grown in a closed hydroponic system. J. Appl. Bot. 76, 153–158.

Seppänen, M., Turakainen, M., Hartikainen, H., 2003. Selenium effects on oxidative stress in potato. Plant Sci. 165, 311–319.

Shanker, A.K., 2006. Countering UV-B stress in plants: does selenium have a role? Plant Soil 282, 21–26.

Shen, X., Li, J., Duan, L., Li, Z., Eneji, A.E., 2009. Nutrient acquisition by soybean treated with and without silicon under ultraviolet-B radiation. J. Plant Nutr. 32, 1731–1743.

Shen, X., Zhou, Y., Duan, L., Li, Z., Eneji, A.E., Li, Z., 2010. Silicon effects on photosynthesis and antioxidant parameters of soybean seedlings under drought and ultraviolet-B radiation. J. Plant Phys. 167, 1248–1252.

Shi, G., Cai, Q., Liu, C., Wu, L., 2010. Silicon alleviates cadmium toxicity in peanut plants in relation to cadmium distribution and stimulation of antioxidative enzymes. Plant Growth Regul. 61, 45–52.

Shi, X.H., Zhang, C.C., Wang, H., Zhang, F.S., 2005. Effect of Si on the distribution of Cd in rice seedlings. Plant Soil 272, 53–60.

Shi, Y., Wang, Y., Flowers, T.J., Gong, H., 2013. Silicon decreases chloride transport in rice (*Oryza sativa* L.) in saline conditions. J. Plant Physiol. 170, 847−853.

Shrift, A., 1973. Selenium compounds in nature and medicine. In: Klayman, D.L., Gunther, W.H.H. (Eds.), Organic Selenium Compounds: Their Chemistry and Biology. Wiley, New York, pp. 763−814.

Sippola, J., 1979. Selenium content of soils and timothy (*Phleum pratense* L.) in Finland. Ann. Agric. Fenn. 18, 182−187.

Smrkolj, P., Germ, M., Kreft, I., Stibilj, V., 2006. Respiratory potential and Se compounds in pea (*Pisum sativum* L.) plants grown from Se-enriched seeds. J. Exp. Bot. 57, 3595−3600.

Sommer, M., Kaczorek, D., Kuzyakov, Y., Breuer, J., 2006. Silicon pools and fluxes in soils and landscapes— a review. J Plant Nutr. Soil Sci. 169, 310−329.

Son, M.S., Song, J.Y., Lim, M.Y., Sivanesan, I., Jeong, B.R., 2011. Effect of silicon on tolerance to high temperatures and drought stress in *Euphorbia pulcherrima* willd. "Ichiban". Proceedings of the 5th International Conference on Silicon in Agriculture, Beijing, China, p. 188.

Song, A., Li, Z., Zhang, J., Xue, G., Fan, F., Liang, Y., 2009. Silicon-enhanced resistance to cadmium toxicity in *Brassica chinensis* L. is attributed to Si-suppressed cadmium uptake and transport and Si-enhanced antioxidant defense capacity. J. Hazard. M. 172, 74−83.

Song, A., Li, P., Li, Z., Fan, F., Nikolic, M., Liang, Y., 2011. The alleviation of zinc toxicity by silicon is related to zinc transport and antioxidative reactions in rice. Plant Soil 344, 319−333.

Sonobe, K., Hattorri, T., An, P., Tsuji, W., Eneji, A.E., Kobayashi, S., et al., 2010. Effect of silicon application on sorghum root responses to water stress. J. Plant Nutr. 34, 71−82.

Sonobe, K., Hattori, T., An, P., Tsuji, W., Eneji, A.E., Kobayashi, S., et al., 2011. Effect of silicon application on sorghum root responses to water stress. J. Plant Nutr. 34, 71−82.

Sors, T.G., Ellis, D.R., Salt, D.E., 2005. Selenium uptake, translocation, assimilation and metabolic fate in plants. Photosynth. Res. 86, 373−389.

Soylemezoglu, G., Demir, K., Inal, A., Gunes, A., 2009. Effect of silicon on antioxidant and stomatal response of two grapevine (*Vitis vinifera* L.) rootstocks grown in boron toxic, saline and boron toxic-saline soil. Sci. Hort 123, 240−246.

Srivastava, M., Ma, L.Q., Rathinasabapathi, B., Srivastava, P., 2009. Effects of selenium on arsenic uptake in arsenic hyperaccumulator *Pteris vittata* L. Bioresour. Technol. 100, 1115−1121.

Sun, H.W., Ha, J., Liang, S.X., Kang, W.J., 2010. Protective role of selenium on garlic growth under cadmium stress. Commun. Soil Sci. Plant Anal. 41, 1195−1204.

Tadina, N., Germ, M., Kreft, I., Breznik, B., Gaberščik, A., 2007. Effects of water deficit and selenium on common buckwheat (*Fagopyrum esculentum* Moench.) plants. Photosynthetica 45, 472−476.

Tahir, M.A., Aziz, T., Farooq, M., Sarwar, G., 2012. Silicon-induced changes in growth, ionic composition, water relations, chlorophyll contents and membrane permeability in two salt-stressed wheat genotypes. Arch. Agron. Soil Sci. 58, 247−256.

Takahashi, E., Ma, J.F., Miyake, Y., 1990. The possibility of silicon as an essential element for higher plants. Comments Agric. Food Chem. 2, 99−122.

Tamai, K., Ma, J.F., 2003. Characterization of silicon uptake by rice roots. New Phytol. 158, 431−436.

Terry, N., Zayed, A.M., de Souza, M.P., Tarun, A.S., 2000. Selenium in higher plants. Annu. Rev. Plant Physiol. Plant Mol. Biol. 51, 401−432.

Tesfay, S.Z., Bertling, I., Bower, J.P., 2011. Effects of postharvest potassium silicate application on phenolics and other antioxidant systems aligned to avocado fruit quality. Postharvest. Biol. Technol. 60, 92−99.

Tuna, A.L., Kaya, C., Higgs, D., Murillo-Amador, B., Aydemir, S., Gergon, A.R., 2008. Silicon improves salinity tolerance in wheat plants. Environ. Exp. Bot. 62, 10−16.

Turakainen, M., 2007. Selenium and its effects on growth, yield and tuber quality in potato. University of Helsinki, Department of Applied Biology, Publication No 30, Helsinki, p 50.

Turakainen, M., Hartikainen, H., Seppänen, M.M., 2004. Effects of selenium treatments on potato (*Solanum tuberosum* L.) growth and concentrations of soluble sugars and starch. J. Agric. Food Chem. 52, 5378–5382.

Turakainen, M., Hartikainen, H., Ekholm, P., Seppänen, M., 2006. Distribution of selenium in different biochemical fractions and raw darkening degree of potato (*Solanum tuberosum* L.) tubers supplemented with selenate. J. Agric. Food Chem. 54, 8617–8622.

Vaculík, M., Landberg, T., Greger, M., Luxová, M., Stoláriková, M., Lux, A., 2012. Silicon modifies root anatomy, and uptake and subcellular distribution of cadmium in young maize plants. Ann. Bot. 110, 433–443.

Valadabadi, S.A., Shiranirad, A.H., Farahani, H.A., 2010. Ecophysiological influences of zeolite and selenium on water deficit stress tolerance in different rapeseed cultivars. J. Ecol. Nat. Environ. 2, 154–159.

Valdez Barillas, J.R., Quinn, C.F., EAH, Pilon-Smits, 2011. Selenium accumulation in plants—phytotechnological applications and ecological implications. Int. J. Phytoremed. 13, 166–178.

Valkama, E., Kivimäenpää, M., Hartikainen, H., Wulff, A., 2003. The combined effects of enhanced UV-B radiation and selenium on growth, chlorophyll fluorescence and ultrastructure in strawberry (*Fragaria × ananassa*) and barley (*Hordeum vulgare*) treated in the field. Agric. For Meteorol. 120, 267–278.

Van Hoest, P.J., 2006. Rice straw, the role of silica and treatments to improve quality. Anim. Feed Sci. Technol. 130, 137–171.

Vorobets, N., Mykiyevich, I., 2000. Single and combined effects of lead and selenium on sunflower seedlings. Scientific Workshop on Horticulture and Vegetable Growing 19, 390.

Wang, C.Q., Xu, H.J., Liu, T., 2011. Effect of selenium on ascorbate–glutathione metabolism during PEG-induced water deficit in *Trifolium repens* L. J. Plant Growth Regul. 30, 436–444.

Wang, Y.X., Stass, A., Horst, W.J., 2004. Apoplastic binding of aluminium is involved in silicon-induced amelioration of aluminium toxicity in maize. Plant Physiol. 136, 3762–3770.

Whanger, P.D., 2002. Review. Selenocompounds in plants and animals and their biological significance. J. Amer. College Nutr. 21, 223–232.

White, P.J., Bowen, H.C., Parmaguru, P., Fritz, M., Spracklen, W.P., Spidy, R.E., et al., 2004. Interactions between selenium and sulphur nutrition in *Astragalus thaliana*. J. Exp. Bot. 55, 1927–1937.

White, P.J., Bowen, H.C., Marshall, B., Broadley, M.R., 2007. Extraordinarily high leaf selenium to sulfur ratios define "Se-accumulator" plants. Ann. Bot. 100, 111–118.

Xiaoqin, Y., Jianzhou, C., Guangyin, W., 2009a. Effects of drought stress and selenium supply on growth and physiological characteristics of wheat seedlings. Acta. Physiol. Plant 31, 1031–1036.

Xiaoqin, Y., Jianzhou, C., Guangyin, W., 2009b. Effects of selenium on wheat seedlings under drought stress. Biol. Trace Elem. Res. 130, 283–290.

Xu, P.L., Guo, Y.K., Bai, J.G., Shang, L., Wang, X.J., 2008. Effects of long-term chilling on ultrastructure and antioxidant activity in leaves of two cucumber cultivars under low light. Physiol. Plant 132, 467–478.

Xue, T., Hartikainen, H., Piironen, V., 2001. Antioxidative and growth-promoting effect of selenium in senescing lettuce. Plant Soil 237, 55–61.

Yamaji, N., Ma, J.F., 2007. Spatial distribution and temporal variation of the rice silicon transporter *Lsi1*. Plant Physiol. 143, 1306–1313.

Yamaji, N., Mitani, N., Ma, J.F., 2008. A transporter regulating silicon distribution in rice shoots. Plant Cell 20, 1381–1389.

Yao, X., Chu, J., Wang, G., 2009. Effects of selenium on wheat seedlings under drought stress. Biol. Trace Elem. Res. 130, 283–290.

Yao, X., Chu, J., Ba, C., 2010a. Responses of wheat roots to exogenous selenium supply under enhanced ultraviolet-B. Biol. Trace Elem. Res. 137, 244–252.

Yao, X.Q., Chu, J.Z., Ba, C.J., 2010b. Antioxidant responses of wheat seedlings to exogenous selenium supply under enhanced ultraviolet-B. Biol. Trace Elem. Res. 136, 96–105.

Ye, J., Yan, C., Liu, J., Lu, H., Liu, T., Song, Z., 2012. Effects of silicon on the distribution of cadmium compartmentation in root tips of *Kandelia obovata* (S., L.) Yong. Environ. Pollut. 162, 369–373.

Yeo, A.R., Flowers, S.A., Rao, G., Welfare, K., Senanayake, N., Flowers, T.J., 1999. Silicon reduces sodium uptake in rice (*Oryza satival* L.) in saline conditions and this is accounted for by a reduction in the transpirational bypass flow. Plant Cell Environ. 22, 559–565.

Yläranta, T., 1983a. Sorption of selenite and selenate added in the soil. Ann. Agric. Fenn. 22, 29–39.

Yläranta, T., 1983b. Selenium in Finnish agricultural soils. Ann. Agric. Fenn. 22, 122–136.

Yläranta, T., 1985. Increasing the selenium content of cereals and grass crops in Finland. Academic dissertation, Agricultural Research Centre, Institute of Soil Science, Jokioinen. Yliopistopaino Helsinki 72.

Zahedi, H., Noormohammadi, G., Rad, A.H.S., Habibi, D., Boojar, M.M.A., 2009. Effect of zeolite and foliar application of selenium on growth, yield and yield component of three canola cultivar under conditions of late season drought stress. Not. Sci. Biol. 1, 73–80.

Zayed, A., Lytle, C.M., Terry, N., 1998. Accumulation and volatilization of different chemical species of selenium by plants. Planta 206, 284–292.

Zeng, F.R., Zhao, F.S., Qiu, B.Y., Ouyang, Y.N., Wu, F.B., Zhang, G.P., 2011. Alleviation of chromium toxicity by silicon addition in rice plants. Agr. Sci. China 10, 1188–1196.

Zhang, C., Wang, L., Nie, Q., Zhang, W., Zhang, F., 2008. Long-term effects of exogenous silicon on cadmium translocation and toxicity in rice (*Oryza sativa* L.). Environ. Exp. Bot. 62, 300–307.

Zhao, D., Hao, Z., Tao, J., Han, C., 2013. Silicon application enhances the mechanical strength of inflorescence stem in herbaceous peony (*Paeonia lactiflora* Pall.). Sci. Hort 151, 165–172.

Zhu, J., Liang, Y.C., Ding, Y.F., Li, Z.J., 2006. Effects of silicon on photosynthesis and its related physiological parameters in two winter wheat cultivars under cold stress. Sci. Agric. Sin. 39, 1780–1788.

Zhu, Z., Wei, G., Li, J., Qian, Q., Yu, J., 2004. Silicon alleviates salt stress and increases antioxidant enzymes activity in leaves of salt-stressed cucumber (*Cucumis sativus* L.). Plant Sci. 167, 527–533.

Herbicides, Pesticides, and Plant Tolerance: An Overview

Qaisar Mahmood, Muhammad Bilal and Sumira Jan

17.1 Introduction

Pesticides are man-made and naturally occurring chemicals that control insects (Khan et al., 2010; Xiao et al., 2010), weeds, fungi, and other pests that destroy crops. Pesticide is a general term that includes a large number of biocidal compounds like fungicides, nematicides, insecticides, molluscicides, rodenticides, herbicides, and plant growth hormones. The most common pesticides like organochlorines (OC) have been widely used to overcome many diseases like malaria and typhus, but due to their recalcitrant nature, their use has been banned or at least reduced during last 50 years in the majority of developed nations. The application of various organophosphates (OP) during the 1960s, carbamates during the 1970s, and pyrethroids during the 1980s, along with the synthesis of other pesticides (1970s–1980s), have effectively reduced the pest hazards and resulted in increased productivity (Aktar et al., 2009).

Pesticides are the chemical species that cause death and avoid or reduce growth of plants or animals that are considered as pests. Herbicides are a class of pesticides that are used to kill weeds and other undesirable life forms in agricultural crops, including insects, while fungicides are employed to restrict the growth of molds and mildew. The use of disinfectants prevents the outbreak of bacteria and is also used to control mice and rats. Due to excessive use of these chemicals in agriculture, humans may be exposed to low concentrations of these biocidal residues through the food chain. The health effects of these pesticide residues are currently being investigated. Surveys have revealed that more prevalent cases of insomnia, dizziness, headaches, hand tremors, fatigue, and other neurological symptoms in children among the farming community are due the use of agricultural insecticides and pesticides. Hepel et al. (2012) established that catechol-containing compounds in the presence of copper(II) or iron(II) ions may induce a Fenton cascade leading to reactive oxygen species (ROS) generation, which are potent enough to cause damage to DNA. Currently, more than half of the world production of catechol is used by the pesticide and herbicide industry (Nurzhanova et al., 2013).

Restrained use of pesticides is critical to meet the food supply for an ever-increasing human population, and to avoid pest attack by ensuring the safety of human health. Crop protection and agricultural production are important factors of food safety, which can be ensured through judicious application of pesticides (Bolognesi, 2003). According to the United Nations Food and Agricultural Organization, the world's potential human food losses are about 55%, which include pre-harvest (35%) and post-harvest (20%) losses. In advanced countries, about 10–30% of crop loss occurs

P. Ahmad (Ed): Emerging Technologies and Management of Crop Stress Tolerance, Volume 1.
DOI: http://dx.doi.org/10.1016/B978-0-12-800876-8.00017-5

due to pests; however, these losses are 75% in developing nations (Ohaya-Mitoko, 1997). Pesticides are crucial to enhance agricultural production, but at the same time they are toxic recalcitrant substances. Almost 90% of food is exposed to different herbicides, which are used in crops (Rosa et al., 2008). The herbicide residues in food are a consequence of their being directly sprayed onto crop plants and less likely due to residues persisting in the soil (Businelli et al., 1992). The presence of pesticide residues in food plants, especially vegetables, represents the most critical threat to human health. Best agricultural management practices can greatly reduce the risk associated with pesticide food chain contamination.

Aktar et al. (2009) have discussed the benefits and hazards of the application of pesticides in agriculture. Overwhelmingly large benefits result from the application of pesticides on agricultural crops and these xenobiotics present an ideal opportunity to study risk assessment. It has been estimated that developing countries need $8 billion on an annual basis. An in-depth comparison of financial gains from the use of pesticides and their subsequent impact on human health is required. The total cost and associated benefits vary among developed and developing countries. To ensure food security, pesticide use in developing nations is more intensive and the risk to human health is sometimes more demanding. However, the approach to pesticide use should be more prudent and rational, and practiced on scientific principles rather than merely for profit. Numerous factors, i.e., sex, race, age, state of health, socioeconomic status, and diet, have to be taken into account to make a proper scientific judgment on the effects of pesticide residues on human health. The long-term exposure to minute quantities of pesticides is greatly influenced by their interaction with other pollutants found in air, water, and food (Aktar et al., 2009).

Pesticide contamination is a greater risk to the sustainability of the environment and non-target organisms like soil flora and fauna, birds, and fishes. However, the cost of these pesticides is ever-increasing and their environmental toxicity is a consequence of their increased use. In reality, the persistent use of herbicides has caused greater damage to the environment. Weedicides are more toxic in elevated concentration. Aktar et al. (2009) recommended the standard procedure of minimizing xenobiotic residues and their harmful impact on the environment through their safer use and utilization of non-chemical pest control strategies. Focus should be centered on the avoidance of serious health consequences and endorsement of health as a lucrative venture for workers to achieve sustainable financial development. Health education should be based on information, capacity, and performance, and communities should be educated on the risks to health of pesticides and how their exposure can be reduced (Aktar et al., 2009).

17.2 Global pesticide use

A dramatic rise in pesticide usage throughout the globe is evident along with the ever-increasing human population and crop production (Zhang et al., 2011). The increasing grave misuse of pesticides is causing critical damage to the environment. The status of pesticide pollution was analyzed on a global scale (especially in China) to protect human health and endangered plant and animal species. During this course, pesticide development and usage were also considered and reviewed. The global pesticide consumption pattern has changed significantly over the last 50 years. Compared to pesticides, herbicide consumption has rapidly escalated, while the proportionate consumption of insecticides and fungicides/bactericides has declined. China is the major herbicide manufacturer and exporter worldwide. Pesticide contamination of the environment and resultant deaths in China remain a serious issue. An alternate option of bio-pesticides has been suggested by Zhang et al. (2011).

Historically, the use of pesticides has been divided into three periods (Zhang et al., 2006): (1) the main period covers the time prior to the 1870s when naturally occurring chemicals like sulfur were used as pesticides; sulfur was used in ancient Greece for pest management; (2) the next period, from the 1870s to 1945, employed inorganic synthetic compounds; the natural compounds found in various plant extracts along with some synthetic inorganics were the main focus of this period; (3) the last period, since 1945, comprises the time when organic synthetic pesticides were used. Over the last 60 years, synthetic pesticides like DDT, 2,4-D, and later HCH and dieldrin replaced inorganic and natural pesticides. Also, a number of synthetic pesticides were prepared called chemical pesticides. The use of synthetic organic herbicides has remained a hallmark for human society, which has greatly facilitated pest control and enhanced agricultural productivity. During the initial years of last period, the main types of organic insecticides, i.e., carbamate, organophosphate, and organochloride, were synthesized. Subsequently, herbicides and fungicides were used as the main pest control strategy. However, the use of insecticides is seen to be gradually declining and herbicides will gain popularity for the future (Zhang et al., 2011).

The United States Environmental Protection Agency reported the consumption of 3 billion kg of herbicides during 2001, which corresponded to 2 kg per capita in the USA (Toxipedia.org 2011). Roughly 400 million kg of active components and 600 different chemicals were listed as herbicides. Agriculture production used around 300 million kg of herbicides and 45 million kg (11.5%) were sprayed on lawns, gardens, and other areas. Besides this, approximately 1 billion kg were sprayed as disinfectants, and 0.3 billion kg were utilized as wood preservatives (Pesticide Use Statistics, 2011). Globally, around 2 billion kg of active pesticide ingredients were sprayed in agriculture production during 2001. Table 17.1 presents use of pesticides in the USA. Pesticide utilization is gradually decreasing in the USA in view of environmental protection. During 1972, the use of DDT and other organo-chlorinated pesticides was banned in the USA. Since 1975, the use

Table 17.1 The Consumption of Pesticides in the USA

Category of Pesticide	Billions of kg	Percent Use
Traditional pesticides like fungicides, herbicides, insecticides, etc.	0.4	17.7
Petrochemicals*	0.14	6.4
Chemicals to protect wood	0.3	16.1
Antimicrobial chemicals	0.15	7.2
Chlorine/hypochlorites for water disinfection	1.25	52.5
TOTAL	2.4	100

*According to EPA: "These pesticides include sulfur and petroleum oil and other chemical ingredients such as sulfuric acid, insect repellants (e.g., DEET), moth control products (e.g., paradichlorobenzene), and related chemicals." (EPA Pesticides Industry Sales and Usage 2000 and 2001 Market Estimates, Table 3.3 (2004).)

of these chemicals decreased by 35% without affecting agricultural productivity (SDNX, 2005). The total cost of pesticides in the USA during 2001 was $11.09 billion of which $7.4 billion was used in the agricultural sector. Globally, the price of pesticides to enhance agriculture productivity during 2001 was US$31.8 billion (Toxipedia.org 2011).

Future pesticides will be extremely proficient with high biological reactivity, which will result in the decline of pesticide usage and will reduce environmental toxicity. These advanced pesticides will be less toxic, pollution free, and thus eco-friendly. Another recent concept is the development of bio-pesticides that employ direct use of either various life forms or their biochemical intermediates under field conditions and synthetic products from genetically modified organisms (GMOs), pathogenic insects, wild plants, or pathogenic microbes (Zhang and Zhang, 1998; Zhu et al., 2002).

The following benefits are associated with bio-pesticides:

1. High-quality control of pests, safe to non-target animals, no residues left, and biodegradability is highly desirable
2. Highly specific to target species
3. Should result in high productivity for greater sustainability
4. Prone to modification through modern biotechnological and fermentation procedures to enhance output and better qualitative features
5. Low pest resistance generation (Yang, 2001)

Hundreds of bio-pesticides currently exist, of which around 30% are synthesized on a commercial basis (Xu, 2008). Forty-four percent of these bio-pesticides are being used in the USA, Canada, and Mexico, while consumption in the rest of the world is 56% (Qin and Kong, 2006).

17.3 Why pesticide-/herbicide-tolerant plants?

The toxicological and environmental effects of synthetic herbicides are an increasing concern for human health and environmental protection agencies (Sunohara et al., 2010). The production of highly specific herbicides at very low concentrations is a prerequisite to controlling the target plants without harming other non-target organisms. The fine toxicological and environmental performances are still highly preferred in spite of the registered 400 herbicides (Sunohara et al., 2010). Due to its frequent use, pesticide resistance assists the ability of plants and other organisms. Such resistance follows the rules of evolution, resulting in the survival of the fittest and induces a heritable alteration. Multiple herbicide selections may be disregarded if a weed has acquired resistance to numerous herbicides (i.e., cross-resistance). Apparently, reducing herbicide choices may result in important economic and environmental consequences to agriculture. The evaluation of herbicide-resistant weeds involves complex and costly procedures. Because of cross-tolerance, consistent efforts should be put forth to overcome herbicide resistance. The herbicide resistance concern has resolutions and it is best to consider them as a resource. Later, plans for resistance prevention may be devised (Gunsolus, 2008).

Besides certain disadvantages, herbicide resistance in plants may be beneficial for certain plant species. The discovery of herbicide-resistant (HR) weeds during the 1970s activated an interest in imitating this involuntary development into crop breeding (Madsen and Streibig, 2013).

The associated growth in biochemistry realized the incorporation of genes responsible for tolerance in vulnerable plants. The conventional route for producing herbicide-resistant crops (HRCs) was initially carried out by traditional breeding methods. However, genetic engineering emerged as the major contributor to producing HRCs, but the technology has been under scrutiny for its benefits and other ethical issues. HRCs had been produced on a commercial scale during last 30 years when the OAC Triton HR biotype was developed and released in Canada. This cultivar was the product of breeding between the HR *Brassica rapa* L. oilseed rape plant (Hall et al., 1996). Genetically modified (GM) HRCs encompassed the bulk of cultivated areas where GM crops were grown (James, 2001). GM HRCs are normally considered as "first generation crops" and their efficacy has been queried due to hypothetical threats to users and the environment. In the case of a few HRCs, herbicides were substituted with a less promising ecological profile. Additionally, costs of weed control programs in glyphosate-tolerant soybean were reduced in conventional and HRCs due to the low cost of herbicides (Madsen and Streibig, 2013).

Glyphosate-tolerant soybean crops offer farmers a vital tool for fighting weeds and are compatible with no-till methods, which help preserve topsoil. As glyphosate has a strong affinity to become adsorbed to the soil particles, there was little chance of having residues on subsequent crops. The number of chemical sprays on soybeans was decreased to 12% for the season 1995−1999. However, an increase was observed in terms of total concentration of active constituents sprayed (Carpenter and Gianessi, 2001). The implementation effects of GM crops cannot be separated from other factors affecting insect killer applications (Heimlich et al., 2000). According to the American Soybean Association, environment protection by HR soybean involves changes in tillage procedures, pesticide application, and enhanced weed control (Anderson, 2001). The commercially available HR rice varieties are other examples. Agronomically, there are two arguments in favor of development of HR rice: (1) improvements of management in the weed related with rice, particularly red rice and other taxa (Gealy and Dilday, 1997; Olofdotter et al., 2000) and (2) provision of an alternative method to control weeds which formerly attained tolerance to specific pesticides, particularly monocots like *Echinochloa* spp. (Wilcut et al., 1996; Olofdotter et al., 2000). HR rice authorizes the usage of substitutes to presently practiced herbicides (Olofdotter et al., 2000).

Soil loss due to farming procedures is troublesome in many parts of the world. Generally, HRCs will be encouraging for environment protection through pest control in relation to conservative measures. Such practice will allow crop growers to use conventional farming methods in order to decrease soil erosion, especially through non-tillage (Duke, 2001). In the case of HRCs, vulnerabilities may be considered as qualitative estimates including the probability and harshness of both instant and outstanding severe impacts on human health, the environment, and the economies of farmers. Such undesirable effects may be caused by many traits such as crop specificity, particularly resistant features, weeds, atmospheric conditions, and farming practices (Madsen et al., 2002).

The best possible weed eradication often needs chronological applications of glyphosate in glyphosate-resistant (GR) crops. The proper schedule for such application is very crucial and the spray application in relation to weed appearance is important (Swanton et al., 2000). A high selection pressure on weed plants is caused by high annual doses of glyphosate. In a span of around 8 years, a shift in GR weed composition is usually observed (Shaner, 2000; Benbrook, 2001), in which case discrete pesticides are required to eradicate these GR weeds (Shaner, 2000). The classical, post-emergence pesticides were anticipated for effective weed control of GR soybean to support the elimination of GR weeds like *Sesbania exaltata* (Raf.) Cory, *Ipomoea* spp., or *Amaranthus rudis*

Sauer (Payne and Oliver, 2000). Growing GR corn and soybean in their rotation will not result in the control of native corn by glyphosate (Shaner, 2000). Flow of genes within similar crop species can be an alternate method to develop weed-resistant crop populations. Steady spray of pesticides with a similar mode of action can lead to selection of the single gene responsible for resistance if detected, which can be further implemented for similar weed control. Such practice will result in conveying pesticide resistance in plants and natives.

Amplified use of pesticides can be a risk in a few parts of the world, while the toxicity of these pesticides to humans or environmental health is not fully characterized. The toxic effects caused by the extensive use of pesticides on ground water and the pesticide residues in different food plants are critical. Extensive application of herbicides for HT crops may be due to increased pest tolerance to respective pesticides, which proved a compelling factor for farmers to employ high concentration pesticide sprays. Farmers considered that pesticides would not exert harmful effects on plants and they would realize greater yields by controlling weeds. Moreover, resistance of weeds or native plants further compels farmers to enhance pesticide usage for effective weed control and increase crop yield.

The use of herbicides affects the biodiversity of a field when pesticides are used in greater quantities to control weeds or wild species. In addition, weeds show differential response to various herbicides and other methods to eradicate undesirable vegetation; this may result in vegetation shifting of a particular field. According to FAO (2001), consequent reduction in the diversity of local field species will be hazardous for growth of HR crops of genetic origin. Further, HR is not predicted to cause genetic diversity fluctuations among wild plant species because herbicide application is also prevalent outside cultivated lands. Moreover, HR traits cannot result in selective advantage excluding plants exposed to herbicides (Poulsen, 1995; Madsen et al., 1998). Thus, a meager loss of genetic diversity among wild plant species is anticipated in natural habitats. Despite major ethical and ecological threats to genetic diversity of native species, few GM crop plants have been considered safe under discreet scientific norms, e.g., HR sugar beet (Madsen and Sandøe, 2001). However, the final release of HR varieties has to surpass several strict scientific assessments (Madsen et al., 2002).

17.4 Mechanisms of Herbicide/pesticide tolerance in plants

Herbicide/pesticide tolerance is a complicated process involving numerous components of a plant. These include phytochromes, plant antioxidant machinery, glycoproteins, and interaction of various metabolic systems (Figure 17.1). Phytoremediation is one potential method for reducing risk from these pesticides. Genetic heterogeneity of wild populations and weedy species growing on pesticide-contaminated soil provides a source of plant species tolerant to these conditions (Nurzhanova et al., 2013). In this section, we will deal with various aspects of herbicide/pesticide resistance mechanisms found in plants.

17.4.1 Cytochrome P450-mediated resistance

The enzymes of cytochrome P450 (Cyt P450) have an important role in the detoxification processes that confer HR through biochemical pathways in plants (Schuler, 1996; Mizutani and Ohta, 2010).

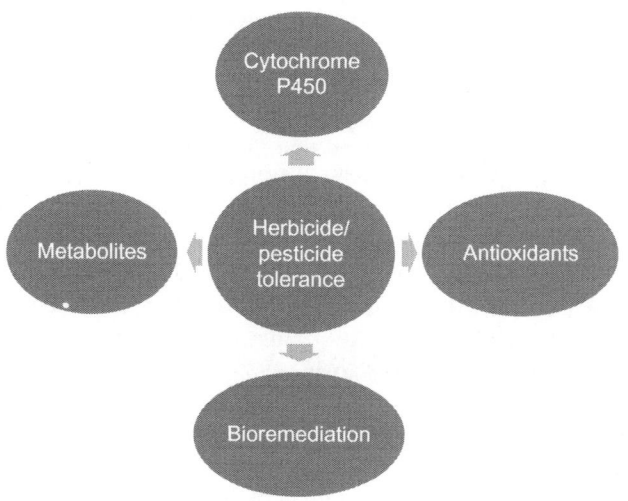

FIGURE 17.1

Various herbicide/pesticide tolerance mechanisms found in plants.

The biochemistry of a particular herbicide detoxicity mediated by enzymes of Cyt P450 has been regarded as an important HR pathway in certain HR plants (Powles and Yu, 2010). Recent research has established that the Cyt P450 might have numerous isoforms exhibiting variable specificity to particular herbicides, thereby defining differential capacity of their metabolism (Siminszky, 2006; Powles and Yu, 2010). However, the biochemical characterization of specific P450 genes that harbor herbicide HR through biochemical pathways is yet to be accomplished. The alteration in P450 substrate specificity caused by mutation, gene regulation, or gene duplication and gene mutation are major research themes that need to be researched (Schuler and Werck-Reichhart, 2003). Resistance to a particular herbicide in the HR *L. rigidum* population (biotype VLR69) is independently governed by the metabolism-based genetic characteristic P450 (Preston, 2003). Another study on HR *Alopecurus myosuroides* (Huds.) suggested the presence of both individual and multiple additive genetic controls on resistance (Letouze and Gasquez, 2001; Petit et al., 2010). The Cyt P450-based monooxygenases represent tremendously significant biochemical control over xenobiotic compounds. These monooxygenases have been found to be involved in the regulation of endogenous substrates like hormones, fatty acids, and steroids, and in the metabolism of exogenous drugs, herbicides, and other toxic substances. These monooxygenases were thought to be prevalent in all aerobic living organisms (Stegeman and Livingstone, 1998). The enzymes found in Cyt P450 in plants have an important function in detoxification, which is quite similar to animal systems. Thus, these enzymes are commercially important both in detoxification and HR perspectives. The metabolism based on monooxygenases is the usual metabolism in many insects conferring them insecticide resistance (Scott, 1999).

Cyt P450 is a heme protein that acts as the final oxidase in monooxygenases. These enzymes are capable of catalyzing oxidation of a variety of substrates and thus carry out an array of functions (Kulkarni and Hodgson, 1980; Rendic and Di Carlo, 1997; Mansuy, 1998). The Cyt P450 enzymes are located in endoplasmic reticulum and mitochondrial membranes. The electron transport of Cyt P450 found in the mitochondria is discreet, exhibiting minor similarity with the primitive prokaryote Cyt P450s (Wilkinson, 1980).

According to Lu and Coon (1968), the pioneer Cyt P450 was extracted in pure form from mammalian liver and its reconstitution study revealed that the smallest requirements of monooxygenase-based oxidation are as P450, NADPH Cyt P450 oxidoreductase, NADPH, and phospholipid. Cyt P450 reductase catalyzes the shift of reducing equivalents from NADPH to Cyt P450. Cyt b5 is concerned with certain monooxygenase pathways, which depend on the P450 and/or concerned substrates (Vatsis et al., 1980; Peterson and Prough, 1986; Pompon, 1987; Epstein et al., 1989; Zhang and Scott, 1996). Complete Cyt P450 was first discovered in an insect in 1967 (Ray, 1967). Subsequently, many investigators noted the persuasive confirmation of numerous Cyt P450 enzymes from different species of insects (Agosin, 1985). More than 100 Cyt P450 enzymes have been found in insects (Nelson et al., 1996).

The most prevalent resistance in insects is basically enzymatic detoxification and insensitivity of the target site (Oppenoorth, 1985; Agosin, 1985; Scott, 1991). Numerous researchers have reported an amplified biochemical detoxification most prevalent resistance system, even though the presence of changed target sites is also common (Wilkinson, 1983; Oppenoorth, 1985; Scott, 1991). It has been speculated that enzymes of Cyt P450 present the most significant basis of herbicide resistance (Hodgson and Kulkarni, 1983; Oppenoorth, 1985; Brattsten et al., 1986; Scott, 1991) followed by esterases (Hemingway and Karunarantne, 1998) and glutathione S-transferases (Yu, 1996). It is also worth mentioning that the extent of monooxygenase-mediated detoxification in vulnerable species considerably confines the toxicity and usefulness of some insecticides, like pyrethrins (Sawicki, 1962), imidacloprid (Wen and Scott, 1997), and carbaryl (Wilkinson, 1967). Moreover, the enzymes of Cyt P450 mediate the activation of several organophosphates (Hodgson et al., 1991). Organophosphates are the most extensively employed pesticides. Monooxygenase detoxification has the potential to confer cross-resistance to numerous poisons independent of their target sites (Wilkinson, 1983; Oppenoorth, 1985; Scott, 1991). During the 1980s, several investigators reported high concentrations of Cyt P450 enzymes in HR plants (Hodgson, 1985). However, the clear-cut evidence in favor of correlation between enzymes of Cyt P450s, HR level, and/or enzymatic activity was scant. However, few reports provided evidence of the presence of many Cyt P450 enzymes in insects and their tolerance attributed to a single Cyt P450 enzyme (Scott, 1991). Later, it was established that the enzymes of Cyt P450 might be regulated and that model substrates were not capable of measuring P450 enzyme activities responsible for tolerance (Wilkinson, 1983). It was also established that these enzymes should be isolated for characterization to obtain further insights into monooxygenases. Amplified intensity of P450 reductases (Vincent et al., 1985) and b5 (Scott and Georghiou, 1986) linked with monooxygenase-mediated resistance against insecticides was originally identified in house flies, and later found in other organisms (Sun et al., 1992; Kotze, 1993; Kotze and Wallbank, 1996; Valles and Yu, 1996). The detection of increased activity of Cyt P450 enzymes (reductases) and b5 linked with monooxygenase-mediated tolerance established that mutations in these enzymes may achieve resistance (Scott and Georghiou, 1986).

Many instances of Cyt P450-based tolerance may result in augmenting detoxification. The majority of organophosphates are detoxified through Cyt P450 monooxygenases to achieve herbicide resistance; however, decreased activation may also be involved in the process and it does not seem to be a prevalent resistance pathway. This clarifies the rationale behind the prevalence of esterase familiarity over monooxygenases in achieving tolerance to organophosphates (Oppenoorth, 1985; Scott, 1991). Detoxification may involve a change in the enzymatic activity of the concerned Cyt P450 enzymes and/or an alteration in their expression intensity (Oppenoorth, 1985). It is generally speculated that herbicide resistance is attained by the increased Cyt P450 activity of enzymes compared to susceptible species whose level is low. Scott et al. (1998) established two criteria as an indication of the involvement of Cyt P450 in acquiring resistance as follows:

1. The Cyt P450 enzymes should exhibit detoxification (or sequestration) of a pesticide to a plant that has attained tolerance; and
2. The tolerant biotype must possess a greater concentration of Cyt P450 enzymes or increased genetic expression exhibited in terms of greater catalytic activity that would result in improved detoxification compared to non-tolerant species.

Liu and Scott (1998) reported that herbicide resistance is achieved by high transcriptional rates of Cyt P450, which result in high gene expression and amplified pesticide detoxification. The gene regulation was proved to involve *cis*- and *trans*-mechanisms (Liu and Scott, 1998). An interesting finding revealed that the factors causing *trans*-regulation of the Cyt P450 enzymes concerned with herbicide resistance were also found to regulate the expression of genes that do not infer resistance especially in house flies. This finding created some ambiguity Cyt P450 in enzymes studies as gene regulation and substrate specificities were found to be variable. The identification of specific factors regulating the function of Cyt P450 enzymes would greatly assist in revealing cross-resistance models. It was confirmed that herbicide resistance occurs as a result of detoxification through a single P450 (i.e., CYP6D1) and the metabolic attack may be restricted to a particular site on the insecticide (Zhang and Scott, 1994). Moreover, b5 is required for P450-interceded detoxification of certain pesticides (Zhang and Scott, 1994; Dunkov et al., 1997), and is implicated in insecticide resistance in some cases (Zhang and Scott, 1996).

It has been proved that various pesticide classes exhibit phase metabolism (Balazs, 2006). The herbicide tolerance mechanism was evident from *in vitro* investigation on plant microsomes exposed to many pesticides in the presence of numerous Cyt P450 inhibitors and activators. Further categorization of Cyt P450 enzymes was established after gene isolation responsible for encoding specific isoforms carrying out pesticide metabolism. Evidence has established the association of increased concentrations of Cyt P450 enzyme activity involving herbicide resistance in weed plants. Herbicide resistance based on increased detoxification is difficult to achieve, as it may include tolerance to several diverse chemical compounds. Further research on the fate of various pesticides, the role of Cyt P450 proteins in plants, and herbicide resistance development is a prerequisite. Current progress in this field has opened new avenues for genetic engineering of herbicide resistance and the biodegradation of pesticides (Balazs, 2006). The herbicide resistance pathway in plants is affected by various components like enzymes and heredity (Busi et al., 2011). Herbicide resistance may be attained by increased herbicide metabolic rates based on Cyt P450 proteins; however, such herbicide resistance in woody plants is poorly interpreted. Busi et al. (2011) studied the hereditary control of Cyt P450 enzyme-mediated herbicide resistance for *Lolium rigidum*. It was concluded that herbicide resistance in this plant species may be accompanied by the build-up of resistant genes (Busi et al., 2011).

Many additive genes were found responsible for Cyt P450-based resistance against chlortoluron or it was based on the non-target site to the acetyl coenzyme A carboxylase (ACCase) herbicide pinoxaden in *Alupercurus myosuroides* (Chauvel, 1991; Petit et al., 2010). Polygenic, quantitative inheritance was reported by Mackenzie et al. (1995) regarding chlorsulfuron tolerance by *Lolium perenne*. However, single Cyt P450 genes were found to be responsible for herbicide resistance against a group of herbicides in *L. rigidum* (Preston, 2003). Neve and Powles (2005) hypothesized that some gene(s) conferring tolerance for minor effect might be augmented under herbicide selection, which exerts a complementary consequence in endurance of HR plant species. It was contradictory to the two-gene segregation model proposed by Wang et al. (1996) for HR *Setaria italica* L. Two-gene segregation states that resistant plants may survive at low pesticide concentrations.

Pan et al. (2012) investigated the herbicide metribuzin tolerance mechanism in narrow-leafed lupin by comparing two induced mutants of higher metribuzin tolerance with the susceptible wild type. It was concluded that the metribuzin tolerance mechanism in lupin mutants was non-target site based, likely involving P450-mediated metribuzin metabolism. Clethodim is the lowest resistance risk ACCase-inhibiting herbicide, with only two of 11 target-site mutations (amino acid substitutions) in weed populations that confer resistance. However, there are no reduced-risk acetolactate synthase/acetohydroxy acid synthase (ALS/AHAS) herbicides or other herbicide classes (Beckie and Francois, 2012). Dayan and Zaccaro (2012) developed a simple three-step assay to test selected herbicides representative of the known herbicide mechanisms of action and a number of natural phytotoxins to determine their effect on photosynthesis as measured by chlorophyll fluorescence. The most active compounds were those interacting directly with photosynthesis (inhibitors of photosystem I and II), those inhibiting carotenoid synthesis, and those with mechanisms of action generating reactive oxygen species and lipid peroxidation (uncouplers and inhibitors of protoporphyrinogen oxidase). Other active compounds targeted lipids (very-long-chain fatty acid synthase and removal of cuticular waxes). Therefore, induced chlorophyll fluorescence is a good biomarker to help identify certain herbicide modes of action and their dependence on light for bioactivity.

17.4.2 Bioremediation and phytoremediation

The understanding of the biochemical pathways of pesticides in various plants and microbes is crucial for efficient clean-up, safe use of these chemicals, and their bioremediation in soil and water. Metabolism or co-metabolism may be involved in the pesticide metabolism, which is a multi-step process. The majority of pesticides undergo widespread breakdown in plants and the environment. The breakdown of these pesticides involves a number of diverse physicochemical rearrangements like redox reactions, hydrolysis, and conjugation (Hoagland et al., 2000). The mechanisms of bioremediation and/or phytoremediation are presented in Figure 17.2. Two major mechanisms of phytoremediation for organic pesticides are rhizodegradation and phytoextraction (Pascal-Lorber and Lauren, 2011). Co-metabolism is a term that describes the biological transformation of various organics but does not act as a source of energy and this process also aids in the biotransformation of the pesticides (Alexander, 1994). During early co-metabolic stages, the chemical reaction decreases the toxicity of the pesticides for both target and non-target plants and increases the susceptibility of pesticides to various biological, biochemical, or physicochemical degradation. Various enzymes like hydrolytic enzymes (esterases, amidases, nitrilases, etc.), transferases

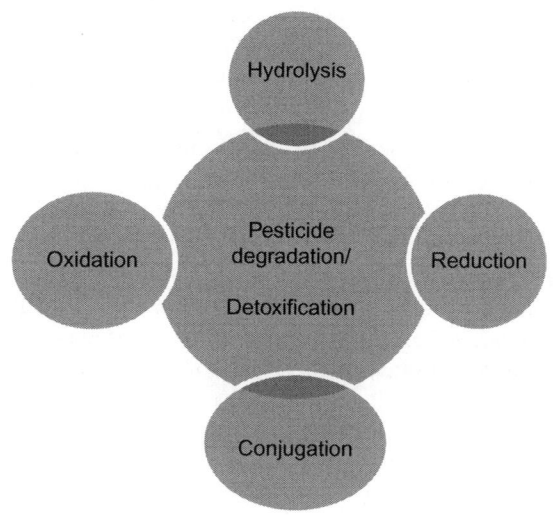

FIGURE 17.2

Various mechanisms involved in the metabolism of pesticides/herbicides through bioremediation and phytoremediation.

(glutathione S-transferase, glucosyl transferases, etc.), oxidases (cytochrome P-450 s, peroxidases, etc.), and reductases (nitroreductases, reductive dehalogenases, etc.) take part in the initial stages of co-metabolism (Hatzios, 1991; Mandelbaum et al., 1995). Plants and microbes have many common enzymes for the detoxification of various synthetic organic toxins. However, there are many differences in the details of the metabolic processes of these pesticides (Hoagland et al., 2000).

Many soil microbes like *Arthrobacter*, *Burkholderia*, *Pseudomonas*, and *Sphingomonas* have the ability to use a variety of xenobiotics. Bumpus (1993) reported that white rot fungi possess various enzymes employed in lignin and other xenobiotic degradation (lignin peroxidase and manganese peroxidase). The pesticide degradation by various microbial species was reported by many workers; however, bacterial species are not capable of pesticide degradation but rare cases of plants are found (Carr et al., 1985; Ramanand et al., 1988; Ryan and Bumpas, 1989; Krueger et al., 1989; Nagasawa et al., 1993; Mandelbaum et al., 1995; Leung et al., 1997). Previously, the 2,4-dichlorophenoxyacetic acid and 2,4,5-trichlorophenoxyacetic acid were reported to be biodegraded by various bacterial cultures (*Alcaligines*, *Arthrobacter*, *Flavobacterium*, and *Pseudomonas*) (Häggbloom, 1992). Certain herbicides were completely biodegraded while serving as a nitrogen source such as atrazine degradation, accomplished by *Pseudomonas* sp. ADP (Mandelbaum et al., 1995); while paraquat was mineralized by the yeast *Lipomyces starkeyi* (Carr et al., 1985).

Dubey and Fulekar (2013) reported rhizoremediation—the biodegradation catalyzed by plant–microbe interaction to mineralize various organic pollutants. Rhizodegradation of pesticides like pyrethroid presents a potentially cheap and promising method of reclamation of polluted sites. Hydrolysis and photolysis are the most effective in degradation of cypermethrin in soil. The degradation route involving the hydrolysis of the ester bonds produces 3-phenoxybenzoic acid (PBA) and cyclopropane carboxylic acid derivatives, especially 3-(2,2-dichlorovinyl)-2,2-dimethyl cyclopropane carboxylic acid (DCVA) (Kaufman et al., 1981). Aerobic microbes accomplish the degradation of cypermethrin under oxic conditions. Factors like the presence of heavy organic soils and reduced microbial metabolism were associated with high prevalence of this pesticide in the soil. The end result of rhizodegradation is always formation of non-toxic end products. The roots of plants directly affect the structure and density of the soil microbes; this is called rhizosphere effect. The structure of root and its secretions have significant bearing on the biochemical reactions and biodegradation occurring in the root zone. Several workers reported pesticide rhizodegradation (Yu et al., 2003; Singh et al., 2004; Sun et al., 2004), metal uptake (Gaur and Adholeya, 2004), and the mineralization of a few other organic compounds through rhizodegradation (Jordahl et al., 1997; Nakamura et al., 2004; Chaudhry et al., 2005; Biryukova et al., 2007). It was also reported that such mineralization was greater in the rhizosphere compared to bulk soils. Dubey and Fulekar (2013) reported the ability of *Pennisetum pedicellatum* to mineralize cypermethrin. The microbes capable of pesticide degradation were found higher in the rhizosphere. The effective rhizodegradation of cypermethrin was reported by *Stenotrophomonas* spp. from the rhizosphere of *P. pedicellatum*. It was suggested that rhizodegradation can serve as an effective tool for the reclamation of natural habitats (Dubey and Fulekar, 2013).

The hereditary potential to generate hydroxyatrazine was formerly credited to a 1.9-kb AvaI DNA section isolated from *Pseudomonas* sp. ADP (de Souza et al., 1996). The study confirmed open reading frame atzA responsible for coding an enzyme that converts atrazine to hydroxyatrazine. The enzyme AtzA was isolated, homogenized, and its structure revealed that it was chlorohydrolase instead of oxygenase. The selection of plant species to accomplish the rhizodegradation is very important. Thus, the intensity of pesticide-degrading bacteria is always high in the rhizosphere of plants inhabiting polluted soils. The physical parameters of root are the most critical in rhizoremediation; these parameters include width of root, depth, surface-to-volume ratio, root biomass, and surface area, and these parameters are variable among various plant species, which results in variable herbicide resistance (Dubey and Fulekar, 2013). Due to the extensive shallow root system in grasses, they are very promising candidates for rhizoremediation. As roots absorb plenty of water from the soil, they help in the movement of pollutants to the rhizosphere (Erickson, 1997). *Pennisetum pedicellatum* resisted large amounts of pesticide cypermethrin (Dubey and Fulekar, 2013). Yu et al. (2003) also reported that diversity and richness of soil microbes were important for pesticide metabolism. An interesting report stated that the excessive pesticide that does not reach a target organism is absorbed by plants like many vegetables and fruits, and also processed foods (González-Rodríguez et al., 2011). Kreuz and Martinoia (1999) concluded that the principle metabolic step in the detoxification of a pesticide is hydrolysis or oxidation, generating metabolites that could be processed further by secondary enzymatic conjugation to endogenous substrates like glutathione (GSH), carbohydrates, or organic acids. The final disposition of such substrates is carried out in vacuoles as conjugates, and/or extracellular disposal. An important constituent in the regulatory metabolism is the transfer of intermediates like GSH or glucosyl conjugates from the cytoplasm by specific ATPase conjugate pumps (Hoagland et al., 2000). It was reported that genetic control of enzymes in higher plants is different from that in prokaryotic organisms. In bacteria, operon clusters possess genes regulating the metabolism of pesticides. The enzymes involved in the metabolism of pesticides can be stimulated or suppressed by the same effect. The eukaryotes are different in this regulation where multiple effectors regulate these enzymes and a gene is regulated by a specific promoter. Such systems have innumerable proteins that influence expression, i.e., specific cytoplasm enzymes (WD proteins) controlling the transcription process, and they are merely restricted to the cytoplasm (Ma, 1994). A thorough understanding of the gene regulation of plant gene expression is yet to be accomplished (Hoagland et al., 2000). Nandula et al. (2008) established glyphosate resistance in a rye grass T1 population partially due to abridged glyphosate absorption and translocation, while in the T2 population it was caused by decreased transport of glyphosate.

17.4.3 Antioxidants

It is a known fact that herbicides like many other pollutants initiate the development of ROS in the cells (Wu and von Tiedemann, 2002; Peixoto et al., 2006; Song et al., 2006). The endogenous defense system combats such oxidative stress in living beings (Kahkonen et al., 1999; Wang et al., 2004). ROS like superoxides ($O_2^{\bullet-}$), hydrogen peroxide (H_2O_2), and the hydroxyl radical ($OH^{\bullet}$) are predictable consequences in all living aerobes due to disturbances in ETC; however, these undesirable reactions are efficiently controlled within cells (Valavanidis et al., 2006). Various abiotic stresses like metal and pesticide exposure are common stimuli for ROS generation and subsequent disturbance in cellular functions (Wang et al., 2004; Peixoto et al., 2006; Valavanidis et al., 2006; Song et al., 2007; Zhou et al., 2007). The plants have many control measures against these ROS stresses, which include enzymatic antioxidants, i.e., superoxide dismutase (SOD), peroxidase (POD), catalase (CAT), and ascorbate peroxidase (APX). The non-enzymatic antioxidant defense comprises polyphenols, ascorbic acid, and carotenoids (Mittler, 2002). These ROS are hard to detect after their production (Dorta et al., 2003; Radetski et al., 2004); normally they are inferred from variations in the antioxidants (Valavanidis et al., 2006). The antioxidant defense can be regarded as biomarkers and the changes in their activities result in an increase of thiobarbituric acid reactive substances (TBARS) (Pang et al., 2001; Wu and von Tiedemann, 2002). The cellular concentrations of glutathione S-transferase (GST) may be stimulated by a number of pollutants (Pascal et al., 1998). The enzyme is responsible for the binding of GSH to many electrophilic substances, is considered a constituent of defense against stress-inducing pollutants, and, hence, is regarded as a biomarker of abiotic stress (Pascal et al., 1998). The consequences of the isoproturon exposures and its consequent changes were investigated in wheat (a significant economic crop throughout the world) by Yin et al. (2008). After exposure to 20 mg/kg of the pesticide, the seedlings failed to grow. The photosynthetic pigments were found to be sensitive and decreased significantly at 2 mg/kg of the pesticide. The concentrations of TBARS were found to be enhanced, indicating stress. Considerable modification in catalysis of various antioxidants like SOD, POD, CAT, and APX were increased under pesticide exposure, CAT activity gradually decreased in leaves (Yin et al., 2008).

Alscher et al. (2002) reported that antioxidants are metalloenzymes present in various isoforms like Cu−Zn-SOD, Mn-SOD, and Fe-SOD. Augmented levels of SOD action may be a consequence of higher levels of superoxides, which result in the up-regulation of gene expression (Foyer et al., 1997; Mishra et al., 2006). Higher SOD activity was found associated with the amplified oxidative stress in wheat (Song et al., 2007). Hydrogen peroxide (H_2O_2) produced as a result of SOD-mediated response is extremely poisonous; thus, it should be immediately fixed in cells. In plants, many enzymes like POD, APX, and CAT convert H_2O_2 to H_2O or at least detoxify it (Zhang and Kirham, 1994). APX is the chief H_2O_2 scavenger that accomplishes its elimination

(De Gara, 2004). Among many antioxidants in plants, POD is another key enzyme whose extra- and intracellular components take part in eliminating H_2O_2 and lignin biosynthesis in its presence (Passardi et al., 2004; Wang et al., 2004; Wang and Yang, 2005). POD utilizes many electron donors like NAD(P)H, while guaiacol is normally used to detect its presence, as guaiacol corresponds to the non-specific activity of POD (Asad 1992; Passardi et al., 2004). CAT found in peroxisomes, glyoxysomes, and mitochondria eradicates the majority of the photo-respiratory and respiratory H_2O_2 (Asad, 1992). Unlike APX, reducing metabolites are not needed in the catabolism of H_2O_2 to H_2O and O_2 by CAT; the reaction is fast but possesses low affinity with the substrate. In contrast, APX exhibits greater substrate affinity and is capable of catalyzing minute H_2O_2 concentrations of (Nakano and Asada 1981; Amako et al., 1994). GST is present in many aerobes and controls the nucleophilic addition GSH to electrophilic centers of different organics (Armstrong 1997). GST has various classes: α, μ, π, θ, δ, ζ, and β that can act as GSH peroxidase, which catalyzes the reduction of fatty acid, hydroperoxides, or thymidine hydroperoxides to the related hydroxy derivatives that result in the formation of GSSG (Mannervik and Danielson, 1988; Bartling et al., 1993). Herbicides may cause the enhancement of cellular GST concentrations (Pascal et al., 1998; Edwards and Cole, 1996).

17.4.4 Metabolites in pesticide/herbicide resistance

From 1997 to 2002, numerous researchers reported the ability of plants to treat various contaminants from soil and water without any apparent mechanism. However, the treatment of recalcitrant organic compounds produces inconsistent results for bio-treatment by plants because of differences in structure of recalcitrant compounds from naturally occurring molecules (Singer et al., 2003). Numerous pesticides usually attack the target enzymes, which are normally inhibited; this may be due to overexpression or overstimulation of the target proteins. The enzymes of non-target organisms and microbial enzymes can also be inhibited by a few pesticides (Frear and Still, 1968; Blake and Kaufman, 1973; Hoagland and Zablotowicz, 1995).

Certain plant species have attained selectivity on the basis of detoxification; the examples of such pesticides include 6-oxidation of phenoxy butyric acids (Wain and Smith, 1976), sulfoxidation of thiocarbamate herbicides (Carringer et al., 1978), and hydrolytic de-esterification of diclofop-methyl {(±)-2-[4-(dichlorophenoxy)phenoxy]propanoic acid} (Shimabukuro et al., 1979). Various metabolic pathways have been discovered in a number of plants for pesticidal detoxification. The enzyme specificity was employed in differential metabolisms of pesticides in crops and weeds (Brown et al., 1991). The metabolic rates of herbicides by a specific plant are also crucial in determining selectivity. Aryl acylamidase activity was found to bring tolerance against Propanil in *Echinochloa crusgalli* and *Echinochloa colona* (Leah et al., 1994). Acquisition of a new GST isozyme resulted in atrazine tolerance in velvet leaf (*Abutilon theophrasti*) (Anderson and Gronwald, 1991). Improved N-dealkylation concentration caused concurrent tolerance to simazine and chlortoluron in rigid ryegrass (*Lolium rigidum*) (Burnet et al., 1993a,b). An explicit cultivar of rigid ryegrass VLR69 was consequently observed to be tolerant to nine classes of herbicides, after exposure to five pesticides for 21 years (Preston et al., 1996). For HR weeds, combinations of various pesticides should be used (herbicide synergistic) to suppress their herbicide resistance (Hoagland et al., 2000).

Current progress in biotechnology has progressed to develop herbicide tolerance in many crops. Previously, classical breeding was tried to accomplish this goal, which was slow but fruitful in producing metribuzin-resistant soybean (Marshall, 1991). Tissue culture was also exploited to produce cellular lines resistant to different herbicides like 2,4-D, picloram, paraquat, chlorsulfuron, and imazaquin (Marshall, 1991). However, selection-based herbicide resistance through tissue culturing was not found inheritable. Currently, cloning and genetic transformation are being used to produce herbicide resistance in the majority of cases. HR crops containing foreign genes have been produced against many herbicides. Two related genes, bar (Block et al., 1987) and pat (Wohlleben et al., 1988; Broer et al., 1989) conifer tolerance to glufosinate through a bacterial acetyl transferase gene, while tolerance to bromoxynil and phenmedipham was attained via bacterial genes for nitrilase (bxn) (Stalker et al., 1988) and carbamate hydrolase (Streber et al., 1994), respectively. In view of the crucial role of Cyt P450 enzymes in herbicide resistance, various workers used inhibitors against these enzymes to effectively control HR weeds of various crops. Such an inhibitor, piperonyl butoxide (PBO), may improve herbicidal action of atrazine and terbutryn in corn (Varsano et al., 1992). PBO improved the efficiency of thiazopyr in barn yard grass, grain sorghum (*Sorghum vulgare*), and redroot pigweed (*Amaranthus retroflexus*) (Rao et al., 1995).

17.5 **The selection process of tolerant plants**

The HR herbaceous plants are privileged via the process of natural selection in polluted habitats because they are better adapted to such conditions compared to non-tolerant plants. The selection pressure in numerous herbs is directed to the natural evolutionary progress of HR genotypes consequent to a broad variety of pesticides. The herbicide-sensitive plants display noticeable impacts on their growth and reproduction upon exposure to various herbicides. The seedlings of the majority of plants are the most susceptible to pesticide toxicity. However, the considerable toxic effects may be displayed during vegetative and reproductive growth periods. Many annual herbs exhibit the process of selection for herbicide resistance under herbicide exposure. The mechanism of tolerance under pesticide exposure in perennial plant species is different from annuals. The response to agrochemical pollution is similar to abiotic stress. However, the mechanism of tolerance in these herbaceous plants is supposed to be linked with their better genetic make-up, which enables them to survive under stressful conditions. How better genetic make-up originated is not clear (Gunsolus, 2008). Herbicides have not been thought to cause gene mutations allowing herbicide resistance. The HR species may be found in low numbers within large plant populations and survive till reproduction when herbicides are applied; while other susceptible species are wiped out. Upon continuous use of herbicides on HR species, the number of such HR plants increases with the passage of time. Selection pressure functions like a sieve, which sorts the HR species from intolerant ones and the end result is the survival of just HR species. Being herb killers, herbicides have the ability to exert selection pressure on the weed populations. The susceptibility of a weed species results in its elimination upon exposure to an herbicide; susceptibility has a linear relation with weed control. Consequently, the rate of selection for herbicide resistance may be quick if the same herbicide is repeatedly used on similar weeds. Thus, even the use of very effective herbicides may increase the number of HR weed species. HR plant species are detected only when present in about 30% of total weed populations. In many instances it was observed that use of the same pesticide resulted in a 1% HR population after many years of pesticide use. Prolonged use of the same herbicide will result in the growth and reproduction of HR weed species. Gunsolus (2008) listed the following features of herbicides that may result in HR species: (1) they may target solitary sites of weeds; (2) spraying the same herbicide for repeated periods of time in the same crop; (3) repeated use for several growing seasons having a single site of action; (4) herbicides employed exclusively for additional weed control programs and sometimes called "stand-alone" pesticides.

Herbicide resistance is more likely to be triggered due to application of a single site of action herbicide, and the mutation in merely one gene is sufficient to accomplish its binding to the target site. It is more likely to develop an HR weed population when dissimilarity in a single gene is needed. The use of herbicides having multiple sites of action will not result in the development of HR plant species. Such resistance is generally against those herbicides to which plants were exposed; there will be no resistance to other herbicides. The presence of many binding sites for a particular site of action may be the reason and such sites of action are specific to each herbicide. Thus, many herbicides may have the ability to bind with the same enzyme but at diverse sites of action. Consequently, the cross-resistance of a herbicide cannot be predicted; apart from the fact that a particular herbicide works at single or multiple sites, it may be transformed by target plants until it reaches its site of action. The speed of herbicide metabolism is crucial in shaping damage to plants and weed management. An example is the change of metabolic rate caused by a single gene mutation in biotypes of atrazine-resistant velvet leaf (*Abutilon theophrasti*). However, the majority of metabolic processes have polygenic control in nature and thus there is a little chance for weed species to be resistant against a particular pesticide based on its improved metabolic functions. As metabolic process influences the action of herbicides with different sites of action, metabolic resistance can be a challenge. The reduction in selection intensity may be the key factor to preventing herbicide resistance. Consequently, weeds also exhibit an ability to adapt various pesticide management programs. When an HR weed dominates a field, two important points should be focused on for their effective control: (1) ability of a weed to reproduce and (2) dispersal of HR weed seeds.

The ability of HR weed spread depends upon its reproductive success to become a dominant weed species. Based on increased viability of certain HR weeds after establishment in a field, it may be difficult to eradicate in spite of costly chemical application. Due to different seed dispersal means, farmers should consider interception of the seed dispersal mechanism as well as employing effective herbicide management strategies. A number of strategies were developed by the North Central Weed Science Society (NCWSS) and Herbicide Resistance Committee to either avoid HR weeds or eradicate them. These include (Gunsolus, 2008):

1. Apply the pesticides only if essential. Their utilization should be based on financial thresholds.
2. Pesticides with alternate modes of action should be used. Avoid a number of successive sprays of herbicides with a similar site of action on the same farm until accompanied by other management methods. Two successive applications might be acceptable for 2 years, or two separate sprays in 1 year.
3. Pesticides with many sites of action should be used. Sequential mixtures of pre-packed herbicides are better. The use of multi-spectral pesticides may be costly but their use may be preferred.
4. Crop rotation involving diverse life cycles should be encouraged.

5. Discourage the use of more than two sprays of herbicides having a single site of action on HR crops.
6. Mechanical weed control should be combined with pesticide use.
7. Primary tillage along with minimal soil erosion potential should be part of the weed control strategy.
8. Explore the farm area on a regular basis to observe and identify troublesome weeds. Act swiftly to variations in weeds to confine their spread.
9. Clean farming equipment prior to use to avoid the spread of HR plants.
10. Influence various departments to avoid the practice of the weed control options leading to selection of HR plant species. HR plants generally spread from small areas to the whole cropland.

17.6 Conclusion and future prospects

Continuous endeavors for greater yields have prompted scientists to invent new herbicides/pesticides. Pesticide application is still the most effective and accepted mode of protection of plants from pests, and has contributed significantly to enhance agricultural productivity and crop yield. Thus, the future work on herbicide/pesticide resistance could be promising if it involves the identification of metabolites responsible for detoxification of these applied xenobiotics, various enzymes, their controlling genes, and the rhizosphere bacterial species conferring herbicide/pesticide tolerance to plants. The quest of biotechnology to identify resistance responsible genes could be promising, which in turn could be transferred to transgenic plants for enhancing resistance. Among major resistance mechanisms, the roles of enzymes like Cyt P450 in plants have a supreme function in the metabolism for the detoxification of herbicides. The detoxification and metabolism of herbicide-catalyzed enzymes of Cyt P450 present an outstanding resistance mechanism found in HR plants. Knowledge of the pesticide resistance mechanism in plants and other microorganisms is crucial for devising a safe plan for its use and biodegradation in contaminated environments.

Secondary plant metabolites also play a crucial role in developing the multitude of enzymes responsible for the breakdown of various organic pollutants. Further research work is required to provide a link between secondary plant metabolites and enzymatic diversity, which can be applied in fields for pest management, bioremediation, and fine chemical production (Singer et al., 2003). Pest management has been challenged by economic and ecological constraints globally. The characterization and synthesis of new and effective insecticides is crucial to combat increasing resistance. The plant extracts having active insecticidal properties seem to be promising to control some of these problems. Thus, continuous efforts are required to explore novel active molecules with innovative mechanisms of action via gene transformation methods.

Recent research on gene shuffling furnished glyphosate resistance in plants, moreover, has resulted in numerous genes being patented, and substantial efforts being put into developing herbicide-resistant transgenic crops. Currently, few commercialized herbicide-resistant crops are available, e.g., glyphosinate- and glyphosate-resistant herbicides. However, transgenes have significantly greater environmental risk than HR crops. Further, wild-type species can be screened for resistance genes and then transferred to susceptible plant species for increasing survival and suitability in the natural environment. However, transgenic plants pose a greater risk than HR crops. Ameliorating the allelopathic mechanism in plants can result in greater resilience and minimize herbicide/pesticide usage. However, this work could be painstaking and the practical outcome would be difficult to achieve. Transforming wild-type species as transgenes can increase their suitability in natural ecosystems; however, transgenes have significantly greater environmental risk than HR crops. Research on the development of allelopathic plants is in progress with the aim of reducing the use of herbicides/pesticides. Even if made successful and safe, this technology will not be available for at least 10 years (Duke et al., 2007).

References

Agosin, M., 1985. Role of microsomal oxidations in insecticide degradation. In: Kerkut, G.A., Gilbert, L.I. (Eds.), Comprehensive Insect Physiology, Biochemistry, and Pharmacology, vol. 12. Pergamon, NY, pp. 647–712.

Aktar, M.W., Sengupta, D., Chowdhury, A., 2009. Impact of pesticides use in agriculture: their benefits and hazards. Interdiscip Toxicol. 2, 1–12.

Alexander, M.A., 1994. Biodegradation and Bioremediation. Academic Press, San Diego, pp. 177–193

Alscher, R.G., Erturk, N., Heath, L.S., 2002. Role of superoxide dismutases (SODs) in controlling oxidative stress in plants. J. Exp. Bot. 53, 1331–1341.

Amako, K., Chen, G.X., Asada, K., 1994. Separate assays for ascorbate peroxidase and guaiacolperoxidase and for the chloroplastic and cytosolic isozymes of ascorbate peroxidase in plants. Plant Cell Physiol. 35, 497–504.

Anderson, M.P., Gronwald, J.W., 1991. Atrazine resistance in velvet leaf (Abutilon theophrasti Medic) biotype due to enhanced glutathione S-transferase activity. Plant Physiol. 96, 104–109.

Anderson, T., 2001. Biotech soybean seed helps growers produce safe and profitable crops. American Soybean Association. Available at: http://www.monsanto.co.uk/news/ukshowlib.phtml?uid=5063. Accessed June 2001.

Armstrong, R.N., 1997. Structure, catalytic mechanism, and evolution of the glutathione-transferases. Chem. Res. Toxicol. 10, 2–18.

Asad, A., 1992. Ascorbate peroxidase: a hydrogen peroxide scavenging enzyme in plants. Physiol. Plant 85, 235–241.

Balazs, S., 2006. Plant cytochrome P450-mediated herbicide metabolism. Phytochem. Rev. 5, 445–458.

Bartling, D., Radzio, R., Steiner, U., Weiler, E.W., 1993. A glutathione S-transferase with glutathione-peroxidase activity from Arabidopsis thaliana. Molecular cloning and functional characterization. Eur. J. Biochem. 216, 579–586.

Beckie, H.J., Francois, J.T., 2012. Herbicide cross resistance in weeds. Crop. Protect. 35, 15–28.

Benbrook, C.M., 2001. Trouble times amid commercial success for Roundup Ready soybeans. AgBioTech. Info Net 3, 6.

Biryukova, O.V., Fedorak, P.M., Quideau, S.A., 2007. Biodegradation of naphthenic acids by rhizosphere microorganisms. Chemistry. 67, 2058–2064.

Blake, J., Kaufman, D.D., 1973. Microbial degradation of several acetamide, acylanilide, carbamate, toluidine and urea pesticides. Soil Biol. Biochem. 5, 297–308.

Block, M.D., Botterman, J., Vandewiele, M., Dockx, J., Thoen, C., Gosselé, V., et al., 1987. Engineering herbicide resistance in plants by expression of a detoxifying enzyme. EMBO J. 6, 2513–2518.

Bolognesi, C., 2003. Genotoxicity of pesticides: a review of human biomonitoring studies. Rev. Mutat. Res. 543, 251–272.

Brattsten, L.B., Holyoke, C.W., Leeper, J.R., Raffa, K.F., 1986. Insecticide resistance: challenge to pest management and basic research. Science 231, 1255−1260.

Broer, I., Arnold, W., Wohlleben, W., Puhler, A., 1989. Proc. Braunschweig Symp. Applied Plant Molecular Biol 240−246.

Brown, H.M., Fuesler, T.P., Ray, T.P., Strachan, S.D., 1991. Pesticide chemistry: advances in international research, development, and legislation. In: Frehse, H.M. (Ed.), Proceedings of the 7th International Congress on Pesticides and Chemicals (IUPAC). VCH Verlagsgsellschaft, Weinhem, Germany, pp. 257−266.

Bumpus, J.A., 1993. Recent advances in the use of fungi in environmental remediation and biotechnology. In: Bollag, J.-M., Stotzky, G. (Eds.), Soil Biochemistry, vol. 8. Marcel Decker, New York, pp. 65−100.

Burnet, M., Loveys, B., Holtum, S., Powles, S., 1993a. A mechanism of chlorotoluron resistance in *Lolium rigidum*. Planta 190, 182−189.

Burnet, M., Loveys, B., Holtum, S., Powles, S., 1993b. Increased detoxification is mechanism of simazine resistance in *Lolium rigidum*. Pest Biochem. Physiol. 46, 207−218.

Busi, R., Vila-Aiub, M.M., Powles, S.B., 2011. Genetic control of a cytochrome P450 metabolism-based herbicide resistance mechanism in *Lolium rigidum*. Heredity 106, 817−824.

Businelli, A., Vischetti, C., Coletti, A., 1992. Validation of Koc approach for modeling the fate of some herbicides in Italian soil. Fres Analyt. Chem. 1, 583−588.

Carpenter, J.E., Gianessi, L.P., 2001. Agricultural Biotechnology: Updated Benefit Estimates. Report from the National Center for Food and Agricultural Policy, Washington DC, pp. 46

Carr, R.J.G., Bilton, R.F., Atkinson, T., 1985. Mechanism of bio-degradation of paraquat by Lipomyces starkeyi. Appl. Environ. Microbiol. 49, 1290−1294.

Carringer, R.D., Rieck, C.E., Bush, L.P., 1978. Effect of R-25788 on EPTC metabolism in corn (*Zea mays*). Weed Sci. 26, 157−160.

Chaudhry, Q., Blom-Zandstra, M., Gupta, S.K., Joner, E., 2005. Utilising the synergy between plants and rhizosphere microorganisms to enhance breakdown of organic pollutants in the environment. Environ. Sci. Pollut. Res. 12, 34−48.

Chauvel, B., 1991. Polymorphisme génétique et sélection de résistance aux urées substituées chez Alopecurusmyosuroides Huds. PhD thesis dissertation. Université de Paris-Sud Centre d, Orsay, Paris, France.

Dayan, F.E., Zaccaro, M.L.M., 2012. Chlorophyll fluorescence as a marker for herbicide mechanisms of action. Pest Biochem. Physiol. 102, 189−197.

De Gara, L., 2004. Class III peroxidases and ascorbate metabolism in plants. Phytochem. Rev. 3, 195−205.

de Souza, M.L., Sadowsky, M.J., Wackett, L.P., 1996. Atrazine chlorohydrolase from Pseudomonas sp. strain ADP: gene sequence, enzyme purification, and protein characterization. J. Bacteriol. 178, 4894−4900.

Dorta, D.J., Leite, S., De Marco, K.C., Prado, I.M.R., Rodrigues, T., 2003. A proposed sequence of events for cadmium-induced mitochondrial impairment. J. Inorg. Biochem. 97, 251−257.

Dubey, K.K., Fulekar, M.H., 2013. Investigation of potential rhizospheric isolate for cypermethrin degradation. Biotechnology. 3, 33−43.

Duke, S.O., 2001. Herbicide-resistant crops. In: Pimentel, D. (Ed.), Encyclopedia of pest management. Marcel Dekker, Inc, NY.

Duke, S.O., Baerson, S.R., Rimando, A.M., Pan, Z., Dayan, F.E., Belz, R.G., 2007. Biocontrol of weeds with allelopathy: conventional and transgenic approaches. In: Vurro, M., Gressel, J. (Eds.), Novel Biotechnologies for Biocontrol Agent Enhancement and Management. Springer, The Netherlands, pp. 75−85.

Dunkov, B.C., Guzov, V.M., Mocelin, G., Shotkoski, F., Brun, A., Amichot, M.R.H., et al., 1997. The Drosophila cytochrome P450 gene Cyp6a2: structure, localization, heterologous expression, and induction by phenobarbital. DNA Cell Biol. 16, 1345−1356.

Edwards, R., Cole, D.J., 1996. Glutathione transferases in wheat (Triticum) species with activity toward fenoxaprop-ethyl and other herbicides. Pest Biochem. Physiol. 54, 96–104.

Epstein, P.M., Curti, M., Jansson, I., Huang, C.K., Schenkman, J.B., 1989. Phosphorylation of cytochrome P450: regulation by cytochrome b5. Arch. Biochem. Biophys. 271, 424–432.

Erickson, L.E., 1997. An overview of research on the beneficial effects of vegetation in contaminated soil. Ann. NY Acad. Sci. 829, 30–35.

FAO, 2001. Draft of Guidelines for Assessment of Ecological Hazards of Herbicide- and Insect-Resistant Crops. Plant Protection Division, Rome (In collaboration with Kathrine H. Madsen, Bernal E. Valverde, Jens C. Streibig of the Royal Veterinary and Agricultural University, Denmark), pp. 18.

Foyer, C.H., Lopez-Delgado, H., Dat, J.F., Scott, I.M., 1997. Hydrogen peroxide and glutathione associated mechanism of acclamatory stress tolerance and signaling. Physiol. Plant 100, 241–254.

Frear, D.S., Still, G.G., 1968. The metabolism of 3,4-dicholoropropionanilide in plants. Partial purification and properties of an aryl acylamidase from rice. Phytochem 7, 913–920.

Gaur, A., Adholeya, A., 2004. Prospects of arbuscular mycorrhizal fungi in phytoremediation of heavy metal contaminated soils. Curr. Sci. 86, 528–534.

Gealy, D.R., Dilday, R.H., 1997. Biology of red rice (*Oryza sativa* L.) accessions and their susceptibility to glufosinate and other herbicides. Weed Sci. Soc. Am. Abstr. 37, 34.

González-Rodríguez, R.M., Rial-Otero, R., Cancho-Grande, B., Gonzalez-Barreiro, C., Simal-Gándara, J., 2011. A review on the fate of pesticides during the processes within the food-production chain. Crit. Rev. Food Sci. Nutr. 51, 99–114.

Gunsolus, J.L., 2008. Herbicide resistant weeds. Extension Agronomist, Weed Science. Department of Agronomy and Plant Genetics, University of Minnesota, USA.

Häggbloom, M.M., 1992. Microbial breakdown of halogenated aromatic pesticides and related compounds. FEMS Microbiol. Rev. 103, 29–72.

Hall, J.C., Donnelly-Vanderloo, M.J., Hume, D.J., 1996. Triazine-resistant crops: the agronomic impact and physiological consequences of chloroplast mutation. In: Duke, S.O. (Ed.), Herbicide-resistant crops. agricultural environmental, economic, regulatory and technical aspects. CRC Press, USA, pp. 107–126.

Hatzios, K.K., 1991. Biotransformations of herbicides in higher plants. In: Grover, R., Cessna, A.J. (Eds.), Environmental Chemistry of Herbicides. CRC Press, Boca Raton, FL, pp. 141–185.

Heimlich, R.E., Fernandez-Cornejo, J.F., McBride, W., Klotz-Ingram, C., Jans, S., Brooks, N. (2000) Adoption of genetically engineered seed in U.S. agriculture. In: Fairbairn, C., Scoles G., McHughen A. (eds.), Proceedings of 6th International Symposium on the Biosafety of GMOs, Saskatoon, Canada, pp. 56–63.

Hemingway, J., Karunarantne, SHPP, 1998. Mosquito carboxylesterases: a review of the molecular biology and biochemistry of a major insecticide resistance mechanism. Med. Vet Entomol. 12, 1–12.

Hepel, M., Stobiecka, M., Peachey, J., Miller, J., 2012. Intervention of glutathione in pre-mutagenic atechol-mediated DNA damage in the presence of copper(II). Mutation Res. 2, 1–11.

Hoagland, R.E., Zablotowicz, R.M., Hall, J.C., 2000. Pesticide metabolism in plants and microorganisms: an overview. In: Hall, J., et al., (Eds.), Pesticide Biotransformation in Plants and Microorganisms. ACS Symposium Series. American Chemical Society, Washington, DC.

Hoagland, R.E., Zablotowicz, R.M., 1995. Rhizobacteria with exceptionally high aryl acylamidaseactivity. Pest Biochem. Physiol. 52, 190–200.

Hodgson, E., 1985. Microsomal mono-oxygenases. In: Kerkut, G.A., Gilbert, L.C. (Eds.), Comprehensive Insect Physiology Biochemistry and Pharmacology, vol. 11. Pergamon Press, Oxford, pp. 647–712.

Hodgson, E., Kulkarni, A.P., 1983. Characterization of cytochrome P-450 in studies of insecticide resistance. In: Georghiou, G.P., Saito, Y. (Eds.), Pest Resistance to Pesticides. Plenum Press, NY.

Hodgson, E., Silver, I.S., Butler, L.E., Lawton, M.P., Levi, P.E., 1991. Metabolism. In: Hayes Jr, W.J., Laws Jr, E. R. (Eds.), General Principles. Handbook of Pesticide Toxicology, vol. 1. Academic Press, NY, pp. 107–168.

James, C. (2001) Global GM Crop Area continues to grow and exceeds 50 million hectares for first time in 2001. Intl Service for the Acquisition of Agri-biotech Applications. Available at http://www.isaaa.org/press percent20release/Global percent20Area_Jan2002.htm

Jordahl, J.L., Foster, L., Schnoor, J.L., Alvarez, P.J.J., 1997. Effect of hybrid poplar trees on microbial populations important to hazardous waste bioremediation. Environ. Toxicol. Chem. 16, 1318–1321.

Kahkonen, M.P., Hopia, A.I., Vuorela, H.J., Rauha, J.P., Pihlaja, K., Kujala, T.S., 1999. Antioxidant activity of plant extracts containing phenolic compounds. J. Agri. Food Chem 47, 3954–3962.

Kaufman, D.D., Russell, B.A., Helling, C.S., Kayser, A.J., 1981. Movement of cypermethrin, decamethrin, permethrin and their degradation products in soil. J. Agric. Food Chem. 29, 239–245.

Khan, N., Muller, J., Khan, S.K., Amjad, S., Nizamani, S., Bhanger, M.I., 2010. Organochlorine pesticides (OCPS) contaminants in sediments from Karachi harbour, Pakistan. J. Chem. Soc. Pak. 32, 542.

Kotze, A.C., 1993. Cytochrome P450 monooxygenases in larvae of insecticide-susceptible and-resistant strains of the Australian sheep blowfly, Luciliacuprina. Pesti Biochem. Physiol. 46, 65–72.

Kotze, A.C., Wallbank, B.E., 1996. Esterase and monooxygenase activities in organophosphate-resistant strains of Oryzaephilussurinamensis (Coleoptera: Cucujidae). J. Econ. Entomol. 89, 571–576.

Kreuz, K., Martinoia, E., 1999. Herbicides metabolism in plants: integrated pathways of detoxification. In: Pesticide Chemistry and Bioscience. Brooks, G.T., Roberts, T.R. (Eds.), Royal Society of Chemistry, Cambridge, UK. pp. 279–297.

Krueger, J.P., Butz, R.G., Atallah, Y.H., Cork, D.J., 1989. Isolation and identification of microorganisms for the degradation of dicamba. J. Agric. Food Chem. 37, 534–538.

Kulkarni, A.P., Hodgson, E., 1980. Metabolism of insecticides by mixed function oxidase systems. Pharmac. Ther. 8, 379–475.

Leah, J.M., Caseley, J.C., Riches, C.R., Valverde, B., 1994. Association between elevated activity of acyl amidase and propanil resistance in jungle rice, *Echinochloa colona*. Pestic. Sci. 42, 281–289.

Letouze, A., Gasquez, J., 2001. Inheritance of fenoxaprop-P-ethyl resistance in a blackgrass (*Alopecurus myosuroides* Huds.) population. Theor. Appl. Gen. 103, 288–296.

Leung, K.T., Cassidy, M.B., Shaw, K.W., Lee, H., Trevors, J.T., Lohmeier-Vogel, E.M., et al., 1997. Isolation and characterization of pentachlorophenol degrading Pseudomonas spp. UG25 and UG30. World J. Microbiol. Biotechnol. 13, 305–313.

Liu, N., Scott, J.G., 1998. Increased transcription of CYP6D1 causes cytochrome P450-mediated insecticide resistance in house fly. Insect Biochem. Mol. Biol. 28, 531–535.

Lu, A.Y.H., Coon, M.J., 1968. Role of hemoprotein P-450 in fatty acid omega-hydroxylation in a soluble enzyme system from liver microsomes. J. Biol. Chem. 243, 1331–1332.

Ma, H., 1994. GTP-binding proteins in plants: new members of an old family. Plant Mol. Biol. 26, 1611–1636.

Mackenzie, R., Mortimer, A.M., Putwain, P.D., Bryan, I.B., Hawkes, T.R., 1995. Brighton Crop Protection Conference—Weeds. British Crop Protection Council Publications, Farnham, UK, The inheritance of chlorsulfuron resistance in perennial ryegrass: strategic implications for management of resistance; pp. 769–774.

Madsen, K.H., Poulsen, G.S., Fredshavn, J.R., Jensen, J.E., Steen, P., Streibig, J.C., 1998. A method to study competitive ability of hybrids between seabeet (Beta vulgaris ssp. maritima) and transgenic glyphosate tolerant sugarbeet (Beta vulgaris ssp. vulgaris). Acta Agricul. Scand B., Soil Plant Sci. 48, 170–174.

Madsen, K.H., Sandøe, P., 2001. Herbicide resistant sugar beets—what is the problem? J. Agricul. Environ. Ethic. 14, 161–168.

Madsen, K.H., Streibig, J.C., 2013. Benefits and risks of the use of herbicide-resistant crops. Available at www.fao.org. Accessed on May 8, 2013.

Madsen, K.H., Valverde, B.E., Jensen, J.E., 2002. Risk assessment of herbicide resistant crops: a Latin American perspective using rice (*Oryza sativa*) as a model. Weed Tech. 16, 215–223.

Mandelbaum, R.T., Allan, D.L., Wackett, L.P., 1995. Isolation and characterization of a *Pseudomonas* spp. that mineralizes the s-triazine herbicide atrazine. Appl. Environ. Microbiol. 61, 1451–1457.

Mannervik, B., Danielson, U.H., 1988. Glutathione transferases—structure and catalytic activity. CRC Crit. Rev. Biochem. 23, 283–337.

Mansuy, D., 1998. The great diversity of reactions catalyzed by cytochromes P450. Comp. Biochem. Physiol. C Pharmacol. Toxicol. Endocrinol. 121, 5–14.

Marshall, G., 1991. In: Caseley, J.C., Cussans, G.W., Atkin, R.K. (Eds.), Herbicide resistance in weeds and crops. Butterworth-Heinemann Ltd, Oxford, UK, pp. 331–341.

Mishra, S., Srivastava, S., Tripathi, R.D., Govindarajan, R., Kuriakose, S.V., Prasad, M.N.V., 2006. Phytochelatin synthesis and response of antioxidants during cadmium stress in *Bacopa monnieri* L. Plant Physiol. Biochem. 44, 25–37.

Mittler, R., 2002. Oxidative stress, antioxidants and stress tolerance Trends. Plant Sci. 7, 405–410.

Mizutani, M., Ohta, D., 2010. Diversification of P450 genes during land plant evolution. Ann. Rev. Plant Biol. 61, 291–315.

Nagasawa, S., Kikuchi, R., Nagat, Y., Takagi, M., Matsuo, M., 1993. Aerobic mineralization of γ-HCH by Pseudomonas paucimobilis UT26. Chemistry. 26, 1719–1728.

Nakamura, T., Motoyama, T., Suzuki, Y., Yamaguchi, I., 2004. Biotransformation of pentachlorophenol by Chinese chive and a recombinant derivative of its rhizosphere competent microorganism, Pseudomonas gladioli M-2196. Soil Biol. Biochem. 36, 787–795.

Nakano, Y., Asada, K., 1981. Hydrogen peroxide is scavenged by ascorbate-specific peroxidase in spinach chloroplasts. Plant Cell Physiol. 22, 867–880.

Nandula, V.K., Reddy, K.N., Poston, D.H., Rimando, A.M., Duke, S.O., 2008. Glyphosate tolerance mechanism in Italian ryegrass (*Lolium multiflorum*) from Mississippi. Weed Sci. 56, 344–349.

Nelson, D.R., Koymans, L., Kamataki, T., Stegeman, J.J., Feyereisen, R., Waxman, D.J., et al., 1996. P450 superfamily: update on new sequences, gene mapping, accession numbers and nomenclature. Pharmacogen 6, 1–42.

Neve, P., Powles, S.B., 2005. Recurrent selection with reduced herbicide rates results in the rapid evolution of herbicide resistance in *Lolium rigidum*. Theor. Appl. Gen. 110, 1154–1166.

Nurzhanova, A., Kalugin, S., Zhambakin, K., 2013. Obsolete pesticides and application of colonizing plant species for remediation of contaminated soil in Kazakhstan. Environ. Sci. Pollut. Res. 20, 2054–2063.

Ohaya-Mitoko, G.J.A., 1997. Occupational pesticide exposure among Kenyan agricultural workers. PhD thesis. Wageningen University.

Olofdotter, M., Valverde, B.E., Madsen, K.H., 2000. Herbicide resistant rice (*Oryza sativa* L.) in a global perspective: implications for weed management. Annal. Appl. Biol. 137, 279–295.

Oppenoorth, F.J., 1985. Biochemistry and genetics of insecticide resistance. In: Kerkut, G.A., Gilbert, L.I. (Eds.), Comprehensive Insect Physiology, Biochemistry, and Pharmacology. Pergamon Press, Oxford, pp. 731–774.

Pan, G., Si, P., Yu, Q., Tu, J.M., Powles, S., 2012. Non-target site mechanism of metribuzin tolerance in induced tolerant mutants of narrow-leafed lupin (Lupinus angustifolius L.). Crop Pasture Sci. 63, 452–458.

Pang, X., Wang, D., Peng, H.A., 2001. Effect of lead stress on the activity of antioxidant enzymes in wheat seedling. Environ. Sci. 22, 108–111.

Pascal-Lorber, S., Lauren, F., 2011. Phytoremediation techniques for pesticide contaminations. In: Lichtfouse, E. (Ed.), Alternative Farming Systems, Biotechnology, Drought Stress and Ecological Fertilisation. Springer, New York, pp. 77–105.

Pascal, S., Debrauwer, L., Ferte, M.P., Anglade, P., Rouimi, P., Scalla, R., 1998. Analysis and characterization of glutathione S-transferase subunits from wheat (Triticumaestivum L.). Plant Sci. 134, 217–226.

Passardi, F., Penel, C., Dunand, C., 2004. Performing the paradoxical: how plant peroxidases modify the cell wall. Trend Plant Sci. 9, 534–540.

Payne, S.A., Oliver, L.R., 2000. Weed control programs in drilled glyphosate-resistant soybean. Weed Tech. 14, 413– 422.

Peixoto, F., Alves-Fernandes, D., Santos, D., Fontánhas-Fernandes, A., 2006. Toxicological effects of oxyfluorfen on oxidative stress enzymes in tilapia Oreochromisniloticus Pestic. Biochem. Physiol. 85, 91–96.

Peterson, J.A., Prough, R.A., 1986. Cytochrome P-450 reductase and cytochrome b5 in cytochrome P-450 catalysis. In Ortiz De Montellano, P.R. (Eds.), Cytochrome P-450 structure, mechanism, and biochemistry. Plenum Press, NY and London, pp. 89

Petit, C., Duhieu, B., Boucansaud, K., Delye, C., 2010. Complex genetic control of non-target-site-based resistance to herbicides inhibiting acetyl-coenzyme A carboxylase and acetolactate-synthase in *Alopecurus myosuroides* Huds. Plant Sci. 178, 501–509.

Pompon, D., 1987. Rabbit liver cytochrome P-450 LM2: roles of substrates, inhibitors and cytochrome b5 in modulating the partition between productive and abortive mechanisms. Biochemistry 26, 6429–6435.

Poulsen, G.S., 1995. Weediness of transgenic oilseed rape—evaluation methods. The Royal Veterinary and Agricultural University, Department of Agricultural Sciences (Weed Science), Denmark (PhD thesis).

Powles, S.B., Yu, Q., 2010. Evolution in action: plants resistant to herbicides. Ann. Rev. Plant Biol. 61, 317–347.

Preston, C., 2003. Inheritance and linkage of metabolism-based herbicide cross-resistance in rigid ryegrass (*Lolium rigidum*). Weed Sci. 51, 4–12.

Preston, C., Tardif, F.J., Christopher, J.T., Powles, S.B.P., 1996. Multiple herbicide resistance to dissimilar herbicide chemistries in a biotype of Loliumrigidum is due to enhanced activity of several degrading enzymes. Pest Biochem. Physiol. 55, 123–134.

Qin, X.F., Kong, F.B., 2006. Prospect and situation of biopesticides. J. Anhui Agricult. Sci. 34, 4024–4057.

Radetski, C.M., Ferrari, B., Cotelle, S., Masfaraud, J.F., Ferrard, J.F., 2004. Evaluation of the genotoxic, mutagenic and oxidant stress potentials of municipal solid waste incinerator bottom ash leachates. Sci. Total Environ. 333, 209–216.

Ramanand, K., Sharmilla, M., Sethunathan, N., 1988. Mineralization of carbofuran by a soil bacterium. Appl. Environ. Microbiol. 54, 2129–2133.

Rao, S.R., Feng, P.C.C., Schafer, D.E., 1995. Enhancement of thiazopyr bioefficacy by inhibitors of monooxygenases. Pestic. Sci. 45, 209–213.

Ray, J.W., 1967. The epoxidation of aldrin by housefly microsomes and its inhibition by carbon monoxide. Biochem. Pharmacol. 16, 99–107.

Rendic, S., Di Carlo, F.J., 1997. Human cytochrome P450 enzymes: a status report summarizing their reactions, substrates, inducers, and inhibitors. Drug Metab. Rev. 29, 413–580.

Rosa, M., González, R., Raquel, R.-O., Beatriz, C.-G., Jesús, S.-G., 2008. Occurrence, fungicide and insecticide residues in trade samples of leafy vegetables. Food Chem. 107, 1342–1347.

Ryan, T.P., Bumpus, T.A., 1989. Biodegradation of 2,4,5-trichloro-phenoxyacetic acid in liquid culture and in soil by the white rot fungus *Phanerochaete chrysosporium*. Appl. Microbiol. Biotechnol. 31, 1387–1398.

Sawicki, R.M., 1962. Insecticidal activity of pyrethrum extract and its four insecticidal constituents against house flies. III. Knock-down and recovery of flies treated with pyrethrum extract with and without piperonylbutoxide. J. Sci. Food Agric. 13, 283–291.

Schuler, M.A., 1996. Plant cytochrome P450 monooxygenases. Crit. Rev. Plant Sci. 15, 235–284.

Schuler, M.A., Werck-Reichhart, D., 2003. Functional genomics of P450s. Annu. Rev. Plant Biol. 54, 629–667.

Scott, J.G., 1991. Insecticide resistance in insects. In: Pimentel, D. (Ed.), Handbook of Pest Management in Agriculture, vol. 2. CRC Press, Boca Raton, p. 663.

Scott, J.G., 1999. Cytochromes P450 and insecticide resistance. Insect Biochem. Mol. Biol. 29, 757–777.

Scott, J.G., Georghiou, G.P., 1986. Mechanisms responsible for high levels of permethrin resistance in the house fly. Pestic. Sci. 17, 195–206.

Scott, J.G., Sridhar, P., Liu, N., 1996. Adult specific expression and induction of cytochrome P450lpr in house flies. Arch. Insect Biochem. Physiol. 31, 313–323.

Scott, J.G., Liu, N., Wen, Z., 1998. Insect cytochromes P450: diversity, insecticide resistance and tolerance to plant toxins. Comp. Biochem. Physiol. 121C, 147–155.

SDNX, 2005. Reduction situation of pesticide applications of some countries. Shandong Pesticide News 11, 34.

Shaner, D.L., 2000. The impact of glyphosate-tolerant crops on the use of other herbicides and on resistance management. Pest Manage. Sci. 56, 320–326.

Shimabukuro, R.H., Walsh, W.C., Hoerauf, R.A.J., 1979. Metabolism and selectivity of diclofop-methyl in wild oat and wheat. Agric. Food Chem. 27, 615–623.

Siminszky, B., 2006. Plant cytochrome P450-mediated herbicide metabolism. Phytochem. Rev. 5, 445–458.

Singer, A.C., Crowley, D.E., Thompson, I.P., 2003. Secondary plant metabolites in phytoremediation and biotransformation. Trends Biotechnol. 21, 123–130.

Singh, B.K., Walker, A., Morgan, J.A.W., Wright, D.J., 2004. Biodegradation of chlorpyrifos by Enterobacter strain B-14 and its use in bioremediation of contaminated soils. Appl. Environ. Microbiol. 70, 4855–4863.

Song, N.H., Yang, Z.M., Zhou, L.X., Wu, X., Yang, H., 2006. Effect of dissolved organic matter on the toxicity of chlorotoluron to Triticumaestivum. J. Environ. Sci. 17, 101–108.

Song, N.H., Yin, X.L., Chen, G.F., Yang, H., 2007. Biological responses of wheat (Triticumaestivum) plants to the herbicide chlorotoluron in soils. Chemistry. 68, 1779–1787.

Stalker, D.M., McBride, K.E., Malyj, L.D., 1988. Herbicide resistance in transgenic plants expressing a bacterial detoxification gene. Science 242, 419–423.

Stegeman, J.J., Livingstone, D.R., 1998. Forms and functions of cytochrome P450. Comp. Biochem. Physiol. 212C, 1–3.

Streber, W.R., Kutschka, U., Thomas, F., Pohlenz, H.D., 1994. Expression of a bacterial gene in transgenic plants confers resistance to the herbicide phenmedipham. Plant Mol. Biol. 25, 977–987.

Sun, C.N., Tsai, Y.C., Chiang, F.M., 1992. Resistance in the diamondback moth to pyrethroids and benzoylphenylureas. In: Mullin, C.A., Scott, J.G. (Eds.), ACS Symposium Series, No. 505. ACS, Washington, DC, pp. 149–167.

Sun, H., Xu, J., Yang, S., Liu, G., Dai, S., 2004. Plant uptake of aldicarb from contaminated soil and its enhanced degradation in the rhizosphere. Chemistry. 54, 569–574.

Sunohara, Y., Shirai, S., Wongkantrakorn, N., Hiroshi Matsumoto, H., 2010. Sensitivity and physiological responses of Eleusine indica and Digitaria adscendens to herbicide quinclorac and 2,4-D. Environ. Exp. Bot. 68, 157–164.

Swanton, C.J., Shrestha, A., Chandler, K., Deen, W, 2000. An economic assessment of weed control strategies in no-till glyphosate-resistant soybean (Glycine max). Weed Tech. 14, 755–763.

Toxipedia.org. (2011) Pesticide use statistics. Available from: http://toxipedia.org/display/toxipedia/Pesticide + Use + Statistics. Accessed November 2013.

Valavanidis, A., Vlahogianni, T., Dassenakis, M., Scoullos, M., 2006. Molecular biomarkers of oxidative stress in aquatic organisms in relation to toxic environmental pollutants. Ecotoxicol. Environ. Saf. 64, 178−189.

Valles, S.M., Yu, S.J., 1996. Detection and biochemical characterization of insecticide resistance in the German cockroach (Dictyoptera: Blattelidae). J. Econ. Entomol. 89, 21−26.

Varsano, R., Rabinowitch, H.D., Rubin, B., 1992. Mode of action of piperonylbutoxide as herbicide synergist of atrazine and terbutryn in maize. Pestic. Biochem. Physiol. 44, 174−182.

Vatsis, K.P., Gurka, D.P., Hollenberg, P.F., 1980. Involvement of cytochrome b5 in the NADPH-dependent regioselective hydroxylation of N-methylcarbazole by cytochrome P-450LM2 and P-450LM4 in a reconstituted liver microsomal enzyme system. In: Gustafsson, J., Carlstedt-Duke, J., Mode, A., Rafter, J. (Eds.), Biochemistry, Biophysics, and Regulation of Cytochrome P-450. Elsevier, New York, pp. 347−350.

Vincent, D.R., Moldenke, A.F., Farnsworth, D.E., Terriere, L.C., 1985. Cytochrome P-450 in insects. 6. Age dependency and phenobarbital induction of cytochrome P-450, P-450 reductase, and monooxygenase activities in susceptible and resistant strains of *Musca domestica*. Pestic Biochem. Physiol. 23, 171−181.

Wain, R.L., Smith, M.S., 1976. In: Audus, L.J. (Ed.), Herbicides: Physiology, Biochemistry, Ecology, vol. 1. Academic Press, London, England, pp. 279−302.

Wang, S.H., Yang, Z.M., Lu, B., LiS, Q., Lu, Y.P., 2004. Copper induced stress and antioxidative responses in roots of *Brassica juncea* L. Bot. Bull. Acad. Sin. 45, 203−212.

Wang, T., Fleury, A., Ma, J., Darmency, H., 1996. Genetic control of dinitroaniline resistance in foxtail millet (*Setaria italica*). J. Heredity 87, 423−426.

Wang, Y.S., Yang, Z.M., 2005. Nitric oxide reduces aluminum toxicity by preventing oxidative stress in the roots of Cassia tora L. Plant Cell Physiol. 46, 1915−1923.

Wen, Z., Scott, J.G., 1997. Cross-resistance to imidacloprid in strains of German cockroach (*Blattella germanica*) and house fly (*Musca domestica*). Pestic. Sci. 49, 367−371.

Wilcut, J.W., Coble, H.D., York, A.C., Monks, D.W., 1996. The niche for herbicide-resistant crops in U.S. agriculture. In: Duke, S.O. (Ed.), Herbicide-Resistant Crops: Agricultural, environmental, economic, regulatory, and technical aspects. CRC Press Inc, Boca Raton, Florida, USA, pp. 213−230.

Wilkinson, C.F., 1967. Penetration, metabolism, and synergistic activity with carbaryl of some simple derivatives of 1,3-benzodioxole in the housefly. J. Agric. Food Chem. 15, 139−147.

Wilkinson, C.F., 1980. The metabolism of xenobiotics: a study in biomedical evolution. In: Witschi, H.R. (Ed.), The Scientific Basis of Toxicity Assessment. Elsevier/North-Holland Biomedical Press, Amsterdam, pp. 251−268.

Wilkinson, C.F., 1983. Role of mixed-function oxidases in insecticide resistance. In: Georghiou, G.P., Saito, Y. (Eds.), Pest Resistance to Pesticides. Plenum Press, New York, pp. 175−206.

Wohlleben, W., Arnold, W., Broer, I., Hillemann, D., Strauch, E., Puhler, A., 1988. Nucleaotide-sequence of the phosphinothricin N-acetyltransferase gene from Streptomyces-viridochomogenes-TU494 and its expression in Nicotiana-tabacum. Gene 70, 25−37.

Wu, X.Y., von Tiedemann, A., 2002. Impact of fungicides on active oxygen species and antioxidant enzymes in spring barley (Hordeumvulgare L.) exposed to ozone. Environ. Pollut. 116, 37−47.

Xiao, Y.M., Wang, J.A., Wang, M.A., Liu, J.P., Yuan, H.Z., Qin, Z.H., 2010. Study on the inclusion complexes of flumorph and dimethomorph with β-cyclodextrin to improve fungicide formulation. J. Chem. Soc. Pak. 32, 363.

Xu, Y.Z., 2008. The application of biological pesticides and industrial development measures. Chin. Agric. Sci. Bull. 24, 402−404.

Yang, J.L., 2001. Green Chemistry and Technology. Beijing University of Posts and Telecommunications, Beijing, China.

Yin, X.L., Jiang, L., Song, N.H., Yang, H., 2008. Toxic reactivity of wheat (Triticum aestivum) plants to herbicide isoproturon. J. Agric. Food Chem. 56, 4825−4831.

Yu, S., 1996. Insect glutathione S-transferases. Zool. Stud. 35, 9–19.

Yu, Y.L., Chen, Y.X., Luo, Y.M., Pan, X.D., He, Y.F., Wong, M.H., 2003. Rapid degradation of butachlor in wheat rhizosphere soil. Chemistry 50, 771–774.

Zhang, H.L., Zhang, H.M., 1998. New direction of pesticide development: biopesticides. J. Mod. Agricult. 4, 9.

Zhang, J.X., Kirham, M.B., 1994. Drought stress-induced changes in activities of superoxide dismutase, catalase and peroxidase in wheat species. Plant Cell Physiol. 35, 785–791.

Zhang, M., Scott, J.G., 1994. Cytochrome b5 involvement in cytochrome P450 monooxygenase activities in house flymicrosomes. Arch. Insect Biochem. Physiol. 27, 205–216.

Zhang, M., Scott, J.G., 1996. Cytochrome b5 is essential for cytochrome P450 6D1-mediated cypermethrin resistance in LPR houseflies. Pestic. Biochem. Physiol. 55, 150–156.

Zhang, W.J., Jiang, F.B., Ou, J.F., 2011. Global pesticide consumption and pollution: with China as a focus. Proc. Intl. Acad. Ecol. Environ. Sci. 1, 125–144.

Zhang, W.J., Qi, Y.H., Zhang, Z.G., 2006. A long-term forecast analysis on worldwide land uses. Environ. Monit. Assess. 119, 609–620.

Zhou, Z.S., Huang, S.Q., Guo, K., Mehta, S.K., Zhang, P.C., Yang, Z.M., 2007. Metabolic adaptations to mercury-induced oxidative stress in roots of Medicago sativa L. J. Inorg. Biochem. 101, 1–9.

Zhu, C.X., Bai, X.S., Zhang, M., 2002. The status quo of development and perspective of biopesticides. Shanghai Environ. Sci. 21, 654–659.

Effects of Humic Materials on Plant Metabolism and Agricultural Productivity

18

Andrés Calderín García, Fernando Guridi Izquierdo and Ricardo Luis Louro Berbara

18.1 Introduction

One of the challenges that the planet faces is the need to avoid increased environmental contamination and improve crop production while providing quality food products. For this purpose, recent studies have focused on the reutilization of easily available materials, such as humic materials (compost and vermicompost), that are innocuous to nature (Bhattacharya et al., 2012; Shafawati and Siddiquee, 2013; De Carvalho et al., 2013). These materials have a variety of environmental functions, such as improving soil quality, increasing plant yield, and interacting with soluble cations (Singh et al., 2013; Doan et al., 2013; Yang et al., 2013).

The fact that these materials have natural effects that are both indirect (on plants, improvement of chemical, physical, and biological soil properties and soil decontamination) and direct (metabolism stimulation) is of extreme environmental importance. Regarding the direct actions, compost materials can improve the agricultural yield of plants by increasing plant nutritional status, improving resistance to stress, and stimulating various metabolic pathways (Gutiérrez-Miceli et al., 2007; Tejada et al., 2009; Lazcano et al., 2010; Srivastava et al., 2012; Azizi et al., 2013; Doan et al., 2013; Singh et al., 2013).

These effects are important aspects of agricultural resource management. Although there is no technology for the use of humic materials, the fact that they reduce the use of synthetic chemicals in the soil makes them an environmentally valuable alternative, particularly because they have a low cost and are easily available and harmless to the environment.

With this general view and without the objective of creating booklets on the use of humic materials for improving crop yields, we will discuss the potential of these substances. We first describe the chemical, physical, and spectroscopic characteristics of humic materials and their fractions to provide context for understanding their properties in the environment. We then discuss the relation between the structural properties and function of humic materials.

P. Ahmad (Ed): Emerging Technologies and Management of Crop Stress Tolerance, Volume 1.
DOI: http://dx.doi.org/10.1016/B978-0-12-800876-8.00018-7

18.2 General aspects of the characteristics of humic materials and their functions

18.2.1 Vermicomposting and its application on soil and plants

Vermicomposting is a process that involves chemical, physical, and biological transformations of solid organic materials (agricultural residues of plant and animal origin) through the use of worms and microorganisms (Garg and Gupta, 2009). The use of worms in the process aids the fragmentation of fresh organic matter, thereby increasing the surface area available for microbial colonization (Domínguez et al., 2010). Utilization of vermicompost (VC) in agricultural and environmental activities is related to the ability of this material to improve the chemical, physical, and biological soil conditions and to directly and indirectly improve the biological and agricultural plant yield. Therefore, VC has been used to remediate contaminated soils, to promote plant regulation, and to stimulate plant growth (García et al., 2013b; Oo et al., 2013).

With respect to the use of VC in soil remediation, it has been shown that its use, when based on olive cake compost, may promote a rapid decrease in the concentration of atrazine herbicides in the soil. This reduction occurs because VC may stimulate microbial activity related to the degradation of these substances. The same results were not obtained when using only olive cake that was not previously composted, which indicates that atrazine degradation depends on the type of organic matter applied to the soil, and humic organic matter is more favorable to the degradation of this contaminant (Delgado-Moreno and Peña, 2009).

Other studies have also shown that the application of VC to soil stimulates the bacterial community related to the degradation of organic chemical species that are contaminants, such as polycyclic aromatic hydrocarbons (PAH). Di Gennaro et al. (2009) documented the use of olive mill VC for the degradation of naphthalene in soil based on gene expression related to biodegradation in the community of autochthonous soil bacteria that are able to degrade PAH. Similarly, authors such as Zhang et al. (2012) discuss inhibition of *in situ* PAH remediation due to the low presence of nutrients and organic carbon in the soil. These authors conclude that the application of organic matter to soil stimulates the activity of dehydrogenase and the microorganisms that degrade PAH.

Humic materials also directly affect physical, chemical, and biological soil properties. The use of composted vinasse residues and vermicomposted plant residues decreased soil loss by 31.2% and increased plant cover by 68.7% (Tejada et al., 2009). Beneficial results have also been obtained when VC inoculated with microorganisms was applied to soil. This VC and microorganism consortium can improve physical conditions of the soil, such as the density and water-retention capacity, as well as chemical properties, such as pH and the levels of organic carbons, N, P, and K (Singh et al., 2013).

Therefore, VC can be an ecological alternative as a regular agricultural practice. However, there are still several questions that must be clarified before establishing phytotechnological recommendations because, as previously discussed, the beneficial effects depend on the type of raw material that was used in the VC, the type of soil to which it will be applied, and the species of plant (Figure 18.1).

18.2.2 Some structural characteristics of vermicomposts

The VC quality depends on the source of the raw materials used. Thus, it is necessary to know the chemical, physical, and biological characteristics before making a recommendation. Table 18.1

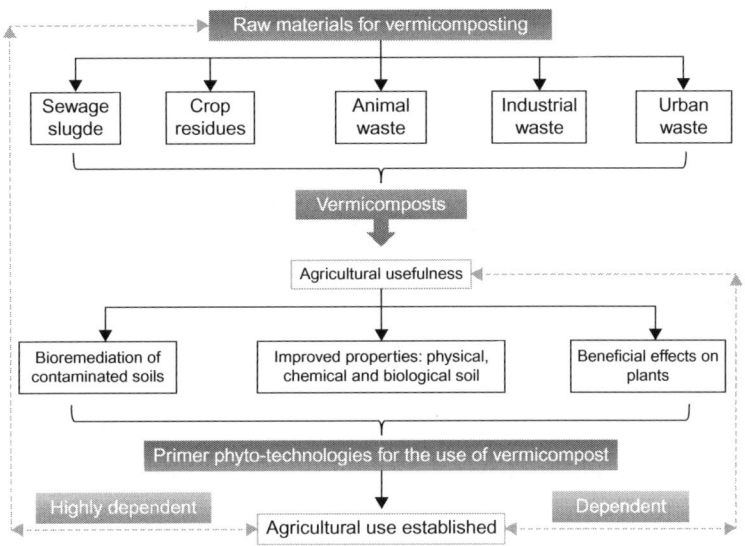

FIGURE 18.1

Schematic summary of some of the current main aspects in the study of VC and their utilization.

Table 18.1 Quantities of Some of the Chemical Parameters Present in Five Different VC Samples

VC	Pep and PPep	C	N	C/N	M.O	C-FHS	C-FAF	C-FAH	HA/FA
1	3.45	21.41	1.85	11.57	36.9	11.71	9.81	1.9	0.19
2	4.37	20.36	1.76	11.57	35.1	12.28	10.58	1.7	0.16
3	2.99	20.00	1.73	11.56	34.5	11.37	9.57	1.8	0.18
4	3.33	22.30	1.92	11.61	38.4	7.16	5.73	1.43	0.24
5	3.22	21.40	1.85	11.57	36.9	12.51	11.16	1.35	0.12

Pep and PPep (peptides and polypeptides): values in g/100 g (VC).
C, N, M.O, C-FAF, C-FAH, and HA/FA: values in %.
(1) VC from cow manure, Red Californian worms and 1 year of maturation, (2) VC from rabbit manure, Red Californian worms and 1 month of maturation, (3) VC from cow manure, Red African worms and 1 year of maturation, (4) VC from goat manure, Red African worms and 3 months of maturation, (5) VC from cow manure, Red African worms and 3 months of maturation.
Data provided by research professor Eduardo Ruiz Vasallo-UNAH-Cuba.

shows some chemical parameters of five different VC that were created from different raw materials with varying types of worms and maturation times. The table shows that an important characteristic of VC is the amount of carbon in the form of humic carbon. This factor could have important consequences on the effects of the VC when applied to soil.

Another important question related to the chemical characteristics of VC is the insoluble solid fraction of VC in aqueous solution. Researchers have documented the ability of this insoluble residual fraction to retain dissolved heavy metals (García et al., 2007, 2013c). This ability increases the

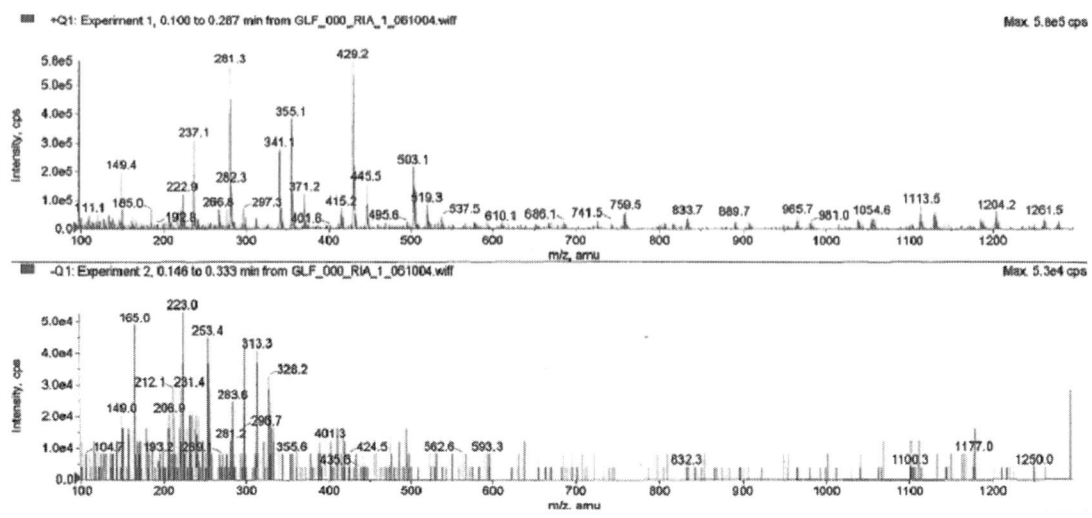

FIGURE 18.2

Mass spectrum obtained by the APCI-MS technique of the insoluble solid fraction in aqueous solution in cow manure VC. Upper spectrum: positive fragments, lower spectrum: negative fragments (authors' data).

continuation of metal contaminants in the soil and is explained by the presence of ionizable functional groups, which can establish interactions of different natures with metal cations.

Figure 18.2 shows a chemical fragmentation technique (atmospheric pressure chemical ionization-mass spectrometry [APCI-MS]) applied to the insoluble solid fraction of cow manure VC. The mass spectrum of the fraction shows the structural variability that is present in one of the VC fractions, for both the positive and negative fractions, but particularly in the latter due to the importance of interactions with positive metallic elements.

The diverse effects that VC have on soil properties can be observed in the ability of these materials to interact with distinct environments. When these VC are applied, they can enrich the soil with different fragments of diverse chemical properties that provide nutrients for both plants and microorganisms. Evidence of this structural diversity is shown in Figure 18.3, which shows how cow manure VC has considerable quantities of soluble fragments in solvents with different polarities. This figure indicates a greater quantity of less hydrophilic substances solubilized in chloroform and more hydrophobic substances solubilized in water. These characteristics are proof of a more stable structural arrangement of humic organic matter, which maintains its structural integrity. These facts may explain some of the action mechanisms of humic substances (HS) in plants and interaction with roots (Nebbioso and Piccolo, 2011; Song et al., 2013).

Regarding the content of the mineral elements, VC have a variable composition, depending on the VC's source and production process. Some of these elements can be released and become part of the soil when applied in the field, which contributes to increased fertility. Figure 18.4 shows an EDX spectrum of a cow manure VC. A greater quantity of the elements C, K, Ca, O, and Si, possibly in the form of salts such as carbonates and silicates (CO_3^{2-}, SiO_4^{4-}), can be observed in this spectrum. Other mineral elements, such as Mg, Fe, and Al, are part of the inorganic pool of this VC.

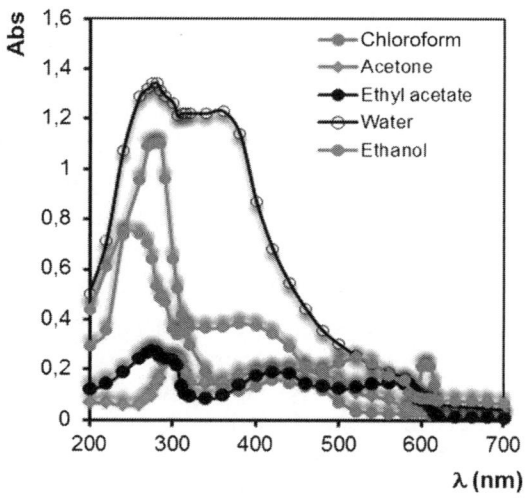

FIGURE 18.3

Ultraviolet-visible spectroscopy (UV-vis) spectra of soluble substance fractions in 1 gram of cow manure VC using 100 mL of solvents with different polarities (authors' data).

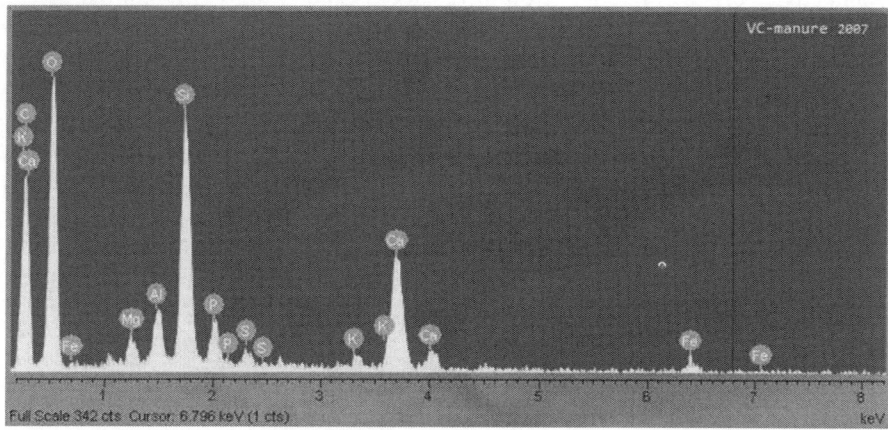

FIGURE 18.4

Energy-dispersive X-ray (EDX) spectrum of cow manure VC produced by Red Californian worms (*Eisenia foetida*) and plant residues (authors' data).

A study that used spectroscopic techniques (diffuse-reflectance infrared Fourier transform spectroscopy [DRIFT]) to determine the quantities and relative abundance of structural fragments (by pyrolysis-gas chromatography-mass spectrometry) during the VC of five materials (cow manure [M]; cow manure and sugarcane bagasse [SB]; cow manure and sunflower oil cake [SC]; cow manure, sugarcane residues, and sunflower-oil cake [SBSC], and filter cake [FC]) concluded that the thermochemolysis of the VC released compounds derived from lignins, carbohydrates, proteins, acids, fatty alcohols, terpenic compounds, and hydrocarbons, whose relative abundances varied with the maturation of the VC, and indicated that the relative variations and abundance of these composts were characteristics of each VC studied (Balmori et al., 2013). An example of the dynamics of the composts presented in this study can be found in Figure 18.5.

18.2.3 Chemical composition of liquid humus

Liquid humus is obtained through the extraction of an extracting dissolution with different chemical characteristics and a VC. The liquid humus obtained from the VC can be used as liquid fertilizers applied to foliage because they have high levels of mineral nutrients and HS that regulate metabolic processes in plants that contribute to improved development (Gutiérrez-Miceli et al., 2008; Singh et al., 2010).

Logically, the variability of VC as a source of raw materials causes the liquid humus to have a variable composition as well, although in a general manner, with some parameters fluctuating within the same range. Liquid humus has a high amount of elements and substances with different chemical natures that vary even with the type of method used for attaining the humus (Terry et al., 2012; Aşik and Katkat, 2013). Caro (2004) obtained a humic acid/fulvic acid (HA/FA) relationship of 0.98, pH of 8.4, and electrical conductivity of 12.82 mS/cm^{-1} from a cow manure VC liquid humus (Liplant$^{®}$), while Hernández (2011) obtained HA/FA values of 1.21, a pH of 8.7, and an electrical conductivity of 5.81 mS/cm^{-1} from this same VC but with a different extractive solution. Gutiérrez-Miceli et al. (2008) reported that leachates obtained from cow manure VC had AH/AF values of 1.6, a pH of 7.8, and an electrical conductivity of 2.6 mS/cm^{-1}. In contrast, Canellas et al. (2013), in their study with soluble HS in basic aqueous solution obtained from cow manure VC, found AH/AF values of 0.98, a pH of 8.67, and an electrical conductivity of 11.7 mS/cm^{-1}.

Arteaga et al. (2007) found in the liquid humus Liplant$^{®}$ a microbial community of bacteria (6.55×10^7 UFC/mL^{-1}), fungi (2.21×10^5 UFC/mL^{-1}) and actinomycetes (5.4×10^3 UFC/mL^{-1}), while Hernández (2011) applied thermal treatments in several of the steps to obtain this liquid

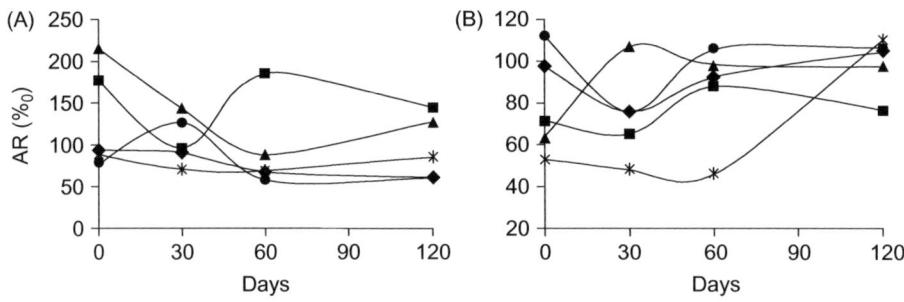

FIGURE 18.5

Relative abundance of carbohydrates (A) and nitrogen compounds (B) in VC with different states of maturation (0, 30, 60, and 120 days); M (♦), SB (■), SC (▲), SBSC (●), and FC (✳).

Modified from Balmori et al., 2013.

humus, changed some of the extracting solution, and obtained a quantity of bacteria (3.4×10^5 UFC/mL^{-1}) and fungi (1.40×10^4 UFC/mL^{-1}), but no actinomycetes were detected.

This evidence indicates that liquid humus is enriched with various types of organic substances (plant hormones, proteins, amino acids, peptides, and fatty acids), mineral elements, microorganisms, and HS and thus constitutes a viable source of organic fertilizer. The scientific literature provides methodologies for ecological and sustainable use (García et al., 2013a).

18.2.4 **Some structural characteristics of humic fractions**

Concern for knowing the structural characteristics of HS and its fractions has led to a group of chemical and physical techniques being used to understand the superstructure of HS. If it is certain that VC HS fractions have spectral differences, it is also possible to observe similarities in spectral "signatures," such as those obtained by the ultraviolet-visible spectroscopy (UV-vis), Fourier-transform infrared spectroscopy (FT-IR), and carbon-13 nuclear magnetic resonance (^{13}C-NMR) techniques (Ke et al., 2013; Lv et al., 2013; Xiao-Song et al., 2013).

A study by Aguiar et al. (2013) demonstrates the spectral characteristics obtained by FT-IR and ^{13}C-CPNMR of humic acids (HA) isolated from four VC of different raw materials. Although there were some differences, the NMR spectra showed the presence of C-alkyl, N-alkyl/methoxyl, O-alkyl, di-O-alkyl, acrylic, O-acrylic, and carboxylic structures in the HA of all of the VC.

Therefore, HA, even when isolated from composts derived from different raw materials, have great similarities regarding the presence of their chemical structures, although they also show differences in spectral peak intensity, which indicate the quantity of structures (Al-Faiyz, 2013).

Humic substances of sewage sludge VC characterized by FT-IR had a low aliphatic character, with structures that came from proteins and polysaccharides, and an elevated presence of functional groups of acids and high aromaticity (Li et al., 2011). In contrast, an extraction of HA from cow manure VC by chromatography showed that larger molecular fragments of HA have predominantly aromatic characteristics, while smaller fragments have a predominance of more oxidized carbons and are less aromatic (Canellas et al., 2010).

Variability in the structure of HA during the composting process has also been studied. Amir et al. (2010) determined the structures of HA in the starting material of a compost at 15, 60, and 135 days of composting. In the first 15 days of composting, the predominant structures in the HA of the VC were *O/N*-alkyl and *C*-carboxyl. In the next 60 days, the predominant structures were *C*-alkyl and *C*-aliphatic, while at 135 days, the structural predominance was *C*-aromatic and *C*-carboxylic, with an aromaticity/aliphaticity ratio of 1.51. Regardless of the composting processes, HA exhibit characteristic structural characteristics based on the source of material. However, if it is certain that the HS and fractions have a high variability depending on the source of origin and production processes, then it is also certain that there are spectral patterns that repeat and homogenize their structural behavior. Thus, the growing use of VC as biofertilizers that promote plant growth is based more on the elevated content of soluble humic substances that these materials possess than on the composition of nutritional elements (Trevisan et al., 2010).

A complement for understanding the similarities and differences of HS and their fractions is the material on this topic on the International Humic Substance Society (IHSS) site (http://www.humic-substances.org/).

18.2.5 Some basic knowledge about the structure—activity relationship of humic substances

The most important previous knowledge for understanding HS properties and functions is a structural complexity. According to studies in "*Humeomics*," introduced by Nebbioso and Piccolo (2011, 2012), HS have a supramolecular structural organization, distributed in large hydrophobic structures and other smaller hydrophilic ones. The hydrophobic fractions are basically composed of humic fractions of linear aliphatic chains and condensed aromatic rings, while the hydrophilic fractions are composed of more irregular humic fractions. Therefore, it is understood that the supramolecular structural arrangement of HS is a result of a non-uniform relationship of heterogeneous humic molecules that interact as a function of their size, form, chemical affinity, and hydrophobicity (Nebbioso and Piccolo, 2012).

This heterogeneity and variability of HS, both organizational and in terms of the diversity of functional groups, allows us to understand the many chemical, physical, and biological processes in which they are involved in the environment. Among the most studied processes are their ability to interact with metal cations and the possibility of interaction with plant root systems, triggering various beneficial events that result in greater plant and agricultural yield. Studies about the role of HS in the environment have created interpretations to understand the modes of action and structure—activity relationship of HS (Nuzzo et al., 2013; Xitao et al., 2013; Wang-Wang et al., 2014).

Regarding interactions between HS and metals, it is known that these interactions are favored, which causes a high presence of ionizable functional groups (carboxyls: COOH, carbonyls: $C = O$, and phenols: OH) that allows the formation of HS-metal organometallic compounds. It has been documented that these functional groups have a greater participation in interactions with the elements Cu^{2+}, Pb^{2+}, and Ca^{2+} for humic and fulvic acids, establishing an order of interaction of $Cu^{2+} > Pb^{2+} > Ca^{2+}$ (Piccolo and Stevenson, 1982). The structural arrangement of the HS also participates in interactions with metals. Nebbioso and Piccolo (2009) reported that in the interaction of HA with the metals Al^{3+} and Ca^{2+}, carboxylic acids participated favorably, and this participation led to an increase in the conformational structural rigidity and the size of the hydrophobic domains.

One of the most discussed questions in the scientific literature today in the area of humic materials is about the action mechanisms of HS in plants. The discussion involves the physiological stimulation of the root system of plants by HS action (Muscolo et al., 2013; Zandonadi et al., 2013). Some studies demonstrate a mimetic action of HA with auxins, which stimulates the emission of secondary roots through a mechanism that stimulates the activity of H^+-ATPases (Zandonadi et al., 2013). Authors such as Canellas et al. (2012) report that HS, regardless of the source of isolation, have a superstructural organization with a superficial hydrophobic domain and a hydrophilic interior, and a break in the superstructure can release hydrophilic components structurally similar to auxins that could cause these mimetic effects.

In turn, Elena et al. (2009) showed that in isolated leonardite HA, there are no substances that are structurally similar to plant hormones, which indicates that the effects of root stimulation could be exerted through mechanisms that are independent of the pathways used by auxins. Humic substances that are soluble in water with a low molecular weight did not show the same type of activity as auxin in *Arabidopsis* plants (Schmidt et al., 2007).

Other studies reported that HA can interact with the root system and modify root functionality. This type of event leads to a decrease of up to 44% in hydraulic conductivity and has been called colloidal stress (Asli and Neumann, 2010). Similar results that support this type of interaction have been reported by García et al. (2012b, 2014), who found that HA from VC can form clusters in the roots of rice plants, promote the production of radical oxygen species, and stimulate some enzymes related to antioxidative defense mechanisms. Another study also reported a stimulation of the enzyme catalase due to the application of humic substances in maize plants (Cordeiro et al., 2011).

18.3 **Use of humic materials for sustainable plant production**

As a result of research performed under controlled environments, the favorable effects of HS on plant growth and development have been documented (Alfonso et al., 2011; Ertani et al., 2013). During the last two decades, many studies were developed to reduce the use of inorganic fertilizers. The theoretical assumption is that there is a possibility of release of mineral elements when the fertilizers are completely mineralized. For example, a dosage of 15 kg/plant VC produced indicators that the given biological and agricultural productivity decreased the demand for mineral fertilizer by 50% in oil palm plantations in Colombia (Garcia et al., 2012). However, in *Cleome gynandra*, the most significant dosages were 11.5 and 15 t ha^{-1} (Ng'etich et al., 2012).

The progressive introduction of VC in conventional agriculture has been studied under different conditions. There has been evidence of positive effects on the microbial biomass and organic carbon fractions in the surface layer of tropical soils, which would translate to important improvements in agricultural productivity over time. Santos et al. (2012) considers that a period of at least 2 years in tropical soils is necessary to ensure adequate transfer to organic agriculture.

The application of plant residue compost, particularly to soil affected by salinity in coastal areas or those with a predominantly sandy texture with low fertility, has produced favorable effects on both development and biological and agricultural productivity in crops (Ayyappan et al., 2013) with *Vigna radiata* L. and Adebayo et al. (2013) with *Abelmoscus esculentus*). In silt clay soils with less than 1% organic matter, the most widely used composts have been those made from manure, sometimes in combination with urban organic residues (the latter only after previous verification that they do not contain persistent organic contaminants or heavy metals). This type of use requires prior application during planting or sowing, mixed with the soil surface layer. Generally, application of VC in agricultural production occurs in three fundamental variations: incorporating the VC into the soil, using the previously extracted soluble fraction in an aqueous solution (neutral or alkaline), or using dissolutions of HA isolated from the soluble fraction.

VC (created from macrophytes mixed with cow manure) incorporated in the first 15 cm of a soil with low organic matter in proportions corresponding to 2, 4, and 6 t ha^{-1} used for tomato crops (*Lycopersicon esculentum* Mill) caused significant increases in biological productivity parameters (height, leaf area, above-ground dry mass, and dry root mass) and agricultural productivity (number of fruits per stem and per plant, average fruit mass, and proportion of commercial quality fruits), which proves that this alternative source of organic matter is ecological and economically viable for agricultural production in this type of soil, especially at a dose of 6 t ha^{-1} (Ahmed and Khan, 2013).

The considerations and examples cited here about the impact of the use of composts on biological and agricultural productivity confirm the viability of this practice in different parts of the world.

The accumulated experience and knowledge that have been cited in the scientific literature are summarized in Box 18.1.

The soils employed in intensive agricultural systems, particularly in tropical and subtropical regions, experience a gradual decline in total organic carbon and alterations in the structural characteristics of the humic fraction (Quintero et al., 2012). These structural modifications in the humic fraction, as previously described, lead to changes in the potential interactions of these substances in the soil—plant system. García et al. (2012a) found that when they applied different procedures for extracting and purifying HA, structural differences were reflected in spectropic studies, as is shown in Figure 18.6.

The above-stated issue is a very important question to consider because another way of applying humic organic matter to stimulate plant development is using extracts of soluble humic substances in aqueous solution (or isolating the HA of these extracts to later dissolve into water), which can be obtained from humic materials (natural or derived from induced humification processes, such as composting and vermicomposting).

This alternative has had practical use in hydroponics through fertigation or direct foliar application. One of the first studies on this topic in the scientific literature discusses the application of humates and fulvates extracted from natural humic material in field conditions with barley (Gonet

---

BOX 18.1 GENERAL CONSIDERATIONS FOR THE USE OF COMPOST IN AGRICULTURAL PRODUCTION

Types of Soil Where Favorable Impact Is Most Significant
- In tropical and subtropical regions.
- Of predominantly sandy texture.
- With low organic matter content.
- Affected by salinity.
- Acidic pH.
- With low fertility (natural or induced by management system).

Raw Material for Creation (Possibly Available in the Same Location)
- Various plant residues.
- Livestock manure.
- Other organic residues (not contaminated with heavy metals or harmful organic substances).

Main Target Crops
- Horticultural crops in general.
- Annual crops.

Main Effects
- Stimulation of the production of above-ground and root biomass.
- Increase in agricultural productivity.
- Decrease in mineral fertilization requirements. Provision of better indicators of the quality of agricultural products.
- Favoring of the development (diversity and activity) of soil microbiota.

Dosages
- Variable.
- Dosages that appear in the literature range from 5 to 20 t ha^{-1}, but it is necessary to define the dosage based on the characteristics of the soil, crop, and compost developed

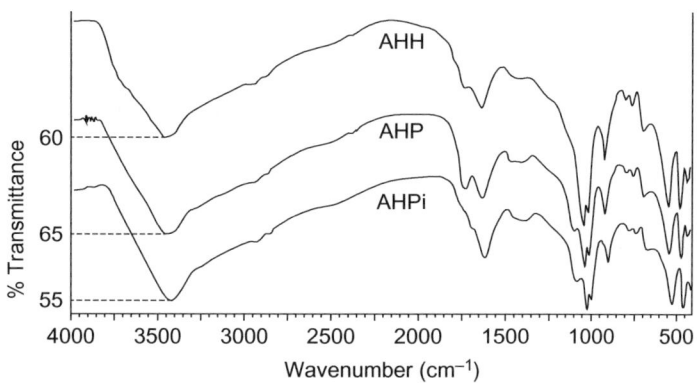

FIGURE 18.6

Infrared spectra (FT-IR) of HA extracted with the IHHS procedure (using NaOH—AHH) with hydrogen peroxide (AHP) and sodium pyrophosphate (AHPi) (García et al., 2012a).

et al., 1996). The application of the humic material preparations achieved positive results for grain production (15 to 20% greater) compared to mineral fertilizer application, and there was also a higher level of total nitrogen in the grains.

Due to the practical fact that HA are relatively easier to isolate than fulvic acids, most studies of soluble humic fractions applied in agricultural production systems have been developed using HA. VC of different materials are one of the most commonly used sources to obtain HA (García et al., 2013c), although it is possible to use composts and other natural humic materials such as leonardites (Mora et al., 2010). Aqueous dissolutions of HA isolated from VC in different concentrations were tested in different crops to evaluate the impact on the biological development of different commercially important plants. Some dissolutions were applied to the soil, while others were sprayed on leaves and seeds before planting. In direct applications to the soil for tomato (*Lycopersicon esculentum* Mill) and cucumber plants (*Cucumis sativus* L.), the addition of up to 500 mg kg^{-1} HA stimulated plant growth, but the intensity of the effect differed according to the source of the VC used in the extraction and isolation of the HA, and there were differences in the optimal concentration value. Very high values caused unfavorable effects, thus confirming the overall behavior equivalent to plant hormones that is attributed to HA (Atiyeh et al., 2002).

Various studies have been performed to associate structural characteristics of HA with their biological effects. Dobbss et al. (2010) induced structural modifications in humic substances contained in VC with reactions of hydrolysis, oxidation, reduction, and methylation. The biological activity in stimulating root growth and the activation of proton pumps were studied in all of the structural derivatives in tomato and corn seedlings. The effects of the derivatives were more intense than those caused by the original humic substances. There was no defined relationship between biological activity and the molecular size of derivatives, but the hydrophobicity index was directly related with proton pump stimulation.

Applications of HA were also used to protect plants from different conditions of abiotic stress, particularly water and salt stress and by chemical contaminants such as heavy metals. In bean plants (*Phaseolus vulgaris* L.) grown in soil contaminated with heavy metals, three concentrations (20, 40, and 60 mg of HA L^{-1}) of HA dissolutions (from cow manure VC) were sprayed, and the biochemical—physiological effects of the stress from the heavy metals were mitigated (Portuondo, 2011). Similarly, different concentrations of HA were applied to potatoes (var. Spunta) on a sandy entisol soil in Egypt under different levels of water stress. The results obtained indicated that applications of up to 120 kg ha^{-1} HA increased growth parameters, tuber production, and biochemical indicators such as the content of chlorophyll, ascorbic acid, starch, and total soluble acids, which proves that HA also mitigate the effects of water stress (Selim et al., 2012).

Another tendency in the development of alternative methods for the application of soluble humic substances to stimulate biological and agricultural productivity has been to combine solutions of HA with suspensions of beneficial microorganisms. In this way, a suspension (10^9 cells mL^{-1}) of diazotrophic endophytic bacteria (*Herbaspirillum seropedicae*) combined or not with a dissolution of 20 mg of C L^{-1} HA extracted from VC of cow manure was used in corn plants (*Zea mays* L.) under field conditions on an ultisol soil in Macaé, Rio de Janeiro (Brazil), with a low total content of carbon (0.6%) and nitrogen (0.04%). The treatments (a suspension of bacteria, an HA dissolution, and a combination of both) were applied via foliar spray one time 45 days after plant germination (hybrid P6875) at a density of 120,000 ha^{-1}. The results obtained in grain production demonstrated that the independent application of the bacterial suspension and HA dissolution increased production by 17 to 20%, while the bacteria—HA combination produced an increase of 65% greater than the control (Canellas et al., 2013).

Over the past few years, different forms of commercial HA have appeared in the agricultural market, and the possible effects of these products on agricultural production are being studied. In the early twenty-first century, the Chemistry Department of the Agrarian University of Havana (*Universidad Agraria de la Habana*) standardized a procedure for obtaining an extract of cow manure VC (called "liquid humus") used in an alkaline aqueous solution, characterized by a high percentage of organic carbon (such as HA, FA, free amino acids, and other compounds with biological activity) and a lower amount of a mineral fraction with both macro- and micronutrients. Caro (2004) demonstrated the physicochemical characterization of HA and FA contained in liquid humus, and dilutions of this extract were sprayed on the leaves of corn plants on a small rural property. Two dilutions (1:20 v:v and 1:30 v-v) of this humus were applied two times with an interval of 25 days. The two dilutions significantly increased indicators of growth and productivity. The 1:20 dilution treatment produced the greatest economic benefits.

The same liquid humus was used to evaluate the impact of foliar application of two aqueous dilutions of the extract on tomato plants (*Lycopersicon esculentum* Mill) for two consecutive years (Arteaga et al., 2006). Two aqueous dilutions (1/30 v:v and 1/40 v-v) were sprayed 7 days after transplant and repeated every 15 days. The biological factors measured were the stem length and diameter, number of leaves, and number of flowers. Productivity indicators included the number and mass of fruits per plant and the yield per unit surface area (t ha^{-1}). The liquid humus dilutions improved the biological development of plants, with indicators that were superior to those of the controls, and thus produced better results for agricultural productivity. The most notable results were obtained with the 1/30 dilution.

Some modifications were recently introduced to the process of obtaining liquid humus, and the effects of the new extract were tested in bean plants (*Phaseolus vulgaris* L.) and lettuce (*Lactuca sativa* L.) (Hernández, 2011). Using these modifications, aqueous dilutions of 1:50, 1:60, and 1:70 (v:v) of the extract were sprayed on the bean plant leaves (primary leaves, trifoliate leaves, and trifoliate tertiary leaves) in CC−25−9 plants planted by conventional planting in a red ferralitic soil (Rhodic Eutrustok according to soil taxonomy). These dilution sprays increased the plant height, dry mass, leaf surface, quantity of pods per plant, number of grains per pod, and yield (Hernández et al., 2012). In urban lettuce crops in Cuba, similar aqueous dilutions (1:50, 1:60, and 1:70 v:v, which corresponds to 1.2; 1.3 and 1.6 mmol C L^{-1}, respectively) were sprayed after 10 and 25 days of transplanting at a dose of 300 L ha^{-1}. Both the biochemical−physiological (leaf composition of carbohydrates and proteins, nitrate-reductase activity, and specific leaf area) and biological and agricultural productivity indicators (number of leaves per plant, leaf length and area, and production cycle length) that were evaluated indicated significantly positive effects for the application of all the dilutions. The 1:60 dilution produced the most notable impacts and reduced the production cycle by 21 days, which greatly benefits producers (Hernández et al., 2013).

In an attempt to summarize the general questions regarding the application of humic materials in agricultural production, we suggest Figure 18.7.

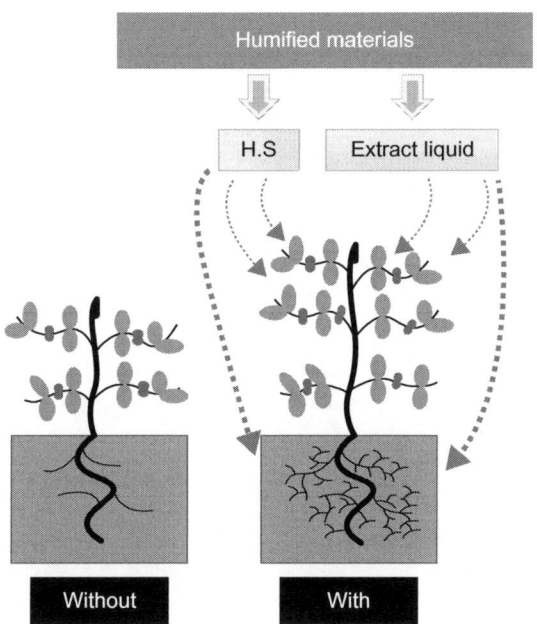

FIGURE 18.7

Schematic representation of alternative uses and effects of the application of humic materials in agricultural production. (H.S. = isolated humic substances from aqueous extracts of humic materials; intact extracts = soluble fraction of humic materials in a neutral or alkaline aqueous solution; arrow with dotted line = foliar spray of aqueous dilutions.)

18.4 Conclusion and future prospects

A general evaluation of the previously cited examples in relation to the use of VC in plant productions allows the formulation of at least the following considerations:

- There are many organic raw materials (separate or combined) that can be subjected to the VC process, the majority of which are subproducts of the agricultural production systems to which they will be applied, which facilitates the practical application of these products.
- The chemical, physical, and biological characteristics of VC depend on the nature of the original materials, temperature, time of maturation, and type of worm.
- VC generally have a higher proportion of organic carbon (particularly soluble humic substances) than composts and a greater variety of organic compounds that are potentially active in the soil–plant system.
- Four ways of using VC (applying whole material directly on the soil, using soluble humic substances previously isolated from the material, spraying aqueous extracts obtained in an alkaline solution, and spraying aqueous leachates) have demonstrated success in increasing the development and biological and agricultural productivity of various commercial crops.
- Conditions for the applicability of different ways of using VC are similar to those of composts, but the dosages are generally lower.

The use of humified materials in agricultural activities has increased. Its acceptance has been due to its advantages regarding environmental safety, improvement of soil properties (chemical, physical, and biological), and growth and yield of agricultural crops. However, further studies are needed on the structural features of these materials targeting its standardization in relation to its effect and purpose. According to the actual studies, one way to ensure greater acceptance is to focus research on finding alternatives phytotechnical which were included the humified materials. These materials, VC or compost, for example, could be used as a substitute for the large consumption of chemicals and fertilizers of synthetic origin. On the other hand, it is necessary for the preparation and training of skilled manpower for the purpose of acquiring knowledge about concepts of organic agriculture and humus carbon.

References

Adebayo, A.G., Shokalu, A.O., Akintoye, H.A., 2013. Effect of compost mixes on vegetative development and fruit yield of okra (*Abelmoscus Esculentus*). J. Agric. Anim. Sci. 3, 42–49.

Aguiar, N.O., Novotny, E.H., Oliveira, A.L., Rumjanek, V.M., Olivares, F.L., Canellas, L.P., 2013. Prediction of humic acids bioactivity using spectroscopy and multivariate analysis. J. Geochem. Explor. 129, 95–102.

Ahmed, I., Khan, A.B., 2013. Effect of vermicompost on growth and productivity of tomato (*Lycopersicon esculentum* Mill) under field conditions. Acta. Biol. Malays. 2, 12–21.

Al-Faiyz, Y.S.S., 2013. CPMAS 13C NMR characterization of humic acids from composted agricultural Saudi waste. Arabian J. Chem. http://dx.doi.org/10.1016/j.arabjc.2012.12.018.

Alfonso, E.T., Padrón, J.R., Peraz, T.T., Escobar, I.R., de Armas, M.M.D., 2011. Lettuce (*Lactuca sativa* L.) crop response to the application of different bioactive products. Cultivos Tropicales 32, 77–82.

Amir, S., Jouraiphy, A., Meddich, A., Gharous, M., Winterto, P., Hafidi, M., 2010. Structural study of humic acids during composting of activated sludge-green waste: elemental analysis, FTIR and ^{13}C NMR. J. Hazard Matter. 177, 524–529.

Arteaga, M., Garcés, N., Guridi, F., Pino, J.A., López, A., Menéndez, L.A., et al., 2006. Evaluación de las aplicaciones foliares de humus líquido en cultivo del tomate (*Lycopersicon esculentum* Mill) var. Amalia en condiciones de producción. Cultivos Tropicales 27, 95–101.

Arteaga, M., Garcés, N., Novo, R., Izquierdo, F.G., Pino, J.A., Acosta, M., et al., 2007. Influencia de la aplicación foliar del bioestimulante liplant sobre algunos indicadores biológicos del suelo [Influence of foliar application of biostimulant liplant on some soil biological indicators]. Rev. Protección. Veg. 22, 110–111.

Asli, S., Neumann, P.M., 2010. Hizosphere humic acid interacts with root cell walls to reduce hydraulic conductivity and plant development. Plant. Soil. 336, 313–322.

Atiyeh, R.M., Lee, S., Edwards, C.A., Arancon, N.Q., Metzger, J.D., 2002. The influence of humic acids derived from earthworm-processed organic wastes on plant growth. Bioresour. Technol. 84, 7–14.

Ayyappan, D., Sanjiviraja, K., Balakrishnan, V., Ravindran, K.C., 2013. Impact of halophytic compost on growth and yield characteristics of *Vigna radiata* L. Afr. J. Agric. Res. 8, 2663–2672.

Azizi, A.B., Lim, M.P.M., Noor, Z.M., Abdullah, N., 2013. Vermiremoval of heavy metal in sewage sludge by utilizing *Lumbricus rubellus*. Ecotox. Environ. Safe. 90, 13–20.

Aşik, B.B., Katkat, A.V., 2013. Determination of effects on solid and liquid humic substances to plant growth and soil micronutrient availability. J. Food Agric. Environ. 11, 1182–1186.

Balmori, D.M., Olivares, F.L., Spaccini, R., Aguiar, K.P., Araújo, M.F., Aguiar, N.O., et al., 2013. Molecular characteristics of vermicompost and their relationship to preservation of inoculated nitrogen-fixing bacteria. J. Anal. Appl. Pyrol. 104, 540–550.

Bhattacharya, S.S., Iftikar, W., Sahariah, B., Chattopadhyay, G.N., 2012. Vermicomposting converts fly ash to enrich soil fertility and sustain crop growth in red and lateritic soils. Resour. Conserv. Recy. 65, 100–106.

Canellas, L.P., Piccolo, A., Dobbss, L.B., Spaccini, L.B., Olivares, F.L., Zandoni, D.B., et al., 2010. Chemical composition and bioactivity properties of size-fractions separated from a vermicompost humic acid. Chemosphere 78, 457–466.

Canellas, L.P., Dobbs, L.B., Oliveira, A.L., Chagas, J.G., Aguiar, N.O., Rumjanek, V.M., et al., 2012. Chemical properties of humic matter as related to induction of plant lateral roots. Europ. J. Soil. Sci. 63, 315–324.

Canellas, L.P., Balmori, D.M., Médici, L.O., Aguiar, N.O., Campostrini, E., Rosa, R.C.C., et al., 2013. A combination of humic substances and Herbaspirillum seropedicae inoculation enhances the growth of maize (*Zea mays* L.). Plant. Soil. 1–2, 119–132.

Caro, I., 2004. Caracterización de algunos parámetros químico—físico del humus líquido obtenido a partir de vermicompost de estiércol vacuno y su evaluación sobre algunos indicadores biológicos y productivos de dos cultivos. Tesis de Máster Science, Departamento de Química, Facultad de Agronomía, Univ. Agraria de la Habana, 75 p.

Cordeiro, F.C., Santa-Catarina, C., Silveira, V., de Souza, S.R., 2011. Humic acid effect on catalase activity and the generation of reactive oxygen species in corn (*Zea mays* L). Biosci. Biotechnol. Biochem. 75, 70–74.

De Carvalho, A.M.X., De Souza, M.E.P., Deliberalia, D.C., Jucksch, I., Brown, G.G., Mendonça, E.S., et al., 2013. Vermicomposting with rock powder increases plant growth. Appl. Soil. Ecol. 69, 56–60.

Delgado-Moreno, L., Peña, A., 2009. Compost and vermicompost of olive cake to bioremediate triazines-contaminated soil. Sci. Total Environ. 407, 1489–1495.

Di Gennaro, P., Moreno, B., Annoni, E., García-Rodríguez, S., Bestetti, G., Benitez, E., 2009. Dynamic changes in bacterial community structure and in naphthalene dioxygenase expression in vermicompost-amended PAH-contaminated soils. J. Hazard. Mater. 172, 1464–1469.

Doan, T.T., Ngo, P.T., Rumpel, C., Nguyen, B.V., Jouquet, P., 2013. Interactions between compost, vermicompost and earthworms influence plant growth and yield: a one-year greenhouse experiment. Sci. Hortic-Amsterdam 160, 148–154.

Dobbss, L.B., Canellas, L.P., Olivares, F.L., Oliveira, N.A., Peres, L.E., Azevedo, M., et al., 2010. Bioactivity of chemically transformed humic matter from vermicompost on plant root growth. J. Agric. Food. Chem. 58, 3681–3688.

Domínguez, J., Aira, M., Gómez-Brandón, M., 2010. Vermicomposting: earthworms enhance the work of microbes. In: Insam, H., et al., (Eds.), Microbes at Work. Springer-Verlag, Berlin, Heidelberg, pp. 93–114.

Elena, A., Diane, L., Eva, B., Fuentes, M., Baigorri, R., Zamarreño, A.M., et al., 2009. The root application of a purified leonardite humic acid modifies the transcriptional regulation of the main physiological root responses to Fe deficiency in Fe-sufficient cucumber plants. Plant Physiol. Bioch. 47, 215–223.

Ertani, A., Pizzeghello, D., Baglieri, A., Cadili, V., Tambone, F., Gennari, M., et al., 2013. Humic-like substances from agro-industrial residues affect growth and nitrogen assimilation in maize (*Zea mays* L.) plantlets. J. Geochem. Explor. 129, 103–111.

García, A.C., Izquierdo, F.G., Nieblas, E.G., Rosado, E., Valdés, R., Pimentel, J.J., et al., 2007. Material de origen natural que retiene cationes de metales pesados. Rev. Iber. Polímeros 3, 204–214.

García, A.C., Lima, W.L., Gómez, E.F., Izquierdo, F.G., Berbara, R.L.L., 2012a. Modificaciones estructurales en ácidos húmicos durante su extracción y purificación: monitoreo espectroscópico FTIR y UV-VIS. Rev. Iber. Polímeros 13, 227–237.

García, A., Angulo, J., Mercedes, M., Gutiérrez, V., 2012b. Effect of phosphate—solubilizing bacteria and compost on the nutritional characteristics of the oil palm crop (*Elaeis guineensis* Jacq.) in Casanare, Colombia. Agron Colomb. 30, 274–281.

García, A.C., Izquierdo, F.G., González, O.L.H., Armas, M.M.D., López, R.H., Rebato, S.M., et al., 2013a. Biotechnology of humified materials obtained from vermicomposts for sustainable agroecological purposes. Afr. J. Biotechnol. 7, 625–634.

García, A.C., Izquierdo, F.G., Souza, L.G.A., Santos, L.A., Berbara, R.L.L., 2013b. Agromateriais humificados. Potencialidades e desafios da sua utilização ecológica. AGROTec 1, 59–61.

García, A.C., Izquierdo, F.G., Sobrino, N.M.B.A., Castro, R.N., Santos, L.A., Souza, L.G.A., et al., 2013c. Humified insoluble solid for efficient decontamination of nickel and lead in industrial effluents. J. Environ. Chem. Eng. 1, 916–924.

García, A.C., Santos, L.A., Izquierdo, F.G., Rumjanek, V.M., Castro, R.N., dos Santos, F.S., et al., 2014. Potentialities of vermicompost humic acids to alleviate water stress in rice plants (*Oryza sativa* L.). J. Geochem. Explor. 136, 48–54.

Garg, V.K., Gupta, R., 2009. Biotechnology for agro-industrial residues utilisation. In: Singh-Nee Nigam, P., Pandey, A. (Eds.), Biotechnology for Agro-industrial Residues Utilisation. Springer Science + Business Media BV, pp. 431–456.

Gonet, S.S., Dziamski, A., Gonet, E., 1996. Application of humus preparations from oxyhumolite in crop production. Environ. Int. 22, 559–562.

Gutiérrez-Miceli, F.A., Santiago-Borraz, J., Molina, J.A.M., Nafate, C.C., Abud-Archila, M., Llaven, M.A.O., et al., 2007. Vermicompost as a soil supplement to improve growth, yield and fruit quality of tomato (*Lycopersicum esculentum*). Bioresour. Technol. 98, 2781–2786.

Gutiérrez-Miceli, F.A., García-Gómez, R.C., Reiner, R.R., Abud-Archila, M., Angela, O.L.M., Cruz, M.J.G., et al., 2008. Formulation of a liquid fertilizer for sorghum (*Sorghum bicolor* (L.) Moench) using vermicompost leachate. Bioresour. Technol. 99, 6174–6180.

Hernández, G., Hernández, O.L., Guridi, F., Arbelo, N., 2012. Influencia de la siembra directa y las aplicaciones foliares de extracto líquido de vermicompost en el crecimiento y rendimiento del frijol (*Phaseolus vulgaris* L.) cv. cc–25–9. Rev. Cie. Téc. Agr. 21, 86–90.

Hernández, O.L., 2011. Modificaciones al proceso de obtención de sustancias húmicas a partir de vermicompost: efectos biológicos. Tesis de Máster Science, Departamento de Química, Facultad de Agronomía, Univ Agraria de la Habana, 78 p.

Hernández, O.L., Huelva, R., Guridi, F., Olivares, F.L., Canellas, L.P., 2013. Humates isolated from vermi-compost as growth promoter in organic Lettuce production. Rev. Cie. Téc. Agr. 22, 70–75.

Ke, W., Weiguang, L., XuJin, G., Yunbei, L., Chuandong, W., Nanqi, R., 2013. Spectral study of dissolved organic matter in biosolid during the composting process using inorganic bulking agent: UVevis, GPC, FTIR and EEM. Int. Biodeter. Biodegr. 85, 617–623.

Lazcano, C., Revilla, P., Malvar, R.A., Domínguez, J., 2010. Yield and fruit quality of four sweet corn hybrids (*Zea mays* L) under conventional and integrated fertilization with vermicompost. J. Sci. Food. Agric. 91, 1244–1253.

Li, X., Xing, M., Yang, J., Huang, Z., 2011. Compositional and functional features of humic acid-like fractions from vermicomposting of sewage sludge and cow dung. J. Hazard. Mater. 185, 740–748.

Lv, B., Xing, M., Yang, J., Qi, W., Lu, Y., 2013. Chemical and spectroscopic characterization of water extractable organic matter during vermicomposting of cattle dung. Bioresour. Technol. 132, 320–326.

Mora, V., Bacaicoa, E., Zamarreño, A.M., Aguirre, E., Garnica, M., Fuentes, M., et al., 2010. Action of humic acid on promotion of cucumber shoot growth involves nitrate-related changes associated with the root-to-shoot distribution of cytokinins, polyamines and mineral nutrients. J. Plant. Physiol. 167, 633–642.

Muscolo, A., Sidari, M., Nardi, S., 2013. Humic substance: relationship between structure and activity. Deeper information suggests univocal findings. J. Geochem. Explor. 129, 57–63.

Nebbioso, A., Piccolo, A., 2009. Molecular rigidity and diffusivity of Al^{3+} and Ca^{2+} humates as revealed by NMR spectroscopy. Environ. Sci. Technol. 43, 2417–2424.

Nebbioso, A., Piccolo, A., 2011. Basis of a humeomics science: chemical fractionation and molecular charac-terization of humic biosuprastructures. Biomacromolecules 12, 1187–1199.

Nebbioso, A., Piccolo, A., 2012. Advances in humeomics: enhanced structural identification of humic mole-cules after size fractionation of a soil humic acid. Anal. Chim. Acta. 720, 77–90.

Ng'etich, O.K., Aguyoh, J.N., Ogweno, J.O., 2012. Effects of composted farmyard manure on growth and yield of spider plant (*Cleome gynandra*). IJSN 3, 514–520.

Nuzzo, A., Sánchez, A., Fontaine, B., Piccolo, A., 2013. Conformational changes of dissolved humic and ful-vic superstructures with progressive iron complexation. J. Geochem. Explor. 129, 1–5.

Oo, A.N., Iwai, C.B., Saenjan, P., 2013. Soil properties and maize growth in saline and nonsaline soils using cassava-industrial waste compost and vermicompost with or without earthworms. Land Degrad. Develop. 10.1002/ldr.2208.

Piccolo, A., Stevenson, F.J., 1982. Infrared spectra of Cu^{2+}, Pb^{2+} and Ca^{2+} complexes of soil humic sub-stances. Geoderma 27, 195–208.

Portuondo, L.B., 2011. Efectos de los ácidos húmicos sobre el estrés por metales pesados en plantas de frijol (*Phaseolus vulgaris* L.). Tesis de Máster Science, Departamento de Química, Facultad de Agronomía, Univ Agraria de la Habana, 71 p.

Quintero, D., Huelva, R., Hernández, O.L., Guridi, F., Berbara, R.L.L., 2012. El sistema de usos de los suelos Ferralíticos modifica la estructura y las propiedades de sus ácidos húmicos. Rev. Cie. Téc. Agr. 21, 55–60.

Santos, V.B., Araújo, A.S.F., Leite, L.F.C., Nunes, LAPL, Melo, W.J., 2012. Soil microbial biomass and organic matter fractions during transition from conventional to organic farming systems. Geoderma 170, 227–231.

Schmidt, W., Santi, S., Pinton, R., Varanini, Z., 2007. Water-extractable humic substances alter root develop-ment and epidermal cell pattern in Arabidopsis. Plant. Soil. 300, 259–267.

Selim, E.M., Shedeed, S.I., Asaad, F.F., El-Neklawy, A.F., 2012. Interactive effects of humic acid and water stress on chlorophyll and mineral nutrient contents of potato plants. J. Appl. Sci. Res. 8, 531–537.

Shafawati, S.N., Siddiquee, S., 2013. Composting of oil palm fibres and Trichoderma spp. as the biological control agent: a review. Int. Biodeterior. Biodegrad. 85, 243–253.

Singh, R., Gupta, R.K., Patil, R.T., Sharma, R.R., Asrey, R., Kumar, A., et al., 2010. Sequential foliar applica-tion of vermicompost leachates improves marketable fruit yield and quality of strawberry (*Fragaria x ana-nassa* Duch.). Sci. Hortic-Amsterdam 124, 34–39.

Singh, R., Singh, R., Soni, S.K., Singh, S.P., Chauhan, U.K., Kalra, A., 2013. Vermicompost from biodegraded distillation waste improves soil properties and essential oil yield of Pogostemon cablin (patchouli) Benth. Appl. Soil. Ecol. 70, 48–56.

Song, X.Y., Spaccini, R., Pan, G., Piccolo, A., 2013. Stabilization by hydrophobic protection as a molecular mechanism for organic carbon sequestration in maize-amended rice paddy soils. Sci. Total Environ. 458–460, 319–330.

Srivastava, P.K., Gupta, M., Upadhyay, R.K., Sharma, S., Shikha, Singh, N., et al., 2012. Effects of combined application of vermicompost and mineral fertilizer on the growth of *Allium cepa* L. and soil fertility. J. Plant. Nutr. Soil. Sci. 175, 101–107.

Tejada, M., García-Martínez, A.M., Parrado, J., 2009. Effects of a vermicompost composted with beet vinasse on soil properties, soil losses and soil restoration. Catena 77, 238–247.

Terry, E., de Armas, M.M., Ruiz, J., Tejeda, T., Zea, M.E., Camacho-Ferre, F., 2012. Effects of different bioactive products used as growth stimulators in lettuce crops (*Lactuca sativa* L.). J. Food Agric. Environ. 10, 386–389.

Trevisan, S., Francioso, O., Quaggiotti, S., Nardi, S., 2010. Humic substances biological activity at the plant-soil interface. From environmental aspects to molecular factors. Plant Signal Behav. 5, 635–643.

Wang-Wang, Guang-Ming, Z., Ji-Lai, G., Jie, L., Piao, X., Chang, Z., et al., 2014. Impact of humic/fulvic acid on the removal of heavy metals from aqueous solutions using nanomaterials: a review. Sci. Total Environ. 468–469, 1014–1027.

Xiao-Song, H., Bei-Dou, X., Yong-Hai, J., Lian-Sheng, H., Dan, L., Hong-Wei, P., et al., 2013. Structural transformation study of water-extractable organic matter during the industrial composting of cattle manure. Microchem. J. 106, 160–166.

Xitao, L., Wenjuan, Z., Ke, S., Chunye, L., Ye, Z., 2013. Immobilization of cadmium onto activated carbon by microwave irradiation assisted with humic acid. J. Taiwan. Inst. Chem. E 44, 972–976.

Yang, J., Zhao, C., Xing, M., Lin, Y., 2013. Enhancement stabilization of heavy metals (Zn, Pb, Cr and Cu) during vermifiltration of liquid-state sludge. Bioresour. Technol. 146, 649–655.

Zandonadi, D.B., Santos, M.P., Busato, J.G., Pereira, L.E.P., Façanh, A.R., 2013. Plant physiology as affected by humified organic matter. Theor. Exp. Plant. Physiol. 25, 12–25.

Zhang, J., Lin, X., Liu, W., Wang, Y., Zeng, J., Chen, H., 2012. Effect of organic wastes on the plant-microbe remediation for removal of aged PAHs in soils. J. Environ. Sci. 8, 1476–1482.

Climate Changes and Potential Impacts on Quality of Fruit and Vegetable Crops

Leonora M. Mattos, Celso L. Moretti, Sumira Jan, Steven A. Sargent, Carlos Eduardo P. Lima and Mariana R. Fontenelle

19.1 Introduction

Fruits and vegetables are detrimentally affected by numerous abiotic stresses during post-harvest including production, handling, storage, and distribution. These stresses can have ambiguous effects resulting in either no loss or rapid improvement in quality. The sternness of the influences of climate change on food production will be inconstant and crop specific (Jackson et al., 2011). However, when the abiotic stress is moderate or severe, quality losses almost always are incurred at market. As a consequence, it is important to understand the nature and sources of abiotic stresses that affect fruits and vegetables. With advancing knowledge, alternatives for better management or resistance have now become accessible (Hodges et al., 2005; Toivonen, 2009). At a global scale, fruit and vegetable production is potentially confronted by the annual climatic variability of particular regions. Since post-harvest stress limits the potential storage and shelf-life of fruits and vegetables, effects of field abiotic stresses, e.g., drought, extreme temperatures, light, and salinity, need to be focused on in detail to gain a greater perspective on post-harvest stress susceptibility (Toivonen, 2005). In this context, this chapter is framed first to describe the pre- and post-harvest abiotic stresses, then to evaluate their significance for product quality and marketing, and finally to investigate the technologies accessible to ameliorate and manage fruit and vegetable sensitivity during post-harvest and the distribution chain.

Rockstrom et al. (2009) defined adaptation to global climate change as one of the vital processes to assure human survival. Recent studies have confirmed the human involvement in climate change on the planet through simulations. Rowlands et al. (2012) predict increases in average global temperature between 1.4°C and 3°C by 2050. The most pessimistic scenario, called the A2, suggests that the average temperature of the planet may reach up to 5.8°C by 2100 (IPCC, 2007). It is expected that this increase in global average temperature will lead to more frequent occurrence of extreme weather and changes in rainfall patterns. Coumou and Rahmstorf (2012) demonstrated associations between already observed increased temperatures and extreme weather events recorded in the last decade, especially in the incidence of heat waves and intense precipitation. Semi-arid and arid regions are undergoing fewer incidences of precipitation, supplementary aridity, and extended periods deprived of precipitation (Mukherjee, 2013).

P. Ahmad (Ed): Emerging Technologies and Management of Crop Stress Tolerance, Volume 1.
DOI: http://dx.doi.org/10.1016/B978-0-12-800876-8.00019-9

Climate change has emerged as the most prominent global environmental issue and there is an urgent need to evaluate its impact on agriculture. Atmospheric carbon dioxide levels have increased beyond that experienced in the past million years and are rapidly approaching levels never before noticed since the Eocene, evidenced by no ice caps, sea levels 100 m above current, and the presence of crocodiles near the North Pole. Alteration of ice sheet size and atmospheric CO_2 intensify the entire Earth system sensitivity by an expanse that hinges on the time scale considered (Hansen et al., 2013). Global temperatures have been rising steadily for nearly 40 years; the eight warmest years on record have occurred since 2001 and the global average land surface temperature for the period January to September 2010 was the second warmest on record after 2007. Climate alteration may affect the degree, scheduling, and incidence of precipitation, river runoff, and flood measures over variations to the land surface, atmosphere, and oceans (Anderson, 2013; Bales, 2013). In addition to rising mean global surface temperatures, climatic disruption is causing other changes to the physical conditions of the planet. Among these are rising sea levels and changes in the pattern of precipitation leading to more frequent droughts, floods, forest fires, and extreme storms. Some of the health consequences of climate change are simple: warmer temperatures, changes in the hydrologic cycle, increased ground-level ozone, and enhanced pollen production will result in heat stress, alter patterns of infectious disease, and compromise air quality. Numerous studies specify that climate change will deleterious potential on the crop yields and profits until the end of this century (Deschenes and Greenstone, 2012; Medellín-Azuara et al., 2011).

19.2 Impacts of climate change on vegetable production systems in Brazil

The impacts of climatic changes on vegetable crop production are certainly perturbing. Even moderate elevations of average temperatures during the day and/or night can affect the yield and quality of vegetables. Expected losses to vegetable crops like spinach, potato, broccoli, and lettuce, mainly due to the increase in temperature, have been reported (United States Global Change Research Program, 2000). Crops adapted to lower temperatures such as carrots, tomatoes, and other brassica species are predicted to be potentially damaged. Production system changes may be needed even for vegetables better adapted to warmer climates due to changes in rainfall patterns and/or the occurrence of pests and diseases. Sato et al. (2006) demonstrated that even moderate elevations of daytime and/or night-time temperatures can have negative effects on tomato yield. The authors reported that there were no major effects on plant development; however, their reproductive capacity was damaged resulting in lower quantity and quality of fruit. These alterations were caused predominantly by less effective pollination, reduction of photosynthetic rate, and increased respiration rates. Climate change poses possibly fewer threats for crop pollination than other constituents of agriculture; modern-day crop pollination methods are now extremely vulnerable because agriculture depends on a single pollinator species—the honey bee (Kremen, 2013).

Gioria et al. (2008) reported that ongoing climate change in Brazil will most likely increase the frequency and gravity of tomato diseases such as powdery mildew (*Leveilula taurica*), early blight (*Alternaria solani*), fusarium wilt (*Fusarium oxysporum* f. sp. *lycopersici*), bacterial wilt (*Ralstonia solanacearum*), spotted wilt (*Tomato spotted wilt virus*—TSWV, *Tomato chlorotic spot virus*—TCSV, *Groundnut ring spot virus*—GRSV, *Chrysanthemum stem necrosis virus*—CSNV), and tomato golden mosaic (several species of geminivirus). Other diseases such as late blight (*Phytophthora infestans*), Verticilium wilt (*Verticilium albo-atrum, V. dahliae*), and Sclerotinia rot (*Sclerotinia sclerotiorum*) will be probably less prevalent and cause less damage, while others like the tomato mosaic virus (*Tomato mosaic virus*—ToMV) will remain unchanged.

Potato is an important vegetable crop in Brazil and expected to be negatively affected by climate change. Lopes et al. (2011) discuss the production of these potatoes, which prefer mild temperatures, are presently grown in regions with a temperature range between 10 and 30°C, optimally between 10°C and 25°C, and are most productive with large amplitude between day- and nighttime temperatures. They observed that high temperatures not only reduce the synthesis of photoassimilates but also inhibit their partition for tubers, which significantly affects production levels at higher temperatures. Excessive strains from pests and diseases may also impair the tuber quality through the appearance of disease-caused anomalies and physiological disorders such as internal cracks and stains.

Lettuce, a wild plant species native to temperate regions of Brazil territory, is also vulnerable to climate change. At temperatures higher than 30°C, seedling malformation, low germination potential, and even death of the seeds can be observed. Besides a decline in plant size, high temperature can also lead to latex accumulation in lettuce making leaves bitter and rigid. Other problems such as loss of apical meristem and browned edges can also occur. Higher temperatures can also be associated with stem cleft resulting from deficiency of calcium and boron (Filgueira, 2007; Kobori et al., 2011). Nevertheless, Kobori et al. (2011) studied smooth, American and curly lettuce cultivars at high temperatures and observed that they maintained high productivity at 26°C. Kobori et al. (2011) evaluated the impacts of climate change on Brazilian lettuce production for the years 2071−2100 and reported that climate change will impact the production of vegetables because diseases caused by *Pythium* spp., *Sclerotiium rolfsii, Rhizoctonia solani, Fusarium oxysporum* f. sp. *lactucae, Thielaviopsis basicola, Cercospora longissima, Erysiphe cichocearum, Pectobacterium carotovorum, Tomato spotted wilt virus, Groundnut ring spot virus* and others will also exaggerate.

The production of *Brassica* species is more concentrated in colder regions across the southeast, south, and areas with altitudes above 800 m (Brunelli et al., 2011). The authors focused on the production of the three most important *Brassica* species produced in Brazil: cabbage, broccoli, and cauliflower. The authors estimated the variation in temperature and precipitation in these regions using climate models, meteorological data, and the scenarios of the Intergovernmental Panel on Climate Change (IPCC). The results reported by Brunelli et al. (2011) demonstrated that almost all the *Brassica* species produced in Brazil, even in the colder months of the year, will be restricted to the use of summer cultivars adapted to high temperatures. They further reported that the most probable distribution of *Brassica* will be affected due to high temperatures in important cultivated regions. The authors further reported the possibility of diseases even in high temperature-resistant varieties, such as stem cracks caused by the slow translocation of boron and calcium, pilosity and redness of cauliflower inflorescences, and bad compaction of cabbage heads throughout the year, except from July to August.

Negative effects of climate change in the production of *Brassica* species have been recorded on a global scale as well. McKeown et al. (2006) and Warland et al. (2006) demonstrated negative correlation between yield and the number of growing days at temperatures above 30°C in cold adapted *Brassica* species including cabbage, cauliflower, rutabaga, and other vegetables. Obviously, these results can be variable if heat-tolerant cultivars are used. In addition, data obtained indicate the negative effect of high temperatures on the yield of *Brassica* species. The family Brassicaceae consists of a group of species from temperate climates, which develops more favorably in cold or mild weather conditions (Filgueira, 2007). Many species have been improved over time, resulting in the development of summer cultivars better adapted to the climatic conditions of Brazil. However, even adapted cultivars have high temperature restrictions such as impairment in seed germination, which can result in the death of the seedlings at temperatures above 35°C.

Hamada et al. (2011), using climate projections, observed that the Brazilian territory must suffer increases in the average air temperature of up to 4.8°C at the end of the 21st century. In general, according to the authors' projections, increases in the average air temperature should be observed for all regions at all times of the year. Average temperatures observed for Brazil should be around 23.8°C in the fall, 21.8°C in the winter, 24.2°C in the spring and 25.0°C in the summer (IPCC, 2007).

19.3 Harvest and post-harvest

Fruit and vegetable crop harvest times are variable throughout the year depending on cultivar, water regime, climate conditions, pest control, cultural practices, exposure to direct sunlight, temperature management, and maturity index. Respiration is the main process to be controlled after the crops have been harvested; however, post-harvest physiologists and food scientists have restricted alternatives that interfere with the respiratory process, since the majority of them depend on the specific characteristics of harvested commodities (Saltveit, 2002). In order to minimize undesirable changes in quality parameters during the post-harvest period, growers and entrepreneurs can adopt a series of techniques to extend the shelf-life of perishable plant products. Post-harvest technology comprises different methods of harvesting, packaging, rapid cooling, storage under refrigeration as well as modified and controlled atmospheres and proper transportation. These strategies can be significant to help growers everywhere withstand the confrontations that climate changes may impose throughout the foreseeable future. The principal key climate change trends that affect post-harvest fruit and vegetable yields are a general increase in temperature and more frequent occurrences of dry spells, droughts, high winds, storms, heavy precipitation, and flooding.

19.4 Effects of temperature

Global temperature increase is an outcome of climate change, and is amplified especially on regional and subregional scales. Numerous crops are classified as vegetables and the wide crop diversity makes it difficult to summarize the potential effects of climate change on growth and yield. Few studies on the impact of climate change on vegetables have been carried out. Moreover, varietal differences may be too vast, making it difficult to generalize the results of some experiments (Nelson et al., 2010). Fruit and vegetable growth and development are influenced by different environmental factors (Bindi et al., 2001; Ilić et al., 2012). Peet and Wolfe (2000) concluded that elevated levels of CO_2 may benefit most crops, provided temperatures are not limiting. However, they suggested that the effects of increasing temperatures are much more complex to predict. During their development, high temperatures can affect photosynthesis, respiration, water relations, and membrane stability as well as plant hormone levels, primary and secondary metabolites (Magan et al., 2011). Seed germination can be reduced or even inhibited by high temperatures depending on the species and stress level (Bewley, 1997; Lobell et al., 2011; Ramos et al., 2013). Springate and Kover (2013) demonstrated early flowering and larger flower size in a set of recombinant inbred lines of *Arabidopsis thaliana* exposed to high temperatures in a simulated global warming treatment under field conditions. Higher winter temperatures could result in reduced snowpack accumulation, which decreases irrigation supplies to agriculture; reduced water accessibility would consequently be an *indirect* temperature change effect (USDA, 2013).

High temperatures may also increase the mortality rate of pollinators and affect pest population, diseases, and natural enemies (Chancellor and Kubiriba, 2006; Chakraborty et al., 2008; Chakraborty and Newton, 2011). Generally, temperature change risk in crops includes: modified phenology (timing) of leafing, flowering, harvest and fruit production; reduced winter chill; and asynchrony between flowering and pollinators (Baldocchi, 2012). These effects may be positive or negative depending on the cropping system. Further elevated temperatures may favor the growth of weed species enabling them to compete more effectively with crops. Warmer spring temperatures also have adverse effects on crop pollen germination, flower, and ovule size, which can end in declined fruit yields in the form of smaller, deformed (double), and fewer fruits (Beppu and Kataoka, 2011; Pope, 2012). The response of crops to temperature stress depends upon several factors. These include the crop growth stage (Porter and Semenov, 2005), the type of plant tissue, and the nature of temperature stress. Some vegetable crops such as tomatoes and peppers are extremely sensitive to high temperatures during the reproductive phase. Tomato plants are particularly vulnerable to heat stress during the period immediately prior to flowering when pollen release and its function are affected (Peet et al., 1998; Li et al., 2012). Amplified spring temperatures bring earlier spring blooms through western states (Pope et al., 2013). The effects of global warming on wine grapes include: extended frost-free phases; increasing degree-days; fewer winter chills and a swing to earlier bud break, bloom, and veraison (onset of ripening) resulting in negative yield quantity and possible quality effects (Battany, 2012).

Most of the temperature effects on plants are mediated by plant biochemistry. For the well-irrigated plants, the Q_{10} factor is very high resulting in vigorous growth. As plants are subjected to water deficit, temperature proves to be a physical facilitator for balancing normal and latent heat exchange at the shoot, which is modulated by relative humidity and wind. Most of the physiological processes proceed normally in temperatures ranging from 0 to 40°C. However, cardinal temperatures for the development of fruit and vegetable crops are lower depending on the species and ecological origin. These temperatures can be shifted towards 0°C for temperate species from cold regions such as carrots and lettuce. Conversely, these temperatures reach 40°C in species from tropical regions, such as cucurbits and cactus species (Went, 1953).

A general temperature effect for plants involves the ratio between photosynthesis and respiration. In order to obtain a high yield, the photosynthesis/respiration ratio should be much higher. At temperatures around 15°C, the above-mentioned ratio is usually higher than 10, which explains better survival of plants in temperate regions than in tropical ones (Went, 1953). Higher than normal temperatures affect the photosynthetic process through the modulation of enzyme activity as well as the electron transport chain (Sage and Kubien, 2007). Moreover, higher temperatures can affect the photosynthetic process increasing leaf temperatures indirectly. This defines the magnitude of the leaf-to-air vapor pressure difference (D), a key factor which can influence stomatal conductance (Lloyd and Farquhar, 2008). Photosynthetic activity is proportional to temperature variations. High temperatures can increase the rate of biochemical reactions catalyzed by different enzymes. However, above a certain temperature threshold, major enzymes lose their function causing potential changes in plant tissue tolerance to heat stresses (Bieto et al., 1996). Temperature is highly significant for establishing a harvest index. The higher the temperature during the growing season, the sooner the crop will mature. Hall et al. (1996) and Wurr et al. (1996) reported that lettuce, celery, cauliflower, and kiwi grown under higher temperatures matured earlier than when grown under lower temperatures. There could be several positive effects and prospects accompanying new temperature regimes due to climate change; for instance, the facility to develop some crops in different areas, all undesirable effects eventually stand to decrease crop quality such as decreased size and yields (Ackerman and Stanton, 2013).

19.4.1 Fruit ripening

Prolonged exposure to direct sunlight causes a rise in surface temperatures in fruit and thus results in rapid ripening and other associated events. One of the classic examples is that of grapes, where berries exposed to direct sunlight ripen faster than those ripened in shaded areas (Kliewer and Lider, 1968). The ripening process of "Hass" avocados exhibited discreet behavior to direct sunlight and shade during growth and development. The "Hass" avocados when exposed to direct sunlight demonstrated a rise in pulp temperature to 35°C and attained ripening 1.5 days longer than those growing in shade (Woolf et al., 1999). "Fuerte" and "Hass" fruits exposed to direct sunlight were firmer than fruits growing in shaded areas. Cell wall enzyme activities of cellulose and polygalacturonase were negatively correlated with fruit firmness indicating that higher temperatures during growth and development can delay ripening. However, this delay did not occur via a direct effect on the enzymes associated with cell wall degradation (Chan and Linse, 1989). Tomatoes grown at temperatures above 36°C for 3 days in direct sunlight resulted in ripening in terms of color development, ethylene evolution, and respiratory climacteric. However, ripening was slower than freshly harvested fruit (Lurie and Klein, 1991). The immediate effects of elevated treatments generally include inhibition in respiration and protein synthesis and increment in protein breakdown and ethylene production (Eaks, 1978; Ferguson et al., 1994; Lurie and Klein, 1990, 1991). However, in vitro apples, treated at 38 and 40°C for 2−6 days, did not exhibit any distinct effects on respiration, although ethylene production was reduced (Liu, 1978; Porritt and Lidster, 1978).

There are some reports on other specific effects and subsequent changes in ripening physiology as result of high temperature that prevail during the growing season. But there is a greater need to bring forth extrapolations procured from post-harvest ripening (Woolf and Ferguson, 2000). High temperatures on fruit surfaces caused by pronounced exposure to sunlight can slow ripening and other associated events. The above studies suggest that changes in ripening behavior possibly occur when fruit and vegetable crops are exposed to higher temperatures prior to harvest. Chan et al. (1981) and Picton and Grierson (1988) observed that high temperature stresses inhibited ethylene production and cell wall softening in papaya and tomato fruits. On the other hand, cucumber fruits showed increased tolerance to high temperature stress (32.5°C) with no change in in vitro aminocyclopropane carboxylate oxidase activity (Chan and Linse, 1989).

19.4.2 Quality parameters

For more than three decades extensive work has been carried out on fruit and vegetable crop quality when exposed to high temperatures during growth and development. High temperature was known to affect the flavor of fruits and vegetables. Apple fruits exposed to direct sunlight had higher sugar content compared to those fruits grown in shaded sites (Brooks and Fisher, 1926). Grapes also had higher sugar content and lower levels of tartaric acid when grown under high temperatures (Kliewer and Lider, 1968, 1970). Coombe (1987) observed that a 10°C increase in temperature caused a 50% reduction in tartaric acid content. Kliewer and Lider (1970) and Lakso and Kliewer (1975) verified greater temperature sensitivity of malic acid than tartaric acid synthesis during growth.

Fruit firmness is also affected by high temperature conditions during growth. "Fuerte" avocados exposed to direct sunlight (35°C) were 2.5 times firmer than those on the shaded side (20°C). Changes in cell wall composition, cell number, and cell turgor properties were postulated as being associated with the observed phenomenon (Woolf et al., 2000). Dry matter content is used as a harvest indicator for avocados due to its direct correlation with oil content, a key quality component (Lee et al., 1983). For example, the state of California produces about 80% of the avocados grown in the USA (Mexican and Guatemalan strains and their hybrids) and requires a minimum oil content of 19 to 25% depending upon the cultivar (Kader and Arpaia, 2002). "Hass" avocados grown under high ambient temperatures (45 ± 2°C) exhibit greater moisture content at harvest than fruit grown under lower temperatures (30 ± 2°C) (Woolf et al., 1999). They also demonstrated higher temperature-influenced oil composition, where the concentration of certain specific fatty acids increased, e.g., palmitic acid by 30% with apparently no effect in oleic acid. Avocados with higher dry matter content take a longer time to ripe, which could pose a serious problem for growers planning to market their fruits immediately after harvest (Woolf and Ferguson, 2000; Woolf et al., 1999). Mineral accumulation was also reported to be effected by high temperatures and/or direct sunlight. "Hass" avocado fruits exposed to direct sunlight showed higher calcium (100%), magnesium (51%), and potassium (60%) contents when compared to fruits grown under shaded conditions (Woolf et al., 1999). The authors suggested that these changes might be related to water movement through the fruit. Thus, fruit and vegetable growers, packers, and shippers must pay close attention to ambient temperatures during growth and development as well as maturity indices to assure harvest at the appropriate time.

Temperature is the most significant factor affecting antioxidant activity in vegetables and fruits. Generally, rise in temperature accelerates initiation reactions, and hence results in decline of existing and augmented antioxidant activities (Pokorný, 1986; Réblová, 2012). Antioxidants in fruit and vegetable crops can also be altered by exposure to high temperatures during the growing season. Wang and Zheng (2001) observed that "Kent" strawberries grown during warmer nights (18−22°C) and warmer days (25°C) had a higher antioxidant activity than berries grown during cooler days (12°C). The investigators also observed that high temperature conditions significantly increased the level of flavonoids and, consequently, antioxidant capacity. Galletta and Bringhurst (1990) verified that higher day and night temperatures had a direct influence on strawberry fruit color. Berries grown under such conditions were redder and darker. McKeown et al. (2006) also addressed the effects of climate changes in functional components. The authors verified that higher temperatures inclined to reduce vitamin content in fruit and vegetable crops. Cordenunsi et al. (2005) observed an increase in soluble sugars, anthocyanin, and vitamin C contents in three strawberry cultivars grown at the same commercial plantation, which were harvested at the ripening stage and stored at high temperatures for 6 days. However, variations in temperature may change the action mechanism of some antioxidants (Yanishlieva, 2001; Brewer, 2011) or vice versa. Temperature can affect particular reactions in which antioxidants participate in reactions with lipid radicals compared to side reactions. The tested compounds act as antioxidants or pro-oxidants (Marinova and Yanishlieva, 1992, 2003; Armando et al., 1998), or the antioxidants that vaporize (Zandi and Ahmadi, 2000; Zhang et al., 2004).

19.4.3 Physiological disorders and tolerance to high temperatures

High temperature exposures in fruit and vegetable crops can result in physiological disorders and other associated internal and external symptoms. Tomato fruits exposed to temperatures above 30°C suppress the majority of parameters involved in fruit ripening including color development, softening, respiration rate, and ethylene production (Buescher, 1979; Hicks et al., 1983). It is established now that exposure of fruit to temperature extremes approaching 40°C can induce metabolic disorders and facilitate fungal and bacterial invasion. Although symptoms of heat injury and disease incidence are easily observed at the end of storage, the incipient incidence of these disorders is often not recognized in time to cope with corrective treatment. In general, visible evidence of heat injury on tomatoes appears as yellowish-white side patches on fruits (Mohammed et al., 1996; Ghini et al., 2011; Pangga et al., 2011). Electrolyte leakage in harvested "Dorado" tomatoes exposed to direct sunlight ($34 \pm 2°C$) for 5 h was 73% higher than fruits held in shaded ($29 \pm 2°C$) conditions. However, no significant changes were recorded for electrolyte leakage and direct sunlight treatments following storage at 20°C for 18 days. But the percentage of infected fruits augmented to 35% when exposed to direct sunlight. Exposure of tomato fruit to elevated temperatures affected quality-determining characteristics. Titratable acidity and soluble solids content were 20% higher and 10% lower, respectively, in those tomatoes exposed to direct sunlight (Mohammed et al., 1996).

Frequent and high temperature exposure of apples to 40°C can result in sunburn, development of a watery core, texture loss, and induce low temperature tolerance in apple trees either close to or at harvest stage (Ferguson et al., 1999). Higher temperatures released as hot spells may delay rather than speed up ripening of wine grapes since berries are extremely sensitive to direct radiation; they are vulnerable to sunburn in extreme temperatures (Matthews, 2012). Avocado fruit grown in New Zealand and exposed to direct sunlight had pulp temperatures that frequently exceeded 35°C (Woolf et al., 1999). During subsequent storage at 0°C (below the recommended temperature), these fruit had lower incidences of chilling injury than fruit harvested from shaded parts of the tree. Brooks and Fisher (1926) and Schroeder (1965) observed increased tolerance to high temperatures after fruit and vegetable crop harvest for apple and avocados, respectively. Woolf et al. (1999) and Woolf and Ferguson (2000) observed that avocados grown under direct sunlight, with pulp temperatures frequently around 40°C, had a significant tolerance to high temperature stresses during post-harvest operations.

The practical effects of climate change have already been experienced in some parts across the globe. For example, increased temperatures in Sambalpur, India, have delayed the onset of winter resulting in the decline of cauliflower yields (Pani, 2008). Earlier growers used to harvest 1 kg of inflorescences head but now weight has reduced to 0.25−0.30 kg each. Such reductions in yield hiked production costs of other vegetables too like tomatoes, radishes, and other native Indian vegetable crops. In Brazil, the Brazilian Agricultural Research Corporation (Embrapa) has predicted a 50% reduction in soybean yield in the center-west region ("cerrado") by 2020, assuming an average increase of 0.3 and 0.5°C per year (unpublished data).

19.5 Effects of carbon dioxide exposure

Earth's atmosphere mainly comprises nitrogen (78.1%), oxygen (20.9%), argon (0.93%), and carbon dioxide (0.031%) (Lide, 2009). Nitrogen and oxygen do not play any substantial role in global warming because both gases are virtually transparent to terrestrial radiation. The greenhouse effect is primarily a combination of the effects of water vapor, CO_2, and trace amounts of other gases like methane, nitrous oxide, and ozone that absorb the radiation leaving Earth's surface (IPCC, 2001). The warming effect is explained by the fact that CO_2 and other gases absorb Earth's infrared radiation, trapping heat. A major part of energy emanating from Earth occurs in the form of infrared radiation resulting in CO_2 concentrations rising and more energy preserves contributing to global warming (Lloyd and Farquhar, 2008; Yin, 2013). Carbon dioxide concentrations in the atmosphere have increased approximately 35% from pre-industrial times to 2005 (IPCC, 2007). Besides industrial activities, agriculture also contributes to the emission of greenhouse gases. A current study reports that urban land consumption in Yolo County, California, had regular emissions of more than 70 times that of watered cropland (van Haden et al., 2013). Methane and nitrous oxide were the primary sources emitted by US agricultural activities (EPA, 2009). As shown in Table 19.1, carbon dioxide, methane, and nitrous oxide emissions are very high because of the extensive use of fuel for soil preparation.

Table 19.1 Carbon Dioxide, Methane, and Nitrous Oxide Emissions from Farm to Table for Vegetables and Fruits

Vegetables and Fruits	Emissions			
	Carbon Dioxide	Nitrous Oxide	Methane	Total
	kg CO_2 Equivalents/kg Product over a 100-y Time Period			
Carrots	0.38	0.04	0	0.42
Potatoes	0.40	0.06	0	0.45
Apples	0.80	0.02	0	0.82
Oranges	1.1	0.10	0	1.2
Vegetables: frozen, overseas by boat, boiled	2.2	0.05	0	2.3
Tropical fruit	11	0.23	0	11

Adapted from *Carlsson-Kanyama and González, 2009.*

19.5.1 Growth and physiological alterations

Changes in CO_2 concentrations in the atmosphere can alter the growth and physiological behavior of plants. The majority of these effects have been studied in detail for some vegetable crops (Bazzaz, 1990; Cure and Acock, 1986; Idso and Idso, 1994). These studies concluded that increased atmospheric CO_2 modifies net photosynthesis, biomass production, sugars and organic acid contents, stomatal conductance, fruit firmness, seed yield, light, water, nutrient use efficiency, and plant water potential. Certainly, CO_2, temperature rise, and changes in precipitation patterns will, singly and together, act to modify crop ranges and productivities. These are a result of direct effects on crop physiology. There will also be associated changes in the movement of crop pests and diseases, thereby changing the dynamics of their interactions (Harrington and Woiwod, 1995; Bale et al., 2002). All crop production is operated through the three biochemical pathways of photosynthesis, C_3 and C_4 plants, and CAM crassulacean acid metabolism, which have evolved depending on the plant environment, and differentiated on the basis of the first product of CO_2 assimilation (Cowie, 2007). CAM evolved independently in plants such as cacti and other succulents, including the pineapples better adapted to semi-arid environments. C_4 plants are more efficient at lower concentrations of atmospheric CO_2; hence any minor change in CO_2 may benefit these plants (Drake et al., 2005; Cowie, 2007). However, assessment of plant photosynthetic response to CO_2 rise is difficult and the results are sometimes contradictory. The combined effects of temperature rise as a consequence of increased atmospheric CO_2 concentrations and changes to plant biochemistry may change the palatability of crops to humans and other consumers, as well as the geographical regions of cultivation. C_3 plants respond with increased photosynthetic rates but with a decline in leaf nitrogen levels when grown under enriched CO_2 levels (Sage and Kubien, 2007). C_3 plants demonstrated increased growth under elevated CO_2 concentrations and ambient temperature compared to C_4 plants (Patterson et al., 1999), but this is less evident at higher temperatures. Most plants are expected to decline in nutritional quality and leaf nitrogen; this might affect insect pests feeding on them. Elevated atmospheric carbon dioxide will result in greater carbon to nitrogen ratios in plant foliage and may stimulate greater feeding activity in some herbivores. The fact that host plant quality is often increased by drought-induced plant stress may result in increased feeding by pests and lead to greater crop damage (Hill and Dymock, 1989; Porter et al., 1991); however, this theme remains controversial. Clark (2004), while working on tropical forests, argued that increasing atmospheric CO_2 has no or little effect on biomass production rates and stressed that the growth of tropical forests is not carbon limited, since higher temperature increase, respiration, and other metabolic processes result in increased atmospheric CO_2, which reduces forest productivity.

19.5.2 Quality parameters

Högy and Fangmeier (2009) studied the effects of high CO_2 concentrations on the physical and chemical quality of potato tubers. They observed that the rise in atmospheric CO_2 (50% higher) increased tuber malformation by approximately 63%, resulting in poor quality, and inferior tuber greening (around 12%). Higher CO_2 levels (550 µmol CO_2/mol) increased the occurrence of common scab by 134% but no significant change in dry matter content, specific gravity, and underwater weight. Higher (550 µmol CO_2/mol) concentrations of CO_2 increased glucose (22%), fructose (21%), and reducing sugar (23%) concentrations. It also declined tuber quality due to increased browning and acryl amide formation in french fries. The authors further reported a decline in proteins, potassium, and calcium levels of tubers exposed to high CO_2 concentrations, indicating loss of nutritional and sensory quality.

Bindi et al. (2001) studied the effects of high atmospheric CO_2 during growth on the quality of wines. These authors observed that elevated atmospheric CO_2 levels had a significant increased fruit dry weight ranging from 40 to 45% at 550 mmol CO_2/mol treatment and from 45 to 50% at 700 mmol CO_2/mol. Tartaric acid and total sugar content increased around 8 and 14%, respectively, because of the rise in CO_2 level to the 700 mmol maximum in the mid ripening season. However, as the grapes reached the maturity stage, the CO_2 effect on both quality parameters almost completely disappeared. Overall wine quality was not significantly affected by elevated CO_2. Furthermore, no significant differences were detectable among plants grown in the two enriched treatments (550 and 700 mmol CO_2/mol) and the effects of elevated CO_2 concentration were similar in the two growing seasons. The absence of any further stimulation of the highest CO_2 treatment (700 mmol/mol) on grapevine growth and yield quality (i.e., grapes and wine) may be explained as a result of transport and/or sink limitations. The researchers concluded that the expected rise in CO_2 concentrations may strongly stimulate grapevine production without causing negative repercussion on grape and wine quality.

19.6 **Effects of ozone exposure**

19.6.1 **Formation and distribution**

Ozone in the troposphere is the result of a series of photochemical reactions involving carbon monoxide (CO), methane (CH_4), and other hydrocarbons in the presence of nitrogen species ($NO + NO_2$) (Schlesinger, 1991). Ozone is formed under high temperature and solar irradiation, normally during summer seasons (Mauzerall and Wang, 2001). It is also formed naturally during other seasons, reaching the peak of natural production in the spring (Singh et al., 1978). However, higher concentrations of atmospheric ozone are found during summer due to increases in nitrogen species and emission of volatile organic compounds (Mauzerall and Wang, 2001). Concentrations are at maximum values in the late afternoon and minimum in the early morning hours particularly in industrialized cities and vicinities. The opposite phenomenon occurs at high latitude sites (Oltmans and Levy, 1994). Another potential source for increased levels of ozone in a certain region is via the movement of local winds or downdrafts from the stratosphere.

19.6.2 **Physiological effects**

The effects of ozone on vegetation have been studied both under laboratory and in field experiments. Stomatal conductance and ambient concentrations are the most important factors associated with ozone uptake by plants. Ozone enters plant tissues through the stomates, causing direct cellular damage, especially in the palisade cells (Mauzerall and Wang, 2001). The damage is probably due to changes in membrane permeability and may or may not result in visible injury, reduced growth, and, ultimately, reduced yield (Krupa and Manning, 1988). Visible injury symptoms of exposure to low ozone concentrations include changes in pigmentation also known as bronzing, leaf chlorosis, and premature senescence (Felzer et al., 2007). Since leafy vegetable crops are often grown in the vicinity of large metropolitan areas, it can be expected that increasing concentrations of ozone will result in increased yellowing of leaves. Ozone-stressed leaf tissue can result in inhibition of the photosynthetic rate, declined biomass production, and ultimately impaired post-harvest quality in terms of overall appearance, color, and flavor compounds.

Additionally, Percy et al. (2003) observed that ozone exposure causes a reduction in photosynthesis and an increased turnover of antioxidant systems. Furthermore, Reich (1987) verified that a reduction in photosynthesis declined growth in conifers by 3%, in hardwoods by 13%, and 30% in other crops. Fuhrer et al. (1997) concluded that ozone concentrations higher than 40 nmol O_3/mol can result in 10% yield reduction in different tree species in southern Europe using modeling tools. In open field studies, two-fold increase in CO_2 concentration caused a 15% increase in soybean yield, whereas a 20% increase in the atmospheric ozone offset the yield-increasing effect of CO_2 (Henson, 2008). Grulke and Miller (1994) and Tjoelker et al. (1995) observed that higher ozone concentrations can affect both the photosynthetic and respiratory processes. These authors further verified that ozone concentrations of 95 nmol O_3/mol (twice-ambient concentrations) in branches within the upper canopy of sugar maple (*Acer saceharum* Marsh.) exhibit reduced light saturation in net photosynthetic rates by 56% and increased dark respiration by 40%. These researchers also observed that ozone reduced net photosynthesis and impaired stomatal function.

The present chapter related to plant responses under elevated ozone exposure reveals that there is considerable variation in species response. The greatest impacts on fruit and vegetable crops may occur from changes in carbon transport. Underground storage organs (e.g., roots, tubers, bulbs) normally accumulate carbon in the form of starch and sugars in both fresh and processed crops. If carbon transport to these structures is restricted, there is great potential to lower quality in important crops such as potatoes, sweet potatoes, carrots, onions, and garlic. Exposure of other crops to elevated concentrations of atmospheric ozone can induce external and internal disorders that occur simultaneously or independently. These physiological disorders can decline post-harvest quality of fruits and vegetables resulting in poor marketing and aesthetic value owing to disagreeable visual quality. Decreased biomass production directly affects the size, appearance, and other important visual quality parameters. Furthermore, impaired stomatal conductance due to ozone exposure can reduce root growth, affecting crops such as carrots, sweet potatoes, and beets (Felzer et al., 2007).

19.6.3 Quality parameters

Skog and Chu (2001) carried out a set of experiments to determine the effectiveness of ozone in preventing ethylene-mediated deterioration and post-harvest decay in both ethylene-sensitive and ethylene-producing commodities stored under optimal and suboptimal temperatures. Mushrooms exhibit no known site of ethylene activity so effects from ozone would be antimicrobial only. Ozone at the concentration of 0.04 μl/L appeared to have potential for extending the storage life of broccoli and seedless cucumbers, both stored at 3°C. When mushrooms were stored at 4°C and cucumbers at 10°C, response to ozone was minimal. Ozone also showed the capability of removing ethylene from the environment inside cold rooms. At concentrations of 0.4 μl/L, ozone was effective at removing ethylene (1.5−2.0 μl/L) from an apple and pear storage room. Apples and pears submitted to ozone-enriched atmospheres showed no difference in the fruit quality. Strawberries cv. camarosa stored for 3 days under refrigerated storage (2°C) in an ozone-enriched atmosphere

(0.35 μl/L) showed a three-fold increase in vitamin C content when compared to berries stored at the same temperature under normal atmosphere and 40% reduction in emissions of volatile esters in ozonized fruits (Perez et al., 1999). On the other hand, Kute et al. (1995) reported that strawberries stored in atmospheres with ozone ranging from 0.3 to 0.7 μl/L showed no effect on ascorbic acid levels after 7 days of storage under refrigerated conditions. Quality attributes and sensory characteristics were evaluated in tomato fruits cv. carousel after ozone exposure (concentration ranging from 0.005 to 1.0 μmol/mol) at 13°C and 95% relative humidity. Soluble sugars (glucose, fructose), fruit firmness, weight loss, antioxidant status, CO_2/HO_2 exchange, ethylene production, citric acid, vitamin C (pulp and seed), and total phenolic content were not significantly affected by ozone treatment when compared to fruits kept under ozone-free air. A transient but inconsistent increase in β-carotene, lutein, and lycopene content was observed in ozone-treated fruit. Sensory evaluation revealed a significant preference for fruits subjected to low-level ozone enrichment (0.15 μmol/mol) (Tzortzakisa et al., 2007). Sugiura et al. (2013) illustrated decline in acid concentration, fruit firmness, and watercore development in apples regardless of the maturity index used for harvest date in response to climate change.

The quality of persimmons (*Diospyros kaki* L. F., cv. Fuyu) harvested at two different harvest dates was evaluated after ozone exposure. Fruits were exposed to 0.15 μmol/mol (vol/vol) of ozone for 30 days at 15°C and 90% relative humidity (RH). Astringency removal treatment (24 h at 20°C, 98% CO_2) was performed and fruits were then stored for 7 days at 20°C (90% RH), imitating commercial conditions. Flesh softening was the most important disorder that appeared when the fruit was transferred from 15°C to commercial conditions. Ozone exposure maintained firmness in harvested fruits even after 30 days at 15°C plus shelf-life. Ozone-treated fruit displayed highest weight loss and maximum electrolyte leakage. However, ozone exposure had no significant effect on color, ethanol, soluble solids, and pH. Furthermore, ozone-treated fruits showed no signs of phytotoxic injuries (Salvador et al., 2006).

19.7 Conclusion and future prospects

Understanding how climate changes will impact mankind in the decades to come is of paramount importance for our survival. Temperature, carbon dioxide, and ozone directly and indirectly affect the production and quality of fruit and vegetable crops grown in different climates around the world. Temperature variation can directly affect crop photosynthesis, and increased global temperatures can be expected to have significant impact on post-harvest quality by altering important quality parameters such as synthesis of sugars, organic acids, antioxidant compounds, and firmness. Rising levels of carbon dioxide also contribute to global warming by entrapping heat in the atmosphere. Prolonged exposure to CO_2 concentrations could induce higher incidences of tuber malformation, increased levels of sugars in potatoes, and declined protein and mineral contents, leading to loss of nutritional and sensory quality. Increased levels of ozone in the atmosphere can lead to detrimental effects on post-harvest quality of fruit and vegetable crops. Elevated levels of ozone can induce visual injury and physiological disorders in different species, as well as significant changes in dry matter, reducing sugars, citric and malic acid, among other important quality parameters.

The focus should now be on the novel combination of temperature manipulation in the field and the use of MAGIC lines as adopted in recent studies, which can determine plasticity and assess the role of plant capability to survive climate changes. Although plants exhibit higher fitness on average under elevated temperatures, all genotypes seldom exhibit similar response resulting in decline of their fitness. Numerous significant changes in the genetic makeup of populations are expected to occur in reaction to impending climate changes. The enormity of this genetic alteration will depend on the population size and the species generation time. Future work should be focused on exploring the genetic mechanisms underlying plastic responses that amplify fitness under climate change. However, most studies maintain the initiative that genotypes that are more phenologically approachable are inclined to thrive more under elevated temperatures, signifying that phenological sensitivity can be a practical indicator of genotypes to be utilized in re-establishment. Moreover, individual studies on the responses of crop species or varieties and their diseases to climate change are indispensable. Furthermore, wide-scale projections of disease threats are crucial with the aim of categorizing research priorities, while industry should be targeted and public guidelines developed to ascertain adaptation procedures and to defer impending food security catastrophes. Only by accomplishing long-term and multidisciplinary studies can we decrease the improbability concerning the effects of climate change on plant diseases.

References

Ackerman, F., Stanton, E., 2013. Climate impacts on agriculture: a challenge to complacency? 13-04. Global Development and Environment Institute Working Paper. Medford MA 02155, USA, Tufts University.

Anderson, M., 2013. Climate change, floods and adaptation. Presented at the California Department of Food and Agriculture Climate Change Adaptation Consortium, January 23, Monterey, CA.

Armando, C., Maythe, S., Beatriz, N.P., 1998. Antioxidant activity of grapefruit seed extract on vegetable oils. J. Sci. Food Agric. 77, 463−467.

Baldocchi, D., 2012. Accumulated winter chill is decreasing in fruit growing regions of California. Presented at the California Department of Food and Agriculture Climate Change Adaptation Consortium, November 28, Modesto, CA.

Bale, J.S., Masters, I.D., Hodkinson, C., Awmack, T.M., Bezemer, V.K., Brown, J., et al., 2002. Herbivory in global climate change research: direct effects of rising temperature on insect herbivores. Glob. Chan. Biol. 8, 1−16.

Bales, R.C., 2013. Managing forests for snowpack storage and water yield. Presented at the California Department of Food and Agriculture Climate Change Adaptation Consortium, January 23, Monterey, CA.

Battany, M., 2012. Climate change in CA: vineyard cultural practices. Presented at the California Department of Food and Agriculture Climate Change Adaptation Consortium, November 28, Modesto, CA.

Bazzaz, F.A., 1990. The response of natural ecosystems to the rising global CO_2 levels. Ann. Rev. Ecol. Syst. 21, 167−196.

Beppu, K., Kataoka, I., 2011. Studies on pistil doubling and fruit set of sweet cherry in warm climate. J. Jpn. Soc. Hort. Sci. 80, 1−13.

Bewley, J.D., 1997. Seed germination and dormancy. Plant Cell 9, 1055−1066.

Bieto, J.A., Talon, M., Fisiologia, Y., 1996. Bioquimica vegetal. Madrid: Interamericana. McGraw-Hill, pp. 581.

Bindi, M., Fibbi, L., Miglietta, F., 2001. Free air CO_2 Enrichment (FACE) of grapevine (*Vitis vinifera* L.): II Growth and quality of grape and wine in response to elevated CO_2 concentrations. Eur. J. Agro. 14, 145−155.

Brewer, M.S., 2011. Natural antioxidants: sources, compounds, mechanisms of action, and potential applications. Comp. Rev. Food Sci. Food Saf. 10, 221−247.

Brooks, C., Fisher, D.F., 1926. Some high temperature effects in apples: contrasts in the two sides of an apple. J. Agric. Res. 23, 1−16.

Brunelli, K.R., Gioria, R., Kobori, R.F., 2011. Impacto potencial das mudanças climáticas sobre as doenças das brássicas no Brasil. In: Ghini, R., Hamada, E., Bettiol, W. (Eds.), Impactos das mudanças climáticas sobre doenças de importantes culturas no Brasil. Embrapa Meio Ambiente, Jaguariúna, pp. 145−160.

Buescher, R.W., 1979. Influence of high temperature on physiological and compositional characteristics tomato fruits. Leben-Wissen Techn. 12, 162−164.

Carlsson-Kanyama, A., González, A.D., 2009. Potential contributions of food consumption patterns to climate change. Am. J. Clin. Nutr. 89, 1704S−1709S.

Chakraborty, S., Newton, A.C., 2011. Climate change, plant diseases and food security: an overview. Plant Pathol. 60, 1−14.

Chakraborty, S., Luck, J., Hollaway, G., Freeman, A., Norton, R., Garrett, K.A., et al., 2008. Impacts of global change on diseases of agricultural crops and forest trees. CAB Rev.: Perspect. Agric. Vet. Sci. Nutr. Nat. Resour. 3, 1−15.

Chan, H.T., Linse, E., 1989. Conditioning cucumbers to increase heat resistance in the EFE system. J. Food Sci. 54, 1375−1376.

Chan, H.T., Tam, S.Y.T., Seo, S.T., 1981. Papaya polygalacturonase and its role in thermally injured ripening fruit. J. Food Sci. 46, 190−197.

Chancellor, T.C.B., Kubiriba, J., 2006. The effects of climate change on infectious diseases of plants. Office of Science and Innovation Foresight project on Infectious Diseases: Preparing for the Future. Available at <http://www.foresight.gov.uk/InfectiousDiseases/t7_2a.pdf>.

Clark, D.A., 2004. Sources or sinks? The responses of tropical forests to current and future climate and atmospheric composition. Phil. Trans. R Soc. B 359, 477−491.

Coombe, B.G., 1987. Influence of temperature on composition and quality of grapes. Acta Hortic. 206, 23−35.

Cordenunsi, B.R., Genovese, M.I., do Nascimento, J.R.O., Hassimotto, N.M.A., dos Santos, R.J., Lajolo, F.M., 2005. Effects of temperature on the chemical composition and antioxidant activity of three strawberry cultivars. Food Chem. 91, 113−121.

Coumou, D., Rahmstorf, S.A., 2012. Decade of weather extremes. Nat. Clim. Change 2, 491−496.

Cowie, J., 2007. Climate Change: Biological and Human Aspects. Cambridge University Press, 487pp.

Cure, J.D., Acock, B., 1986. Crop responses to carbon dioxide doubling: a literature survey. Agric. For. Meteorol. 38, 127−145.

Deschenes, O., Greenstone, M., 2012. The economic impacts of climate change: evidence from agricultural output and random fluctuations in weather: reply. Am. Econ. Rev. 102, 3761−3773.

Drake, B.G., Hughes, L., Johnson, E.A., Seibel, B.A., Cochrane, M.A., Fabry, V.J., et al., 2005. Synergistic effects. In: Lovejoy, T.E., Hannah, L. (Eds.), Climate Change and Biodiversity. Tale University Press, New Haven.

Eaks, I.L., 1978. Ripening, respiration and ethylene production of "Hass" avocado fruit at 20−40°C. J. Am. Soc. Hort. Sci. 103, 576−578.

Environmental Protection Agency (EPA), 2009. Available at: <http://www.epa.gov/climatechange/emissions/downloads09/Agriculture.pdf> (accessed February 2012).

Felzer, B.S., Cronin, T., Reilly, J.M., Melillo, J.M., Wang, X., 2007. Impacts of ozone on trees and crops. Comp. Ren. Geosci. 339, 784−798.

Ferguson, I., Lurie, S., Bowen, J.H., 1994. Protein synthesis and breakdown during heat shock of cultured pear (Pyrus communis L.) cells. Plant Physiol. 104, 1429−1437.

Ferguson, I., Volz, R., Woolf, A., 1999. Preharvest factors affecting physiological disorders of fruit. Postharvest Biol. Technol. 15, 255−262.

Filgueira, F.A.R., 2007. In: Viçosa, M.G. (Ed.), Novo Manual de Olericultura: agrotecnologia moderna na produção e comercialização de hortaliças, third ed. UFV, p. 421.

Fuhrer, J., Skarby, L., Ashmore, M.R., 1997. Critical levels for ozone effects on vegetation in Europe. Environ. Poll. 97, 91−106.

Galletta, G.J., Bringhurst, R.S., 1990. Strawberry management. In: Galletta, G.J., Bringhurst, R.S. (Eds.), Small Fruit Crop Management. Prentice-Hall, Englewood Cliffs, pp. 83−156.

Ghini, R., Bettiol, W., Hamada, E., 2011. Diseases in tropical and plantation crops as affected by climate changes: current knowledge and perspectives. Plant Pathol. 60, 122−132.

Gioria, R., Brunelli, K.R., Kobori, R.F., 2008. Impacto potencial das mudanças climáticas sobre as doenças do tomate no Brasil. In: Ghini, R., Hamada, E. (Eds.), Mudanças climáticas: impactos sobre doenças de plantas no Brasil. Emb Inform Tecn, Brasília, DF, pp. 95−128.

Grulke, N.E., Miller, P.R., 1994. Changes in gas exchange characteristics during the life span of giant sequoia: implications for response to current and future concentrations of atmospheric ozone. Tree Physiol. 14, 659−668.

Hall, A.J., McPherson, H.G., Crawford, R.A., Seager, N.G., 1996. Using early season measurements to estimate fruit volume at harvest in kiwifruit. New Zea. J. Crop Hort. Sci. 24, 379−391.

Hamada, E., Ghini, R., Marengo, J.A., Thomaz, M.C., 2011. Projeções de mudanças climáticas para o Brasil no final do século XXI. In: Ghini, R., Hamada, E., Bettiol, W. (Eds.), Impactos das mudanças climáticas sobre doenças de importantes culturas no Brasil. Embrapa Meio Ambiente, Jaguariúna, pp. 41−74.

Hansen, J., Mki, S., Russell, G., Kharecha, P., 2013. Climate sensitivity, sea level, and atmospheric carbon dioxide. Phil. Trans. R Soc. A 371, 2012−2094.

Harrington, R., Woiwod, I.P., 1995. Insect crop pests and the changing climate. Weather 50, 200−208.

Henson, R., 2008. The Rough Guide to Climate Change. second ed. Penguin Books, London, pp. 384

Hicks, J.R., Manzano-Mendez, J., Masters, J.F., 1983. Temperature extremes and tomato ripening. Proc. Fourth Tomato Qual. Workshop 4, 38−51.

Hill, M.G., Dymock, J.J., 1989. Impact of Climate Change: Agriculture/Horticulture Systems. DSIR Entomology Division Submission to the New Zealand Climate Change Programme. Department of Scientific and Industrial Research, Auckland, New Zealand.

Hodges, D.M., Lester, G.E., Munro, K.D., Toivonen, P.M.A., 2005. Oxidative stress: importance for postharvest quality. Hort. Sci. 39, 924−929.

Högy, P., Fangmeier, A., 2009. Atmospheric CO_2 enrichment affects potatoes: 2 tuber quality traits. Eur. J. Agro. 30, 85−94.

Idso, K.E., Idso, S.B., 1994. Plant responses to atmospheric CO_2 enrichment in the face of environmental constraints: a review of the past 10 years' research. Agric. For. Meteorol. 69, 153−203.

Ilić, Z.S., Trajković, R., Perzelan, Y., 2012. Influence of 1-methylcyclopropene (1-MCP) on postharvest storage quality in green bell pepper fruit. Food Biopro. Technol. 5, 2758−2767.

IPCC, 2001. Climate change, 2001: Impacts, adaptations and vulnerability. In: McCarthy, J.J., Canziani, O.F., Leary, N.A., Dokken, D.J., White, K.S. (Eds.), Contribution of Working Group II to the Third Assesment Report of the Intergovernmental Panel on Climate Change. Cambridge University Press, Cambridge.

IPCC, 2007. Climate change, 2007: The physical science basis. In: Solomon, S., Qin, D., Manning, M., Chen, Z., Marquis, M., Averyt, K.B., et al., (Eds.), Contribution of Working Group I to the Fourth Assessment Report of the Intergovernmental Panel on Climate Change. Cambridge University Press, Cambridge.

Jackson, L.E., Wheeler, S.M., Hollander, A.D., O'Geen, A.T., Orlove, B.S., Six, J., et al., 2011. Case study on potential agricultural responses to climate change in a California landscape. Clim. Change 109, 407−427.

Kader, A.A., Arpaia, M.L., 2002. Postharvest handling systems: subtropical fruits. In: Kader, A.A. (Ed.), Postharvest Technology of Horticultural Crops, third ed. University of California Agricultural and Natural Resources, Oakland, pp. 375−384.

Kliewer, M.W., Lider, L.A., 1968. Influence of cluster exposure to the sun on the composition of Thompson seedless fruit. Am. J. Eno. Viticul. 19, 175−184.

Kliewer, M.W., Lider, L.A., 1970. Effects of day temperature and light intensity on growth and composition of *Vitis vinifera* L. fruits. J. Am. Soc. Hort. Sci. 95, 766−769.

Kobori, R.F., Brunelli, K.R., Gioria, R., 2011. Impacto potencial das mudanças climáticas sobre as doenças da alface no Brasil. In: Ghini, R., Hamada, E., Bettiol, W. (Eds.), Impactos das mudanças climáticas sobre doenças de importantes culturas no Brasil. Embrapa Meio Ambiente, Jaguariúna, pp. 129−144.

Kremen, C., 2013. Integrated crop pollination for resilience against climate change and other problems. Presented at the California Department of Food and Agriculture Climate Change Adaptation Consortium, March 20, American Canyon, CA.

Krupa, S.V., Manning, W.J., 1988. Atmospheric ozone: formation and effects on vegetation. Environ. Poll. 50, 101−137.

Kute, K.M., Zhou, C., Barth, M.M., 1995. The effect of ozone exposure on total ascorbic acid activity and soluble solids contents in strawberry tissue. In: Proceedings of the Annual Meeting of the Institute of Food Technologists (IFT). LA, New Orleans, pp. 82.

Lakso, A.N., Kliewer, W.M., 1975. The influences of temperature on malic acid metabolism in grape berries I. Enzyme responses. Plant Physiol. 56, 370−372.

Lee, S.K., Young, R.E., Shiffman, P.M., Coggins, C.W., 1983. Maturity studies of avocado fruit based on picking date and dry weight. J. Am. Soc. Hort. Sci. 108, 390−394.

Li, Z., Palmer, W.M., Martin, A.P., Wang, R., Rainsford, F., Jin, Y., et al., 2012. High invertase activity in tomato reproductive organs correlates with enhanced sucrose import into, and heat tolerance of, young fruit. J. Exp. Bot. 63, 1155−1166.

Lide, D.R., 2009. CRC Handbook of Chemistry and Physics. 90th ed. CRC Press, Boca Raton, pp. 2804.

Liu, F.W., 1978. Effects of harvest date and ethylene concentration in controlled atmosphere storage on the quality of McIntosh apples. J. Am. Soc. Hort. Sci. 103, 388−392.

Lloyd, J., Farquhar, G.D., 2008. Effects of rising temperatures and [CO_2] on the physiology of tropical forest trees. Phil. Trans. Royal Soc. Bio. Sci. 363, 1811−1817.

Lobell, D.B., Schlenker, W., Costa-Roberts, J., 2011. Climate trends and global crop production since 1980. Science 333, 616−620.

Lopes, C.A., Silva, G.O., Cruz, E.M., Assad, E., Pereira, A.S., 2011. Uma análise do efeito do aquecimento global na produção de batata no Brasil. Horti. Bras. 29, 7−15.

Lurie, S., Klein, J.D., 1990. Heat treatment of ripening apples: differential effects on physiology and biochemistry. Physiol. Plant 78, 181−186.

Lurie, S., Klein, J.D., 1991. Acquisition of low-temperature tolerance in tomatoes exposed to high-temperature stress. J. Am. Soc. Hort. Sci. 116, 1007−1012.

Magan, N., Medina, A., Aldred, D., 2011. Possible climate-change effects on mycotoxin contamination of food crops pre- and postharvest. Plant Pathol. 60, 150−163.

Marinova, E.M., Yanishlieva, N.V., 1992. Effect of temperature on the antioxidative action of inhibitors in lipid autoxidation. J. Sci. Food Agric. 60, 313−318.

Marinova, E.M., Yanishlieva, N.V., 2003. Antioxidant activity and mechanism of action of some phenolic acids at ambient and high temperatures. Food Chem. 81, 189−197.

Matthews, M., 2012. Temperature—grapevine development and fruit composition. Presented at the California Department of Food and Agriculture Climate Change Adaptation Consortium, November 28, Modesto, CA.

Mauzerall, D.L., Wang, X., 2001. Protecting agricultural crops from the effects of tropospheric ozone exposure: reconciling science and standard setting in the United States, Europe, and Asia. Ann. Rev. Energy Environ. 26, 237–268.

McKeown, A.W., Warland, J., McDonald, M.R., 2006. Long-term climate and weather patterns in relation to crop yield: a mini review. Can. J. Bot. 84, 1031–1036.

Medellín-Azuara, J., Howitt, R.E., Duncan, J., MacEwan, Lund, J.R., 2011. Economic impacts of climate-related changes to California agriculture. Clim. Change 109, 387–405.

Mohammed, M., Wilson, L.A., Gomes, P.I., 1996. Influence of high temperature stress on postharvest quality of processing and non-processing tomato cultivars. J. Food Qual. 19, 41–55.

Mukherjee, M., 2013. Irrigated agricultural adaptation to water and climate variability. Presented at the California Department of Food and Agriculture Climate Change Adaptation Consortium, January 23, Monterey, CA.

Nelson, V., Morton, J., Chancellor, T., Burt, P., Pound, B., 2010. Climate change, agricultural adaptation and fairtrade: identifying the challenges and opportunities. Nat. Res. Ins. 1, 1–45.

Oltmans, S.J., Levy, H., 1994. Surface ozone measurements from a global network. Atmos. Environ. 28, 9–24.

Pangga, I.B., Hannan, J., Chakraborty, S., 2011. Pathogen dynamics in a crop canopy and their evolution under changing climate. Plant Pathol. 60, 70–81.

Pani, R.K., 2008. Climate change hits vegetable crop. Indian Express. Available from: <http://www.express-buzz.com>.

Patterson, D.T., Westbrook, J.K., Joyce, R.J.V., Lingren, P.D., Rogasik, J., 1999. Weeds, insects and diseases. Clim. Change 43, 711–727.

Peet, M.M., Wolfe, D., 2000. Vegetable crop responses to climatic change. In: Reddy, K.R., Hodges, H.F. (Eds.), Climate Change and Global Crop Productivity. CABI Publishing, Wallingford, UK, pp. 213–243.

Peet, M.M., Sato, S., Gardner, R.G., 1998. Comparing heat stress effects on male-fertile and male-sterile tomatoes. Plant Cell Environ. 21, 225–231.

Percy, K.E., Legge, A.H., Krupa, S.V., 2003. Troposphere ozone: a continuing threat to global forests? In: Karnosky, D.F.E.A. (Ed.), Air Pollution: Global Change and Forests in the New Millennium. Elsevier Ltd, pp. 85–118.

Perez, A.G., Sanz, C., Rios, J.J., Olias, R., Olias, J.M., 1999. Effects of ozone treatment on postharvest strawberry quality. J. Agric. Food Chem. 47, 1652–1656.

Picton, S., Grierson, D., 1988. Inhibition of expression of tomato-ripening genes at high temperature. Plant Cell Environ. 11, 265–272.

Pokorný, J., 1986. Addition of antioxidants for food stabilization to control oxidative rancidity. Czech J. Food Sci. 4, 299–307.

Pope, K.S., 2012. Climate change adaptation: temperate perennial crops. Presented at the California Department of Food and Agriculture Climate Change Adaptation Consortium, November 28, Modesto, CA.

Pope, K.S., Dose, V., Da Silva, D., Brown, P.H., Leslie, C.A., DeJong, T.M., 2013. Detecting nonlinear response of spring phenology to climate change by Bayesian analysis. Glob. Chan. Biol. 19, 1518–1525.

Porritt, S.W., Lidster, P.D., 1978. The effect of prestorage heating on ripening and senescence of apples during cold storage. J. Am. Soc. Hort. Sci. 103, 584–587.

Porter, J.H., Parry, M.L., Carter, T.R., 1991. The potential effects of climatic change on agricultural insect pests. Agric. For. Meteorol. 57, 221–240.

Porter, J.R., Semenov, M.A., 2005. Crop responses to climatic variation. Philos. Trans. R. Soc. Lond. B Biol. Sci. 360, 2021–2035.

Ramos, B., Miller, F.A., Brandão, T.R.S., Teixeira, P., Silva, C.L.M., 2013. Fresh fruits and vegetables—an overview on applied methodologies to improve its quality and safety. Innovative Food Sci. Emerg. Technol. 20, 1–15.

Réblová, Z., 2012. Effect of temperature on the antioxidant activity of phenolic acids. Czech J. Food Sci. 30, 171–177.

Reich, P.B., 1987. Quantifying plant response to ozone: a unifying theory. Tree Physiol. 3, 63–91.

Rockstrom, J., Steffen, W., Noone, K., Persson, A., Chapin, F.S., Lambin, E.F., et al., 2009. A safe operating space for humanity. Nature 461, 472–475.

Rowlands, D.J., Frame, D.J., Ackerley, D., Aina, T., Booth, B.B.B., Christensen, C., et al., 2012. Broad range of 2050 warming from an observationally constrained large climate model ensemble. Nat. Geosci. 5, 256–260.

Sage, R.F., Kubien, D., 2007. The temperature response of C_3 and C_4 photosynthesis. Plant Cell Environ. 30, 1086–1106.

Saltveit, M.E., 2002. Respiratory metabolism, in the commercial storage of fruits, vegetables, and florist and nursery crops. In: Gross, K.C., Wang, C.Y., Saltveit M.E. (Eds.), United States Department of Agriculture, Agriculture Handbook 66, Available at <http://www.ba.ars.usda.gov/hb66/index.html>.

Salvador, A., Abad, I., Arnal, L., Martinez-Javegam, J.M., 2006. Effect of ozone on postharvest quality of persimmon. J. Food Sci. 71, 443–446.

Sato, S., Kamiyama, M., Iwata, T., Makita, N., Furukawa, H., Ikeda, H., 2006. Moderate increase of mean daily temperature adversely affects fruit set of *Lycopersicon esculentum* by disrupting specific physiological processes in male reproductive development. Ann. Bot. 97, 731–738.

Schlesinger, W.H., 1991. Biogeochemistry: An Analysis of Global Change. Academic Press, New York, 443p.

Schroeder, C.A., 1965. Temperature relationships in fruit tissues under extreme conditions. Proc. Am. Soc. Hort. Sci. 87, 199–203.

Singh, H.B., Ludwig, F.L., Johnson, W.B., 1978. Tropospheric ozone: concentrations and variabilities in clear remote atmospheres. Atmos. Environ. 12, 2185–2196.

Skog, L.J., Chu, C.L., 2001. Effect of ozone on qualities of fruits and vegetables in cold storage. Can. J. Plant Sci. 81, 773–778.

Springate, D.A., Kover, P.X., 2013. Plant responses to elevated temperatures: a field. Glob. Chan. Biol. 10.1111/gcb.12430.

Sugiura, T., Ogawa, H., Fukuda, N., Moriguchi, T., 2013. Changes in the taste and textural attributes of apples in response to climate change. Sci. Rep. . 10.1038/srep02418.

Tjoelker, M.G., Volin, J.C., Oleksyn, J., Reich, P.B., 1995. Interaction of ozone pollution and light effects on photosynthesis in a forest canopy experiment. Plant Cell Environ. 18, 895–905.

Toivonen, P.M.A., 2005. Postharvest storage procedures and oxidative stress. Hort. Sci. 39, 938–942.

Toivonen, P.M.A., 2009. Benefits of combined treatment approaches to maintaining fruit and vegetable quality. Fresh Prod. 3, 58–64.

Tzortzakisa, N., Borlanda, A., Singletona, I., Barnes, J., 2007. Impact of atmospheric ozone-enrichment on quality-related attributes of tomato fruit. Postharvest Biol. Technol. 45, 317–325.

United States Global Change Program Research, 2000. Global climate change impacts in the United States: Agriculture. Washington, DC, pp. 71–78.

USDA, 2013. Economic research service: overview. Retrieved March 2013. Available at <http://www/ers.usda.gov/data-products/feed-grains-database.aspx#.UVpoYaXWGoc>.

Van Haden, R., Dempseya, M., Wheelerb, S., Salasc, W., Jacksona, L.E., 2013. Use of local greenhouse gas inventories to prioritise opportunities for climate action planning and voluntary mitigation by agricultural stakeholders in California. J. Environ. Plan. Manage. 56, 553–571.

Wang, S.Y., Zheng, W., 2001. Effect of plant growth temperature on antioxidant capacity in strawberry. J. Agric. Food Chem. 49, 4977–4982.

Warland, J., McKeown, A.W., McDonald, M.R., 2006. Impact of high air temperatures on Brassicaceae crops in southern Ontario. Can. J. Plant Sci. 86, 1209–1215.

Went, F.W., 1953. The effect of temperature on plant growth. Ann. Rev. Plant Physiol. 4, 347–362.

Woolf, A.B., Ferguson, I.B., 2000. Postharvest responses to high fruit temperatures in the field. Postharvest Biol. Technol. 21, 7–20.

Woolf, A.B., Bowen, J.H., Ferguson, I.B., 1999. Preharvest exposure to the sun influences postharvest responses of "Hass" avocado fruit. Postharvest Biol. Technol. 15, 143–153.

Woolf, A.B., Wexler, A., Prusky, D., Kobiler, E., Lurie, S., 2000. Direct sunlight influences postharvest temperature responses and ripening of five avocado cultivars. J. Am. Soc. Hort. Sci. 125, 370–376.

Wurr, D.C.E., Fellows, J.R., Phelps, K., 1996. Investigating trends in vegetable crop response to increasing temperature associated with climate change. Sci. Hort. 66, 255–263.

Yanishlieva, N.V., 2001. Inhibiting oxidation. In: Pokorný, J., Yanishlieva, N.V., Gordon, H. (Eds.), Antioxidants in Food: Practical Applications. Woodhead Publishing, Cambridge, pp. 22–70.

Yin, X., 2013. Improving ecophysiological simulation models to predict the impact of elevated atmospheric CO_2 concentration on crop productivity. Ann. Bot. 112, 465–475.

Zandi, P., Ahmadi, L., 2000. Antioxidant effect of plant extracts of Labiatae family. J. Lipid Sci. Technol. 37, 436–439.

Zhang, C.X., Wu, H., Weng, X.C., 2004. Two novel synthetic antioxidants for deep frying oils. Food Chem. 84, 219–222.

Interplays of Plant Circadian Clock and Abiotic Stress Response Networks

Agnieszka Kiełbowicz-Matuk and Jagoda Czarnecka

20.1 Introduction

Plants having a sedentary lifestyle are constantly exposed to various environmental factors, which often cause limited and uneven germination leading to a significant reduction in their productivity. In the course of evolution, plants have developed strategies to ensure their survival among stressors. Recently, one of the most intensively investigated mechanisms that allow organisms to adapt to changing environmental conditions by predicting the environmental variations that appear during the day—night shift are circadian rhythms. The occurrence of circadian rhythms is a common phenomenon in the world of living organisms (Yerushalmi and Green, 2009). Many aspects of metabolism, physiology, and behavior vary during the day and night. A characteristic feature of circadian rhythms is that they persist under constant environment conditions (without external stimulation) with the period length of approximately 24 hours.

Plant circadian rhythms were first described about 300 years ago. Circadian rhythms are controlled by a special mechanism of the molecular clock, synchronizing the performance of plants with daily cycles of light and temperature to maintain homeostasis (Wijnen and Young, 2006).

Smooth operation of the circadian clock mechanism is very important for the proper functioning of organisms. Since the expression of at least one-third of *Arabidopsis* genes is controlled at the transcriptional level by the circadian clock, it can be assumed that the clock is the potential regulator of circadian rhythms for nearly all physiological and metabolic processes occurring in plants (Covington et al., 2008; Doherty and Kay, 2010; Kinmonth-Schultz et al., 2013). Therefore, the clock functions as a control unit that manages the complex cellular networks of the organisms. Numerous studies revealed that the circadian clock affects various biological phenomena such as photosynthesis, leaf movements, hypocotyl growth, time of flowering, and above all biotic and abiotic stress responses (Locke et al., 2005; Nozue et al., 2007; Yakir et al., 2007; Kant et al., 2008; Nakamichi et al., 2009; Sanchez et al., 2011; Lai et al., 2012; Maibam et al., 2013). Thus far, little is known about the role of the clock in the modulation of response to biotic factors and plant defense responses (Goodspeed et al., 2012; Shin et al., 2012; Bolouri Moghaddam and Van den

P. Ahmad (Ed): Emerging Technologies and Management of Crop Stress Tolerance, Volume 1.
DOI: http://dx.doi.org/10.1016/B978-0-12-800876-8.00020-5

Ende, 2013), whereas its involvement in the regulation of plant response to abiotic environmental factors is extensively studied and largely known.

This review focuses on multidimensional mutual relationships among the clock and the complicated network of plant abiotic stresses, investigates the existence of multiple feedback loops among these components, and highlights the prominent role of the clock in modulating abiotic responses utilizing a gating phenomenon.

20.2 Molecular basis of the circadian clock function in plants

In plants, the molecular mechanism underpinning the circadian clock and its rhythmicity has been mainly investigated in the model plant *Arabidopsis thaliana*. However, studies on the extension of the *Arabidopsis* clock mechanism for other plant species, particularly those which are important in agriculture, including rice, maize, soybean, *Brassica rapa*, and tobacco, is still ongoing (Nagy et al., 1988; Murakami et al., 2007; Liu et al., 2009, Xu et al., 2010; Wang et al., 2011; Yon et al., 2012; McClung, 2013).

20.2.1 Transcriptional regulation

Several biochemical and genetic studies have revealed that the key measure in the circadian clock is an oscillator that generates rhythms. It was originally specified by a feedback loop between morning- and evening-phased elements, which constitute the core of the internal circulation. This mechanism consists of two morning MYB transcription factor proteins, CIRCADIAN CLOCK-ASSOCIATED1 (CCA1) and LATE ELONGATED HYPOCOTYL (LHY), and one evening protein TIMING OF CAB EXPRESSION1/PSEUDORESPONSE REGULATOR1 (TOC1/PRR1) (Schaffer et al., 1998; Wang and Tobin, 1998). At the onset of dawn, TOC1 protein activates the transcription of *CCA1* and *LHY* genes whose protein products are then transported to the nucleus, where they bind to the EVENING ELEMENT (EE) motif within the promoter of the *TOC1* gene, and directly repress its expression. Progressive reduction in TOC1 protein levels leads to the inhibition of the *CCA1* and *LHY* genes expression, which, in turn, results in the initiation of *TOC1* gene transcription, and the whole cycle starts again (Alabadi et al., 2001; Mizoguchi et al., 2002; Kolmos et al., 2008). A precise control of transcriptional events is, therefore, crucial for appropriate circadian function (Hemmes et al., 2012).

Even though the roles of the three core elements has been proven previously, significantly greater complexity of this process was revealed by research based on mathematical analysis. According to this model, it was found that the molecular mechanism of the clock in plants is based on the action of a number of proteins that form three interconnected loops that operate in a positive and negative feedback (Locke et al., 2006; Shin and Davis, 2010).

In a three-loops model, evening TOC1 protein acts indirectly by activating an unidentified X component that supports the transcription of *CCA1* and *LHY* genes (Locke et al., 2006). Additionally, TOC1 protein probably interacts with CCA1 HIKING EXPEDITION (CHE) protein that acts as a repressor of *CCA1* gene expression, and antagonizes it through the direct protein—protein interaction. This causes the activation of the expression of the *CCA1* gene (Pruneda-Paz et al., 2009). The second loop (morning loop) of the clock builds the CCA1 and LHY proteins that act as transcriptional activators of the genes *PSEUDO RESPONSE REGULATOR 7* and *9* (*PRP7* and *PRP9*), whose maximum expression falls on the morning (Farré et al., 2005; Zeilinger et al., 2006). In addition, the PRP7—PRP9 protein complex inhibits the expression of *CCA1* and *LHY* genes, using a previously unknown mechanism, thereby closing the cycle (Farré et al., 2005). Current data suggest that members of the plant Groucho/TUP1 corepressor family, TOPLESS/ TOPLESS RELATED PROTEINs (TPL/TPRs), specifically interact with three members of the PRR family (PRR5, PRR7, PRR9) and together inhibit the expression of two core circadian transcription factors, *CCA1* and *LHY* (Wang et al., 2013). The third loop (evening loop) is based on the action of two elements, the TOC1 protein and an unidentified Y component. The first acts as a repressor of the second component, which feeds back by inducing expression of its predecessor. It is believed that one of the proteins that may play the role of a Y factor is GIGANTEA (GI), whose expression is cyclic (Locke et al., 2005).

Recent reports indicate that many regulatory mechanisms acting at the post-transcriptional, translational, post-translational, and metabolic levels are required for efficient and precise operation of the circadian clock (Troncoso-Ponce and Mas, 2012; Henriques and Mas, 2013; Rund-Kangisser et al., 2013).

20.2.2 Involvement of alternative splicing in the action and control of the circadian clock

In addition to the transcriptional relationship, substantial progress has been made towards understanding the post-transcriptional regulation of the circadian gene networks in plants (Staiger and Koster, 2011; Bertoni, 2012). Among them, alternative splicing (AS) seems to be the most widespread and complex mechanism that plays an important role in the regulation of the circadian clock and its output (Filichkin et al., 2010; James et al., 2012; Syed et al., 2012). The consequence of its action is the formation of different splice variants from a single gene, resulting in the synthesis of structurally and functionally diverse proteins. In plants, AS can occur at various stages of development. Abiotic stresses and types of specific tissue may affect its frequency. Furthermore, AS of transcription factors and key regulatory proteins in signaling pathways is a major contributor to the magnitude of response to abiotic stresses.

Analysis based on the ultra-high-throughput RNA sequencing revealed that about 42% of genes containing intron in *Arabidopsis* are subjected to the AS mechanisms (Filichkin et al., 2010). The data contained in the *Arabidopsis* Information Resource (TAIR) indicate that for the core clock genes there exist several mRNA splice variants, for example five for *LHY*, two transcripts for *CCA1* and *PRR9* each, and a single splice variant for *TOC1*, *PRR7*, *GI*, and *CHE*. In addition, other splice variants have been reported for *CCA1* (intron 4 retention), for *LHY* (intron 1 retention and the additional exon 5a), for *TOC1* (intron 4 retention), and *PRR9* (intron 3 retention) (Hazen et al., 2009; Sanchez et al., 2010; Filichkin et al., 2010). Current data suggest that the number of the AS events for the core clock genes that supposedly contribute to the creation of various splice variants in *Arabidopsis* is much greater than the number of AS events available in the TAIR (James et al., 2012). Based on the high-resolution RT-PCR (HR RT-PCR) system, 63 different AS events for 10 clock genes have been reported (James et al., 2012). This shows how important is the AS mechanism in the proper operation and regulation of the circadian timekeeping.

Unfortunately, current knowledge regarding the molecular mechanisms that connect the clock to the AS regulation is very limited in plants. In *Arabidopsis*, a crucial role in proper mRNA AS of the circadian clock components has been attributed to PROTEIN ARGININE METHYL TRANSFERASE 5 (PRMT5), which transfers methyl groups to arginine residues of certain spliceosomal proteins (Deng et al., 2010). *PRMT5* expression exhibits daily and circadian rhythms, therefore it seems that PRMT5 protein can control the circadian expression and AS of many clock genes, functioning as a platform between environmental fluctuations and the amount of unproductive splicing isoforms of the clock genes. An example can be significant changes in the alternative splicing of the core clock gene *PRR9* observed in *Arabidopsis prmt5* mutant (Hong et al., 2010; Sanchez et al., 2010; Petrillo et al., 2011).

20.2.3 The impact of individual stress on the type of AS of the clock genes

In plants, different types of AS have been identified. Among them, a significant quantitative advantage is the intron retention event, often associated with abiotic stress response (Palusa et al., 2007). This type of AS leads to the formation of numerous alternative isoforms with premature termination codons (PTC). Cold-induced intron retention was observed for *Arabidopsis IDD14* transcription factor gene (Seo et al., 2011). Low temperature-induced intron retention was also noticed for two early cold-regulated genes from *Triticum durum*, putatively encoding for a ribokinase and a C3H2C3 RING-finger protein (Mastarangelo et al., 2005).

The relationship between a particular splicing event and an individual stress condition has been demonstrated for many of the circadian clock-associated genes (Filichkin et al., 2010; Filichkin and Mockler, 2012). Low temperature suppressed the retention of Intron 4 in the *Arabidopsis CCA1* core clock gene, while the accumulation of a high level of PCT isoforms is promoted in high light conditions (Filichkin et al., 2010).

Another example is the clock-regulated *SUPPRESSOR OF OVEREXPRESSIONOF CO1* (*SOC1*) gene for which alternatively spliced mRNA with a retained sixth intron appears to be dependent on the specific abiotic conditions (Filichkin et al., 2010).

An interesting example of regulation of circadian clock genes expression by AS is negative autoregulation. It has been observed for *Arabidopsis GRP7* (*CCR2*) and *GRP8* (*CCR1*) (GLYCINE-RICH PROTEIN 7 and 8, COLD CIRCADIAN REGULATED 2 and 1) genes, elements of the circadian oscillator (Staiger et al., 2003; Schoning et al., 2008). After reaching the abundance limit, GRP7 and GRP8 proteins bind to their own pre-mRNA to initiate AS-producing transcripts with premature termination codons (Schoning et al., 2007, 2008). Interestingly, both proteins create output pathway controlling responses to environmental stresses (Kim et al., 2008; Schmidt et al., 2011).

20.2.4 Temperature affects AS of the key clock genes

Under natural conditions, plants are exposed to sudden and frequent temperature changes. In order to survive rapid temperature and weather fluctuations, plants have evolved a variety of adaptation strategies that enable them to quickly receive temperature stimuli and initiate a defense response in extreme temperatures. Temperature compensation is one of the factors that defines the circadian rhythms and is integrally connected with the clock. This phenomenon helps maintain the length period of the circadian rhythm (under constant conditions) over a large range of temperature entrainments. Recent reports underline the primary role of *Arabidopsis* CCA1, LHY, PRR7, and PRR9 proteins in this process (Salomé et al., 2010).

However, despite the temperature compensation phenomenon, rapid external temperature changes can lead to disturbances in the operation of the clock by entering it into a new phase. Therefore, study of the interactions between the circadian clock and temperature variations is the subject of several research groups (McClung and Davis, 2010). It is known that low temperatures can alter the periodic expression pattern of the clock components, and that the clock (the time of day) can affect the cold-induced expression of numerous genes, including genes encoding transcription factors (Fowler et al., 2005; Espinoza et al., 2008). However, precise function of the molecular clock during low temperatures has not yet been determined in plants.

The impact of temperature alteration on the AS events for the clock genes has been studied intensively. Originally, AS was noticed for the circadian-associated genes *Per* and *frq* in *Drosophila melanogaster* and *Neurospora crassa*, respectively, and the products of its actions are involved in the determining phase and periodicity of the clock (Majercak et al., 1999; Diernfellner et al., 2007; Akman et al., 2008; Kojima et al., 2011).

A major effect of temperature change on the AS event was observed for *CCA1* and *LHY* transcripts. While studying changes in AS patterns at dawn in two groups of plants, half day adapting and already adapted to 20°C, 12°C, and 4°C, differences of cold influence on AS events for *CCA1* and *LHY* genes were noticed (James et al., 2012). The authors observed that transient and long-term exposure of plants in the cold leads to a significant reduction in amount of fourth intron-containing *CCA1* splice variant (*CCA1β*) compared to those grown at 20°C. Cold treatment resulted in the accumulation of the fully spliced (functional) *CCA1α* variant and CCA1 protein level significantly increased (James et al., 2012; Henriques and Mas, 2013). Interestingly, temperature-dependent differential splicing was also reported for *LHY*. In this case, there was a significant reduction in the fully spliced *LHY* form encoding functional LHY protein (James et al., 2012).

In plants, AS of *CCA1* is a conservative phenomenon that leads to the longer (*CCA1β*) and shorter (*CCA1α*) splice isoforms, with the latter supplemented with the MYB DNA Binding Domain (Filichkin et al., 2010). Recent reports indicate that the presence of the *CCA1α* isoform contributes to plants' freezing tolerance, since 35 S:*CCA1α* transgenic *Arabidopsis* plants are characterized by increased resistance to freezing (Seo et al., 2012). However, the smaller protein isoform (CCA1β) acts as a negative regulator of the CCA1α isoform by suppressing its activity. Structurally, the CCA1β isoform looks like a small competitive inhibitor, which interacts with CCA1α isoform and LHY and prevents the formation of functional dimers, such as CCA1α-CCA1α, CCA1α-LHY, LHY-LHY (Seo et al., 2012). Interaction of CCA1β with CCA1α leads to the formation non-functional heterodimers and prevents the binding of CCA1α to DNA. Transgenic plants overexpressing an aberrant *CCA1β* isoform exhibited altered rhythmicity and reduced periods of circadian oscillations, like the *cca1-lhy* double mutant. Moreover, in the 35 S:*CCA1β* transgenic plants, the rhythmic expression pattern of cold response pathway genes under the control of the circadian clock is disrupted (Seo et al., 2012).

Thus, the temperature-dependent AS of the *CCA1* gene has a major impact on the functioning of its protein products, in the regulation of the circadian clock and the plant freezing tolerance.

20.3 Molecular basis of the interaction between the clock components and cold response

Low temperature exerts dramatic effects on plant survival and affects agricultural productivity. Plants respond to low temperature by running the adaptive response known as cold acclimation, which involves reprogramming the wide range of cellular processes and activates mechanisms leading to an increase in freezing tolerance.

Understanding the biological effects of cold on changes in the cyclic expression of the clock components and output genes is fascinating. Cold-dependent damage or inhibition of circadian oscillations may in turn cause particular contributions of affected genes in response to this stress factor. Recent studies have emphasized the adverse effect of cold on rhythmic expression of circadian clock genes in *Arabidopsis*, chestnut, and poplar, pointing to the differences between species (Ramos et al., 2005; Bieniawska et al., 2008; Ibáñez et al., 2008, 2010). These authors revealed that under diurnal light—dark conditions, cold disrupts the cyclic expression of the clock genes only in chestnut, while for *Arabidopsis* clock components, the cold just reduces the amplitude of cycles (Bieniawska et al., 2008; Ibáñez et al., 2008). Interestingly, in such conditions, the cycles of standard circadian output marker genes in *Arabidopsis* are suppressed or disrupted (Bieniawska et al., 2008).

One significant aspect of the interaction between cold response and circadian clock that has been reported previously is "gating" cold acclimation. Participation of circadian clock in cold signaling in *Arabidopsis* was first identified when cold-dependent transcript accumulation varied considerably in the time of day at which the stressor was applied (Bieniawska et al., 2008). Strong evidence suggests that that circadian clock gates the low temperature induction of more than 50 transcription factors including *C-REPEAT BINDING FACTORS* (*CBFs* also known as *DREB*), *ZAT12*, and *RAV1* in *Arabidopsis* and alters low temperature Ca^{2+} signals (Fowler et al., 2005; Dodd et al., 2006). The CBF1-3 transcription factors are key components of the canonical cold response pathway (known as CBF regulon), which activates the expression of more than 100 genes known as *COLD REGULATED* (*COR*) leading to increased tolerance to freezing (Fowler and Thomashow, 2002; Maruyama et al., 2004; Van Buskirk and Thomashow, 2006). Clock modulates *CBF* genes expression through CCA1 and LHY MYB transcription factors, which bind directly to *CBF* promoters and up-regulate their transcription in the morning, thereby inducing cold tolerance (Dong et al., 2011). Transgenic plants overexpressing the *CCA1* gene exhibited abolished gating effects on *CBF* genes expression. Extensive analysis of the promoter sequences of cold-inducible genes revealed an abundance of *cis*-Evening Elements (EE\EEL) often conjugated to ABA-responsive elements (ABRE) (Mikkelsen and Thomashow, 2009). Previous studies have shown that the EE motif, AAAATATCT, is a key element in circadian-regulated gene expression, the presence of which is required to induce peak expression in the evening (Harmer et al., 2000). Thus, the EE motif is the "connecting point" for configuring cold and circadian transcriptomes (Mikkelsen and Thomashow, 2009).

Despite numerous studies, the role of particular clock components in plants at low temperature has not been fully demonstrated. As mentioned before, alternative splicing of the *CCA1* clock core gene plays an important role in plant adaptation to cold conditions (James et al., 2012; Seo et al., 2012). During cold acclimation, constitutive expression of *CCA1α* reduces the gating effects of the clock and also activates the *CBF* and *GI* genes to enhance freezing tolerance (Fowler et al., 2005). Further, *Arabidopsis* plants with mutation in clock genes exhibit diminished cold acclimation (Cao et al., 2005; Nakamichi et al., 2009; Espinoza et al., 2010; Dong et al., 2011). For example, *cca1* single and *cca1/lhy* double mutants are more susceptible to freezing than wild-type plants (Espinoza et al., 2010; Dong et al., 2011). Similarly, the GIGANTEA deficient *gi* mutant showed enhanced sensitivity to freezing and impaired cold acclimation ability (Cao et al., 2005). These authors indicated existence of additional signal pathways (CBF independent) for the *GI* gene, which are involved in the regulation of freezing tolerance.

20.4 Crosstalk between the circadian clock and ABA transcriptional networks

Finding the relationship between the circadian clock and abscisic acid (ABA) is the subject of interest to many researchers. The main reason for this interest is the involvement of both in the regulation of many cellular processes occurring at developmental, physiological, and metabolic levels as well as in the response of plants to environmental stress factors (Finkelstein et al., 2002; Zhu, 2002). Many genes for which the changes in expression profiles were observed under the influence of cold, drought, and salinity are up- and down-regulated by both the clock and the ABA (Finkelstein et al., 2002; Nakashima et al., 2009). Some ABA up-regulated genes such as *RD29A (RESPONSE TO DESICATION)*, *COR15A* and *B (COLD-REGULATED)*, *ERD10*, and *7 (EARLY RESPONSE TO DEHYDRATATION)*, the transcription of which is activated during the day, may be cited as examples. The microarray analysis showed that a significant proportion of genes involved in ABA precursors and biosynthesis as well as ABA signal transduction are under the control of the clock (Covington and Harmer, 2007; Dodd et al., 2007; Mizuno and Yamashino, 2008). In addition, the clock can affect not only the expression of ABA-related genes, but also act as the controller of various cellular processes regulated by ABA. By exercising control over these processes, the clock can change the plant response to ABA, which in turn can modulate its activity.

The concentration of ABA increases rapidly due to change in water conditions and availability in the cells caused by drought, low temperature, and salinity. Evoked by these stress factors, even a small decrease in the water potential in the roots and/or leaf cells stimulates the synthesis of this hormone. Contribution of ABA in the regulation of gene expression in plant responses to the stressors is currently extensively studied. Recently, the mechanism linking the clock to the drought stress response has been identified. It was demonstrated that in the middle of the day when the transpiration rate is high, above the level of the water supply, causing moderate drought stress, plants trigger a mechanism that allows them to sensitize towards the prevailing conditions (Figure 20.1) (Legnaioli et al., 2009; Castells et al., 2010; Pokhilko et al., 2013). A dry environment promotes the production of ABA that, in turn, induces the expression of a key clock gene, *TOC1*, in a clock-dependent manner. TOC1 protein acts as a negative regulator of the ABA-related gene, *ABAR/CHLH/GUN5*, encoding magnesium-protoporphyrin IX chelatase by binding to its promoter and inhibiting its circadian expression; which, in turn, is essential for the ABA-dependent activation of *TOC1* (Figure 20.1). It has been shown that the H subunit of ABAR may function as the ABA

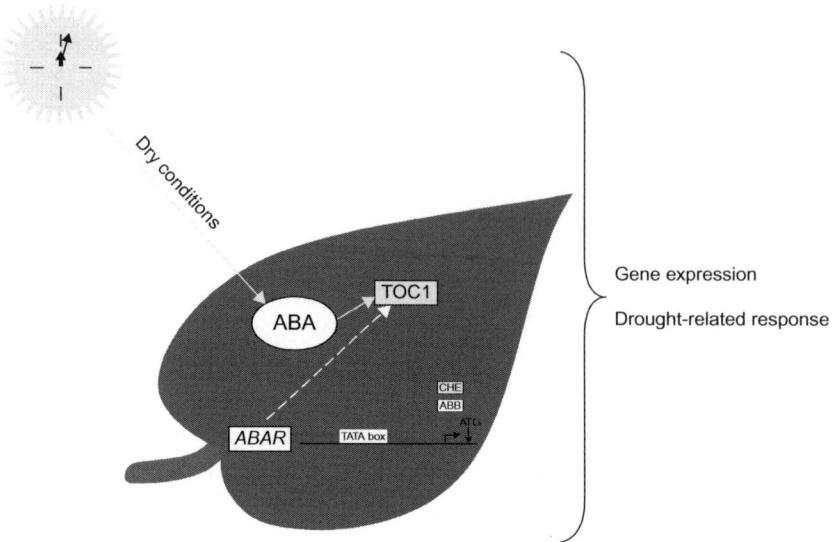

FIGURE 20.1

Graphical representation illustrating the mutual interaction between the clock transcriptional regulator TOC1 and *ABAR* in plant response to drought. In the middle of the day when the transpiration rate is high, above the level of the water supply, causing moderate drought stress, plants promote the production of ABA that, in turn, induces the expression of *TOC1*. TOC1 protein inhibits circadian expression of the ABA-related gene *ABAR* by binding to its promoter, which, in turn, is essential for activation of *TOC1* by ABA. The involvement of other factors such as CHE or ABI3 for TOC1 binding to the promoter appears to be likely.

receptor, wherein specificity of the ABA binding to a particular ABAR subunit has been confirmed (Zhang et al., 2002; Shen et al., 2006; Wu et al., 2009). Interestingly, TOC1 binding to the *ABAR/CHLH/GUN5* promoter is highly dependent on both the clock and the ABA, and can vary during the day (Legnaioli et al., 2009; Castells et al., 2010). Studies carried out on plants expressing the luciferase (*LUC*) gene under a *TOC1* promoter control have shown that ABA strongly induces the expression of *TOC1* only during the subjective day. It was also speculated that in the middle of the day, ABA activity is gated by the circadian clock that precisely establishes the time of TOC1 connecting to the *ABAR* promoter (Castells et al., 2010). It seems likely that TOC1 binding to this promoter can be influenced by other factors such as a TCP clock-associated transcription factor (CHE) or ABA signaling factor (ABI3), whose physical interaction with TOC1 protein has been demonstrated (Kurup et al., 2000; Pruneda-Paz et al., 2009). The validity of *TOC1* expression in the regulation of ABA-mediated response to drought has been demonstrated for *TOC1* and *ABAR* overexpressing and *Arabidopsis RNAi* plants, where significant changes in drought-related response, stomatal function, and transpiration rates have been observed when compared to wild-type plants (Legnaioli et al., 2009).

There is evidence that indicates that other crucial components of the circadian system such as PPR9, 7, and 5 (PSEUDORESPONSE REGULATOR) can regulate the production of ABA, acting as a negative regulator of its biosynthetic pathway (Fukushima et al., 2009). In the triple-knockout plants of *PRR9,7,5 (d975)*, a significant increase in the expression of genes involved in the ABA biosynthesis and a further significant increase in ABA levels were observed.

Looking for links between the clock and ABA signaling pathway in plants, it is worth noting the evidence that points towards the overlap among the circadian transcriptome and ABA-induced cyclic adenosine diphosphate ribose (cADPR), which forms the feedback loop within the clock (Dodd et al., 2007). Studies have indicated an important role of cADPR in molecular ABA responses by inducing a group of ABA-responsive genes (Sánchez et al., 2004). Both cADPR and ABA up-regulated the expression of a number of genes encoding transcription factors associated with the plant responses to ABA and stress such as *DREB1A* and *2 A (DEHYDRATION-RESPONSIVE ELEMENT-BINDING PROTEIN), MYC2, MYB2*, as well as many ABA and stress-related genes including *COR15A* and *B (COLD-REGULATED), RD20, RD22*, and RD29a *(RESPONSE TO DESICATION), ERD15 (EARLY RESPONSE TO DEHYDRATATION)*, and *LT16 (LOW TEMPERATURE)* (Sánchez et al., 2004).

20.5 **Light inputs to the clock**

The best-studied input signals entraining the circadian clock to environmental cycles are light and temperature, wherein the influence of light on the central oscillator components appears to be predominant. Because photosynthesis is dependent on light, it is obvious that the light is the main environmental factor controlling different metabolic pathways in plants, including functioning of the molecular mechanism of the clock. The first stage of the light signal transduction to the oscillator is its perception by the special molecule chromoproteins performing the functions of the photoreceptors. In plants, there are two basic types of photoreceptor proteins: phytochromes, such as phyA, phyB, phyC, phyD, phyE, which are sensitive to red and far-red light, and cryptochromes including CRY1 and CRY2, which are sensitive to blue light (Somers et al., 1998; Kami et al., 2010; Wenden et al., 2011). Currently, a very important role in the transduction of light to clock is attributed to the ZEITLUPE (ZTL) protein acting as a blue-light photoreceptor (Kim et al., 2007).

Although the role of photoreceptors in the clock coordination with the external environmental conditions is well known, the mechanism of the light signal transduction into the central oscillator is still poorly understood. The present data indicate that perception of the light signal and subsequent transfer to the central oscillator has a very large impact on the periodicity of the clock. Further, transcriptional and post-translational mechanisms participate in the regulation of this signaling pathway. The onset of dawn initiates the interaction between the clock components, CCA1 and LHY, and three proteins performing the function of positive regulators in PHY-A pathways such as FAR-RED ELONGATED HYPOCOTYLS3 (FHY3), FAR-RED IMPAIRED RESPONSE 1 (FAR1), and ELONGATED HYPOCOTYL5 (HY5), resulting in inhibition of EARLY FLOWERING 4 (ELF4), the clock central component (Li et al., 2011). In turn, the clock can modulate the expression pattern of the circadian photoreceptors and ultimately gates phytochrome signaling as part of the clock resetting onset (Tóth et al., 2001). Previous studies led to the identification of the proteins EARLY FLOWERING 3 (ELF3), TIME FOR COFFEE (TIC), and FAR-RED ELONGATED HYPOCOTYLS3 (FHY3), which are components of photoreceptors' gating signaling pathway (Covington et al., 2001; Hall et al., 2003; Allen et al., 2006). Interestingly, the first two appear to gate the light signal relayed from both phytochromes and cryptochromes during the subjective night, whereas FHY3 protein specifically regulates the light signal from the phytochrome receptors during the subjective day.

Originally, the primary role in photo-transduction was attributed to the basic photoreceptors, which may interact with each other to supervise entrance of light to the oscillator (Martínez-García et al., 2000; Devlin and Kay, 2000; Tóth et al., 2001). However, studies using quadruple and quintuple photoreceptor mutants, *phyA/phyB/cry1/cry2* and *phyA/phyB/phyC/phyD/phyE*, respectively, challenged this concept. It was demonstrated that light entrance to the clock may occur without the participation of the classical photoreceptors (Yanovsky et al., 2000; Strasser et al., 2010). A hypothetical candidate that may be responsible for light transduction to the oscillator is ZTL, mentioned above.

20.6 Relationships between ROS transcriptional network and the circadian timekeeping system

For several years, an increasing number of experimental data indicate that the common response of plant cells to various environmental stress factors may be an imbalance between the production of toxic oxygen products known as reactive/active oxygen species (ROS/AOS) such as singlet oxygen ($^1O_2$), superoxide anion ($O_2^{\cdot-}$), hydrogen peroxide (H_2O_2), and hydroxyl radical ($^\cdot HO$), and their removal by the appropriate antioxidant systems, known as free radical enzymatic and non-enzymatic scavengers (Oelze et al., 2008).

ROS are characteristic of normal plant metabolism, and one of the most important metabolic processes leading to the generation of ROS in photosynthesis (Sanchez et al., 2011). The increased accumulation of ROS also occurs as a result of action of many factors including biotic and abiotic stresses, atmospheric pollution, and herbicides. Oxidative stress is the state in which there is an overproduction of free radicals and the imbalance between production of ROS and antioxidants (Moller, 2001; Jaspers and Kangasjärvi, 2010). ROS production is dependent on rate of photosynthesis, which, in turn, is dependent on the energy source, i.e., light. It means that expression of ROS-responsive genes may be subject to a diurnal fluctuation (Lai et al., 2012). By studying the potential impact of the circadian clock on the transcriptional network of ROS, it was determined that the clock plays a primary role in maintaining ROS cellular homeostasis and transcriptional response, through the supervision of generation, scavenging, and transcriptional regulations of ROS-responsive genes (Lai et al., 2012). Available data indicate that the clock can regulate 34% of ROS-signaling genes, for which the time-of-day-dependent fluctuation of expression has been demonstrated (Covington et al., 2008). A key role in this regulation is performed by the CIRCADIAN CLOCK-ASSOCIATED 1 (CCA1) component of the central oscillator, which can interact *in vivo* with the Evening Element (EE) in promoters of ROS-signaling genes. It was proved that the mutations in *CCA1* affect the time-of-day-specific phasing of expression of ROS genes leading to disturbances in ROS homeostasis and response to oxidative stress (Lai et al., 2012). Interestingly, only at dawn, when CCA1 level is high, can it control the expression of ROS genes (Lai et al., 2012). It appears that the expression of ROS genes is gated by the circadian clock, which clearly indicates the time of CCA1 binding to ROS gene promoters. In turn, ROS may feed back to interfere with the processes under the control of the clock, for example, by changing the expression of some clock-regulated genes, such as *FKF1* (Lai et al., 2012).

In summary, plants have evolved a complex system to deal with excessive ROS production and scavenging. Although there is a link between the clock and ROS homeostasis and oxidative stress response, the details of reciprocal interactions await elucidation and further research.

20.7 **Contribution of cellular metabolism to circadian network**

In higher plants, output signals generated by the central oscillator lead to the formation rhythms, associated with the various cellular metabolic processes including photosynthesis, redox homeostasis, sugar metabolism, nutrient assimilation, or secondary metabolism (Dodd et al., 2005; Gutiérrez et al., 2008; Graf et al., 2010; Kerwin et al., 2011; Feugier and Satake, 2012; Lai et al., 2012; Haydon et al., 2013). Emerging evidence indicates a close connection of the clock components with the maintenance of metabolic homeostasis through temporal compartmentation of inconsistent metabolic processes (Hotta et al., 2007). Numerous studies have focused on both the precise determination of role for the circadian clock in the supervision of metabolic processes (Fukushima et al., 2009), as well as the participation of metabolism in regulating clock timing.

Abiotic stresses are the main causes of changes in the metabolic processes. As mentioned previously, they are associated with impaired redox homeostasis by significantly increasing ROS levels. In *Arabidopsis*, more than one-third of ROS-related transcripts are under the control of the clock (Covington et al., 2008; Lai et al., 2012). Searching for connectors among metabolism, generation of ROS and response to abiotic stresses, and then their contribution to the clock, researchers have focused on the influence of stress factors on a metabolic co-enzyme nicotinamide adenine dinucleotide (NAD^+), whose rhythmical synthesis has been documented in mammals (Ramsey et al., 2009). It has been reported that NAD^+ and its derivative NADP perform many important functions in cells, as an energy source contributing to redox reaction, and in systems regulating adaptation to abiotic stresses such as salinity, heat shock, and drought (Mittler et al., 2004; Chai et al., 2005, 2006). In addition, NAD^+ is the substrate for ADP-ribosylation and the precursor for cyclic ADP-ribose, which participates in the formation of cADPR (Hashida et al., 2009).

As mentioned above, in plants, the occurrence of an additional loop in cytosol, which is associated with cADPR signaling, has been revealed. ADPR cyclase is involved in the management of $[Ca^{2+}]_{cyt}$ circadian oscillation (Dodd et al., 2007). Inhibition of cADPR by nicotinamide abolished the diurnal oscillations of $[Ca^{2+}]_{cyt}$, and extended the period of expression of the core clock genes such as *CCA1*, *LHY*, and *TOC1* (Sethi et al., 1996). In turn, the circadian oscillations of cADPR are driven by transcriptional feedback loops of the central oscillator (Dodd et al., 2007). It has previously been demonstrated that $[Ca^{2+}]_{cyt}$ concentration and cADPR play an important role in the transmission of signals from ABA and abiotic stresses (Sánchez et al., 2004). The results obtained by Sánchez et al. (2004) indicated that ABA stirs cADPR, which, in turn, induces the expression of ABA-related genes. Microarray data have suggested that cADPR functions as a "bridge" between metabolism and the clock (Dodd et al., 2007).

Further, the action of stress changes in poly(ADP)ribosylation also indicates a connection between metabolic input to the clock and response to external factors. Available evidence suggests

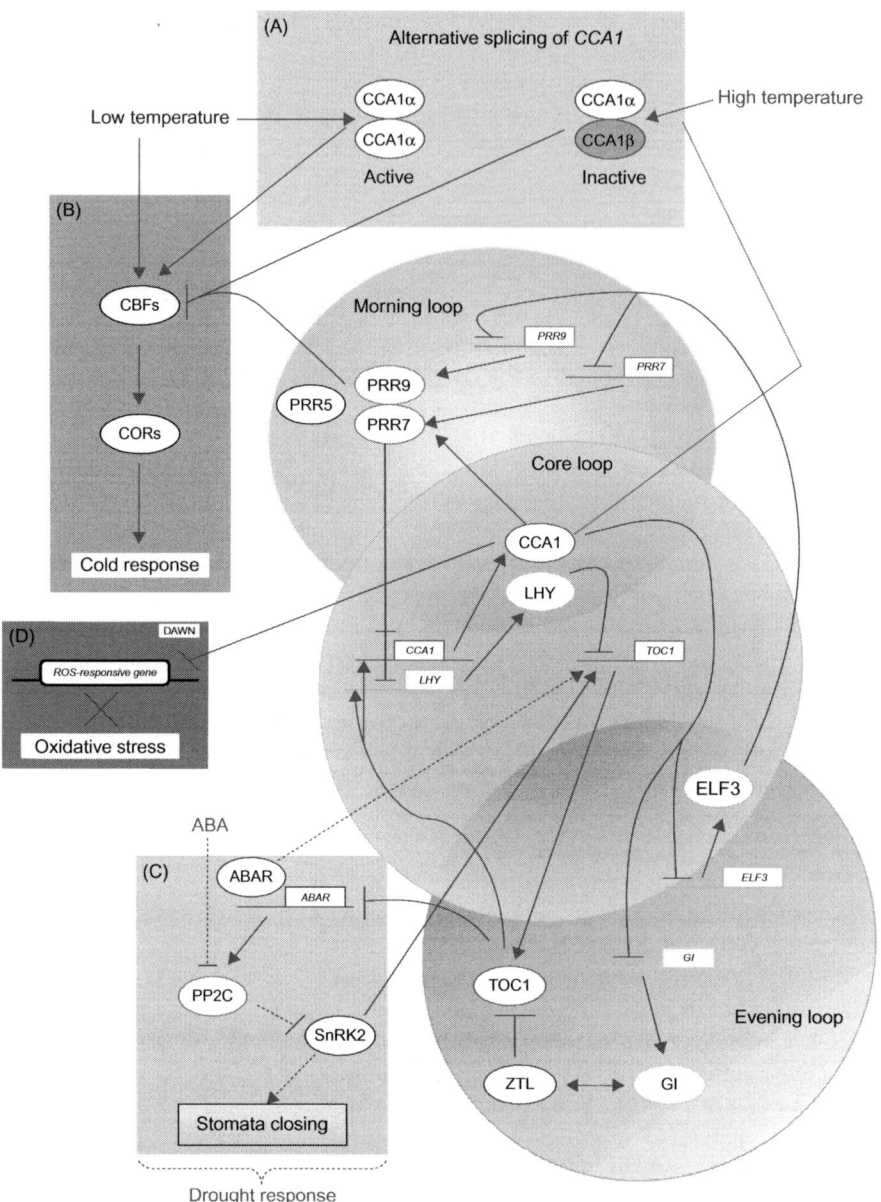

FIGURE 20.2

Schematic representation depicting multidimensional reciprocal relationships among the circadian clock components and the complicated network of plant abiotic stresses (low and high temperature, drought, oxidative stress) and ABA. Positive regulation is represented by solid arrows and negative regulation is

(*Continued*)

that a deficiency of poly(ADP-ribose) polymerases (PARP) causes increased tolerance of plants to abiotic stresses (De Block et al., 2005; Vanderauwera et al., 2007). However, different reasons for increased plant stress tolerance including a reduced consumption of energy (NAD and ATP) and an induction of ABA response have been suggested (De Block et al., 2005; Vanderauwera et al., 2007). Despite numerous studies to determine the relationships among energy homeostasis, the circadian clock, and a changing environment, the total mechanism responsible for these interactions has not been fully elucidated. Understanding the connection between abiotic stress and energy requires multiple additional researches.

20.8　Conclusion and future perspectives

The circadian clock as a time-keeping mechanism is responsible for the translation of external information such as light and temperature and coordination of physiological and metabolic response of any organisms to its ambient environment. In the last decade, there has been a large breakthrough in research into the understanding of the molecular-genetic basis of clock operation. The precise mechanism of action of the clock based on the multiple interlocking feedback loops has been decoded. This allowed the actual model of transcription and translation control. Accomplishment is the knowledge of molecular function of the clock protein components and of the interactions between them in complex networks.

Our knowledge of understanding of the potential crosstalk among the circadian clock and the responses to various abiotic stresses has dramatically improved in the last decade (Figure 20.2). Varied impact of stress on circadian output genes and its unique effect on the components of the central oscillator is well known. On the other hand, the clock by gating supervises the reaction of plants to a wide range of environmental stimuli and allows replay only to favorable at the moment.

Researchers are looking for answers to significant question about how plants integrate the abiotic stress and the circadian clock mechanisms to ensure the appropriate conditions for their growth and development. Detailed analysis of the transcriptome, proteome, and metabolome gives us the possibility of finding an answer to this question, and indicates the relationship between the endogenous timekeeper and the response of plants to environmental variables.

◀ marked as perpendicular lines. (A, B) Temperature-dependent alternative splicing of *CCA1* transcriptional regulator. Under low temperature, a proper splice variant *CCA1α* accumulates. This leads to an increase in the level of the fully functional CCA1α/CCA1α dimer, which in turn binds to the *cis*-regulatory element and initiates transcription of cold-induced genes such as *CBFs*. Conversely, the high temperature leads to the formation of an aberrant splice form *CCA1β* that encodes CCA1β protein lacking the DNA binding MYB domain. CCA1β isoform interacts with CCA1α and prevents the formation of the functional dimer. CCA1α/CCA1β dimer exhibits a limited DNA binding activity and is unable to initiate transcription of downstream genes. (C) Mutual regulation between *TOC1* and *ABAR* expression (detailed explanation is described in the caption to Figure 20.1). (D) Gating of ROS-related gene expression by the key circadian component CCA1. With the coming of dawn, CCA1 transcription factor binds to the Evening Element (EE) in promoters of ROS-signaling genes and inhibits their transcription. CCA1 seems to play a primary role in maintaining ROS cellular homeostasis and transcriptional response.

References

Akman, O.E., Locke, J.C., Tang, S., Carré, I., Millar, A.J., Rand, D.A., 2008. Isoform switching facilitates period control in the *Neurospora crassa* circadian clock. Mol. Syst. Biol. 4, 164.

Alabadi, D., Oyama, T., Yanovsky, M.J., Harmon, F.G., Mas, P., Kay, S.A., 2001. Reciprocal regulation between TOC1 and LHY/CCA1 within the *Arabidopsis* circadian clock. Science 293, 880–883.

Allen, T., Koustenis, A., Theodorou, G., Somers, D.E., Kay, S., Whitelam, G.C., et al., 2006. *Arabidopsis* FHY3 specifically gates phytochrome signaling to the circadian clock. Plant Cell 18, 2506–2516.

Bertoni, G., 2012. Circadian rhythms require proper RNA splicing. Plant Cell 24, 3856.

Bieniawska, Z., Espinoza, C., Schlereth, A., Sulpice, R., Hincha, D.K., Hannah, M.A., 2008. Disruption of the *Arabidopsis* circadian clock is responsible for extensive variation in the cold-responsive transcriptome. Plant Physiol. 147, 263–279.

Bolouri Moghaddam, M.R., Van den Ende, W., 2013. Sweet immunity in the plant circadian regulatory network. J. Exp. Bot. 64, 1439–1449.

Cao, S., Ye, M., Jiang, S., 2005. Involvement of *GIGANTEA* gene in the regulation of the cold stress response in *Arabidopsis*. Plant Cell Rep. 24, 683–690.

Castells, E., Portolés, S., Huang, W., Mas, P., 2010. A functional connection between the clock component TOC1 and abscisic acid signaling pathways. Plant Signal Behav. 5, 409–411.

Chai, M.F., Chen, Q.J., An, R., Chen, Y.M., Chen, J., Wang, X.C., 2005. NADK2, an *Arabidopsis* chloroplastic NAD kinase, plays a vital role in both chlorophyll synthesis and chloroplast protection. Plant Mol. Biol. 59, 553–564.

Chai, M.F., Wei, P.C., Chen, Q.J., An, R., Chen, J., Yang, S., et al., 2006. NADK3, a novel cytoplasmic source of NADPH, is required under conditions of oxidative stress and modulates abscisic acid responses in *Arabidopsis*. Plant J. 47, 665–674.

Covington, M.F., Harmer, S.L., 2007. The circadian clock regulates auxin signaling and responses in *Arabidopsis*. PLoS Biol. 5, e222.

Covington, M.F., Maloof, J.N., Straume, M., Kay, S.A., Harmer, S.L., 2008. Global transcriptome analysis reveals circadian regulation of key pathways in plant growth and development. Genome. Biol. 9, R130.

Covington, M.F., Panda, S., Liu, X.L., Strayer, C.A., Wagner, D.R., Kay, S.A., 2001. ELF3 modulates resetting of the circadian clock in *Arabidopsis*. Plant Cell 13, 1305–1315.

De Block, M., Verduyn, C., De Brouwer, D., Cornelissen, M., 2005. Poly(ADP-ribose) polymerase in plants affects energy homeostasis, cell death and stress tolerance. Plant J. 41, 95–106.

Deng, X., Gu, L., Liu, C., Lu, T., Lu, F., Lu, Z., et al., 2010. Arginine methylation mediated by the *Arabidopsis* homolog of PRMT5 is essential for proper pre-mRNA splicing. Proc. Natl. Acad. Sci. USA 107, 1919–19114.

Devlin, P.F., Kay, S.A., 2000. Cryptochromes are required for phytochrome signaling to the circadian clock but not for rhythmicity. Plant Cell 12, 2499–2510.

Diernfellner, A., Colot, H.V., Dintsis, O., Loros, J.J., Dunlap, J.C., Brunner, M., 2007. Long and short isoforms of *Neurospora* clock protein FRQ support temperature-compensated circadian rhythms. FEBS Lett. 581, 5759–5764.

Dodd, A.N., Salathia, N., Hall, A., Kévei, E., Tóth, R., Nagy, F., et al., 2005. Plant circadian clocks increase photosynthesis, growth, survival, and competitive advantage. Science 309, 630–633.

Dodd, A.N., Jakobsen, M.K., Baker, A.J., Telzerow, A., Hou, S.W., Laplaze, L., et al., 2006. Time of day modulates low-temperature Ca signals in *Arabidopsis*. Plant J. 48, 962–973.

Dodd, A.N., Gardner, M.J., Hotta, C.T., Hubbard, K.E., Dachau, N., Love, J., et al., 2007. The *Arabidopsis* circadian clock incorporates a cADPR-based feedback loop. Science 318, 1789–1792.

Doherty, C.J., Kay, S.A., 2010. Circadian control of global gene expression patterns. Annu. Rev. Genet. 44, 419–444.

Dong, M.A., Farré, E.M., Thomashow, M.F., 2011. Circadian clock-associated 1 and late elongated hypocotyl regulate expression of the C-repeat binding factor (CBF) pathway in *Arabidopsis*. Proc. Natl. Acad. Sci. USA 108, 7241–7246.

Espinoza, C., Bieniawska, Z., Hincha, D.K., Hannah, M.A., 2008. Interactions between the circadian clock and cold-response in *Arabidopsis*. Plant Signal Behav. 3, 593–594.

Espinoza, C., Degenkolbe, T., Caldana, C., Zuther, E., Leisse, A., Willmitzer, L., et al., 2010. Interaction with diurnal and circadian regulation results in dynamic metabolic and transcriptional changes during cold acclimation in *Arabidopsis*. PLoS One 5, e14101.

Farré, E.M., Harmer, S.L., Harmon, F.G., Yanovsky, M.J., Kay, S.A., 2005. Overlapping and distinct roles of PRR7 and PRR9 in the *Arabidopsis* circadian clock. Curr. Biol. 15, 47–54.

Feugier, F.G., Satake, A., 2012. Dynamical feedback between circadian clock and sucrose availability explains adaptive response of starch metabolism to various photoperiods. Front Plant Sci. 3, 305.

Filichkin, S.A., Mockler, T.C., 2012. Unproductive alternative splicing and nonsense mRNAs: a widespread phenomenon among plant circadian clock genes. Biol. Direct 7, 20.

Filichkin, S.A., Triest, H.D., Givan, S.A., Shen, R., Bryant, D.W., Fox, S.E., et al., 2010. Genome-wide mapping of alternative splicing in *Arabidopsis thaliana*. Genome. Res. 20, 45–58.

Finkelstein, R.R., Gampala, S.S., Rock, C.D., 2002. Abscisic acid signaling in seeds and seedlings. Plant Cell 14 (Suppl.), S15–S45.

Fowler, S., Thomashow, M.F., 2002. *Arabidopsis* transcriptome profiling indicates that multiple regulatory pathways are activated during cold acclimation in addition to the CBF cold response pathway. Plant Cell 14, 1675–1690.

Fowler, S.G., Cook, D., Thomashow, M.F., 2005. Low temperature induction of *Arabidopsis* CBF1, 2, and 3 is gated by the circadian clock. Plant Physiol. 137, 961–968.

Fukushima, A., Kusano, M., Nakamichi, N., Kobayashi, M., Hayashi, N., Sakakibara, H., et al., 2009. Impact of clock-associated *Arabidopsis* pseudo-response regulators in metabolic coordination. Proc. Natl. Acad. Sci. USA 106, 7251–7256.

Goodspeed, D., Chehab, E.W., Min-Venditti, A., Braam, J., Covington, M.F., 2012. *Arabidopsis* synchronizes jasmonate-mediated defense with insect circadian behavior. Proc. Natl. Acad. Sci. USA 109, 4674–4677.

Graf, A., Schlereth, A., Stitt, M., Smith, A.M., 2010. Circadian control of carbohydrate availability for growth in *Arabidopsis* plants at night. Proc. Natl. Acad. Sci. USA 107, 9458–9463.

Gutiérrez, R.A., Stokes, T.L., Thum, K., Xu, X., Obertello, M., Katari, M.S., et al., 2008. Systems approach identifies an organic nitrogen-responsive gene network that is regulated by the master clock control gene *CCA1*. Proc. Natl. Acad. Sci. USA 105, 4939–4944.

Hall, A., Bastow, R.M., Davis, S.J., Hanano, S., McWatters, H.G., Hibberd, V., et al., 2003. The *TIME FOR COFFEE* gene maintains the amplitude and timing of *Arabidopsis* circadian clocks. Plant Cell 15, 2719–2729.

Harmer, S.L., Hogenesch, J.B., Straume, M., Chang, H.S., Han, B., Zhu, T., et al., 2000. Orchestrated transcription of key pathways in *Arabidopsis* by the circadian clock. Science 290, 2110–2113.

Hashida, S.N., Takahashi, H., Uchimiya, H., 2009. The role of NAD biosynthesis in plant development and stress responses. Ann. Bot. 103, 819–824.

Haydon, M.J., Hearn, T.J., Bell, L.J., Hannah, M.A., Webb, A.A.R., 2013. Metabolic regulation of circadian clocks. Semi. Cell Dev. Biol. 24, 414–421.

Hazen, S.P., Naef, F., Quisel, T., Gendron, J.M., Chen, H., Ecker, J.R., et al., 2009. Exploring the transcriptional landscape of plant circadian rhythms using genome tiling arrays. Genome. Biol. 10, R17.

Hemmes, H., Henriques, R., Jang, I.C., Kim, S., Chua, N.H., 2012. Circadian clock regulates dynamic chromatin modifications associated with *Arabidopsis* CCA1/LHY and TOC1 transcriptional rhythms. Plant Cell Physiol. 53, 2016–2029.

Henriques, R., Mas, P., 2013. Chromatin remodeling and alternative splicing: Pre- and post-transcriptional regulation of the *Arabidopsis* circadian clock. Semin. Cell Dev. Biol. 24, 399–406.

Hong, S., Song, H.R., Lutz, K., Kerstetter, R.A., Michael, T.P., McClung, C.R., 2010. Type II protein arginine methyltransferase 5 (PRMT5) is required for circadian period determination in *Arabidopsis thaliana*. Proc. Natl. Acad. Sci. USA 107, 21211–21216.

Hotta, C.T., Gardner, M.J., Hubbard, K.E., Baek, S.J., Suhita, D., Dodd, A.N., et al., 2007. Modulation of environmental responses of plants by circadian clocks. Plant Cell Environ. 30, 333–349.

Ibáñez, C., Ramos, A., Acebo, P., Contreras, A., Casado, R., Allona, I., et al., 2008. Overall alteration of circadian clock gene expression in the chestnut cold response. PLoS One 3, e3567.

Ibáñez, C., Kozarewa, I., Johansson, M., Ogren, E., Rohde, A., Eriksson, M.E., 2010. Circadian clock components regulate entry and affect exit of seasonal dormancy as well as winter hardiness in *Populus* trees. Plant Physiol. 153, 1823–1833.

James, A.B., Syed, N.H., Bordage, S., Marshall, J., Nimmo, G.A., Jenkins, G.I., et al., 2012. Alternative splicing mediates responses of the *Arabidopsis* circadian clock to temperature changes. Plant Cell 24, 961–981.

Jaspers, P., Kangasjärvi, J., 2010. Reactive oxygen species in abiotic stress signaling. Physiol Plantarum 138, 405–413.

Kami, C., Lorrain, S., Hornitschek, P., Fankhauser, C., 2010. Light-regulated plant growth and development. Curr. Top. Dev. Biol. 91, 29–66.

Kant, P., Gordon, M., Kant, S., Zolla, G., Davydov, O., Heimer, Y.M., et al., 2008. Functional-genomics-based identification of genes that regulate *Arabidopsis* responses to multiple abiotic stresses. Plant Cell Environ. 31, 697–714.

Kerwin, R.E., Jimenez-Gomez, J.M., Fulop, D., Harmer, S.L., Maloof, J.N., Kliebenstein, D.J., 2011. Network quantitative trait loci mapping of circadian clock outputs identifies metabolic pathway-to-clock linkages in *Arabidopsis*. Plant Cell 23, 471–485.

Kim, J.S., Jung, H.J., Lee, H.J., Kim, K.A., Goh, C.H., Woo, Y., et al., 2008. Glycine-rich RNA-binding protein7 affects abiotic stress responses by regulating stomata opening and closing in *Arabidopsis thaliana*. Plant J. 55, 455–466.

Kim, W.Y., Fujiwara, S., Suh, S.S., Kim, J., Kim, Y., Han, L., et al., 2007. ZEITLUPE is a circadian photoreceptor stabilized by GIGANTEA in blue light. Nature 449, 356–360.

Kinmonth-Schultz, H.A., Golembeski, G.S., Imaizumi, T., 2013. Circadian clock-regulated physiological outputs: dynamic responses in nature. Semin. Cell Dev. Biol. 24, 407–413.

Kojima, S., Shingle, D.L., Green, C.B., 2011. Post-transcriptional control of circadian rhythms. J. Cell Sci. 124, 311–320.

Kolmos, E., Schoof, H., Plümer, M., Davis, S.J., 2008. Structural insights into the function of the core–circadian factor TIMING OF CAB2 EXPRESSION 1 (TOC1). J. Circadian Rhythms 6, 3.

Kurup, S., Jones, H.D., Holdsworth, M.J., 2000. Interactions of the developmental regulator ABI3 with proteins identified from developing *Arabidopsis* seeds. Plant J. 21, 143–155.

Lai, A.G., Doherty, C.J., Nueller-Roeber, B., Kay, S.A., Schippers, J.H.M., Dijkwel, P.P., 2012. CIRCADIAN CLOCK-ASSOCIATED 1 regulates ROS homeostasis and oxidative stress responses. PNAS 109, 17129–17134.

Legnaioli, T., Cuevas, J., Mas, P., 2009. TOC1 functions as a molecular switch connecting the circadian clock with plant responses to drought. EMBO J. 28, 3745–3757.

Li, G., Siddiqui, H., Teng, Y., Lin, R., Wan, X.Y., Li, J., et al., 2011. Coordinated transcriptional regulation underlying the circadian clock in *Arabidopsis*. Nat. Cell Biol. 13, 616–622.

Liu, H., Wang, H., Gao, P., Xü, J., Xü, T., Wang, J., et al., 2009. Analysis of clock gene homologs using unifoliolates as target organs in soybean (*Glycine max*). J. Plant Physiol. 166, 278−289.

Locke, J.C.W., Southern, M.M., Kozma-Bognar, L., Hibberd, V., Brown, P.E., Turner, M.S., et al., 2005. Extension of a genetic network model by iterative experimentation and mathematical analysis. Mol. Syst. Biol. 1 (2005), 0013.

Locke, J.C.W., Kozma-Bognár, L., Gould, P.D., Fehér, B., Kevei, E., Nagy, F., et al., 2006. Experimental validation of a predicted feedback loop in the multi-oscillator clock of *Arabidopsis thaliana*. Mol. Syst. Biol. 2, 59.

Maibam, P., Nawkar, G.M., Park, J.H., Sahi, V.P., Lee, S.Y., Kang, C.H., 2013. The influence of light quality, circadian rhythm, and photoperiod on the CBF-mediated freezing tolerance. Int. J. Mol. Sci. 14, 11527−11543.

Majercak, J., Sidote, D., Hardin, P.E., Edery, I., 1999. How a circadian clock adapts to seasonal decreases in temperature and day length. Neuron 24, 219−230.

Martínez-García, J.F., Huq, E., Quail, P.H., 2000. Direct targeting of light signals to a promoter element-bound transcription factor. Science 288, 859−863.

Maruyama, K., Sakuma, Y., Kasuga, M., Ito, Y., Seki, M., Goda, H., et al., 2004. Identification of cold-inducible downstream genes of the *Arabidopsis* DREB1A/CBF3 transcriptional factor using two microarray systems. Plant J. 38, 982−993.

Mastrangelo, A.M., Belloni, S., Barilli, S., Ruperti, B., Di Fonzo, N., Stanca, A.M., et al., 2005. Low temperature promotes intron retention in two *e-cor* genes of durum wheat. Planta 221, 705−715.

McClung, C.R., 2013. Beyond *Arabidopsis*: the circadian clock in non-model plant species. Semin. Cell Dev. Biol. 24, 430−436.

McClung, C.R., Davis, S.J., 2010. Ambient thermometers in plants: from physiological outputs towards mechanisms of thermal sensing. Curr. Biol. 20, 1086−1092.

Mikkelsen, M.D., Thomashow, M.F., 2009. A role for circadian evening elements in cold-regulated gene expression in *Arabidopsis*. Plant J. 60, 328−339.

Mittler, R., Vanderauwera, S., Gollery, M., Van Breusegem, F., 2004. Reactive oxygen gene network of plants. Trends Plant Sci. 9, 490−498.

Mizoguchi, T., Wheatley, K., Hanzawa, Y., Wright, L., Mizoguchi, M., Song, H.R., et al., 2002. LHY and CCA1 are partially redundant genes required to maintain circadian rhythms in *Arabidopsis*. Dev. Cell 2, 629−641.

Mizuno, T., Yamashino, T., 2008. Comparative transcriptome of diurnally oscillating genes and hormone-responsive genes in *Arabidopsis thaliana*: insight into circadian clock-controlled daily responses to common ambient stresses in plants. Plant Cell Physiol. 49, 481−487.

Moller, I.M., 2001. Plant mitochondria and oxidative stress: electron transport, NADPH turnover, and metabolism of reactive oxygen species. Annu. Rev. Plant Physiol. Plant Mol. Biol. 52, 561−591.

Murakami, M., Tago, Y., Yamashino, T., Mizuno, T., 2007. Comparative overviews of clock associated genes of *Arabidopsis thaliana* and *Oryza sativa*. Plant Cell Physiol. 48, 110−121.

Nagy, F., Kay, S.A., Chua, N.H., 1988. A circadian lock regulates transcription of the wheat *Cab-1* gene. Genes. Dev. 2, 376−382.

Nakamichi, N., Kusano, M., Fukushima, A., Kita, M., Ito, S., Yamashino, T., et al., 2009. Transcript profiling of an *Arabidopsis* PSEUDO RESPONSE REGULATOR arrhythmic triple mutant reveals a role for the circadian clock in cold stress response. Plant Cell Physiol. 50, 447−462.

Nakashima, K., Ito, Y., Yamaguchi-Shinozaki, K., 2009. Transcriptional regulatory networks in response to abiotic stresses in *Arabidopsis* and Grasses. Plant Physiol. 149, 88−95.

Nozue, K., Covington, M.F., Duek, P.D., Lorrain, S., Fankhauser, C., Harmer, S.L., et al., 2007. Rhythmic growth explained by coincidence between internal and external cues. Nature 448, 358−361.

Oelze, M.L., Kandlbinder, A., Dietz, K.J., 2008. Redox regulation and overreduction control in the photo-synthesizing cell: complexity in redox regulatory networks. Biochim. Biophys. Acta 1780, 1261–1272.

Palusa, S.G., Ali, G.A., Reddy, A.S.N., 2007. Alternative splicing of pre-mRNAs of *Arabidopsis* serine/argi-nine-rich proteins: regulation by hormones and stresses. Plant J. 49, 1091–1107.

Petrillo, E., Sanchez, S.E., Kornblihtt, A.R., Yanovsky, M.J., 2011. Alternative splicing adds a new loop to the circadian clock. Commun. Integr. Biol. 4, 284–286.

Pokhilko, A., Mas, P., Millar, A.J., 2013. Modelling the widespread effects of TOC1 signalling on the plant circadian clock and its outputs. BMC Syst. Biol. 7, 23.

Pruneda-Paz, J.L., Breton, G., Para, A., Kay, S.A., 2009. A functional genomics approach reveals CHE as a novel component of the *Arabidopsis* circadian clock. Science 323, 481–1485.

Ramos, A., Pérez-Solís, E., Ibáñez, C., Casado, R., Collada, C., Gómez, L., et al., 2005. Winter disruption of the circadian clock in chestnut. Proc. Natl. Acad. Sci. USA 102, 7037–7042.

Ramsey, K.M., Yoshino, J., Brace, C.S., Abrassart, D., Kobayashi, Y., Marcheva, B., et al., 2009. Circadian clock feedback cycle through NAMPT-mediated NAD + biosynthesis. Science 324, 651–654.

Rund-Kangisser, S., Yakir, E., Green, R., 2013. Proteasomal regulation of CIRCADIAN CLOCK ASSOCIATED 1 (CCA1) stability is part of the complex control of CCA1. Plant Signal Behav. 8, e23206.

Salomé, P.A., Weigel, D., McClung, C.R., 2010. The role of the *Arabidopsis* morning loop components CCA1, LHY, PRR7 and PRR9 in temperature compensation. Plant Cell 22, 3650–3661.

Sanchez, A., Shin, J., Davis, S.J., 2011. Abiotic stress and the plant circadian clock. Plant Signal Behav. 6, 223–231.

Sánchez, J.P., Duque, P., Chua, N.H., 2004. ABA activates ADPR cyclase and cADPR induces a subset of ABA-responsive genes in *Arabidopsis*. Plant J. 38, 381–395.

Sanchez, S.E., Petrillo, E., Beckwith, E.J., Zhang, X., Rugnone, M.L., Hernando, C.E., et al., 2010. A methyl transferase links the circadian clock to the regulation of alternative splicing. Nature 468, 12–116.

Schaffer, R., Ramsay, N., Samach, A., Corden, S., Putterill, J., Carré, I.A., et al., 1998. The late elongated hypocotyl mutation of *Arabidopsis* disrupts circadian rhythms and the photoperiodic control of flowering. Cell 93, 1219–1229.

Schmidt, F., Marnef, A., Cheung, M.K., Wilson, I., Hancock, J., Staiger, D., et al., 2011. A proteomic analysis of oligo(dT)-bound mRNP containing oxidative stress-induced *Arabidopsis thaliana* RNA-binding proteins ATGRP7 and ATGRP8. Mol. Biol. Rep. 37, 839–845.

Schoning, J.C., Streitner, C., Meyer, I.M., Gao, Y., Staiger, D., 2008. Reciprocal regulation of glycine-rich RNA-binding proteins via an interlocked feedback loop coupling alternative splicing to nonsense-mediated decay in *Arabidopsis*. Nucleic Acids Res 36, 6977–6987.

Schoning, J.C., Streitner, C., Page, D.R., Hennig, S., Uchida, K., Wolf, E., et al., 2007. Auto-regulation of the circadian slave oscillator component AtGRP7 and regulation of its targets is impaired by a single RNA rec-ognition motif point mutation. Plant J. 52, 119–1130.

Seo, P.J., Kim, M.J., Ryu, J.Y., Jeong, E.Y., Park, C.M., 2011. Two splice variants of the IDD14 transcription factor competitively form nonfunctional heterodimers which may regulate starch metabolism. Nat. Commun. 2, 303.

Seo, P.J., Park, M.J., Lim, M.H., Kim, S.G., Lee, M., Baldwin, I.T., et al., 2012. A self-regulatory circuit of CIRCADIAN CLOCK-ASSOCIATED1 underlies the circadian clock regulation of temperature responses in *Arabidopsis*. Plant Cell 24, 2427–2442.

Sethi, J.K., Empson, R.M., Galione, A., 1996. Nicotinamide inhibits cyclic ADP-ribose-mediated calcium sig-naling in sea urchin eggs. Biochem. J. 319, 613–617.

Shen, Y.Y., Wang, X.F., Wu, F.Q., Du, S.Y., Cao, Z., Shang, Y., et al., 2006. The Mg-chelatase H subunit is an abscisic acid receptor. Nature 443, 823–826.

Shin, J., Davis, S.J., 2010. Recent advances in computational modeling as a conduit to understand the plant circadian clock. F1000 Biol. Rep. 2, 49.

Shin, J., Heidrich, K., Sanchez-Villarreal, A., Parker, J.E., Davis, S.J., 2012. TIME FOR COFFE represses accumulation of the MYC2 transcription factor to provide time-of-day regulation of jasmonate signaling in *Arabidopsis*. Plant Cell 24, 2470−2482.

Somers, D.E., Devlin, P.F., Kay, S.A., 1998. Phytochromes and cryptochromes in the entrainment of the *Arabidopsis* circadian clock. Science 282, 1488−1490.

Staiger, D., Koster, T., 2011. Spotlight on post-transcriptional control in the circadian system. Cell Mol. Life Sci. 68, 71−83.

Staiger, D., Allenbach, L., Salathia, N., Fiechter, V., Davis, S.J., Millar, A.J., 2003. *Arabidopsis SRR1* gene mediates phyB signaling and is required for normal circadian clock function. Genes. Dev. 17, 256−268.

Strasser, B., Sánchez-Lamas, M., Yanovsky, M.J., Casal, J.J., Cerdán, P.D., 2010. *Arabidopsis thaliana* life without phytochromes. Proc. Natl. Acad. Sci. USA 107, 4776−4781.

Syed, N.H., Kalyna, M., Marquez, Y., Barta, A., Brown, J.W., 2012. Alternative splicing in plants-coming of age. Trends Plant Sci. 17, 616−623.

Tóth, R., Kevei, E., Hall, A., Millar, A.J., Nagy, F., Kozma-Bognár, L., 2001. Circadian clock-regulated expression of phytochrome and cryptochrome genes in *Arabidopsis*. Plant Physiol. 127, 1607−1616.

Troncoso-Ponce, M.A., Mas, P., 2012. Newly described components and regulatory mechanisms of circadian clock function in *Arabidopsis thaliana*. Mol. Plant 5, 545−553.

Van Buskirk, H.A., Thomashow, M.F., 2006. *Arabidopsis* transcription factors regulating cold acclimation. Physiol. Plantarum 126, 72−80.

Vanderauwera, S., De Block, M., Van de Steene, N., van de Cotte, B., Metzlaff, M., Van Breusegem, F., 2007. Silencing of poly(ADP-ribose) polymerase in plants alters abiotic stress signal transduction. Proc. Natl. Acad. Sci. USA 104, 15150−15155.

Wang, L., Kim, J., Somers, D.E., 2013. Transcriptional corepressor TOPLESS complexes with pseudoresponse regulator proteins and histone deacetylases to regulate circadian transcription. Proc. Natl. Acad. Sci. USA 110, 761−766.

Wang, W., Barnaby, J.Y., Tada, Y., Li, H., Tor, M., Caldelari, D., et al., 2011. Timing of plant immune responses by a central circadian regulator. Nature 470, 110−114.

Wang, Z.Y., Tobin, E.M., 1998. Constitutive expression of the CIRCADIAN CLOCK ASSOCIATED 1 (CCA1) gene disrupts circadian rhythms and suppresses its own expression. Cell 93, 1207−1217.

Wenden, B., Kozma-Bognár, L., Edwards, K.D., Hall, A.J., Locke, J.C., Millar, A.J., 2011. Light inputs shape the *Arabidopsis* circadian system. Plant J. 66, 480−491.

Wijnen, H., Young, M.W., 2006. Interplay of circadian clocks and metabolic rhythms. Annu. Rev. Genet. 40, 409−448.

Wu, F.Q., Xin, Q., Cao, Z., Liu, Z.Q., Du, S.Y., Mei, C., et al., 2009. The magnesium-chelatase H subunit binds abscisic acid and functions in abscisic acid signaling: new evidence in *Arabidopsis*. Plant Physiol. 150, 1940−1954.

Xu, X., Xie, Q., McClung, C.R., 2010. Robust circadian rhythms of gene expression in *Brassica rapa* tissue culture. Plant Physiol. 153, 841−850.

Yakir, E., Hilman, D., Hassidim, M., Green, R.M., 2007. CIRCADIAN CLOCK ASSOCIATED1 transcript stability and the entrainment of the circadian clock in *Arabidopsis*. Plant Physiol. 145, 925−932.

Yanovsky, M.J., Whitela, G.C., Casal, J.J., 2000. hy3-1 retains inductive responses of phytochrome A. Plant Physiol. 123, 235−242.

Yerushalmi, S., Green, R.M., 2009. Evidence for the adaptive significance of circadian rhythm. Ecol. Lett. 12, 970−981.

Yon, F., Seo, P.J., Ryu, J., Ch, Park, Baldwin, I.T., Kim, S.G., 2012. Identification and characterization of circadian clock genes in a native tobacco, *Nicotina attenuate*. BMC Plant Biol. 12, 172.

Zeilinger, M.N., Farré, E.M., Taylor, S.R., Kay, S.A., Doyle 3rd, F.J., 2006. A novel computational model of the circadian clock in *Arabidopsis* that incorporates PRR7 and PRR9. Mol. Syst. Biol. 2, 58.

Zhang, D.P., Wu, Z.Y., Li, X.Y., Zhao, Z.X., 2002. Purification and identification of a 42-kilodalton abscisic acid-specific-binding protein from epidermis of broad bean leaves. Plant Physiol. 128, 714–725.

Zhu, J.K., 2002. Salt and drought stress signal transduction in plants. Annu. Rev. Plant Biol. 53, 247–273.

Development of Water Saving Techniques for Sugarcane (*Saccharum officinarum* L.) in the Arid Environment of Punjab, Pakistan

Abdul Gaffar Sagoo, Abdul Hannan, Muhammad Aslam, Ejaz Ahmed Khan, Amir Hussain, Imam Bakhsh, Muhammad Arif and Muhammad Waqas

21.1 Introduction

Sugarcane, one of the most important cash crops, plays a vital role in the national economy of Pakistan. It is an important source of income and employment for farmers, particularly during the winter season. The challenge of increasing sugar from sugarcane has been erratic, and there has been insufficient rainfall distribution to support the entire crop cycle (Pires et al., 2008; Cheavegatti-Gianotto et al., 2011). Sugarcane requires about 1200 mm of annual rainfall equally distributed throughout the entire growing period. The sugarcane crop requires 88–118 kg water/kg cane and 884–1157 kg water/kg sugar to produce plant and ratoon crops (Ashok et al., 2011). In Pakistan, it has been estimated that rain meets only 25% of the water needs of the crop; the remaining 75% of water is met through additional irrigation, but water is not always available in the desired quantities and crops frequently suffer from lack of available soil moisture. The prevalence of drought during the peak growth period of sugarcane results in a considerable reduction in its yield ability. Drought remains a research topic of high priority for agricultural scientists (Medici et al., 2014) causing reduction in growth and development of plants due to the cessation of cell expansion directly related to carbon assimilation and photosynthetic rates (Benesová et al., 2012; Zingaretti et al., 2012). In order to realize the full benefits of the land and environmental resources, it is necessary to place the plants in the field in such a pattern that there is the least competition among them for essential growth factors.

Proper orientation of plants in the field plays a significant role in the development and functioning of vital plant organs. Planting geometry provides the right direction of plants in the field in the form of multi-row strips, which facilitate efficient and expeditious intercultural operations, conserve irrigation water with saving in labor, and permit systematic planting and handling of intercrops.

P. Ahmad (Ed): Emerging Technologies and Management of Crop Stress Tolerance, Volume 1.
DOI: http://dx.doi.org/10.1016/B978-0-12-800876-8.00021-7

Plant spacing accordingly affects the number of plants, plant height, plant health, and plant canopy, which are directly associated with plant yield. The present study, therefore, was designed to investigate the impact of soil moisture depletion and planting geometry on the yield and quality of sugar cane under arid climatic conditions in Pakistan on two soils: silty clay and sandy loam.

21.2 Methodology

Sugarcane (*Saccharum officinarum* L.) is a tropical crop of long duration and requires abundant water. Increasing urban growth and environmental concerns are limiting the amount of water available for agricultural use. In Pakistan, especially in Punjab and Kyber Pakhtunkhwa, the sugarcane lies in areas where its successful growth needs artificial irrigation. Water is not always available in desired quantities during the long growth period of the crop and drought conditions result in considerable loss of yield. To overcome drought stress against the sugarcane crop, a factorial arrangement was laid out in randomized complete block design (RCBD) with four replications. The net plot size was 6×4 m^2. Each year the crop was planted during the first week of September and harvested in the first week of December the following year. The seed was used at the rate of 70,000 double-budded setts ha^{-1}. A cane cultivar, "HSF 240," was used as test variety. Chemical fertilizer was applied at the rate of 200, 200, and 100 kg NPK ha^{-1} on the farm of urea, P$_2$O$_5$, and K$_2$O, respectively. All the phosphorus, potassium, and one-quarter of total nitrogen were applied at the time of sowing while the remaining nitrogen was applied in two equal splits each at completion of germination at the end of February and at the start of cane formation at the end of March. The crop was kept free of weeds. All agronomic practices were kept normal and uniform for all the treatments. The experiment comprised the treatments shown in Table 21.1. Water was applied to the respective plots as soon as the desired available soil moisture depletion level was reached in the soil within the crop root zone.

21.2.1 Soil sampling and determination of moisture content

The gravimetric procedure of direct soil water measurement was applied to determine the water contents in the soil. Soil sampling for soil moisture measurement was carried out regularly on alternate days keeping in view the weather conditions from the first week of May to the last week of

Table 21.1 Treatment Undertaken in the Experiment

Available Soil Moisture Depletion Levels (ASMDL)		Planting Geometries	
ASMDL$_1$	20%	G-60	60 cm spaced single row strips
ASMDL$_2$	40%	G-75	75 cm spaced single row strips
ASMDL$_3$	60%	G-30/90	30/90 cm spaced paired row strips
ASMDL$_4$	80%	G-30/120	30/120 cm spaced paired row strips

September. Composite soil samples at depth intervals of 30 cm up to 150 cm were taken from randomly located sites per plot for moisture determination. The soil samples were dried in an oven at $105-110°C$ for 24 hours.

21.2.2 Moisture percentage on volume basis

The moisture content percentage on volume basis was computed with the following details:

$$O_w = \frac{(WS_1 - WS_2) \times 100}{WS_2} \qquad (21.1)$$

where:

O_W = soil moisture percentage on dry weight basis
WS_1 = soil sample weight before oven drying
WS_2 = soil sample weight after oven drying

The bulk density of soil was calculated as below:

$$P_b = \frac{W_d}{V_i} \qquad (21.2)$$

where:

P_b = bulk density of soil in g/cm^3
W_d = weight of oven dry soil core in g
V_i = total volume of soil and voids in cm^3

$$O_v\% = \frac{P_b \times O_W}{P_W} \qquad (21.3)$$

where:

O_v = soil moisture content on a volume basis in percent
P_b = bulk density of soil in g/cm^3
P_W = density of water in g/cm^3

$$ASMDL = \frac{RAW \times 100}{AW} \qquad (21.4)$$

where:

ASMDL = available soil moisture depletion level in percent
RAW = readily available water in cm
AW = available water in cm; and

AW and RAW are defined as under:

$$AW = \frac{D_{r \cdot z}(F_c - P_{w \cdot p})}{100} \qquad (21.5)$$

where:

$D_{r.z}$ = depth of root zone in cm
F_c = field capacity in percent by volume
$P_{w.p}$ = permanent wilting point in percent by volume

$$RAW = \frac{D_{r.z}(F_c - O_c)}{100} \qquad (21.6)$$

where O_c = critical soil water content in percent by volume.

By combining Eqs. 21.4, 21.5, and 21.6 we get critical soil water contents in percent by volume for different ASMD levels. Critical soil water contents in percent by volume basis (O_c) were calculated by the above-mentioned formulae and are given in Table 21.2.

21.2.3 Measurement of depth of irrigation water applied

The irrigation requirements of each plot were predetermined on the basis of soil sampling, and irrigation was applied at four different ASMD levels, being 20, 40, 60, and 80% depletion of available moisture in the root zone to bring back the soil moisture to field capacity (F_c); a measured quantity of water was applied to each treatment. A cut throat flume ($3' \times 8''$) was installed in the water course at the field entrance to measure irrigation water applied to each plot. Water was allowed to flow up to 10 minutes in the non-experimental area to maintain constant flow of water. The depth of irrigation water applied to each plot (Table 21.3) was calculated from the pre-irrigation soil moisture content in the root zone by using the following relationships:

$$d_w = \frac{D_{r.z}(F_c - O_i)}{100}$$

Table 21.2 Critical Soil Water Contents in Percent by Volume for Different ASMD levels

Available Soil Moisture Depletion Levels (ASMDL)	Critical Soil Water Contents Percent by Volume Basis (O_c) Calculated by Above Formulae			
	Location-I		Location-II	
	2003–2004	2004–2005	2003–2004	2004–2005
ASMDL1 = 20%	21.4	21.8	14.0	16.0
ASMDL2 = 40%	18.9	19.3	12.3	14.1
ASMDL3 = 60%	16.5	16.8	10.6	12.2
ASMDL4 = 80%	14.0	14.3	8.9	10.3

Irrigation was applied to respective plots as soon as the desired available soil moisture depletion level reached in the soil in the crop root zone.

where:

d_w = depth of water to be applied in mm
$D_{r.z}$ = depth of root zone in mm
F_c = field capacity in percent by volume
O_i = soil moisture content before irrigation in percent by volume

The time required to supply the required depth of irrigation water to each plot was calculated with the help of following equation:

$$t = \frac{d(a)}{q}$$

where:

t = time in minutes
d = depth of water in cm
a = area in m^2
q = discharge of irrigation water in liter/sec

In a well-leveled field, a border strip irrigation system was adopted to check conveyance losses and conserve application and distribution efficiencies.

Table 21.3 Water Applied to Different Treatments During Both Years at Both Locations

Available Soil Moisture Depletion Levels (ASMDL)	Location-I (Silty Clay Soil) 2003–2004			Location-II (Sandy Loam Soil) 2003–2004		
	Total Number of Irrigations Applied	Depth of Irrigation (mm)	Irrigation + Rainfall (mm)	Total Number of Irrigations Applied	Depth of Irrigation (mm)	Irrigation + Rainfall (mm)
ASMDL1 = 20%	55	36.93	2359.15	93	25.71	2820.76
ASMDL2 = 40%	16	73.86	1509.76	28	51.42	1869.49
ASMDL3 = 60%	15	110.80	1990	22	77.13	2126.59
ASMDL4 = 80%	12	147.72	2100.64	16	102.84	2075.17
Total rainfall received = 328.00 mm				Total rainfall received = 373.23mm		
	Location-I 2004–2005			Location-II 2004–2005		
ASMDL1 = 20%	50	37.17	2442.5	87	28.92	2896.28
ASMDL2 = 40%	14	74.33	1624.62	25	57.84	1826.24
ASMDL3 = 60%	13	111.50	2033.5	20	86.76	2115.44
ASMDL4 = 80%	11	148.65	2219.15	14	115.68	1999.76
Total rainfall received = 584.00 mm				Total rainfall received = 380.24 mm		

21.2.4 Soil analyses of experimental sites

Soil analyses of experimental sites were determined using procedures of all three experiments and are presented in Table 21.4.

Table 21.4 Physicochemical Characteristics of Soils at Experimental Sites

Particular		Location-I		Location-II	
		2003–2004	2004–2005	2003–2004	2004–2005
Soil Texture	Unit	Silty Clay	Silty Clay	Sandy Loam	Sandy Loam
N	%	0.03	0.04	0.044	0.049
P	ppm	8	8.5	3.55	4.75
K	ppm	80	92.5	55	60
Field capacity	% by volume	23.85	24.24	15.71	17.97
Bulk density	g/cm^{-3}	1.3	1.35	1.4	1.38
Permanent wilting point	% by volume	11.54	11.85	7.14	8.33
pH (1:5)	1–14	8	8.1	7.7	8
EC	$dS\ m^{-1}$	4.6	5.2	1	1

21.2.5 Observations

The following observations were recorded through standard procedures:

- Germination percentage
- Number of shoots (m^{-2})
- Crop growth rate ($g\ m^{-2}\ day^{-1}$)
- Leaf area index
- Total leaf area duration (days)
- Net assimilation rate ($g\ m^{-2}\ day^{-1}$)
- Plant height (m)
- Weight per stripped cane (kg)
- Mean stripped cane length (m)
- Mean cane diameter (cm)
- Unstrapped cane yield ($t\ ha^{-1}$)
- Stripped cane yield ($t\ ha^{-1}$)
- Harvest index percentage
- Sugar recovery percentage
- Sugar yield ($t\ ha^{-1}$)
- Net field benefit ($Rs\ ha^{-1}$)

The weather data of two locations are shown in Figures 21.1–21.3.

21.2.6 Statistical analysis

The observations were recorded using standard procedures. The data were analyzed statistically using Fisher's analysis of variance technique. A least significant difference (LSD) test at $0.05p$ was

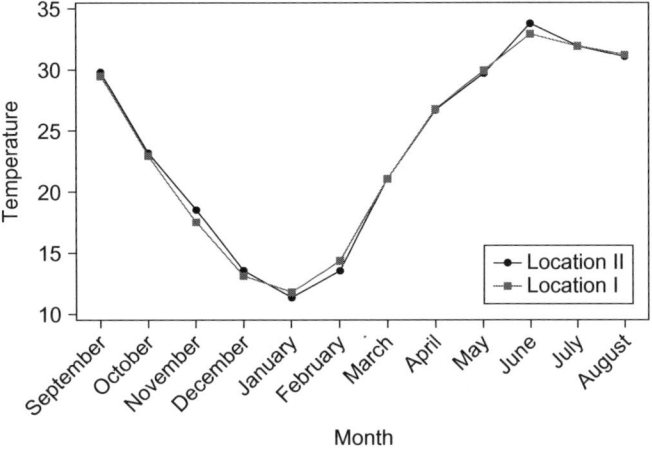

FIGURE 21.1

Temperature data of two locations for 2003–2004 and 2004–2005 (average of 2 years).

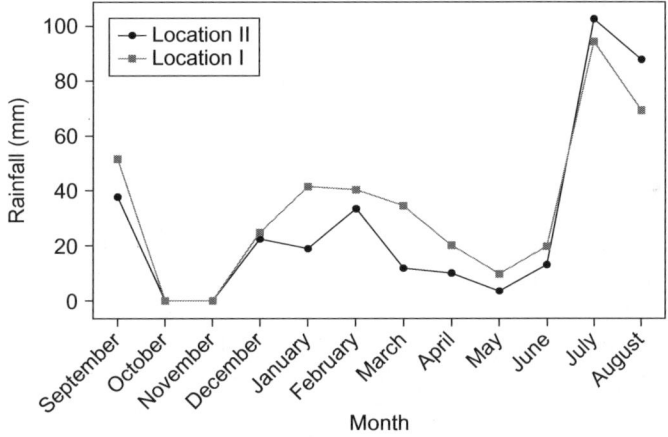

FIGURE 21.2

Rainfall data of two locations for 2003–2004 and 2004–2005 (average of 2 years).

employed to compare the differences among the treatment means (Steel et al., 1997). In the tables of results, the test of significance is presented in the conventional way:

*$P < 0.05$
**$P < 0.01$
NS. Non-significant

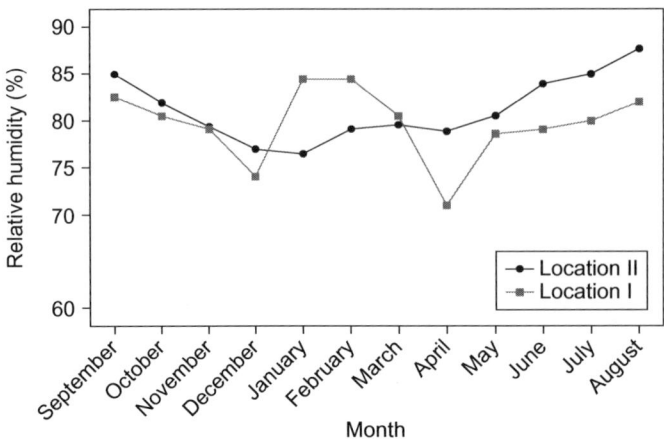

FIGURE 21.3

Relative humidity data of two locations for 2003–2004 and 2004–2005 (average of 2 years).

21.3 Results and discussion

21.3.1 Germination percentage

Germination data for 2 years were pooled for Location-I and Location-II and are presented in Table 21.5. The results showed that the effect of different ASMD levels and planting patterns were non-significant on germination percentage at both locations. All the depletion levels showed the same germination percentage, which can be attributed to the even supply of irrigation water until completion of germination. The subsequent irrigations that were applied according to soil moisture deficit levels showed the effects thereafter. Malik et al. (1992) stated that germination at various moisture regimes was found to be the same. They further revealed that for the study of the effect of different moisture regimes on the growth and development of sugarcane crop, optimum moisture should be applied until the completion of germination. Malik and Ali (1990) found slightly better germination at a closer spacing of 1.0 m than that at a wider spacing (1.5 m). They further advocated that the negative effect in wider spacing was due to more seed per unit area. Row spacing of 100 and 125 cm did not affect germination percentage (Ali et al., 1999). Interactive effects of ASMD levels and planting patterns (ASMDL × G) on germination percentage were non-significant. The same germination trend at both locations was due to similar environmental conditions of temperature and moisture during germination. Other stress aspects may be studied particularly at germination stage to quantify extent of factors (Dourado et al., 2013).

21.3.2 Number of shoots (m^{-2})

The pooled data of 2 years (Location-I and Location-II) on number of shoots m^{-2} is given in Table 21.6. The effects of ASMD levels and planting patterns on number of shoots m^{-2} were

Table 21.5 Pooled Average Data of 2003–2004 and 2004–2005 Regarding Germination Percentage of Autumn Sugarcane Influenced by Different Available Soil Moisture Depletion Levels and Planting Patterns on soils of Location-II and Location-I Under Arid Conditions

Location	Available Soil Moisture Depletion Levels (ASMDL)	Planting Patterns (G)				Means
		G-60	G-75	G-30/90	G-30/120	
Location-II	ASMDL$_1$ (20%)	47.78a	48.00a	48.43a	47.16a	**47.84a**
	ASMDL$_2$ (40%)	48.67a	48.78a	50.02a	47.73a	**48.80a**
	ASMDL$_3$ (60%)	47.82a	48.20a	48.83a	47.38a	**48.06a**
	ASMDL$_4$ (80%)	47.54a	47.54a	48.00a	47.01a	**47.52a**
	Means	**47.95a**	**48.13a**	**48.82a**	**47.32a**	
Location-I	ASMDL$_1$ (20%)	47.02a	48.07a	48.39a	47.17a	**47.66a**
	ASMDL$_2$ (40%)	48.32a	48.90a	50.26a	47.97a	**48.86a**
	ASMDL$_3$ (60%)	48.12a	48.17a	49.29a	47.67a	**48.31a**
	ASMDL$_4$ (80%)	47.26a	47.60a	48.02a	45.81a	**47.17a**
	Means	**47.68a**	**48.19a**	**48.99a**	**47.16a**	
		Location-II			**Location-I**	
LSD$_{0.05}$ ASMDL		3.60			1.80	
LSD$_{0.05}$ Planting patterns		3.81			3.99	
LSD$_{0.05}$ ASMDL $\times$ planting pattern		7.20			1.80	
CV (%)		22.41			10.52	

Mean in the respective category does not differ significantly at 5% level of probability according to LSD test.

significant at both locations. Maximum number of shoots m^{-2} (30.19 and 34.22) were recorded at Location-II and Location-I, respectively, at the interaction of 40% ASMDL $\times$ 30/90 cm and minimum of 13.31 and 13.15 were recorded by 80% ASMDL $\times$ 30/120 cm. It was observed that 126.30 and 159.11% higher number of shoots m^{-2} at Location-II and Location-I, respectively, were recorded by the interaction of 40% ASMDL $\times$ 30/90 cm compared to an 80% ASMDL $\times$ 30/120 cm. This increase in number of shoots m^{-2} might be ascribed to the fact that optimum use of water increased nutrient availability, and the complementary effect of increased nutrient availability and improved air circulation and light interception with 40% ASMDL $\times$ 30/90 cm spaced paired row strip planting pattern reduced shoot mortality. Malik et al. (1996) recorded significantly more shoots per plant at wider row spacing than at closer spacing. It was also noted that too big an increase in inter-strip spacing, as in 30/120 cm paired row strip planting pattern number of strips per plot, decreased and number of plants per unit area increased by decreasing plant-to-plant distance to maintain optimum plant population, due to inter-plant competition increasing and causing adverse effects on number of shoots m^{-2}.

Table 21.6 Pooled Average Data of 2003–2004 and 2004–2005 Regarding Total Number of Shoots m^{-2} of Autumn Sugarcane Influenced by Different Available Soil Moisture Depletion Levels and Planting Patterns on soils of Location-II and Location-I Under Arid Conditions

Location	Available Soil Moisture Depletion Levels (ASMDL)	Planting Patterns (G)				Means
		G-60	G-75	G-30/90	G-30/120	
Location-II	ASMDL$_1$ (20%)	21.76bcdef	22.95bcdef	25.83abcd	16.52fg	**20.77c**
	ASMDL$_2$ (40%)	27.72abc	28.61ab	30.19a	20.78cdef	**26.83a**
	ASMDL$_3$ (60%)	24.94abcde	26.13abcd	27.42abc	19.20defg	**23.42b**
	ASMDL$_4$ (80%)	17.31fg	18.10efg	20.18defg	13.31g	**17.23d**
	Means	**19.93c**	**22.95b**	**25.91a**	**17.45d**	
Location-I	ASMDL$_1$ (20%)	22.89ef	23.99de	27.23cd	16.64ij	**22.69c**
	ASMDL$_2$ (40%)	29.73bc	31.54ab	34.22a	22.65ef	**29.54a**
	ASMDL$_3$ (60%)	27.22cd	28.77bc	29.90bc	20.76efg	**26.66b**
	ASMDL$_4$ (80%)	17.11hi	18.32ghi	20.38fgh	13.15j	**17.24d**
	Means	**24.24c**	**25.66b**	**27.93a**	**18.3d**	
		Location-II		**Location-I**		
LSD$_{0.05}$ ASMDL		3.6		1.8		
LSD$_{0.05}$ Planting patterns		3.8		3.9		
LSD$_{0.05}$ ASMDL × planting pattern		7.2		1.8		
CV (%)		22.41		10.52		

Mean in the respective category does not differ significantly at 5% level of probability according to LSD test.

21.3.3 Crop growth rate (g m^{-2} day^{-1})

Crop growth rate shown in Table 21.7 revealed that it was significantly dissimilar under different ASMD levels and planting patterns at both locations. The maximum CGR of 8.75 and 10.08 g m^{-2} day^{-1} was recorded at Location − II and Location-I, respectively, at 40% ASMD level. It was observed that a 25.86, 17.98, and 12.23% increase in CGR was recorded at Location-II and 35.48, 27.15, and 15.73% at Location-I with 40, 60, 20, and over 80% ASDM. The increase in CGR at 40% ASMD level might be ascribed to an increased nutrient availability which in turn improved the crop growth. Sheu et al. (1992) studied the effect of different moisture regimes on the growth rate of various sugarcane varieties and found that crop growth rate was reduced with moisture stress because of a reduction in translocation of photosynthates. Soares et al. (2004) and Wiedenfeld and Juan Enciso (2008) reported an increase in stalk growth rate with increase in water application during the grand growth period.

Maximum CGR of 8.58 and 9.74 g m^{-2} day^{-1} at Location-II and Location-I, respectively, was recorded with a 30/90 cm spaced paired row strip planting pattern followed by a 75 and 60 cm spaced single row planting pattern. It was observed that an increase of 22.75, 16.02, and 13.30% and 27.15, 19.45, and 15.40% in CGR at Location-II and Location-I, respectively, was recorded with 30/90 cm

Table 21.7 Pooled Average Data of 2003–2004 and 2004–2005 Regarding Total Crop Growth Rate (g^{-2} day^{-1}) of Autumn Sugarcane Influenced by Different Available Soil Moisture Depletion Levels and Planting Patterns on Soils of Location-II and Location-I Under Arid Conditions

Location	Available Soil Moisture Depletion Levels (ASMDL)	Planting Patterns (G)				Means
		G-60	G-75	G-30/90	G-30/120	
Location-II	$ASMDL_1$ (20%)	7.80cdef	8.02bcde	8.56abc	6.82fg	**7.80c**
	$ASMDL_2$ (40%)	8.91ab	9.08a	9.38a	7.61cdef	**8.75a**
	$ASMDL_3$ (60%)	8.39abcd	8.62abc	7.32ef	7.32ef	**8.20b**
	$ASMDL_4$ (80%)	6.96fg	7.11efg	7.50def	6.21g	**6.95d**
	Means	**7.92c**	**8.11b**	**8.58a**	**6.99d**	
Location-I	$ASMDL_1$ (20%)	8.65cde	8.90cd	9.58bc	7.30fg	**8.61c**
	$ASMDL_2$ (40%)	10.12ab	10.51ab	11.09a	8.60cde	**10.08a**
	$ASMDL_3$ (60%)	9.58bc	9.92b	10.16ab	8.19def	**9.46b**
	$ASMDL_4$ (80%)	7.41fg	7.67ef	8.12def	6.56g	**7.44d**
	Means	**8.84c**	**9.15b**	**9.74a**	**7.66d**	

		Location-II	Location-I
$LSD_{0.05}$ ASMDL		1.13	0.51
$LSD_{0.05}$ Planting patterns		0.55	0.59
$LSD_{0.05}$ ASMDL × planting pattern		1.01	1.14
CV (%)		8.93	7.98

Mean in the respective category does not differ significantly at 5% level of probability according to LSD test.

paired row and 75 and 60 cm spaced single row planting patterns, respectively, compared to a 30/120 cm spaced paired row strip planting pattern. This increase in shoot dry weight with a 30/90 cm spaced paired row strip planting pattern might be due to more available space for air circulation and light interception, which increased photosynthetic efficiency and improved growth. Gill (1995) stated that a significant effect was observed on crop growth rate under various planting systems. The utmost CGR of 9.38 and 11.09 g m^{-2} day^{-1} at Location-II and Location-I, respectively, was recorded with a 40% ASMDL × 30/90 cm spaced paired row strip planting pattern followed by a 40% ASMDL × 75 cm, 40% ASMDL × 60 cm, 40% ASMDL × 30/120 cm spaced paired row strip planting pattern. The increase of 51.05 and 69.05% in CGR was recorded at Location-II and Location-I, respectively, with a 40% ASMDL × 30/90 cm planting pattern compared to an 80% ASMDL × 30/120 cm planting pattern. It was observed that an increase in CGR with a 40% ASMDL × 30/90 cm planting pattern was due to the complementary effect of increased nutrient availability and improved air circulation and light interception, which enhanced photosynthetic efficiency.

21.3.4 Leaf area index

Leaf area index (LAI) is an indicator of the size of assimilatory surface of a crop. The analysis of 2 years of pooled data of both locations (Location-I and Location-II) regarding leaf area index given

Table 21.8 Pooled Average Data of 2003–2004 and 2004–2005 Regarding Maximum Leaf Area Index of Autumn Sugarcane Influenced by Different Available Soil Moisture Depletion Levels and Planting Patterns on Soils of Location-II and Location-I Under Arid Conditions

Location	Available Soil Moisture Depletion Levels (ASMDL)	Planting Patterns (G)				Means
		G-60	G-75	G-30/90	G-30/120	
Location-II	ASMDL₁ (20%)	5.69a	5.89a	6.01a	5.59a	**5.70c**
	ASMDL₂ (40%)	6.10a	6.26a	6.43a	6.03a	**6.21a**
	ASMDL₃ (60%)	6.01a	6.12a	6.33a	5.93a	**6.00b**
	ASMDL₄ (80%)	5.93a	5.74a	5.84a	5.51a	**5.57d**
	Means	**5.85a**	**6.00a**	**6.15a**	**5.77a**	
Location-I	ASMDL₁ (20%)	5.95a	6.22a	6.29a	5.85a	**5.98c**
	ASMDL₂ (40%)	6.37a	6.66a	6.81a	6.24a	**6.52a**
	ASMDL₃ (60%)	6.24a	6.15a	6.37a	6.02a	**6.16b**
	ASMDL₄ (80%)	5.87a	6.01a	6.12a	5.78a	**5.80d**
	Means	**6.06a**	**6.32a**	**6.43a**	**5.95a**	

		Location-II		Location-I
LSD $_{0.05}$ ASMDL		0.51		0.55
LSD $_{0.05}$ Planting patterns		0.58		0.59
LSD $_{0.05}$ ASMDL × planting pattern		1.01		1.08
CV (%)		11.95		11.47

Mean in the respective category does not differ significantly at 5% level of probability according to LSD test.

in Table 21.8 revealed that the cane LAI was significantly affected by different ASMD levels than by different planting patterns. The maximum LAI of 6.21 and 6.52 was recorded at Location-II and Location-I, respectively, at 40% ASMD level followed by 60 and 20%. It was observed that LAI increased by 55.72, 35.93, 20.55 and 71.35, 54.64, 31.61% at Location-II and Location-I, at 40, 60, 20, and 80%, respectively. Interactive effects of ASMD levels and planting patterns on LAI were also not significantly different.

21.3.5 Total leaf area duration (days)

As shown in Table 21.9, the greatest total leaf area duration (TLAD) of 847.41 and 980.92 days at Location-II and Location-I, respectively was recorded with a 40% ASMDL × 30/90 cm spaced paired row strip planting pattern followed by a 40% ASMDL × 75 cm, 40% × 60 cm, 40% × 30/120 cm planting pattern spaced. Minimum TLAD of 621.34 and 681.87days was recorded at Location-II and Location-I, respectively, in an 80% ASMDL × 30/120 cm spaced paired row strip planting pattern. It was seen that a 36.38 and 43.86% increase in TLAD at Location-II and Location-I, respectively, was recorded with a 40% ASMDL × 30/90 cm spaced paired row strip planting pattern compared to an 80% ASMDL × 30/120 cm spaced paired row strip planting pattern. The increase in TLAD with a 40% ASMDL × 30/90 cm spaced paired row strip planting pattern might be

Table 21.9 Pooled Average Data of 2003–2004 and 2004–2005 Regarding Total Leaf Area Duration (days) of Autumn Sugarcane Influenced by Different Available Soil Moisture Depletion Levels and Planting Patterns On Soils of Location-II and Location-I Under Arid Conditions

Location	Available Soil Moisture Depletion Levels (ASMDL)	Planting Patterns (G)				Means
		G-60	G-75	G-30/90	G-30/120	
Location-II	$ASMDL_1$ (20%)	663.74gh	718.38de	738.98cd	640.43hi	**690.38c**
	$ASMDL_2$ (40%)	761.16c	803.68b	847.41a	744.43cd	**789.17a**
	$ASMDL_3$ (60%)	740.28cd	767.45c	821.06ab	719.74de	**762.13b**
	$ASMDL_4$ (80%)	640.43hi	674.19fg	699.50ef	621.34i	**658.87d**
	Means	**701.40c**	**740.92b**	**776.74a**	**681.49d**	
Location-I	$ASMDL_1$ (20%)	749.55ij	819.42efg	836.75de	723.35jk	**782.27c**
	$ASMDL_2$ (40%)	859.02d	938.55b	980.92a	824.08ef	**900.64a**
	$ASMDL_3$ (60%)	800.16fg	857.87d	896.16c	767.61hi	**830.45b**
	$ASMDL_4$ (80%)	705.17kl	765.05hi	791.97gh	681.87l	**736.01d**
	Means	**778.48c**	**845.22b**	**876.45a**	**749.23d**	

		Location-II	Location-I
$LSD_{0.05}$ ASMDL		15.17	15.27
$LSD_{0.05}$ Planting patterns		15.19	15.25
$LSD_{0.05}$ ASMDL × planting pattern		30.40	30.96
CV (%)		2.94	2.62

Mean in the respective category does not differ significantly at 5% level of probability according to LSD test.

due to the complementary effect of increased nutrient availability and improved air circulation and light interception, which enhanced photosynthetic efficiency and ultimately TLAD.

21.3.6 Net assimilation rate (g m^{-2} day^{-1})

Both excessive and deficient irrigations for sugarcane crop are equally harmful. Due to frequent irrigation, soil remains continuously wet, while due to the arid climate at Location-I and Location-II, frequent wind storms in summer season caused lodging of crop during the growth period. The highest net assimilation rate (NAR) of 1.93 and 2.02 g m^{-2} day^{-1} at Location-II and Location-I, respectively, was recorded with a 40% ASMDL × 30/90 cm planting pattern followed by a 40% ASMDL × 75 cm, 40% ASMDL × 60 cm, 40% ASMDL × 30/120 cm spaced paired row strip planting pattern. NAR was minimum at 1.25 and 1.48 g^{-2} day^{-1} at Location-II and Location-I, respectively, with an 80% ASMDL × 30/120 cm planting pattern. It was found that a 35.23 and 36.49% increase with a 40% ASMDL × 30/90 cm planting pattern at Location-II and Location-I, respectively, was recorded compared to an 80% ASMDL × 30/120 cm spaced paired row strip planting pattern. It was also observed that optimum NAR with a 40% ASMDL × G-30/90 was due to enhanced LAI, TLAD, and CGR attributes. These results are in agreement with those of Bashir (1997) and Ali (1999) who reported that planting patterns had a significant effect on NAR.

Table 21.10 Pooled Average Data of 2003–2004 and 2004–2005 Regarding Net Assimilation Rate (g m^{-2} day^{-1}) Planted Sugarcane Influenced by Different Available Soil Moisture Depletion Levels and Planting Patterns on Soils of Location-II and Location-I Under Arid Conditions

Location	Available Soil Moisture Depletion Levels (ASMDL)	Planting Patterns (G)				Means
		G-60	G-75	G-30/90	G-30/120	
Location-II	ASMDL$_1$ (20%)	1.45efgh	1.81abcd	1.84abcd	1.31gh	**1.50c**
	ASMDL$_2$ (40%)	1.72abcde	1.91ab	1.93a	1.58cdefg	**1.79a**
	ASMDL$_3$ (60%)	1.56defgh	1.84abcd	1.89abc	1.40fgh	**1.57b**
	ASMDL$_4$ (80%)	1.34fgh	1.60bcdefg	1.65abcdef	1.25h	**1.46d**
	Means	**1.42c**	**1.69b**	**1.83a**	**1.29d**	
Location-I	ASMDL$_1$ (20%)	1.51de	1.82abcd	1.95abc	1.51de	**1.64c**
	ASMDL$_2$ (40%)	1.92abc	1.99a	2.02a	1.80abcd	**1.93a**
	ASMDL$_3$ (60%)	1.80abcd	1.95abc	1.98ab	1.66cde	**1.75b**
	ASMDL$_4$ (80%)	1.59de	1.73abcde	1.81abcd	1.48e	**1.55d**
	Means	**1.65c**	**1.77b**	**1.94a**	**1.51d**	
			Location-II		**Location-I**	
LSD$_{0.05}$ ASMDL			0.28		0.30	
LSD$_{0.05}$ Planting patterns			0.16		0.19	
LSD$_{0.05}$ ASMDL × planting pattern			0.31		0.36	
CV (%)			13.46		12.24	

Mean in the respective category does not differ significantly at 5% level of probability according to LSD test.

The maximum NAR was recorded in pit planting system compared to the minimum from the crop sown with a 60 cm spaced single rows of sugarcane crop. It was also noted that too much of an increase in inter-strip spacing, as in a 30/120 cm spaced paired row strip planting pattern, decreased the number of strips per plot and the number of plants per unit area was increased to maintain equal plant population, due to increased inter-plant competition causing adverse effects on LAI, TLAD, and CGR (see Table 21.10).

21.3.7 Plant height (m)

A larger plant height of 4.07 and 4.46 m at Location-II and Location-I, respectively, was documented (Table 21.11) with a 40% AMSDL × 30/90 cm planting pattern, and a minimum of 1.79 and 1.71 m at Location-II and Location-I, respectively, with interaction of 80% AMSDL × 30/120 cm planting patterns. It was observed that a 127.37 and 160.82% higher plant height at Location-II and Location-I, respectively, was recorded in the interaction of a 40% AMSDL × 30/90 cm planting pattern followed by 56.02 and 58.39% at Location-II and Location-I, respectively, in a 40% AMSDL × 75 cm spaced planting pattern compared to the interaction of 80% AMSDL × 30/120 cm planting patterns. The maximum plant height at 40% × 30/90 cm planting pattern might be

Table 21.11 Pooled Average Data of 2003–2004 and 2004–2005 Regarding Plant Height (m) of Autumn Sugarcane Influenced by Different Available Soil Moisture Depletion Levels and Planting Patterns on Soils of Location-II and Location-I Under Arid Conditions

Location	Available Soil Moisture Depletionlevels (ASMDL)	Planting Patterns (G)				
		G-60	G-75	G-30/90	G-30/120	Means
Location-II	ASMDL₁ (20%)	2.93defgh	3.09cdefg	3.48abcde	2.22ij	**2.93b**
	ASMDL₂ (40%)	3.73abc	3.85ab	4.07a	2.80efghi	**3.61a**
	ASMDL₃ (60%)	3.36bcdef	3.52abcd	2.59ghi	2.59ghi	**3.29a**
	ASMDL₄ (80%)	2.33hij	2.44ghij	2.72fghi	1.79j	**2.32c**
	Means	**3.09b**	**3.23ab**	**3.49a**	**2.35c**	
Location-I	ASMDL₁ (20%)	2.98def	3.13cde	3.55bcd	2.17gh	**2.96c**
	ASMDL₂ (40%)	3.88ab	4.11ab	4.46a	2.95def	**3.85a**
	ASMDL₃ (60%)	3.55bcd	3.75bc	3.75bc	2.71efg	**3.48b**
	ASMDL₄ (80%)	2.23gh	2.39fgh	2.66efg	1.71h	**2.32c**
	Means	**3.16b**	**3.35ab**	**3.64a**	**2.39c**	

		Location-II			Location-I
$LSD_{0.05}$ ASMDL		0.35			0.39
$LSD_{0.05}$ Planting patterns		0.38			0.37
$LSD_{0.05}$ ASMDL × planting pattern		0.69			0.73
CV (%)		15.98			15.49

Mean in the respective category does not differ significantly at 5% level of probability according to LSD test.

ascribed to better CGR. Pandian et al. (1992) concluded that cane height was taller at 1.1 IW/CPE than at 0.9 IW/CPE. Naik et al. (1993) observed the effect of moderate and severe water stress and evaluated that plant height was 117.3 and 105.7 cm at moderate and sever water stress, respectively, against control where a 148 cm plant height was found. Lal (1988) in India compared conventional planting at 90 cm row spacing with 60 cm clumps (also at 90 cm row spacing) and pit plantation (4167 pits ha^{-1}) with up to 20 setts per pit and a row spacing of 2 m. He found that plant height increased with increase in space around the cane clumps. Domini and Plana (1989) reported that planting patterns had a significant effect on plant height. Nisachon et al. (2012) found that drought significantly reduced stalk diameter, but it did not significantly affect relative rate of height growth.

21.3.8 Weight per stripped cane (kg)

The reciprocal effects of ASMD levels and planting patterns on weight per stripped cane were found significant at $P = 0.05$ (Table 21.12). Maximum individual stripped cane weight of 0.95 and 1.22 kg at Location-II and Location-I, respectively, was recorded in the interaction of a 40% AMSDL × 30/90 cm planting pattern. Minimum weight per stripped cane of 0.42 and 0.47 kg at

Table 21.12 Pooled Average Data of 2003—2004 and 2004—2005 Regarding Individual Stripped Cane Weight (kg) of Autumn Sugarcane Influenced by Different Available Soil Moisture Depletion Levels and Planting Patterns on Soils of Location-II and Location-I Under Arid Conditions

Location	Available soil moisture depletion levels (ASMDL)	Planting patterns (G)				Means
		G-60	G-75	G-30/90	G-30/120	
Location-II	ASMDL$_1$ (20%)	0.69bcdefg	0.73abcdefg	0.82abcde	0.52gh	**0.69b**
	ASMDL$_2$ (40%)	0.88abc	0.9ab	0.95a	0.66cdefg	**0.85a**
	ASMDL$_3$ (60%)	0.79abcdef	0.83abcde	0.87abcd	0.61efgh	**0.78ab**
	ASMDL$_4$ (80%)	0.55gh	0.57fgh	0.64defgh	0.42h	**0.55c**
	Means	**0.63b**	**0.76a**	**0.82a**	**0.50c**	
Location-I	ASMDL$_1$ (20%)	0.82defg	0.86cdef	0.97bcde	0.59gh	**0.81b**
	ASMDL$_2$ (40%)	1.06abc	1.13ab	1.22 a	0.81defg	**1.06a**
	ASMDL$_3$ (60%)	0.97bcde	1.03abcd	1.07abc	0.74efg	**0.95a**
	ASMDL$_4$ (80%)	0.61gh	0.65fgh	0.73fg	0.47h	**0.62c**
	Means	**0.77b**	**0.82a**	**1.00a**	**0.60c**	
			Location-II		**Location-I**	
LSD$_{0.05}$ ASMDL			0.12		0.18	
LSD$_{0.05}$ Planting patterns			0.15		0.14	
LSD$_{0.05}$ ASMDL × planting pattern			0.23		0.28	
CV (%)			22.77		18.96	

Mean in the respective category does not differ significantly at 5% level of probability according to LSD test.

Location-II and Location-I, respectively, was recorded with the interaction of an 80% AMSDL × 30/120 cm planting pattern. It was further observed that a 126.19 and 159.57% increase in individual stripped cane weight at Location-II and Location-I, respectively, was traced by the interaction of a 40% AMSDL × 30/90 cm planting pattern compared to the interaction of an 80% AMSDL × 30/120 cm planting pattern.

21.3.9 Stripped cane length (m)

Significant improvement was seen by mutual effects of ASMD levels and planting patterns on individual cane length (Table 21.13) at both locations and maximum individual cane length 2.38 and 2.59 m at Location-II and Location-I was recorded with a 40% AMSDL × 30/90 cm planting pattern. Individual cane length was a minimum of 1.0 m at both locations with an 80% AMSDL × 30/120 cm planting pattern. It was observed that a 138 and 159 % higher individual cane length at Location-II and Location-I, respectively, was achieved by a 40% AMSDL × 30/90 cm planting pattern compared to an 80% AMSDL × 30/120 cm planting pattern. Ingram (1986) also stated a 40 to 50% greater stalk length in 75 cm spaced rows than in 150 cm spaced rows.

Table 21.13 Pooled Average Data of 2003–2004 and 2004–2005 Regarding Mean Stripped Cane Length (m) of Autumn Sugarcane Influenced by Different Available Soil Moisture Depletion Levels and Planting Patterns on Soils of Location-II and Location-I Under Arid Conditions

Location	Available Soil Moisture Depletion Levels (ASMDL)	Planting Patterns (G)				Means
		G-60	G-75	G-30/90	G-30/120	
Location-II	$ASMDL_1$ (20%)	1.72cdefg	1.81bcdef	2.04abcd	1.3gh	1.72b
	$ASMDL_2$ (40%)	2.19abc	2.26ab	2.38a	1.64defg	2.12a
	$ASMDL_3$ (60%)	2.06bcd	2.18abcd	2.27abc	1.57ef	1.93ab
	$ASMDL_4$ (80%)	1.3fg	1.39efg	1.55ef	1.00g	1.36c
	Means	**1.81a**	**1.89a**	**2.04a**	**1.38b**	
Location-I	$ASMDL_1$ (20%)	1.74def	1.82cde	2.06bcd	1.26fg	1.72b
	$ASMDL_2$ (40%)	2.25abc	2.39ab	2.59a	1.72def	2.24a
	$ASMDL_3$ (60%)	2.06bcd	2.18abcd	2.27abc	1.57ef	2.02a
	$ASMDL_4$ (80%)	1.3fg	1.39efg	1.55ef	1.00g	1.31c
	Means	**1.84b**	**1.95ab**	**2.12a**	**1.39c**	
		Location-II			Location-I	
$LSD_{0.05}$ ASMDL		0.24			0.27	
$LSD_{0.05}$ Planting patterns		0.28			0.29	
$LSD_{0.05}$ ASMDL × planting pattern		0.49			0.53	
CV (%)		19.24			18.8	

Mean in the respective category does not differ significantly at 5% level of probability according to LSD test.

21.3.10 Cane diameter (cm)

A combination of ASMDL and planting pattern on individual cane diameter was identified by a 40% ASMDL × 30/90 cm planting pattern, which resulted in individual cane diameter of 7.83 and 8.11 cm at Location-II and Location-I, respectively (Table 21.14). Lesser values of 3.46 and 3.13 cm at Location-II and Location-I, respectively, were noticed with an 80% ASMDL × 30/120 cm planting pattern. An increase of up to 126.30 and 159.11% higher individual cane diameter was recorded in the interaction of a 40% ASMDL × 30/90 cm planting pattern compared to an 80% ASMDL × 30/120 cm planting pattern at Location-II and Location-I, respectively. Naidu and Venkataramana (1989) reported that cane girth was decreased as compared to 80% ASMD level, when the crop was subjected to water stress. Sharma and Gupta (1990) investigated the effect of different moisture regimes (0.75, 1.0, and 1.5 IW/CPE) and stated that a greater cane girth of 9.29 and 9.95 cm was obtained at 1.0 and 1.51 IW/CPE, respectively, as compared to 8.38 cm obtained from 0.75 IW/CPE.

21.3.11 Unstripped cane yield (t ha^{-1})

Different ASMD levels exhibited different unstripped cane yield at both the locations (Table 21.15). The highest unstripped cane yield of 180.88 and 215.75 t ha^{-1} at Location-II and

Table 21.14 Pooled Average Data of 2003–2004 and 2004–2005 Regarding Mean Cane Diameter (cm) of Autumn Sugarcane Influenced by Different Available Soil Moisture Depletion Levels and Planting Patterns on Soils of Location-II and Location-I Under Arid Conditions

Location	Available Soil Moisture Depletion Levels (ASMDL)	Planting patterns (G)				Means
		G-60	G-75	G-30/90	G-30/120	
Location-II	ASMDL$_1$ (20%)	5.66fg	5.97ef	6.70cd	4.29i	**5.66c**
	ASMDL$_2$ (40%)	7.19abc	7.42ab	7.83a	5.39fgh	**6.96a**
	ASMDL$_3$ (60%)	6.48de	6.78bcd	7.11bcd	4.98ghi	**6.34b**
	ASMDL$_4$ (80%)	4.50i	4.71hi	5.25gh	3.46j	**4.48d**
	Means	**5.96b**	**6.22b**	**6.72a**	**4.53c**	
Location-I	ASMDL$_1$ (20%)	5.44de	5.71d	6.46c	3.94g	**5.39c**
	ASMDL$_2$ (40%)	7.05bc	7.48ab	8.11a	5.38de	**7.01a**
	ASMDL$_3$ (60%)	6.46c	6.83bc	7.09bc	4.93ef	**6.33b**
	ASMDL$_4$ (80%)	4.07g	4.36fg	4.85ef	3.13h	**4.10d**
	Means	**5.76b**	**6.10b**	**6.63a**	**4.35c**	

		Location-II		Location-I	
LSD$_{0.05}$ ASMDL		0.35		0.38	
LSD$_{0.05}$ Planting patterns		0.37		0.40	
LSD$_{0.05}$ ASMDL × planting pattern		0.69		0.73	
CV (%)		8.29		8.51	

Mean in the respective category does not differ significantly at 5% level of probability according to LSD test.

Location-I, respectively, was recorded in a 40% ASMDL followed by 60 and 20% ASMD levels. The lowest unstripped cane yield was 126 and 159.13 t ha^{-1} at Location-II and Location-I, respectively, in an 80% ASMD level. About 43.56, 34.92, 23.02% and 35.58, 28.83 and 12.25% higher unstripped cane yield was obtained at Location-II and Location-I in 40, 60, and 20% ASMDL over an 80% ASMD level, respectively. The increased unstripped cane yield at 40% ASMD level might be ascribed to increased nutrient availability, which resulted in greater number of stalks per square meter.

The highest unstripped cane yield of 176.62 and 212.13 t ha^{-1} at Location − II and Location-I, respectively, was recorded in 30/90 cm (G-30/90) followed by a 75 cm (G-75), 60 cm spaced single row planting pattern (G-60), and the lowest stripped cane yield of 126.50 and 153.13 t ha^{-1} at Location-II and Location-I, respectively, in 30/120 cm spaced paired row planting pattern (G-30/120). The increase was to the tune of 39.62, 32.32, 27.57% and 38.53, 30.37, and 26.45% at Location-II and Location-I, respectively, with G-30/90, G-75, and G-60 over G-30/120. It was observed that improvement in unstripped cane in the 30/90 cm spaced paired row planting pattern was due to accelerated CGR, more NAR, longer TLAD, and greater LAI attributes.

The ceiling of the unstripped cane yield of 197.50 and 238 t ha^{-1} at Location-II and Location-I, respectively, was obtained with 40% ASMDL × G-30/90 followed by 40% ASMDL × G-75 and

Table 21.15 Pooled Average Data of 2003–2004 and 2004–2005 Regarding Unstripped Cane Yield (t ha^{-1}) of Autumn Sugarcane Influenced by Different Available Soil Moisture Depletion Levels and Planting Patterns on Soils of Location-II and Location-I Under Arid Conditions

Location	Available Soil Moisture Depletion Levels (ASMDL)	Planting Patterns (G)				Means
		G-60	G-75	G-30/90	G-30/120	
Location-II	ASMDL$_1$ (20%)	157.50fg	164.00ef	177.50cde	121.00j	**155.00c**
	ASMDL$_2$ (40%)	187.50abc	192.50ab	197.50a	146.00gh	**180.88a**
	ASMDL$_3$ (60%)	173.00de	180.50bcd	187.50abc	144.00gh	**170.00b**
	ASMDL$_4$ (80%)	127.50ij	132.50hij	144.00gh	100.00k	**126.00d**
	Means	**161.38b**	**167.38b**	**176.62a**	**126.50c**	
Location-I	ASMDL$_1$ (20%)	182.50e	186.50e	205.50d	140.00i	**178.63c**
	ASMDL$_2$ (40%)	221.00bc	227.50ab	238.00a	176.50efg	**215.75a**
	ASMDL$_3$ (60%)	210.50cd	218.50bcd	223.50bc	167.50fgh	**205.00b**
	ASMDL$_4$ (80%)	160.50h	166.00gh	181.50ef	128.50i	**159.13d**
	Means	**193.63b**	**199.63b**	**212.13a**	**153.13c**	

		Location-II		**Location-I**	
LSD$_{0.05}$ ASMDL		7.04		7.12	
LSD$_{0.05}$ Planting patterns		7.09		7.19	
LSD$_{0.05}$ ASMDL × planting pattern		14.08		14.15	
CV (%)		6.26		5.21	

Mean in the respective category does not differ significantly at 5% level of probability according to LSD test.

40% ASMDL × G-60, respectively. The minimum unstripped cane yield of 100 and 128.5 t ha^{-1} at Location-II and Location-I, respectively, was calculated in ASMDL$_4$ × G-30/120 treatment. It was noteworthy that 97.50 and 85.21% higher unstripped cane yield was recorded in the interaction of a 40% ASMDL × 30/90 cm planting pattern compared to an 80% ASMDL × 30/120 cm planting pattern at Location-II and Location-I, respectively. The optimum unstripped cane yield in 40% ASMDL × G-30/90 might be due to accelerated CGR, more NAR, longer TLAD, and greater LAI.

21.3.12 Stripped cane yield (t ha^{-1})

The challenges faced by the agricultural sector under the climate change scenarios are to provide food security for an increasing world population while protecting the environment and the functioning of its ecosystems (Rosenzweig et al., 2012). Sugarcane crop is one of those challenges. The highest stripped cane yield of 135.50 and 171.38 t ha^{-1} at Location-II and Location-I, respectively, was taken with a 40% ASMD level followed by 60 and 20% ASMDL (Table 21.16), while the lowest cane yield of 87.06 and 100.25 t ha^{-1} was observed at Location-II and Location-I, respectively, with an 80% ASMD level. It was further observed that 55.64, 41.72, 26.35% at Location-II and 70.95, 54.36, and 31.42% higher stripped cane yield at Location-I, were recorded in 40, 60, and 20%

Table 21.16 Pooled Average Data of 2003–2004 and 2004–2005 Regarding Stripped Cane Yield (t ha^{-1}) of Autumn Sugarcane Influenced by Different Available Soil Moisture Depletion Levels and Planting Patterns on Soils of Location-II and Location-I Under Arid Conditions

Location	Available Soil Moisture Depletion Levels (ASMDL)	Planting Patterns (G)				Means
		G-60	G-75	G-30/90	G-30/120	
Location-II	ASMDL$_1$ (20%)	110.00ef	116.00de	130.50bc	83.50h	**110.00c**
	ASMDL$_2$ (40%)	140.00abc	144.50ab	152.50a	105.00efg	**135.50a**
	ASMDL$_3$ (60%)	126.00cd	132.00bc	138.50abc	97.00fgh	**123.38b**
	ASMDL$_4$ (80%)	87.50h	91.50gh	102.00efg	67.25i	**87.06d**
	Means	**105.88b**	**121.00b**	**130.87a**	**88.19c**	
Location-I	ASMDL$_1$ (20%)	133.00ef	139.50e	158.00d	96.50i	**131.75c**
	ASMDL$_2$ (40%)	172.50bc	183.00b	198.50a	131.50efg	**171.38a**
	ASMDL$_3$ (60%)	158.00d	167.00cd	173.50bc	120.50fgh	**154.75b**
	ASMDL$_4$ (80%)	99.50i	106.50hi	118.50gh	76.50j	**100.25d**
	Means	**140.75c**	**149.00b**	**162.12a**	**106.25d**	

	Location-II	Location-I
LSD$_{0.05}$ ASMDL	7.04	7.13
LSD$_{0.05}$ Planting patterns	7.15	7.16
LSD$_{0.05}$ ASMDL × planting pattern	14.08	14.25
CV (%)	8.67	7.09

Mean in the respective category does not differ significantly at 5% level of probability according to LSD test.

ASMD levels, respectively, over an 80% ASMD level. The increased stripped cane yield at the 40% ASMD level was ascribed to greater weight per cane, cane length, and cane diameter. It was found that excess or deficient irrigation water was equally harmful for sugarcane crops in the case of irrigation at the 20% ASMD level (excess irrigation). Due to the longer intervals between successive irrigations the soil remained deficient in available soil moisture content and both of these conditions adversely affected the yield components and ultimately stripped cane yield of sugarcane crop at both locations. Similar results were also reported by Sharma and Gupta (1990), Pandian et al. (1992), Malik et al. (1992), and Ghugare et al.(1995), whereas Ali (1996) studied the effect of two moisture levels, i.e., 45% ASM (available soil moisture) and 15% ASM, and found no significant effect on cane yield at both the levels. Siddique et al. (2008) obtained the highest yield of sugarcane as well as cabbage as an inter-crop at a pan ratio of 1 with 450 mm irrigation water in addition to 176.43 mm of rainfall. Siddique et al. (2008) also reported in another study that varieties intolerant to water stress gave better response to optimum soil moisture than varieties tolerant to water stress.

The maximum stripped cane yield of 130.87 and 162.12 t ha^{-1} at Location-II and Location-I, respectively, was recorded in a 30/90 cm spaced paired row strip planting pattern, while the lowest stripped cane yield of 88.19 and 106.25 t ha^{-1} at Location-II and Location-I, respectively, was obtained in a 30/120 cm spaced paired row strip planting pattern. It was found that 48.40, 37.20,

20.06% as well as 52.58, 40.24, and 32.47 % higher stripped cane yield at Location-II and Location-I, respectively, was recorded in 30/90, 75, and 60 cm planting patterns over a 30/120 cm planting pattern. It was also noted that too much increase in inter-strip spacing, as in the 30/120 cm spaced paired row strip planting, decreased the number of strips per plot and the number of plants per unit area were increased (by decreasing plant-to-plant distance to maintain equal plant population in all the treatments) due to inter-plant increased competition causing adverse effects on yield components and ultimately on yield of sugarcane crop. Higher stripped cane yield in a 30/90 cm spaced paired row planting pattern was attributed to improved weight cane^{-1}, and increased cane length, and accelerated CGR. The maximum stripped cane yield of 152.50 and 198.50 t ha^{-1} at Location-II and Location-I, respectively, could be seen with 40% × 30/90 cm followed by 40% × 75 cm, 40% × 60 cm, and 40% × 30/120 cm planting patterns. Minimum stripped cane yield of 67.25 and 76.50 t ha^{-1} at Location-II and Location-I, respectively, was entered with an 80% ASMDL × 30/120 cm planting pattern. It was also observed that 126.77 and 159.48% higher stripped cane yield at Location-II and Location-I, respectively, was recorded in the interaction of a 40% × 30/90 cm planting pattern compared to an 80% × 30/120 cm planting pattern. Similar results were reported by El-Geddawy et al. (2002). Khan et al. (2004) reported that crop planted at an inter-row spacing of 0.72 m produced maximum stripped cane yield compared to wider inter-row spaces of 0.90 and 1.20 m. Whereas, Singh et al. (2006) reported significantly higher cane yield at 45 cm spacing followed by 60 and 75 cm in ratoon crop.

21.3.13 Harvest index percentage

The analysis of 2 years of pooled average data of Location-I and Location-II regarding harvest index (HI), shown in Table 21.17, revealed that HI was not significantly different under different ASMD levels at Location-II. Whereas at Location-I, a maximum HI of 79.08 was recorded at the 40% ASMD level, which was statistically similar to 60% ASMDL and 20% ASMDL. HI was a minimum of 62.56 at the 80% ASMD level. HI was not significantly affected by different planting patterns at Location-II. Whereas at Location-I the maximum (HI) of 75.76 was recorded in a 30/90 cm spaced paired row strip planting pattern followed by 75 and 60 cm spaced single row planting patterns. A minimum HI of 68.66 was calculated in a 30/120 cm spaced paired row strip planting pattern. The variation in HI values between the two sites might be due to the difference in local soil and climatic conditions. Interactive effects of ASMD levels and planting patterns regarding HI were not significantly different at Location-II, whereas a maximum HI of 83.38 was recorded in a 40% ASMDL × 30/90 cm planting pattern followed by 40% ASMDL × 75 cm, 40% ASMDL × 60 cm, and 40% ASMDL × 30/120 cm spaced paired row strip planting patterns, while a minimum of 59.36 at Location-I was recorded in an 80% ASMDL × 30/120 cm planting pattern. The variation between HI values of the two sites might be due to the difference in soils and microclimate of both locations. A literature survey indicated straw yields of sugarcane in the range of 7.4−24.3 mg ha^{-1} (dry basis), and a straw-to-stalk ratio ranging from 9.7 to 29.5%. The averages were, respectively, 14.1 mg ha^{-1} and 18.2% (Hassuani et al., 2005).

21.3.14 Sugar recovery percentage

Although the quality components of sugarcane are mainly genetic characters, they are also influenced by various agronomic practices. The analysis of 2 years of pooled average data of

Table 21.17 Pooled Average Data of 2003–2004 and 2004–2005 Regarding Harvest Index (%) of Autumn Sugarcane Influenced by Different Available Soil Moisture Depletion Levels and Planting Patterns on Soils of Location-II and Location-I Under Arid Conditions

Location	Available Soil Moisture Depletion Levels (ASMDL)	Planting Patterns (G)				Means
		G-60	G-75	G-30/90	G-30/120	
Location-II	ASMDL$_1$ (20%)	69.79	70.67	72.47	68.99	**70.48**
	ASMDL$_2$ (40%)	74.67	75.06	77.21	71.90	**74.71**
	ASMDL$_3$ (60%)	72.79	73.11	73.85	69.78	**72.38**
	ASMDL$_4$ (80%)	68.63	68.99	70.83	67.32	**68.943**
	Means	**71.47**	**71.96**	**73.59**	**69.50**	
Location-I	ASMDL$_1$ (20%)	72.73abcde	74.60abcd	76.90abc	68.93bcde	**73.29a**
	ASMDL$_2$ (40%)	78.04abc	80.41ab	83.38a	74.47abcd	**79.08a**
	ASMDL$_3$ (60%)	75.07abcd	76.42abc	77.68abc	71.89abcde	**75.27a**
	ASMDL$_4$ (80%)	61.79de	63.99cde	65.08cde	59.36e	**62.56b**
	Means	**71.91ab**	**73.86ab**	**75.76a**	**68.66b**	

	Location-II	**Location-I**
LSD$_{0.05}$ ASMDL	7.04	7.11
LSD$_{0.05}$ Planting patterns	4.10	7.16
LSD$_{0.05}$ ASMDL × planting pattern	14.08	14.16
CV (%)	13.8	13.63

Mean in the respective category does not differ significantly at 5% level of probability according to LSD test.

Location-II and Location-I regarding sugar recovery percentage shown in Table 21.18 revealed that sugar recovery percentage was significantly affected by different ASMD levels at Location-I rather than at Location-II. Sugar recovery percentage was 9.60% higher at the 80% ASMD level as compared to 20% ASMDL. However, it was statistically similar in 40 and 60% ASMD levels with a minimum of 8.40% in the 20% ASMD level at Location-I. It was noted that higher sugar percentage recovery of 7.7, 4.88, and 2.71% and 14.29, 10.12, and 6.07% was recorded at Location-II and Location-I, respectively, in 80, 60, and 40% ASMD levels, respectively, over the 20% ASMD level. In India, Sharma and Gupta (1990) concluded that juice extraction was increased but juice purity, commercial cane sugar (CSS%), and sucrose content decreased drastically as moisture level increased from 0.75 to 1.5 IW/CPE. Ghugare et al. (1995) studied the effect of 50, 75, 100, and 125 CPE against canal rotation as control and found non-significant effects in sucrose content in juice as well as in purity percentage. Ali (1996) reported that irrigation applied at 15 and 45% ASM did not affect the CCS percentage.

With respect to different planting patterns, the maximum sugar recovery of 10.39 and 10.17% at Location-II and Location-I, respectively, was recorded in a 30/120 cm spaced paired row strip planting pattern followed by a 30/90 and 75 cm spaced planting pattern. Minimum sugar recovery

Table 21.18 Pooled Average Data of 2003–2004 and 2004–2005 Regarding Sugar Recovery Percentage of Autumn Sugarcane Influenced by Different Available Soil Moisture Depletion Levels and Planting Patterns on Soils of Location-II and Location-I Under Arid Conditions

Location	Available Soil Moisture Depletion Levels (ASMDL)	Planting Patterns (G)				Means
		G-60	G-75	G-30/90	G-30/120	
Location-II	ASMDL$_1$ (20%)	8.44d	9.15bcd	9.19bcd	10.09abc	**9.22a**
	ASMDL$_2$ (40%)	8.73cd	9.25abcd	9.68abcd	10.21abc	**9.47a**
	ASMDL$_3$ (60%)	8.83cd	9.45abcd	9.92abcd	10.48ab	**9.67a**
	ASMDL$_4$ (80%)	9.16bcd	9.62abcd	10.16abc	10.78a	**9.93a**
	Means	**8.79c**	**9.37bc**	**9.74ab**	**10.39a**	
Location-I	ASMDL$_1$ (20%)	7.28g	8.44defg	8.44defg	9.45abcde	**8.40b**
	ASMDL$_2$ (40%)	7.80fg	9.25abcd	9.26bcdef	9.99abc	**8.91ab**
	ASMDL$_3$ (60%)	8.14efg	9.03bcdef	9.47abcde	10.36ab	**9.25a**
	ASMDL$_4$ (80%)	8.63cdefg	9.06bcdef	9.84abcd	10.87a	**9.60a**
	Means	**7.96c**	**8.78b**	**9.25b**	**10.17a**	

	Location-II	Location-I
LSD$_{0.05}$ ASMDL	0.77	0.97
LSD$_{0.05}$ Planting patterns	0.72	0.75
LSD$_{0.05}$ ASMDL × planting pattern	0.53	0.61
CV (%)	11.88	11.23

Mean in the respective category does not differ significantly at 5% level of probability according to LSD test.

percentage of 8.79 and 7.96% at Location-II and Location-I, respectively, was observed in a 60 cm spaced planting pattern. It was noted that 18.20, 10.81, 6.60% and 27.76, 16.21, and 10.30% higher sugar recovery was recorded in 30/120, 30/90, and 75 cm spaced planting patterns over a 60 cm planting pattern. These results are in line with those of Kathirisan and Narayanasamy (1991), who reported that sugarcane grown in widely spaced rows had higher sucrose content than that grown in narrow spaced rows. On the contrary, Vains et al. (2000) reported that sucrose content in cane juice was not significantly affected by different spatial arrangements and plantation methods. Such differential impact of row spacing on sucrose content might be due to the difference in soil and climatic conditions under which the various experiments were conducted.

Regarding interactive effects, the maximum sugar recovery of 10.78 and 10.87% at Location-II and Location-I, respectively, was recorded in an 80% ASMDL × 30/120 cm planting pattern followed by 80% ASMDL × 30/90 cm, 80% ASMDL × 75 cm, and 80% ASMDL × 60 cm spaced planting patterns, respectively, while a minimum of 8.44 and 7.28% at Location-II and Location-I, respectively, was recorded in a 20% ASMDL × 60 cm planting pattern. It was logical that 27.73 and 49.31% higher sugar recovery percentage at Location-II and Location-I, respectively, was recorded in the interaction of an 80% × 30/120 cm planting pattern compared to a 20% × 60 cm planting pattern.

Table 21.19 Pooled Average Data of 2003–2004 and 2004–2005 Regarding Sugar Yield (t ha^{-1}) of Autumn Sugarcane Influenced by Different Available Soil Moisture Depletion Levels and Planting Patterns on Soils of Location-II and Location-I Under Arid Conditions

Location	Available Soil Moisture Depletion Levels (ASMDL)	Planting Patterns (G)				Means
		G-60	G-75	G-30/90	G-30/120	
Location-II	ASMDL$_1$ (20%)	9.14f	10.53de	11.93c	8.40gh	10.00c
	ASMDL$_2$ (40%)	12.20c	13.35b	14.74a	10.70de	12.75a
	ASMDL$_3$ (60%)	11.07d	12.44c	13.72b	10.16e	11.85b
	ASMDL$_4$ (80%)	7.95h	8.72fg	10.31e	7.25i	8.56d
	Means	**10.09c**	**11.26b**	**12.68a**	**9.13d**	
Location-I	ASMDL$_1$ (20%)	9.69h	11.77g	13.30e	9.13hi	10.97c
	ASMDL$_2$ (40%)	13.45e	15.72c	18.36a	13.12e	15.16a
	ASMDL$_3$ (60%)	12.85ef	15.04d	16.42b	12.47f	14.20b
	ASMDL$_4$ (80%)	8.57ij	9.62h	11.63g	8.31j	9.53d
	Means	**11.14c**	**13.04b**	**14.93a**	**10.76d**	

		Location-II		Location-I	
LSD$_{0.05}$ ASMDL		0.32		0.38	
LSD$_{0.05}$ Planting patterns		0.36		0.35	
LSD$_{0.05}$ ASMDL × planting pattern		0.64		0.7	
CV (%)		4.18		3.62	

Mean in the respective category does not differ significantly at 5% level of probability according to LSD test.

21.3.15 Sugar yield (t ha^{-1})

Sugarcane is a highly productive crop plant with the capacity of storing large amounts of sucrose (Monica et al., 2011). Sugar yield (t ha^{-1}) is the product of stripped cane yield (t ha^{-1}) and sugar recovery percentage. As is evident from the data given in Table 21.19, the highest sugar yield of 12.75 and 15.16 t ha^{-1} at Location-II and Location-I, respectively, was recorded in 40% ASMD level followed by 60 and 20% ASMD levels, while the lowest sugar yield of 8.56 and 9.53 t ha^{-1} at Location-II and Location-I, respectively, was recorded in 80% ASMD level. It was also observed that 48.95, 38.43, 16.82 and 59.08, 49, and 15.11% higher sugar yield at Location-II and Location-I, respectively, was recorded in 40, 60, and 20% ASMD levels, respectively, over the 80% ASMD level. The higher sugar yield at the 40% ASMD level is ascribed to higher stripped cane yield. Saini and Chakor (1992) stated that irrigated plots produced significantly higher commercial cane sugar and extraction percentage over unirrigated plots with no effects on quality parameters. Pandian et al. (1992) obtained 12.93 and 12.75 t ha^{-1} sugar yields at 0.9 and 1.1 IW/CPE, respectively. Ali (1996) reported that irrigation applied at 15% ASM and 45% ASM did not affect the CCS%; however, total sugar was significantly reduced at 45% ASM due to decrease in cane yield at this moisture level.

Maximum sugar yield of 12.68 and 14.93 t ha^{-1} at Location-II and Location-I, respectively, was recorded in a 30/90 cm spaced paired row strip planting pattern. The lowest sugar yield of 9.13 and 10.76 t ha^{-1} at Location-II and Location-I, respectively, was obtained in a 30/120 cm spaced paired row strip planting pattern. It was found that 38.88, 23.33, 10.51 and 38.75, 21.19, and 3.53% higher sugar yield at Location-II and Location-I, respectively, was recorded in 30/90, 75, and 60 cm planting patterns over the 30/120 cm planting pattern. Higher stripped cane yield in a 30/90 cm spaced paired row planting pattern was attributed to higher stripped cane yield. Similar results were reported by El-Geddawy et al. (2002).

The maximum stripped cane yield of 14.74 and 18.36 t ha^{-1} at Location-II and Location-I, respectively, was recorded in the interaction of 40% $\times$ 30/90 cm followed by 40% $\times$ 75 cm, 40% $\times$ 60 cm, and 40% $\times$ 30/120 cm planting patterns. Whereas the minimum sugar yield of 7.25 and 8.31 t ha^{-1} at Location-II and Location-I, respectively, was recorded in an 80% $\times$ 30/120 cm planting pattern. It was also observed that 102.76 and 121.20% higher sugar yield at Location-II and Location-I, respectively, was recorded in the treatments interaction of a 40% ASMDL $\times$ 30/90 cm planting pattern compared to an 80% ASMDL $\times$ 30/120 cm planting pattern. The increase in sugar yield in the interaction of the 40% $\times$ 30/90 cm planting pattern was ascribed to complementary effects of increased nutrient availability at the 40% ASMD level and improved air circulation and light interception in the 30/90 cm paired row strip planting pattern, which resulted in higher stripped cane yield.

21.3.16 Net field benefit (Rs ha^{-1})

Water for irrigation is a limited resource and its effective management is critical, not only in reducing wasteful usage, but also in reducing production costs and sustaining productivity (Qureshi and Afghan, 2005). Total water requirement (cm) for sugarcane ranges from 140 to 240 cm through irrigation (Narendra, 2008). To have this amount, farmers have to resort to artificial irrigation through tubewells, therefore the demand for available water resource is fast exceeding the economic supply (Namara et al., 2007). Net field benefit (NFB) was calculated on the basis of 2-year pooled data of Location-I and Location-II (Table 21.20). Regarding ASMD levels, the crop receiving irrigation at the 40% ASMD level gave more NFB of Rs 70,917 ha^{-1}, followed by 60 and 20% ASMD levels and the lowest NBF of Rs 25022 ha^{-1} in an 80% ASMDL. Sugarcane grown in different planting patterns generated different NFB. The crop grown in the 30/90 cm spaced paired row strips gave the maximum NFB of Rs 65,266 ha^{-1}, followed by 75 and 60 cm spaced planting patterns with Rs 55,871 and Rs 45,561 ha^{-1}, respectively, while a minimum NFB of Rs 26,241 ha^{-1} was obtained from the crop grown in a 30/120 cm spaced paired row strip planting pattern. Higher NFB in the 30/90 cm spaced paired row strip planting pattern was ascribed to added stripped cane yield than the other three planting patterns.

Different treatment combinations resulted in different NFB. The combination of 40% $\times$ 30/90 cm planting pattern gave the maximum NFB of Rs 88,577, followed by 40% $\times$ G-75, 60% $\times$ G-30/90, 40% $\times$ G-60, with NFB of Rs 79,277, 74,127, and 73,327 ha^{-1}, respectively, against the minimum of Rs 8392 ha^{-1} with 80% $\times$ G-30/120 combination.

21.4 Conclusion and future prospects

It was concluded that maximum sugar yield was obtained from autumn sugarcane by irrigating it at 40% ASMD level and planting it at 30/90 cm spaced paired row strips. Thus, 10 and 24% irrigation

Table 21.20 Pooled Average Data of Location-II and Location-I Khan for the Years 2003–2004 and 2004–2005 Regarding Net Field Benefit (Rs ha^{-1}) of Autumn Sugarcane Influenced by Available Soil Moisture Levels And Planting Patterns Under Arid Conditions

Treatments	Stripped Cane Yield (ton)	Variable Cost (Rs)*	Total Cost (Rs)**	Gross Cost (Rs)***	Gross Income (Rs)	Net Field Benefit (Rs)
A. N: P: K. doses (kg ha^{-1})						
ASMDL$_1$ = 20%	24,175	26,025	53,238	79,263	120,875	41,612
ASMDL$_2$ = 40%	30,688	31,800	50,723	82,523	153,440	70,917
ASMDL$_3$ = 60%	27,813	29,200	49,473	78,673	139,065	60,392
ASMDL$_4$ = 80%	18,731	20,750	47,883	68,633	93,655	25,022
Planting patterns (G)						
G-60 = 60 cm	24,663	27,425	50,329.3	77,754.3	123,315	45,561
G-75 = 75 cm	27,000	28,800	50,329.3	79,129.3	135,000	55,871
G-30/90 = 30/90 cm	29,299	30,900	50,329.3	81,229.3	146,495	65,266
G-30/120 = 30/120 cm	19,444	20,650	50,329.3	70,979.3	97,220	26,241
(ASMDL × G)						
ASMDL$_1$ × G-60	24,300	26,800	53,238	80,038	121,500	41,462
ASMDL$_1$ × G-75	25,550	28,100	53,238	81,338	127,750	46,412
ASMDL$_1$ × G-30/90	28,850	30,200	53,238	83,438	144,250	60,812
ASMDL$_1$ × G-30/120	18,000	19,000	53,238	72,238	90,000	17,762
ASMDL$_2$ × G-60	31,250	32,200	50,723	82,923	156,250	73,327
ASMDL$_2$ × G-75	32,750	33,800	50,723	84,523	163,750	79,227
ASMDL$_2$ × G-30/90	35,100	36,200	50,723	86,923	175,500	88,577
ASMDL$_2$ × G-30/120	23,650	25,000	50,723	75,723	118,250	42,527
ASMDL$_3$ × G-60	28,400	30,000	49,473	79,473	142,000	62,527
ASMDL$_3$ × G-75	29,900	31,400	49,473	80,873	149,500	68,627
ASMDL$_3$ × G-30/90	31,200	32,400	49,473	81,873	156,000	74,127
ASMDL$_3$ × G-30/120	2,750	23,000	49,473	72,473	108,750	36,277
ASMDL$_4$ × G-60	18,700	20,700	47,883	68,583	93,500	24,917
ASMDL$_4$ × G-75	19,800	21,900	47,883	69,783	99,000	29,217
ASMDL$_4$ × G-30/90	22,050	24,800	47,883	72,683	110,250	37,567
ASMDL$_4$ × G-30/120	14,375	15,600	47,883	63,483	71,875	8,392

Govt. rate of stripped cane for the years 2003–2004 and 2004–2005 was Rs 1000 per 1000 kg.
Gross income (stripped cane yield in tons × rate of one tone of stripped cane).
*Variable cost (output cost) is the cost of harvesting stripping loading and carriage of crop at the rate of Rs 200 per ton.
**Total cost is the cost incurred on purchase of inputs and cultural operations (input cost).
***Gross cost is the sum of variable cost and total cost (input cost + output cost).

water was saved on sandy loam and silty clay soils, respectively, as compared to the traditional method of irrigating autumn sugarcane crop through flooding.

In developing countries like Pakistan and others, more determined effort is needed to bridge the gap between science and the application of this knowledge for sugarcane on the farm in view of decreasing surface water resources augmented with high growth population rate. We need to develop simple and practicable monitoring tools and engage irrigators in a process of adaptive learning for the promotion of existing technologies related to direct measurement of volumetric soil−water content coupled with advances in remote data access through the internet or via cellular networks.

References

Ali, F.G., 1999. Impact of Moisture Regime and Planting Pattern on Bio-Economic Efficiency Of Spring-Planted Sugarcane (*Saccharum officinarum* L) Under Different Nutrient and Weed Management Strategies. PhD thesis. Department Agronomy University Agriculture, Faisalabad, Pakistan.

Ali, F.G., Iqbal, M.A., Chatha, A.A., 1999. Cane yield response towards spacings and methods of irrigation under Faisalabad conditions. Pak. Sugar J. 14, 8−10.

Ali, S.A., 1996. Response of varying doses of nitrogen and soil moisture regimes on sugarcane crop. Bharatiya Sugar 21, 7−10.

Bashir, S., 1997. Planting Patterns and Population Dynamics Effects on Bio-Economic Efficiency of Autumn Sugarcane (*Saccharum officiarum* L.). PhD thesis. Department Agronomy University Agriculture, Faisalabad, Pakistan.

Benesová, M., Holá, D., Fischer, L., Jedelsky, P.L., Hnilicka, F., Wilhelmová, N., et al., 2012. The physiology and proteomics of drought tolerance in maize: early stomatal closure as a cause of lower tolerance to short-term dehydration? PLoS ONE 7, 38017.

Cheavegatti-Gianotto, A., Abreu, H.M.C., Arruda, P., Filho, J.C.B., Burnquist, W.L., Creste, S., et al., 2011. Sugarcane (Saccharum X officinarum): a reference study for the regulation of genetically modified cultivars in Brazil. Trop. Plant Biol. 4, 62−89.

Domini, M.E., Plana, R., 1989. Effect of planting density on sugarcane stalk growth and yield. Cultiaos Trop. 11, 67−73.

Dourado, M.N., Martins, P.F., Quecine, M.C., Piotto, F.A., Souza, L.A., Franco, M.R., et al., 2013. Burkholderia sp. SCMS54 reduces cadmium toxicity and promotes growth in tomato. Ann. Appl. Biol. 163, 494−507.

El-Geddawy, I.H., Darweish, D.G., Ael-Sherbing, A., Eldin, Abd El-Hddy E, 2002. Effect of row spacing and number of buds/seed setts on yield component of ratoon crops for some sugarcane cultivars ratoon. Pak. Sugar J. 17, 2−8.

Ghugare, R.V., More, D.V., Kenjale, S.T., 1995. Effect of levels of irrigation and nitrogen on yield and quality of ratoon sugarcane (CO.7219). Bharatiya Sugar 21, 44−49.

Gill, M.B., 1995. Physio-Agronomic Studies on Flat Versus PIT Plantation of Autumn and Spring Planted Sugarcane (Saccharumofficinrum L). PhD thesis. Department Agronomy University Agriculture, Faisalabad, Pakistan.

Hassuani, S.J., MRLV, Leal, Macedo, I.C., 2005. Biomass power generation: sugarcane bagasse and trash. Série Caminhos para Sustentabilidade. Piracicaba PNUD-CTC.

Ingram, K.T., 1986. Agronomic improvement of sugarcane seed production and quality. Proc. 19th Cong. ISSCT 35−43.

Kathirisan, G., Narayanasamy, R., 1991. Fixing the spacing, seed rate and nitrogen levels for COC 8201, sugarcane cultivar. Indian Sugar 41, 241−242.

Khan, M.Z., Bashir, S., Bajwa, M.A., 2004. Performance of promising sugarcane varieties in response of inter row spacings towards stripped cane and sugar yield. Pak. Sugar J. 19, 15−18.

Lal, B., 1988. Yield, quality and root distribution pattern of sugarcane Co 1148 as influenced by methods of planting. Bharatyia Sugar 14, 57−59.

Malik, K.B., Ali, F.G., 1990. Cane yield response to seed density and row spacing in spring and autumn planting. Proc 27th Ann. Cov. Pak. Soc. Sugar Tech. 239−243.

Malik, K.B., Ali, F.G., Zahid, A.R., 1992. Effect of different irrigation regimes on yield of some sugarcane varieties. Pak. J. Soil Sci. 7, 7−10.

Malik, K.B., Ali, F.G., Khaliq, A., 1996. Effect of plant population and row spacing on cane yield of spring-planted cane. J. Agric. Res. 34, 389−395.

Medici, L.O., Reinert, F., Carvalho, D.F., Kozak, M., Azevedo, R.A., 2014. What about keeping plants well watered? Environ. Exp. Bot. 99, 38−42.

Monica, S., Surekha, B., Suresh, K.B., 2011. Sucrose accumulation in sugarcane: a potential target for crop improvement. Acta. Physiol. Plant 33, 1571−1583.

Naidu, K.M., Venkataramana, S., 1989. Sugar yield and harvest index in water stressed cane varieties. Sugarcane 6, 5−7.

Naik, G.R., Somaskekhar, R., Hiremeth, S.M., 1993. Effect of water stress on growth and stomatal characteristics in sugarcane cultivars. Indian Sugar 43, 645−649.

Namara, R.E., Nagar, R.K., Upadhyay, B., 2007. Economics, adoption determinants, and impacts of micro-irrigation technologies: empirical results from India. Irrigation Sci. 25, 283−297.

Narendra, S., 2008. Is sugarcane a water guzzler (II)—irrigation water requirement for sugarcane and alternate crop sequences in India—an analysis. Indian Sugar 58, 35−49.

Nisachon, J., Sompong, T., Prasit, Patcharin S, 2012. Effects of drought and recovery from drought stress on above ground and root growth, and water use efficiency in sugarcane (Saccharum officinarum L.). Aust. J. Crop Sci. 6, 1298−1304.

Pandian, B.J., Muthukrishnan, P., Rajasekaran, S., 1992. Efficiency of different irrigation methods and regimes in sugarcane. Indian Sugar 42, 215−219.

Pires, R.C.M., Arruda, F.B., Sakai, E., 2008. Irrigação e drenagem. In: Dinardo-Miranda, L.L., Vasconcelos, A.C.M., Landell, M.G.A. (Eds.), Cana-de-açúcar. Instituto Agronômico de Campinas, Campinas, p. 882.

Qureshi, M.A., Afghan, S., 2005. Sugarcane cultivation in Pakistan. Sugar Book Pub. Pakistan Society of Sugar Technologist, Lahore.

Rosenzweig, C., Jones, J.W., Hatfield, J.L., Ruane, A.C., Boote, K.J., Thorburn, P., et al., 2012. The Agricultural Model Intercomparison and Improvement Project (AgMip): Protocols and pilot studies. Forest. Agr. Meteorol. 170, 166−182.

Saini, J.P., Chakor, I.S., 1992. Effect of irrigation and weed management on quality parameters of sugarcane under mid Hill conditions of H.P. Indian Sugar 42, 843−847.

Sharma, O.L., Gupta, P.C., 1990. Effect of levels of irrigation, nitrogen and method of planting on growth and yield of sugarcane. Indian Sugar 40, 313−315.

Sheu, Y.S., Yang, P.C., Li, S.W., 1992. Effect of soil moisture on growth and maturity of recent sugarcane varieties in Taiwan. Taiwan Sugar.

Shrivastava, A.K., Srivastava, A.K., Solomon, S., 2011. Sustaining sugarcane productivity underdepleting water resources. Curr. Sci. 101, 748−754.

Siddique, M.A.B., Karim, K.M.R., Rahman, M.A., Mahmud, K., 2008. Relative response of sugarcane varieties at the same irrigation level. Pak. Sugar J. 13, 9–11.

Singh, A.K., Lal, M., Prasad, S.R., 2006. Effect of row spacing and nitrogen on ratoonability of early maturing high sugar genotypes of sugarcane (Saccharum spp) hybrids. Indian J. Agric. Sci. 76, 3.

Soares, R.A.B., Oliveira, P.F.M., Cardoso, H.R., Vasconcelos, A.C.M., Landell, M.G.A., Rosenfeld, U., 2004. Efeiro da irrigacao sobre o desenvolvimento e a produtividade de duas variedades de cana-de-acucar colhidas eminicio de safra. STABAc.esubp. 22, 38–41.

Steel, R.G.D., Torrie, J.H., Dicky, D., 1997. Principles and Procedures of Statistics. McGraw Hill Book Co Inc, New York.

Vains, S.N., Ahmad, R., Jabbar, A., Mahmood, T., Ahmad, A., 2000. Biological traits of autumn sugarcane as influenced by spatial arrangement and plantation method. Pak. Sugar J. 15, 15–19.

Wiedenfeld, B., Enciso, J., 2008. Sugarcane responses to irrigation and nitrogen in semiarid South Texas. Agron. J. 100, 7.

Zingaretti, S.M., Aparecida Rodrigues, F., Perez da Graça, J., de Matos Pereira, L., Lourenço, M.V., 2012. Sugarcane responses at water deficit conditions, water stress. In: Prof. Ismail Md. Mofizur Rahman (Ed.), InTech, pp. 255–276.

Index

Note: Page numbers followed by "*f*", and "*t*" refers to figures and tables respectively.

537

CPI Antony Rowe
Eastbourne, UK
November 19, 2018